Canadian Edition

ENVIRONMENTAL SCIENCE

A Global Concern

William P. Cunningham
University of Minnesota

Mary Ann Cunningham
Vassar College

Barbara Woodworth Saigo
St. Cloud State University

Robert Bailey
University of Western Ontario

Dan Shrubsole
University of Western Ontario

McGraw-Hill Ryerson

Toronto Montréal Boston Burr Ridge, IL Dubuque, IA Madison, WI New York San
Francisco St. Louis Bangkok Bogotá Caracas Kuala Lumpur Lisbon London Madrid
Mexico City Milan New Delhi Santiago Seoul Singapore Sydney Taipei

McGraw-Hill
Ryerson

ENVIRONMENTAL SCIENCE
A GLOBAL CONCERN
CANADIAN EDITION

Statistics Canada information is used with the permission of the Minister of Industry, as Minister responsible
for Statistics Canada. Information on the availability of the wide range of data from Statistics Canada can be
obtained from Statistics Canada's Regional Offices, its World Wide Web site at http://www.statcan.ca, and its
toll-free access number 1-800-263-1136.

ISBN-13: 978–0–07–091664–7
ISBN-10: 0–07–091664–0

2 3 4 5 6 7 8 9 10 TCP 0 9 8 7

Printed and bound in Canada

Vice President, Editorial and Media Technology: *Patrick Ferrier*
Sponsoring Editor: *Leanna MacLean*
Developmental Editor: *Suzanne Simpson Millar*
Sales Manager: *Megan Farrell*
i-Learning Marketing Specialist: *Patricia Gibson*
Manager, Editorial Services: *Kelly Dickson*
Supervising Editor: *Joanne Limebeer*
Copy Editor: *Erin Moore*
Production Coordinator: *Paula Brown*
Editorial Coordinator: *Carole Harfst*
Formatter: *SR Nova Pvt Ltd., Bangalore, India*
Interior Design: *Sharon Lucas*
Cover Design: *Sharon Lucas*
Cover Image Credit: © *Pat O'Hara/Corbis*
Printer: *Transcontinental Printing Group*

National Library of Canada Cataloguing in Publication

Environmental science : a global concern / William P. Cunningham ... [et al.]. — Canadian ed.

Includes index.
ISBN 0–00–70916640
1. Environmental sciences–Textbooks. I. Cunningham, William P.

GE105.E57   2004   363.7   C2004-904581-4

About the Canadian Authors

A GLOBAL CONCERN

ROBERT BAILEY

Robert Bailey began his undergraduate career at the University of Toronto, knowing that he was into science but not really having any idea how he was going to make science his life's work. Intrigued by the blend of ecology and environmental problems presented as part of his first year biology class, Bob decided to move to the University of Guelph to complete his BSc in Ecology, and stayed on to complete a Masters in Zoology with aquatic biologist Gerry Mackie. Bob then returned to the University of Toronto to work for acidification researcher Pamela (Stokes) Welborn at the Institute of Environmental Studies, working in the beautiful lakes on the Canadian Shield, including the important Experimental Lakes Area in northwestern Ontario (see Chapter 3: Matter, Energy, and Life).

Inspired to make a career of Evironmental Science, Bob went to the University of Western Ontario to do a PhD with Aquatic Ecologist and Environmental Statistician Roger Green. He stayed on at Western as a faculty member, and is now a Professor of Biology and Environmental Science there, as well as Director of Environmental Research Western, an institute dedicated to building collaborations among environmental researchers of all types at Western, and bridges between the university's environmental researchers and opportunities for partnerships with industry and government.

For the last decade, Bob's research has been mainly concerned with environmental assessment of freshwater ecosystems. He does field work in streams from the Yukon River Basin in northwestern Canada to ponds and wetlands in southern Ontario, and is often called on to contribute to panels and workshops where government agencies in Canada and the U.S. try to determine the best strategies to use in assessing and protecting their aquatic resources. He has taught Introductory Ecology at Western for several years, and is launching a new Environmental Science course for first year students this year.

DAN SHRUBSOLE

Dan Shrubsole is an Associate Professor of Geography at the University of Western Ontario, and has published many papers on his research on the institutional aspects of resource management, particularly water resource management in Canada. He has co-authored a book entitled *Ontario Conservation Authorities: Myth and Reality* (1992).

National perspectives on water management are provided in two other books: *Practising Sustainable Water Management: Canadian and International Experiences* (1997), and *Canadian Water Management: Visions for Sustainability* (1994). He co-edited a book on water conservation entitled *Every Drop Counts* (1994).

He has also published journal articles on floodplain management, wetland management, water conservation, and watershed management in *Environmental Management, Journal of Environmental Management, Canadian Water Resources Journal, Applied Geography,* and *Geoforum.* In 1995/96, Dan was a visiting research fellow with Australia's Commonwealth Research and Scientific Organization (CSIRO) and participated in research concerning catchment management and the environmental management in the sugarcane growing industry.

Contents in Brief

Contents

Preface

A GLOBAL CONCERN

Kananaskis Valley

As I write this, I am sitting at the Kananaskis Field Station in the southern Alberta Rocky Mountains. A wonderful part of my job as a Professor of Environmental Science and Biology at the University of Western Ontario is teaching undergraduate students in a field environment. I am here helping my colleague, Jack Millar, teach a field course in Alpine Ecology by showing the students how to experience and learn about the ecology of mountain streams. When I experience places like Kananaskis, whether in my teaching or research, I feel very keenly my love of the science of the natural world, and my urge to engage students in this passion.

I do not consider myself an "environmentalist." Rightly or wrongly, this label has the same effect on me as "industrialist" or "entrepreneur" or "fundamentalist." It implies decisions and values based on faith rather than knowledge. My approach in this Canadian revision of Cunningham, Cunningham, and Saigo's book is a reflection of this view. Throughout this text I have

tried to integrate natural sciences and the human dimensions of environmental issues with the goal of inspiring and educating environmentally-informed readers who can then make up their own minds about the critical environmental issues facing us. I want to help students bridge the gap between environmental concern and action, which I feel requires knowledge to fill. I feel very strongly that this knowledge and inspiration should be accessible to all students, whether their main interests are in the natural sciences, engineering, social sciences, or the arts.

When I was asked to write a Canadian adaptation of the fine U.S. textbook authored by William Cunningham, Mary Ann Cunningham, and Barbara Saigo, my one request was that I would be free to modify any aspect of the book, not just mention more Canadian examples and data. I wanted this not out of any disrespect for the American edition—I feel it is a fine text that gets significantly better with each edition released—but more because I feel it imperative to give students the tools they need to engage in discussion about an issue, rather than a particular point of view. I have no desire to make up their minds for them, so a more objective approach to issues ranging from nuclear power to mining is used. I hope what we (for definition of the "we" see the Acknowledgements) achieve is close to that goal. That said, many topics in Environmental Science are contentious, so I encourage those teaching Environmental Science with this book as a resource (and especially those taking a course that uses this book), to send me the feedback I need to make the book even better. If you email me (drbob@uwo.ca), I will not always agree, but I will listen, and I will respond!

Robert Bailey

Developments in environmental science are evolving at an astounding pace. Technological advances in our ability to measure the quality and quantity of substances in the environment, and in the areas of geographic information systems and remote sensing continue at an ever increasing rate. So, too, are human responses to environmental issues. New laws and policies, user fees and/or taxes, and governance arrangements are introduced on a very regular basis in Canada and abroad. Thus, dealing with changes in the environment, in technology, and in the laws and policies that guide environmental management, will be an ongoing challenge for environmental scientists and students.

In reading and using this text, I hope students become better grounded in the principles of environmental science. At the same time, you should become aware that other perspectives, such as law, political science, psychology, geography, economics, and sociology, also contribute to our understanding and resolving environmental problems. Having the right breadth and depth of courses that support your environmental education is a difficult task. However, it can also be challenging and rewarding. I hope you enjoy the journey and its destinations.

Dan Shrubsole

AUDIENCE

This book is intended for use in a one or two semester course in Environmental Science in first or second year university or college. Because many students who will use it will be non-science majors, we have tried to make the text readable and accessible without technical jargon or a presumption of much of a prior science background. At the same time, we have tried to provide enough theory, data, and depth to make the book a valuable reference. Periodically, we will point you to more detailed treatments of an issue (see **Want More?** boxes). When technical details seem to overwhelm the more general, important points on an issue, the instructor or student is invited to intellectually skim as appropriate in crafting their definition of what one needs to know to be educated and informed in Environmental Science.

SUSTAINABILITY

An overarching theme in this book is sustainability: Can we find ways to meet our present needs without compromising the ability of future generations to meet their own needs? Can we live on renewable energy sources and the surplus produced by biogeochemical cycles, without damaging the productive capacity of our environment? The concepts of inherent values, ethical rights, stewardship, and equity between generations and between people living under different conditions now all play important roles in our consideration of how natural resources should be managed. Consequently, ethics, philosophy, and environmental worldviews are among the first topics we discuss in this book.

CRITICAL THINKING

Critical thinking is another central theme in this book. Environmental science is a complex field, one in which a large number of special interests, contradictory data, and conflicting interpretations

battle for our attention. How can we decide what to believe when apparently equally eminent experts hold diametrically opposed opinions on controversial topics? Perhaps the most valuable skill any student can gain from the study of Environmental Science is the ability to think purposively, analytically, and clearly about evidence. To understand the complexity and conflicting interpretations of environmental problems, students need a number of skills. They need to be able to identify and evaluate biases, recognize and assess assumptions, and understand conceptual frameworks. They must also learn to acknowledge and clarify uncertainties, equivocations, and contradictions in arguments. Reaching satisfactory conclusions about environmental dilemmas isn't just a matter of logic and rationality; we also need open-mindedness, skepticism, independence, and an ability to empathize with others. We discuss these skills and then model their application in boxed readings, case studies, and questions at the end of each chapter.

BALANCED VIEW

In every edition of this textbook, we have tried to pull together and summarize the most important current environmental information, and to explain the context and significance of scientific evidence. There's a temptation, in discussing environmental conditions to focus on extremes. While acknowledging problems, we also are careful to describe good news, progress towards sustainability, and the many ways individuals can make positive contributions toward environmental protection. Because science is always conditional, and there can be many ways to interpret data, we also present a balanced view that recognizes uncertainties and conflicting interpretations. At the same time, we stress that scientific consensus does emerge on major issues. We feel it is essential students understand the need for differing interpretations of evidence and also recognize the value of general agreement among scholars.

We hope you will find this book a valuable source of information about our global environment, as well as an inspiration for solutions to the dilemmas we face. Everyone has a role to play in this endeavour. Whether as students, educators, researchers, activists, or consumers, each of us can find ways to contribute in solving our common problems.

GLOBAL CONCERN

We live in an increasingly interconnected world. An awareness of international events, population trends, health conditions, and environmental quality are essential for educated citizens. The coal mining in Nova Scotia, the nuclear waste dumped in the ocean by Russia, or the pesticides used on farms in Central America affects all of us. This text incorporates a worldview of environmental issues into each chapter with discussions in the text, photos, examples used, boxed readings, and data.

Acknowledgements

Many reviewers from across Canada who teach Environmental Science have helped to shape this first Canadian edition of *Environmental Science: A Global Concern*. I appreciate their thorough reading of our initial draft, and productive challenges and ideas that significantly enhanced the book. I thank my friends at McGraw-Hill Ryerson, who did so much to launch and bring to fruition this project. Cathy Koop initially intrigued me enough about the idea of the adaptation to get me involved. Sandra de Ruiter was my Development Editor in the early stages, and helped me throughout the initial drafting of the text. Brendan Bailey helped immensely during this stage as my researcher, seeking out data, documents, and photos on a moment's notice, and often suggesting ideas for additions and further modifications to the text. Suzanne Simpson Millar came on in the latter stages of the project as my Development Editor, and she did valiant service as a collector and synthesizer of reviews, gently prodding me to deadlines but always with an eye on the quality and value of the project. Joanne Limebeer, as Supervising Editor of Editorial, Design, and Production, and Erin Moore, as Copy Editor, oversaw the final stages of the book's creation, with a careful eye to detail. Finally, I thank my friend and colleague Dan Shrubsole, who extensively revised and rewrote several sections of the text with his extensive knowledge of the human side of Environmental Science, and did so at a time when facing many personal challenges.

Robert Bailey

Authors owe many debts of thanks to the people who encourage and nurture them. Friend and colleague Bob Bailey provided the invitation to participate. The staff at McGraw-Hill have been very supportive and responsive to my many questions and unpredictable schedule. Suzanne Simpson Millar made life easier with her ability to obtain photographs and other graphics, provide constructive comments and proofread effectively. Leanna MacLean and Jennifer Matyczak gave excellent instructions on how to adapt the book. Joanne Limebeer was able to discern my scribbles and etchings during the proofing stage. I appreciate their professionalism and friendship. Finally, thanks to Connie, Evan, and Ethan.

Dan Shrubsole

We also gratefully acknowledge the constructive criticism of the many colleagues who provided reviews of this new Canadian edition. They include:

Michal Bardecki *Ryerson University*

Stephen Bocking *Trent University*

Lyle G. Courtney *Kwantlen University College*

Susan Dakin *University of Lethbridge*

Mrinal K. Das *University of Alberta*

Tim Elkin *Camosun College*

Isobel W. Heathcote *University of Guelph*

Herbert J. Kronzucker *University of Western Ontario/University of Toronto*

Lawrence Licht *York University*

Linda Lusby *Acadia University*

Tim Patterson *Carleton University*

Anthony G. Price *University of Toronto (Scarborough)*

Hilary Sandford *Camosun College*

Kathleen Smith *Confederation College*

R. Greg Thorn *University of Western Ontario*

Karen Tomic *University of Alberta*

Michael C. Wilson *Douglas College*

Ann P. Zimmerman *University of Toronto*

LEARNING AIDS

This text is designed to be useful as a self-education tool for students. Full-page chapter openers set the stage with a vibrant photo. To facilitate studying and encourage high-level thinking, these pages include a set of **Objectives** based on major concepts that students should master. In addition, a new box entitled **WebQuest** includes a list of keywords for students to use when searching for more information online—avoiding broken links and out-of-date material.

To begin each chapter, an **opening vignette** introduces some of the key factors that will be covered in the chapter. Several new examples have been incorporated into this Canadian edition, and all are real stories from around the world that will help open up discussion to the specific chapter topic. A new box entitled **Want More?** is incorporated into various vignettes and boxes. These boxes provide full references of the subject, allowing students to research original data, theories, and information.

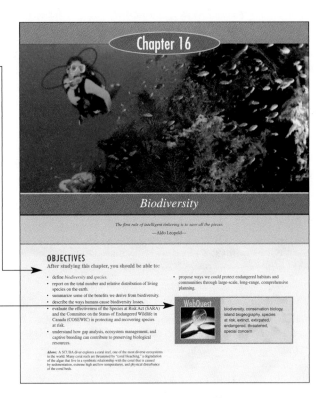

Chapter 16

Biodiversity

The first rule of intelligent tinkering is to save all the pieces.
—Aldo Leopold—

OBJECTIVES
After studying this chapter, you should be able to:

- define *biodiversity* and *species*.
- report on the total number and relative distribution of living species on the earth.
- summarize some of the benefits we derive from biodiversity.
- describe the ways humans cause biodiversity losses.
- evaluate the effectiveness of the Species at Risk Act (SARA) and the Committee on the Status of Endangered Wildlife in Canada (COSEWIC) in protecting and recovering species at risk.
- understand how gap analysis, ecosystem management, and captive breeding can contribute to preserving biological resources.

- propose ways we could protect endangered habitats and communities through large-scale, long-range, comprehensive planning.

biodiversity, conservation biology, island biogeography, species at risk, extinct, extirpated, endangered, threatened, special concern

Above: A SCUBA diver explores a coral reef, one of the most diverse ecosystems in the world. Many coral reefs are threatened by "coral bleaching," a degradation of the algae that live in a symbiotic relationship with the coral that is caused by sedimentation, extreme high and low temperatures, and physical disturbance of the coral beds.

The Cough Heard Round the World

Early in 2003, news began to trickle out of China that a very infectious "atypical pneumonia" was spreading rapidly in hospitals around Guangzhou (formerly known as Canton) in Guangdong Province. Symptoms included fever, chills, headaches, muscle pains, and a dry cough, but in many patients, especially the elderly, the disease would quickly turn into a deadly pneumonia. In February 2003, a doctor from Guangzhou, who had contracted this disease from his patients, travelled to Hong Kong. There he passed the infection to other travellers, who carried what is now known as **Severe Acute Respiratory Syndrome** (SARS) to Beijing, Canada, Taiwan, Singapore, and Vietnam.

Within six months, SARS spread to 31 countries around the world, where more than 8,500 probable cases and 812 deaths were reported. Fear of SARS travelled even faster and farther than the disease itself. Rumours multiplied across the Internet as conferences and sporting events were cancelled, factories closed, and tourism to China fell by as much as 85 percent after the epidemic was revealed. The economic impact of SARS is estimated to be at least $30 billion (U.S.) in 2003 alone. By July 2003, the World Health Organization (WHO) declared the outbreak contained, but suggested that the world remain vigilant for further infections.

The rapid transmission of this disease shows how interconnected we all are. A virus can travel in just a few days anywhere a plane flies. One flight attendant is thought to have been the source of infection for 160 people in seven countries. SARS also points to the need for better communication and identification of new diseases. Because the Chinese government hid the extent and

FIGURE 20.2 Often cited for his love of Toronto, lead singer of the Rolling Stones Mick Jagger helped Toronto overcome the shadow of SARS by performing a free concert in the city. The Concert for SARS Relief was held in July 2003 and included 15 other acts. More than 500,000 people filled Downsview Park to see the concert, seen by many as a light at the end of the dark tunnel that SARS had put Canada in.

severity of the disease for several months, patients with this highly infectious disease mingled with the general hospital population, spreading the illness. Medical professionals, not knowing how serious the infection was, treated patients without wearing protective clothing. Major hospitals in Beijing, Taipei, Hong Kong, and Singapore were closed because so many staff members were sick, making treatment of the contagion even more difficult.

While globalization helped spread SARS, it also helped in rapid recognition and treatment of the disease. It took centuries to discover the cause of cholera. Identifying the virus that causes AIDS took two years. But within weeks after the WHO issued its first warning about SARS, an electron microscopist at the Centers for Disease Control in Atlanta found a new coronavirus in cell cultures infected with tissue from SARS patients. Less than a month later, labs from Vancouver to Singapore were sequencing its RNA. In May 2003, Scientists at Hong Kong University announced they had found coronavirus nearly identical to those from SARS patients in civets, badgers, and raccoon dogs being sold in Guangdong meat markets.

Wild species had been suspected as a source of the SARS virus since some of the first infections occurred among chefs and animal merchants. Exotic animals are regarded as delicacies in southern China, where they are featured at banquets and dinners at expensive restaurants. In April of 2003, Chinese police raided animal markets and hotels and restaurants, seizing 838,500 animals including many rare and endangered species. The emergence of SARS may reduce animal smuggling, but the existence of a reservoir of this virus in the wild may mean that it will be impossible to completely eradicate the disease.

FIGURE 20.1 In spring 2003, Canada was in the midst of what could be described as its worst disease epidemic in modern times. The World Health Organization advised against travel to Toronto and Vancouver and health care workers who treated the sick were at high risk of contracting SARS themselves. Masks were the first response to the feared airborne virus that caused SARS, especially among health care workers in Vancouver and Toronto.

Case Studies, **What Do You Think?** essays, some with **Ethical Considerations** attached, will also give students real-life examples to evaluate. All of these features are carefully planned to build upon chapter content and encourage students to practise critical thinking skills and formulate reasoned opinions.

what do you think?

The Yukon Placer Authorization

In 1898, the famous Klondike Gold Rush began an era of placer (pronounced *plasser*) gold mining in the Yukon that has continued ever since. Placer mining is essentially "panning for gold" with the help of bulldozers and other large equipment. In some areas, sediment originally deposited by rivers contains gold particles (and a few larger nuggets). Miners dredge, dig up, or dislodge the sediment with water, dump it into a large, rotating cylinder to filter out the larger rocks and stones, and then separate out the gold by gravity selection (gold is heavier than most other minerals in the sediment). All of these steps require large amounts of water, which is usually found in nearby streams and pumped to the filter system.

There are three main effects on the environment of such mining. The first is the construction of access roads and the site itself for the bulldozers and other infrastructure needed to operate a mine. Next is the obvious disruption of terrestrial ecosystems as the surface sediment and soil is scraped off for processing in the mine and extraction of the gold. Finally, the waste water from the process of gold recovery goes back into the stream ecosystem; even with sometimes large networks of settling ponds there is often a considerable increase in the suspended material in the stream, with resulting direct (toxicity) and indirect (habitat) effects on the biota in the stream.

The Canadian Fisheries Act, which has been in force since Confederation, "... prohibits the deposit of deleterious (toxic or harmful) substances into fish-frequented waters or in a place or under any conditions where it may enter fish-frequented waters." This means that it is unlawful to deposit a harmful substance either directly into a fish-bearing stream, or into a place like the top of a bank or a storm drain that leads to a fish-bearing stream. On the face of it, placer mining seems contrary to the Fisheries

Act, and in 1993, the Yukon Placer Authorization (YPA) formally recognized this contradiction. This authorization specifies the classification of streams, the allowable sediment discharges under which the placer mining industry can operate in different types of streams, and plans for inspecting and monitoring. It also provides for the temporary deferment of water quality standards on streams.

In 2002, the Minister of Fisheries and Oceans decided that the placer mining industry was not being subjected to the same rigorous standards of environmental protection as other industries. The Minister announced that the YPA would be phased out over four years and that during that time, any new mines would have to meet environmental protection standards, which the industry claimed were economically

unachievable. In the Yukon, passionate objections to this position came from all sides ... from local communities to the territorial government. Conservationists, who had lobbied hard for increased environmental protection, continued to press the Minister to go ahead with his plan. Placer miners, and the businesses and communities that depended on them, fought equally hard for concessions and retention of the YPA. Placer-mined gold production amounts to about $40 million (Cdn.) per year in the Yukon. What would you recommend to the Minister of Fisheries and Oceans in this situation? Should the same regulations apply to placer mining as other industries? Or is placer mining special because of the historical significance and the fact that it is usually a small or medium sized business, similar to family farming?

The **What Can You Do?** boxes throughout the text bring the global subject of environmental science, protection, and economics to a level that students can not only understand, but can implement in their own lives. These include examples of various Canadian environmental policy bodies as well as the United Nations, World Bank, and many more. Students will be eager to implement these ideas (recycling, electricity usage, etc.) into their own lives.

In Depth boxes expand a topic of the chapter into a more in-depth look at the subject. It will enhance student knowledge as well as jump-start class discussion.

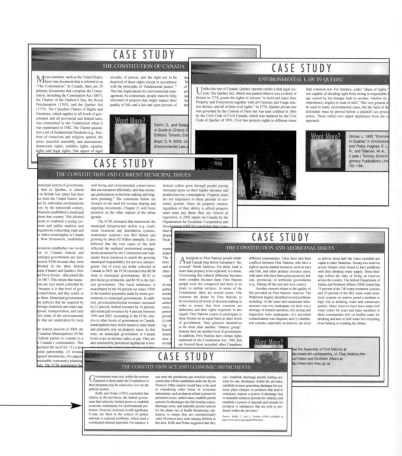

New Chapters 19 and 21 on urbanization and policy were written specifically for Canadian classrooms and feature Canadian content. **Chapter 19: Environmental Law: Its Role in Guiding Governing Instruments** opens with the following **Objectives:**

- differentiate between statute and common law, and know the basis for each in Canadian law.
- identify federal and provincial responsibilities for resource and environmental management as defined in Canada's Constitution.
- understand the five approaches that have characterized the evolution of environmental law and policy in Canada, and remain in use today.
- consider the reasons that international treaties have not been successful.
- appreciate the importance of wicked problems, resilience and adaptive management in environmental planning.
- scrutinize collaborative, community-based planning methods.

Chapter 21: Urbanization and Sustainable Cities "Canada's Urban Centres in Transition," explores the role urbanization has played in the development of a strong and stable economy and society. It also discusses the issues that still need to be addressed: infrastructure built in the 1960s, traffic and gridlock, Aboriginal poverty as well as a shift from resource-based economy to a knowledge-based economy. It also discusses the Caucus Task Force on Urban Issues established by the government in 2001.

The Sydney Tar Ponds are explored in detail, taking us from the advent of the coal-mining economy of Nova Scotia in the 1880s, through to the termination of the proposed cleanup in 1994, and the government's renewed funding for cleanup in 2004.

As well, urban issues such as Ontario's decision to privatize its Highway 407 is debated as a way to ease the cash-strapped economy, as is the recent Vancouver Agreement, signed to help alleviate drug problems plaguing the city's downtown east side.

At the end of each text section is a **Profile** of a specifically chosen Canadian who works in the fields that have been discussed. Each profile includes photos of the subject at work and/or at play (which are often the same place!). Not only a teaching tool for students, these profiles will show them the myriad of ways in which these individuals got involved in environmental science, and what their day-to-day work entails.

New **Three-Dimensional Art** has transformed certain chapters in this edition and raised it to a new standard, providing students with images that are more realistic and identifiable. For example, life-like images of wolves, hares, and other organisms involved in the arctic food web allow the students to more accurately visualize the connections between these various components.

INSTRUCTOR RESOURCES

Integrated Learning

Your Integrated Learning Sales Specialist is a McGraw-Hill Ryerson representative who has the experience, product knowledge, training, and support to help you assess and integrate any of the below-noted products, technology, and services into your course for optimum teaching and learning performance. Whether it's using our test bank software, helping your students improve their grades, or putting your entire course online, your *i*-Learning Sales Specialist is there to help you do it. Contact your local *i*-Learning Sales Specialist today to learn how to maximize all of McGraw-Hill Ryerson's resources!

i-Learning Services Program

McGraw-Hill Ryerson offers a unique iServices package designed for Canadian faculty. Our mission is to equip providers of higher education with superior tools and resources required for excellence in teaching. For additional information visit **www.mcgrawhill.ca/highereducation/eservices/.**

PageOut

Visit **www.mhhe.com/pageout** to create a Web page for your course using our resources. PageOut is the McGraw-Hill Ryerson website development centre. This Web page-generation software is free to adopters and is designed to help faculty create an online course, complete with assignments, quizzes, links to relevant websites, and more—all in a matter of minutes.

In addition, content cartridges are available for the course management systems **WebCT** and **Blackboard.** These platforms provide instructors with user-friendly, flexible teaching tools. Please contact your local McGraw-Hill Ryerson *i*-Learning Sales Specialist for details.

e-Instruction's Classroom Performance System (CPS)

Bring interactivity to the classroom or lecture hall. CPS is a student response system using wireless connectivity. It gives instructors and students immediate feedback from the entire class. The response pads are remotes that are easy to use and engage students.

- CPS helps you to increase student preparation, interactivity, and active learning so you can receive immediate feedback and know what students understand.
- CPS allows you to administer quizzes and tests, and provide immediate grading.
- With CPS you can create lecture questions that can be multiple-choice, true/false, and subjective. You can even create questions on the fly as well as conduct group activities.
- CPS not only allows you to evaluate classroom attendance, activity, and grading for your course as a whole, but CPSOnline allows you to provide students with an immediate study guide. All results and scores can easily be imported into Excel and can be used with various classroom management systems.

Please contact your *i*-Learning Sales Specialist for more information on how you can integrate CPS into your Environmental Science classroom.

Primis Online

Professors can now create their unique custom eBook or printed text online. With Primis Online you can select, view, review your table of contents, fill out your custom cover and shipping information, and then have the opportunity to approve a complimentary sample book! Once you have approved your sample eBook or printed book, your students can purchase it—either through McGraw-Hill's Primis eBookstore or your local campus bookstore. Start creating your own customized text through Primis Online by going to **www.mcgrawhill.ca/highereducation/primis+online.**

Online Learning Centre (OLC)

Visit the site at **www.mcgrawhill.ca/college/cunningham**. In addition to all services available to students (see below), you'll also receive:

- Questions for eInstruction (CPS)
- Answers to Web exercises
- Additional case studies
- PowerPoint lecture slides
- Answers to critical thinking questions
- PageOut

Digital Content Manager (DCM) CD-ROM. This multimedia collection of visual resources allows instructors to utilize artwork from the text in multiple formats to create customized classroom presentations, visually based tests and/or quizzes, dynamic course website content, or attractive printed support materials.

Instructor's Testing and Resource CD-ROM. This cross-platform CD-ROM provides a computerized test bank using Brownstone Diploma© testing software to quickly create customized exams. The user-friendly program allows instructors to search for questions by topic, format, or difficulty level; edit existing questions or add new ones; and scramble questions and answers keys for multiple versions of the same test.

Transparencies. A set of 100 transparencies is available to users of the text. These acetates include key figures from the text, including new art from this edition.

STUDENT RESOURCES

Online Learning Centre (OLC)

Visit the site at **www.mcgrawhill.ca/college/cunningham**. Everything you need in one place:

- "How to Study" chapter
- How to study tips
- Practice quizzing
- Web links
- Web exercises
- Guide to electronic research
- Regional Perspectives (case studies)
- Key term flashcards
- Interactive maps
- How to contact your elected officials
- Additional readings
- Metric equivalents and conversion tables

Online **LearningCentre**

Chapter 1

The Science of Our Environment

There are no passengers on spaceship earth. We are all crew.

—Marshall McLuhan—

OBJECTIVES

After studying this chapter, you should be able to:

- define the terms *environment* and *environmental science* and identify some important environmental concerns that we face today.
- discuss the history of environmental science and how it has changed in scale.
- briefly describe some current environmental dilemmas and issues.

WebQuest

deformed frogs, environmental science, environmentalism, industrialism, Rachel Carson

Above: Fishing boat lights glisten on the Li River in Guilin, China.

Deformed Frogs

In 1995, school children on a summer field trip to a marsh in southern Minnesota discovered dozens of frogs with malformed, missing, or extra legs (Fig. 1.1). The students' questions about these deformities led to studies that seem to show a sudden, widespread increase in developmental abnormalities in amphibians. Subsequent investigations in California, Iowa, Kansas, Missouri, New York, Oregon, Texas, Vermont, and Quebec have found as many as 60 percent of some frog or salamander species in certain areas have abnormal limbs, digits, or eyes. When dissected, many of these animals were found to have internal problems as well, including nonfunctioning digestive systems, grossly distended bladders, and abnormal reproductive organs.

What could be causing these problems? The widespread use of pesticides and herbicides in agriculture, together with industrial toxins released into the air and water, make toxic synthetic chemicals likely suspects. Since frogs spend most of their lives in water, they easily absorb toxins through their thin skin. And there seems to be more deformities in ponds near agricultural areas, compared to relatively "pesticide-free" areas, but these studies are not yet conclusive. We *do* know, however, that a number of synthetic chemicals disrupt endocrine hormone functions critical in regulating normal growth and development of many species.

Other factors could also play a role in these deformities. Retinoids (such as vitamin A) are known to regulate key processes in embryo development. They have been found in ponds with deformed frogs, perhaps from pesticide residues. Metals such as arsenic, mercury, selenium, and cadmium that are released by industry or carried in agricultural runoff also may be involved. Ultraviolet radiation from the sun is increasing in many areas because of damage to the stratospheric ozone shield and frog eggs and young are known to be sensitive to UV radiation. Parasites such as trematodes (flukes) are known to burrow into tadpoles during early development, forming cysts that interfere with normal growth and development. Unfortunately, there may be no single, simple cause for all the diverse problems observed at different places and times. Varying combinations of some or all of these factors may be at work.

What should we make of this apparently sudden increase in deformed animals? We don't have good historic data against which we can compare recent surveys; but scientists with many years of field experience report they have never seen the numbers of aberrant frogs found recently. Some people see this as an ominous warning that human activities are disrupting nature in ways that threaten not only amphibians but us as well. They see frogs as an indicator species, much like the canaries that early miners took into the mines to warn of life-threatening conditions.

Skeptics argue that frog abnormalities have probably always existed. Perhaps we're finding more now simply because we are looking for them. Just because the deformed frogs look pathetic, they contend, we shouldn't let our emotions run wild. Their condition may not mean anything for us. Clearly we need more careful research and reliable data before we know how serious the problem may be.

Deformed limbs are not the only threatening condition facing frogs and their amphibian kin today. Over the past 20 or 30 years, populations of frogs, toads, salamanders, and newts have diminished or even vanished entirely from wetlands around the world where they

FIGURE 1.1 An alarming number of deformed frogs have been found in recent years. What causes these abnormalities is not yet known. Are they a warning of pollution or other serious environmental problems?

once were common. Wetlands are drained for agriculture and urban expansion, forests are cleared by lumber operations, and erosion fills streams and ponds with silt. Water pollution, acid rain, pesticide runoff, and other sources of toxic chemicals are known to be highly damaging to aquatic species. Naturally occurring diseases appear to be more virulent because hormone-disrupting chemicals damage immune system protection. Weather changes, perhaps caused by global warming, may be altering habitat.

These situations introduce a number of important themes in environmental science. What are we doing to other species and what threats do our activities expose *us* to as well? What can we do to reduce our negative impact and live in harmony with our environment? In this book we will survey a variety of global environmental issues and offer some suggestions for what we might do about them. In many cases—such as the example of deformed frogs—the problems are complex and only partially understood. Environmental science is a new and evolving field. It will require active participation on your part to think critically about these topics and to work out what you personally believe is correct. Sometimes, the answers to your questions will be more questions, and more ideas for the research that will bring us closer to solutions. You are undertaking a journey of discovery that will continue, we hope, long after you finish the course for which you bought this book.

Want More?

Burkhart, J.G., Ankley, G., Bell, H., Carpenter, H., Fort, D., Gardiner, D., Gardner, H., Hale, R., Helgen, J.C., Jepson, P., Johnson, D., Lannoo, M., Lee, D., Lary, J., Levey, R., Magner, J., Meteyer, C., Shelby, M.D., Lucier, G. 2000. "Strategies for Assessing the Implications of Malformed Frogs for Environmental Health." *Environmental Health Perspectives*. 108 (1): 83–90.

WHAT IS ENVIRONMENTAL SCIENCE?

Humans have always inhabited two worlds. One is the natural world of plants, animals, soils, air, and water that preceded us by billions of years and of which we are a part. The other is the world of social institutions and artifacts that we create for ourselves using science, technology, and political organization. Both worlds are essential to our lives, but integrating them successfully causes enduring tensions.

Where earlier people had limited ability to alter their surroundings, we now have power to extract and consume resources, produce wastes, and modify our world in ways that threaten both our continued existence and that of many organisms with which we share the planet. To ensure a sustainable future for ourselves and future generations, we need to understand something about how our world works, what we are doing to it, and what we can do to protect and improve it.

Environment (from the French *environner:* to encircle or surround) can be defined as the circumstances or conditions that surround an organism or group of organisms. Humans (and a few other animal species!) have a physical and chemical environment, but also a social and cultural environment (Fig. 1.2).

Environmental science, then, is the scientific study of the environment as we have just defined it. At first glance, it may seem that environmental science is "the science of everything" … after all, what is really *not* part of the environment? But it is at the interface of these different aspects of the environment that the work of environmental scientists really shines. To put it simply, we can think of environmental science as answering questions (or solving problems) about the environment that are of concern to people.

Because of the complex interplay of the natural and human environment (Fig. 1.2), environmental scientists often work in interdisciplinary teams, and with a broader, more holistic perspective than scientists who are "just" physicists or ecologists or chemists. An engineer may develop a revolutionary industrial wastewater treatment technology, but before it is applied, the available resources for treatment must be determined, and cost of the new technology must be weighed against its benefits of better treatment. Conservation biologists may know of an ecosystem or species that merits protection, but conservation priorities need to be considered in the context of societal values, including fish and wildlife or forestry management, and treaty rights of aboriginal people.

Chapter 2 looks more closely at science as a way of knowing, environmental ethics, and other tools that help us analyze and understand the world around us. For the remainder of this chapter, we'll complete our overview with a short history of environmental science and a survey of some important current issues that face us.

A BRIEF HISTORY OF ENVIRONMENTAL SCIENCE

It may seem to you like environmental science was "invented" recently, as many people became more concerned about the environment and our effect on it. Taking a longer view, you might think that the counter-culture of the 1960s as experienced by your parents and grandparents motivated much of the awareness and activism you see today. But human awareness of the environment, and particularly its degradation in the face of all sorts of human stressors, goes back a very long time. We'll take a brief look at the history of our observations of the environment and environmental degradation, and the science that was spawned by these observations. Rather than taking a strictly historical approach, we will look at the increasing scale of environmental problems and what the changing scale has meant for research of, and solutions to, the problems.

Local Problems: In Your Back Yard

The late Tip O'Neill, a popular American politician in his day, used to say, "All politics is local." This is often true of environmental problems as well, where people don't seem to get really concerned until they can observe the problem and are affected by it in their day-to-day lives. Many of the earliest concerns about the environment were very local.

Management of human waste, water and air quality, as well as maintenance of sustainable agricultural, are typical of issues that have concerned people for thousands of years and still perplex us today. As noted by Noel Hynes of the University of Waterloo, even prehistoric, nomadic hunting humans fouled their local environments with middens (garbage dumps). The overgrazing that degraded the ancient Greek landscape saddened Plato

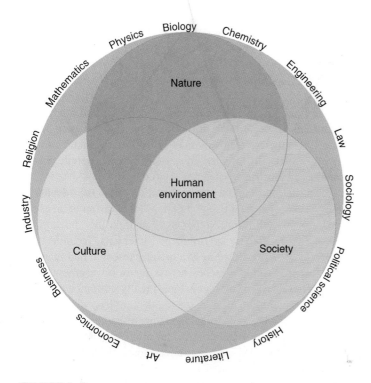

FIGURE 1.2 The intersections of the natural world with the social and cultural worlds encompass our environment. Many disciplines contribute to environmental science and help us understand how our worlds intertwine as well as our proper role in them.

FIGURE 1.3 Nearly 2,500 years ago, Plato lamented land degradation that denuded the hills of Greece. Have we learned from history's lessons?

FIGURE 1.4 Rachel Carson's book *Silent Spring* was a landmark in modern environmental history. She alerted readers to the dangers of indiscriminate pesticide use.

(Fig. 1.3). His student, Aristotle, remarked on the white colour (sewage fungus) and red threads (sludge worms) in the technologically remarkable sewer system of ancient Athens. England's King Edward I (reigned 1239–1307) imposed the death penalty upon anyone caught burning soft coal because it was believed to give off "poisonous odours." In the eighteenth and nineteenth centuries, human sewage and other domestic waste in cities and large towns were managed by burial or dumping it into the street gutter.

With the awareness of and subsequent research into local environmental problems came local solutions. Sewer systems facilitated the removal of human waste from where it was created to nearby waterways and away from the city. Taller smokestacks moved airborne pollutants "somewhere else." This practice continued well into the twentieth century: Inco's "super stack" was built in Sudbury, Ontario, in 1972, with the express goal of moving the sulphur dioxide generated in nickel production away from the city and thus improving the city's environment. Only later were scrubber technologies developed that somewhat reduced the emissions headed for Quebec and beyond, causing long-lasting acidification problems where they eventually deposited.

Regional and Continental Problems: Upscale Problems Need Upscale Solutions

Lake Erie in the early 1960s was dead. Or that's what the newspapers said, anyway. Every summer a massive algal bloom would form and then later die, causing all sorts of problems as it rotted at the bottom of the lake. Recognition of Lake Erie's death, or cultural eutrophication (Chapter 11) led to coordinated large-scale studies and action (legislative, water treatment) on the part of the national (Canada, U.S.) and provincial/state (Ontario, Michigan, Ohio, New York) governments. In Canada, some pioneering work by scientists in the Experimental Lakes Area of northwestern Ontario showed clearly that phosphorus control was the key to recovering lakes like Lake Erie that were degraded by having too many mineral nutrients and too much primary production (Chapter 3). By the late 1990s there were many indications that the lake was alive and well, albeit still affected by the millions of people whose personal and industrial waste ends up in the lake.

Regional and continental problems such as the "death" of Lake Erie, loss of coastal fish populations, or invasion of an exotic species are, by definition, bigger in scale than the "back yard" problems first tackled by environmental scientists. To understand and hope to solve them, we need a broader research and management effort that often stretches across many government jurisdictions. These sorts of problems didn't really get much attention until after the middle of the twentieth century, when people started to develop the tools for recognizing, researching, and sometimes solving regional and continental problems. Rachel Carson (Fig. 1.4) was one of the pioneer environmental scientists of this period. She started her career as a "pure" scientist and educator, but ended up as one of the key figures in the birth of the twentieth century environmental movement. Her concern was about the proliferation of pesticides in the environment and how they affected humans, birds, and other organisms.

Global Problems: First You Have to Agree There's a Problem

By the late 1980s, many scientists accepted the hypothesis that increased concentrations of so-called greenhouse gases (such as carbon dioxide (CO_2)), were causing climate change at a global scale. Technology had progressed such that it was certain that CO_2 was increasing, and most (but not all) modellers thought this was causing global warming by the greenhouse effect (Chapter 8). Recognition of a global scale environmental problem (climate change) and the presumed cause of the problem (greenhouse gases) suggested a strategy (reducing greenhouse gases) that required coordinated action at an international scale. The text of the United Nations Framework Convention on Climate Change (UNFCCC) was adopted at the United Nations Headquarters, New York, in May 1992. On December 11, 1997, the text of the Protocol to the UNFCCC was adopted in Kyoto, Japan, and thus became known as the Kyoto Protocol. By late 2002, almost 100 countries, including Canada, had signed the accord, but it will not come into force until countries responsible for at least 55 percent of total 1990 emissions have signed. The agreement commits Canada and other signatories to reducing greenhouse gas emissions to six percent below 1990 rates by 2012. In Canada, that would require a cut of 20 to 30 percent from current levels.

The history of the science and political response to the issue of climate change presents us with several lessons. First of all, environmental problems at a global scale have only been recognized very recently, when we have had the means to measure the hypothesized causes and effects involved (Fig. 1.5). For example, it is not enough to say with false certainty that pesticides are "everywhere" in the world … we must have the means to sample and measure them in remote ecosystems (e.g., polar bear tissue in the Arctic), and then determine what effect such contaminants have on the organisms and environment, and what needs to be done to reduce the cause of the effects we see on the environment.

Another lesson from the climate change issue is more political. Two people can debate or even argue about the merits of a particular hypothesis, and usually resolve their differences. When several countries are involved in an environmental issue, inevitably the scientific debate gets mixed in with other concerns such as what price industry will pay, and how will this translate into jobs lost. Environmental science at a global scale is much more complex both from the perspective of the science itself and in the application of the science to policy and action.

Environmental Science, Environmentalism, and Industrialism: Building Healthy Relationships

As we've seen, environmental science has evolved from the smallest scale of an individual recognizing the degradation of their living environment, to the point where we can actually detect change in both the environment and the organisms living in it (including humans) at a global scale. At each step of the way there has often

FIGURE 1.5 It is only in the last couple of decades that we have observed human effects on the environment at a global scale.

been three distinct perspectives. Many times, environmentalists are the first to "sound the alarm" about issues, and continue raising the consciousness of the public and vigorously encouraging politicians to change policy and legislation to solve or at least address the problem. Industrialists often respond to such alarms defensively, since they are supporters of the economic and industrial growth that often contributes to the perceived problem. Although we have stereotyped these two groups here (there are environmentalists who work hand in hand with industrialists creating environmentally "friendly" industry), more often than not they disagree.

What is the role of the environmental scientist in such a debate? We will see in Chapter 2 that approaching environmental issues objectively is easier said than done, but it is something that environmental scientists must try their best to do. Throughout this book, we will challenge you with issues and problems to try and respond with a critical, objective analysis. You can't approach a problem believing industry is inherently evil and environmentalists are saints (or *vice versa*!). We hope you will learn to put the pieces to a given issue together and make an informed judgement about it.

CURRENT CONDITIONS

As you probably already know, many environmental problems now face us. Before surveying them in the following section, we should pause for a moment to consider the extraordinary natural world that we inherited and that we hope to pass on to future generations in as good—perhaps even better—a condition than when we arrived.

FIGURE 1.6 We are fortunate to live in a beautiful, prolific, agreeable world. It will take knowledge, care, and hard work to keep it this way.

A Marvelous Planet

Imagine that you are an astronaut returning to Earth after a long trip to the moon or Mars. What a relief it would be to come back to this beautiful, bountiful planet (Fig. 1.6) after experiencing the hostile, desolate environment of outer space. Although there are dangers and difficulties here, we live in a remarkably hospitable world that is, as far as we know, unique in the universe. Compared to the conditions on other planets in our solar system, temperatures on the earth are mild and relatively constant. Plentiful supplies of clean air, fresh water, and fertile soil are regenerated endlessly by biological, geological, and chemical cycles (discussed in Chapters 3 and 5).

Perhaps the most amazing feature of our planet is the rich diversity of life that exists here. Millions of species populate the earth that are sustained by and help sustain a habitable environment. This vast multitude of life creates complex, interrelated communities where towering trees and huge animals live together with, and depend upon, tiny life-forms such as viruses, bacteria, and fungi. Together all these organisms make up delightfully diverse, self-sustaining communities, including dense, rainforests, vast sunny savannas, and richly colourful coral reefs. From time to time, we should pause to remember that, in spite of the challenges and complications of life on earth, we are incredibly lucky to be here. We should ask ourselves: what is our proper place in nature? What *ought* we do and what *can* we do to protect the irreplaceable habitat that produced and supports us? These are some of the central questions of environmental science.

Environmental Dilemmas

While there are many things to appreciate and celebrate about the world in which we live, many pressing environmental problems cry out for our attention. Human populations have grown at alarming rates in this century. More than 6 billion people now occupy the earth and we are adding about 85 million more each year. In the next decade, our numbers will increase by almost as many people as are now alive in India. Most of that growth will be in the poorer countries where resources and services are already strained by present populations.

FIGURE 1.7 Three-quarters of the world's poorest nations are in Africa. Millions of people lack adequate food, housing, medical care, clean water, and safety. The human suffering engendered by this poverty is tragic.

Some demographers believe that this unprecedented growth rate will slow in this century and that the population might eventually drop back below its present size. Others warn that the number of humans a century from now could be four or five times our present population if we don't act quickly to bring birth rates into balance with death rates. Whether there are sufficient resources to support 6 billion humans—let alone 25 billion—on a sustainable basis is one of the most important questions we face. How we might stabilize population and what level of resource consumption we and future generations can afford are equally difficult parts of this challenging equation.

Food shortages and famines already are too familiar in many places and may increase in frequency and severity if population growth, soil erosion, and nutrient depletion continue at the same rate in the future as they have in the past (Fig. 1.7). We are coming to realize, however, that food security often has more to do with poverty, democracy, and equitable distribution than it does with the amount of food available. Water deficits and contamination of existing water supplies threaten to be critical environmental issues in the future for agricultural production as well as for domestic and industrial uses. Many countries already have serious water shortages and more than 1 billion people lack access to clean water or adequate sanitation. Violent conflicts over control of natural resources may flare up in many places if we don't learn to live within nature's budget.

How we obtain and use energy is likely to play a crucial role in our environmental future. Fossil fuels (oil, coal, and natural gas) presently supply about 80 percent of the energy used in industrialized countries (Fig. 1.8). Supplies of these fuels are diminishing at an alarming rate and problems associated with their acquisition and use—air and water pollution, mining damage, shipping accidents, and political insecurity—may limit where and how we use remaining reserves. Cleaner renewable energy resources—solar power, wind, and biomass—together with conservation, may replace environmentally destructive energy sources if we invest in appropriate technology in the next few years.

As we burn fossil fuels, we release carbon dioxide and other heat-absorbing gases that cause global warming and may bring about sea-level rises and catastrophic climate changes. Acids formed in the air as a result of fossil fuel combustion already have caused extensive damage to building materials and sensitive ecosystems in many places (Fig. 1.9). Continued fossil fuel use without better pollution control measures could cause even more extensive damage. Chlorinated compounds, such as the chlorofluorocarbons used in refrigeration and air conditioning, also contribute to global warming, as well as damaging the stratospheric ozone that protects us from cancer-causing ultraviolet radiation in sunlight.

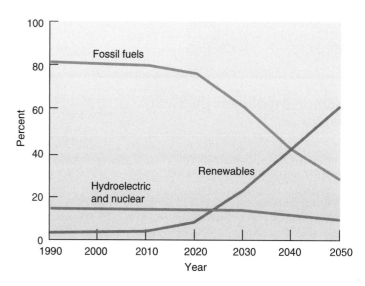

FIGURE 1.8 Fossil fuel supplies are dwindling at an alarming rate. They could be replaced by renewables such as solar power, wind, and biomass energy that could be more equitably distributed and would be less polluting than our current energy sources.

Sources: Data from World Bank estimates; D. Anderson and C.D. Bird, "Carbon Accumulations and Technical Progress—A Simulation Study of Costs," *Oxford Bulletin of Economics & Statistics.* 54 (1): 1–29.

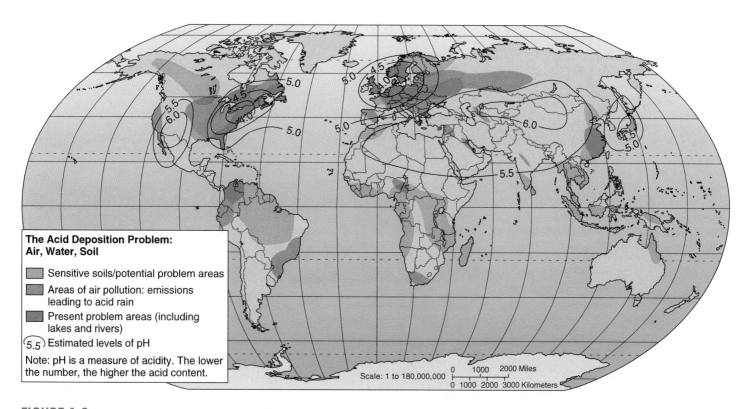

FIGURE 1.9 Sulphuric and nitric acids formed in the air by burning fossil fuels and other industrial activities threaten crops, water quality, building materials, and living organisms, especially in areas where soils have little buffering capacity.

Source: John Allen, *Student Atlas of Environmental Issues,* Dushkin/McGraw-Hill, 1997.

Cultural Whaling in the Pacific Northwest

On May 17, 1999, for the first time in more than 70 years, members of the Makah tribe from Washington's Olympic peninsula struck and killed a grey whale. Using a mixture of modern technology and ancient traditions, nine men paddling a handmade cedar canoe drew close enough to harpoon the 20-tonne whale, after which high-speed motorboats closed in to kill the wounded animal with a .50 calibre anti-tank rifle.

For months before the hunt, native people, animal rights groups, sea conservation societies, and members of the whale-watching industry had engaged in shouting matches, racial slurs, political maneuvering, media campaigns, and dangerous motorboat confrontations as each side tried to promote its own views. At issue is a complex mixture of international environmental policy, aboriginal culture and hunting rights, ethical values, commercial interests, and politics that is not easy to resolve. This case study presents a good opportunity for you to practise critical-thinking skills.

For at least 1,500 years the Makah people collected fish and shellfish and hunted for seals, whales, and sea otters along this section of fog-shrouded coast. The 1855 Treaty of Neah Bay recognized the right of the Makah people to hunt whales, the only such covenant in United States history. By the beginning of the twentieth century, however, commercial exploitation had driven nearly into extinction the whales, seals, otters, and fish on which the Makah once depended. Fewer than 2,000 Pacific grey whales survived out of a historic population of about 30,000. By 1920, most Makah had given up their traditional subsistence culture and turned to farming, logging, or factory jobs.

Since their protection under the Endangered Species Act and the International Whaling Ban, grey whales have made a remarkable and encouraging recovery. The population today is thought to be around 26,000 animals and is growing at about 2.5 percent per year. It may well have reached the carrying capacity of its habitat. In 1999 scientists estimated that about 800 whales died of starvation and disease during their 4,000-mile migration from Baja California to the Arctic. A sustainable harvest might actually be beneficial for the whale population. When the species was removed from the U.S. endangered species list in 1994, the Makah announced their intention to resume whaling under their treaty rights. They consider this an important step in recovering their traditional culture and combating the poverty, drugs, alcoholism, violence, and despair that have afflicted their community. Reviving the rituals, discipline, and pride of whaling, they believe, could lead to a cultural renaissance for their people.

To anti-hunting groups, killing magnificent, intelligent, social animals like whales is tantamount to murder. In their view, calling for a sustainable harvest of whales is equivalent to establishing a sustainable level of genocide or torture. This isn't subsistence hunting, they claim, because the Makah haven't depended on whale meat for many years and don't need it now. To the Makah and their supporters from whaling nations such as Japan and Norway, on the other hand, it is hypocritical to condemn hunting marine mammals when most Americans happily eat the millions of cattle, pigs, sheep, and other animals slaughtered every year in commercial packing plants.

How would you approach this complex question? What information would you need to determine whether native claims for cultural traditions should take precedence over ethical rights of wild animals? Does the fact that native people have suffered economically and socially give them special standing? Should subsistence hunting be limited to traditional tools and people who have a demonstrated nutritional need? What about whale hunting by Europeans and Americans that went on for hundreds of years? What information sources would you regard as reliable and authoritative in this situation? How would you decide whom or what to believe?

As you study environmental science, you will find that many of the dilemmas we face, like this question of hunting whales by native people, require a combination of scientific data and contextual sensitivity and empathy. Because humans are involved, questions of history, culture, ethics, and politics also come into play. How would you resolve this clash in world views? Can you suggest a compromise that might defuse this tense situation?

Members of a Makah whaling crew set out in their handmade cedar canoe on a practice run. Whale hunting is central in the Makah culture, but is opposed by animal rights groups.

Destruction of tropical forests, coral reefs, wetlands, and other biologically rich landscapes is causing an alarming loss of species and a reduction of biological variety and abundance that could severely limit our future options. Many rare and endangered species are threatened directly or indirectly by human activities (Fig. 1.10). In addition to practical values, aesthetic and ethical considerations suggest that we should protect these species and the habitat necessary for their survival. On the other hand, sometimes cultural or economic interests are directly at odds with biodiversity conservation (What Do You Think? above).

FIGURE 1.10 This emperor tamarin is a marmoset from South America. Although we hear a great deal about destruction of tropical *rainforests,* the *dry* tropical forests that once lined the Atlantic coast of South America and the Pacific coast of Central America have been destroyed to a much greater extent. More than 99 percent of the once-extensive dry tropical forests of Latin America have been destroyed, threatening a host of interesting and important species with extinction.

Toxic air and water pollutants, along with mountains of solid and hazardous wastes, are becoming overwhelming problems in industrialized countries. We produce hundreds of millions of tonnes of these dangerous materials annually, and much of it is disposed of in dangerous and irresponsible ways. No one wants this noxious stuff dumped in their own back yard, but too often the solution is to export it to someone else's. We may come to a political impasse where our failure to decide where to put our wastes or how to dispose of them safely will close down industries and result in wastes being spread everywhere. The health effects of pollution, toxic wastes, stress, and the other environmental ills of modern society have become a greater threat than infectious diseases for many of us in industrialized countries.

These and other similarly serious problems illustrate the importance of environmental science for everyone. What we are doing to our world, and what that may mean for our future and that of our children is of paramount concern. We trust that's why you're reading this book.

Signs of Hope

The dismal litany of problems we have just reviewed seems pretty overwhelming, doesn't it? But is there hope that we may find solutions to these problems? We think so. As you will see in subsequent chapters, progress has been made in many areas in controlling air and water pollution and reducing wasteful resource uses. Many cities are cleaner and less polluted than they were a generation or so ago. Population has stabilized in most industrialized countries and even some very poor countries where social security and democracy have been established. Over the last 20 years, the average number of children born per woman worldwide has decreased from 6.1 to 3.4. This is still above the zero population growth rate of 2.1 children per couple, but it is an encouraging improvement. If this rate of progress continues in the next 20 years as it has in the past 20 the world population could stabilize early in this century.

The incidence of life-threatening infectious diseases has been reduced sharply in most countries during the nineteenth century, while the average life-expectancy has nearly doubled. Many new resources have been discovered and more efficient ways of using existing supplies have been invented that allow us to enjoy luxuries and conveniences that would have seemed miraculous only a few generations ago. Although modern life has many stresses and strains, few people would willingly return to conditions that existed 10,000, 1,000, or even 100 years ago. Would you?

Still, we can do much more, both individually and collectively, to protect and restore our environment. Being aware of the problems we face is the first step toward finding their solutions. Increased media coverage has brought environmental issues to public attention. In a poll by the Environics Research Group just prior to the 2004 federal election in Canada, the "environment" ranked in the Top 10, similarly to "law and order," "relations with the U.S.," and "deficit and debt" as an important issue to the public, and well behind "health care" and "government accountability and scandals." This growing understanding and concern are themselves hopeful signs. Young people today may be in a unique position to address these issues because, for the first time in history, we now have resources, motivation, and knowledge to do something about our environmental problems. Unfortunately, if we don't act now, we may not have another chance to do so.

NORTH/SOUTH: A DIVIDED WORLD

We live in a world of haves and have-nots; a few of us live in increasing luxury while many others lack the basic necessities for a decent, healthy, productive life. The World Bank estimates that more than 1.3 billion people—about one-fifth of the world—live in **acute poverty** with an income of less than $1 (U.S.) per day. These poorest of the poor generally lack access to an adequate diet, decent housing, basic sanitation, clean water, education, medical care, and other essentials for a humane existence. Seventy percent of those people are women and children. In fact, four out of five people in the world live in what would be considered poverty in Canada or the United States. The plight of these poor people is not just a humanitarian concern. Policymakers are becoming aware that eliminating poverty and protecting our common environment are inextricably interlinked (Fig. 1.11).

The world's poorest people have become both the victims and the agents of environmental degradation. The poorest people are too often forced to meet short-term survival needs at the cost of long-term sustainability. Desperate for croplands to feed themselves and their families, many move into virgin forests or cultivate steep, erosion-prone hillsides where soil nutrients are exhausted after only a few years. Others migrate to the grimy, crowded slums and ramshackle shantytowns that now surround most major cities in the developing world. With no way to dispose of wastes, the residents often foul their environment further and contaminate the air they breathe and the water on which they depend for washing and drinking.

The cycle of poverty, illness, and limited opportunities can become a self-sustaining process that passes from one generation to another. People who are malnourished and ill can't work

FIGURE 1.11 While many of us live in luxury, more than a billion people lack access to food, housing, clean water, sanitation, education, medical care, and other essentials for a healthy, productive life. Often the poorest people are both the victims and agents of environmental degradation as they struggle to survive. Helping them meet their needs is not only humane, it is essential to protect our mutual environment.

productively to obtain food, shelter, or medicine for themselves or their children, who also are malnourished and ill. About 250 million children—mostly in Asia and Africa and some as young as 4 years old—are forced to work under appalling conditions weaving carpets, making ceramics and jewellery, or in the sex trade. Growing up in these conditions leads to mental and developmental deficits that condemn these children to perpetuate this cycle.

Faced with immediate survival needs and few options, these unfortunate people often have no choice but to overharvest resources; in doing so, however, they diminish not only their own options, but also those of future generations. And in an increasingly interconnected world, the environments and resource bases damaged by poverty and ignorance are directly linked to those on which we depend. It is in our own self-interest to help everyone find better ways to live.

Rich and Poor Countries

Where do the rich and poor live? About one-fifth of the world's population lives in the 20 richest countries where the average per capita income is above $25,000 (U.S.) per year (Fig. 1.12). Almost every country, however, even the richest, such as Canada and the United States, has poor people. No doubt everyone reading this book knows about homeless people or other individuals who lack resources for a safe, productive life. The Canadian Association of Food Banks reports that more than 800,000 Canadians turn to food banks each month for more than 2.3 million meals.

The other four-fifths of the world live in middle- or low-income countries where nearly everyone is poor by our standards. More than 3.5 billion people live in the poorest nations where the average per capita income is below $750 (U.S.) per year. Nearly two-thirds of those people live in China and India, the largest countries in the world. Among the 41 other nations in this category, 33 are in sub-Saharan Africa. All the other lowest-income nations, except Haiti, are in Asia. Although poverty levels in countries such as China and Indonesia have fallen in recent years, most countries in sub-Saharan Africa and much of Latin America have made little progress. The destabilizing and impoverishing effects of earlier colonialism continue to play important roles in the ongoing problems of these unfortunate countries. Meanwhile, the relative gap between rich and poor has increased dramatically (Fig. 1.13).

The 10 poorest countries in the world, in 2002, were (in ascending order from very poorest): Sierra Leone, Tanzania, Malawi, Democratic Republic of Congo, Democratic Republic of Burundi, Congo, Yemen, Zambia, Mali, and Nigeria. According to the Population Reference Bureau 2003 data, each of these countries has annual per capita Gross National Product (GNP) of less than $220 (U.S.) per year. They also have low levels of food security, social welfare, and quality of life as indicated by Table 1.1.

By contrast, each of the 10 richest countries in the world—Luxemborg, United States, Switzerland, Norway, Iceland, Denmark, French Polynesia, Netherlands, Ireland, and Canada (in

PART ONE   Fundamental Principles of Ecology and Environmental Science

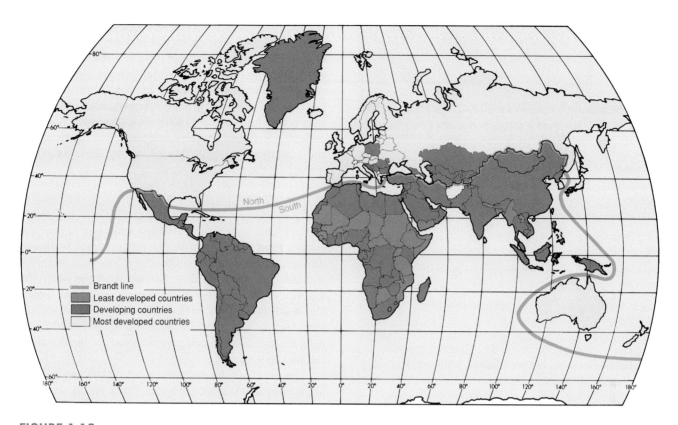

FIGURE 1.12 The world of underdevelopment. The Brandt line, denoted by the Independent Commission on International Development Issues, separates the richer, industrialized nations of Europe, North America, Japan, Australia, and New Zealand from their southern neighbours in the "developing" world. The 42 "least developed" countries, according to the United Nations Development Program, are mostly in Africa and Asia. Note that since this map was drawn, conditions have worsened in parts of the former USSR and Africa but have often improved elsewhere.

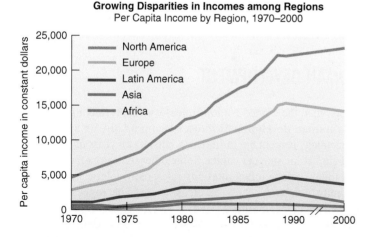

FIGURE 1.13 Although the percentage of the world's population living in poverty has decreased slightly over the past 30 years, the relative gap between rich and poor nations has increased sharply.

Source: United Nations (U.N.), *Critical Trends: Global Change and Sustainable Development* (U.N., New York, 1997), p. 58.

descending order from richest)—has an annual per capita GNP more than one hundred times that of the poorest countries. As you can see in Table 1.1, other conditions in the rich countries reflect this wide disparity in wealth.

The gulf between rich and poor is even greater at the individual level. The richest 200 people in the world have a combined wealth of $1 trillion. This is more than the total wealth of the poorest half—some 3 billion people—of the world's population.

A Fair Share of Resources?

The affluent lifestyle that many of us in the richer countries enjoy consumes an inordinate share of the world's natural resources and produces a shockingly high proportion of pollutants and wastes. The United States, for instance, is the richest country in the world. With less than 5 percent of the total population, it consumes about one-quarter of most commercially traded commodities and produces a quarter to half of most industrial wastes.

To get an average American through the day takes about 450 kg of raw materials, including 18 kg of fossil fuels, 13 kg of other minerals, 12 kg of farm products, 10 kg of wood and

TABLE 1.1 Average Indicators of Quality of Life for the 10 Richest and Poorest Countries[1]

INDICATOR	POOR COUNTRIES	RICH COUNTRIES
GNP/capita	$175	$35,000
Life expectancy	46 years	77 years
Infant mortality[2]	102	6
Child deaths[3]	194	7.5
Safe drinking water	42%	NA[4]
Female literacy	38%	97%
Birth rate[5]	42	12

[1] Averaged as a group
[2] per 1,000 live births
[3] per 1,000 children before age 5
[4] not available, but close to 100 percent
[5] per 1,000 people

Source: Popluation Reference Bureau, 2000 World Data Sheet.

paper, and 450 litres of water. Every year we throw away some 160 million tonnes of garbage, including 50 million tonnes of paper, 67 billion cans and bottles, 25 billion styrofoam cups, 18 billion disposable diapers, and 2 billion disposable razors.

This profligate resource consumption and waste disposal strains the life-support systems of the earth on which we all depend. If everyone in the world tried to live at consumption levels approaching ours, the results would be disastrous. Unless we find ways to curb our desires and produce the things we truly need in less destructive ways, the sustainability of human life on our planet is questionable.

North/South Division

Notice in the lists of rich and poor countries discussed earlier that the wealthiest countries tend to be in the north, while the poorest countries tend to be located closer to the equator. It is common to speak of a **North/South division** of wealth and power, even though many poorer nations such as India and China are in the Northern Hemisphere and some relatively rich nations like Australia and New Zealand are in the Southern. Another way of describing the rich nations is that they tend to be highly industrialized and generally are members of the Organization for Economic Cooperation and Development (OECD), made up of Western Europe, North America, Japan, Korea, Turkey, Australia, and New Zealand. This group makes up about 20 percent of the world's population but consumes more than half of most resources. The other, poorer four-fifths of humanity are sometimes called the majority world because they make up the majority of the population.

Since poverty and social welfare are closely linked with environmental degradation, there will be many instances in this book in which we will want to distinguish between conditions in groups of countries. In fact, income, social services, human development, and other indicators of quality of life don't always fit neatly in just two categories. Nations actually spread across a spectrum—or perhaps a three-dimensional matrix—of these indicators, making it difficult to establish inclusive categories. The following categories and corresponding acronyms are among those we will refer to throughout the rest of this book.

Political Economies

Countries are sometimes classified according to their economic system. **First World** describes the industrialized, market-oriented, democracies of Western Europe, North America, Japan, Australia, New Zealand, and their allies. **Second World** originally described the centrally planned, socialist countries, such as the former Soviet Union and its Eastern European allies. Several Asian socialist countries, such as China, Mongolia, North Korea, and Vietnam, also once belonged in this category but most are rapidly changing to market economies.

In the 1960s, nonaligned, ex-colonial nations in Latin America, Africa, and Asia were labelled **Third World** countries. Economist Alfred Sauvy coined the term in an article in the French magazine *The Observer* of August 14, 1952. It became popular during the Cold War of the 1950s to 1980s, when many poorer nations adopted the category to describe themselves as neither being aligned with NATO nor with the USSR, but instead composing a non-aligned "Third World."

The amazing political and economic upheavals of the past decade have thrown all of these categories into a turmoil. Few countries any more are either purely socialist or capitalist. Nearly every government plans centrally and intervenes in its economy to some extent, and nearly every nation has at least some market-orientation. These categories have decreasing significance as old political alliances break down, forcing us to look for other ways to describe nations and peoples.

HUMAN DEVELOPMENT

Every year, the United Nations releases a report ranking countries by a **human development index** based primarily on average life expectancy, percentage of literate adults, mean years of schooling, and annual income per capita. Some other indicators factored into this index include infant mortality rates, daily calorie supply, child malnutrition, and access to clean water. The highest possible human development index (HDI) is 1.0; the lowest is 0.0. In 2000, Canada was highest in the world with an HDI of 0.96. Although it slipped to 8th in 2003 with a value of 0.94, Canada, along with other industrialized nations, is clearly a country with exceptionally good living conditions. As you might expect, there is a close correlation between wealth and human development. In 2000, all the top 20 nations (all above 0.92) were in North America or

Western Europe, except Japan and Singapore. The lowest HDI was Sierra Leone at 0.19. In fact, all of the 20 lowest rankings (all below 0.35) were in Africa, except Haiti and Bhutan.

Developmental Discrepancies

Of course aggregate numbers such as these hide many important issues. One of these is gender inequities. Men generally fare better than women on almost every socioeconomic indicator. For the few countries that keep such data, Japan had the lowest female-to-male wage ratio (51 percent), while Sweden had the highest (90 percent). Similarly, the female-to-male ratio in nonagricultural work varies from a low of 22 percent in Bahrain to 89 percent in Finland. If the HDI is weighed for these gender discrepancies, egalitarian Scandinavian countries like Sweden, Finland, and Norway move into top place.

Geographical origin (often referred to as race), is another variable that determines socioeconomic status in many countries. If white South Africa were a separate country, it would rank 24th in the world in human development (just after Spain). Black South Africa, on the other hand, would rank 123rd in the world (just above the Congo). In some countries, regional or ethnic differences create disparities. In Nigeria, the state of Bendel, with a HDI of 0.66 is equivalent to Cuba, while the poor state of Borno, with a HDI of 0.156 is lower than any country in the world. Similarly, there are significant differences in HDI between southern provinces in Canada such as Ontario and Quebec, and our largely aboriginal, northern territories.

The United Nations states that these vast discrepancies and the grinding poverty experienced by the poorest of the poor are the greatest threats to political stability, social cohesion, and environmental health on our planet. Some suggested strategies for poverty reduction and social justice include:

- basic social services, especially basic education and primary health care;
- agrarian reform for more equitable land distribution;
- credit to tide those without resources through tough times;
- employment to ensure a secure livelihood for all;
- civil rights that enable everyone to participate in planning and management decisions;
- a social safety net to catch those whom markets exclude;
- economic growth that benefits the poor; and
- sustainable resource use that reduces material-intensive and energy-intensive lifestyles and turns, instead, to renewables.

Good News and Bad News

Over the past 50 years, human ingenuity and enterprise have brought about a breathtaking pace of technological innovations and scientific breakthroughs. The world's gross domestic product increased more than tenfold during that period, from $2 trillion to $22 trillion per year. While not all that increased wealth was

applied to human development, there has been significant progress in increasing general standard of living nearly everywhere. In 1960, for instance, nearly three-quarters of the world's population lived in abysmal conditions (HDI below 0.5). Now, less than one-third are still at this low level of development.

Since World War II, average real income in developing countries has doubled; malnutrition declined by almost one-third; child death rates have been reduced by two-thirds; average life expectancy increased by 30 percent; and the percentage of rural families with access to safe drinking water has risen from less than 10 percent to almost 75 percent. Overall, poverty rates have decreased more in the last 50 years than in the previous 500. Nonetheless, while general welfare has increased, so has the gap between rich and poor worldwide. In 1960, the income ratio between the richest 20 percent of the world and the poorest 20 percent was 30 to 1. In 2000, this ratio was 100 to 1. Because perceptions of poverty are relative, people may feel worse off compared to their rich neighbours than development indices suggest they are.

Sustainable Development

By now, it is clear that security and living standards for the world's poorest people are inextricably linked to environmental protection. One of the most important questions in environmental science is how we can continue improvements in human welfare within the limits of the earth's natural resources. A possible solution to this dilemma is **sustainable development,** a term popularized by *Our Common Future,* the 1987 report of the World Commission on Environment and Development, chaired by Norwegian Prime Minister Gro Harlem Brundtland (and consequently called the Brundtland Commission). In the words of this report, sustainable development means "meeting the needs of the present without compromising the ability of future generations to meet their own needs." Of course, meeting the needs of the present is a goal that is really a moving target, since as population increases, more resources will be needed to just maintain humans in their present state.

Another way of saying this is that development means improving people's lives. Sustainable development, then, means progress in human well-being that can be extended or prolonged over many generations rather than just a few years. To be truly enduring, the benefits of sustainable development must be available to all humans rather than to just the members of a privileged group.

Many industry people see economic growth as the only path to a more sustainable society with a high quality of life for all. But economic growth is not sufficient in itself to meet all essential needs. As the Brundtland Commission pointed out, political stability, democracy, and equitable economic distribution are needed to ensure that the poor will get a fair share of the benefits of greater wealth in a society.

Can Development Be Truly Sustainable?

Many ecologists regard "sustainable" growth of any sort as impossible in the long run because of the limits imposed by nonrenewable resources and the capacity of the biosphere to absorb our wastes. Using ever-increasing amounts of goods and services to make human life more comfortable, pleasant, or agreeable must inevitably interfere with the survival of other species and, eventually, of humans themselves in a world of fixed resources. But, supporters of sustainable development assure us, both technology and social organization can be managed in ways that meet essential needs and provide long-term—but not infinite—growth within natural limits, if we use ecological knowledge in our planning.

While economic growth makes possible a more comfortable lifestyle, it doesn't automatically result in a cleaner environment. As Figure 1.14 shows, people will purchase clean water and sanitation if they can afford to do so. For low-income people, however, more money tends to result in higher air pollution because they can afford to burn more fuel for transportation and heating. Given enough money, people will be able to afford both convenience *and* clean air. Some environmental problems, such as waste generation and carbon dioxide emissions, continue to rise sharply with increasing wealth because their effects are diffuse and delayed. If we are able to sustain economic growth, we will need to develop personal restraint or social institutions to deal with these problems.

Some projects intended to foster development have been environmental, economic, and social disasters. Large-scale hydro-power projects, like that in the James Bay region of Quebec or the Brazilian Amazon that were intended to generate valuable electrical power, also displaced indigenous people, destroyed wildlife, and poisoned local ecosystems with acids from decaying vegetation and heavy metals leached out of flooded soils. Similarly, introduction of "miracle" crop varieties in Asia and huge grazing projects in Africa financed by international lending agencies crowded out wildlife, diminished the diversity of traditional crops, and destroyed markets for small-scale farmers.

Other development projects, however, work more closely with both nature and local social systems. Socially conscious businesses and environmental, nongovernmental organizations sponsor ventures that allow people in developing countries to grow or make high-value products—often using traditional techniques and designs—that can be sold on world markets for good prices (Fig. 1.15). Pueblo to People, for example, is a nonprofit organization that buys textiles and crafts directly from producers in Latin America. It sells goods in America, with the profits going to community development projects in Guatemala, El Salvador, and Peru. It also informs customers in wealthy countries about the conditions in the Third World.

As the economist John Stuart Mill wrote in 1857, "It is scarcely necessary to remark that a stationary condition of capital and population implies no stationary state of human improvement. There would be just as much scope as ever for all kinds of mental culture and moral and social progress; as much room for improving the art of living and much more likelihood of its being improved when minds cease to be engrossed by the art of getting on." Somehow, in our rush to exploit nature and consume resources, we have forgotten this sage advice.

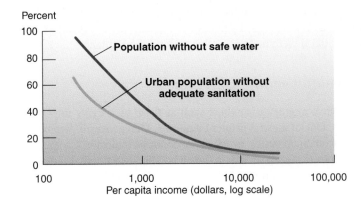

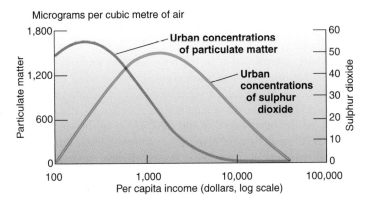

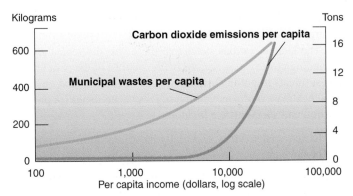

FIGURE 1.14 Environmental indicators show different patterns as average national income increases. When people have more money they invariably will purchase clean water and better sanitation. Rising income may temporarily produce increased urban air pollution (for example, particulates and sulphur dioxide) as people burn more fuel; eventually, however, people can afford both clean air and the benefits of technology. Some environmental problems such as waste generation and carbon dioxide emissions rise sharply with increasing wealth because of increased demands for goods and services without recognition of adverse environmental consequences.

Source: Data from World Bank, *World Development Report,* 1996.

Our Ecological Footprint

Many environmental scientists feel that it is not sufficient (and may in fact be misleading) to base a country's resource use on population numbers alone. If we think about the resources that

PART ONE   Fundamental Principles of Ecology and Environmental Science

FIGURE 1.15 A Mayan woman from Guatemala weaves on a backstrap loom. A member of a women's weaving cooperative, she sells her work to nonprofit organizations in the United States at much higher prices than she would get at the local market.

Want More?

Our Ecological Footprint; Reducing Human Impact on the Earth, by Mathis Wackernagel and William Rees. Illustrated by Phil Testemale. New Society Publishers, Gabriola Island, BC. Canada and Philadelphia, PA USA. 6" × 9". 176 pages Illustrated. Charts. Resources. Canada Paperback ISBN 1-55092-251-3. Hardcover ISBN 1-55092-250-3. USA Paperback ISBN 0-86571-312-X.

FIGURE 1.16 We only have one globe, but if everyone on Earth used it the way Canadians do, it would take three or four to meet our resource demands.

individuals consume, and the waste they generate, we realize that not everyone is equal when it comes to the environment. Imagine if everyone on the planet lived like North Americans do today! Two Canadian environmental scientists, Mathis Wackernagel and William Rees, have quantified the different aspects of human life that use resources or generate waste, and translated this into an individual's *ecological footprint*. The average Canadian requires about 3.5 ha of land to sustain themselves. If everyone on earth lived like the average North American, it would require at least three earths to provide all the material and energy she or he currently uses.

Preliminary estimates show that the ecological footprint of today's consumption in food, forestry products, and fossil fuels alone might already exceed global carrying capacity by roughly 30 percent. About 3/4 of the current consumption goes to the 1.1 billion people who live in affluence, while 1/4 of the consumption remains for the other 4.6 billion people. This demonstrates the ethical implications of the sustainability dilemma and questions economic expansionism as a remedy for poverty.

The beauty of the ecological footprint concept is that it translates individual actions into the environmental consequences of those actions at the scale of populations (Fig. 1.16). You can even calculate your own ecological footprint by answering a few simple questions (see Web Exercises, p. 19).

ENVIRONMENTAL PERSPECTIVES

In *A Christmas Carol* by Charles Dickens, Scrooge questions the Ghost of Christmas Future after seeing the disparity between the rich and poor in London. "Answer me one question," says Scrooge, "are these the shadows of the things that will be or are they shadows of the things that may be only?" We could ask something similar today. Are the problems and dilemmas outlined in this book warnings of what *will* be, or only what *may* be if we fail to take heed and adjust our course of action?

What will be our environmental future and what can we do to shape it? These are, perhaps, the most important questions in environmental science. There are many interpretations of what the data show and how we should interpret them. Think about the following worldviews and tactical positions as you read the subsequent chapters in this book.

Pessimism and Outrage

You will find much in this book that justifies pessimism. A number of very serious environmental problems threaten us. Many environmental scientists see our world as one of scarcity and

IN DEPTH. . . . . .

Getting to Know Our Neighbours

Imagine the world as a village of 1,000 people: who are our neighbours?

- Ethnically, this global village has 333 East Asians, 274 South Asians, 132 Africans, 120 Europeans, 86 Latin Americans, 50 from Canada and the United States, and 5 from Oceania (Australia, New Zealand, and the Pacific Islands).

- About half the residents speak one or more of the six major world languages (Mandarin, English, Hindi/Urdu, Spanish, Russian, and Arabic—in that order), while the rest speak some of the other 6,000 known languages.

- Children under age 15 make up 31 percent of the village, while those over age 65 make up 7 percent. The number of senior citizens is increasing rapidly while the number of children is falling slowly. At some point in the future, there may be more retirees than workers in the village, bankrupting the social security system.

- Every year, 22 children are born in the village and 9 people die. One-third of yearly deaths are children under age 5, mostly from infectious diseases aggravated by malnutrition. One person dies from heart disease, one from cancer, two from chronic or acute lung diseases, and two from violence, accidents, and various infectious diseases.

- The average life expectancy is 68 years for women and 64 for men, but there is considerable variation in longevity. In the richest families, women live to an average age of 80 and men to an average of 78 years, while in the poorest families the life expectancy is less than 48 years for both men and women.

- The average annual income for the village is $4,890 (U.S.), but this average also obscures great discrepancies. Nearly 60 percent of the village lives in households where the annual income is less than $650 per person, while the richest 100 citizens (mostly Canadians, Americans, Europeans, and Japanese) enjoy annual incomes over $35,000 each.

- As a consequence of this income gap, the richest 200 villagers own or control

Nine out of ten people in the world now have access to television. There may be only one TV per village, as in this scene from Nigeria, but it opens a window to the wider world. Most of what everyone watches is made in the United States and glorifies consumerism and commidification.

80 percent of the resources and consume 80 percent of all products sold in the marketplace, while the other 800 people make do with 20 percent of the goods and merchandise available for sale.

- Half of the 620 adults in the village are illiterate. Lacking an education, most of these people work as day labourers or seasonal farm workers. Among the poorer families, girls are half as likely as boys to attend school.

- Women and girls make up slightly more than half the village population. They do two-thirds of all manual labour, receive one-tenth of the wages, and own less than one-hundredth of the property. Seventy percent of the poorest members of the population are women and children.

- About 400 villagers suffer from ill health at any given time. Much of that illness is related to lack of clean water, sanitation, and food. Some 250 of our neighbours don't have clean water to drink or adequate sanitation. About 150 are chronically hungry, lacking the calories and nutrients needed for normal growth and development in children or a healthy, productive life for adults.

- Only 452 people actually live in the village itself; the other 548 live in the surrounding countryside. Sixty percent of the rural families are landless or have too little land to subsist. They make up a majority of those who lack clean water, sanitation, food, housing, and health care.

- Generally the worst pollution problems are borne by those who live in the poorest parts of the village, where air and water pollution, noise, congestion, and toxic wastes are most common. Those who live in the better parts of town actually enjoy a cleaner, safer environment than their parents did only a few decades ago.

- In the past, people didn't often travel from one part of the village to another. Today, travel is easier and cheaper than ever before. Furthermore, 90 percent of the population has access to television so that the lower class is exposed to both consumer pressures and news about how the upper class lives.

- How long do you suppose these great discrepancies in opportunity and quality of life will persist? What might we do to reduce them?

PART ONE   Fundamental Principles of Ecology and Environmental Science

competition in which too many people fight for too few resources. This viewpoint is often called **neo-Malthusian** after Thomas Malthus, who predicted a dismal cycle of overpopulation, misery, vice, and starvation as a result of human fallibility. We will discuss Malthus and his predictions in Chapter 4 as part of our discussion of population growth and resource economics.

This grim view of human nature and resources persuades many environmentalists to issue dire warnings of impending doom unless we make immediate, drastic changes in our way of life. While it is easy to feel moral outrage about the excesses and abuses that have occurred, shock, shame, and fear are generally poor motivation for positive action. How helpful is it in getting you to abandon your bad habits to have someone harp about how guilty and vile you are? Pioneering ecologist Aldo Leopold said: "I have no hope for a conservation based on fear." Furthermore, when the predicted disasters fail to occur, the public may assume that nothing was wrong and that there is no reason to change anything. In 1969, prominent biologist and environmental activist Paul Ehrlich predicted that England would not exist by the year 2000. How many times can we cry wolf before people stop paying attention?

Hopeful Optimism

Science and technology have provided many benefits to humanity; they also have caused many difficulties, as you will see in the course of this book. **Technological optimists** believe that human ingenuity and enterprise will find cures for all our problems. They see the world as one of abundance and opportunity. Geographer Martin Lewis calls this "**Promethean environmentalism**" after the Greek Titan who stole fire from the gods and gave it to humans. He fails to mention, however, that Prometheus suffered eternal torment for this sin of pride. Critics describe this optimistic outlook as "**cornucopian fallacy**" (after the mythical horn of plenty) and see it as either wishful thinking or deliberate denial. Optimists argue, however, that they merely expect historic patterns of progress to continue in the future as they have in the past.

While emphasizing only good news is popular and comfortable, it also can mislead the public. Blind faith in technology can be merely an excuse for business as usual. By telling us that everything is just fine, optimists may lull us into complacency and apathy. What do you think is the most appropriate and useful balance between Malthusian and cornucopian worldviews? How can we use environmental information to bring about positive change?

Pragmatic Realism

Perhaps the best compromise between these two extreme positions—dismal pessimism and unfettered optimism—is a realistic pragmatism. By looking honestly at both our opportunities and our mistakes, and by applying the principles of environmental science, we can strive to understand our world and our place in it. Author Wes Jackson points out that our alienation from nature has diminished both our knowledge of the world around us and our connection to it. Our challenge, he says, is to learn once again how to become native to the place where we live. We should, he urges reestablish a relationship with the land and a sense of stewardship for it. By studying environmental science, we all can learn how to make more informed decisions about issues and thus become better environmental citizens. As Aldo Leopold said, "All history consists of successive excursions from a single starting point to which man returns again and again to organize yet another search for a durable scale of values."

SUMMARY

Humans always have inhabited two worlds: one of nature and another of human society and technology. Environmental science is the systematic study of the intersection of these worlds. An interdisciplinary field, environmental science draws from many areas of inquiry to help us understand the worlds in which we live and our proper role in them.

The most amazing features of our planet may be the self-sustaining ecological systems that make life possible and the rich diversity of life that is part of, and dependent upon, those ecological processes. In spite of the many problems that beset us, the earth is wonderfully bountiful and beautiful.

Concerns about pollution and land degradation date back at least 2,500 years. Clearly, we have pragmatic interests in conserving resources and preserving a habitable environment. There also are ethical reasons to believe that nature has a right to continue to exist for its own sake. Unprecedented population growth, food shortages, scarce energy supplies, air and water pollution, and destruction of habitats and biological resources are all serious threats to our environment and our way of life.

As our ability to measure the causes and effects of environmental degradation at a large scale has improved, we realize that these problems encompass our whole planet and require global cooperation to find solutions. Still, there is good news. Pollution has been reduced and population growth has slowed in many places. Perhaps we can extend these advances to other areas as well.

The 20 percent of us in the world's richest countries consume an inordinate amount of resources and produce a shocking amount of waste and pollution. More than 1.3 billion people live in acute poverty and lack access to an adequate diet, decent housing, basic sanitation, clean water, education, medical care, and other essentials for a humane existence.

Concern for the poor is more than a humanitarian issue. Faced with immediate survival needs, these desperate people often have no choice but to overharvest resources and reduce long-term sustainability for themselves and their children. Since we share the same environment it is in our own self-interest to help them find better options than they currently have.

There are several ways of describing the economic and developmental status of different countries. The First World is generally industrialized and more highly developed. Many Third World countries have made encouraging progress in improving the quality of life for residents, but much remains to be done.

There are valid reasons to be pessimistic about our environmental conditions, but we must be careful that dire predictions don't overwhelm us and become self-fulfilling prophecies. Many people find an optimistic outlook a better motivation than fear, but blind faith in technological progress can be simply an excuse for business as usual. Although we still have far to go in protecting our environment, some heartening progress already has been made toward building a just and sustainable world.

QUESTIONS FOR REVIEW

1. Define environment and environmental science.

2. List six environmental dilemmas that we now face and describe how each concerns us.

3. Describe the differences between the North/South or rich/poor or more developed/less developed nations. What do we mean by First, Second, and Third World?

4. Compare some indicators of quality of life between the richest and poorest nations.

5. Why should we be concerned about the plight of the poor? How do they affect us?

6. Give some reasons for pessimism and optimism about our environmental future and summarize how you feel personally about the major environmental problems that we face.

4. What would it take for human development to be really sustainable? What does sustainable mean to you?

5. Are there enough resources in the world for 8 or 10 billion people to live decent, secure, happy, fulfilling lives? What do those terms mean to you? Try to imagine what they mean to others in our global village.

6. What responsibilities do we have to future generations? What have they done for us? Why not use whatever resources we want right now?

7. Do you believe that outrage, optimism, or pragmatism is the most appropriate attitude about current environmental conditions? Why?

QUESTIONS FOR CRITICAL THINKING

1. How could we determine whether the deformity of frogs is linked to stressors controlled by humans such as contaminants?

2. Some people argue that we can't afford to be generous, tolerant, fair, or patient. There isn't enough to go around as it is, they say. What questions would you ask such a person?

3. Others claim that we live in a world of bounty. They believe there would be plenty for all if we just shared equitably. What questions would you ask such a person?

KEY TERMS

acute poverty   9
cornucopian fallacy   17
environment   3
environmentalism   17
environmental science   3
First World   12
human development
   index (HDI)   12
neo-Malthusian   17
North/South division   12
Promethean
   environmentalism   17
Second World   12
sustainable development   13
technological optimists   17
Third World   12

UNDERSTANDING THE HUMAN DEVELOPMENT INDEX

Go to http://www.undp.org/ to find more about the United Nations Development Program and the Human Development Index. The most recent summary of the Index available online (at the time this was written) was from the 1998 report at http://infomanage.com/international/98hdi.htm. Go there to evaluate the factors that make up this important measure of human welfare. In which categories does Canada rank higher than the United States? How might you explain these differences? Now compare the United States or Canada with one of the countries at the bottom of the index. In which categories do you find the greatest differences? Which of these measures is most closely related to environmental quality? If you were going to invest in improving the quality of life for residents of one of the countries with the lowest HDI, where would you start? What factors do you think have the greatest impact on this ranking?

You can find much more detailed information about particular countries in databases maintained by the World Bank (http://www.worldbank.org/data/countrydata/countrydata.html), EMAN, the federal government's Ecological Monitoring and Assessment Network (eman-rese.ca/eman), the U.S. Library of Congress (http://memory.loc.gov/frd/cs/cshome.html), and the U.S. Central Intelligence Agency (http://www.cia.gov/cia/publications/factbook/indexgco.html). Go to one of these sources to understand the factors that lead to higher or lower quality of life in a specific country.

SIZING YOUR ECOLOGICAL FOOTPRINT

We talked about the ecological footprint, a concept developed by researchers at the University of British Columbia. Now it's time to measure your own footprint. Go to www.earthday.net/footprint/info.asp and answer a few simple questions about your lifestyle. You'll be told the size of your ecological footprint!

Now, vary some of the answers that you've given to get a sense of how much (or how little) they affect your ultimate footprint. Discuss the advice presented to individuals, municipalities, and industry. Do you think it is unbiased? Look at the conservation question ... how much space and resources should be set aside for other species? How does this affect our total ecological footprint?

INVESTIGATING GLOBAL ENVIRONMENTAL CONDITIONS

The World Resources Institute in Washington, D.C., collects a wide range of valuable environmental data in cooperation with the United Nations Development Program (UNDP), the United Nations Environment Program (UNEP), and the World Bank. The biennial World Resources Report presents a summary of this data in the form of extensive databases, maps, and narrative summaries. You can find current assessments of this information on-line at http://www.wri.org/wri/wr2000/.

1. In the left frame on the opening page, choose "Global topics" to find a list of 10 major environmental issues. Summarize, in a few words, what you think the most important global problem is in each of these issues.

2. Back on the home page for World Resources 2000–2001, find the page for "Pilot analysis of global ecosystems (PAGE)." Look at the individual pages for the five major ecosystems assessed in this analysis. What is the most important risk (in your estimation) for each of these ecosystems? (Hint: look at the executive summary for each page. You may have to download Adobe Acrobat to view the PDF document where you'll find this information.)

3. Which of the global environmental issues listed by WRI do you find mentioned in the first chapter in your textbook? Look at the table of contents of the textbook and your class syllabus; will you be covering most of these topics in your class?

Women in New Delhi, India, draw water from a shallow, communal well.

Chapter 2

Environmental Ethics and Philosophy

The highest good is ... the knowledge of the union that the mind has with the whole of nature.

—Benedict Spinoza—

OBJECTIVES

After studying this chapter, you should be able to:

- understand some principles of environmental ethics and philosophy.
- compare and contrast how different ethical perspectives shape our view of nature and our role in it.
- realize how your own worldview and core values shape your perceptions of nature.
- explain anthropocentrism, biocentrism, ecocentrism, utilitarianism, and ecofeminism, and what each says about human/nature relationships.
- summarize the methods, applications, and limitations of the scientific method.
- discuss the role of technology in *causing* environmental problems as well as helping us solve them.

WebQuest

environmental ethics, environmental philosophy, anthropocentrism, ecocentrism, utilitarianism, ecofeminism, scientific method

Above: When human activity and nature are face to face, such as these birds in front of the steel mills of Hamilton Harbour, Ontario, it reminds us of the ethical and philosophical challenges of environmental science.

Playing God in the Laboratory

Early in 2001, a baby Rhesus monkey was born at the Oregon Regional Primate Research Center (Fig. 2.1). In itself, the birth of a baby monkey isn't particularly newsworthy, but this was a very special newborn. ANDi (an acronym for "inserted DNA" spelled backward) is the first transgenic primate born with foreign genes. These genes are part of ANDi's genetic material and can be passed on to his offspring. More alarming to some critics, ANDi's extra gene is borrowed from a jellyfish that makes flourcescent green protein that glows when exposed to ultraviolet light. In fact, ANDi doesn't glow visibly—at least not yet—but his birth raises a host of questions about the limits to modern science and the ethics of tinkering with the genetic makeup and reproduction of organisms.

Molecular recombination and "assisted reproduction" are not new techniques. *In vitro* reproduction of humans has been possible for more than three decades, and the first genetically modified mammal—a mouse—was created in 1980 by inserting a new gene into bone-marrow cells and then transplanting those cells to a living animal. In 1997, Dolly the sheep was born, the first mammal to be cloned from an adult cell. To produce Dolly, scientists destroyed the nucleus of a normal sheep egg cell and replaced it with another nucleus taken from a mature intestinal cell. The egg was then fertilized in vitro (in a glass dish), cultured for a few days in an incubator, and the resulting embryo was implanted in a surrogate mother. This successful result was a breakthrough because it had been thought that the genome (the collection of DNA) in a mature cell was somewhat altered so that it could no longer support the development of an entire animal.

To create ANDi, another set of bioengineering steps was required. The flourcescent green protein gene was first extracted from a jellyfish genome and amplified (copied) by molecular techniques. Using recombinant-DNA technology, scientists then spliced the gene into a modified virus that would carry it into monkey eggs and insert it into their DNA. These techniques aren't very efficient. Of 224 monkey eggs infected with the virus and fertilized with normal monkey sperm, only about half resulted in living embryos. Most of these embryos either died or were destroyed because of abnormalities. Forty of the apparently healthiest embryos were implanted into 20 surrogate mothers, resulting in five successful pregnancies. In the end, only three baby monkeys were born alive in this experiment, and only ANDi carries the foreign gene in all his cells.

This collection of techniques opens up a multitude of opportunities and questions. Would it be a good idea to allow human parents to artificially design their offspring? If they want to insert synthetic genes to make their children bigger, stronger, more beautiful, or more intelligent, should they be allowed to do so? If the process remains as hit and miss as it is now, what would we do with the hundreds of abnormal individuals created to make one desired copy? Suppose a totalitarian government wants to create a

FIGURE 2.1 ANDi, the baby Rhesus monkey, is the first primate born carrying foreign genes inserted by molecular engineering techniques into an egg cell.

whole race of super soldiers or slow-witted drones; would that be acceptable?

There also are conservation implications for this technology. At about the same time that ANDi was born in Oregon, a baby gaur (a wild ox from South Asia) named Noah was born in Iowa. Like Dolly, Noah resulted from the combination of a nucleus from a skin cell inserted into an enucleated egg. The artificially fertilized embryo was then implanted in a surrogate mother—in this case, an ordinary domestic cow, because guars are rare—and carried to birth. Proponents of this research claim it could be used to augment rare and endangered species or even to recreate already extinct organisms. Giant panda and white rhinoceros populations, for instance, might be expanded by cloning. And when the last remaining Spanish bucardo (a wild goat) was killed by a falling tree in 2000, scientists immediately began trying to recreate it with nuclei transplanted into the eggs of related species.

All these developments raise questions about the ethics of tinkering with life and our environment as well as about how science has gotten us to this point. This chapter presents a brief introduction to environmental ethics and a review of how science works to give you some understanding of these important topics.

Want More?

Chan, A.W.S., Chong, K.Y., Martinovich, C., Simerly, C., Schatten, G., "Transgenic Monkeys Produced by Retroviral Gene Transfer into Mature Oocytes." *Science* 291 (5502): 309–312 January 12, 2001.

ENVIRONMENTAL ETHICS AND PHILOSOPHY

Ethics is a branch of philosophy concerned with **morals** (the distinction between right and wrong) and **values** (the ultimate worth of actions or things). Ethics evaluates the relationships, rules, principles, or codes that require or forbid certain conduct. Most Western ethicists consider the roots of their field to be the famous questions posed by Socrates and the Greek philosophers 2,500 years ago: "What is the good life? How ought we, as moral beings, behave?"

Environmental ethics asks about the moral relationships between humans and the world around us. Do we have special duties, obligations, or responsibilities to other species or to nature in general? Are there ethical principles that constrain how we use resources or modify our environment? If so, what are the foundations of those constraints and how do they differ from principles governing our relations to other humans? How are our obligations and responsibilities to nature weighed against human values and interests? Do some interests or values supercede others?

Are There Universal Ethical Principles?

The first question considered in ethics is whether *any* moral laws are objectively valid and independent of cultural context, history, or situation. How the question is answered depends largely on the philosophical disposition of the respondent. **Universalists,** such as Plato and Kant, assert that the fundamental principles of ethics are universal, unchanging, and eternal (Fig. 2.2). These rules of right and wrong are valid regardless of our interests, attitudes, desires, or preferences. Some believe these rules are revealed by God, while others maintain they can be discovered through reason and knowledge.

Relativists, such as Plato's opponents, the Sophists, claim that moral principles always are relative to a particular person, society, or situation. Although there may be right and wrong—or at least better and worse—things to do, relativists assert that no transcendent, absolute principles apply regardless of circumstances. In this view, ethical values always are contextual. Friedrich Nietzsche's famous aphorism, "there are no facts, only interpretations" has been the banner for recent generations of relativists.

Nihilists, such as Schopenhauer, claim that the world makes no sense at all. Everything is completely arbitrary, and there is no meaning or purpose in life other than the dark, instinctive, unceasing struggle for existence. According to this view, there is no reason to behave morally. Only power, strength, and sheer survival matter. "Might is right; eat or be eaten." There is no such thing as a "good" life: we live in a world of uncertainty, pain, and despair. Nevertheless, Schopenhauer enjoyed living in a comfortable, civilized society where rules of normative behaviour and good conduct prevailed.

Utilitarians hold that an action is right that produces the greatest good for the greatest number of people. This philosophy is usually associated with the English philosopher Jeremy Bentham (1748–1832); but something very similar was suggested by Plato,

FIGURE 2.2 Plato and Aristotle debate moral philosophy in a painting by Raphael. Plato (*left*) motions upward, indicating a transcendent, universal moral truth, while Aristotle (*right*) motions downward to suggest grounded, situational ethics.

Socrates, Aristotle, and others. Bentham was an eccentric genius and a hedonist who equated goodness with happiness, and happiness with pleasure. He regarded pleasure as the only thing worth having in its own right. Thus, the good life is one of maximum pleasure. Insofar as people are moral animals, in his view, we should act to produce the greatest pleasure for the greatest number. To do so is good; not to do so is wrong.

Utilitarianism was modified and made less hedonistic by Bentham's brilliant protégé, John Stuart Mill (1806–1873). Mill believed that pleasures of the intellect are superior to pleasures of the body. He held that the greatest pleasure is to be educated and to act according to enlightened, humanitarian principles. This enlightened form of utilitarianism inspired the early conservationists (see Chapter 1), who argued that the purpose of conservation is to protect resources for the "greatest good for the greatest number *for the longest time.*"

Although utilitarianism remains widely popular today, it has drawbacks. It can, for instance, be used to justify reprehensible acts. If 10,000 Romans greatly enjoyed watching a few Christians being eaten by lions, did that make it the right thing to do? Does the pleasure of the tormentors outweigh the suffering of victims? Most of us would conclude that it does not. Justice, freedom, morals, and loyalty take precedence over pleasure, or even happiness, although it could be argued that furthering moral ends and right action ought to bring the greatest happiness in the long run.

Modernism and Postmodernism

Much of the modern, Western worldview is based on theories about reason and progress of seventeenth-century Enlightenment philosophers and scientists such as René Descartes, Francis Bacon, and Isaac Newton. These positivists hoped to develop universal laws of morality through objective science that allows control of nature as well as an understanding of the world and of ourselves. The inevitable outcome, they believed, would be moral progress, universal justice, and, ultimately, the happiness of all humans.

The experiences of the twentieth century, especially the horrors of world wars, have led to a kind of cultural despair among many philosophers who see the quest for individual liberty, self-realization, and experience leading to hedonism, narcissism, and social disintegration. Poststructuralists and postmodernists, such as Jacques Derrida, Jean-Francois Lyotard, and Michel Foucault, claim there is no grand narrative of history and no universal philosophy. Focusing on marginalized, disempowered groups and local, contingent knowledge, they cast doubt on classical notions of truth, reality, and meaning. According to this perspective, reality is constructed by our discourses rather than revealed by them; we never can know anything unequivocally.

How, you may be wondering, does this apply to environmental ethics? From a postmodernist perspective, nature—or at least our perception of it—is an arbitrary, ever-shifting social construction. If you ask a diverse group of people what nature is, you will most likely get many different answers. To a city dweller, nature might be a park with a few trees and squirrels; to a farmer, it may mean a productive field; and to a backpacker, it probably means a pristine wilderness. On the other hand, a logger would see that same wilderness as merely a supply of lumber. From a postmodernist view, no single viewpoint is inherently better than any other.

While relativism recognizes unique differences in the human condition, it can lead to paralysis of will. If every opinion or interpretation is equally valid, how can we choose whom to believe or what to do? What do you think? Is nature whatever we believe it is? Can we replace original nature with a human-constructed "second nature" (Fig. 2.3) or virtual reality whenever we like? Are there ethical absolutes in our relations with nature or is everything relative and subjective?

Values, Rights, and Obligations

For many philosophers, only humans are **moral agents,** beings capable of acting morally or immorally and who can—and

FIGURE 2.3 Is nature only a social creation? Can we replace nature with a "built" environment?

should—accept responsibility for their acts. Capacities that enable humans to form moral judgements include moral deliberation, the resolve to carry out decisions, and the responsibility to hold oneself answerable for failing to do what is right.

Of course, not all humans have all these capacities all the time. Children and those who are intellectually disabled are not regarded as moral agents. If a child murders someone, we don't hold the child responsible. Nonetheless, she/he still has rights. Children are considered **moral subjects,** beings who are not moral agents themselves but who have moral interests of their own and can be treated rightly or wrongly by others.

Historically, the idea that weaker members of society deserve equal treatment with those who are stronger was not a universally held opinion. In many societies, women, children, outsiders, serfs, and others were treated as property by those who were more powerful. The Greek philosophers, for example, accepted slavery as natural and necessary. Gradually, we have come to believe that all humans have certain inalienable rights: life, liberty, and the pursuit of happiness, for example. No one can ethically treat another human as a mere object for their own pleasure, gratification, or profit. This gradually widening definition of whom we consider ethically significant is called **moral extensionism.**

Do Other Organisms Have Rights?

Perhaps the most important question in environmental ethics is whether moral extensionism encompasses nonhumans. Do other species have rights as well? Are they moral agents or at least moral subjects? For many philosophers, the answer is no. Reason and consciousness—or at least a potential thereof—are essential for moral considerability in this view. René Descartes (1596–1650), for instance, claimed that animals are mere automata (machines) and can neither reason nor feel pain (Fig. 2.4).

Most pet owners disagree, claiming that while animals may not have the same self-consciousness as humans, they are intelligent and clearly have feelings. As sentient (perceptive) beings, they deserve ethical treatment. But what about nonsentient beings? Should we extend moral consideration to trees, rocks, landscapes? Some people think so. Let's see why.

Intrinsic and Instrumental Values

Rather than couch ethics strictly in terms of rights, some philosophers prefer to consider values. Value is a measure of the worth of something. But value can be either inherent or conferred. All humans, we believe, have **inherent value**—an intrinsic or innate worth—simply because they are human. They deserve moral consideration no matter who they are or what they do. Tools, on the other hand, have conferred, or **instrumental value.** They are

worth something only because they are valued by someone who matters. If I hurt you without good reason, I owe you an apology. If I borrow your car and smash it into a tree, however, I don't owe the car (or the tree!) an apology. I owe *you* the apology—or reimbursement—for ruining your car. The car is valuable only because you want to use it. It doesn't have inherent values or rights of itself.

Do Nonsentient Things Have Inherent Value?

How does this apply to nonhumans? Domestic animals clearly have an instrumental value because they are useful to their owners. But some philosophers would say they also have inherent values and interests. By living, breathing, struggling to stay alive, the animal carries on its own life independent of its usefulness to someone else. Some would draw a line of moral considerability at the limit of sentience (the ability to perceive sensations such as pain). Others argue that it is in the best interest of bugs, worms, and plants to be treated well, even if they aren't aware of it. In this perspective, just being alive gives things inherent value.

Some people believe that even nonliving things also have inherent worth. Rocks, rivers, mountains, landscapes, and certainly the earth itself, have value. These things were in existence before we came along and we couldn't recreate them if they are altered or destroyed. In a landmark 1969 court case, the Sierra Club sued the Disney Corporation on behalf of the trees, rocks, and wildlife of Mineral King Valley in the Sierra Nevada Mountains (Fig. 2.5) where Disney wanted to build a ski resort. The Sierra Club argued that it represented the interests of beings that could not speak for themselves in court.

A legal brief entitled *Should Trees Have Standing?,* written for this case by Christopher D. Stone, proposed that organisms as well as ecosystems and processes should have standing (or rights) in court. After all, corporations—such as Disney—are treated as persons and given legal rights even though they are really only figments of our imagination. Why shouldn't nature have similar standing? The case went all the way to the Supreme Court but was overturned on a technicality. In the meantime, Disney lost interest in the project and the ski resort was never built. What do you think? Where would you draw the line of what deserves moral considerability? Are there ethical limits on what we can do to nature?

PHILOSOPHICAL PERSPECTIVES ON THE ENVIRONMENT

Our beliefs about our proper roles in the world are deeply conditioned by our ethical perspectives (What Do You Think? p. 28). As historian Lynn White, Jr., said, "What people do about their ecology depends on what they think about themselves in relation to the things around them." In this section, we will look at how some different moral philosophies reflect our attitudes toward nature.

FIGURE 2.4 Do other organisms have inherent values and rights? These chickens are treated as if they are merely egg-laying machines. Many people argue that we should treat them more humanely.

FIGURE 2.6 Do humans have a right to use or destroy other species or natural objects in any way we choose? Do we have duties, obligations, or responsibilities toward nature?

FIGURE 2.5 Mineral King Valley at the southern border of Sequoia National Park was the focus of an important environmental law case in 1969. The Disney Corporation wanted to build a ski resort here, but the Sierra Club sued to protect the valley on behalf of the trees, rocks, and native wildlife.

Humanism and Anthropocentrism

Throughout history, many cultures have regarded humans as particularly important. Secular humanists claim that we are unique because of our intelligence, foresight, and creativity. According to this view, achievements in art, music, science, and other areas give humans special rights and values. Religious humanists believe that we have a unique place in creation according to God's plan. Some people regard the special attributes of humans as justification for our domination of nature (Fig. 2.6).

In an influential 1967 paper entitled *The Historic Roots of Our Ecological Crisis,* Lynn White, Jr. traced this tradition to the biblical injunction to "Be fruitful, and multiply, and replenish the earth, and subdue it: and have dominion over the fish of the sea, and over the fowl of the air, and over every living thing that moveth

upon the earth" (Genesis 1:28). In antiquity, White claimed, "every tree, every spring had its guardian spirit. Before one cut a tree, dammed a brook, or killed an animal, it was important to placate the spirit in charge of that particular situation. By destroying pagan animism, Christianity made it possible to exploit nature in a mood of indifference to feelings of natural objects." This view of humans as more important than any other species (Fig. 2.6) is termed **anthropocentric,** or human-centred.

Although many people agree with this analysis, others argue that this passage is translated and interpreted inaccurately. Genesis is really intended, they claim, to teach us to love and nurture creation rather than dominate and exploit it. White himself pointed out that Judeo–Christian culture has a long tradition of caring for nature. He recommended St. Francis of Assisi as an inspiration and patron saint of the environment.

In recent years, religious groups have played important roles in nature protection. The National Religious Partnership for the Environment is made up of representatives of most major denominations in North America. "Although we may disagree on how the earth's creatures were created," they say, "we really need to work together to save what's left." Acting on this belief, a coalition of evangelical Christians has been instrumental in opposing U.S. Congressional attacks on the Endangered Species Act. The Central Conference of American Rabbis declared that desecration of the Headwaters Redwood Forest in California breaks our covenant with the Creator. And Orthodox Patriarch Bartholomew has said that "to commit a crime against nature is a sin."

Stewardship

Many tribal or indigenous people, as well as agricultural and hunting federations have a strong sense of **stewardship** or responsibility to manage and care for a particular place. As custodians of resources, they see their proper role as working together with

PART ONE   Fundamental Principles of Ecology and Environmental Science

FIGURE 2.7 For many people a productive, domesticated landscape such as this mosaic of farmland and sugarbush presents the ideal perspective. With careful stewardship—including a balance of population density and environmental resources—the land can be stable, harmonious, and fruitful.

human and nonhuman forces to sustain life. Humility and reverence are essential in this worldview, where humans are seen as partners in natural processes rather than masters—not outside of nature but part of it.

This attitude also is held by many modern farmers (Fig. 2.7) or others in close contact with nature. Authors René Dubos and Wendell Berry have written eloquently about the need to nurture and sustain both the rural landscape and culture. In their view, humans can improve the world and make it a better place, both for themselves and for other organisms. As Voltaire said in *Candide,* "This may be the best of all possible worlds, but we must tend our garden."

Biocentrism, Animal Rights, and Ecocentrism

Many environmentalists criticize both stewardship and dominion as being too anthropocentric. They favour, instead, the **biocentric** (life-centred) egalitarianism of John Muir or Aldo Leopold, who claimed that all living organisms have intrinsic values and rights regardless of whether they are useful to us. Leopold wrote, "Of the 22,000 higher plants and animals native to Wisconsin, it is doubtful whether more than 5 percent can be sold, fed, eaten, or otherwise put to economic use. Yet these creatures are members of the land community, and if (as I believe) its stability depends on its integrity, they are entitled to continuance ... A thing is right when it tends to preserve the integrity, stability, and beauty of the biotic community. It is wrong when it tends otherwise." For many biocentrists, biodiversity is the highest ethical value in nature. Species and populations, as the basic units of biodiversity, are the locus of inherent value.

Some animal rights advocates question the importance of species or populations, claiming that each individual organism is

of value. They point out that the individual lives, reproduces, and experiences pleasure or pain, not the group. Many ecologists, in contrast, view larger-scale ecological processes such as evolution, adaptation, and the grand biogeochemical cycles as the most important aspects of nature. In this view, which is described as **ecocentric** (ecologically centred) because it claims moral values and rights for ecological processes and systems, the whole is considered more important than its individual parts. If you kill an individual organism, you deny it a few months or years of life, but if you eliminate an entire species or a whole landscape, you have destroyed something that took millions of years to create.

In the broadest ecocentric view, individuals could be seen to have only instrumental value while abiotic resources, ecological cycles, and the whole earth possess inherent value (Table 2.1). Nature doesn't seem to care about individuals. Vastly more offspring are born than can ever survive. Even species come and go. What seems to have longevity in nature are processes like photosynthesis and evolution.

Ecofeminism

Many feminists believe that none of these philosophies is sufficient to solve environmental problems or to tell us how we ought to behave as moral agents. They argue that all these philosophies come out of a patriarchal system based on domination and duality that assigns prestige and importance to some things but not others. In a patriarchal worldview, men are superior to women, minds are better than bodies, and culture is higher than nature. Feminists contend that domination, exploitation, and mistreatment of women, children, minorities, and nature are intimately connected and mutually reinforcing. They reject all "isms" of domination: sexism, racism, classism, heterosexism, and speciesism.

Ecofeminism, a pluralistic, nonhierarchical, relationship-oriented philosophy that suggests how humans could reconceive themselves and their relationships to nature in nondominating

TABLE 2.1	Worldviews and Ethical Perspectives—A Comparison		
PHILOSOPHY	INTRINSIC VALUE	INSTRUMENTAL VALUE	ROLE OF HUMANS
Anthropocentric	Humans	Nature	Masters
Stewardship	Humans and Nature	Tools	Caretakers
Biocentric	Species	Abiotic nature	One of many
Animal rights	Individuals	Processes	Equals
Ecocentric	Processes	Individuals	Destroyers
Ecofeminist	Relationships	Roles	Caregivers

Worldviews and Values

Worldviews are sets of basic beliefs, images, and values that make up a way of looking at and making sense of the world around us. They shape how we interpret and interact with our environment and other people. Most worldviews are learned early in life and are not easily changed. In cases where we encounter evidence that doesn't fit our worldview, we often reject the evidence and cling to prior beliefs. You have probably heard someone say, "I couldn't believe my eyes." What does that imply?

Our basic beliefs impact not only the way we think about ourselves and our place in the world, but they also determine what questions are valid to ask and who has a right to ask them. Worldviews are not usually arrived at rationally, nor are they necessarily accurate, but they strongly influence our behaviour. People often cannot explain why they believe what they do. It is difficult to verbalize the underlying reasons for a particular set of ideas. When you find yourself saying, "I can't explain this, it's just the way things are," you are probably describing your worldview.

Values are a measure of the ultimate worth of a thing or an action. They can represent principles, standards, or rules of conduct derived from one's worldview. As you read this book, you will find that many of the issues discussed are highly contentious and have no single clear answer. What each of us thinks about these issues depends on our particular set of values.

Each of us has deeply held, core values that we generally learn as very young children or develop—often without being conscious of it—from life experiences. Think about the following list of attributes: friendship, safety, popularity, security, wisdom, frugality, beauty, creativity, fame, fulfilling duties, accomplishment, happiness, individuality, knowledge, kindness. If everything else were stripped away, by what two or three basic characteristics would you want to be remembered? Perhaps this list doesn't really capture who you think you are or who you'd like to be. Try completing the following sentence: "I'd give up everything if I could be —." That's probably your real core value.

A principle of cooperative education is that you never really understand your own position on a particular issue until you understand the beliefs of someone who has a different point of view from your own. Environmental science provides an excellent opportunity to examine a variety of ways of looking at the world. This doesn't mean that you have to agree with everyone around you. But it's helpful to examine others' fundamental values as a way of gaining insight into your own. In discussions about environmental issues, try to see problems from the perspective of someone from an entirely different socioeconomic, religious, or political background from yours. You may find that your own values and worldviews become clearer.

In evaluating what you believe and why you believe what you do, think about your position on the following statements. Note that these are generally in opposing pairs but there is no intended association between any particular combination of statements. You may find that you agree only partially with some or all of the positions. The point of this exercise is to invite you to think about why you agree or don't agree with some of all of these propositions and what that says about your worldview. What social experiences, cultural beliefs, family background, or other personal influences may have shaped your views on these issues? If you were rewording some or all of these statements, how would you state them?

1. Humans are distinct and separate from nature. Our role is to subdue, command, and use other organisms and natural resources for our own benefit.

2. Every organism has rights and values of its own, regardless of its usefulness to us. We are merely part of nature but we may have special responsibilities to care for our neighbours because of our unique abilities.

---

ways, is proposed as an alternative to patriarchal systems of domination. It is concerned not so much with rights, obligations, ownership, and responsibilities as with care, appropriate reciprocity, and kinship. This worldview promotes a richly textured understanding or sense of what human life is and how this understanding can shape people's encounters with the natural world (Fig. 2.8). Among ecofeminist leaders are Karen Warren, Vandana Shiva, Carolyn Merchant, Rosemary Ruether, and Ynestra King.

According to ecofeminist philosophy, when people see themselves as related to others and to nature, they will see life as bounty rather than scarcity, as cooperation rather than competition, and as a network of personal relationships rather than isolated egos. Like the postmodernists, ecofeminists reject the view of a single, ahistoric, context-free, neutral observation stance. Instead, they favour multiple understandings, complex relationships, and "embodied objectivity."

In *Healing the Wounds* Ynestra King wrote, "We can use [ecofeminism] as a vantage point for creating a different kind of culture and politics that would integrate intuitive, spiritual, and rational forms of knowledge, embracing both science and magic insofar as they enable us to transform the nature-culture distinction and to envision and create a free, ecological society."

SCIENCE AS A WAY OF KNOWING

Science (from "knowing" in Latin) is a way of exploring and explaining the world around us. Ideally, it provides an orderly, methodical approach to investigating ideas and phenomena, and suggests testable explanations of why things work as they do. Science requires analytical, logical, purposeful thinking; and since it provides an understanding of processes and problems, scientific knowledge can also be applied to making predictions and finding solutions to problems.

Science is often an exciting and rewarding enterprise that requires creativity, skill, and insight. It takes many forms and is practised in a variety of ways by widely diverse people. Science

3. There are some things that are always right or wrong regardless of the circumstances or situation.

4. Whether an act is right or wrong depends on how, when, why, and by whom it was done. There are shades of grey when it comes to ethics and morals.

5. The world is characterized by scarcity and competition for limited resources. Further growth of either human populations or standard of living will be disastrous.

6. The world is one of opportunity and potential. Continued economic and population growth will have more good than bad effects.

7. Technology has caused most of our environmental problems. Returning to a simpler, nontechnological way of life is the only way to solve these problems.

8. Further technological advances are not only beneficial but essential for solving the problems we now face.

9. Problems that occur in other parts of the globe are not our concern. We should pay attention to the issues in our own back yard.

10. Whether we like it or not, we are all part of a global system. What happens anywhere in the world affects us in some way.

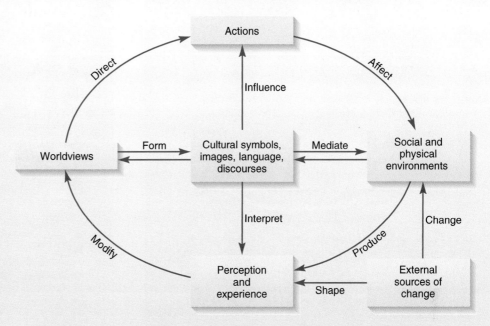

Each of us has a worldview that is shaped by, and in turn informs and interprets, our actions and experiences. We often aren't aware of our own worldview, only that others seem peculiar and obviously wrong. Note the similarity between this diagram and the concept maps in the Introduction.

11. I regard friendship, family, a sense of belonging, and fulfilling my duties as the highest or core values in my life.

12. I regard individuality, creativity, excitement, adventure, and accomplishment as the highest or core values in my life.

and technology have become such important factors in our lives, that everyone needs to know something about how the scientific process works (Fig. 2.9).

The Scientific Worldview

Science assumes that the world is knowable, and that systematic, rigorous investigations can yield meaningful insights about our environment. Science demands evidence; it depends on objective, reproducible results from careful observations or controlled experiments. Of course it is difficult to be completely neutral and unbiased. The process of making observations often changes the observed as well as the observer. Still, scientists attempt to identify and eliminate bias as far as possible, and to adjust for unavoidable errors.

Parsimony (keeping things as simple as possible) is an important principle in science. Parsimony says that an explanation for unknown phenomena should first be attempted in terms of what is already known. And where two competing explanations appear equally plausible, we should accept the simpler one. In other words, don't make things more complex than they need to be.

Science is a complex social activity that can't operate in secrecy. Through the give and take of open, honest discussion, scientists compare data, debate ideas, and forge an agreement on how to interpret evidence. Because it's difficult to be absolutely sure we've considered all relevant factors, scientific conclusions are almost always conditional. We must keep in mind that our findings may be incomplete and that even theories that seem firmly settled may be overturned in the future by new evidence.

Inductive and Deductive Reasoning

In 1919, Sir Arthur Eddington led an expedition to Príncipe Island off the coast of West Africa to photograph stars during a solar eclipse. These photographs represented the first successful empirical test of Einstein's theory of general relativity. Einstein's calculations predicted that photons of light should be affected by gravity.

FIGURE 2.8 Ecofeminism re-examines our relations with other humans and with nature. It calls for a new ethic of care, reciprocity, and kinship.

FIGURE 2.9 Researchers use a directional antenna to track radio-collared wildlife.

Thus, light from a distant star should be deflected if it passes close to the sun. These stars should appear shifted slightly in position during the day, when the sun is present, compared to the night, when it is not. Normally you can't make such an observation because stars close to the sun are invisible in the daytime. During an eclipse, however, it's possible to photograph stars and compare their apparent day and night locations. This is what Eddington did, and his corroboration of Einstein's prediction was a triumphant moment in modern physics.

This series of logical steps is **deductive reasoning.** Starting with a general principle (the theory of general relativity), a testable prediction is derived about a specific case: that certain stars should appear closer together in the daytime than at night. You might think of this as "top-down" reasoning—proceeding from the general to the specific. Although deduction is often regarded as a scientific ideal, it is difficult to do in fields such as ecology or environmental

science where there are few universal laws on which we all agree. What we do more often in these fields is "bottom-up" **inductive reasoning** that attempts to infer general principles by careful examination of specific cases.

Consider the deformed frogs found in many places. There is no general theory that would explain why this may be occuring. Studying a problem such as this is like a detective investigation. We gather all available evidence and then sift through the data looking for possible clues and connections. By examining many different examples and making careful observations, we hope to infer some overarching environmental principles and perhaps even find some universal truths. Interpreting inductive evidence requires caution, however, because correlation doesn't prove causation. We may, for example, find pesticide residues in ponds where deformed frogs live, but that doesn't prove that pesticides are causing the problem. There may be some other completely unrelated factors such as UV radiation or diseases that are really responsible.

Hypotheses and Scientific Theories

Whether our science is inductive or deductive, we examine the validity of our suppositions by forming a **hypothesis;** that is, a provisional explanation that can be supported or falsified by further investigation. Notice that while you can show a hypothesis to be wrong, you can almost never show it to be unquestionably true. The philospher Ludwig Wittgenstein gives the following example to illustrate this principle. Suppose all the swans you have ever seen are white. You might form the hypothesis that *all* swans are white. You could test this hypothesis by examining a large number of swans. If you never find a black one, you might tentatively conclude that your hypothesis is right. But you could never be sure that you've see all the swans in the world. Even if you've looked at a million white swans, there still may be a black one somewhere that proves your hypothesis wrong. Thus, refutability is possible, but positive proof is elusive.

When a hypothesis is supported by our tests, we consider it to be provisionally true until some additional test uncovers a flaw in it. If an explanation has been supported by a large number of tests, and a majority of experts in a given field have reached a general consensus that it is the best description available, we call it a **scientific theory.** Note that the use of this term by scientists is very different from the way it is understood by the general public. To many people, a theory is speculative and unsupported by facts. To a scientist, it means just the opposite; while all explanations are contingent, one that counts as a scientific theory is supported by an overwhelming body of data and experience, and is widely accepted in the scientific community.

Using the Scientific Method

You may already be using scientific technique without being aware of it. Suppose your flashlight doesn't work. The problem could be in the batteries, the bulb, or the switch—or all of them could be faulty. How can you distinguish between the possibilities? If you change all the components at once, you might have a working flashlight, but you won't know which was the faulty one. A series of methodical steps to test each component can be helpful. First, you might try new batteries. If that doesn't help, replace the bulb with one that you know works. If neither of these steps solves your problem, perhaps the flashlight has a faulty switch. You could try the original battery and bulb in a different flashlight. By testing the variables one at a time, you should be able to identify the problem.

What you have just employed in this example is an orderly cycle of observations, methodical testing, and interpretation of results known formally as the **scientific method.** The general flow of an experimental scientific study is shown in Figure 2.10. We start with observations: in this case, my flashlight doesn't work. From this we formulate a testable hypothesis: the batteries must be dead. From this hypothesis, we make a prediction: if we replace

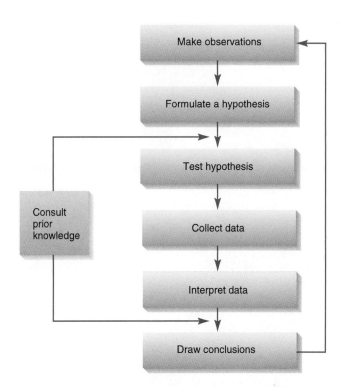

FIGURE 2.10 Ideally, scientific investigation follows a series of logical, orderly steps to formulate and test hypotheses.

the dead batteries with fresh ones, we should see some light. Finally, we carry out the experiment, collect data, and draw conclusions. As described previously, the results of one experiment may give us information that leads to further hypotheses and additional experiments. If the batteries are okay, then we suspect the bulb is burned out and design a way to test that hypothesis. In every case, prior knowledge and experience help us design experiments and interpret results. Eventually, with evidence from a group of related investigations, we may be able to formulate a theory to explain a set of general principles.

Descriptive and Interpretive Science

Not every field of science is amenable to direct experimentation. We don't regard it as acceptable, for instance, to study the effects of toxins on human health by deliberately poisoning people, no matter how useful that information might be. Although geologists and ecologists can do certain kinds of experiments, they can hardly build entire mountain ranges in the laboratory or recreate the conditions over millions of years that gave rise to different species. Scientists in these fields often test their ideas and explanations indirectly by looking for historical evidence or natural experiments that support or contradict their hypotheses. In addition, scientists often use mathematical models as virtual representations that can be tested. The degree to which such simulations represent reality varies, however, with the quality of the information used to construct the model or interpret historical events. If the model is

FIGURE 2.11 A wildlife biologist examines an endangered Blandings turtle.

in research. Many important discoveries were made not because of superior scientific method or objective detachment, but because the investigators were passionately interested in their topic and pursued hunches and intuitions that appeared reasonable to fellow scientists. A good example is Barbara McClintock, the geneticist who discovered transposable elements or "jumping genes" in corn. Her years of experience in corn breeding and an uncanny ability to recognize patterns in what seemed to be random variation, gave her an intuitive understanding that took other investigators years to accept.

Paradigms and Scientific Consensus

The process of reaching agreement on scientific knowledge is as much a process of consensus-building as it is about conducting tests and gathering data. The historian of science, Thomas Kuhn, describes science as long periods of puzzle-solving (what he calls "normal science") followed by the dramatic change of scientific revolutions. In normal science, most research is based on commonly shared **paradigms:** models that provide a framework for interpreting results and developing theories. These paradigms determine which phenomena are worth investigating, and how we understand them. Much of what scientists do during this phase is to measure and very accurately remeasure previously known facts.

Eventually, however, anomalies and contradictions to generally accepted paradigms build up to a point at which they cannot be ignored. These contradictions result in resistance, confusion, and uncertainty among practitioners. A crisis looms as previous beliefs become increasingly problematic until a few unusually bold or visionary individuals suggest that the old paradigms must be discarded entirely. Sometimes this revolution is accomplished rather quickly. Quantum mechanics and Einstein's theory of relativity, for example, overturned classical physics in only about 30 years. In other cases, it may take a century or more for new ideas to supplant old ones. Occasionally two mutually exclusive paradigms can coexist in the same field for many decades. They may even be used to explain contradictory evidence from a single study. Typically, new paradigms attract new generations of scholars; those who are more established tend to resist change. Kuhn claims that new scientific truth doesn't triumph by convincing its opponents, but rather because the old guard eventually dies, and a new generation grows up for whom the new paradigm is familiar and non-threatening.

As you study this book and think about environmental science, try to identify some of the paradigms that guide our investigations and explanations today. This is one of the skills involved in critical thinking and in becoming a reflective thinker about environmental dilemmas.

Technology and Progress

For the past three centuries, a central tenet of Western culture has been an almost religious faith in progress: an inevitable

flawed, so will be the results it produces. But sometimes we learn a lot about how a system works by looking carefully at the "mistakes" in our results.

Sometimes simply knowing a biological community really well and being able to make careful observations and comparisons of the organisms that make it up can yield significant insights about patterns and relationships (Fig. 2.11). Biogeographers and taxonomists, for example, can gather important information about the variety of species and where they live that help us understand the natural world. Where it may be impossible or unethical to do deliberate experiments on rare or endangered species, simply describing where and how they live can be valuable. Furthermore, while experiments under controlled conditions allow scientists to study factors one at a time, they can result in such artifical systems that they bear little resemblance to real nature.

Many people—even scientists—fail to recognize the role that human factors such as creativity, insight, aesthetics, and luck play

PART ONE   Fundamental Principles of Ecology and Environmental Science

cause social and environmental crises. Whether the root causes of these problems are in technology or human nature, technology clearly allows us to make mistakes faster and on a larger scale than ever before.

A nineteenth-century English backlash against the excesses of industrialization led by Ned Ludd gave rise to the term *Luddites* for opponents of rampant technology. The Luddites smashed power looms and other machines that were threatening the craft guilds, cottage industries, and village networks that sustained traditional rural communities.

In the twentieth century, even more dangerous technologies, such as nuclear weapons, biological warfare, and the petrochemical industry, along with the problems caused by earlier technologies, such as biodiversity losses, global climate changes, and destruction of stratospheric ozone, have led many intellectuals and young people to question whether progress is either possible or desirable. **Neo-Luddites** now assert that all large-scale human endeavours eventually fail, that science and technology cause more problems than they solve, and that our only hope is to abandon modern life and go back to a low-tech pastoral or hunting-and-gathering society.

Some neo-Luddites resort to terrorist bombings and sabotage to try to bring down mainstream culture. Others flee to end-of-the-road refuges where they attempt to recreate a simpler, agrarian life. Rural life can be more resource intensive than urban living, however, especially if you demand all the modern conveniences. We probably couldn't all live off the land in remote places without very destructive environmental effects.

Appropriate Technologies

As historian Lewis Mumford pointed out, technology consists of more than machines. It includes all the techniques, knowledge, and organization that we use to accomplish tasks. When you build a fire, or use a rock as a tool, you are employing some of the oldest and most revolutionary human technologies. Whether our technology is destructive or constructive depends, in part, on the tools themselves, but even more on our worldview about how and why we use them. As sustainable-energy expert Amory Lovins points out, some technologies such as nuclear power, genetic engineering, and nanotechnology might be benign in the hands of a "wise, far-seeing, and incorruptible people." Unfortunately that seems not to describe most of us.

In 1973, British economist E. F. Schumacher published a widely popular book entitled *Small Is Beautiful.* It introduced the concept of **appropriate technology,** which promotes machines and approaches suitable for local conditions and cultures. The appropriate-technology movement attempts to design productive facilities in places where people now live, not in urban areas. It looks for products that are affordably made by simple production methods from local materials for local use. It advocates safe, creative, environmentally sound, emotionally satisfying work conducted in conditions of human dignity and freedom that creates social bonds rather than breaking them down.

FIGURE 2.12 Technology has brought us benefits but has also had costs.

march of human betterment. Originally formulated during eighteenth-century enthusiasm over the American and French Revolutions, this theory seemed to be proven by the increase in material wealth and standard of living provided by the Industrial Revolution. But while technology and development brought many benefits, they had a darker side as well (Fig. 2.12). Pollution, rapid urbanization, inhumane working conditions for many, and vast disparities in wealth and power between classes still

FIGURE 2.13 Appropriate technology, such as this multipurpose tractor/utility cart, is designed to suit local needs and conditions in developing countries. Small, simple, and affordable machines that are locally manufactured and repaired can reduce drudgery and improve productivity without reducing the number of jobs available.

Rather than to try to convert local economies and tastes into copies of Western culture, appropriate technologists try to work with indigenous people to create sustainable livelihoods suitable for prevailing conditions (Fig. 2.13). They hope that appropriate technology can help us avoid future environmental damage and to repair mistakes made in the past. Some advances have been made toward these goals, but the promise and power of the dominant Western paradigm are very seductive. Perhaps technology that effectively reduces the size of a society's ecological footprint (see Chapter 1) while improving the quality of life of its members is what really defines "appropriate technology."

SUMMARY

Are there universal, eternally valid ethical principles or moral laws? Universalists think so. Nihilists disagree. Relativists believe that everything is contextual. Utilitarians hold that something is good that brings the greatest good to the greatest number. Postmodernists argue that everything is socially constructed. Ecofeminists contend that patriarchal systems of domination and duality cause both environmental degradation and social dysfunction. They call for a more pluralistic, non-hierarchical, caring treatment of both nature and other people. Whichever of these worldviews you hold shapes your views about nature and our place in it.

Anthropocentrists claim that the world was made for our domination, and that only humans have inherent or intrinsic rights and values. Nature, from this perspective may be only a source of materials for humans. Many people who live close to the land, on the other hand, often feel a sense of stewardship or responsibility to care for creation. They see themselves as caretakers rather than dominators. Biocentrists consider all living things to have inherent value. We are merely one of many species. Animal rights advocates place their emphasis on individual animals. Ecocentrics maintain that ecological processes such as evolution, adaptation, and biogeochemical cycles are the most important parts of nature. In their view, individuals don't count for much and humans are mostly a negative influence.

Science is a methodical, meticulous study of the natural world. It takes many different forms and is practised in various ways by diverse people, but observation, hypothesis formation, testing, analysis, and reevaluation of hypotheses in light of new data form the core of positivist scientific methodology. Descriptive or interpretive sciences take a more holistic, inductive approach to knowledge.

Technology brings us many benefits, but it also creates pollution, consumes resources, despoils nature, and allows us to separate individuals, classes, and nations into those who have and those who do not. It gives us the power to make mistakes faster and on a larger scale than ever before. Then again, appropriate technology may also provide options to avoid environmental damage in the future or to repair mistakes made in the past.

QUESTIONS FOR REVIEW

1. Who was ANDi and why was he remarkable?

2. Define universalist, nihilist, relativist, utilitarian, postmodern, and ecofeminist ethics.

3. Explain how resource exploiters and social justice advocates use these ethical positions to support their causes.

4. Describe the differences between moral agents, moral subjects, and those who fail to qualify for moral considerability.

5. What are inherent and instrumental values? Who has them? Why?

6. Compare and contrast stewardship, anthropocentric, biocentric, ecocentric, and environmental justice worldviews. Which is closest to your own views?

7. What did Lynn White, Jr., say about the role of religion in environmental problems?

8. Draw a diagram showing the scientific method and describe, in your own words, what it means.

9. Explain the idea of a *paradigm* and why it is important.

10. Not every science is experimental. Explain how geologists or evolutionary biologists might test their hypotheses.

QUESTIONS FOR CRITICAL THINKING

1. Review a topic in this chapter—ecofeminism or animal rights for example—and arrange the arguments for or against this idea as a series of short statements. Determine whether each is a statement of fact or opinion.

2. How would you verify or disprove fact statements given in question 1? Do the opinions or conclusions reached in this argument *necessarily* follow from the facts given? What alternatives could be proposed?

3. Reflect on the preconceptions, values, beliefs, contextual perspective that you bring to the discussion above. Does it coincide with those in the textbook? What different interpretation would your perspective impose on the argument?

4. What is your environmental ethic? What experiences, cultural background, education, religious beliefs help shape your worldview?

5. Try some role-playing with a classmate, friend, or family member. Take the position of a universalist, biocentrist, ecocentrist, postmodernist, or ecofeminist and debate the merits of cutting trees in old growth forests. On what points would you agree and where would you disagree in this issue?

KEY TERMS

anthropocentric   26
appropriate technology   33
biocentric/biocentrism   27
deductive reasoning   30
ecocentric   27
ecofeminism   27

environmental ethics   23
hypothesis   31
inductive reasoning   30
inherent value   25
instrumental value   25
moral agents   24
moral extensionism   25
moral subjects   24

morals   23
neo-Luddites   33
nihilists   23
paradigms   32
parsimony   29
relativists   23
science   28
scientific method   31

scientific theory   31
stewardship   26
universalists   23
utilitarians   23
values   23
worldviews   28

Web Exercises

EXPLORING ENVIRONMENTAL ETHICS

Go to Ethics Updates, a survey of selected resources on environmental ethics maintained by the University of San Diego at http://ethics.acusd.edu/environmental_ethics.html. On this site you'll find a wealth of resources related to environmental ethics, including audio and video files, discussion groups, home pages for environmental groups with ethical interests, and governmental and legal resources on the environment. Follow a few of these links to sample current issues and debates about environmental ethics. Analyze the positions you

find on these web pages with respect to the categories described in Table 2.1 in your textbook. Which philosophical worldviews do you find represented? Do the assignments of intrinsic and instrumental values and roles of humans described in this book fit with what you find in material on the Internet? Why or why not, in your opinion? Do these authors seem to be universalists, relativists, nihilists, or utilitarians? What underlying view of environmental ethics do you find to be most common in this collection?

Chapter 3

Matter, Energy, and Life

Nature does nothing uselessly.

—Aristotle—

OBJECTIVES

After studying this chapter, you should be able to:

- describe matter, atoms, and molecules and give simple examples of the role of four major kinds of organic compounds in living cells.
- define *energy* and explain the difference between kinetic and potential energy.
- understand the principles of conservation of matter and energy and appreciate how the laws of thermodynamics affect ecosystems.
- know how photosynthesis captures energy for life and how respiration releases that energy to do useful work.
- define *species, populations, communities,* and *ecosystems* and understand the ecological significance of these levels of organization.
- discuss food chains, food webs, and trophic levels in biological communities and explain why there

are pyramids of productivity in the trophic levels of an ecosystem.
- compare the ways that carbon, nitrogen, sulphur, and phosphorus cycle within ecosystems and globally.

WebQuest

atoms, molecules, and compounds; carbon dating; stable isotopes; cell biology; Conservation of Matter; earth impact database; Laws of Thermodynamics; photosynthesis; food webs; trophic levels; biogeo-chemistry

Above: When scientists added phosphorus to this northwestern Ontario lake on one side of the curtain, they caused a huge increase in primary productivity of the phytoplankton.

The Mystery of Lake Laberge
Toxins Abound in a Boreal Lake

Lake Laberge is a narrow finger lake (5 km wide by 50 km long) just north of the town of Whitehorse in the headwaters of the fabled Yukon River (Fig. 3.1). Beginning about a decade ago, scientists collected water and biological samples from the lake to use as controls in a study of uptake of toxic industrial chemicals. The intent was to compare toxin levels in organisms from this remote and supposedly pristine watershed with samples collected closer to industrial and agricultural areas. Surprisingly, the top predator fish from Lake Laberge turned out to have some of the highest levels of PCBs, toxaphene, DDT, and other industrial and agricultural chemicals found anywhere in Canada.

Located about 1,600 km north of Vancouver, B.C., Lake Laberge would seem to be well removed from major pollution sources. Far from being pure and pristine, however, the lake water was found to have up to 100 times the concentration of toxic organic chemicals of lakes much further south. And trout and pike, the top predators in the lake, accumulate these compounds at levels thousands of times that of their counterparts in other ecological systems. How could this be? The answer is that volatile organic chemicals are being transported thousands of kilometres by air currents and are falling into the lake in the form of contaminated rain and snow. The transport apparently is not in a single leap but rather a series of shorter hops. These compounds evaporate from water and soil in warmer climates. They are carried aloft by rising air currents where they cool, condense, and precipitate out. On reaching the surface, the most volatile chemicals reevaporate, and with each cycle, they move further north until they reach very cold places, such as Lake Laberge, where they accumulate to extraordinary levels. It's a sort of distillation process on a continental scale.

Contaminants in the water are taken up first by tiny free-floating microorganisms known as plankton. Insects and crustaceans prey on plankton and store toxins they contain in tissues and organs. Minnows and small fish concentrate the toxins further. Finally, long-living top predators in the lake accumulate the highest levels of all.

This process of biomagnification is especially intense in Lake Laberge, where the clear, cold waters are low in mineral nutrients.

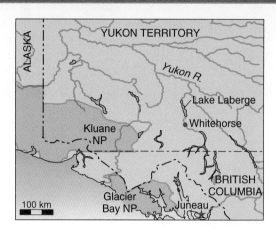

FIGURE 3.1 The Yukon River flows through Lake Laberge just north of the historic town of Whitehorse. This was the route to the Klondike during the gold rush of 1898.

This causes a low growth rate of the particular species of algae adapted to these conditions, and similarly slow-growing populations of consumer organisms at each of the many steps in the food chain. At the top of the food chain, the many steps of biomagnification have resulted in "toxic top predator" trout and pike.

If you ever have a chance to visit Lake Laberge, you wouldn't be advised to make a meal of the fish you might catch there. We'll discuss the health effects and some of the possible political solutions for this contamination in subsequent chapters of this book. In this chapter, we'll examine the biogeochemical cycles and ecological processes that are responsible for the movement of materials through our environment and into and through living food chains.

Want More?

Kidd, K.A., Schindler, D.W., Muir, D.C.G., Lockhart, W.L., and Hesslein, R.H. 1995. "High Concentrations of Toxaphene in Fishes from a Sub-Arctic Lake." *Science.* 269: 240–242.

FROM ATOMS TO CELLS

The endless cycle of materials between the living and nonliving parts of our environment is an important part of **ecology,** the scientific study of relationships between organisms and their environment. Modern biology covers a wide range of inquiry, ranging from molecules to ecosystems to the entire planet. Ecology examines the life histories, distribution, and behaviour of individual organisms, as well as the structure and function of natural systems at the level of populations, communities, ecosystems, and landscapes. Ecology encourages us to think holistically and to see interconnections that make whole systems more than just the sum of their individual parts.

In this chapter, we will survey some fundamental aspects of energy flow and material recycling within ecosystems. We will look at why living organisms need a constant supply of these essential ingredients, how energy and materials are transferred from one organism to another, and how humans benefit from, and also disrupt, those ecological relationships.

PART ONE Fundamental Principles of Ecology and Environmental Science

In a sense, every organism is a chemical factory that captures matter and energy from its environment and transforms them into structures and processes that make life possible. To understand how these processes work, we need to understand something about the fundamental properties of matter and energy. This section presents a survey of some of these principles.

Atoms, Molecules, and Compounds

Everything that takes up space and has mass is **matter.** Matter exists in three distinct states—solid, liquid, and gas—due to differences in the arrangement of its constitutive particles. Water, for example, can exist as vapour (gas), fluid (liquid), or ice (solid). Matter also exists in unique chemical forms we call **elements,** which are substances that cannot be broken down into simpler forms by ordinary chemical reactions. Just four elements—oxygen, carbon, hydrogen, and nitrogen—are responsible for more than 96 percent of the mass of most living organisms. Each of the 115 known elements (92 natural, plus 23 created under special conditions) has distinct chemical characteristics.

All elements are composed of discrete units called **atoms,** which are the smallest particles that exhibit the characteristics of the element. Atoms are tiny units of matter composed of positively charged protons, negatively charged electrons, and electrically neutral neutrons. Protons and neutrons, which have approximately the same mass, are clustered in the nucleus in the centre of the atom (Fig. 3.2). Electrons, which are tiny in comparison to the other particles, spin around the nucleus at high speed.

Each element has a characteristic number of protons per atom, called its **atomic number.** The number of neutrons in different atoms of the same element can vary within certain limits. Thus, the atomic mass, which is the sum of the protons and neutrons in each nucleus, also can vary. We call forms of a single element that differ in atomic mass **isotopes.** For example, hydrogen, the lightest element, normally has only one proton (and no neutrons) in its nucleus. A small percentage of hydrogen atoms have one proton and one neutron. We call this isotope deuterium (^{2}H). An even smaller percentage of natural hydrogen called tritium (^{3}H) has one proton plus two neutrons.

Some isotopes, such as ^{14}C, radioactively **decay** on a known schedule once an organism dies, and are therefore useful for establishing the age of organic matter (**carbon dating**). **Stable isotopes,** on the other hand, do not decay over time. Different stable isotopes of the same element (e.g., ^{2}H vs. ^{1}H, ^{13}C vs. ^{12}C, ^{18}O vs. ^{16}O, and ^{34}S vs. ^{32}S) are differentially involved in chemical reactions or biological processes depending on the environment (e.g., temperature) of the reactions. This makes the relative proportions of stable isotopes a useful "fingerprint" of the pathway of matter as it moves in and around the ecosystem, and the environments it has experienced.

Atoms can join together to form **molecules.** A **compound** is a molecule containing different kinds of atoms (Fig. 3.3). Water, for example, is a compound composed of two atoms of hydrogen attached to a single oxygen atom, shown by the formula H_2O. Sometimes, two atoms of the same element combine to form a molecule. Hydrogen gas (H_2), molecular oxygen (O_2), and molecular nitrogen (N_2) consist of such diatomic molecules. Most molecules are incredibly small, but some can be relatively large. The genetic information in your cells, for instance, is contained in deoxyribonucleic acid (DNA) molecules, each of which contains billions of atoms and can be several metres long. Tightly wound upon themselves at the time of cell division, DNA molecules become visible as the chromosomes in the cell nucleus.

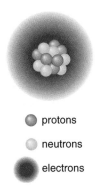

protons

neutrons

electrons

FIGURE 3.2 As difficult as it may be to imagine when you look at a solid object, all matter is composed of tiny, moving particles, separated by space and held together by energy. It is hard to capture these dynamic relationships in a drawing. This model represents carbon-12, with a nucleus of six protons and six neutrons; the six electrons are represented as a fuzzy cloud of potential locations rather than as individual particles.

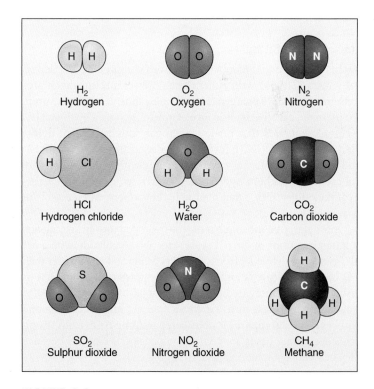

FIGURE 3.3 These common molecules, with atoms held together by covalent bonds, are important components of the atmosphere or important pollutants.

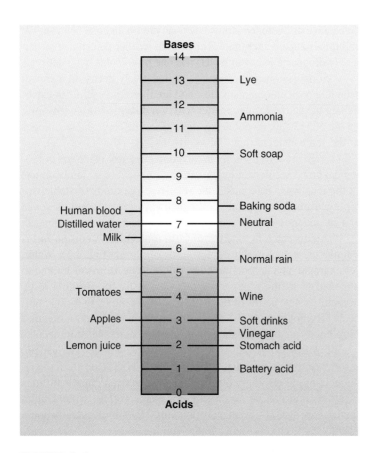

FIGURE 3.4 The pH scale. The numbers represent the negative logarithm of the hydrogen ion concentration in water.

Ions, Acids, and Bases

Atoms with an equal number of electrons and protons are electrically neutral. Those with either more or fewer electrons than protons are electrically charged and are called **ions.** A hydrogen atom, for example, can give up its sole electron to become a hydrogen ion (H^+). Compounds that release hydrogen ions are called **acids.** Hydrochloric acid (HCl), for instance, dissociates in water into H^+ and Cl^- ions. Substances that bond readily with hydrogen ions are called **bases.** Sodium hydroxide (NaOH), for example, releases hydroxide ions (OH^-) that bind to hydrogen ions when dissolved in water.

We describe the number of free hydrogen ions in an aqueous solution by means of the **pH** scale (Fig. 3.4). A solution with a pH of 7 is neutral, while a pH less than 7 is acidic and one greater than 7 is basic. The scale is logarithmic, which means that a pH 6 represents 10 times more hydrogen ions in solution than pH 7. A substance with a pH less than 3 is considered a strong acid, while one with a pH greater than 12 is considered a strong base. Both strong acids and strong bases are considered dangerous caustics because they can damage materials and kill living cells.

Chemical Bonds

The forces holding atoms together in molecules and compounds are called **chemical bonds.** Each bond represents a certain amount of chemical energy. Creating a bond requires an energy input; breaking that bond usually results in an energy release. Covalent bonds involve the sharing of electrons between adjacent atoms. A simple example of a covalent bond is the joining of two hydrogen atoms in a hydrogen molecule. Each hydrogen atom shares its single electron with its partner so that the pair of electrons orbit the two hydrogen nuclei equally, joining the two atoms. Carbon is able to form covalent bonds simultaneously with four other atoms, making it possible to create the complex structures of organic compounds. The main function of photosynthesis, which we will discuss later in this chapter, is the creation of energy-rich, carbon-carbon bonds in organic molecules.

Ionic bonds form between positively charged cations and negatively charged anions. For example, the bond between sodium and chloride is an ionic bond. Both covalent and ionic bonds can be very strong and can form highly stable compounds. Several types of weak bonds also are important in creating biological structures. Perhaps the most important of these is the hydrogen bond. The attraction between adjacent water molecules is a good example of hydrogen bonding. The oxygen atom in water attracts shared electrons much more strongly than do the hydrogen atoms. This makes the hydrogens partially positive, while the oxygen is partially negative. Consequently, the hydrogen atoms in one water molecule have a weak attraction for the oxygen atoms in an adjacent molecule. Although these bonds have very little energy individually and typically last only for a very short time, on average, the vast number of them in a body of water at any given instant is responsible for many of the remarkable qualities of liquid water.

Organic Compounds

Organisms use some elements in abundance, others in trace amounts, and others not at all. Certain vital substances are concentrated within cells, while others are actively excluded. Carbon is a particularly important element because chains and rings of carbon atoms form the skeletons of **organic compounds,** the material of which biomolecules, and therefore living organisms, are made.

The four major categories of bioorganic compounds are lipids, carbohydrates, proteins, and nucleic acids. Most lipids belong to a family of molecules called hydrocarbons because they are composed of chains of carbon atoms, most of which have two hydrogens attached. Methane and propane are simple examples of hydrocarbons (Fig. 3.5*a*). Some common examples of lipids are fats and oils. They make up an important part of the membranes that surround cells as well as their internal organelles. Carbohydrates, as their name suggests, are composed of carbon, hydrogen, and oxygen $(CH_2O)_n$, where *n* represents many repeated units arranged in rings or linear series. Glucose (Fig. 3.5*b*) is an example of a simple sugar. Starch and cellulose are made of long strands of conjoined sugar molecules.

Proteins are composed of chains of subunits called amino acids (Fig. 3.5*c*). Folded into complex three-dimensional shapes, proteins make up much of both the structural and functional components of cells. Enzymes, for instance, are proteins. Nucleic acids

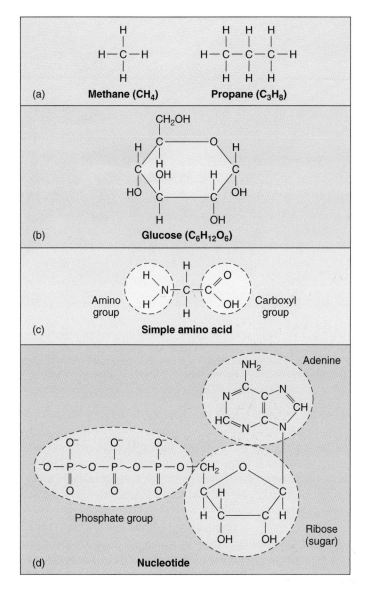

FIGURE 3.5 The four major groups of organic molecules are based on repeating subunits of these carbon-based structures. Basic structures are shown for (*a*) butyric acid (a building block of lipids) and a hydrocarbon, (*b*) a simple carbohydrate, (*c*) a protein, and (*d*) a nucleic acid.

are composed of combinations of a sugar molecule, a nitrogen-containing ring structure, and a phosphate bridge that can link many subunits (known as nucleotides (Fig. 3.5*d*)) into long molecules such as DNA and RNA. Nucleic acids store the genetic information that directs the processes of life.

Cells: The Fundamental Units of Life

All living organisms are composed of one or more **cells,** minute compartments within which the processes of life are carried out (Fig. 3.6). Microscopic organisms such as bacteria and protists are composed of single cells. Animals, plants, and fungi are multicellular, usually with many different cell varieties. Your body, for

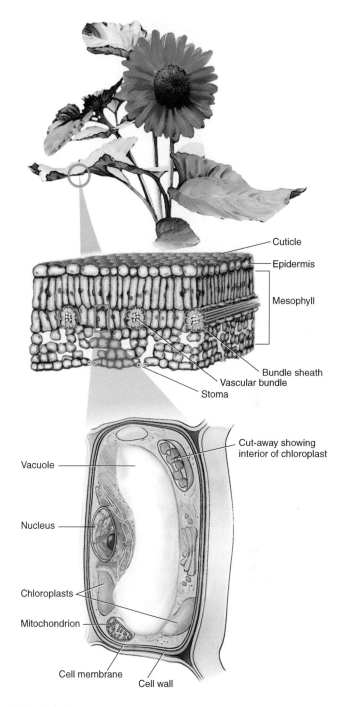

FIGURE 3.6 Plant tissues and a single cell's interior. Cell components include a cellulose cell wall, a nucleus, a large, empty vacuole, and several chloroplasts, which carry out photosynthesis.
Source: © E. H. Newcomb.

instance, is composed of several trillion cells of about 200 distinct types. Every cell is surrounded by a thin but dynamic membrane of lipid and protein that receives information about the exterior world and regulates the flow of materials between the cell and its environment. Inside, cells are subdivided into tiny organelles and subcellular particles that provide the machinery for life. Some of these organelles store and release energy. Others manage and distribute information. Still others create the internal structure that gives the cell its shape and allows it to fulfill its role.

All of the chemical reactions required to create these various structures, provide them with energy and materials to carry out their functions, dispose of wastes, and perform other functions of life at the cellular level are carried out by a special class of proteins called **enzymes.** Enzymes are molecular catalysts, that is, they regulate chemical reactions without being used up or inactivated in the process. Think of them as tools: like hammers or wrenches, they do their jobs without being consumed or damaged as they work. There are generally thousands of different kinds of enzymes in every cell, all necessary to carry out the many processes on which life depends. Altogether, the multitude of enzymatic reactions performed by an organism is called its **metabolism.**

ENERGY AND MATTER

Energy and matter are essential constituents of both the universe and living organisms. Matter, of course, is the material of which things are made. Energy provides the force to hold structures together, tear them apart, and move them from one place to another. In this section we will look at some fundamental characteristics of these components of our world.

Energy Types and Qualities

Energy is the ability to do work such as moving matter over a distance or causing a heat transfer between two objects at different temperatures. Energy can take many different forms. Heat, light, electricity, and chemical energy are examples that we all experience. The energy contained in moving objects is called **kinetic energy.** A rock rolling down a hill, the wind blowing through the trees, water flowing over a dam (Fig. 3.7), or electrons speeding around the nucleus of an atom are all examples of kinetic energy. **Potential energy** is stored energy that is latent but available for use. A rock poised at the top of a hill and water stored behind a dam are examples of potential energy. **Chemical energy,** like the energy stored in the food that you eat and the gasoline that you put into your car, is potential energy that can be released to do useful work. Energy is usually measured in **joules** (J). One joule is the work done when one kg is accelerated 1 m per second ($1J = 1 Kg \cdot m^2/s^2$). When people count calories in the food they eat, they are actually talking about kilocalories (1,000 calories), where one calorie is the amount of heat energy needed to raise 1 g of water's temperature from 14.5° C to 15.5° C. Sometimes it helps to think of the energy we talk about in the environment in more familiar units, so we will use the **doughnut,** where one glazed yeast ring doughnut is 199 kilocalories, which is about 835 kilojoules (kJ).

Work is the transfer of energy from one mechanical system to another, so it can also be measured in joules or doughnuts. **Power** is the rate that work is done and is usually measured in watts

FIGURE 3.7 Water stored behind this dam represents potential energy. Water flowing over the dam has kinetic energy, some of which is converted to heat. This kinetic energy is sometimes translated into electricity at hydro generating stations, an important source of power in Canada.

(W = J per second). A 50W light bulb is transforming 50 joules per second (or about one bite of doughnut per hour) of electrical energy into light energy and a fair amount of heat. **Heat** is a *process* rather than a *substance*. It is the transfer of energy between two objects due to temperature differences, whereas **temperature** is a property that is directly proportional to the kinetic energy of the substance under examination.

Energy that is diffused, dispersed, and low in temperature is considered low-quality energy because it is difficult to gather and use for productive purposes. The heat stored in the oceans, for instance, is low-quality. Conversely, energy that is intense, concentrated, and high in temperature is high-quality energy because of its usefulness in carrying out work. The intense flames of a very hot fire or high-voltage electrical energy are examples of high-quality forms that are valuable to humans. These distinctions are important, because many of our most common energy sources are low-quality and must be concentrated or transformed into high-quality before they are useful to us.

Conservation of Matter

Under ordinary circumstances, matter is neither created nor destroyed but rather is recycled over and over again. Some of the molecules that make up your body probably contain atoms that once made up the body of a dinosaur and most certainly were part of many smaller prehistoric organisms, as chemical elements are used and reused by living organisms. Matter is transformed and combined in different ways, but it doesn't disappear; everything goes somewhere. These statements paraphrase the physical principle of **conservation of matter.**

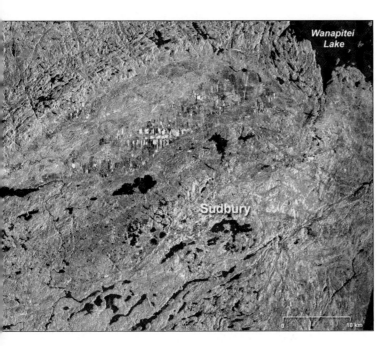

FIGURE 3.8 The Sudbury, Ontario, impact crater from a meteorite is almost 2 billion years old and 250 km across.

Conservation of Matter
Although it can be moved or transformed, matter cannot be created or destroyed.

At a global scale, the ramifications of this physical law are pretty clear. We can move matter (e.g., solid waste, fossil fuels) around, but we can't get rid of it or create it. Later, when we discuss biogeochemical cycles, we will see how humans usually affect them by accelerating the movement of matter or altering its pathway. In rare instances, matter enters the global ecosystem as large meteorites. Although small relative to the total mass of the earth, sometimes the addition can have major consequences, such as when economically and environmentally important metals landed in the Sudbury, Ontario, area hundreds of millions of years ago (Fig. 3.8).

Thermodynamics and Energy Transfers

Organisms use gases, water, and nutrients and then return them to the environment in altered forms as by-products of their metabolic processes. Year after year, century after century, the same atoms find endless reincarnation in new molecules synthesized by succeeding organisms as they feed, grow, and die. This exchange and continuity are made possible, however, by something that cannot be recycled: energy. *Energy must be supplied from an external source to keep biological processes running.* Energy flows in a one-way path through biological systems and eventually into some low-temperature sink such as outer space. It can be used to accomplish work as it flows through the system, and it can be stored temporarily in the chemical bonds of organic molecules, but eventually it is released and dissipated.

The study of thermodynamics deals with how energy is transferred in natural processes. More specifically, it deals with the rates of flow and the transformation of energy from one form or quality to another. Thermodynamics is a complex, quantitative discipline, but you don't need a great deal of math to understand some of the broad principles that shape our world and our lives.

The **first law of thermodynamics** states that energy is *conserved;* that is, it is neither created nor destroyed under normal conditions. It may be transferred from one place or object to another, but the total amount of energy remains the same. Similarly, energy may be transformed, or changed from one form to another (for example, from the energy in a chemical bond to heat energy), but the total amount is neither diminished nor increased.

The **second law of thermodynamics** states that, with each successive energy transfer or transformation in a system, less energy is available to do work. This is not a contradiction of the first law; the energy is not lost or destroyed, merely degraded or dissipated from a higher-quality form to a lower-quality form. We can think of this process as an energy "expenditure," or the "cost" in terms of useful energy of doing work.

The conservation of matter and energy laws got a bit more slippery after Einstein's time. It is possible, as we see from nuclear reactors and bombs, to simultaneously create energy and destroy matter ($E = mc^2$). But for most practical purposes in environmental science, the two laws hold. The second law of thermodynamics is often harder for people to understand, but it plays out in many important ways in the natural world. The water flowing over a dam (Fig. 3.7) will have a higher temperature than the water stored behind the dam, but some of this kinetic energy heats the air over top of the river. The transformation from potential to kinetic energy has not been 100 percent efficient. Some of the fuel put into the bus you ride to school, or that you eat with your supper, will end up not doing useful work such as moving the bus down the road or building and replacing tissue in your body. It will be lost to the bus and your body as waste heat.

Thermodynamics
1. Although it can be moved or transformed, energy cannot be either created or destroyed.
2. When energy is moved or transformed, some is lost as heat.

ENERGY FOR LIFE

Ultimately, most organisms depend on the sun for the energy needed to create structures and carry out life processes. A few rare biological communities live in or near hot springs or thermal vents in the ocean where hot, mineral-laden water provides energy-rich chemicals that form the basis for a limited and unique way of life. For most of us, however, the sun is the ultimate energy source. In this section, we will look at how primary producers (plants and photosynthesizing bacteria and protists) capture solar energy and use it to create organic molecules that are essential for life.

Solar Energy: Warmth and Light

Our sun is a star, a fiery ball of exploding hydrogen gas. Its thermonuclear reactions emit powerful forms of radiation, including potentially deadly ultraviolet and shortwave radiation (Fig. 3.9), yet life here is nurtured by, and dependent upon, this searing, energetic source. Solar energy is essential to life for two main reasons.

First, the sun provides warmth. Most organisms survive within a relatively narrow temperature range. In fact, each species has its own range of temperatures within which it can function normally. At very high temperatures, biomolecules break down or become distorted and nonfunctional. At very low temperatures, the chemical reactions of metabolism occur too slowly to enable organisms to grow and reproduce. The earth's water and atmosphere help to moderate, maintain, and distribute the sun's heat.

Second, organisms depend on solar radiation for life-sustaining energy, which is captured by green plants, algae, and some bacteria in a process called **photosynthesis.** Photosynthesis converts radiant energy into useful, high-quality chemical energy in the bonds that hold together organic molecules.

How much of the available solar energy is actually used by organisms? The amount of incoming, extraterrestrial solar radiation is enormous, about 1,372 W/m^2, or about 1 doughnut/m^2 every 10 minutes, at the top of the atmosphere. However, not all of this radiation reaches the earth's surface. More than half of the incoming sunlight may be reflected or absorbed by atmospheric clouds, dust, and gases. In particular, harmful, short wavelengths are filtered out by gases (such as ozone) in the upper atmosphere; thus, the atmosphere is a valuable shield, protecting life-forms from harmful doses of ultraviolet and other forms of radiation. Even with these energy reductions, however, the sun provides much more energy than biological systems can harness.

Of the solar radiation that does reach the earth's surface, about 10 percent is ultraviolet, 45 percent is visible, and 45 percent is infrared. Most of that energy is absorbed by land or water or is reflected into space by water, snow, and land surfaces. Seen from outer space, the earth shines about as brightly as Venus.

Fortunately for us, some radiation is captured by organisms through photosynthesis. Even then, however, the amount of energy that can be used to build organic molecules is further reduced. Photosynthesis can use only certain wavelengths of solar energy that are within the visible light range of the electromagnetic spectrum. These wavelengths are in the ranges we perceive as red and blue light. Furthermore, half, or more, of the light energy absorbed by leaves is consumed as water evaporates. Consequently, only about 1 to 2 percent of the sunlight falling on plants is captured for photosynthesis. This small percentage represents the energy base for virtually all life in the biosphere!

How Does Photosynthesis Capture Energy?

Photosynthesis occurs in tiny membranous organelles called chloroplasts that reside within the cells of most photosynthesizers

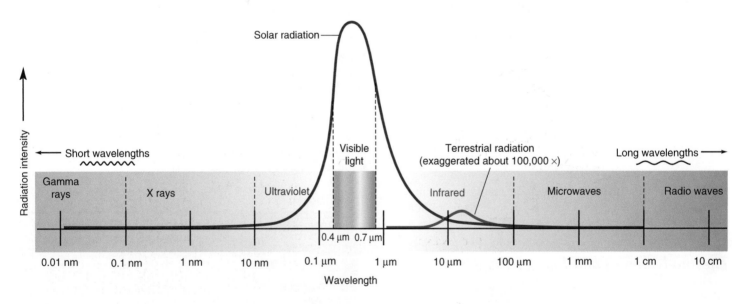

FIGURE 3.9 The electromagnetic spectrum. Our eyes are sensitive to light wavelengths, which make up nearly half the energy that reaches the earth's surface (represented by the area under the curve). Photosynthesizing plants use the most abundant solar wavelengths (light and infrared). The earth reemits lower-energy, longer wavelengths, mainly the infrared part of the spectrum.

PART ONE   Fundamental Principles of Ecology and Environmental Science

(see Fig. 3.6). The most important key to this process is chlorophyll, a unique green molecule that can absorb light energy and use it to create high-energy chemical bonds in compounds that serve as the fuel for all subsequent cellular metabolism. Chlorophyll doesn't do this important job all alone, however. It is assisted by a large group of other lipid, sugar, protein, and nucleotide molecules. Together these components carry out two interconnected cyclic sets of reactions (Fig. 3.10).

Photosynthesis begins with a series of steps called light-dependent reactions, because they occur only while light is being received by the chloroplast. During these reactions, water molecules are split, releasing molecular oxygen. This is the source of all the oxygen in the atmosphere on which all animals, including humans, depend for life. The other products of the light reactions are small, mobile, high-energy molecules that serve as the fuel or energy source for the next set of processes, the light-independent reactions. These reactions, as their name implies, can occur in the chloroplast after the light has been turned off. The enzymes in this complex use energy captured from light to add a carbon atom (from carbon dioxide) to a small sugar molecule.

In most temperate-zone plants, photosynthesis can be summarized in the following equation:

$$6H_2O + 6CO_2 + \text{solar energy} \xrightarrow{\text{chlorophyll}} C_6H_{12}O_6 \text{ (sugar)} + 6O_2$$

We read this equation as "water plus carbon dioxide plus energy produces sugar plus oxygen." The reason the equation uses six water and six carbon dioxide molecules is that it takes six carbon atoms to make the sugar product. If you look closely, you will see that all the atoms in the reactants balance with those in the products. This is consistent with the law of conservation of matter, as discussed above.

You might wonder how making a simple sugar benefits the plant. Obviously, we couldn't live on a diet of just sugar, and photosynthesizers are no different. But glucose is an energy-rich compound that serves as the central, primary fuel for all metabolic processes of cells. The energy in its chemical bonds—the ones created by photosynthesis—can be released by other enzymes and used to make other molecules (lipids, proteins, nucleic acids, or other carbohydrates), or it can drive kinetic processes such as movement of ions across membranes, transmission of messages, changes in cellular shape or structure, or movement of the cell itself in some cases. This process of releasing chemical energy, called **cellular respiration,** involves splitting carbon and hydrogen atoms from the sugar molecule and recombining them with oxygen to recreate carbon dioxide and water. The net chemical reaction, then, is the reverse of photosynthesis:

$$C_6H_{12}O_6 + 6O_2 \longrightarrow 6H_2O + 6CO_2 + \text{released energy}$$

Note that in photosynthesis, energy is *captured,* while in respiration, energy is *released.* Similarly, photosynthesis *consumes* water and carbon dioxide to *produce* sugar and oxygen, while respiration does just the opposite. In both sets of reactions, energy is stored temporarily in chemical bonds, which constitute a kind of energy currency for the cell.

We animals don't have chlorophyll and can't carry out photosynthetic food production. We do have the components for cellular respiration, however. In fact, this is how we get all our energy for life. We eat plants—or other animals that have eaten plants—and break down the organic molecules in our food through cellular respiration to obtain energy. In the process, we also consume oxygen and release carbon dioxide, thus completing the cycle of photosynthesis and respiration. Later in this chapter we will see how these feeding relationships work.

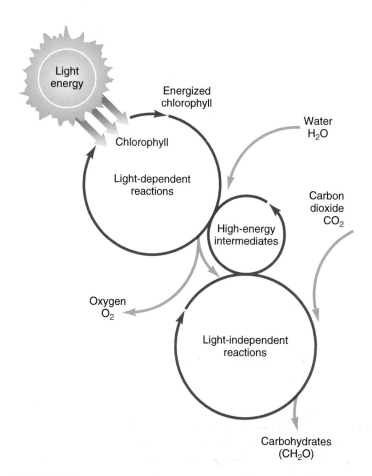

FIGURE 3.10 Photosynthesis represents the capture of light energy by chlorophyll molecules and the use of that energy to synthesize energy—rich chemical compounds that can be used by cells to drive the processes of life. Carbon dioxide and water are consumed in these reactions while oxygen and carbohydrates are produced.

FROM INDIVIDUALS TO ECOSYSTEMS

While cellular and molecular biologists study life processes at the microscopic level, ecologists study interactions at the individual organism, population, biotic community, or ecosystem level.

Individual organisms interact with their environment and other individuals as they take up resources, grow, reproduce, and eventually die. Every individual organism, whether animal, plant, fungi, protist, or bacteria is a member of a particular **species.** The definition of a species is hotly debated among biologists, but most agree a close degree of genetic relatedness through shared ancestry is necessary.

A **population** consists of all the members of a species living in a given area at the same time. Chapter 4 deals further with population growth and dynamics. All of the populations of organisms living in a particular area make up a **biological community.** What populations make up the biological community of which you are a part? The population sign marking your city limits announces only the number of humans who live there, disregarding the other populations of animals, plants, fungi, and microorganisms that are part of the biological community within the city's boundaries. Characteristics of biological communities are discussed in more detail in Chapter 5.

An ecological system, or **ecosystem,** is composed of a biological community and its environment. The environment includes abiotic factors (nonliving components), such as climate, water, minerals, and sunlight, as well as biotic factors, such as organisms, their products (secretions, wastes, and remains), and effects in a given area. Broadening our perspective from the biological community to the community plus its surroundings fosters study of the ways in which energy and materials are obtained, processed, stored, or cycled between components of the ecosystem. This systems approach tends to focus more on roles played by various members of the community rather than on the unique life stories of the individuals themselves. Looking at the organization and functions of a system can give us valuable insights into how it works.

For simplicity's sake, we think of ecosystems as fixed ecological units with distinct boundaries. If you look at a patch of woods surrounded by farm fields, for instance, a relatively sharp line separates the two areas, and conditions such as light levels, wind, moisture, and shelter are quite different in the woods than in the fields around them. Because of these variations, distinct populations of plants and animals live in each place. By studying each of these areas, we can make important and interesting discoveries about who lives where and why and about how conditions are established and maintained there.

The division between the fields and woods may not be as definite and impenetrable as we might imagine at first glance, however. Air, of course, moves freely from one to another, and the runoff after a rainfall may carry soil, leaf litter, and even live organisms between the areas. Birds may feed in the field during the day but roost in the woods at night, giving them roles in both places. Are they members of the woodland community or the field community? Is the edge of the woodland ecosystem where the last tree grows, or does it extend to every place that has an influence on the woods? Depending on how tenuous the connections you are willing to consider, the whole world might be part of an ecosystem.

As you can see, it may be difficult to draw clear boundaries around communities and ecosystems. To some extent we define these units by what we want to study and how much information we can handle. Thus, an ecosystem might be as large as a whole watershed or as small as a pond or even the surface of your skin. Even though our choices may be somewhat arbitrary, we still can make important discoveries about how organisms interact with each other and with their environment within these units. The woods are, after all, significantly different from the fields around them.

Like the woodland we just considered, most ecosystems are open, in the sense that they exchange materials and organisms with other ecosystems. A stream ecosystem is an extreme example. Water, nutrients, and organisms enter from upstream and are lost downstream. The species and numbers of organisms present at a particular spot in the stream may be relatively constant, but they are made up of continually changing individuals. Some ecosystems are relatively closed, in the sense that very little enters or leaves them. A balanced aquarium is a good example of a closed ecosystem. Aquatic plants, animals, and decomposers can balance material cycles in the aquarium if care is taken to balance their populations. Because of the second law of thermodynamics, however, every ecosystem must have a constant inflow of energy and a way to dispose of heat. Thus, at least with regard to energy flow, every ecosystem is open.

Many ecosystems have mechanisms that maintain composition and functions within certain limits. A forest tends to remain a forest for the most part and to have forestlike conditions if it isn't disturbed by outside forces. Some ecologists suggest that ecosystems—or perhaps all life on the earth—may function as superorganisms (What Do You Think? p. 48).

Food Chains, Food Webs, and Trophic Levels

Photosynthesis is the base of the energy economy of all but a few special ecosystems, and ecosystem dynamics are based on how organisms share food resources. In fact, one of the major properties of an ecosystem is its **productivity,** the amount of energy (accumulated material) and other **biomass** in a given area during a given period of time. Photosynthesis is described as *primary productivity* because it is the basis for almost all other growth in an ecosystem. Manufacture of biomass by organisms like us that eat energy-rich biomass rather than photosynthesizing, is termed *secondary productivity.*

Productivity, whether primary or secondary, is the *rate* of accumulation of energy. It is usually reported in units of energy per unit area (or volume) per unit time (e.g., doughnuts/m^2/year), although sometimes rather than energy, biomass or even carbon is reported. It is important to distinguish productivity from **standing crop,** the amount of energy or biomass sitting in an ecosystem at a given time. Sometimes an ecosystem can have high productivity and low standing crop, or the reverse. There are some lakes where the phytoplankton have a high rate of primary production, but they are being eaten so quickly by microcrustaceans that there is never much biomass in the water at a given time. An unmowed pasture can have lots of standing crop of grass present in the dog days of August, but a very low rate of primary production.

Think about what you have eaten today and trace it back to its photosynthetic source. If you have eaten an egg, you can trace it back to a chicken, which ate corn. This is an example of a

PART ONE Fundamental Principles of Ecology and Environmental Science

FIGURE 3.11 Each time an organism feeds, it becomes a link in a food chain. In an ecosystem, food chains become cross-connected when predators feed on more than one kind of prey, and all organisms are eaten by more than one kind of consumer, thus forming a food web. The arrows in this diagram and in Figure 3.12 indicate the direction in which matter and energy are transferred through feeding relationships. Only a few representative relationships are shown here. What others might you add?

food chain, a linked feeding series. Now think about a more complex food chain involving you, a chicken, a corn plant, and a grasshopper. The chicken could eat grasshoppers that had eaten leaves of the corn plant. You also could eat the grasshopper directly—some humans do. Or you could eat corn yourself, making the shortest possible food chain. Humans have several options as to where we fit into food chains.

In ecosystems, some consumers feed on a single species, but most consumers have multiple food sources. In this way, individual food chains become interconnected to form a **food web.** Figure 3.11 shows feeding relationships among some of the larger organisms in a woodland and lake community. If we were to add all the insects, worms, and microscopic organisms that belong in this picture, however, we would have overwhelming complexity. Perhaps you can imagine the challenge ecologists face in trying to quantify and interpret the precise matter and energy transfers that occur in a natural ecosystem!

An organism's feeding status in an ecosystem can be expressed as its **trophic level** (from the Greek *trophe,* food). In our first example, the corn plant is at the **producer** level; it transforms solar energy into chemical energy, producing food molecules.

Other organisms in the ecosystem are **consumers** of the chemical energy originally harnessed by producers. An organism that eats producers is a primary consumer. An organism that eats primary consumers is a secondary consumer, which may, in turn, be eaten by a tertiary consumer, and so on. Most terrestrial food chains are relatively short (seeds → mouse → owl), but aquatic food chains may be quite long (microscopic algae → copepod → minnow → crayfish → bass → osprey). The length of a food chain also may reflect the physical characteristics of a particular ecosystem. A harsh arctic landscape has a much shorter food chain than a temperate or tropical one (Fig. 3.12).

Organisms can be identified both by the trophic level at which they feed and by the *kinds* of food they eat (Fig. 3.13). **Herbivores** are plant eaters, **carnivores** are flesh eaters, and **omnivores** eat both plant and animal matter. What are humans? We are natural omnivores, by history and by habit. Tooth structure is an important clue to understanding animal food preferences, and humans are no exception. Our teeth are suited for an omnivorous diet, with a combination of cutting and crushing surfaces that are not highly adapted for one specific kind of food, as are the teeth of a wolf (carnivore) or a horse (herbivore).

Chaos or Stability in Ecosystems?

The questions we ask are shaped by our conceptual frameworks or paradigms, models that serve as general explanations of how things work. You may be surprised to learn that not all ecologists agree about something as central to their science as the basic properties of biological communities and ecosystems.

One paradigm for nature is that of pioneer biogeographer F. E. Clements (1874–1945), who argued that biological communities behave as if they are superorganisms. Species and populations, in his view, work together something like the organs of the body to maintain a stable, well-balanced set of functions and composition. Clements borrowed the physiological term **homeostasis** (from the Greek *homeo*, same, and *stasis*, stationary) to describe the equilibrium achieved by nature when left undisturbed.

Clements also saw a similarity between the growth and development of an individual and the regular developmental stages of an ecosystem from bare ground or fallow field to mature forest (see Chapter 5). He claimed that every landscape has a characteristic "climax" community toward which it will develop if free from external interference. The climax community, he believed, represents the maximum state of complexity and stability possible for a given set of environmental conditions.

Many people are attracted by this purposeful, deterministic view of nature. Others, however, believe that assuming intentional design or plan in nature is not substantiated by observation. H. A. Gleason, for instance, who was a contemporary of Clements, regarded biological communities as merely chance associations of species able to migrate into and live in a specific place at a particular time. The presence or absence of any one species, Gleason argued, is independent of all others. He suggested that we see ecosystems as stable and uniform only because our lifetimes are so short and our perspective so limited.

Clements' views were very influential in the early days of ecology, and for many years homeostasis generated by diversity was regarded as the most important principle in nature. More recently, however, the individualistic, probabilistic views of Gleason have attracted more supporters. Nevertheless, both of these viewpoints continue to have proponents.

Recent research on cybernetics and planetary systems has awakened a new interest in the self-maintaining, equilibrating features of biological systems. The **Gaia** hypothesis, named by biophysicist James Lovelock after the ancient Greek goddess of the earth, originated in speculations about the unique environmental conditions of the earth. In contrast to our neighbouring planets, which are either much too hot or much too cold for life as we know it, most of the earth is just right for our existence (some people call Earth the Goldilocks planet).

This fortunate set of circumstances is more than a happy accident, according to the Gaia hypothesis. Rather it is the result of active participation by the living biota to create and sustain a livable environment. In this view, the whole earth operates as if it were a single superorganism. This homeostasis depends on active feedback processes operated automatically and unconsciously by the biota. Although Lovelock dissociates himself from any mystical or religious implications of his theory, many people believe it suggests design and meaning in the world.

The Gleasonian view of nature as individualistic and unpredictable also has modern supporters. Paleobiogeography (the study of distribution of plants and animals through history based on analysis of fossils and sediments) shows a much greater variability in landscapes than Clements thought possible. What appears to be a stable forest or prairie community may have had dramatically different species composition and ecosystem characteristics over the millennia. From this perspective, patchiness, variability, and randomness of species distribution in communities are more important and characteristic than homogeneity. Constant change seems to be the rule rather than the exception for much of nature.

What triggers these changes? Random, often catastrophic events appear to be major forces in shaping our world. Erratic climate change, irregular volcanic eruptions, giant asteroids crashing into the earth, and the slow but inexorable processes of continental drift, mountain building, and erosion modify landscapes in complex ways. The science of chaos and catastrophe allows us to find patterns—or at least limits—in what had hitherto seemed merely noise. An important revelation from this view is that complex systems can exhibit emergent properties, characteristics that could not have been predicted from a study of individual components.

Both of these contrasting views of nature have interesting implications. From the Gaia hypothesis we might infer that even though nature has self-correcting mechanisms, there may be thresholds of disruption beyond which it may not be able to recover. If nature has a plan for survival, it may not include us. From chaos and catastrophe theory, we realize that there are great uncertainties in nature. If we can't predict what may happen next, we might want to leave a margin of safety in how much we disturb nature.

Perhaps even more useful is the illustration of the importance of conceptual frameworks in shaping our views. If you look for examples of stability and order, you will find them. At the same time, if you look for change, you will find many examples of that as well. Probably neither of these paradigms is exclusively right, but both give useful insights. Which is closest to your worldview? Do you see the world in terms of harmony and stability? Or do you see more evidence of randomness and constant change? What observations or experiments would enable us to pick the paradigm that better describes our world?

Want More?

Barlow, C. (editor) 1991. *From Gaia to Selfish Genes: Selected Writings in the Life Sciences.* The MIT Press.

FIGURE 3.12 Harsh environments tend to have shorter food chains than environments with more favourable physical conditions. Compare the arctic food chains depicted here with the longer food chains in the food web in Figure 3.11.

The **detrital food chain** is a critically important but sometimes overlooked part of the food web (Fig. 3.13). Every trophic level contributes dead organisms to its base. The consumers of this dead material vary from **scavengers,** such as crows, jackals, and vultures, which eat carcasses of large, dead animals, to the **decomposer** bacteria and fungi that **remineralize** matter so nutrients can once again be taken up by primary producers.

Ecological Pyramids

If we arrange the organisms in a food chain according to trophic levels, with a block for each trophic level proportional to its productivity, they form a pyramid, with a broad base representing primary producers and only a small rate of production in the highest trophic levels (Fig. 3.14).

Why is productivity of trophic levels a pyramid shape? Not all of the primary production gets consumed; some of it ends up in the detrital food chain. Some of what gets consumed doesn't get assimilated; this energy also ends up at the base of the detrital food chain. Finally, some of the energy consumed and assimilated is not accumulated as biomass, but respired in doing the work of metabolism and lost as heat. So each block in the pyramid of productivity is smaller than that of the next lower trophic

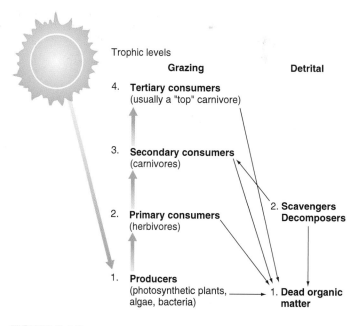

FIGURE 3.13 Organisms in an ecosystem may be identified by how they obtain food for their life processes (producer, herbivore, carnivore, omnivore, scavenger, decomposer) or by consumer level (producer; primary, secondary, or tertiary consumer) or by trophic level (1st, 2nd, 3rd, 4th).

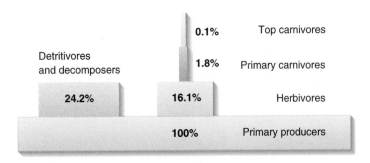

FIGURE 3.14 A classic example of an energy pyramid from Silver Springs, Florida. The numbers in each bar show the percentage of the energy captured in the primary producer level that is incorporated in the biomass of each succeeding level. Detritivores and decomposers feed at every level but are shown attached to the producer bar because this level provides most of their energy.

Source: Data from Howard T. Odum, "Trophic Structure and Productivity of Silver Springs, Florida," in *Ecological Monographs,* 27:55–112, 1957, Ecological Society of America.

level (Fig. 3.15). Each trophic level contributes to the productivity of the level above it as well as the **detritivores** and decomposers (who also contribute to their own production, since they die too).

The **10 percent rule** of ecosystem productivity is that each trophic level in the **grazing food chain** has about 10 percent of the productivity of the next lower level. This helps us understand that the number of trophic levels in any ecosystem must be limited. If we try to build pyramids of standing crop or numbers in a trophic level, sometimes it can be inverted. Think of the total biomass of a herd of sheep chewing their pasture down to a few millimetres of grass. Imagine thousands of beetles chewing on the leaves of just one tree!

MATERIAL CYCLES AND LIFE PROCESSES

To our knowledge, the earth is the only planet in our solar system that provides a suitable environment for life as we know it. Even our nearest planetary neighbours, Mars and Venus, do not meet

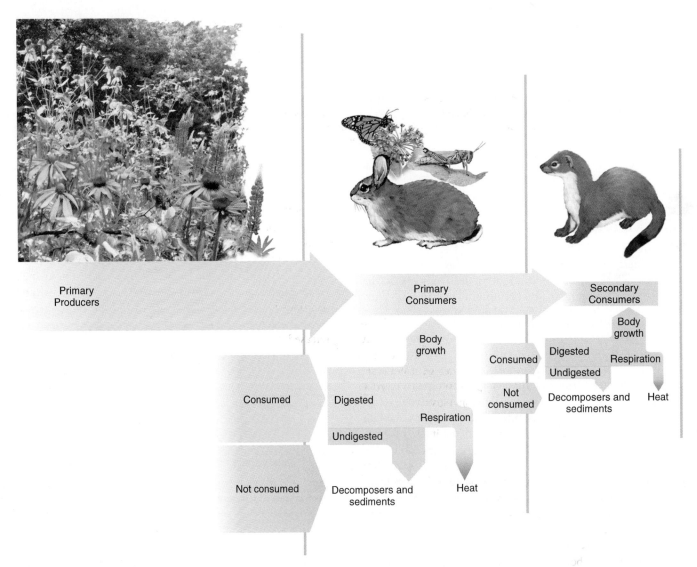

FIGURE 3.15 The productivity pyramid is understood more clearly if we trace the pathway of energy and nutrients through the trophic levels.

PART ONE   Fundamental Principles of Ecology and Environmental Science

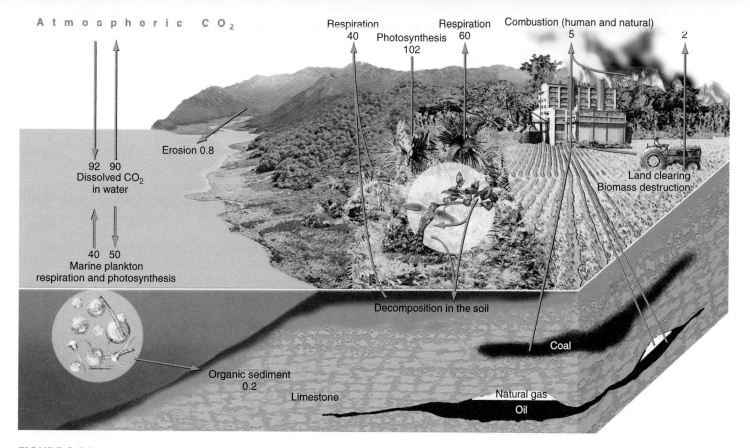

FIGURE 3.16 Atmospheric carbon dioxide is the "source" of carbon in the carbon cycle. Numbers indicate approximate exchange of carbon in giga-tonnes (Gt) per year. It passes into ecosystems through photosynthesis and is captured in the bodies and products of living organisms. It is released to the atmosphere by weathering, respiration, and combustion. Carbon may be locked up for long periods in both organic (coal, oil, gas) and inorganic (limestone, dolomite) geological formations, which are, therefore, referred to as carbon "sinks."

Source: Intergovernmental Panel for Climate Change (IPCC), 2002.

these requirements. Maintenance of these conditions depends on a constant recycling of materials between the biotic (living) and abiotic (nonliving) components of ecosystems, known as **bio-geochemical cycles.**

The Carbon Cycle

Carbon serves a dual purpose for organisms: (1) it is a structural component of organic molecules, and (2) the energy-holding chemical bonds it forms represent energy "storage." The **carbon cycle** begins with the intake of carbon dioxide (CO_2) by photosynthetic organisms (Fig. 3.16). Carbon (and hydrogen and oxygen) atoms are incorporated into sugar molecules during photosynthesis. Carbon dioxide is eventually released during respiration, closing the cycle.

The path followed by an individual carbon atom in this cycle may be quite direct and rapid, depending on how it is used in an organism's body. Imagine for a moment what happens to a simple sugar molecule you swallow in a glass of fruit juice. The sugar molecule is absorbed into your bloodstream where it is made available to your cells for cellular respiration or for making more complex biomolecules. If it is used in respiration, you may exhale the same carbon atom as CO_2 the same day.

Alternatively, that sugar molecule can be used to make larger organic molecules that become part of your cellular structure. The carbon atoms in it could remain a part of your body until it decays

after death. Similarly, carbon in the wood of a thousand-year-old tree will be released only when the wood is digested by fungi and bacteria that release carbon dioxide as a by-product of their respiration.

Can you think of examples where carbon may not be recycled for even longer periods of time, if ever? Fossil fuels are the compressed, chemically altered remains of plants (coal) or microorganisms (oil) that lived millions of years ago. Their carbon atoms (and hydrogen, oxygen, nitrogen, sulphur, etc.) are not released until the coal and oil are burned. Enormous amounts of carbon also are locked up as calcium carbonate ($CaCO_3$), used to build shells and skeletons of marine organisms from tiny proto-zoans to corals. These shells and skeletons in turn become components of limestone, mostly deposited on the ocean floor of continental shelves or platforms. The world's extensive surface limestone deposits are biologically formed calcium carbonate from ancient oceans, exposed by geological events. The carbon in lime-stone has been locked away for millennia, which is the likely fate of carbon currently being deposited in ocean sediments. Eventually, even the deep ocean deposits are recycled as they are drawn into deep molten layers and released via volcanic activity. Geologists estimate that every carbon atom on the earth has made about 30 such round trips over the last 4 billion years.

How does tying up so much carbon in the bodies and by-products of organisms affect the biosphere? Favourably. It helps balance CO_2 generation and utilization. Carbon dioxide is one of the so-called greenhouse gases because it absorbs heat radiated from the earth's surface, retaining it instead in the atmosphere. This phenomenon is discussed in more detail in Chapter 8. Photosynthesis and deposition of $CaCO_3$ remove atmospheric carbon dioxide; therefore, vegetation (especially large forested areas such as the boreal forests) and the oceans are very important **carbon sinks** (storage deposits). Cellular respiration and combustion both release CO_2, so they are referred to as carbon sources of the cycle.

Presently, natural fires and human-created combustion of organic fuels (mainly wood, coal, and petroleum products) release huge quantities of CO_2 at rates that seem to be surpassing the pace of CO_2 removal. Scientific concerns over the linked problems of increased atmospheric CO_2 concentrations, massive deforestation, and reduced productivity of the oceans due to pollution are discussed in Chapters 8 and 9.

The Nitrogen Cycle

Organisms cannot exist without amino acids, peptides, nucleic acids, and proteins, all of which are organic molecules containing nitrogen. The nitrogen atoms that form these important molecules are provided by producer organisms. Using some of the energy collected via photosynthesis, primary producers assimilate (take up) inorganic nitrogen from the environment and use it to build their own protein molecules, which are eaten by consumer organisms, digested, and used to build their bodies. Even though nitrogen is the most abundant gas (about 78 percent of the atmosphere), however, plants cannot use N_2, the stable diatomic (2-atom) molecule in the air.

Where and how, then, do photosynthesizers get *their* nitrogen? The answer lies in the most complex of the gaseous cycles, the **nitrogen cycle.** Figure 3.17 summarizes the nitrogen cycle. The key natural processes that make nitrogen available are carried out by nitrogen-fixing bacteria (including some blue-green algae or cyanobacteria). These organisms have a highly specialized ability to "fix" nitrogen, meaning they change it to less mobile, more useful forms by combining it with hydrogen to make ammonia (NH_3).

Nitrite-forming bacteria combine the ammonia with oxygen, forming nitrites, which have the ionic form NO_2^-. Another group of bacteria then convert nitrites to nitrates, which have the ionic form NO_3^-, that can be absorbed and used by green plants. After nitrates have been absorbed into plant cells, they are reduced

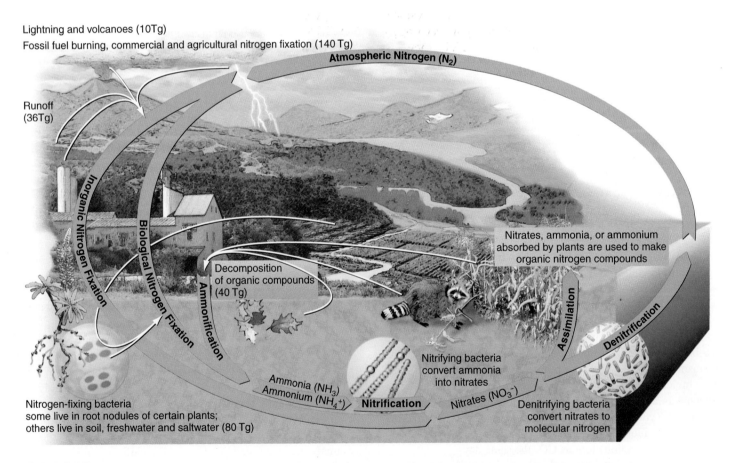

FIGURE 3.17 Nitrogen is incorporated into ecosystems when plants and bacteria use it to build their own amino acids and is released from ecosystems by bacterial decomposition. Both natural and human interactions with the nitrogen cycle are depicted here. Units are teragrams (Tg) nitrogen per year.
Source: Galloway and Cowling, 2002.

PART ONE   Fundamental Principles of Ecology and Environmental Science

to ammonium (NH_4^+), which is used to build amino acids that become the building blocks for peptides and proteins.

Members of the bean family (legumes) and a few other kinds of plants are especially useful in agriculture because they have nitrogen-fixing bacteria actually living *in* their root tissues (Fig. 3.18). Legumes and their associated bacteria enrich the soil, so interplanting and rotating legumes with crops such as corn that use but cannot replace soil nitrates are beneficial farming practices that take practical advantage of this relationship.

Nitrogen reenters the environment in several ways. The most obvious path is through the death of organisms. Their bodies are decomposed by fungi and bacteria, releasing ammonia and ammonium ions, which then are available for nitrate formation. Organisms don't have to die to donate proteins to the environment, however. Plants shed their leaves, needles, flowers, fruits, and cones; animals shed hair, feathers, skin, exoskeletons, pupal cases, and silk. Animals also produce excrement and urinary wastes that contain nitrogenous compounds. Urinary wastes are especially high in nitrogen because they contain the detoxified wastes of protein metabolism. All of these by-products of living organisms decompose, replenishing the nitrogen supply for primary producers.

How does nitrogen reenter the atmosphere, completing the cycle? Denitrifying bacteria break down nitrates into N_2 and nitrous oxide (N_2O), gases that return to the atmosphere; thus, denitrifying bacteria compete with plant roots for available nitrates. However, denitrification occurs mainly in waterlogged soils that have low oxygen availability and a high amount of decomposable organic matter. These are suitable growing conditions for many wild plant species in swamps and marshes, but not for most cultivated crop species, except for rice, a domesticated wetlands grass.

In recent years, humans have profoundly altered the nitrogen cycle. By using synthetic fertilizers, cultivating nitrogen-fixing crops, and burning fossil fuels, we now convert more nitrogen to ammonia and nitrates than all natural land processes combined. However, this excess nitrogen input is causing serious loss of soil nutrients such as calcium and potassium, acidification of rivers and lakes, and rising atmospheric concentrations of the greenhouse gas, nitrous oxide. It also encourages the spread of weeds into areas such as prairies occupied by native plants adapted to nitrogen-poor environments. Blooms of toxic algae and dinoflagellates ("red tides") in coastal areas may also be linked to excess nitrogen carried by rivers.

The Phosphorus Cycle

Minerals become available to organisms after they are released from rocks. Two mineral cycles of particular significance to organisms are phosphorus and sulphur. Why do you suppose phosphorus is a primary ingredient in fertilizers? At the cellular level, energy-rich, phosphorus-containing compounds are primary participants in energy-transfer reactions, as we have discussed. The amount of available phosphorus in an environment can, therefore, have a dramatic effect on primary productivity. Abundant phosphorus stimulates lush plant and algal growth, making it a major contributor to water pollution.

The **phosphorus cycle** (Fig. 3.19) begins when phosphorus compounds are leached from rocks and minerals over long periods of time. Because phosphorus has no atmospheric form, it is usually transported in aqueous form. Inorganic phosphorus is taken up by producer organisms, incorporated into organic molecules, and then passed on to consumers. It is returned to the environment by decomposition. An important aspect of the phosphorus cycle is the very long time it takes for phosphorus atoms to pass through it. Deep sediments of the oceans are significant phosphorus sinks of extreme longevity. Phosphate ores that now are mined to make detergents and inorganic fertilizers represent exposed ocean sediments that are millennia old. You could think of our present use of phosphates, which are washed out into the river systems and eventually the oceans, as an accelerated mobilization of phosphorus from source to sink. Aquatic ecosystems often are dramatically affected in the process because excess phosphates can stimulate explosive growth of algae and photosynthetic bacteria populations, upsetting ecosystem stability. Notice also that in this cycle, as in the others, the role of organisms is only one part of a larger picture.

The Sulphur Cycle

Sulphur plays a vital role in organisms, especially as a minor but essential component of proteins. Sulphur compounds are important determinants of the acidity of rainfall, surface water, and soil.

FIGURE 3.18 The roots of this adzuki bean plant are covered with bumps called nodules. Each nodule is a mass of root tissue containing many bacteria that help to convert nitrogen in the soil to a form the bean plants can assimilate and use to manufacture amino acids.

FIGURE 3.19 The phosphorus cycle carries this essential mineral from its source in rocks and soil, through plants and animals, and back to sediments as wastes and carrion. Units are teragrams (Tg) phosphorus per year.

In addition, sulphur in particles and tiny air-borne droplets may act as critical regulators of global climate. Most of the earth's sulphur is tied up underground in rocks and minerals such as iron disulphide (pyrite) or calcium sulphate (gypsum). This inorganic sulphur is released into air and water by weathering, emissions from deep seafloor vents, and by volcanic eruptions (Fig. 3.20).

The **sulphur cycle** is complicated by the large number of oxidation states the element can assume, including hydrogen sulphide (H_2S), sulphur dioxide (SO_2), sulphate ion (SO_4^{-2}), and sulphur, among others. Inorganic processes are responsible for many of these transformations, but living organisms, especially bacteria, also sequester sulphur in biogenic deposits or release it into the environment. Which of the several kinds of sulphur bacteria prevail in any given situation depends on oxygen concentrations, pH, and light levels.

Human activities also release large quantities of sulphur, primarily through burning fossil fuels. Total yearly anthropogenic sulphur emissions rival those of natural processes, and acid rain caused by sulphuric acid produced as a result of fossil fuel use is a serious problem in many areas (see Chapter 9). Sulphur dioxide and sulphate aerosols cause human health problems, damage buildings and vegetation, and reduce visibility. They also absorb UV radiation and create cloud cover that cools cities and may be offsetting greenhouse effects of rising CO_2 concentrations.

Interestingly, biogenic sulphur emissions by oceanic phytoplankton may play a role in global climate regulation. When ocean water is warm, tiny, single-celled organisms release dimethlysulphide (DMS) that is oxidized to SO_2 and then to SO_4 in the atmosphere. Acting as cloud droplet condensation nuclei, these sulphate aerosols increase the earth's albedo (reflectivity) and cool the earth (Fig. 3.21). As ocean temperatures drop because less sunlight gets through, phytoplankton activity decreases, DMS production falls, and clouds disappear. Thus DMS, which may account for half of all biogenic sulphur emissions, could be a feedback mechanism that keeps temperature within a suitable range for all life. This hypothesized feedback mechanism is an important part of the Gaia model of the earth as a super organism. (see What Do You Think?, p. 48).

PART ONE   Fundamental Principles of Ecology and Environmental Science

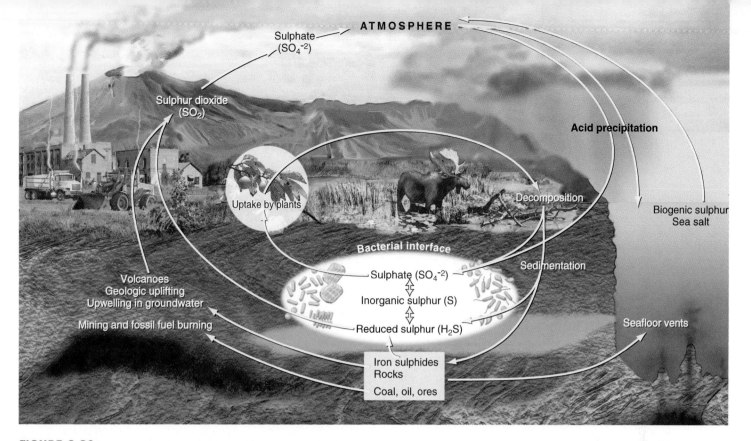

FIGURE 3.20 Sulphur is present mainly in rocks, soil, and water. It cycles through ecosystems when it is taken in by organisms. Combustion of fossil fuels causes increased levels of atmospheric sulphur compounds, which create problems related to acid precipitation.

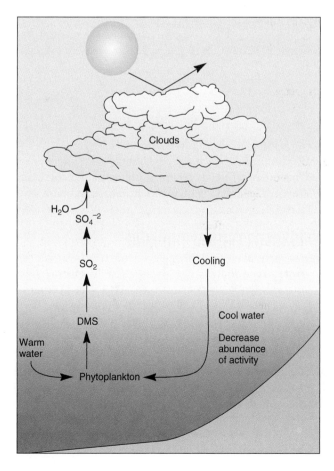

FIGURE 3.21 Dimethylsulphide (DMS), produced by oceanic phytoplankton (microscopic single-celled plants) may act as a governor, or feedback control, for global climate. As oceans warm, more DMS is produced, causing clouds to block sunlight. Less solar energy means a cooler earth, less active phytoplankton, and, thus, less DMS.

SUMMARY

Certain conditions, including the availability of required chemical elements, a steady influx of solar energy, mild surface temperatures, the presence of liquid water, and a suitable atmosphere, are essential for life on earth. Only the earth in our solar system has all these conditions. Life requires and makes possible highly organized exchanges of matter and energy between organisms and their environments.

Matter is the observable material of which the universe is composed. It generally exists in three interchangeable phases: gas, liquid, and solid. Matter is made up of atoms, which are composed of particles called protons, neutrons, and electrons. The energy of bonds that holds atoms together forms the basis for energy transfers in the bodies of living organisms and, therefore, in the biosphere.

Solar radiation ultimately provides the light and much of the heat energy needed to support life in the biosphere. Ecosystem dynamics are ultimately governed by physical laws, including the law of conservation of matter and the first and second laws of thermodynamics. The recycling of matter is the basis of the cycles of elements that occur in ecosystems and globally. Unlike matter, energy is not cycled. Energy always flows through systems in a one-way process in which some energy is converted from a high-quality, concentrated form to a lower-quality, less useful, dispersed form. We describe this increase in disorder as entropy, a fundamental limit to life. In ecosystems, solar energy enters the system and is converted to chemical energy by the process of photosynthesis. The chemical energy stored in the bonds that hold food molecules together is available for metabolism of organisms.

A species is all the organisms of the same kind that are genetically similar enough to breed in nature and produce live, fertile offspring. The populations of different species that live and interact within a particular area at a given time make up a biological community. An ecosystem is composed of a biological community together with all the biotic and abiotic factors that make up the environment in a defined area. Although ecosystem boundaries may be rather arbitrary, the holistic or systems approach to biology has provided rich insights into who lives where, when, how, and why.

Matter and energy move through the trophic levels of an ecosystem via food webs. At each trophic level, there is a lower rate of energy and matter accumulation because not all of that produced by the next lower trophic level is consumed, assimilated, or accumulated. If we build a tower of trophic levels with the blocks proportional to productivity rates, the result is, therefore, a pyramid.

The biosphere is a source of large quantities of essential elements. In a given ecosystem, these elements are constantly used and reused by living organisms. Water, carbon, nitrogen, sulphur, and phosphorus, for instance, are recycled in ecosystems through complex biogeochemical cycles.

QUESTIONS FOR REVIEW

1. Define *atom* and *element*. Are these terms interchangeable?

2. Your body contains vast numbers of carbon atoms. How is it possible that some of these carbon atoms may have been part of the body of a prehistoric creature?

3. In the biosphere, matter follows a circular pathway while energy follows a linear pathway. Explain.

4. The oceans store a vast amount of heat, but (except for climate moderation) this huge reservoir of energy is of little use to humans. Explain the difference between high-quality and low-quality energy.

5. Ecosystems require energy to function. Where does this energy come from? Where does it go? How does the flow of energy conform to the laws of thermodynamics?

6. Heat is released during metabolism. How is this heat useful to a cell and to a multicellular organism? How might it be detrimental, especially in a large, complex organism?

7. Photosynthesis and cellular respiration are complementary processes. Explain how they exemplify the laws of conservation of matter and thermodynamics.

8. What do we mean by carbon-fixation or nitrogen-fixation? Why is it important to humans that carbon and nitrogen be "fixed"?

QUESTIONS FOR CRITICAL THINKING

1. When we say that there is no "away" where we can throw things we don't want anymore, are we stating a premise or a conclusion? If you believe this is a premise, supply the appropriate conclusion. If you believe it is a conclusion, supply the appropriate premises. Does the argument change if this statement is a premise or a conclusion?

2. Suppose one of your classmates disagrees with the statement above, saying, "Of course there is an 'away.' It's anywhere out of *my* ecosystem." How would you answer?

3. A few years ago most laundry detergents contained phosphates for added cleaning power. Can you imagine any disadvantages to adding soluble phosphate to household products?

4. The first law of thermodynamics is sometimes summarized as "you can't get something for nothing." The second law is summarized as "you can't even break even." Explain what these phrases mean. Is it dangerous to oversimplify these important concepts?

5. The ecosystem concept revolutionized ecology by introducing holistic systems thinking as opposed to individualistic life history studies. Why was this a conceptual breakthrough?

6. Why is it important to recognize that ecosystems often are open and that boundaries may be fuzzy? Do these qualifications diminish the importance of the ecosystem study?

7. The holistic or systems approach to biology has sometimes been criticized as "black box" engineering. It allows us to make broad generalizations about what goes into or comes out of a system without knowing the precise details of how the system works. What do you think are the benefits and limitations of this approach?

8. Compare and contrast the views of F. E. Clements and H. A. Gleason concerning the concept of biological communities as superorganisms. How could these eminent biogeographers study the same communities and reach opposite interpretations? What evidence would be necessary to settle this question? Is lack of evidence the problem?

9. The DMS feedback control of global climate is offered by some people as evidence for the Gaia hypothesis. Why might they take this position?

KEY TERMS

acids   40
atoms   39
atomic number   39
bases   40
biogeochemical cycle   51
biological community   46
biomass   46
carbon cycle   51
carbon dating   39
carbon sinks   52
carnivores   47
cells   41
cellular respiration   45
chemical bonds   40
chemical energy   42

compound   39
conservation of matter   42
consumers   47
decay   39
decomposers   49
detrital food chain   49
detritivores   50
doughnut   42
ecology   38
ecosystem   46
elements   39
energy   42
enzymes   42
food chain   47
food web   47
Gaia hypothesis   48
grazing food chain   50

heat   42
herbivores   47
homeostasis   48
ions   40
isotopes   39
joules   42
kinetic energy   42
matter   39
metabolism   42
molecules   39
nitrogen cycle   52
omnivores   47
organic compounds   40
pH   40
phosphorus cycle   53
photosynthesis   44
population   46

potential energy   42
power   42
producers   47
productivity   46
remineralize   49
scavengers   49
species   46
stable isotope   39
standing crop   46
sulphur cycle   54
temperature   42
thermodynamics, first law   43
thermodynamics, second law   43
trophic level   47
10 percent rule   50
work   42

Web Exercises

ASSESSING NET PRIMARY PRODUCTIVITY

To explore the concept of net primary productivity and to see how it varies in different biological communities, go to http://www-eosdis.ornl.gov/NPP/html_docs/npp_site.html, a research page maintained by the Oak Ridge National Laboratory. This clickable map shows 55 sites around the world where long-term ecological data are being collected. Click on a tropical forest, a hot desert, a grassland, and a temperate forest to go to their individual pages. At the bottom of each site page, find highlighted links such as "NPP data" or "treatments" that take you to NPP measurements. Look at the range of data over time and compare the maximum amount of aboveground biomass—generally the seventh column from the left—for each of these sites. How would you account for the variation you observe between different seasons of the year and different sites in this key ecological measurement?

Chapter 4

The Dynamics of Populations

Before 2050, 80 percent of the (world) population will be projected to have below-replacement fertility.

—The Population Division of the United Nations Department of Economic and Social Affairs (2002)—

OBJECTIVES

After studying this chapter, you should be able to:

- appreciate the potential of exponential growth.
- draw a diagram of J and S curves and explain what they mean.
- explain who Thomas Malthus was and what he said about population growth.
- describe environmental resistance and discuss how it can lead to logistic or stable growth.
- define *fecundity, fertility, birth rates, life expectancy, death rates,* and *survivorship.*
- compare and contrast density-dependent and density-independent population processes.
- trace the history of human population growth.
- summarize Malthusian and Marxist theories of limits to growth as well as why technological optimists and supporters of social justice oppose these theories.

- explain the process of demographic transition and why it produces a temporary population surge.
- understand how changes in life expectancy, infant mortality, women's literacy, standards of living, and democracy affect population changes.
- evaluate pressures for and against family planning in traditional and modern societies.

Above: Whether it's a lively group of Canada geese in an urban park or a fast growing city, the dynamics of biological populations have important consequences for their environments.

WebQuest

population growth, exponential growth, logistic growth, density dependence, density independence, Thomas Malthus

The Saga of Easter Island

One of the most remote habitable places on the earth, Easter Island, lies about 3,200 km west of South America, the nearest continent, and more than 2,000 km from the closest occupied island (Pitcairn). With a mild climate and fertile volcanic soils, Easter Island should have been a tropical paradise, but when it was "discovered" by Dutch explorer Jacob Roggeveen in 1722, it resembled a barren wasteland more than a paradise. Covered by a dry grassland, the island had no trees and few bushes more than a metre tall. No animals inhabited the island except humans, chickens, rats, and a few insects.

The 2,000 people living on the island at the time eked out a pitiful existence. Having no seaworthy canoes, they couldn't venture out on the ocean to fish. With no trees to provide building materials or firewood, the island's cool, wet, windy winters were miserable; meagre gardens hardly produced enough food for subsistence.

And yet, scattered throughout the island were hundreds of immense stone heads, some as large as 30 metres tall, weighing more than 200 tonnes (Fig. 4.1). How could such a small population have carved, moved, and erected these enormous effigies? Was there once a larger and more advanced civilization on the island? If so, where did they go?

Historical studies have shown that conditions on the island were once very different than they are now. Until about 1,500 years ago, the island was covered with a lush subtropical forest and the soil was deep and fertile. Polynesian people apparently reached Easter Island about A.D. 400. Anthropological and linguistic evidence suggests they sailed from the Marquesas Islands 3,500 kilometres to the northwest. Excavations of archeological sites show that the early settlers' diet consisted mainly of porpoises, land-nesting seabirds, and garden vegetables. Populations soared, reaching as much as 20,000 on an island only about 15 km across.

By A.D. 1400 the forest appears to have disappeared completely—cut down for firewood and to make houses, canoes, and rollers for transporting the enormous statues. Without a protective forest cover, soil washed off steep hillsides. Springs and streams dried up, while summer droughts made gardens less productive. All wild land birds became extinct and seabirds no longer nested on the island. Lacking wood to build new canoes, the people could no longer go offshore to fish. Statues carved at this time show sunken cheeks and visible ribs suggesting starvation.

At this point, chaos and warfare seem to have racked the land. The main bones found in fireplaces were those of rats and humans. Cannibalism apparently was rampant as the population decreased by 90 percent. The few remaining people cowered in caves, a pitiful remnant of a once impressive civilization. When we try to imagine how people reached this condition, we wonder why they didn't control their population and conserve their resources. What were their thoughts as they cut down the last trees, stranding themselves on this island of diminishing possibilities?

FIGURE 4.1 Giant heads stare stonily at the sea on Easter Island. What happened to the people who built these mysterious artifacts?

Does this story have lessons for us? Is Easter Island an example of what could happen to the rest of us if our population grows and we use up our store of resources? The debate over the carrying capacity of the earth for humans remains one of the most contentious and important issues in environmental science. Some demographers warn that we are headed for a disaster similar to that of Easter Island. Others hope that we will be more clever and perceptive than the unhappy people who destroyed the resource base on which they depended. What do you think? How will we recognize and respond to excess population and consumption levels?

Rainbird P. "A Message for Our Future? The Rapa Nui (Easter Island) Ecodisaster and Pacific Island Environments." *World Archaeology.* 33 (3): 436–451 February 2002.

DYNAMICS OF POPULATION GROWTH

Why do some populations grow to enormous size while others do not? What causes abrupt population increases or decreases? How many individuals does it take to make a viable population? What roles do competition, genetic diversity, and ecological adaptations play in maintaining or reducing wild species? These are some of the questions asked by biologists interested in how and why population size changes through time (population dynamics). In this chapter, we will survey the factors that lead to population growth and decline. We also will consider the mechanisms that regulate total population size, distribution, and age structure. One of the most important concepts in population biology is that both density-dependent and density-independent factors play roles in determining how big a particular population in a given habitat can be.

Exponential Growth and Doubling Times

All species have the capacity for **exponential growth,** or growth at a constant *percentage* per unit of time. It is called exponential because the rate of increase can be expressed as a constant *fraction,* or exponent, by which the existing population is *multiplied.* This pattern also is called **geometric growth** because the sequence of growth follows a geometric pattern of increase, such as 2, 4, 8, 16, and so on. By contrast, a pattern of growth that increases at a constant *number* per unit of time is called **arithmetic growth.** The sequence in this case might be 1, 2, 3, 4 or 1, 3, 5, 7. Notice in these examples that a constant amount is *added* to the population (Fig. 4.2).

Organisms are able to grow exponentially, given an appropriate environment, because of the amplifying power of biological reproduction.

As you can see in Figure 4.2, the number of individuals added to a population at the beginning of a geometric growth curve is rather small. But within a very short time, the numbers begin to increase quickly because a given percentage becomes a much larger amount as the population gets bigger. The growth curve produced by this constant rate of unfettered growth is called a **J curve** because of its shape. At 1 percent per year, a population of humans or ragweed plants doubles in roughly 70 years.

The concept of exponential growth might be easiest to grasp if you imagine being offered a choice. Either you receive a cheque for $1 million today, or you get one penny on the first square of a chess board, two pennies on the second square a week later, four pennies on the third square a week after that, and so on, until the 64-square board is covered with your money after a little more than a year (63 weeks). It might at first seem wise to take the million dollars and run! But if you wait while the money put on the chess board increases exponentially (doubling every week), by the 63rd week you will get about 92×10^{15}, for a total of 184×10^{15}!

Biotic Potential

As shown in Table 4.1, many species have amazingly high reproductive rates that give them the potential to produce enormous

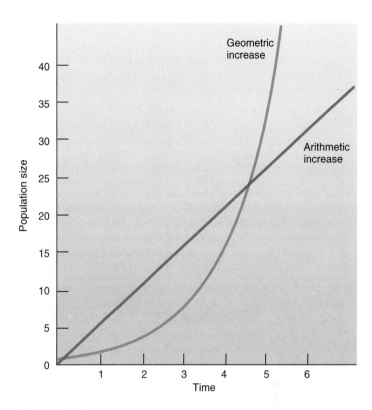

FIGURE 4.2 Arithmetic and geometric growth curves. Note that the geometric or exponential curve grows more slowly at first, but then accelerates past the arithmetic curve, which grows at a steady incremental pace throughout.

TABLE 4.1 Doubling Times of Various Organisms in an Ideal Environment

SPECIES	POPULATION DOUBLING TIME
T phage virus	3.3 minutes
Flour beetle	6.9 days
Brown rat	47 days
Southern beech tree	25 years

Source: From N. J. Gotelli, *A Primer of Ecology.*

populations very quickly, given unlimited resources and freedom from limiting factors (Fig. 4.3). We call the maximum reproductive rate of an organism its **biotic potential.** Table 4.2 shows the potential number of offspring that a single female housefly and her offspring could produce in a year. The result is astounding. Each female fly lays an average of 120 eggs in each generation. The eggs hatch and mature into sexually active adults and lay their own eggs in 56 days. In one year (seven generations), the average female could be the ancestor of 5.6 *trillion* flies. If this rate of reproduction

FIGURE 4.3 Reproduction gives many organisms the potential to expand populations explosively. The cockroaches in this kitchen could have been produced in only a few generations. A single female cockroach can produce up to 80 eggs every six months. This exhibit is in the Smithsonian Institute's National Museum of Natural History.

TABLE 4.2	Biotic Potential of Houseflies in One Year

Assuming that:
—a female lays 120 eggs per generation
—half of these eggs develop into females
—there are seven generations per year

GENERATION

1	120
2	7,320
3	446,520
4	27,237,720
5	1,661,500,920
6	101,351,520,120
7	6,182,442,727,320

Source: Data from E. J. Kormondy, *Concepts of Ecology,* 3d ed., © 1985 Harper & Row Publishers, Inc.

continued for 10 years, the whole earth would be covered several metres deep with houseflies! Fortunately, this has not happened because of factors that limit the reproductive success of houseflies. This example, however, illustrates the potential for biological populations to increase rapidly.

Population Oscillations and Irruptive Growth

In the real world there are limits to growth. When a population exceeds the carrying capacity of its environment or some other limiting factor comes into effect, death rates begin to surpass birth rates. The growth curve becomes negative rather than positive, and the population decreases as fast as, or faster than, it grew.

The extent to which a population exceeds the carrying capacity of its environment is called **overshoot,** and the severity of the dieback is generally related to the extent of the overshoot. This pattern of **population explosion** followed by a **population crash** is called **irruptive.**

Populations may go through repeated oscillating cycles of exponential growth and catastrophic crashes, as shown in Figure 4.4. These cycles may be very regular if they depend on a few simple factors, such as the seasonal light- and temperature-dependent bloom of algae in a lake, or the development time of individuals in the population. They also may be very irregular if they depend on complex environmental and biotic relationships that control cycles, such as the population explosions of migratory locusts in the desert or tent caterpillars and spruce budworms in northern forests.

Growth to a Stable Population

Not all biological populations go through these cycles of irruptive population growth and catastrophic decline. The growth rates of many species are regulated by both internal and external factors so that they come into equilibrium with their environmental resources. These species may grow exponentially when resources are unlimited, but their growth slows as they approach the carrying capacity of the environment. This pattern is called **logistic growth.**

Together, factors that tend to reduce population growth rates are called **environmental resistance.** In later sections of this chapter, we will look in more detail at these factors and how they limit

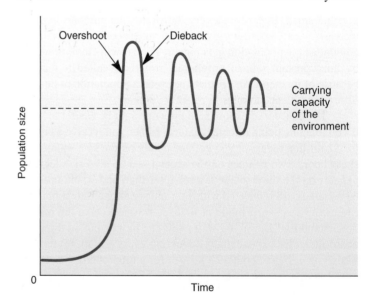

FIGURE 4.4 Population oscillations. Some species demonstrate a pattern of cyclic overshoot and dieback.

PART ONE Fundamental Principles of Ecology and Environmental Science

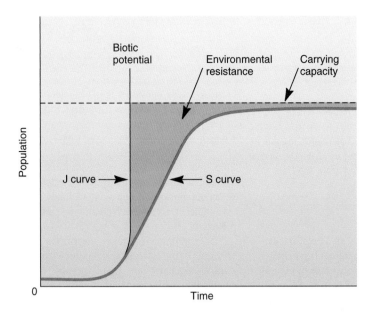

FIGURE 4.5 J and S population curves. The vertical J represents theoretical unlimited growth. The S represents population growth and stabilization in response to environmental resistance.

growth and regulate population size. First, we will see how logistic growth compares to the exponential growth and what kinds of organisms we are talking about in these different patterns.

How does the growth curve of a logistically growing population differ from the J curve of an exponentially growing population? Figure 4.5 shows an idealized comparison between exponential and logistic growth. The J curve on the left in this figure represents the growth without restraint, where numbers rise rapidly. The curve to the right represents logistic growth. We call this pattern an **S curve** because of its shape. It is also called a sigmoidal curve (for the Greek letter sigma). The area between these curves is the cumulative effect of environmental resistance. Note that the resistance becomes larger and the rate of logistic growth becomes smaller as the population approaches the carrying capacity of the environment.

Chaotic and Catastrophic Population Dynamics

Real populations under natural conditions rarely follow the smooth growth curves shown in Figures 4.4 and 4.5. Apparently haphazard fluctuations often characterize population dynamics. Population ecology has begun to incorporate ideas from nonlinear mathematics to describe these patterns as being chaotic or catastrophic.

Chaotic systems exhibit variability that, while not necessarily random, can be of a complexity whose pattern is not observable over a normal human time scale. In these systems, minute differences in initial conditions can lead to dramatically different outcomes after many iterations.

A famous metaphor for chaotic dynamics is the "butterfly effect," first described by meteorologist Edward Lorenz in 1979. Lorenz observed that seemingly insignificant variations in the

beginning parameters of computerized climate models resulted in totally different final results. It is, he said, as if a butterfly flapping its wings in Rio de Janeiro could, through a long series of unpredictable multiplying effects in weather systems, eventually result in a tornado in Texas.

Catastrophe theory was first developed by mathematicians as a purely hypothetical exercise, but some biologists regard it as a good explanation of population dynamics displaying abrupt discontinuities. As conditions change, a **catastrophic system** may jump abruptly from one seemingly steady state to another without any intermediate stages. Balancing a chair on two legs is a good model of catastrophe. You may be stable for a long time with relatively minor corrective input, but, if you pass a certain threshold, suddenly find yourself flat on your back on the floor—and, once again, stable. Among populations, the sudden outbursts and diebacks of spruce budworm in the forests of eastern North America or of lemmings in the arctic tundra often are mentioned as examples of catastrophic dynamics.

Is there any practical difference between chaos or catastrophe and simple random behaviour? Perhaps so. In a chaotic system, although you can't predict behaviour on a short time scale with any certainty, you may be able to identify outer boundary conditions beyond which the system will not go. Knowing these limits, you may be able to avoid pushing a population close to the edge of an unstable flip point.

Strategies of Population Growth

There appear to be evolutionary advantages, as well as disadvantages, in both exponential and logistic growth patterns. Although we should avoid implying intention in natural systems where the controlling forces may be entirely mechanistic, it sometimes helps us see these advantages in terms of "strategies" of adaptation and "logic" in different modes of reproduction.

Exponential Growth "Strategies"

Organisms with exponential growth patterns often tend to occupy low trophic levels in their ecosystems (Chapter 3), or to be pioneers in succession (Chapter 5). As generalists or opportunists, they move quickly into disturbed environments, grow rapidly, mature early, and produce many offspring. They usually do little to care for their offspring or protect them from predation. They depend on sheer numbers and effective dispersal mechanisms to ensure that some offspring survive to adulthood (Table 4.3). They have little investment in individual offspring, using their energy to produce vast numbers instead.

Many insects, rodents, marine invertebrates, parasites, and annual plants (especially the ones we consider weeds) follow this reproductive strategy. Their numbers generally are limited by predators or other controlling factors in the environment. They reproduce at the maximum rate possible to offset these losses. If the external factors that normally control their populations are inoperative, they tend to rapidly overshoot the carrying capacity of the environment and then die back catastrophically, as we have just seen. Among this group are the weeds, pests, or other species we

What Is Earth's Carrying Capacity for Humans?

Human numbers are doubling about every 40 years. At that rate there would be an astounding 170 quadrillion people 600 years from now. That would put one person on each square metre over the entire land surface of the earth. **Carrying capacity** reflects the limits imposed on population growth by finite space and finite resources. Our consideration of ecological footprint in Chapter 1 tells us there is no way one earth would be able to support such a huge number of humans. So, what is the carrying capacity of the earth for humans? That question cannot be meaningfully addressed without clarifying a basic assumption: the type of lifestyle on which to develop the estimates.

Different lifestyles have different resource requirements. Are people to be vegetarians or will meat be a significant part of the diet? Will the earth's resources be counted upon to provide additional amenities beyond food? The answers to those questions have a profound effect on the numbers of people the earth can sustain. For illustration purposes imagine a miniature planet we could call Terrabase. Consider the following scenarios to see how lifestyle affects carrying capacity.

Scenario 1

In this scenario, the humans living on Terrabase are vegetarians, and 100 percent of the planet's space is devoted to raising human food. Let's assume that under these conditions 1,000 people can be sustained. Carrying capacity equals 1,000.

Scenario 2

Most people enjoy meat in the diet. Assume that the people of Terrabase do not live as vegetarians, but obtain half their calories by eating herbivores such as beef cattle. This would require a significant amount of plant mass to be fed to animals. But, as you learned in Chapter 3, much energy is lost in transfer between trophic levels. It takes about 10 calories of plant food to produce one calorie of beef. After doing the calculations, it turns out that under these conditions, Terrabase will produce enough food to sustain only 180 people. Simply by eating meat, the carrying capacity has been reduced by over 80 percent.

Scenario 3

In both scenarios 1 and 2, all of the land has been used exclusively for human food production.

But people have consumer needs as well: cars, parking spaces, televisions, washers, dryers, clothing, shopping centres, and much more, all of which require space that would have to be subtracted from that used to produce food. Recreational space is also important to us. We want athletic fields, golf courses, bird sanctuaries, nature preserves, and hunting lands, but these all divert even more land away from food production.

Assume the residents of Terrabase had the high living standard of industrialized nations, requiring the immense and continuous input of chemicals, energy, paper, and other raw materials, as well as requiring land for waste disposal. How much of Terrabase would be devoted to these uses? If it is 20 percent, the carrying capacity is reduced to 140.

Wild organisms also play ecological roles important to our well-being. Leaving space for wild nature further reduces land available to produce human food.

"What is the earth's carrying capacity for humans?" is not a meaningful question until the cultural context within which people are to live has been clarified.

---

consider nuisances that reproduce profusely, adapt quickly to environmental change, and survive under a broad range of conditions.

Logistic Growth Strategies

While environmental resistance is a factor in controlling population growth in all species, those exhibiting logistic growth tend to grow more slowly and are more likely to be regulated by intrinsic characteristics than those with exponential patterns. These organisms are usually larger, live longer, mature more slowly, produce fewer offspring in each generation, and have fewer natural predators than the species below them in the ecological hierarchy. Some typical examples of this strategy are wolves, elephants, whales, and primates. Each of these species provides more care and protection for its offspring than do "lower" organisms.

Elephants, for instance, are not reproductively mature until they are 18 to 20 years old. During youth and adolescence, a young elephant is part of a complex extended family that cares for it, protects it, and teaches it how to behave. A female elephant normally conceives only once every four or five years after she matures. The gestation period is about 18 months; thus, an elephant herd doesn't

produce many babies in a given year. Since they have few enemies (except humans) and live a long life (often 60 or 70 years), however, this low reproductive rate produces enough elephants to keep the population stable, given appropriate environmental conditions.

An important underlying question to much of the discussion in this book is which of these strategies humans follow. Do we more closely resemble wolves and elephants in our population growth, or does our population growth pattern more closely resemble that of moose and rabbits? Will we overshoot the carrying capacity of our environment (or are we already doing so), or will our population growth come into balance with our resources? (See What Do You Think? above.)

FACTORS THAT INCREASE OR DECREASE POPULATIONS

Now that you have seen population dynamics in action, let's focus on what happens *within* populations, which are, after all, made up

TABLE 4.3 Characteristics of Contrasting Reproductive Strategies

EXPONENTIAL GROWTH	LOGISTIC GROWTH
1. Short life	1. Long life
2. Rapid growth	2. Slower growth
3. Early maturity	3. Late maturity
4. Many small offspring	4. Fewer large offspring
5. Little parental care or protection	5. High parental care and protection
6. Little investment in individual offspring	6. High investment in individual offspring
7. Adapted to unstable environment	7. Adapted to stable environment
8. Pioneers, colonizers	8. Later stages of succession
9. Niche generalists	9. Niche specialists
10. Prey	10. Predators
11. Regulated mainly by extrinsic factors	11. Regulated mainly by intrinsic factors
12. Low trophic level	12. High trophic level

of individuals. In this section, we will discuss how new members are added to and old members removed from populations. We also will examine the composition of populations in terms of age classes and introduce terminology that will apply in subsequent chapters.

Natality, Fecundity, and Fertility

Natality is the production of new individuals by birth, hatching, germination, or cloning, and is the main source of addition to most biological populations. Natality is usually sensitive to environmental conditions so that successful reproduction is tied strongly to nutritional levels, climate, soil or water conditions, and—in some species—social interactions between members of the species. The maximum rate of reproduction under ideal conditions varies widely among organisms and is a species-specific characteristic. We already have mentioned, for instance, the differences in natality between several different species.

Fecundity is the physical ability to reproduce, while **fertility** is a measure of the actual number of offspring produced. Because of lack of opportunity to mate and successfully produce offspring, many fecund individuals may not contribute to population growth. Human fertility often is determined by personal choice of fecund individuals.

Immigration

Organisms are introduced into new ecosystems by a variety of methods. Seeds, spores, and small animals may be floated on winds or water currents over long distances. This is a major route of colonization for islands, mountain lakes, and other remote locations. Sometimes organisms are carried as hitchhikers in the fur, feathers, or intestines of animals travelling from one place to another. They also may ride on a raft of drifting vegetation. Some animals travel as adults—flying, swimming, or walking. In some ecosystems, a population is maintained only by a constant influx of immigrants. Salmon, for instance, must be important predators in some parts of the ocean, but their numbers are maintained only by recruitment from mountain streams thousands of kilometres away.

Mortality and Survivorship

An organism is born and eventually it dies; it is mortal. **Mortality,** or death rate, is determined by dividing the number of organisms that die in a certain time period by the number alive at the beginning of the period.

Mortality is often expressed in terms of **survivorship** (the percentage of a cohort that survives to a certain age) or **life expectancy** (the probable number of years of survival for an individual of a given age). Table 4.4 shows Canadian life expectancies by sex at birth for different time periods in the twentieth century. Presumably, improved medical care and nutrition is responsible for the 15–20 year gain in life expectancy over the 70 years shown. As well, your life expectancy changes as you age. If a female born in 1950 has "made it" to 2003 (age 53), she can expect to live far longer than the 71 years of age predicted at her birth, since she has avoided the mortality factors that affected others of her cohort. Obviously the balance between natality and mortality determine whether a population grows or declines. To understand how these factors interact, we need to define some more population terms.

TABLE 4.4 Life Expectancies in Canada at Birth

TIME PERIOD	BOTH SEXES	MALES	FEMALES
1920–22	**59**	59	61
1930–32	**61**	60	62
1940–42	**65**	63	66
1950–52	**69**	66	71
1960–62	**71**	68	74
1970–72	**73**	69	76
1980–82	**75**	72	79
1990–92	**78**	75	81
2000–2001	**80**	77	82

Source: Statistics Canada "Guide to Health Statistics," Catalogue 82-573, October 11, 2000. Data available at http://www.statcan.ca/Daily/English/030925/d030925c.htm

Life span is the longest period of life reached by a given type of organism. The process of living entails wear and tear that eventually overwhelm every organism, but maximum age is dictated primarily by physiological aspects of the organism itself. There is an enormous difference in life span between different species. Some microorganisms live their whole life cycles in a matter of hours or minutes. Bristlecone pine trees in the mountains of California, on the other hand, have life spans up to 4,600 years.

Most individuals in a population do not live anywhere near the maximum life span for their species. The major factors in early mortality are predation, parasitism, disease, accidents, fighting, and environmental influences, such as climate and nutrition. Important differences in relative longevity among various types of organisms are reflected in the survivorship curves shown in Figure 4.6.

Four general patterns of survivorship can be seen in this idealized figure. Curve (*a*) is the pattern of organisms that tend to live their full physiological life span if they reach maturity and then have a high mortality rate when they reach old age. This pattern is typical of many large mammals, such as whales, bears, and elephants (when not hunted by humans), as well as humans in developed countries. Interestingly, some very small organisms, including predatory protozoa and rotifers (small, multicellular, freshwater animals), have similar survivorship curves even though

their maximum life spans may be hundreds or thousands of times shorter than those of large mammals. In general, curve (*a*) is the pattern for top consumers in an ecosystem, although many annual plants have a similar survivorship pattern.

Curve (*b*) represents the survivorship pattern for organisms for which the probability of death is generally unrelated to age. Their mortality rate is generally constant with age, and thus their survivorship curve is a straight line.

Curve (*c*) is characteristic of many songbirds, rabbits, members of the deer family, and humans in less-developed countries. They have a high mortality early in life when they are more susceptible to external factors, such as predation, disease, starvation, or accidents. Adults in the reproductive phase have a high level of survival. Once past reproductive age, they become susceptible again to external factors and the number of survivors falls quite rapidly.

Curve (*d*) is typical of organisms at the base of a food chain or those especially susceptible to mortality early in life. Many tree species, fish, clams, crabs, and other invertebrate species produce a very large number of highly vulnerable offspring, few of which survive to maturity. Those individuals that do survive to adulthood, however, have a very high chance of living most of the maximum life span for the species.

Emigration

Emigration, the movement of members out of a population, is the second major factor that reduces local population size. The dispersal factors that allow organisms to migrate into new areas are important in removing surplus members from the source population. Emigration can even help protect a species. For instance, if the original population is destroyed by some catastrophic change in their environment, their genes still are carried by descendants in other places. Many organisms have very specific mechanisms to facilitate migration of one or more of each generation of offspring.

Age Structure

An outcome of the interaction between mortality and natality is that growing or declining populations usually will have very different proportions of individuals in various age classes. Figure 4.7 shows three hypothetical population profiles that are distinguished by differences in distribution among prereproductive, reproductive, and postreproductive age classes. Pattern (*a*) is characteristic of a rapidly expanding population. Young make up a large proportion of the population. This large number of prereproductive individuals represent **population momentum** because there is potential for rapid increase in natality once the youngsters reach reproductive age. The age structure of human populations in rapidly growing, developing countries generally resembles pattern (*a*).

When natality comes into balance with mortality, the population enters a stationary phase having an age structure pattern similar to (*b*). This pattern is characteristic of nations with a stable population size, such as many European countries.

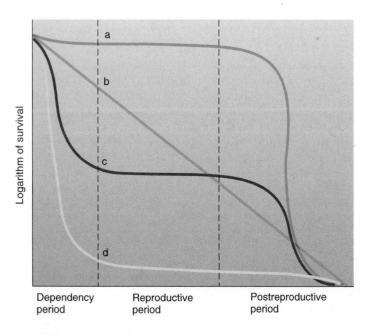

FIGURE 4.6 Four basic types of survivorship curves for organisms with different life histories. Curve (*a*) represents organisms such as humans or whales, which tend to live out the full physiological life span if they survive early growth. Curve (*b*) represents organisms such as sea gulls, for which the rate of mortality is fairly constant at all age levels. Curve (*c*) represents such organisms as white-tailed deer, moose, or robins, which have high mortality rates in early and late life. Curve (*d*) represents such organisms as clams and redwood trees, which have a high mortality rate early in life but live a full life if they reach adulthood. A steeper negative slope means a higher mortality rate.

PART ONE Fundamental Principles of Ecology and Environmental Science

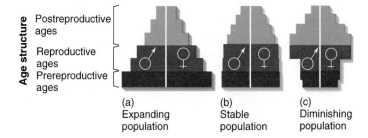

Age structure

Postreproductive ages

Reproductive ages

Prereproductive ages

(a) Expanding population

(b) Stable population

(c) Diminishing population

FIGURE 4.7 Typical age structure diagrams for hypothetical populations that are expanding, stable, or decreasing. Different numbers of individuals in each age class create these distinctive shapes. In each diagram the left half represents males, while the right half represents females.

The age structure shown by pattern (c) is characteristic of a diminishing population, where natality has fallen to a level lower than the replacement number. The bulge in the upper age classes represents adults still living but reproducing at a lower rate. When those individuals die, the population will be much smaller.

FACTORS THAT REGULATE POPULATION GROWTH

So far, we have seen that differing patterns of natality, mortality, life span, and longevity can produce quite different rates of population growth. The patterns of survivorship and age structure created by these interacting factors not only show us how a population is growing but also can indicate what general role that species plays in its ecosystem. They also reveal a good deal about how that species is likely to respond to disasters or resource bonanzas in its environment. But what factors *regulate* natality, mortality, and the other components of population growth? In this section, we will look at some of the mechanisms that determine how a population grows.

Various factors regulate population growth, primarily by affecting natality or mortality, and can be classified in different ways. They can be *intrinsic* (operating within individual organisms or between organisms in the same species) or *extrinsic* (imposed from outside the population). Factors can also be either **biotic** (caused by living organisms) or **abiotic** (caused by nonliving components of the environment). Finally, the regulatory factors can act in a *density-dependent* manner (effects are stronger or a higher proportion of the population is affected as population density increases) or *density-independent* manner (the effect is the same or a constant proportion of the population is affected regardless of population density).

In general, biotic regulatory factors tend to be density-dependent, while abiotic factors tend to be density-independent. There has been much discussion about which of these factors is most important in regulating population dynamics. In fact, it probably depends on the particular species involved, its tolerance levels, the stage of growth and development of the organisms involved, the specific ecosystem in which they live, and the way

combinations of factors interact. In most cases, density-dependent and density-independent factors probably exert simultaneous influences. Depending on whether regulatory factors are regular and predictable or irregular and unpredictable, species will develop different strategies for coping with them.

Density-Independent Factors

Weather (conditions at a particular time) and climate (conditions over a longer period) are often the most important density-independent factors affecting natality and mortality in populations. Extreme cold or even moderate cold at the wrong time of year, high heat, drought, excess rain, severe storms, and geologic hazards—such as volcanic eruptions, landslides, and floods—can have devastating impacts on particular populations.

Abiotic factors can have beneficial effects as well, as anyone who has seen the desert bloom after a rainfall can attest. Fire is a powerful shaper of many biomes. Grasslands, savannas, and some montane and boreal forests often are dominated—even created—by periodic fires. Some species, such as jack pine and Kirtland's warblers, are so adapted to periodic disturbances in the environment that they cannot survive without them.

In a sense, these density-independent factors don't really regulate population *per se,* since regulation implies a homeostatic feedback that increases or decreases as density fluctuates. By definition, these factors operate without regard to the number of organisms involved. They may have such a strong impact on a population, however, that they completely overwhelm the influence of any other factor and determine how many individuals make up a particular population at any given time.

Density-Dependent Factors

Density-dependent mechanisms tend to reduce population size by decreasing natality or increasing mortality as the population size increases. Most of them are the results of interactions *between* populations of a community (especially predation and disease), but some of them are based on interactions *within* a population.

Interspecific Interactions

As we will discuss in Chapter 5, a predator feeds on—and usually kills—its prey species. While the relationship is one-sided with respect to a particular pair of organisms, the prey species as a whole may benefit from the predation. For instance, the moose that gets eaten by wolves doesn't benefit individually, but the moose *population* is strengthened because the wolves tend to kill old or sick members of the herd. Their predation helps prevent population overshoot, so the remaining moose are stronger and healthier.

Sometimes predator and prey populations oscillate in a sort of offset synchrony with each other as is shown in Figure 4.8, which shows the number of furs brought into Hudson's Bay Company trading posts in Canada between 1840 and 1930. As you can see, the number of lynx fluctuates on about a 10-year cycle that is slightly out of phase with that of the snowshoe hare. This is an often cited example of a link between a predator (the lynx) and its

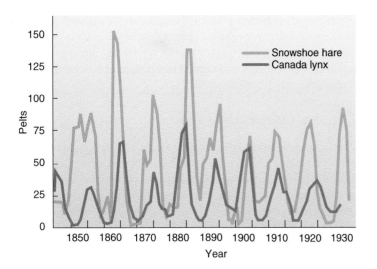

FIGURE 4.8 Oscillations in the populations of snowshoe hare and lynx in Canada suggest the close interdependency of this prey-predator relationship. These data are based on the number of pelts received by the Hudson's Bay Company. Both predator and prey show about a 10-year cycle in population growth and decline.

Source: Data from D. A. MacLulich, *Fluctuations in the Numbers of the Varying Hare (Lepus americus)*. Toronto: University of Toronto Press, 1937, reprinted 1974.

FIGURE 4.9 Animals often battle over resources. This conflict can induce stress and affect reproductive success.

prey (the hare), where an abundance of hare leads to an abundance of lynx, which leads to a decline in hare, and so on. However, further studies of these two populations have proposed other hypotheses, including sunspot activity causing variation in the abundance of the hare's food (herbaceous vegetation), which causes variation in the hare population, which causes variation in the lynx population. So there may or may not be the sort of feedback loop first proposed in the Lotka-Volterra model of predator-prey dynamics in the early twentieth century.

Not all interspecific interactions are harmful to one of the species involved. Mutualism and commensalism, for instance, are interspecific interactions that are beneficial or neutral in terms of population growth (Chapter 5).

Intraspecific Interactions

Individuals within a population also compete for resources if they are in limited supply. When population density is low, resources are likely to be plentiful and the population growth rate will approach the maximum possible for the species, assuming that individuals are not so dispersed that they cannot find mates and they are not limited in numbers by weather or other physical conditions. As population density approaches the carrying capacity of the environment, however, one or more of the vital resources becomes limiting. The stronger, quicker, more aggressive, more clever, or luckier members get a larger share, while others get less and then are unable to reproduce successfully or survive.

Territoriality is one principal way many animal species control access to environmental resources. The individual, pair, or group that holds the territory will drive off rivals if possible, either by threats, displays of superior features (colours, size, dancing ability), or fighting equipment (teeth, claws, horns, antlers). Members of the opposite sex are attracted to individuals that are able to seize and defend the largest share of the resources. From a selective point of view, these successful individuals presumably represent superior members of the population and the ones best able to produce offspring that will survive.

Stress and Crowding

Stress and crowding also are density-dependent population control factors. When population densities get very high, organisms often exhibit symptoms of what is called **stress shock** or **stress-related diseases.** These terms describe a loose set of physical, psychological, and/or behavioural changes that are thought to result from the stress of too much competition and too close proximity to other members of the same species. There is a considerable controversy about what causes such changes and how important they are in regulating natural populations. The strange behaviour and high mortality of arctic lemmings or hares during periods of high population density may be a manifestation of stress shock (Fig. 4.9). On the other hand, they could simply be the result of malnutrition, infectious disease, or some other more mundane mechanism at work.

Some of the best evidence for the existence of stress-related disease comes from experiments in which laboratory animals, usually rats or mice, are grown in very high densities with plenty of food and water but very little living space (Table 4.5). A variety of symptoms are reported, including reduced fertility, low resistance to infectious diseases and aggressive behaviour. Dominant animals seem to be affected least by crowding, while subordinate animals—the ones presumably subjected to the most stress in intraspecific interactions—seem to be the most severely affected.

TABLE 4.5	The Influence of Density on Fecundity in the House Mouse			
	SPARSE	MEDIUM	DENSE	VERY DENSE
Average number/m^3	34	118	350	1,600
Average percentage pregnant	58.3	49.4	51.0	43.4
Average number per litter	6.2	5.7	5.6	5.1

Source: Data from C. Southwick, "Population Characteristics of House Mice Living in English Corn Ricks: Density Relationships," *Proc. Zool. Soc. London,* 131:163–175, 1958, The Zoological Society of London.

FIGURE 4.10 A Mayan family in Guatemala with four of their six living children. Decisions on how many children to have are influenced by many factors, including culture, religion, need for old age security for parents, immediate family finances, household help, child survival rates, and power relationships within the family. Having many children may not be in the best interest of society at large, but may be the only rational choice for individual families.

HUMAN POPULATION GROWTH

Every second, on average, four or five children are born somewhere on the earth. In that same second, two other people die. This difference between births and deaths means a net gain of nearly 2.5 more humans per second in the world population. This means we are growing at a little less than 9,000 per hour, 214,000 per day, or about 77 million more people per year. By mid-2001, the world population stood at about 6.2 billion, making us, along with rats and other pest species we've spread, among the most numerous vertebrate species on the planet. For the families to whom these children are born, this may well be a joyous and long-awaited event (Fig. 4.10). But is a continuing increase in humans good for the planet in the long run?

Many people worry that overpopulation will cause—or perhaps already is causing—resource depletion and environmental degradation that threaten the ecological life-support systems on which we all depend. These fears often lead to demands for immediate, worldwide birth control programs to reduce fertility rates and to eventually stabilize or even shrink the total number of humans.

Others believe that human ingenuity, technology, and enterprise can extend the world carrying capacity and allow us to overcome any problems we encounter. From this perspective, more people may be beneficial rather than disastrous. A larger population means a larger workforce, more geniuses, more ideas about what to do. Along with every new mouth comes a pair of hands. Proponents of this worldview—many of whom happen to be economists—argue that continued economic and technological growth can both feed the world's billions and enrich everyone enough to end the population explosion voluntarily. Not so, counter many ecologists. Growth is the problem; we must stop both population and economic growth.

Yet another perspective on this subject derives from social justice concerns. From this worldview, there are sufficient resources for everyone. Current shortages are only signs of greed, waste, and oppression. The root cause of environmental degradation, in this view, is inequitable distribution of wealth and power rather than population size. Fostering democracy, empowering women and minorities, and improving the standard of living of the world's poorest people are what are really needed. A narrow focus on population growth only fosters racism and an attitude that blames the poor for their problems while ignoring the deeper social and economic forces at work.

Whether human populations will continue to grow at present rates and what that growth would imply for environmental quality and human life are among the most central and pressing questions in environmental science. In this chapter, we will look at some causes of population growth as well as how populations are measured and described. Family planning and birth control are essential for stabilizing populations. The number of children a couple decides to have and the methods they use to regulate fertility, however, are strongly influenced by culture, religion, politics, and economics, as well as basic biological and medical considerations. We will examine how some of these factors influence human demographics.

Human Population History

For most of our history, humans have not been very numerous compared to other species. Studies of hunting and gathering societies suggest that the total world population was probably only a few million people before the invention of agriculture and the domestication of animals around 10,000 years ago. The larger and more secure food supply made available by the agricultural revolution allowed the human population to grow, reaching perhaps 50 million people by 5000 B.C. For thousands of years, the number of humans increased very slowly. Archaeological evidence and historical descriptions suggest that there were only about 300 million people on Earth 2,000 years ago (Table 4.6).

Until the Middle Ages, human populations were held in check by diseases, famines, and wars that made life short and uncertain for most people (Fig. 4.11). Furthermore, there is evidence that many early societies regulated their population size through cultural taboos and practices such as infanticide. Among the most destructive of natural population controls were bubonic plagues that periodically swept across Europe between 1348 and 1650. During the worst plague years (between 1348 and 1350), it is estimated that at least one-third of the European population perished. Notice, however, that this did not retard population growth for very long. In 1650, at the end of the last great plague, there were about 600 million people in the world.

As you can see in Figure 4.11, human populations began to increase rapidly after A.D. 1600. Many factors contributed to this rapid growth. Increased sailing and navigating skills stimulated commerce and communication between nations. Agricultural developments, better sources of power, and better health care and hygiene also played a role. We are now in an exponential or J curve pattern of growth (What Do You Think? p. 71). In fact, some researchers say we are in a period of "double exponential" growth, where the percentage increase is itself going up by a constant percentage every year. With this increased agricultural and transportation technology came exploration and, ultimately, the takeover of indigenous populations around the globe. The main imports were diseases that devastated many of these populations.

It took all of human history to reach 1 billion people in 1804, but little more than 150 years to reach 3 billion in 1960. To go from 5 to 6 billion took only 12 years. Another way to look at population growth is that the number of humans tripled during the twentieth century. Will it do so again in this century? If it does, will we overshoot the carrying capacity of our environment and experience a catastrophic dieback similar to those described earlier in this chapter? As you will see later in this chapter, there is evidence that population growth already is slowing, but whether we will reach equilibrium soon enough and at a size that can be sustained over the long term remains a difficult but important question.

TABLE 4.6	World Population Growth and Doubling Times	
DATE	POPULATION	DOUBLING TIME
5000 B.C.	50 million	?
800 B.C.	100 million	4,200 years
200 B.C.	200 million	600 years
A.D. 1200	400 million	1,400 years
A.D. 1700	800 million	500 years
A.D. 1900	1,600 million	200 years
A.D. 1965	3,200 million	65 years
A.D. 2000	6,100 million	51 years
A.D. 2050 (estimate)	9,300 million	140 years

Source: Data from Population Reference Bureau and United Nations Population Division.

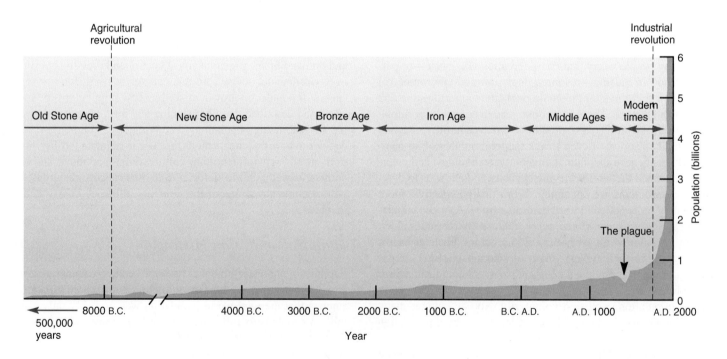

FIGURE 4.11 Human population levels through history. Since about A.D. 1000, our population curve has assumed a J shape. Are we on the upward slope of a population overshoot? Will we be able to adjust our population growth to an S curve? Or can we just continue the present trend indefinitely?

PART ONE Fundamental Principles of Ecology and Environmental Science

First Impressions: Same Data, Different Message

Graphs, pictorial representations of information, can help others understand what we have to say. They are particularly helpful in conveying patterns of relationship. Why are graphs used so widely and why do they make such powerful impressions on us?

Not only are images expressive, but graphs come equipped with substantiating data. It is one thing to read that "the human population is rising explosively," but it is considerably more compelling to see the J curve pictured together with concrete numbers as evidence. The combination of picture and numbers often conveys powerful impressions beyond the numbers themselves.

An important question addressed in this chapter is: How serious is the worldwide population problem? Does it demand action? The following graphs present information relevant to human population studies, yet the impressions left with the reader vary considerably, depending on the type of data used and the design of the graph.

Reexamine Figure 4.11, which presents an historical perspective on the growth in human numbers. What conclusions do you draw from it? Does it suggest we are experiencing an explosive rise in human numbers or that the current increase is unprecedented in our species history? Does it also imply, perhaps more indirectly, that the earth's population problem is serious, that there is an urgency to respond? In light of concepts of environmental resistance, carrying capacity, and overshoot and dieback, many people would conclude that the current growth, highlighted by the graph, is unsustainable.

Now examine Figure 1.

It plots the same variables as Figure 4.11, but suggests that population is rising at a modest rate. It does not produce the sense of explosiveness and urgency as Figure 4.11. How can the impressions be so different?

Notice that the graph in Figure 1 covers a much shorter time period. This greatly changes the time interval lengths on the horizontal scale. In Figure 4.11, one millimetre represents about 50 years, but less than one year in Figure 1. This changes the line's slope from nearly vertical in Figure 4.11 to a modest incline in Figure 1. Slope impacts the visual impression created by a graph, and therefore on its interpretation. How could you change the time axis in Figure 1 to flatten the slope even more?

Next examine Figure 2. This graph plots the stabilization ratio rather than population size over time. This graph reveals that the rate of growth for most regions except Africa has been in decline since about 1970. By itself, does the graph's downward-trending line suggest there really isn't much of a population problem, or at least not much reason for concern?

All three graphs present information relevant to a discussion of worldwide population growth, yet each gives a different impression of the seriousness of the problem. So, how can we analyze graphs and their messages in a thoughtful way? There is no single formula applicable to all situations, certainly, but a few questions are worth keeping in mind.

Is the time frame of reference appropriate or is it too restricted to allow a valid, comprehensive assessment of the issue? Are the unit intervals on the graph appropriately sized? Is the impression created by the graph real or simply an artifact of the graph's format? Do the data shown in a graph present only a partial, perhaps misleading, view of the whole?

Because graphs can create powerful impressions, they need to be interpreted with care.

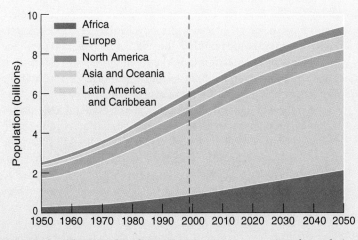

Figure 1. The U.N. Population Division projects continued population growth, although at a gradually slowing rate over the next 50 years.
Source: World Resources Institute, *World Resources 1998–99.*

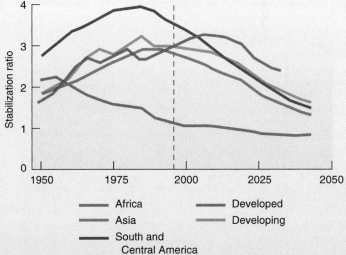

Figure 2. The stabilization ratio is measured by dividing crude birth rate by crude death rate. A ratio of 1 indicates zero population growth.
Source: United Nations (U.N.) Population Division, *World Population Prospects, 1950–2050 (The 1996 Revision),* on diskette (U.N., New York, 1996).

LIMITS TO GROWTH: SOME OPPOSING VIEWS

As with many topics in environmental science, people have widely differing opinions about human population size and resources. Some believe that population growth is the ultimate cause of poverty and environmental degradation. Others argue that poverty, environmental degradation, and overpopulation are all merely symptoms of deeper social and political factors. In this section, we will examine some opposing worldviews and their implications.

Malthusian Checks on Population

In 1798, the Rev. Thomas Malthus wrote *An Essay on the Principle of Population* to refute the views of progressives and optimists—including his father—who were inspired by the egalitarian principles of the French Revolution to predict a coming utopia. The younger Malthus argued that human populations tend to increase at an exponential or compound rate while food production either remains stable or increases only slowly. The result, he predicted, is that human populations inevitably outstrip their food supply and eventually collapse into starvation, crime, and misery. Malthus's theory might be summarized by the graphic in Figure 4.12.

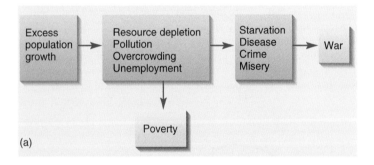

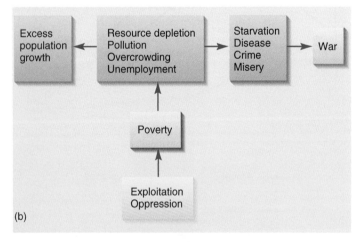

FIGURE 4.12 (*a*) Thomas Malthus argued that excess population growth is the ultimate cause of many other social and environmental problems. (*b*) Karl Marx argued that oppression and exploitation are the real causes of poverty and environmental degradation. Population growth in this view is a symptom or result of other problems, not the source.

According to Malthus, the only ways to stabilize human populations are "positive checks," such as disease or famines that kill people, or "preventative checks," including all the factors that prevent human birth. Among the preventative checks he advocated were "moral restraint," including late marriage and celibacy until a couple can afford to support children. Many social scientists and biologists have been influenced by Malthus. Charles Darwin, for instance, derived his theories about the struggle for scarce resources and survival of the fittest after reading Malthus's essay.

If Malthus's views of the consequences of exponential population growth were dismal, the corollary he drew was even more bleak. He believed that most people are too lazy and immoral to regulate birth rates voluntarily. Consequently, he opposed efforts to feed and assist the poor in England because he feared that more food would simply increase their fertility and thereby perpetuate the problems of starvation and misery.

Not surprisingly, Malthus's ideas provoked a great social and economic debate. Karl Marx was one of his most vehement critics, claiming that Malthus was a "shameless sycophant of the ruling classes." According to Marx, population growth is a symptom rather than a root cause of poverty, resource depletion, pollution, and other social ills. The real causes of these problems, he believed, are exploitation and oppression (Fig. 4.12*b*). Marx argued that workers always provide for their own sustenance given access to means of production and a fair share of the fruits of their labour. According to Marxists, the way to slow population growth *and* to alleviate crime, disease, starvation, misery, and environmental degradation is through social justice.

Malthus and Marx Today

Both Marx and Malthus had ideas about human population growth that rose to prominence and controversy during the nineteenth century, when understanding of the world, technology and society were much different than they are now. Still, the questions they raised are relevant today. While the evils of racism, classism, and other forms of exploitation that Marx denounced still beset us, it is also true that at some point available resources must limit the numbers of humans that the earth can sustain.

Those who agree with Malthus, that we are approaching—or may already have surpassed—the carrying capacity of the earth are called **neo-Malthusians.** In their view, we should address the issue of surplus population directly by making birth control our highest priority. An extreme version of this worldview is expressed by Cornell University entomologist David Pimentel, who claims that the "optimum human population" would be about 2 billion, or about the number living in 1950. He believes this would allow everyone to enjoy a standard of living equal to the average European today.

Neo-Marxists, on the other hand, believe that only eliminating oppression and poverty through technological development and social justice will solve population problems. Claims of resource scarcity, they argue, are only an excuse for inequity and exclusion. If distribution of wealth and access to resources were more fair, they believe, there would be plenty for everyone.

PART ONE Fundamental Principles of Ecology and Environmental Science

Perhaps a compromise position between these opposing viewpoints is that population growth, poverty, and environmental degradation all are interrelated. No factor exclusively causes any other, but each influences and, in turn, is influenced by the others.

Can Technology Make the World More Habitable?

Technological optimists argue that Malthus was wrong in his predictions of famine and disaster 200 years ago because he failed to account for scientific progress. In fact, food supplies have increased faster than population growth since Malthus's time. There have been terrible famines in the past three centuries, but some say they were caused more by politics and economics than lack of resources or sheer population size. Whether this progress will continue remains to be seen, but technological advances have increased human carrying capacity more than once in our history.

The burst of growth of which we are a part, was stimulated by the scientific and industrial revolutions. Progress in agricultural productivity, engineering, information technology, commerce, medicine, sanitation, and other achievements of modern life has made it possible to support approximately 1,000 times as many people per unit area as was possible 10,000 years ago.

Much of our growth in the past 300 years has been based on availability of easily acquired natural resources, especially cheap, abundant fossil fuels. Whether we can develop alternative, renewable energy sources in time to avert disaster when current fossil fuels run out is a matter of great concern (Chapter 12).

Can More People Be Beneficial?

There can be benefits as well as disadvantages in larger populations. More people mean larger markets, more workers, and efficiencies of scale in mass production of goods. Greater numbers also provide more intelligence and enterprise to overcome problems such as underdevelopment, pollution, and resource limitations. Human ingenuity and intelligence can create new resources through substitution of new materials and new ways of doing things for old materials and old ways. For instance, utility companies are finding it cheaper and more environmentally sound to finance insulation and energy-efficient appliances for their customers rather than build new power plants. The effect of saving energy that was formerly wasted is comparable to creating a new fuel supply.

Economist Julian Simon was one of the most outspoken champions of this rosy view of human history. People, he argued, are the "ultimate resource" and there is no evidence that pollution, crime, unemployment, crowding, the loss of species, or any other resource limitations will worsen with population growth. This outlook is shared by leaders of many developing countries who insist that instead of being obsessed with population growth, we should focus on the inordinate consumption of the world's resources by people in richer countries. What constitutes a resource and which resources might limit further human population growth are questions we will return to in subsequent chapters in this book. For now, we will move on to discuss how people are counted.

HUMAN DEMOGRAPHY

Demography is derived from the Greek words demos (people) and graphos (to write or to measure). It encompasses vital statistics about people, such as births, deaths, and where they live, as well as total population size. In this section, we will survey ways human populations are measured and described, and discuss demographic factors that contribute to population growth.

How Many of Us Are There?

On October 12, 1999, the United Nations officially declared that the human population had reached 6 billion. The U.S. Census Bureau, however, had put the date for this landmark three months earlier on July 19. Even in this age of information technology and communication, counting the number of people in the world is like shooting at a moving target. Some countries have never even taken a census, and those that have been done may not be accurate. Governments may overstate or understate their populations to make their countries appear larger and more important or smaller and more stable than they really are. Individuals, especially if they are homeless, refugees, or illegal aliens, may not want to be counted or identified.

We really live in two very different demographic worlds. One of these worlds is poor, young, and growing rapidly. It is occupied by the vast majority of people who live in the less-developed countries of Africa, Asia, and Latin America. These countries represent 80 percent of the world population but more than 90 percent of all projected growth (Fig. 4.13). In countries such as Uganda and Nigeria, the average age is less than 15, the current doubling time is only 23 years, and the average person can expect fewer than 30 years of reasonably good health.

Some countries in the developing world have experienced amazing growth rates and are expected to reach extraordinary

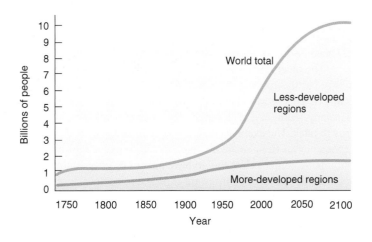

FIGURE 4.13 Estimated human population growth, 1750–2100, in less-developed and more-developed regions. More than 90 percent of all growth in this century and projected for the next is in the less-developed countries.

Source: World Resources Institute, 2000.

TABLE 4.7	10 Most Populous Countries in 2000 and 2050		
MOST POPULOUS IN 2000		**MOST POPULOUS IN 2050**	
COUNTRY	POPULATION (IN MILLIONS)	COUNTRY	POPULATION* (IN MILLIONS)
China	1,200	India	1,600
India	1,000	China	1,300
United States	281	United States	403
Indonesia	212	Indonesia	312
Brazil	170	Nigeria	304
Pakistan	151	Pakistan	285
Russia	145	Brazil	244
Bangladesh	128	Bangladesh	211
Japan	127	Ethiopia	188
Nigeria	123	Congo	182

*Estimate

Source: Population Reference Bureau, 2001.

population sizes by the middle of this century. Table 4.7 shows the 10 largest countries in the world, arranged by their projected size in 2050. Note that, while China reached 1.26 billion people in 2001, India also passed 1 billion and is expected to have the largest population in a few decades because its population control programs have been less successful than China's. Nigeria, which had only 33 million residents in 1950, is forecast to have more than 300 million in 2050. Ethiopia, with about 18 million people 50 years ago, is likely to grow at least tenfold over a century. In many of these countries, rapid population growth is a serious problem. Overall, the population of less developed countries is projected to rise from 5 billion in 2001 to 8.2 billion in 2050. Just six countries (India, China, Pakistan, Nigeria, Bangladesh, and Indonesia) account for almost half this growth.

The other demographic world is made up of the richer countries of North America, Western Europe, Japan, Australia, and New Zealand. This world is wealthy, old, and mostly shrinking. Italy, Germany, Hungary, and Japan, for example, all have negative growth rates. The average age in these countries is now 40, and life expectancy of their residents is expected to exceed 90 by 2050. With many couples choosing to have either one or no children, the populations of these countries are expected to decline significantly over this century. Japan, which has 126 million residents now, is expected to shrink to about 100 million by 2050. Europe, which now makes up about 12 percent of the world population, will constitute less than 7 percent in 50 years, if current trends continue. Even Canada and the United States would have nearly stable populations if immigration were stopped.

It isn't only wealthy countries that have declining populations. Russia, for instance, is now declining by nearly 1 million

people per year as death rates have soared and birth rates have plummeted. A collapsing economy, hyperinflation, crime, corruption, and despair have demoralized the population. Horrific pollution levels left from the Soviet era, coupled with poor nutrition and health care, have resulted in high levels of genetic abnormalities, infertility, and infant mortality. Abortions are twice as common as live births, and the average number of children per woman is now 1.3, one of the lowest in the world. Death rates, especially among adult men, have risen dramatically. According to some medical experts, four out of five Russian men are drunk when they die, and male life expectancy dropped from 68 years in 1990 to 58 years in 2000. After having been the fourth largest country in the world in 1950, Russia is expected to have a smaller population than Vietnam, the Philippines, or the Democratic Republic of Congo by 2050.

The situation is even worse in many African countries, where AIDS and other communicable diseases are killing people at a terrible rate. In Zimbabwe, Botswana, Zambia, and Namibia, for example, up to 36 percent of the adult population have AIDS or are HIV positive. Health officials predict that more than two-thirds of the 15-year-olds now living in Botswana will die of AIDS before age 50. Many of these countries are soon expected to have declining populations. Overall, however, because of high fertility rates, Africa is expected to grow by at least 1.5 billion people over the next century.

Figure 4.14 shows human population distribution around the world. Notice the high densities supported by fertile river valleys of the Nile, Ganges, Yellow, Yangtze, and Rhine Rivers and the well-watered coastal plains of India, China, and Europe. Historic factors, such as technology diffusion and geopolitical power, also play a role in geographic distribution.

Fertility and Birth Rates

Fecundity is the physical ability to reproduce, while fertility describes the actual production of offspring. Those without children may be fecund but not fertile. The most accessible demographic statistic of fertility is usually the **crude birth rate,** the number of births in a year per thousand persons. It is statistically "crude" in the sense that it is an estimate for a one-year period for 1,000 people of all ages, when both the number of people and the age structuring is always changing.

The **total fertility rate** is the number of children born to an average woman in a population during her entire reproductive life. Upper-class women in seventeenth- and eighteenth-century England, whose babies were given to wet nurses immediately after birth and who were expected to produce as many children as possible, often had 25 or 30 pregnancies. The highest recorded total fertility rates for working-class people is among some Anabaptist agricultural groups in North America who have averaged up to 12 children per woman. In most tribal or traditional societies, food shortages, health problems, and cultural practices limit total fertility to about six or seven children per woman even without modern methods of birth control.

PART ONE   Fundamental Principles of Ecology and Environmental Science

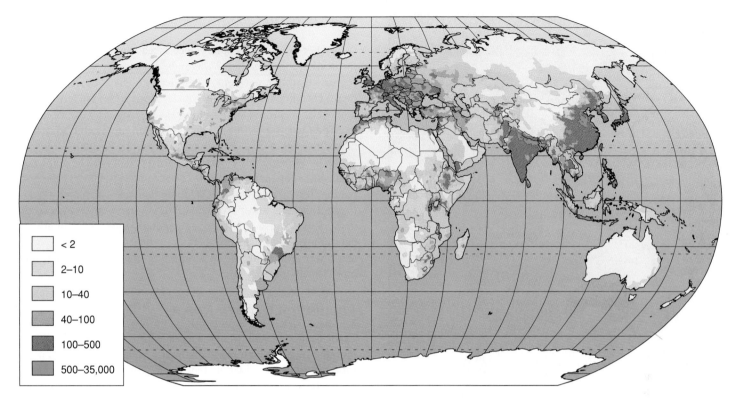

FIGURE 4.14 Population density in persons per square kilometre.
Source: World Bank, 2000.

Fertility is usually calculated as births per woman because, in many cases, it is difficult to establish paternity. Some demographers argue, nevertheless, that we should pay more attention to birth rates per male, because in some cultures men have far more children, on average, than do women. In Cameroon, for instance, due to multiple marriages, extramarital affairs, and a high rate of female mortality, men are estimated to have 8.1 children in their lifetime, while women average only 4.8. In terms of demographics, however, women are the limiting factor in determining the overall number of newborns added to the population.

Zero population growth (ZPG) occurs when births plus immigration in a population just equal deaths plus emigration. It takes several generations of replacement level fertility (where people just replace themselves) to reach ZPG. Where infant mortality rates are high, the replacement level may be five or more children per couple. In the more highly developed countries, however, this rate is usually about 2.1 children per couple because some people are infertile, have children who do not survive, or choose not to have children.

Fertility rates have declined dramatically in every region of the world except Africa over the past 50 years (Fig. 4.15). Only a few decades ago, total fertility rates above 6 were common in many countries. The average family in Mexico in 1975, for instance, had 7 children. By 2000, however, the average Mexican woman had only 2.5 children. According to the World Health Organization,

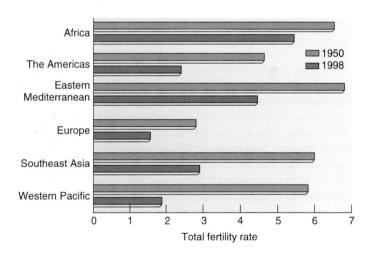

FIGURE 4.15 Declines in fertility rates by region, 1950 and 1998.
Source: Data from World Health Organization, *World Health Report 1999.*

61 out of the world's 190 countries are now at or below a replacement rate of 2.1 children per couple. The greatest fertility reduction has been in Southeast Asia, where rates have fallen by more than half. Most of this decrease has occurred in just the past few decades and, contrary to what many demographers expected, some of the

FIGURE 4.16 China's one-child-per-family policy has been remarkably successful in reducing birth rates. It may, however, have created a generation of "little emperors," since parents and grandparents focus all their attention on an only child.

poorest countries in the world have been remarkably successful in lowering growth rates. Bangladesh, for instance, reduced its fertility rate from 6.9 in 1980, to only 3.1 children per woman in 1998.

China's one-child-per-family policy decreased the fertility rate from 6 in 1970 to 1.8 in 1990. This policy, however, has sometimes resulted in abortions, forced sterilizations, and even infanticide. Another adverse result is that the only children (especially boys) allowed to families may grow up to be spoiled "little emperors" who have an inflated impression of their own importance (Fig. 4.16). Furthermore, there may not be enough workers to maintain the army, sustain the economy, or support retirees when their parents reach old age.

Although the world as a whole still has an average fertility rate of 2.7, growth rates are now lower than at any time since World War II. If fertility declines like those in Bangladesh and China were to occur everywhere in the world, our total population could begin to decline by the end of this century. Interestingly, Spain and Italy, although predominately Roman Catholic, have the lowest reported fertility rates (1.2 children per woman) of any country. And Iran, which has an Islamic government, has seen birth rates fall by more than half in less than 20 years (see Case Study, p. 77).

Mortality and Death Rates

A traveller to a foreign country once asked a local resident, "What's the death rate around here?" "Oh, the same as anywhere," was the reply, "about one per person." In demographics, however, **crude death rates** (or crude mortality rates) are expressed in terms of the number of deaths per thousand persons in any given year. Countries in Africa where health care and sanitation are limited may have mortality rates of 20 or more per 1,000 people. Wealthier countries generally have mortality rates around 10 per 1,000. The number of deaths in a population is sensitive to the age structure of the population. Rapidly growing, developing countries such as Belize or Costa Rica have lower crude death rates (4 per 1,000) than do the more-developed, slowly growing countries, such as Denmark (12 per 1,000). This is because there are proportionately more youths and fewer elderly people in a rapidly growing country than in a more slowly growing one.

Population Growth Rates

Crude death rate subtracted from crude birth rate gives the **natural increase** of a population. We distinguish natural increase from the **total growth rate,** which includes immigration and emigration, as well as births and deaths. Both of these growth rates are usually expressed as a percent (number per 100 people) rather than 1,000. A useful rule of thumb is that if you divide 70 by the annual percentage growth, you will get the approximate doubling time in years. Afghanistan, for example, which is growing 5.3 percent per year, is doubling its population every 13 years. Canada and the United States, which have natural increase rates of 0.8 percent per year, are doubling in 87.5 years. Actually, because of immigration, Canadian total growth is considerably faster than natural increase. Spain and the United Kingdom, with natural increase rates of 0.1 percent, are doubling in about 700 years. Most countries in eastern Europe have negative growth rates and declining populations. The fastest decline currently is Latvia, which at –1.1 percent per year will lose half its population in 64 years. The world growth rate is now 1.4 percent, which means that the population will double in about 50 years if this rate persists.

Life Span and Life Expectancy

Life span is the oldest age to which a species is known to survive. Although there are many claims in ancient literature of kings living for 1,000 years or more, the oldest age that can be certified by written records was that of Jeanne Louise Calment of Arles, France, who was 122 years old at her death in 1997. The aging process is still a medical mystery, but it appears that cells in our bodies have a limited ability to repair damage and produce new components. At some point they simply wear out, and we fall victim to disease, degeneration, accidents, or senility.

Life expectancy is the average age that a newborn infant can expect to attain in any given society. It is another way of expressing the average age at death. For most of human history, we believe that life expectancy in most societies has been between 35 and 40 years.

CASE STUDY

FAMILY PLANNING IN IRAN

After the Islamic Revolution in 1979, Iran had one of the world's highest population growth rates. In spite of civil war, large-scale emigration, and economic austerity, the country surged from 34 million to 63 million in just 20 years. A crude birth rate of 43.4/1,000 people and a total fertility rate of 5.1 per woman during this time resulted in annual population growth of 3.9 percent and a doubling time of less than 18 years. Religious authorities exhorted couples to have as many children as possible. Any mention of birth control or family planning (other than to have as many children as possible) was forbidden, and the marriage age for girls was dropped to 9 years old. When a devastating war with Iraq in the 1980s killed at least 1 million young soldiers, producing more children to rebuild the army became a civic as well as religious duty.

In the late 1990s, however, the Iranian government became aware of the costs of such rapid population growth. With religious moderates gaining greater political power, public policy changed abruptly. Now, the Iranian government is spending millions of dollars to lower birth rates. Couples must pass a national family planning course before they are allowed to marry. While it took a few years to convince people that this change will be long-lasting, most Iranian citizens are now eager for access to birth-control information. Family planning classes are sought out both by engaged couples and those already married. A wide range of birth control methods are available. Implantable or injectable slow-release hormones, condoms, intrauterine devices (IUDs), pills, and male or female sterilization are free to all. Billboards, newspapers, television, and even water towers advertise this national program. Religious leaders have issued a *fatwah*, or command, that all faithful Muslims participate in family planning.

As a consequence, Iran has been remarkably successful in stemming its population growth. Between 1986 and 1996, the fertility rates for urban residents dropped almost by half, to less than three children per woman, and the crude birth rate dropped from 43 to 18 per 1,000 people. By 2000, the average annual growth rate had fallen to 1.4 percent. While the population is still increasing, another decade of such progress would bring the country to a stable or even declining rate of growth.

Several societal changes have contributed to this rapid birth reduction. While the minimum marriage age has been returned to 15, couples are encouraged to wait until at least age 20 to begin their families. The educational benefits of concentrating the family resources on just one or two children are being promoted. Although women's roles are still highly restricted in the Islamic Republic, greater gender equity has given women more control over their reproductive lives. Access to modern, information-age jobs gives people an incentive to seek out education both for themselves and their children.

The demographic transition hasn't spread to all levels of Iranian society. Rural families, ethnic minorities, and some urban poor still tend to have many children. Still, this example of how quickly both ideals of the perfect family size and information about modern birth control can spread through a society—even a highly religious, fundamentalist one—is encouraging for what might be accomplished worldwide in a surprisingly short time.

This doesn't mean that no one lived past age 40, but rather that so many deaths at earlier ages (mostly early childhood) balanced out those who managed to live longer.

Declining mortality, not rising fertility, is the primary cause of most population growth in the past 300 years. Crude death rates began falling in western Europe during the late 1700s. Most of this advance in survivorship came long before the advent of modern medicine and is due primarily to better food and better sanitation.

The twentieth century saw a global transformation in human health unmatched in history. This revolution saw dramatic increases in life expectancy in most places. Worldwide, the average life expectancy rose from about 40 to 65.5 years over the twentieth century. Table 4.8 shows gains in some selected countries. Globally, the number of people over 60 years old is expected to triple, increasing from 600 million today to nearly 2 billion in 2050. The oldest old (over 80 years) is projected to grow five-fold to about 400 million in that same period.

The greatest progress in life expectancy has been in developing countries. Take the case of Chile, for example. In 1900, the

TABLE 4.8	Life Expectancy at Birth, Selected Countries, around 1910 and in 1998			
	AROUND 1910		1998	
COUNTRY	MALES	FEMALES	MALES	FEMALES
Australia	56	60	75	81
Chile	29	33	72	78
Italy	46	47	75	81
Japan	43	43	77	83
New Zealand	60	63	74	80
Norway	56	59	75	81
United States	49	53	73	80
Canada	56	60	75	81

Source: World Health Organization, *World Health Report 1999.*

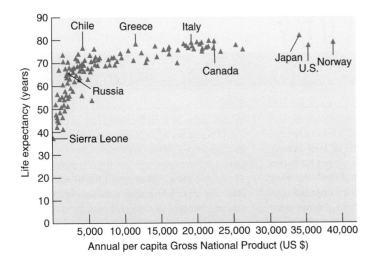

FIGURE 4.17 As incomes rise, so does life expectancy up to about $4,000 (US). Russia is an exception, with a life expectancy nearly 20 years less than that of Chile, even though their GNP is about the same.
Source: The World Bank, *World Development Indicators 2002,* the World Bank, Washington, D.C., 1997.

average Chilean man could expect to live only 29 years, while the average woman would reach just 33 years. By 1998, although Chile had an annual per capita income less than $4,000 (US), the average life expectancy for both men and women had more than doubled and was very close to that of countries with 10 times its income level. Longer lives were due primarily to better nutrition, improved sanitation, clean water, and education rather than miracle drugs or high-tech medicine. While the gains were not as great for the already industrialized countries, residents of Canada, the United States, Italy, and Japan, for example, now live about half-again as long as they did at the beginning of the century.

As Figure 4.17 shows, there is a good correlation between annual income and life expectancy up to about $4,000 (US) per person. Beyond that level—which is generally enough for adequate food, shelter, and sanitation for most people—life expectancies level out at about 75 years for men and 80 for women. Russia is a striking exception to this pattern. With a GNP per person near $5,000 (US), Russian life expectancy is only 58 years for men and 71 for women. Russian men now live about 14 years less, on average, than they did before the breakup of the USSR. As mentioned earlier, disastrous economy, alcoholism, poor nutrition, and substandard medical care all contribute to this decline.

Large discrepancies in how benefits of modernization and social investment are distributed within countries are revealed in differential longevity of various groups. Life expectancy at birth in Canada varies from 66 years for men and 70 years for women in Nunavut, to 78 years for B.C. men and 83 for B.C. females. Differences in medical services and quality of life in the primarily aboriginal Nunavut clearly have significant ramifications for life and health. Even those that reach age 65 in Nunavut have fewer expected senior years when compared to their southern Canada counterparts.

Some demographers believe that life expectancy is approaching a plateau, while others predict that advances in biology and medicine might make it possible to live 150 years or more. If our average age at death approaches 100 years, as some expect, society will be profoundly affected. If workers continue to retire at 65, half of the population could be unemployed, and retirees might be facing 35 or 40 years of retirement. We may need to find new ways to structure and finance our lives.

Living Longer: Demographic Implications

A population that is growing rapidly by natural increase has more young people than does a stable population. One way to show these differences is to graph age classes in a histogram as shown in Figure 4.18. In Kenya, which is growing at a rate of 2.5 percent per year, 42 percent of the population is in the prereproductive category (below age 15). Even if total fertility rates were to fall abruptly, the total number of births, and population size, would continue to grow for some years as these young people enter reproductive age. This phenomenon is called *population momentum.*

A population that has recently entered a lower growth rate pattern, such as Canada, will have a bulge in the age classes for the last high-birth rate generation. Notice that there are more females than males in the older age group because of differences in longevity between the sexes. Canada has a high percentage of retired people because of long life expectancy.

Both rapidly growing countries and slowly growing countries can have a problem with their **dependency ratio,** or the number of nonworking compared to working individuals in a population. In Mexico, for example, each working person supports a high number of children. In Canada and the United States, by contrast, a declining working population is now supporting an ever larger number of retired persons and there are dire predictions that the social security system will soon be bankrupt. This changing age structure and shifting dependency ratio are occurring worldwide (Fig. 4.19). By 2050 the U.N. predicts there will be two older persons for every child in the world.

Emigration and Immigration

The more developed regions are expected to gain about 2 million immigrants per year for the next 50 years. Without migration, the population of the wealthiest countries would already be declining and would be more than 126 million less than the current 1.2 billion by 2050.

Immigration is a controversial issue in many countries. "Guest workers" often perform heavy, dangerous, or disagreeable work that citizens are unwilling to do. Many migrant workers are of a different racial or ethnic background than the majority in their new home. They generally are paid low wages and given substandard housing, poor working conditions, and few rights. Local residents often complain that immigrants take away jobs, overload social services, and ignore established rules of behaviour or social values.

Some countries encourage, or even force, internal mass migrations as part of a geopolitical demographic policy. In the

PART ONE   Fundamental Principles of Ecology and Environmental Science

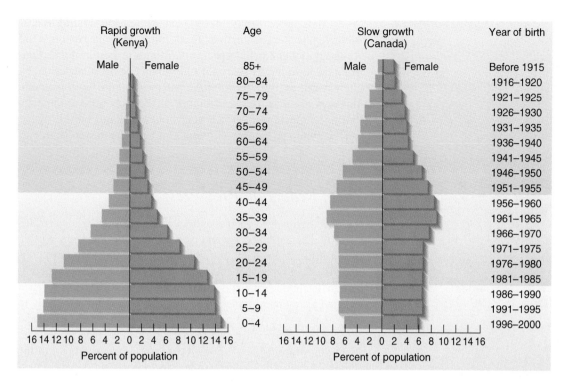

FIGURE 4.18 Population by age and sex in rapidly growing and slowly growing countries. Middle regions represent individuals of reproductive ages. Note the high proportion of children in the rapidly growing population and the high proportion of elderly in the slowly growing population.
Sources: Data from the United Nations, Statistics Canada and the Population Reference Bureau.

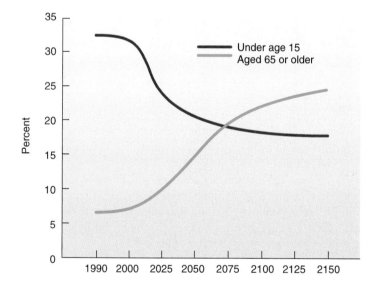

FIGURE 4.19 Changing age structure of world population. In this century, children under 15 years of age will make up a smaller percentage of world population, while people over 65 years old will make up a rapidly rising share of the population.

1970s, Indonesia embarked on an ambitious "transmigration" plan to move 65 million people from the overcrowded islands of Java and Bali to relatively unpopulated regions of Sumatra, Borneo, and New Guinea. Attempts to turn rainforest into farmland had disastrous environmental and social effects, however, and this plan was greatly scaled back. In 2001, native Dayak tribesmen, resentful at the intrusion of outsiders on traditional lands, massacred hundreds of Madurese transmigrants.

POPULATION GROWTH: OPPOSING FACTORS

A number of social and economic pressures affect decisions about family size, which in turn affects the population at large. In this section we will examine both positive and negative pressures on reproduction.

Pronatalist Pressures

Factors that increase people's desires to have babies are called **pronatalist pressures.** Raising a family may be the most enjoyable and rewarding part of many people's lives. Children can be a source of pleasure, pride, and comfort. They may be the only source of support for elderly parents in countries without a social security system. Where infant mortality rates are high, couples may need to have many children to be sure that at least a few will survive to take care of them when they are old. Where there is little opportunity for upward mobility, children give status in society, express parental creativity, and provide a sense of continuity and accomplishment otherwise missing from life. Often children are valuable to the family not only for future income, but even more as a source of current income and help with household chores. In much of the developing world, children as young as 6 years old tend domestic

FIGURE 4.20 Children in rural areas can help with many household chores, such as tending livestock or caring for younger children.

animals and younger siblings, fetch water, gather firewood, and help grow crops or sell things in the marketplace (Fig. 4.20). Parental desire for children rather than an unmet need for contraceptives may be the most important factor in population growth in many cases.

Society also has a need to replace members who die or become incapacitated. This need often is codified in cultural or religious values that encourage bearing and raising children. In some societies, families with few or no children are looked upon with pity or contempt. The idea of deliberately controlling fertility may be shocking, even taboo. Women who are pregnant or have small children are given special status and protection. Boys frequently are more valued than girls because they carry on the family name and are expected to support their parents in old age. Couples may have more children than they really want in an attempt to produce a son.

Male pride often is linked to having as many children as possible. In Niger and Cameroon, for example, men, on average, want 12.6 and 11.2 children, respectively. Women in these countries consider the ideal family size to be only about one-half that desired by their husbands. Even though a woman might desire fewer children, however, she may have few choices and little control over her own fertility. In many societies, a woman has no status outside of her role as wife and mother. Without children, she has no source of support.

Birth Reduction Pressures

In more highly developed countries, many pressures tend to reduce fertility. Higher education and personal freedom for women often result in decisions to limit childbearing. The desire to have children is offset by a desire for other goods and activities that compete with childbearing and childrearing for time and money. When women have opportunities to earn a salary, they are less likely to stay home and have many children. Not only are the challenge and variety of a career attractive to many women, but the money that they can earn outside the home becomes an important part of the family budget. Thus, education and socioeconomic status are usually inversely related to fertility in richer countries. In developing countries, however, fertility is likely to increase as educational levels and socioeconomic status rise. With higher income, families are better able to afford the children they want; more money means that women are likely to be healthier, and therefore better able to conceive and carry a child to term.

In less-developed countries where feeding and clothing children can be a minimal expense, adding one more child to a family usually doesn't cost much. By contrast, raising a child in Canada and the United States can cost hundreds of thousands of dollars by the time the child is through school and is independent. Under these circumstances, parents are more likely to choose to have one or two children on whom they can concentrate their time, energy, and financial resources.

Figure 4.21 contrasts Canadian and Kenyan birth rates between 1950 and 2050. It highlights the baby boom in post-World War II Canada during the 50s and 60s, followed by a steady decline in birth rate. Kenya is on a slower track to lower birth rates, as we saw in looking at its age structure (Figure 4.18).

DEMOGRAPHIC TRANSITION

In 1945, demographer Frank Notestein pointed out that a typical pattern of falling death rates and birth rates due to improved living conditions usually accompanies economic development. He called this pattern the **demographic transition** from high birth and death rates to lower birth and death rates. Figure 4.22 shows an idealized model of a demographic transition. This model is often used to explain connections between population growth and economic development.

PART ONE   Fundamental Principles of Ecology and Environmental Science

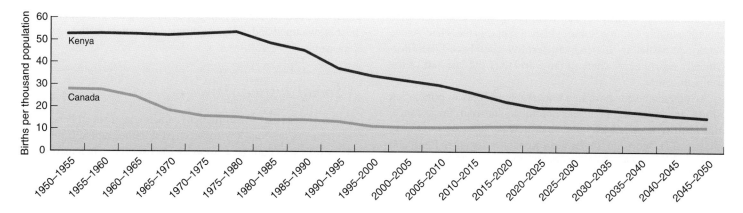

FIGURE 4.21 Historical and projected birth rates in Canada and Kenya, reflecting the "baby boom" and its echo in Canada, and a decline in both countries into this century, but a much higher birth rate in Kenya than Canada.
Source: Statistics Canada.

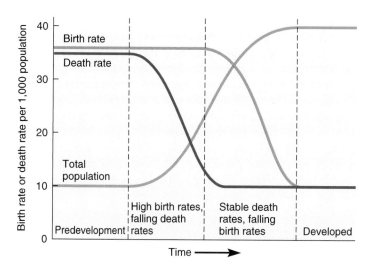

FIGURE 4.22 Theoretical birth, death, and population growth rates in a demographic transition accompanying economic and social development. In a pre-development society, birth and death rates are both high, and total population remains relatively stable. During development, death rates tend to fall first, followed in a generation to two by falling birth rates. Total population grows rapidly until both birth and death rates stabilize in a fully developed society.

Development and Population

The left-hand side of Figure 4.22 represents the conditions in a pre-modern society. Food shortages, malnutrition, lack of sanitation and medicine, accidents, and other hazards generally keep death rates in such a society around 30 per 1,000 people. Birth rates are correspondingly high to keep population densities relatively constant. As economic development brings better jobs, medical care, sanitation, and a generally improved standard of living, death rates often fall very rapidly. Birth rates may actually rise at first as more money and better nutrition allow people to have the children they always wanted. Eventually, however, birth rates fall as people see that all their children are more likely to survive and that the whole family benefits from concentrating more resources on fewer children. Note that populations grow rapidly during the time that death rates have already fallen but birth rates remain high. Depending on how long it takes to complete the transition, the population may go through one or more rounds of doubling before coming into balance again.

The right-hand side of each curve in Figure 4.22 represents conditions in developed countries, where the transition is complete and both birth rates and death rates are low, often, a third or less than those in the predevelopment era. The population comes into a new equilibrium in this phase, but at a much larger size than before. Most of the countries of northern and western Europe went through a demographic transition in the nineteenth or early twentieth century similar to the curves shown in Figure 4.22.

Many of the most rapidly growing countries in the world, such as Kenya, Yemen, Libya, and Jordan, now are in the middle phase of this demographic transition. Their death rates have fallen close to the rates of the fully developed countries, but birth rates have not fallen correspondingly. In fact, both their birth rates and total population are higher than those in most European countries when industrialization began 300 years ago. The large disparity between birth and death rates means that many developing countries now are growing at 3 to 4 percent per year. Such high growth rates in the Third World could boost total world population to 9 billion or more before the end of this century. This raises what may be the two most important questions in this entire chapter: Why are birth rates not yet falling in these countries, and what can be done about it?

An Optimistic View

Some demographers claim that a demographic transition already is in progress in most developing nations. Problems in taking censuses and a normal lag between falling death and birth rates may hide this for a time, but the world population should stabilize sometime in this century. Some evidence supports this view. As we mentioned earlier in this chapter, fertility rates have fallen dramatically nearly everywhere in the world over the last half of the twentieth century.

Some countries have had remarkable success in population control. In Thailand, Indonesia, Colombia, and Iran, for instance, total fertility dropped by more than half in 20 years (see Case Study, p. 77). Morocco, Dominican Republic, Jamaica, Peru, and Mexico all have seen fertility rates fall between 30 percent and 40 percent in a single generation. The following factors contribute to stabilizing populations:

- Growing prosperity and social reforms that accompany development reduce the need and desire for large families in most countries.

- Technology is available to bring advances to the developing world much more rapidly than was the case a century ago, and the rate of technology transfer is much faster than it was when Europe and North America were developing.

- Less-developed countries have historic patterns to follow. They can benefit from our mistakes and chart a course to stability more quickly than they might otherwise do.

- Modern communications (especially television) have caused a revolution of rising expectations that act as a stimulus to spur change and development.

A Pessimistic View

Economist Lester Brown of the Worldwatch Institute takes a more pessimistic view. He warns that many of the poorer countries of the world appear to be caught in a "demographic trap" that prevents them from escaping from the middle phase of the demographic transition. Their populations are now growing so rapidly that human demands exceed the sustainable yield of local forests, grasslands, croplands, or water resources. The resulting resource shortages, environmental deterioration, economic decline, and political instability may prevent these countries from ever completing modernization. Their populations may continue to grow until catastrophe intervenes.

Many people argue that the only way to break out of the demographic trap is to immediately and drastically reduce population growth by whatever means are necessary. They argue strongly for birth control education and bold national policies to encourage lower birth rates. Some agree with Malthus that helping the poor will simply increase their reproductive success and further threaten the resources on which we all depend. Author Garret Hardin described this view as lifeboat ethics. "Each rich nation," he said, "amounts to a lifeboat full of comparatively rich people. The poor of the world are in other much more crowded lifeboats. Continuously, so to speak, the poor fall out of their lifeboats and swim for a while, hoping to be admitted to a rich lifeboat, or in some other way to benefit from the goodies on board . . . . We cannot risk the safety of all the passengers by helping others in need. What happens if you share space in a lifeboat? The boat is swamped and everyone drowns. Complete justice, complete catastrophe."

A Social Justice View

A third view is that **social justice** (a fair share of social benefits for everyone) is the real key to successful demographic transitions. The world has enough resources for everyone, but inequitable social and economic systems cause maldistributions of those resources. Hunger, poverty, violence, environmental degradation, and overpopulation are symptoms of a lack of social justice rather than a lack of resources. Although overpopulation exacerbates other problems, a narrow focus on this factor alone encourages racism and hatred of the poor. A solution for all these problems is to establish fair systems, not to blame the victims. Small nations and minorities often regard calls for population control as a form of genocide. Figure 4.23 expresses the opinion of many people in less-developed countries about the relationship between resources and population.

An important part of this view is that many of the rich countries are, or were, colonial powers, while the poor, rapidly growing countries were colonies. The wealth that paid for progress and security for developed countries was often extracted from colonies, which now suffer from exhausted resources, exploding populations, and chaotic political systems. Some of the world's poorest countries such as India, Ethiopia, Mozambique, and Haiti had rich resources and adequate food supplies before they were impoverished by colonialism. Those of us who now enjoy abundance may need to help the poorer countries not only as a matter of justice but because we all share the same environment.

Infant Mortality and Women's Rights

Survival of children is one of the most critical factors in stabilizing population. When infant and child mortality rates are high, as they are in much of the developing world, parents tend to have high numbers of children to ensure that some will survive to adulthood. There has never been a sustained drop in birth rates that was not first preceded by a sustained drop in infant and child mortality. One of the most important distinctions in our demographically divided world is the high infant mortality rates in the less-developed countries. Better nutrition, improved health care, simple oral rehydration therapy, and immunization against infectious diseases (Chapter 22) have brought about dramatic reductions in child mortality rates, which have been accompanied in most regions by falling birth rates. It has been estimated that saving 5 million children each year from easily preventable communicable diseases would avoid 20 or 30 million extra births.

Increasing family income does not always translate into better welfare for children since men in many cultures control most financial assets. Often the best way to improve child survival is to ensure the rights of mothers. Opportunities for women's education, for instance, as well as land reform, political rights, opportunities

FIGURE 4.23 Controlling our population and resources—there may be more than one side to the issue.
Used with permission of the Asian Cultural Forum on Development.

to earn an independent income, and improved health status of women often are better indicators of family welfare than rising GNP (Fig. 4.24).

FAMILY PLANNING AND FERTILITY CONTROL

Family planning allows couples to determine the number and spacing of their children. It doesn't necessarily mean fewer children—people may use family planning to have the maximum number of children possible—but it does imply that the parents will control their reproductive lives and make rational, conscious decisions about how many children they will have and when those children will be born, rather than leaving it to chance. As the desire for smaller families becomes more common, birth control becomes an essential part of family planning in most cases. In this context, **birth control** usually means any method used to reduce births, including celibacy, delayed marriage, contraception, methods that prevent implantation of embryos, and induced abortions.

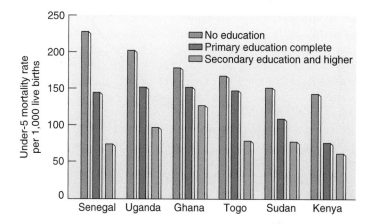

FIGURE 4.24 Childhood deaths drop as female education rises. Under-5 mortality rate is shown for selected African countries compared to level of female education.
Source: World Resources Institute, *World Resources* 1998–99. Reprinted by permission of World Resources Institute.

THE FUTURE OF HUMAN POPULATIONS

How many people will be in the world a century from now? Most demographers believe that world population will stabilize sometime during this century. The total number of humans, when we reach that equilibrium, is likely to be somewhere around 8 to 10 billion people, depending on the success of family planning programs and the multitude of other factors affecting human populations. Figure 4.25 shows four scenarios projected by the U.N. Population Division in its 2000 revision. The optimistic (low) projection shows that world population might stabilize at about 8 billion in the middle of this century. The medium projection suggests that growth might continue to around 9.3 billion. The most pessimistic projection assumes a constant rate of growth (no change from present) to 13 billion people by 2050.

Which of these scenarios will we follow? As you have seen in this chapter, population growth is a complex subject. To accomplish a stabilization or reduction of human populations will require substantial changes from business as usual.

An encouraging sign is that worldwide contraceptive use has increased sharply in recent years. About half of the world's married couples used some family planning techniques in 1999, compared to only 10 percent 30 years earlier, but another 300 million couples say they want but do not have access to family planning. Contraceptive use varies widely by region, with high levels in Latin America and East Asia but relatively low use in much of Africa.

Figure 4.26 shows the unmet need for family planning among married women in selected countries. When people in developing countries are asked what they want most, men say they want better jobs, but the first choice for a vast majority of women is family planning assistance. In general, a 15 percent increase in contraceptive use equates to about one fewer birth per woman per lifetime. In Ethiopia, for example, where only 4 percent of all women use contraceptives, the average fertility is seven children per woman. In South Africa, by contrast, where 53 percent of all women use contraceptives, the average fertility is 3.3.

The World Health Organization estimates that nearly 1 million conceptions occur daily around the world as a result of some 100 million sex acts. At least half of those conceptions are unplanned or unwanted. Still, birth rates already have begun to fall in East Asia and Latin America (Fig. 4.27). Similar progress is expected in South Asia in a few years. Only Africa will probably continue to grow in this century.

Deep societal changes are often required to make family planning programs successful. Among the most important of these are (1) improved social, educational, and economic status for women (birth control and women's rights are often interdependent); (2) improved status for children (fewer children are born as parents come to regard them as valued individuals rather than possessions); (3) acceptance of calculated choice as a valid element in life in general and in fertility in particular (belief that we have

FIGURE 4.25 Estimated and projected world population, 1950 to 2050, with different fertility levels.
Source: United Nations Population Division 2001.

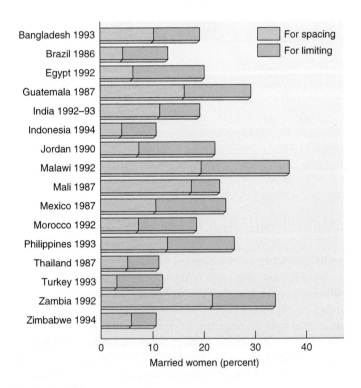

FIGURE 4.26 Unmet need for family planning among married women in selected countries. Globally, at least 300 million couples cannot afford or do not have access to family planning and modern contraception that they desire.

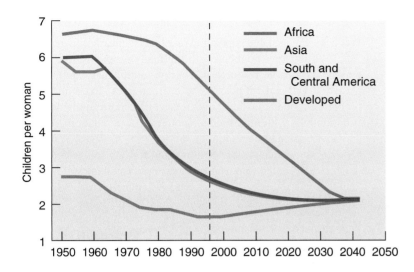

FIGURE 4.27 Fertility declines, real and projected, 1950–2050.
Source: Data from U.N. Population Division, *World Population Prospects, 1996.*

no control over our lives discourages a sense of responsibility); (4) social security and political stability that give people the means and the confidence to plan for the future; (5) knowledge, availability, and use of effective and acceptable means of birth control.

Concerted efforts to bring about these changes can be effective. Twenty years of economic development and work by voluntary family planning groups in Zimbabwe, for example, have lowered total fertility rates from 8.0 to 5.5 children per woman on average. Surveys showed that desired family sizes have fallen nearly by half (9.0 to 4.6) and that nearly all women and 80 percent of men in Zimbabwe use contraceptives. If similar progress can be sustained elsewhere, human populations may be restrained after all. The choices all of us make—both in our private lives and in public policies—will determine our future course.

SUMMARY

Population dynamics play an important role in determining how ecosystems work. Biological organisms generally have the ability to produce enough offspring so populations can grow rapidly when resources are available. Given sufficient resources, populations grow exponentially; that is, they grow at a constant percentage so that the population size doubles in some regular interval of time. If conditions appropriate for this kind of growth persist and other factors don't intervene to reduce abundance, exponential growth can produce astronomical numbers of organisms. We describe the rapidly rising curve of an exponentially growing population as a J curve.

Some populations will grow exponentially until they deplete resources of the environment. Mortality rates rise as resources become limited and the population may crash, often dying back as rapidly as it rose. Other species have lower growth rates as their numbers increase because of competition for limiting resources, resulting in an equilibrium at or near the carrying capacity of the environment.

The most important components of population dynamics are natality, fertility, fecundity, life span, longevity, mortality, immigration, and emigration. The sum of all additions to and subtractions from the population determines the net rate of growth. Mortality rates and longevity are often expressed as survivorship rates that reveal much about a species' place in its ecosystem and the kinds of hazards that eliminate members from the population.

The factors that regulate population dynamics can be either intrinsic or extrinsic to the population. They can be caused by biotic or abiotic forces, and they can act on the population in either a density-dependent or density-independent fashion. The most important abiotic regulatory factors are usually climate and weather. The most important biological factors are usually competition (both interspecific and intraspecific), predation, and disease.

Human populations have grown at an unprecedented rate over the past four centuries. By 2001, the world population stood at 6.2 billion people. If the current growth rate of 1.4 percent per year persists, the population will double in 51 years. Most of that growth will occur in the less-developed countries of Asia, Africa, and Latin America. There is a serious concern that the number of humans in the world and our impact on the environment will overload the life-support systems of the earth.

The crude birth rate is the number of births in a year divided by the average population. A more accurate measure of growth is the general fertility rate, which takes into account the age structure and fecundity of the population. The crude birth rate minus the crude death rate gives the rate of natural increase. When this rate reaches a level at which people are just replacing themselves, zero population growth is achieved.

In the more highly developed countries of the world, growth has slowed or even reversed in recent years so that without immigration from other areas, populations would be declining. The change from high birth and death rates that accompanies industrialization is called a demographic transition. Many developing countries have already begun this transition. Death rates have fallen, but birth rates remain high. Some demographers believe that as infant mortality drops and economic development progresses so that people in these countries can be sure of a secure future, they will complete the transition to a stable population. Others fear that excessive population growth and limited resources will catch many of the poorer countries in a demographic trap that could prevent them from ever achieving a stable population or a high standard of living.

While larger populations bring many problems, they also may be a valuable resource of energy, intelligence, and enterprise that will make it possible to overcome resource limitation problems. A social justice view argues that a more equitable distribution of wealth might reduce both excess population growth and environmental degradation.

We have many more options now for controlling fertility than were available to our ancestors. Some techniques are safer than those available earlier; many are easier and more pleasant to use. Sometimes it takes deep changes in a culture to make family planning programs successful. Among these changes are improved social, educational, and economic status for women; higher values on individual children; accepting responsibility for our own lives; social security and political stability that give people the means and confidence to plan for the future; and knowledge, availability, and use of effective and acceptable means of birth control.

QUESTIONS FOR REVIEW

1. What is the difference between exponential and arithmetic growth?

2. What is the difference between fertility and fecundity?

3. Describe the four major types of survivorship patterns and explain what they show about the role of the species in an ecosystem.

4. What are the main interspecific population regulatory interactions? How do they work?

5. At what point in history did the world population pass its *first* billion? What factors restricted population before that time, and what factors contributed to growth after that point?

6. Where will most population growth occur in the next century? What conditions contribute to rapid population growth in some countries?

7. Define *crude birth rate, total fertility rate, crude death rate,* and *zero population growth.*

8. What is the difference between life expectancy and longevity?

9. Describe the conditions that lead to a demographic transition.

QUESTIONS FOR CRITICAL THINKING

1. Are humans subject to environmental resistance in the same sense that other organisms are? How would you decide whether a particular factor that limits human population growth is ecological or social?

2. There obviously are vast differences in birth and death rates, survivorship, and life spans among species. There must be advantages and disadvantages in living longer or reproducing more quickly. Why hasn't evolution selected for the most advantageous combination of characteristics so that all organisms would be more or less alike?

3. Abiotic factors that influence population growth tend to be density-independent, while biotic factors that regulate population growth tend to be density-dependent. Explain.

4. What do you think is the optimum human population? The maximum human population? Are the numbers different? If so, why?

5. Some people argue that technology can provide solutions for environmental problems; others believe that a "technological fix" will make our problems worse. What personal experiences or worldviews do you think might underlie these positions?

6. Karl Marx called Thomas Malthus a "shameless sycophant of the ruling classes." Why would the landed gentry of the eighteenth century be concerned about population growth of the lower classes? Are there comparable class struggles today?

7. In 1968, biologist Garret Hardin wrote a very controversial article in the journal *Science* called "The Tragedy of the Commons." In it, he proposed that the freedom to reproduce as much as one wanted would eventually lead to the destruction of the planet, because people are ultimately self-centred. He advocated legislated limits on reproduction similar to those we now see in China. Discuss Hardin's views. Is the world ready for widespread, legislative controls on reproduction?

KEY TERMS

Web Exercises

EXPONENTIAL GROWTH

Go to http://www.jump.net/~otherwise/population/exponent.html to run animations that will help you understand exponential growth in a biological population. First run the java applet (see box below second paragraph on the web page) in habitat mode (so you can see the fish). Do a series of simulations in this mode using birth rates of 1.0, 1.2, 1.4, 1.6, 1.8, and 2.0 to see how many generations it takes to completely fill the screen with fish. Then change to Graph view to follow the instructions for Experiment 1. How many generations does it take to get to 1,600 fish at each of these birth rates? Are you surprised at how rapidly the population grows after about 10 generations at a birth rate of 1.5? Do all these birth rates give you a J curve?

Next, follow the instructions for Experiment 2. Does the decline at a negative growth rate mirror that at a positive rate? Why or why not?

If you have time, follow the links on this page to Logistic Growth (or growth to a stable population). How does this pattern differ from exponential growth? What types of animals or plants might follow this pattern? (*Hint:* consult your textbook.)

EXPLORING GROWTH FACTORS IN POPULATION DATA

How and why populations grow are key questions in environmental science. In this exercise, you examine and graph current world population data to explore which factors most strongly correlate with birth rates. Go to www.mcgrawhill.ca/college/global.

There you will find an Excel data file named popdata.xls. Double-click on the file name to copy it to your hard disk. If you have Excel on your computer, you should be able to open the data file by double-clicking on it. (Other spreadsheet programs can also read this file, but you must open it from within your program, not by double-clicking.)

1. This file contains population data for the countries of the world, sorted by the United Nations Human Development Index (HDI) rank. First look at the top 20 countries. Where are they? What is the range of income levels (in GNP per capita) of the top 20? What is the range of income for the bottom 20 countries?

2. Now make an X,Y scatter graph of Adult Literacy and Birth Rate. (Detailed instructions for making graphs in Excel are included at the far right side of the spreadsheet page [column N].) How would you describe the relationship between these variables? How would you explain this relationship? Keep this graph in your spreadsheet while you make three more scatter graphs: (1) GNP per Capita and Birth Rate, (2) Life Expectancy and Birth Rate, and (3) Infant Mortality and Birth Rate. Describe the trends you observe, and explain what they mean.

3. How would you compare the relative amount of scatter in each of your graphs? Why do some curves slope from right to left, while others slope in the opposite direction? If you draw a line through the middle of the dot cluster, some curve smoothly, while others seem to have a break or inflection point. How would you interpret these patterns?

4. Try changing the shape of your graphs. (See instructions on the right side of the spreadsheet to do this.) How does making the graphs taller or wider affect the way your trends look? How could you deliberately manipulate the graph shape to affect other people's interpretation of the data? Is this ethical? Have you ever seen it done?

5. Now make a dot graph of GNP Per Capita and Adult Literacy. Is there a linear relationship between the two variables? Why or why not?

Chapter 5

Factors Controlling the Distribution and Abundance of Organisms

Nothing succeeds like excess.

—Oscar Wilde's definition of natural selection—

OBJECTIVES

After studying this chapter, you should be able to:

- describe how environmental factors influence which species live in a given ecosystem and where or how they live.
- understand how random genetic variation and natural selection lead to evolution, adaptation, niche specialization, and partitioning of resources in biological communities.
- compare and contrast interspecific predation, competition, symbiosis, commensalism, mutualism, and coevolution.
- discuss productivity, diversity, complexity, and structure of biological communities and how these characteristics might be connected to resilience and stability.

- explain how ecological succession is changed through time in an ecosystem and the biological communities that occupy it.
- give some examples of exotic species introduced into biological communities and describe the effects such introductions can have on indigenous species.

WebQuest ecological niche, evolution, natural selection, productivity, diversity, complexity, predation, competition, mutualism, symbiosis, commensalism, ecological succession, exotic species

Above: The bears, birds, and fish of the McNeil River, Alaska, form an interconnected biological community together with terrestrial and aquatic plants and invertebrates.

Orcas, Otters, Urchins, and Kelp
Disrupting a Marine Food Web

What's happening to sea otters on the northern Pacific coast? The sudden disappearance of thousands of sea otters along the Aleutian Island chain in Alaska in recent years has surprised marine biologists and alerted them to a complex and disturbing ecological mystery story.

Once nearly driven into extinction in the northern Pacific by commercial fur hunters, sea otters made a remarkable recovery after Russia, Japan, Canada, and the United States finally agreed to stop hunting them in 1911. From only a few scattered individuals throughout their former range between Siberia and southern California, the otter population had rebounded to more than 100,000 in the Aleutians alone by the end of the twentieth century. The wreck of the *Exxon Valdez* in Prince William Sound in 1989 killed at least 1,000 otters, but the population quickly recovered once the spilled crude oil dissipated. Why, wildlife specialists wondered, are these charismatic animals suddenly vanishing now over much of their former range? The answers to this puzzle seem to involve two highly popular marine mammals, over-fishing by humans, and a complex ecological food web that affects a whole biological community.

Few animals are as cute and cuddly as sea otters. Their round fuzzy faces, dark button eyes, and huge walrus mustaches seem to express perpetual curiosity and good nature (Fig. 5.1). They are beautiful swimmers, supple and swift in the water. Highly skilled hunters, they live mainly on shellfish and sea urchins from the ocean floor. Their thick, luxurious fur protects them from the cold ocean and was the prize sought by hunters who once nearly extirpated them entirely. With few natural enemies other than man, the otters generally seemed safe once hunting was outlawed. But changes in the arctic food web seem now to have changed the otter's future dramatically.

The other charismatic marine mammal in this story is the orca. Once known as killer whales, these bold, black and white animals have become widely popular, especially since the release of the movie *Free Willy*. Whale watchers flock to the British Columbia and Alaskan coasts where they marvel at the graceful power of swimming orcas and listen to underwater recordings of their complex calls. Orcas inhabiting bays and estuaries eat mostly salmon, while those living around outer islands and the open ocean feed primarily on seals and sea lions. In both cases, overfishing by humans has seriously depleted food supplies on which these top predators depend.

Beginning in the late 1980s, seal and sea lion populations over much of the northern Pacific crashed, presumably because there weren't enough fish to sustain them. Although orcas prefer the high fat content of seals and sea lions, apparently some hungry whales have now turned to hunting otters. One piece of evidence for this is that otter populations in bays and inlets that are inaccessible to orcas have remained high, while those in exposed areas have declined by as much as 90 percent. Because otters are much smaller than seals and sea lions, it takes far more otters to sustain

FIGURE 5.1 Almost driven into extinction a century ago by fur hunters, sea otters now face a threat from hungry killer whales.

an orca. Researchers estimate that a single orca would need to eat 1,825 otters a year to satisfy all its nutritional needs. If all the orcas in the northern Pacific were to turn exclusively to otters, the population would be decimated in just a few years.

Loss of sea otters seriously impacts many other organisms. The giant kelp "forests" in which the otters live (see Fig. 5.19) are highly dependent on otters for protection. When otters are removed, sea urchin populations explode. Urchins eat kelp and too many of them can completely destroy a kelp bed and the whole biological community to which it offers shelter, food, and substrate. Without otters to control the urchins, up to 90 percent of the kelp has been eliminated in some areas of Alaska. Denuded urchin "barrens" support few fish, sea birds, or other normal inhabitants. Perhaps if we stop taking so many fish the whole system will return to normal, but it may take many years to restore original relationships. And what if the whales that have learned to hunt otters don't go back to their original prey? Have we permanently disrupted this prolific marine community?

How do we account for the amazing and complex interrelationships in an ecosystem such as this? How do organisms such as sea otters, orcas, kelp, and sea urchins become adapted to their particular environment and to each other? How can we understand the intricate interactions between various members of biological communities so as to avoid introducing other disastrous disruptions in the future? In this chapter we will examine competition, predation, symbiosis, tolerance limits, evolution, and other features of populations, species, and communities that affect both the organisms themselves and the ecosystems of which they are a part.

Want More?

Doroff, A.M., Estes, J.A., Tinker, M.T., Burn, D.M., Evans, T.J. "Sea Otter Population Declines in the Aleutian Archipelago." *Journal of Mammalogy.* 84 (1): 55–64 February 2003.

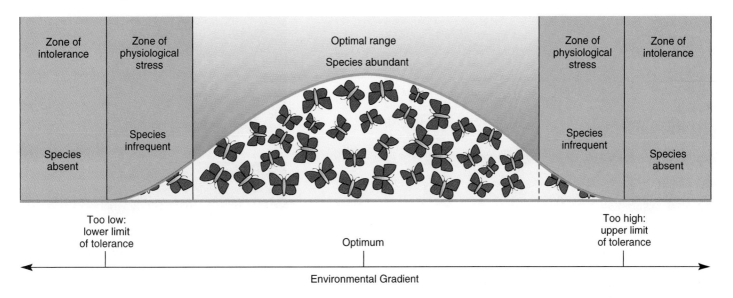

FIGURE 5.2 The principle of tolerance limits states that for every environmental factor, an organism has both maximum and minimum levels beyond which it cannot survive. The greatest abundance of any species along an environmental gradient is around the optimum level of the critical factor most important for that species. Near the tolerance limits, abundance decreases because fewer individuals are able to survive the stresses imposed by limiting factors.

WHO LIVES WHERE, AND WHY?

When we consider carefully the distribution of organisms around the globe, the first obvious conclusion we can draw is everything isn't everywhere. There are no polar bears in Mexico (except perhaps at a zoo), and no palm trees in Alaska (except perhaps on a cruise ship). So ecologists spend a lot of their time figuring out why a given species is one place, and not the other. Sometimes, it's as simple as a problem of individuals "getting there;" they have to be able to disperse to a piece of available and appropriate habitat in order to live there. But leaving the problem of getting there aside, a species has to be able to survive and thrive (reproduce) in the physical and chemical environment (we call this "being there"), and, moreover, it has to be able to maintain a population in the face of interactions with other organisms (we call this "staying there").

In this section we will look at some specific ways in which organisms are limited as to where they can be by both "being there" (physical and chemical conditions) and "staying there" (interactions with other organisms) factors. We'll put these together in trying to understand why groups of species (biological communities) are structured the way they are, and how they change over time.

Being There: Critical Factors and Tolerance Limits

Every living organism has limits to the environmental conditions it can endure. Temperatures, moisture levels, nutrient supply, soil and water chemistry, living space, and other environmental factors must be within appropriate levels for life to persist. In 1840, Justus von Liebig proposed that the single factor in shortest supply relative to demand is the critical determinant in the distribution of that species. Ecologist Victor Shelford later expanded this principle of limiting factors by stating that each environmental factor has both minimum and maximum levels, called **tolerance limits,** beyond which a particular species cannot survive (Fig. 5.2) or is unable to reproduce. The single factor closest to these survival limits, he postulated, is the critical limiting factor that determines where a particular organism can live.

At one time, ecologists accepted this concept so completely that they called it Liebig's or Shelford's Law and tried to identify unique factors limiting the growth of every population of plants and animals. For many species, however, we find that the interaction of several factors working together, rather than a single limiting factor, affects biogeographical distribution. If you have ever explored the rocky coasts of New England or the Pacific Northwest, for instance, you probably have noticed that mussels and barnacles endure extremely harsh conditions but generally are sharply limited to an intertidal zone where they grow so thickly that they often completely cover the substrate. No single factor affects this distribution. Instead, a combination of temperature extremes, drying time between tides, salt concentrations, competitors, and food availability limits the number and location of these animals.

For other organisms, there may be a specific *critical factor* that, more than any other, determines the abundance and distribution of that species in a given area. A striking example of cold intolerance as a critical factor is found in the giant saguaro cactus, which grows in the dry, hot Sonoran desert of southern Arizona and northern Mexico (Fig. 5.3). Saguaros are extremely sensitive to low temperatures. A single exceptionally cold winter night with temperatures below freezing for 12 hours or more will kill growing tips on the branches. Young saguaros are more susceptible to

FIGURE 5.3 Saguaro cacti, symbolic of the Sonoran desert, are an excellent example of distribution controlled by a critical environmental factor. Extremely sensitive to low temperatures, saguaros are found only where minimum temperatures never dip below freezing for more than a few hours at a time.

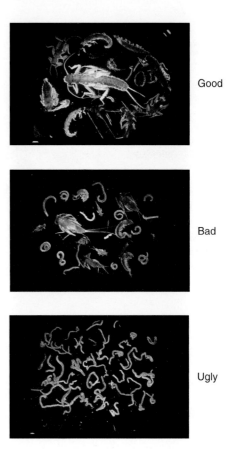

Good

Bad

Ugly

FIGURE 5.4 The particular species of invertebrates present in a stream can tell us whether or not the stream is degraded by pollution.

frost damage than adults, but seedlings typically become established under the canopy of small desert trees such as mesquite that shield the young cacti from the cold night sky. Unfortunately, the popularity of grilling with mesquite wood has caused extensive harvesting of the nurse trees that once sheltered small saguaros, adversely affecting reproduction of this charismatic species.

Animal species, too, exhibit tolerance limits that often are more critical for the young than for the adults. The desert pupfish, for instance, occurs in small isolated populations in warm springs in the northern Sonoran desert. Adult pupfish can survive temperatures between 0°C and 42°C (a remarkably high temperature for a fish) and are tolerant to an equally wide range of salt concentrations. Eggs and juvenile fish, however, can only live between 20°C and 36°C and are killed by high salt levels. Reproduction,

therefore, is limited to a small part of the range of adult fish, which is often restricted anyway by the size of the small springs and desert seeps in which the species lives.

Sometimes the requirements and tolerances of species are useful indicators of specific environmental characteristics. The presence or absence of such species can tell us something about the community and the ecosystem as a whole. Locoweeds, for example, are small legumes that grow where soil concentrations of selenium are high. Because selenium is often found with uranium deposits, locoweeds have an applied economic value as **environmental indicators.** Such indicator species also may demonstrate the effects of human activities. Lichens and eastern white pine are less restricted in habitat than locoweeds, but are indicators of air pollution because they are extremely sensitive to sulphur dioxide and acid precipitation. Bull thistle is a weed that grows on disturbed soil but is not eaten by cattle; therefore, an abundant population of bull thistle in a pasture is a good indicator of overgrazing. Aquatic ecologists can tell whether or not a stream is degraded by pollution just by examining the invertebrate species living in the stream (Fig. 5.4).

Natural Selection, Adaptation, and Evolution

How is it that mussels have developed the ability to endure pounding waves, daily exposure to drying sun and wind, and seasonal threats of extreme cold or hot temperatures? What enables desert pupfish to tolerate hot, mineral-laden springs? How does the saguaro survive in the harsh temperatures and extreme dryness of the desert? We commonly say that each of these species is "adapted" to its special set of conditions, but what does that mean? In this section, we will examine one of the most important concepts in biology: how species acquire traits that allow them to live in unique ways in particular environments.

We use the term *adapt* in two ways. One is a limited range of *physiological modifications* (called acclimation) available to individual organisms. If you keep house plants inside all winter, for example, and then put them out in full sunlight in the spring, they get sunburned. If the damage isn't too severe, your plants will probably grow new leaves with a thicker cuticle and denser pigments that protect them from the sun. They can adapt to some degree, but the change isn't permanent. Another winter inside will make them just as sensitive to the sun as before. Furthermore, the changes they acquire are not passed on to their offspring. Although the potential to acclimate is inherited, each generation must develop its own protective epidermis.

Another type of adaptation operates at the population level and is brought about by inheritance of specific genetic traits that allow a species to live in a particular environment. This process is explained by the theory of **evolution,** developed by Charles Darwin and Alfred Wallace. According to this theory, species change gradually through **natural selection,** a process in which members of a population best adapted to a particular set of environmental conditions will be more likely to survive and produce more offspring

than the less adapted. If their adaptations are at least partly inherited, then future generations will have a greater proportion of individuals with these adaptations.

Natural selection acts on preexisting genetic diversity originally created by a series of small, random mutations (changes in genetic material) that occur spontaneously in every population. These mutations, and the different combinations and recombinations of them that occur in subsequent generations, produce a variety of traits, some of which are more advantageous than others in a given situation. Where resources are limited or environmental conditions place some selective pressure on a population, individuals with those advantageous traits become more abundant in the population, and the species gradually evolves or becomes better suited to that particular environment. Although each change may be very slight, many mutations over a very long time have produced the incredible variety of different life-forms that we observe in nature.

The variety of finches observed by Charles Darwin on the Galápagos Islands is a classic example of speciation driven by availability of different environmental opportunities (Fig. 5.5). Originally derived from a single seed-eating species that somehow crossed the thousands of kilometres from the mainland, the finches have evolved into more than a dozen distinct species that differ markedly in appearance, food preferences, and habitats they occupy. Fruit eaters have thick parrot-like bills; seed eaters have heavy, crushing bills; insect eaters have a variety of bill forms depending on their prey. One of the most unusual species is the woodpecker finch, which pecks at tree bark for hidden insects.

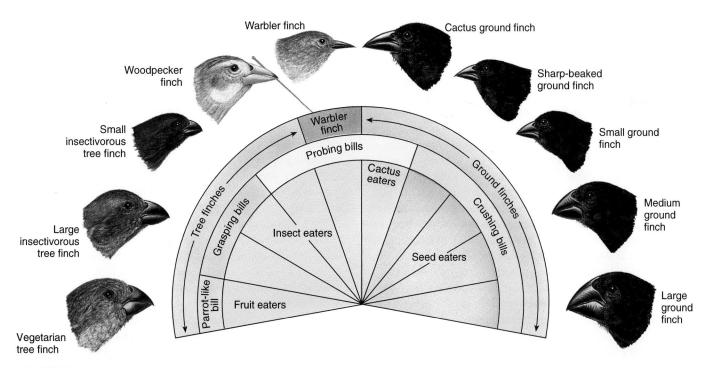

FIGURE 5.5 Some species of Galápagos Island finches. Although all are descendents of a common ancestor, they now differ markedly in appearance, habitat, and feeding behaviour. Ground finches (*lower right*) eat cactus leaves; warbler finches (*upper left*) eat insects; others eat seeds or have mixed diets. The woodpecker finch (*upper left*) pecks tree bark as do woodpeckers, but lacks a long tongue. Instead, it uses cactus spines as tools to extract insects.
Source: From Peter H. Raven and George B. Johnson, *Biology,* 4th edition. Copyright © 1996 McGraw Hill Company, Inc. All rights reserved. Reprinted by permission.

Lacking the woodpecker's long tongue, however, the finch uses a cactus spine as a tool to extract bugs.

The amazing variety of colours, shapes, and sizes of dogs, cats, rabbits, fish, flowers, vegetables, and other domestic species is evidence of deliberate selective breeding. The various characteristics of these organisms arose through mutations. We simply kept the ones we liked. Note that sexual reproduction helps to redistribute genetic material in new and novel combinations that greatly increase the variation and diversity we see in both wild and domestic species. Organisms that reproduce asexually can evolve, but often do so very slowly.

What environmental factors cause selective pressure and influence fertility or survivorship in nature? They include (1) physiological stress due to inappropriate levels of some critical environmental factor, such as moisture, light, temperature, pH, or specific nutrients; (2) predation, including parasitism and disease; (3) competition; and (4) luck. In some cases the organisms that survive environmental catastrophes or find their way to a new habitat where they start a new population may simply be lucky rather than more fit or better suited to subsequent environmental conditions than their less fortunate contemporaries.

Be sure you understand that while selection affects individuals, evolution and adaptation work at the population level. Individuals don't evolve; species do. Each individual is locked in by genetics to a particular way of life. Most plants, animals, or microbes have relatively limited ability to modify their physical makeup or behaviour to better suit a particular environment. Over time, however, random genetic changes and natural selection can change an entire population.

Given enough geographical isolation or selective pressure, the members of a population become so different from their ancestors that they may be considered an entirely new species that has replaced the original one. Alternatively, isolation of population subsets by geographical or behavioural factors that prevent exchange of genetic material can result in branching off of new species that coexist with their parental line. Suppose that two populations of the same species become separated by a body of water, a desert, or a mountain range that they cannot cross. Over a very long time—often millions of years—random mutations and different environmental pressures may cause the populations to evolve along such dissimilar paths that they can no longer interbreed successfully even if the opportunity to do so arises. They have now become separate species as in the case of the Galápagos finches. The barriers that divide subpopulations are not always physical. In some cases, behaviours such as when and where members of a population feed, sleep, or mate—or how they communicate—may separate them sufficiently for divergent evolution and speciation to occur even though they occupy the same territory.

Natural selection and adaptation can cause organisms with a similar origin to become very different in appearance and develop different habits over time, but they can also result in unrelated organisms coming to look and act very much alike. We call this latter process *convergent* evolution. The cactus-eating Galápagos finches (Fig. 5.5), for example, look and act very much like parrots even though they are genetically very dissimilar. The features that enable parrots to eat fruit successfully work well for these finches also.

A common mistake is to believe that organisms develop certain characteristics because they want or need them. This is incorrect. A duck doesn't have webbed feet because it wants to swim or needs to swim in order to eat; it has webbed feet because some ancestor happened to have a gene for webbed feet that gave it some advantage over other ducks in its particular pond and because those genes were passed on successfully to its offspring. A variety of different genetic types are always present in any population, and natural selection simply favours those best suited for particular conditions. Whether there is a purpose or direction to this process is a theological question rather than a scientific one and is beyond the scope of this book.

The Ecological Niche

Habitat describes the place or set of environmental conditions in which a particular organism lives. A more functional term, the **ecological niche,** is a description of either the role played by a species in a biological community or the total set of environmental factors that determine species distribution. Niches as community roles—describing how a species obtains food, what relationships it has with other species, and the services it provides its community, for example—were first described by the British ecologist, Charles Elton in 1927. Thirty years later, the American limnologist G. E. Hutchinson proposed a more biophysical definition of this concept. Every species, he pointed out, has a range of physical and chemical conditions (temperature, light levels, acidity, humidity, salinity, etc.) as well as biological interactions (predators and prey present, defences, nutritional resources available, etc.) within which it can exist. Figure 5.2, for example, shows the abundance of a hypothetical species along a single factor gradient. If it were possible to graph simultaneously all of the factors that affect a particular species, a multidimensional space would result that describes the ecological niche available to that species.

Some species, like raccoons or coyotes, are generalists that eat a wide variety of food and live in a broad range of habitats (including urban areas). Others, such as the panda (Fig. 5.6), are specialists that occupy a very narrow niche. Specialists often tend to be rarer than generalists and less resilient to disturbance or change.

A few species such as elephants, chimpanzees, and baboons learn how to behave from their social group and can invent new ways of doing things when presented with new opportunities or challenges. Most organisms, however, are limited by genetically determined physical structure and instinctive behaviour to established niches.

Over time, though, niches can evolve, just as physical characteristics do. The law of competitive exclusion states that no two species will occupy the same niche and compete for exactly the same resources in the same habitat for very long. Eventually, one group will gain a larger share of resources while the other will either migrate to a new area, become extinct, or change its behaviour or physiology in ways that minimize competition. We call this

PART ONE   Fundamental Principles of Ecology and Environmental Science

FIGURE 5.6 The giant panda feeds exclusively on bamboo. Although its teeth and digestive system are those of a carnivore, it is not a good hunter, and has adapted to a vegetarian diet. In the 1970s, huge acreages of bamboo flowered and died, and many pandas starved.

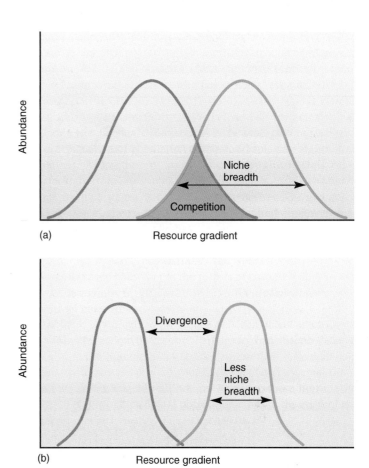

FIGURE 5.7 Resource partitioning and niche specialization caused by competition. Where niches of two species overlap along a resource gradient, competition occurs (shaded area in (*a*)). Individuals occupying this part of the niche are less successful in reproduction so that characteristics of the population diverge to produce more specialization, narrower niche breadth, and less competition between species (*b*).

latter process of niche evolution **resource partitioning** (Fig. 5.7). It can produce high levels of specialization that allow several species to utilize different parts of the same resource and coexist within a single habitat (Fig. 5.8).

Niche specialization also can create behavioural separation that allows subpopulations of a single species to diverge into separate species. Why doesn't this process continue until there is an infinite number of species? The answer is that a given resource can be partitioned only so far. Populations must be maintained at a minimum size to avoid genetic problems and to survive bad times. This puts an upper limit on the number of different niches—and therefore the number of species—that a given community can support.

Perhaps you haven't thought of time as an ecological factor, but niche specialization in a community is a 24-hour phenomenon. Swallows and insectivorous bats both catch insects, but some insect species are active during the day and others at night, providing noncompetitive feeding opportunities for day-active swallows and night-active bats.

STAYING THERE: SPECIES INTERACTIONS

We have seen how predation and competition for scarce resources are major factors in evolution and adaptation that can affect the size and nature of the ecological niche of a species. Not all biological interactions are competitive, however. Organisms also cooperate with, or at least tolerate, members of their own species as well as individuals of other species in order to survive and reproduce. In this section, we will look more closely at the different interactions within and between species that shape biological communities.

Predation

All organisms need food to live. Producers make their own food, and consumers eat organic matter created by other organisms. In most communities, as we saw in Chapter 3, photosynthetic

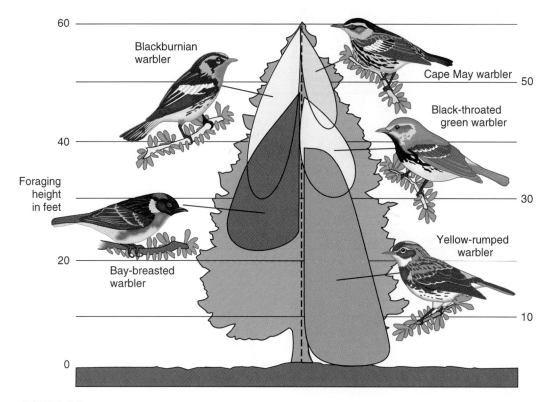

FIGURE 5.8 Resource partitioning and the concept of the ecological niche are demonstrated by several species of wood warblers that use different strata of the same forest. This is a classic example of the principle of competitive exclusion.
Source: Original observation by R. H. MacArthur.

organisms are the producers. Consumers include herbivores, carnivores, omnivores, scavengers, and decomposers. With which of these categories do you associate the term *predator?* Ecologically, the term has a much broader meaning than you might expect. A **predator** in an ecological sense, is an organism that feeds directly upon another living organism, whether or not it kills the prey to do so (Fig. 5.9). By this definition herbivores, carnivores, and omnivores that feed on live prey are predators, but scavengers, detritivores, and decomposers that feed on dead things are not. In this broad sense, **parasites** (organisms that feed on a host organism or steal resources from it without killing it) and even **pathogens** (disease-causing organisms) might also be considered predatory organisms.

Predation is a potent and complex influence on the population balance of communities involving (1) all stages of the life cycles of predator and prey species, (2) many specialized food-obtaining mechanisms, and (3) specific prey-predator adaptations that either resist or encourage predation (Case Study, p. 98).

Predation throughout the life cycle is very pronounced in marine intertidal animals, for example. Many crustaceans, mollusks, and worms release eggs directly into the water, and the eggs and free-living larval and juvenile stages are part of the floating community, or **plankton** (Fig. 5.10). Planktonic animals feed upon each other and are food for successively larger carnivores, including small fish. As prey species mature, their predators change. Barnacle larvae are planktonic and are eaten by fish. Adult barnacles,

FIGURE 5.9 Insect herbivores are predators as much as are lions and tigers. In fact, insects consume the vast majority of biomass in the world. Complex patterns of predation and defense have often evolved between insect predators and their plant prey.

on the other hand, build hard shells that protect them from fish but can be crushed by whelks and other mollusks. Predators also may change their feeding targets. Adult frogs, for instance, are carnivores, but the tadpoles of most species are grazing herbivores. Sorting out the trophic levels in these communities can be very difficult.

FIGURE 5.10 Microscopic plants and animals form the basic levels of many aquatic food chains and account for a large percentage of total world biomass. Many oceanic plankton are larval forms that have very different habitats and feeding relationships than their adult forms.

Predation is an important factor in evolution. Predators prey most successfully on the slowest, weakest, least fit members of their target population, thus reducing competition, preventing excess population growth, allowing successful traits to become dominant in the prey population, and making the prey population stronger and healthier. As the poet Robinson Jeffers said, "What but the wolf's tooth whittled so fine/The fleet limbs of the antelope?".

Prey species have evolved many protective or defensive adaptations to avoid predation. In plants, for instance, this often takes the form of thick bark, spines, thorns, or chemical defences. Animal prey may become very adept at hiding, fleeing, or fighting back against predators. Predators, in turn, evolve mechanisms to overcome the defences of their prey. This process in which species exert selective pressure on each other is called **coevolution.**

Competition

Competition is another kind of antagonistic relationship within a community. For what do organisms compete? To answer this question, think again about what all organisms need to survive: energy and matter in usable forms, space, and specific sites for life activities (What Do You Think? p. 100). Plants compete for growing space for root and shoot systems so they can absorb and process sunlight, water, and nutrients. Animals compete for living, nesting, and feeding sites, as well as for food, water, and mates.

Competition among members of the same species is called **intraspecific competition,** whereas competition between members of different species is called **interspecific competition.**

If you look closely at a patch of weeds growing on good soil early in the summer, you likely will see several types of interspecific competition. First of all, many weedy species attempt to crowd out their rivals by producing prodigious numbers of seeds. After the seeds germinate, the plants race to grow the tallest, cover the most ground, and get the most sun. You may observe several strategies to do this. For example, vines don't build heavy stems of their own; they simply climb up over their neighbours to get to the light.

We often think of competition among animals as a bloody battle for resources. A famous Victorian description of the struggle for survival was "nature red in tooth and claw." In fact, a better metaphor is a race. Have you ever noticed that birds always eat fruits and berries just before they are ripe enough for us to pick? Having a tolerance for bitter, unripe fruit gives them an advantage in the race for these food resources. Many animals tend to avoid fighting if possible. It's not worth getting injured. Most confrontations are more noise and show than actual fighting.

Intraspecific competition can be especially intense because members of the same species have the same space and nutritional requirements; therefore, they compete directly for these environmental resources. How do plants cope with intraspecific competition? The inability of seedlings to germinate in the shady conditions created by parent plants acts to limit intraspecific competition by favouring the mature, reproductive plants. Many plants have adaptations for dispersing their seeds to other sites by air, water, or animals. Undoubtedly you've seen dandelion seeds on their tiny parachutes (Fig. 5.11) or had sticky or burred seeds become attached to your clothing. Some plants secrete substances that inhibit the growth of seedlings near them, including their own and those of other species. This strategy is particularly significant in deserts where water is a limiting factor.

FIGURE 5.11 Dandelions and other opportunistic species generally produce many highly mobile offspring.

Every June, some 2,200 amateur ornithologists and bird watchers across Canada and the United States join in an annual bird count called the Breeding Bird Survey. Organized in 1966 by the U.S. Fish and Wildlife Service to follow bird population changes, this survey has discovered some shocking trends. While birds such as robins, starlings, and blackbirds that prosper around humans have increased their number and distribution over the past 30 years, many of our most colourful forest birds have declined severely. The greatest decreases have been among the true songbirds such as thrushes, orioles, tanagers, catbirds, vireos, buntings, and warblers. These long-distance migrants nest in northern forests but spend the winters in South or Central America or in the Caribbean Islands. Scientists call them neo-tropical migrants.

In many areas of Canada and the eastern United States, three-quarters or more of the neotropical migrants have declined significantly since the survey was started. Some that once were common have become locally extinct. Rock Creek Park in Washington, D.C., for instance, lost 75 percent of its songbird population and 90 percent of its long-distance migrant species in just 20 years. Nationwide, cerulean warblers, American redstarts, and ovenbirds declined about 50 percent in the single decade of the 1970s. Studies of radar images from National Weather Service stations in Texas and Louisiana suggest that only about half as many birds fly across the Gulf of Mexico each spring now compared to the 1960s. This could mean a loss of about half a billion birds in total.

What causes these devastating losses? Destruction of critical winter habitat is clearly a major issue. Birds often are much more densely crowded in the limited areas available to them during the winter than they are on their summer range. Unfortunately, forests throughout Latin America are being felled at an appalling rate. Central America, for instance, is losing about 1.4 million hectares (2 percent of its forests or an area about the size of Yellowstone National Park) each year. If this trend continues, there will be essentially no intact forest left in much of the region in 50 years.

But loss of tropical forests is not the only threat. Recent studies show that fragmentation of breeding habitat and nesting failures in Canada and the United States may be just as big a problem for woodland songbirds. Many of the most threatened species are adapted to deep woods and need an area of 10 hectares or more per pair to breed and raise their young. As our woodlands are broken up by roads, housing developments, and shopping centres, it becomes more and more difficult for these highly specialized birds to find enough contiguous woods to nest successfully.

Predation and nest parasitism also present a growing threat to many bird species. Raccoons, opossums, crows, bluejays, squirrels, and house cats thrive in human-dominated landscapes. They are protected from larger predators like wolves or owls and find abundant supplies of food and places to hide. Their numbers have increased dramatically, as have their raids on bird nests. A comparison of predation rates in the Great Smoky Mountain National Park and in small rural and suburban woodlands shows how devastating predators can be. In a 1,000-hectare study area of mature, unbroken forest in the national park, only one songbird nest in 50 was raided by predators. By contrast, in plots of 10 hectares or less near cities, up to 90 percent of the nests were raided.

Nest parasitism by brown-headed cowbirds is one of the worst threats for woodland songbirds. Rather than raise their young themselves, cowbirds lay their eggs in the nests of other species. The larger and more aggressive cowbird young either kick their foster siblings out of the nest, or claim so much food that the others starve to death. Well adapted to live around humans, there are now about 150 million cowbirds in the United States.

A study in southern Wisconsin found that 80 percent of the nests of woodland species were raided by predators and that three-quarters of those that survived were invaded by cowbirds.

This thrush has been equipped with a lightweight radio transmitter and antenna so that its movements can be followed by researchers.

Another study in the Shawnee National Forest in southern Illinois found that 80 percent of the scarlet tanager nests contained cowbird eggs and that 90 percent of the wood thrush nests were taken over by these parasites. The sobering conclusion of this latter study is that there probably is no longer any place in Illinois where scarlet tanagers and wood thrushes can breed successfully.

What can we do about this situation? Elsewhere in this book, we discuss sustainable forestry and economic development projects that could preserve forests at home and abroad. Preserving corridors that tie together important areas also will help. In areas where people already live, clustering of houses protects remaining woods. Discouraging the clearing of underbrush and trees from yards and parks leaves shelter for the birds.

Could we reduce the number of predators or limit their access to critical breeding areas? Would you accept fencing or trapping of small predators in wildlife preserves? How would you feel about a campaign to keep house cats inside during the breeding season?

Want More?

Schmidt, K.A. "Nest Predation and Population Declines in Illinois Songbirds: A Case for Mesopredator Effects." *Conservation Biology.* 17 (4): 1141–1150 August 2003.

Animals also have developed adaptive responses to intraspecific competition. Two major examples are varied life cycles and territoriality. The life cycles of many invertebrate species have juvenile stages that are very different from the adults in habitat and feeding. Compare a leaf-munching caterpillar to a nectar-sipping adult butterfly or a planktonic crab larva to its bottom-crawling adult form. In these examples, the adults and juveniles of each species do not compete because they occupy different ecological niches.

You may have observed robins chasing other robins during the mating and nesting season. Robins and many other vertebrate species demonstrate **territoriality,** an intense form of intraspecific competition in which organisms define an area surrounding their home site or nesting site and defend it, primarily against other members of their own species. Territoriality helps to allocate the resources of an area by spacing out the members of a population. It also promotes dispersal into adjacent areas by pushing grown offspring outward from the parental territory.

Territory size depends on the size of the species and the resources available. A pair of robins might make do with a suburban yard, but a large carnivore like a tiger may need thousands of square kilometres.

Symbiosis

In contrast to predation and competition, symbiotic interactions between organisms can be nonantagonistic. **Symbiosis** is the intimate living together of members of two or more species. **Commensalism** is a type of symbiosis in which one member clearly benefits and the other apparently is neither benefited nor harmed. Cattle often are accompanied by cattle egrets, small white shore birds who catch insects kicked up as the cattle graze through a field. The birds benefit while the cattle seem indifferent. Many of the mosses, bromeliads, and other plants growing on trees in the moist tropics are also considered to be commensals (Fig. 5.12). These epiphytes get water from rain and nutrients from leaf litter and dust fall, and often neither help nor hurt the trees on which they grow. In a way, the robins and sparrows that inhabit suburban yards are commensals with humans, since the birds benefit from the human modification of the landscape but the humans don't benefit or suffer from the presence of the birds.

Lichens are a combination of a fungus and a photosynthetic partner, either an alga or a cyanobacterium. Their association is a type of symbiosis called **mutualism,** in which both members of the partnership benefit (Fig. 5.13). Some ecologists believe that cooperative, mutualistic relationships may be more important in evolution than we have commonly thought. Aggressive interactions often are dangerous and destructive, while cooperation and compromise may have advantages that we tend to overlook. Survival of the fittest often may mean survival of those organisms that can live best with one another.

FIGURE 5.12 Plants compete for light and growing space in this Indonesian rainforest. Epiphytes, such as the ferns and bromeliads shown here, find a place to grow in the forest canopy by perching on the limbs of large trees. This may be a commensal relationship if the epiphytes don't hurt their hosts. Sometimes, however, the weight of epiphytes breaks off branches and even topples whole trees.

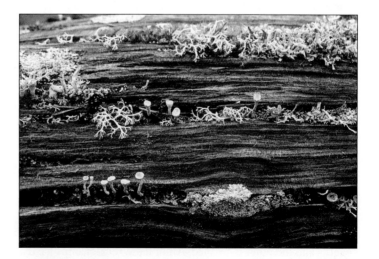

FIGURE 5.13 Lichens, such as the various species growing on this log, are a combination of algae and fungi in a classic example of mutualistic symbiosis.

Understanding Competition

Ecology is a relatively young science. Consequently, many ecological processes are incompletely understood. How a community comes to have its particular organization is one area of uncertainty. Some ecologists feel that physical factors are the most important determinants in community organization, while others feel that interspecific competition is most important.

How can we figure out which force is more powerful in shaping communities, or if the important determinants depend on the particular environmental context? Ecologists employ the scientific method, as described in Chapter 2, to better understand community dynamics. This process is mostly refined common sense and its basic elements can be useful in everyday life.

Once ecologists have decided on the concept to be investigated, they look for a specific situation that can either be observed or manipulated to provide relevant information. For example, ecologist Richard Karban was interested in how competition affected a community. He learned that larvae of two insect species, the meadow spittlebug and the calendula plume moth, both feed and develop on the seaside daisy, a common beach plant on the American west coast. The specific question to be investigated was: Does competition affect these two insect species, therefore impacting community organization?

Competition might reduce survival rate, larval growth, or both. Karban's procedure involved setting up four groups of plants at Bodega Bay, CA: one got both spittlebugs and moths, another got only spittlebugs, another only moths, and a fourth had neither. He compared survival rates of spittlebugs and moths when competitors were present and absent.

There are three important general considerations in designing scientific investigations:

1. Things need to be organized in such a way that the outcome can clearly be linked to a particular cause. In other words, differences in insect survival rates need to be clearly attributable to competition and not to other factors. Karban accomplished this by making his plant/insect groups as uniform as possible, except for the presence or absence of competitors. He eliminated genetic differences between plants by using plants from the same clone. He was careful to put the same numbers of insects on each plant to eliminate animal density as a factor, and so on.

2. The data collected must be a reliable representation of the larger situation and not simply the result of chance. This is usually accomplished by replicating the procedure many times. Instead of setting up just a few plants with one or both insects present, Karban set up 30 plants with each treatment. The procedure was repeated a second year. This gave him a cumulative total of 60 plants that had just spittlebugs, 60 plants having just moths, and 60 plants each having both or neither spittlebugs and moths. With such a large number of replications it was highly likely that differences in survival rates were, in fact, the result of competition and not simply chance occurrences.

3. Finally, conclusions must be justified by the data. Karban's statistical analysis revealed that spittlebug persistence was nearly 40 percent higher when the plume moths were absent. Plume moth persistence was not significantly affected by spittlebug presence, however.

His overall conclusion was:

Evidence from this and other studies supports the contention that interspecific competition can play an important role in influencing densities of plant-feeding insects.

Notice the caution expressed in these words. He did not claim to have proven anything. Instead, his study "*supports* the contention." Second, he states competition "*can* play an important role," instead of using stronger language. And finally, he restricts these conclusions to plant-feeding insects. Karban carefully avoids drawing conclusions beyond the realm supported by his data.

Based on a healthy skepticism, clarity of language, critical evaluation of relationships and information, and caution in coming to judgement, critical thinking in science has been a very successful tool in enhancing understanding.

Spittlebugs produce mounds of foam under which they hide from predators while feeding on host plants.

Parasitism, described earlier as a form of predation, also could be considered a type of symbiosis, where one species benefits and the other is harmed. All of these relationships have a bearing on such ecological issues as resource utilization, niche specialization, diversity, predation, and competition. Symbiotic relationships often enhance the survival of one or both partners.

Symbiotic relationships often entail some degree of coadaptation or coevolution of the partners, shaping—at least in part—their structural and behavioural characteristics. An interesting case of mutualistic coadaptation is seen in Central and South American swollen thorn acacias and their symbiotic ants. Acacia ant colonies live within the swollen thorns on the acacia tree branches and feed on two kinds of food provided by the trees: nectar produced in glands at the leaf bases and special protein-rich structures produced on leaflet tips. The acacias thus provide shelter and food for the ants. Although they spend energy to provide these services, the trees are not physically harmed by ant feeding.

What do the acacias get in return and how does the relationship relate to community dynamics? Ants tend to be aggressive defenders of their home areas, and acacia ants are no exception. They drive off herbivorous insects that attempt to feed on their home acacia, thus reducing predation. They also trim away vegetation that grows around their home tree, thereby reducing competition. This is a fascinating example of how a symbiotic relationship fits into community interactions. It is also an example of coevolution based on mutualism rather than competition or predation.

Defensive Mechanisms

Many species of plants and animals have toxic chemicals, body armor, and other ingenious defensive adaptations to protect themselves from competitors or predators. Arthropods, amphibians, snakes, and some mammals, for instance, produce noxious odours or poisonous secretions to induce other species to leave them alone. Plants also produce a variety of chemical compounds that make them unpalatable or dangerous to disturb. Perhaps you have brushed up against poison ivy or stinging nettles in the woods or you have encountered venomous insects or snakes and appreciate the wisdom of leaving them alone. Often, species possessing these chemical defences will evolve distinctive colours or patterns to warn potential enemies (Fig. 5.14).

Sometimes species that actually are harmless will evolve colours, patterns, or body shapes that mimic species that are unpalatable or poisonous. This is called **Batesian mimicry** after the English naturalist H. W. Bates, who described it in 1857. Wasps, for example, often have bold patterns of black and yellow stripes to warn off potential predators. The rarer longhorn beetle

FIGURE 5.14 Poison arrow frogs of the family Dendrobatidae use brilliant colours to warn potential predators of the extremely toxic secretions from their skin. Native people in Latin America use the toxin on blowgun darts.

(Fig. 5.15), although it has no stinger, looks and acts much like wasps and thus avoids predators as well. Another form of mimicry, called **Müllerian mimicry,** named for the German biologist Fritz Müller, who described it in 1878, involves two species, both of which are unpalatable or dangerous and have evolved to look alike. When predators learn to avoid either species, both benefit.

Species also evolve amazing abilities to avoid being discovered. You very likely have seen examples of insects that look exactly like dead leaves or twigs to hide from predators. Predators also use camouflage to hide as they lie in wait for their prey. The scorpion fish (Fig. 5.16) blends in remarkably well with its surroundings as it waits for smaller fish to come within striking distance. Not all cases of mimicry are to avoid or carry out predation, however. Some tropical orchids have evolved flower structures that look exactly like female flies. Males attempting to mate unwittingly carry away pollen.

FIGURE 5.15 An example of Batesian mimicry. The dangerous wasp (*left*) has bold yellow and black bands to warn predators away. The much rarer longhorn beetle (*right*) has no poisonous stinger, but looks and acts like a wasp and thus avoids predators as well.

FIGURE 5.16 This highly camouflaged scorpion fish lays in wait for its unsuspecting prey. Natural selection and evolution have created the elaborate disguise seen here.

COMMUNITY AND ECOSYSTEM PROPERTIES

The processes and principles that we have studied thus far in this chapter—tolerance limits, species interactions, resource partitioning, evolution, and adaptation—play important roles in determining the characteristics of populations and species. In this section we will look at some fundamental properties of biological communities (sets of species in the same area at the same time) and the ecosystems (the biotic and abiotic components of a given place) that they occupy.

Productivity

A community's **primary productivity** is the rate of biomass production, an indication of the rate of solar energy conversion to chemical energy (Chapter 3). The energy left after respiration is net primary production. Photosynthetic rates are regulated by light levels, temperature, moisture, and nutrient availability. Figure 5.17 shows approximate productivity levels for some major ecosystems.

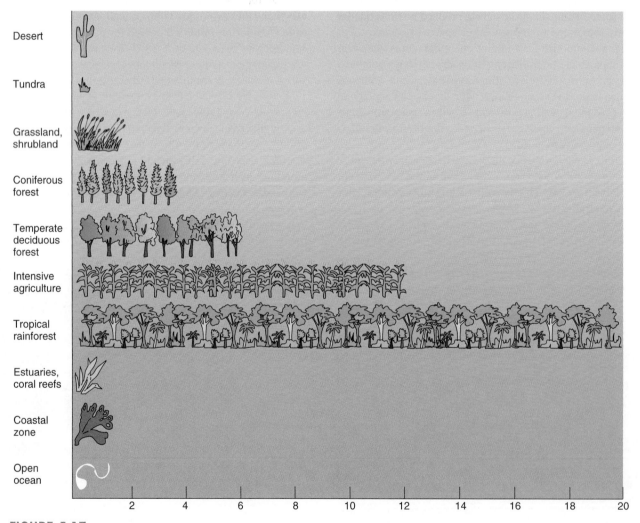

FIGURE 5.17 Net primary productivity of major world ecosystems. This is the ultimate limit on secondary (animal, bacteria, protist, fungi) productivity in an ecosystem.

PART ONE   Fundamental Principles of Ecology and Environmental Science

As you can see, tropical forests, coral reefs, and estuaries (bays or inundated river valleys where rivers meet the ocean) have high levels of productivity because they have abundant supplies of all these resources. In deserts, lack of water limits photosynthesis. On the arctic tundra or in high mountains, low temperatures inhibit plant growth. In the open ocean, low nutrient concentration reduces the ability of algae to make use of plentiful sunshine and water.

Some agricultural crops such as corn (maize) and sugar cane grown under ideal conditions in the tropics approach the productivity levels of tropical forests. Because shallow water ecosystems such as coral reefs, salt marshes, tidal mud flats, and other highly productive aquatic communities are relatively rare compared to the vast extent of open oceans—which are effectively biological deserts—marine ecosystems are much less productive on average than terrestrial ecosystems.

Even in the most photosynthetically active ecosystems, only a small percentage of the available sunlight is captured and used to make energy-rich compounds. Between one-quarter and three-quarters of the light reaching plants is reflected by leaf surfaces. Most of the light absorbed by leaves is converted to heat that is either radiated away or dissipated by evaporation of water. Only 0.1 to 0.2 percent of the absorbed energy is used by chloroplasts to synthesize carbohydrates.

In a temperate-climate oak forest, only about half the incident light available on a midsummer day is absorbed by the leaves. Ninety-nine percent of this energy is used to evaporate water. A large oak tree can transpire (evaporate) several thousand litres of water on a warm, dry, sunny day while it makes only a few kilograms of sugars and other energy-rich organic compounds.

Abundance and Diversity

Abundance is an expression of the total number of organisms in a biological community, while **diversity** is a measure of the number of different species, ecological niches, or genetic variation present. The abundance of a particular species often is inversely related to the total diversity of the community. That is, communities with a very large number of species often have only a few members of any given species in a particular area. As a general rule, diversity decreases but abundance within species increases as we go from the equator toward the poles. The arctic has vast numbers of insects such as mosquitoes, for example, but only a few species. The tropics, on the other hand, have vast numbers of species—some of which have incredibly bizarre forms and habits—but often only a few individuals of any particular species in a given area.

Consider bird populations. Greenland is home to 56 species of breeding birds, while Colombia, which is only one-fifth the size of Greenland, has 1,395. Why are there so many species in Colombia and so few in Greenland?

Climate and history are important factors. Greenland has such a harsh climate that the need to survive through the winter or escape to milder climates becomes the single most important critical factor that overwhelms all other considerations and severely limits the ability of species to specialize or differentiate into new forms.

Furthermore, because Greenland was covered by glaciers until about 10,000 years ago, there has been little time for new species to develop. It is also relatively isolated from potent dispersers.

Many areas in the tropics, by contrast, have relatively abundant rainfall and warm temperatures year-round so that ecosystems there are highly productive. The year-round dependability of food, moisture, and warmth supports a great exuberance of life and allows a high degree of specialization in physical shape and behaviour. Coral reefs are similarly stable, productive, and conducive to proliferation of diverse and amazing life-forms. The enormous abundance of brightly coloured and fantastically shaped fish, corals, sponges, and arthropods in the reef community is one of the best examples we have of community diversity.

Productivity is related to abundance and diversity, both of which are dependent on the total resource availability in an ecosystem as well as the reliability of resources, the adaptations of the member species, and the interactions between species. You shouldn't assume that all communities are perfectly adapted to their environment. A relatively new community that hasn't had time for niche specialization, or a disturbed one where roles such as top predators are missing, may not achieve maximum efficiency of resource use or reach its maximum level of either abundance or diversity.

Complexity and Connectedness

Community complexity and connectedness generally are related to diversity and are important because they help us visualize and understand community functions. **Complexity** in ecological terms refers to the number of species at each trophic level and the number of trophic levels in a community. A diverse community may not be very complex if all its species are clustered in only a few trophic levels and form a relatively simple food chain.

By contrast, a complex, highly interconnected community (Fig. 5.18) might have many trophic levels, some of which can be compartmentalized into subdivisions. In tropical rainforests, for instance, the herbivores can be grouped into "guilds" based on the specialized ways they feed on plants. There may be fruit eaters, leaf nibblers, root borers, seed gnawers, and sap suckers, each composed of species of very different size, shape, and even biological kingdom, but that feed in related ways. A highly interconnected community such as this can form a very elaborate food web.

Keystone Species

A **keystone species** is a species or group of species whose impact on its community or ecosystem is much larger and more influential than would be expected from mere abundance. Originally, keystone species were thought to be top predators, such as wolves, whose presence limits the abundance of herbivores and thereby reduces their grazing or browsing on plants. Recently, it has been recognized that less conspicuous species also play essential community roles. Certain tropical figs, for example, bear during seasons when no other fruit is available for frugivores (fruit-eating animals). If these figs were removed, many animals

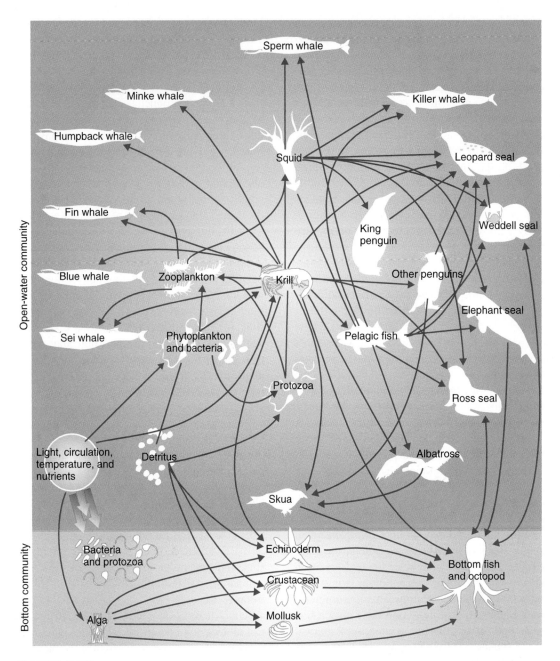

FIGURE 5.18 A complex and highly interconnected community can have many species at each trophic level and many relationships, as illustrated by this Antarctic marine food web. Complexity and diversity would be even greater than this in equatorial communities.

would starve to death during periods of fruit scarcity. With those animals gone, many other plant species that depend on them at other times of the year for pollination and seed-dispersal would disappear as well.

Even microorganisms can play vital roles. In some forest ecosystems, mycorrhizae (fungi associated with tree roots) are essential for mineral mobilization and absorption. If the fungi die, so do the trees and many other species that depend on a healthy forest community. Rather than being a single species,

mycorrhizae are actually a group of species that together fulfill a keystone function.

Often a number of species are intricately interconnected in biological communities so that it is difficult to tell which is the essential key. Consider the Pacific kelp forests and sea otters that opened this chapter (Fig. 5.19). Giant kelp provide shelter for a number of fish and shellfish species and so could be regarded as the key to community structure. Sea urchins, however, feed on the kelp and determine their number and distribution while sea otters

PART ONE   Fundamental Principles of Ecology and Environmental Science

FIGURE 5.19 Giant kelp is a massive alga that forms dense "forests" off the Pacific coast of California. It is a keystone species in that it provides food, shelter, and structure essential for a whole community. Removal of sea otters allows sea urchin populations to explode. When the urchins destroy the kelp, many other species suffer as well.

regulate urchins and kelp provides a resting place for dozing otters. Which of these species is the most important? Each depends on and affects the others. Perhaps we should think in terms of a "keystone set" of organisms in some ecosystems.

Resilience and Stability

Many biological communities tend to remain relatively stable and constant over time. An oak forest tends to remain an oak forest, for example, because the species that make it up have self-perpetuating mechanisms. We can identify three kinds of stability or resiliency in ecosystems: *constancy* (lack of fluctuations in composition or functions), *inertia* (resistance to perturbations), and *renewal* (ability to repair damage after disturbance).

In 1955, Robert MacArthur, who was then a graduate student at Yale, proposed that the more complex and interconnected a community is, the more stable and resilient it will be in the face of disturbance. If many different species occupy each trophic level, some can fill in if others are stressed or eliminated by external forces, making the whole community resistant to perturbations and able to recover relatively easily from disruptions. This theory has been controversial, however. Some studies support it, while others do not. For example, Minnesota ecologist David Tilman, in studies of native prairie and recovering farm fields, found that plots with high diversity were better able to withstand and recover from drought than those with only a few species.

On the other hand, in a diverse and highly specialized ecosystem, removal of a few keystone members can eliminate many other associated species. Eliminating a major tree species from a tropical forest, for example, may destroy pollinators and fruit distributors as well. We might replant the trees, but could we replace the whole web of relationships on which they depend? In this case, diversity has made the forest less resilient rather than more.

Edges and Boundaries

An important aspect of community structure is the boundary between one habitat and its neighbours. We call these relationships **edge effects.** Sometimes, the edge of a patch of habitat is relatively sharp and distinct. In moving from a woodland patch into a grassland or cultivated field, you sense a dramatic change from the cool, dark, quiet forest interior to the windy, sunny, warmer, open space of the field or pasture (Fig. 5.20). In other cases, one habitat type intergrades very gradually into another, so there is no distinct border.

Ecologists call the boundaries between adjacent communities **ecotones.** A community that is sharply divided from its neighbours is called a closed community. In contrast, communities with gradual or indistinct boundaries over which many species cross are called open communities. Often this distinction is a matter of degree or perception. As we saw earlier in this chapter, birds might feed in fields or grasslands but nest in the forest. As they fly back and forth, the birds interconnect the ecosystems by moving energy and material from one to the other, making both systems relatively open. Furthermore, the forest edge, while clearly different from the open field, may be sunnier and warmer than the forest interior, and may have a different combination of plant and animal species than either field or forest "core."

Depending on how far edge effects extend from the boundary, differently shaped habitat patches may have very dissimilar amounts of interior area (Fig. 5.21). In Douglas fir forests of the Pacific Northwest, for example, increased rates of blowdown, decreased humidity, absence of shade-requiring ground cover, and

FIGURE 5.20 Ecologists call the sharp edge, or boundary between the woods and sunny glade seen in this picture, an ecotone. Edge effects such as increased sunshine, wind, and temperature, as well as decreased humidity may extend a considerable distance into the forest. When fragmented into many small patches, the forest may become all edge, with no true interior core at all.

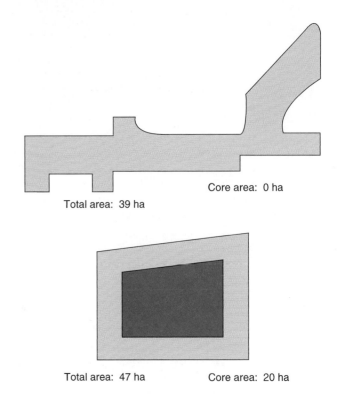

Total area: 39 ha   Core area: 0 ha

Total area: 47 ha   Core area: 20 ha

FIGURE 5.21 Shape can be as important as size in small preserves. While these areas are close to the same size, no place in the top figure is far enough from the edge to have characteristics of core habitat, while the bottom patch has a significant core.

Source: From Verner, Jared, Wildlife 2000. Copyright © 1996. Reprinted by permission of the University of Wisconsin Press.

other edge effects can extend as much as 200 m into a forest. A 400-metre square surrounded by clear-cut would have essentially no true core habitat at all.

Many popular game animals, such as white-tailed deer and pheasants that are adapted to human disturbance, often are most plentiful in boundary zones between different types of habitat. Game managers once were urged to develop as much edge as possible to promote large game populations. Today, however, most wildlife conservationists recognize that the edge effects associated with habitat fragmentation are generally detrimental to biodiversity. Such habitats are known as **evolutionary traps,** since what was once a beneficial habitat (e.g., edge) is now a dangerous or degraded environment. Preserving large habitat blocks and linking smaller blocks with migration corridors may be the best ways to protect rare and endangered species (see Chapter 22).

COMMUNITIES IN TRANSITION

So far our view of communities has focused on the day-to-day interactions of organisms with their environments, set in a context of survival and selection. In this section, we'll step back to look at some transitional aspects of communities, including where communities meet and how communities change over time.

Ecological Succession

Biological communities have a history in a given landscape. The process by which organisms occupy a site and gradually change environmental conditions by creating soil, shade, shelter, or increasing humidity is called ecological succession or development. **Primary succession** occurs when a community begins to develop on a site previously unoccupied by living organisms, such as an island, a sand or silt bed, a body of water, or a new volcanic flow (Fig. 5.22). **Secondary succession** occurs when an existing community is disrupted and a new one subsequently develops at the site. The disruption may be caused by some

PART ONE   Fundamental Principles of Ecology and Environmental Science

Developing a Sense for Where You Live

Although we often feel insignificant and powerless in the face of the big problems facing society, there are things that individuals can do to help improve our environment. One of the first steps toward becoming a more effective environmental citizen is to learn something about the place where you live.

- What natural ecosystems and biological communities existed in your area before European settlement?

- What impact, if any, did indigenous people have on the flora, fauna, or topography of your locality?

- What are the dominant species (besides humans) in your neighbourhood now and where did they originate?

- How much rain falls in your region in a given year? Is there generally more precipitation in one month or season than another? What role does moisture availability play in determining who lives where and how?

- What are the seasonal high, low, and average temperatures where you live? How do native plants and animals adapt to severe weather events and seasonal variations?

- Is there a keystone species or group of species especially important in determining the structure and functions of your local ecosystems? What factors might threaten those keystone components?

- Where do your drinking water, food, and energy come from? What local and regional environmental impacts are caused by production, use, and disposal of those resources?

- Is there a park or wildlife refuge near where you live? Does it contain any rare, threatened, or endangered species? What makes them rare, threatened, or endangered?

- Are there opportunities for volunteer work to improve your local environment: planting trees, cleaning up a river or lake, restoring a wetland, recycling trash, helping to maintain a refuge or park?

natural catastrophe, such as fire or flooding, or by a human activity, such as deforestation, plowing, or mining. Both forms of succession usually follow an orderly sequence of stages as organisms modify the environment in ways that allow one species to replace another.

In primary succession on a terrestrial site, the new site first is colonized by a few hardy **pioneer species,** often microbes, mosses, and lichens that can withstand harsh conditions and lack of resources. Their bodies create patches of organic matter in which protists and small animals can live. Organic debris accumulates in pockets and crevices, providing soil in which seeds can become lodged and grow. We call this process of environmental

modification by organisms **ecological development** or facilitation. The community of organisms often becomes more diverse and increasingly competitive as development continues and new niche opportunities appear. The pioneer species gradually disappear as the environment changes and new species combinations replace the preceding community. In a global sense, the gradual changes brought about by living organisms have created many of the conditions that make life on earth possible. You could consider evolution to be a very slow, planetwide successional and developmental process.

Examples of secondary succession are easy to find. Observe an abandoned farm field or clear-cut forest (Fig. 5.23) in a temperate climate. The bare soil first is colonized by rapidly growing annual plants (those that grow, flower, and die the same year) that have light, wind-blown seeds and can tolerate full sunlight and exposed soil. They are followed and replaced by perennial plants (those that live for several to many years), including grasses, various nonwoody flowering plants, shrubs, and trees. As in primary succession, plant species progressively change the environmental conditions. Biomass accumulates and the site becomes richer, better able to capture and store moisture, more sheltered from wind and climate change, and biologically more complex. Species that cannot survive in a bare, dry, sunny, open area find shelter and food as the field turns to prairie or forest.

Eventually, in either primary or secondary succession, a community often develops that resists further change. Ecologists call this a **climax community** because it appears to be the culmination of the successional process. An analogy is often made between community succession and organism maturation. Beginning with a primitive or juvenile state and going through a complex developmental process, each progresses until a complex, stable, and mature form is reached. It's dangerous to carry this analogy too far, however, because no mechanism is known to regulate communities in the same way that genetics and physiology regulate development of the body.

As mentioned in Chapter 3, the concept of succession to a climax community was first championed by the pioneer biogeographer F. E. Clements. He viewed this process as being like a parade or relay, in which species replace each other in predictable groups and in a fixed, regular order, and as being driven almost entirely by climate. This community-unit theory was opposed by Clements's contemporary, H. A. Gleason, who saw community history as a much more individualistic and random process driven by many environmental factors. He argued that temporary associations are formed according to the conditions prevailing at a particular time and the species available to colonize a given area. You might think of the Gleasonian model as a time-lapse movie of a busy railroad station. Passengers come and go; groups form and then dissipate. Patterns and assemblages that seem significant to us may not mean much in the long run.

The process of succession may not be as deterministic as we once thought, yet mature or highly developed ecological communities may tend to be resilient and stable over long periods of time because they can resist or recover from external disturbances. Many are characterized by high species diversity, narrow niche

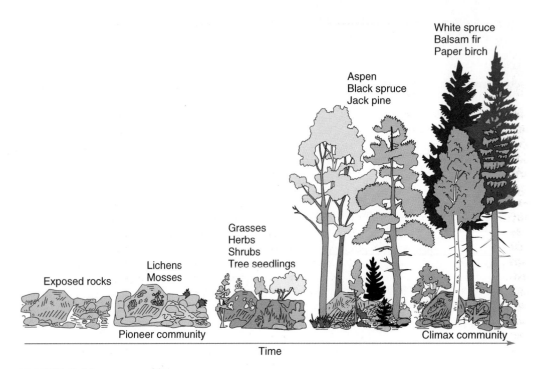

White spruce
Balsam fir
Paper birch

Aspen
Black spruce
Jack pine

Grasses
Herbs
Shrubs
Tree seedlings

Lichens
Mosses

Exposed rocks

Pioneer community

Climax community

Time

FIGURE 5.22 A generalized example of primary succession on a terrestrial site is shown in five stages (*left to right*), beginning with rocks that are initially colonized by a pioneer community of lichens and mosses and ending with a climax forest community.

FIGURE 5.23 This area was once a cool, shady black spruce stand. The forest floor was covered by a deep, moist layer of sphagnum moss. Clear-cutting and burning have turned it into a dry, sunny, barren ground on which few of the former residents can survive. Secondary succession will probably restore previous conditions if the climate doesn't change and further disturbance is prevented.

specialization, well-organized community structure, good nutrient conservation and recycling, and a large amount of total organic matter. Community functions, such as productivity and nutrient cycling, tend to be self-stabilizing or self-perpetuating. What once were regarded as "final" climax communities, however, may still be changing. It's probably more accurate to say that the rate of succession is so slow in a climax community that, from the perspective of a single human lifetime, it appears to be stable.

Some landscapes never reach a stable climax in the traditional sense because they are characterized by, and adapted to, periodic disruption. They are called **equilibrium communities** or **disclimax communities.** Grasslands, the chaparral shrubland of California, and some kinds of coniferous forests, for instance, are shaped and maintained by periodic fires that have long been a part of their history. They are, therefore, often referred to as **fire-climax communities** (Fig. 5.24). Plants in these communities are adapted to resist fires, reseed quickly after fires, or both. In fact, many of the plant species we recognize as dominants in these communities require fire to eliminate competition, to prepare seedbeds for germination of seedlings, or to open cones or thick seed coats. Without fire, community structure may be quite different.

Introduced Species and Community Change

Succession requires the continual introduction of new community members and the disappearance of previously existing species. New species move in as conditions become suitable; others die or move out as the community changes. New species also can be

FIGURE 5.24 This lodgepole pine forest was once thought to be a climax forest, but we now know that this forest must be constantly renewed by periodic fire. It is an example of an equilibrium, or disclimax, community.

FIGURE 5.25 Mongooses were released in Hawaii in an effort to control rats. The mongooses are active during the day, however, while the rats are night creatures, so they ignored each other. Instead, the mongooses attacked defenseless native birds and became as great a problem as the rats.

introduced after a stable community already has become established. Some cannot compete with existing species and fail to become established. Others are able to fit into and become part of the community, defining new ecological niches. If, however, an introduced species preys upon or competes more successfully with one or more populations that are native to the community, the entire nature of the community can be altered.

Human introductions of Eurasian plants and animals to non-Eurasian communities often have been disastrous to native species because of competition or overpredation. Oceanic islands offer classic examples of devastation caused by rats, goats, cats, and pigs liberated from sailing ships. All these animals are prolific, quickly developing large populations. Goats are efficient, nonspecific herbivores; they eat nearly everything vegetational, from grasses and herbs to seedlings and shrubs. In addition, their sharp hooves are hard on plants rooted in thin island soils. Rats and pigs are opportunistic omnivores, eating the eggs and nestlings of seabirds that tend to nest in large, densely packed colonies, and digging up sea turtle eggs. Cats prey upon nestlings of both ground- and tree-nesting birds. Native island species are particularly vulnerable because they have not evolved under circumstances that required them to have defensive adaptations to these predators.

Sometimes we introduce new species in an attempt to solve problems created by previous introductions but end up making the situation worse. In Hawaii and on several Caribbean Islands, for instance, mongooses were imported to help control rats that had escaped from ships and were destroying indigenous birds and devastating plantations (Fig. 5.25). Since the mongooses were diurnal (active in the day), however, and rats are nocturnal, they tended to ignore each other. Instead, the mongooses also killed native birds and further threatened endangered species. Our lessons from this and similar introductions have a new technological twist. Some of the ethical questions currently surrounding the release of genetically engineered organisms are based on concerns that they are novel organisms, and we might not be able to predict how they will interact with other species in natural ecosystems—let alone how they might respond to natural selective forces. It is argued that we can't predict either their behaviour or their evolution.

SUMMARY

The principle of limiting factors states that for every physical factor in the environment, there are both maximum and minimum tolerance limits beyond which a given species cannot survive. Sometimes, the factor in shortest supply or closest to the tolerance limit for a particular species at a particular time is the critical factor that will determine the abundance and distribution of that species in that ecosystem but more often it is a complex interplay of factors that determines if a species can make a living in a given environment. Random genetic variations create diversity in a population that gives some individuals advantages in a given set of circumstances. The best-suited organisms will survive and reproduce more successfully than the ill-suited ones. Eventually, the genes for these successful characteristics predominate in the population and the species becomes adapted to its environment and to a particular role. This process leads to evolution of a species either through gradual replacement of the original parental type or a splitting of a population into two species.

Habitat describes the place in which an organism lives; niche describes either the role an organism plays or the total set of conditions that control its distribution. Natural selection often leads to niche specialization and resource partitioning that reduce competition between species. Organisms interact within communities in many ways. Predation—feeding on another organism— involves pathogens, parasites, and herbivores as well as carnivorous predators. Competition is another kind of antagonistic relationship in which organisms vie for space, food, or other resources. Symbiosis is the intimate living together of two species. Mutualism means that both species benefit; commensalism means that one species benefits while the other is indifferent.

Some fundamental properties of biological communities are productivity, diversity, complexity, resilience, stability, and structure. Productivity is a measure of the rate at which photosynthesis produces biomass made of energy-rich compounds. Tropical rainforests are generally the most productive of all terrestrial communities; coral reefs and estuaries are generally the most productive aquatic communities. Diversity is a measure of the number of different species in a community, while abundance is the total number of individuals. Often the most productive and stable communities are highly diverse and profusely populated, but sometimes a high degree of specialization makes an ecosystem more, rather than less, susceptible to disturbance. Ecological complexity refers to the number of species at each trophic level as well as the total number of trophic levels in a community. Structure concerns the patterns of organization, both spatial and functional, in a community. Often a keystone species, or group of species, plays an unusually important role in determining community structure, composition, or function. All of these characteristics are affected both by physical and chemical factors as well as biological interactions between the organisms that make up the community. Edge effects at the boundaries between different habitat types can be important where landscapes are fragmented into isolated patches.

Ecological succession and development are processes by which organisms alter the environment in ways that allow some species to replace others. Primary succession starts with a previously unoccupied site. Secondary succession occurs on a site that has been disturbed by external forces. Often succession proceeds until a mature, diverse, climax community is established. These mature communities may have self-perpetuating processes that make them resistant to change and resilient to disturbance. Whether diversity always leads to stability, however, is controversial. Communities that are disrupted regularly by fires or other natural disasters sometimes establish dynamic equilibrium or disclimax communities dependent on constant renewal. Introduction of new species by natural processes, such as opening of a land bridge, or through human intervention can upset the natural relationships in a community and cause catastrophic changes for indigenous species.

QUESTIONS FOR REVIEW

1. Explain how tolerance limits to environmental factors determine distribution of a highly specialized species such as the desert pupfish. Compare this to the distribution of a generalist species such as cowbirds or starlings.

2. Productivity, diversity, complexity, resilience, and structure are exhibited to some extent by all communities and ecosystems. Describe how these characteristics apply to the ecosystem in which you live.

3. Describe the general niche occupied by a bird of prey, such as a hawk or an owl. How can hawks and owls exist in the same ecosystem and not adversely affect each other?

4. Define keystone species and explain their importance in community structure and function.

5. All organisms within a biological community interact with each other. The most intense interactions often occur between individuals of the same species. What concept discussed in this chapter can be used to explain this phenomenon?

6. Relationships between predators and prey play an important role in the energy transfers that occur in ecosystems. They also influence the process of natural selection. Explain how predators affect the adaptations of their prey.

This relationship also works in reverse. How do prey species affect the adaptations of their predators?

7. Competition for a limited quantity of resources occurs in all ecosystems. This competition can be interspecific or intraspecific. Explain some of the ways an organism might deal with these different types of competition.

8. Each year fires burn large tracts of forestland. Describe the process of succession that occurs after a forest fire destroys an existing biological community. Is the composition of the final successional community likely to be the same as that which existed before the fire? What factors might alter the final outcome of the successional process? Why may periodic fire be beneficial to a community?

9. Explain the concept of climax community. Why does the climax community often exhibit a higher level of stability than that found in other successional stages?

10. Discuss the dangers posed to existing community members when new species are introduced into ecosystems. What type of organism would be most likely to survive and cause problems in a new habitat?

ecosystem with which you are familiar and decide whether it has a keystone species or keystone set.

5. Some scientists look at the boundary between two biological communities and see a sharp dividing line. Others looking at the same boundary see a gradual transition with much intermixing of species and many interactions between communities. Why are there such different interpretations of the same landscape?

6. The absence of certain lichens is used as an indicator of air pollution in remote areas such as national parks. How can we be sure that air pollution is really responsible? What evidence would be convincing?

7. We tend to regard generalists or "weedy" species as less interesting and less valuable than rare and highly specialized endemic species. What values or assumptions underlie this attitude?

8. What part of this chapter do you think is most likely to be challenged or modified in the future by new evidence or new interpretations?

QUESTIONS FOR CRITICAL THINKING

1. Ecologists debate whether biological communities have self-sustaining, self-regulating characteristics or are highly variable, accidental assemblages of individually acting species. What outlook or worldview might lead scientists to favor one or the other of these theories?

2. The concepts of natural selection and evolution are central to how most biologists understand and interpret the world, and yet the theory of evolution is contrary to the beliefs of many religious groups. Why do you think this theory is so important to science and so strongly opposed by others? What evidence would be required to convince opponents of evolution?

3. What is the difference between saying that a duck has webbed feet because it needs them to swim and saying that a duck is able to swim because it has webbed feet?

4. The concept of keystone species is controversial among ecologists because most organisms are highly interdependent. If each of the trophic levels is dependent on all the others, how can we say one is most important? Choose an

KEY TERMS

abundance 103
Batesian mimicry 101
climax community 107
coevolution 97
commensalism 99
complexity 103
disclimax communities 108
diversity 103
ecological development 107
ecological niche 94
ecotones 105
edge effects 105
environmental indicators 92
equilibrium
    communities 108
evolution 93
evolutionary traps 106
fire-climax communities 108

habitat 94
interspecific competition 97
intraspecific competition 97
keystone species 103
Müllerian mimicry 101
mutualism 99
natural selection 93
parasites 96
pathogens 96
pioneer species 107
plankton 96
predator 96
primary productivity 102
primary succession 106
resource partitioning 95
secondary succession 106
symbiosis 99
territoriality 99
tolerance limits 91

PROJECT FEEDERWATCH

The FeederWatch Program coordinated by the Cornell Laboratory of Ornithology is an excellent example of citizen science. Thousands of volunteers collect data on bird frequency and distribution from back yard feeders throughout winter months. The data are displayed on innovative animated maps that allow you to view dynamic information about a given species in a particular region over time. Go to: http://birds.cornell.edu/PFWMaproom/pfwmaproom.html to find a species and location that interests you; then consider the following questions:

1. Does it surprise you that this species does or doesn't occur in your area?

2. How would you account for the patterns you see on the map? Is it possible that the results show a bias in data collection rather than a real variation in distribution of the species?

3. Some species display seasonal movements. Can you detect a pattern in changing distribution of the species you've chosen during the time shown? How would you account for the pattern (or the lack of a pattern) you observe?

TROPHIC CASCADES IN AQUATIC FOOD WEBS

Ecological relationships can affect physical qualities in our environment. To understand how this occurs, go to http://www.mcgrawhill.ca/college/environmentalscience. Click on the cover photo for your textbook to take you to the online learning centre, and then click on the student centre. You'll see a revolving globe that leads to Regional Perspectives in Environmental Issues. Scroll down to the North region to find a case study titled Food Web Control of Primary Production in Lakes.

Read the text and study the graphics to answer the following questions.

1. Explain the three graphs. Why does an increase in game fish (piscivores) cause a decrease in phytoplankton (algae) in a lake?

2. If you were designing a test of this hypothesis, how would you regulate piscivore biomass experimentally?

3. What would you use as a control in your study?

4. What do the authors mean by top down and bottom up controls?

5. Why do they call this a trophic cascade?

ALIEN INVADERS

With all the media attention invasive species such as zebra mussels and purple loosestrife have received, some people don't realize just how many "exotic species" there are in some parts of Canada. Go to http://www.great-lakes.net/envt/flora-fauna/invasive/invasive.html to learn about invasive species in the Great Lakes. Pick a species to learn more about, following some of the great links at this site. Try to answer the following questions about "your" species.

1. When did it arrive and how fast did it spread?

2. What about its biology made it a good invader?

3. What negative impacts has it caused in the Great Lakes ecosystem?

4. Have there been any benefits to the invasion of this species?

5. How do you think society determines whether or not a species is "good" or "bad" for the ecosystem?

Floating mats of water hyacinth clog the Chao Phrang River in Bangkok, Thailand.

Chapter 6

Biomes

Nature never did betray/The heart that loved her.

—William Wordsworth—

OBJECTIVES

After studying this chapter, you should be able to:

- recognize the characteristics of major aquatic and terrestrial biomes and understand the most important factors that determine the distribution of each type.
- describe ways in which humans disrupt or damage each of these ecosystem types.
- summarize the overall patterns of human disturbance of world biomes as well as some specific, important examples of losses obscured by broad aggregate categories.
- explain the principles and practices of landscape ecology and ecosystem management.

WebQuest

biome, tundra, boreal forest, temperate forest, mountains, landscape ecology

Above: A gallery of Canadian biomes, including (clockwise from top right) tundra, boreal forest, temperate forest, and temperate rainforest.

Disappearing Butterfly Forests

Every fall, in one of the most remarkable spectacles in nature, somewhere around a quarter of a billion monarch butterflies flutter southward from Canada and the United States into the mountains of central Mexico, where they spend the winter in cool, high-altitude, oyamel fir forests (Fig. 6.1). Incredibly, some of these delicate insects travel up to 4,000 km to reach their winter refuge. Discovered by entomologists only 25 years ago, about a dozen small patches of cloud-forest in the mountains west of Mexico City offer exactly the right conditions for monarch hibernation. With night temperatures just above freezing, and cool, foggy days, these forests allow insects to conserve energy and avoid desiccation as they await spring. The average colony contains about 20 million monarchs; some larger ones are thought to have three times this many.

From November to March, tall trees in the mariposa (butterfly) forests are completely covered with bright orange and black blankets as millions of butterflies cling to tree bark and to each other, often breaking off branches with their collective weight. On sunny days when the butterflies awaken and flutter down to streams to drink, the air is filled with floating specks of colour. As temperatures warm in the spring, more and more butterflies become active. Sometime toward the end of March, golden streams of insects pour down out of the Sierra Transvolcanica to begin their journey north once again.

As they move northward through Mexico and then spread out from the Rocky Mountains to the Atlantic Coast, the monarchs follow spring flower emergence, sipping nectar and laying their eggs on milkweed plants, which provide toxins that protect caterpillars and adults from predators. Summer offspring of spring migrants live only about a month. Incredibly, four generations of monarchs hatch, lay eggs, and die during the summer months, but those hatched in the fall somehow sense the need to fly south, over a route they have never seen, to the cloud forests where they will survive until the next spring.

Unfortunately, the oyamel forests on which this whole cycle depends are rapidly disappearing. Representing less than 2 percent of all Mexican forests, this unique ecosystem is one of the rarest and the most endangered in the whole country. Wood harvesting and fires—both accidental and deliberately set to clear land for agriculture—are the greatest threats. In 1986 a presidential decree created the "Reserva de la Biosfera Mariposa Monarca" protecting 161,100 ha including five of the 12 known monarch overwintering areas.

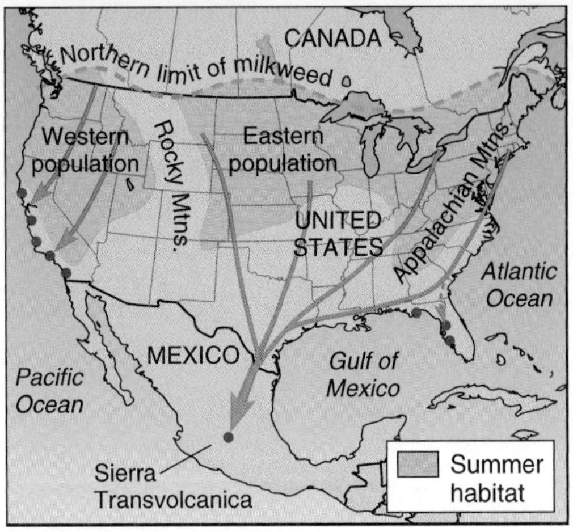

FIGURE 6.1 Fall migration routes and wintering sites for monarch butterflies. Some monarchs winter in Florida and along the Gulf Coast, but most of the population east of the Rockies migrates to Mexico. Western monarchs winter in California.
Source: Data from Brower and Malcolm, *Animal Migration: Endangered Phenomenon.*

Only a small core area is legally protected from logging, however. Most of the land from which the reserve was created was "ejido" or communal property, and local ejido members were never compensated properly for lost income where the logging ban was enforced. Consequently, logging—both legal and illegal—has continued in the sanctuary. Increasing numbers of ecotourists, who come to see the amazing monarch concentrations, provide some income for local communities, but tourism lasts only for about five winter months. Some conservationists are undertaking economic development projects to increase local income. The Monarch Butterfly Sanctuary Foundation, for instance, is paying an ejido with land rights in the Sierra Chincua sanctuary not to log in the reserve buffer zone.

In March 2001, 22 million wintering monarchs—more than 10 percent of the entire population—died suddenly. A sudden cold snap was blamed for the deaths. It appears that loggers encroaching on the butterfly forests have created forest openings that allow temperatures to drop unusually fast. This event demonstrates the vulnerability of populations that rely on just a few remaining fragments of habitat.

In this chapter, we will survey the range of habitats and ecosystems around Canada and the globe so that later (in Chapter 17) we can thoughtfully consider how biomes like those important to the monarch butterfly are changing with human activity.

TERRESTRIAL BIOMES

Many places on the earth share similar climatic, topographic, and soil conditions, and roughly comparable biological communities have developed in response to analogous conditions in widely separated locations. These broad types of communities are called **biomes.** Although, as we pointed out in Chapter 5, there can be considerable variation in the individual species that make up biological communities at two similar sites or even at a single site over time, some broad landscape categories exist. Recognizing these categories gives us insights about the general kinds of plants and

PART ONE   Fundamental Principles of Ecology and Environmental Science

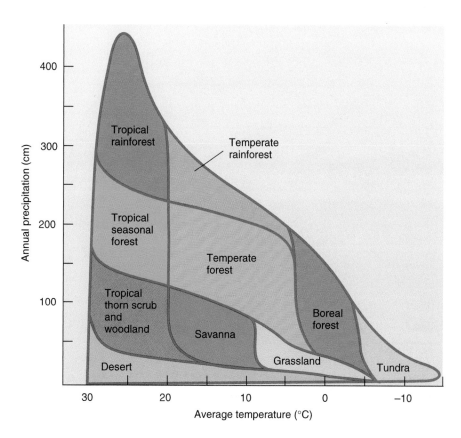

FIGURE 6.2 Biomes most likely to occur in the absence of human disturbance or other disruptions, according to average annual temperature and precipitation. *Note:* this diagram does not consider soil type, topography, wind speed, or other important environmental factors. Still, it is a useful general guideline for biome location.
From *Communities and Ecosystems,* 2/e by R. H. Whitaker, © 1975. Reprinted by permission of Prentice Hall, Upper Saddle River, New Jersey.

animals that we might expect to find there and how they may be adapted to their particular environment.

Temperature and precipitation are among the most important determinants in biome distribution. If we know the general temperature range and precipitation level, we can predict what kind of biological community is likely to develop on a particular site if that site is free of disturbance for a sufficient time (Fig. 6.2). Biome distribution also is influenced by the prevailing landforms of an area, which is in turn a function of its geological history.

Figure 6.3 shows the distribution of major terrestrial biomes around the world. Because of its broad scope, this map ignores the many variations present within each major category. Most terrestrial biomes are identified by the dominant plants of their communities (for example, grassland or deciduous forest). The characteristic diversity of animal life and smaller plant forms within each biome is, in turn, influenced both by the physical conditions and the dominant vegetation.

Deserts

Deserts are characterized by low moisture levels and precipitation that is both infrequent and unpredictable from year to year. With little moisture to absorb and store heat, daily and seasonal temperatures can fluctuate widely. Deserts that have less than 2.5 cm of measurable precipitation support almost no vegetation. Deserts with 2.5 to 5 cm annual precipitation have sparse vegetation (less than 10 percent of the ground is covered), and plants in

this harsh climate need a variety of specializations to conserve water and protect tissues from predation. Seasonal leaf production, water-storage tissues, and thick epidermal layers help reduce water loss. Spines and thorns discourage predators while also providing shade (Fig. 6.4).

Warm, dry, descending air (see Chapter 8) creates broad desert bands in continental interiors at about 30° north and south in the American Southwest, North and South Africa, China, and Australia. These descending air currents also help create desert strips along the west coast of South America and Africa that are among the driest regions in the world. Although we think of deserts as hot, barren, and filled with sand dunes, those at high latitudes or high elevations are often cool or even cold, and sand dunes are actually rather rare away from coastal areas. Most deserts around the world are gravelly or rocky scrubland, where 5 to 10 cm of annual precipitation supports a sparse but often species-rich community dominated by shrubs or small trees. In rare years when winter rains are adequate, a breathtaking profusion of spring ephemerals can carpet the desert with flowers.

Animals of the desert have both structural and behavioural adaptations to meet their three most critical needs: food, water, and heat survival. Many desert animals escape the main onslaught of daytime heat by hiding in burrows or rocky shelters from which they emerge only at night. Pocket mice and kangaroo rats (and their Old World counterparts, gerbils) get most of the moisture they need from the seeds and grains they eat. They have many adaptations to conserve water such as producing highly concentrated urine and

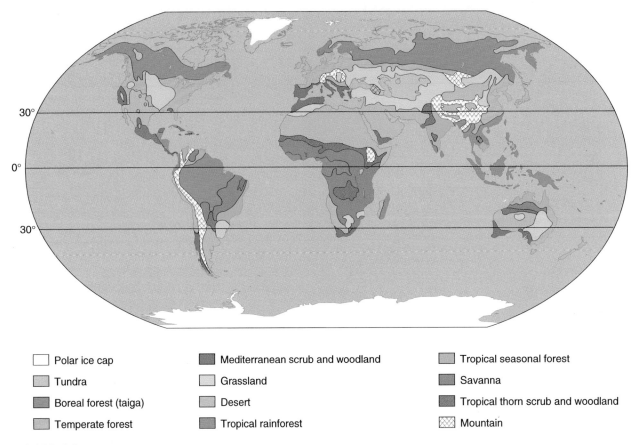

FIGURE 6.3 Major world biomes. Compare this map to Figure 6.2 for general climate conditions, and to Figure 6.11 for a satellite image of primary biological productivity.
Source: Enger & Smith 6/e.

Legend:
- Polar ice cap
- Tundra
- Boreal forest (taiga)
- Temperate forest
- Mediterranean scrub and woodland
- Grassland
- Desert
- Tropical rainforest
- Tropical seasonal forest
- Savanna
- Tropical thorn scrub and woodland
- Mountain

FIGURE 6.4 Joshua trees in the Joshua Tree National Park are really members of the lily family. Like other plants in this hot, dry, rocky landscape, they are adapted to conserve water and repel enemies.

nearly dry feces that allow them to eliminate body wastes without losing precious moisture.

Deserts may seem formidable, but they are more vulnerable than you might imagine. Desert soils are easily disturbed by human activities and are slow to recover because the harsh desert climate severely reduces the ability of desert communities to recover from damage. Tracks left by army tanks practising in the California desert during World War II, for instance, are still visible.

Many dry areas have been overgrazed, mainly by domestic livestock. Other areas are being converted to agricultural land in spite of uncertain future availability of water to sustain crops. Often after humans have degraded and abandoned desert areas they remain wastelands unsuitable for native vegetation or wildlife for a very long time.

Grasslands and Savannas

The moderately dry continental climates of the Great Plains of central North America, the broad Russian steppes, the African veldt, and the South American pampas support **grasslands,** rich biological communities of grasses, seasonal herbaceous flowering plants, and open savannas. Seasonal cycles for temperature and precipitation contribute to abundant vegetative growth that both protects and enriches the soils of these prairies and plains, making them among the richest farmlands in the world.

Grasslands have few trees because inadequate rainfall, large daily and seasonal temperature ranges, and frequent grass fires kill woody seedlings. Exceptions are the narrow gallery forests that form corridors along rivers and streams through the grasslands.

In many parts of the world, grasslands are artificially created or maintained by native people using fire. They do this to improve hunting and ease of travel.

In contrast, fire suppression and conversion of the rich prairie soil to farmland have greatly reduced native grasslands elsewhere. The most productive North American prairie lands now grow wheat, corn, oats, sunflowers, flax, and other cultivated crops. Lands that are less suited for crops because of limited water availability are used mainly as rangeland, where overgrazing reduces natural diversity of the community and contributes to soil erosion.

Before humans converted North America's grasslands to croplands and rangelands, they contained herds of wildlife that rivalled those of Africa's Serengeti Plain. Vast herds of bison roamed the plains along with wolves, deer, elk, and pronghorn antelopes. Shorebirds and migratory waterfowl occupied the millions of ponds, potholes, and marshy spots that once dotted our grasslands. Hunting, fencing, wetland drainage, introduction of alien species, and other human modifications to the land have greatly diminished most wildlife populations since the nineteenth century.

Tundra

Climates in high mountain areas or at far northern or southern latitudes often are too harsh for trees. This treeless landscape, called **tundra,** is characterized by a very short growing season, cold, harsh winters, and the potential for frost any month of the year. Although water may be abundant on the tundra, for much of the year it is locked up in ice or snow and therefore unavailable to plants. As far as plants are concerned, the tundra is a very cold desert.

The *arctic* tundra is a biome of low productivity, low diversity, and low resilience. Winters are long and dark. Only the top several centimetres of the soil thaw out in the summer, and the lower soil is permafrost. This permanently frozen layer prevents snowmelt water from being absorbed into the soil, so the surface soil is waterlogged during the summer. Try to imagine the difficulties encountered by plants in this kind of soil. Most of the year it is completely frozen, and even during the brief growing season, the permafrost is an impenetrable barrier to deep root growth. In addition, the top layer buckles and heaves in response to cycles of freezing and thawing, toppling plants and disrupting root systems.

The *alpine* tundra (Fig. 6.5) differs from the arctic tundra in several ways. Plants of the alpine tundra face different challenges than those of the arctic. The thin mountain air permits intense solar bombardment, especially by ultraviolet radiation; thus, many alpine plants have deep pigmentation that shields their inner cells. The glaring summer sun also causes very hot daytime ground temperatures, even though the night temperatures may return to freezing. Alpine soil is windswept and often gravelly or rocky. The sloping

FIGURE 6.5 The harsh physical conditions of alpine tundra ecosystems are the result of altitude and slope. The growing season is short, ultraviolet radiation in the thin air is greater than at lower altitudes, and extreme temperature fluctuations are possible even during the summer. Air currents and the thin, rocky soil also contribute to the arid conditions of alpine tundra.

terrain causes moisture to drain quickly. Due to this combination of sun, soil, slope, and air currents, drought is a problem—as opposed to the wet conditions in the arctic tundra.

Although the tundra may swarm with life during the brief summer growing season, only a few species are able to survive the harsh winters or to migrate to warmer climates. Dominant tundra plants are dwarf shrubs, sedges, grasses, mosses, and lichens. Its larger life-forms, such as arctic musk ox and caribou, or alpine mountain goats and mountain sheep, must be adapted to survive the harsh climate and sparse food supply. Many animals migrate or hibernate during winter. Flocks of migratory birds nest on the abundant summer arctic wetlands, which also nurture hordes of bloodsucking insects that feed upon the summer flocks and herds (and tourists!).

Damage to the tundra is slow to heal. At present, the greatest threat to this distinctive biome is oil and natural gas wells in the Arctic and mineral excavation in mountain regions. Because plants grow slowly during the brief summer at high altitudes or latitudes, truck ruts and bulldozer tracks on the tundra landscapes may take centuries to heal. Furthermore, some of the most promising sites for oil exploration or mining are summer feeding and breeding grounds for animals such as caribou, grizzly bears, or mountain sheep. Even seemingly minor incursions into this fragile ecosystem can cause great damage.

Boreal Forest and Taiga

Several distinctive biomes are dominated by **conifer** (cone-bearing) trees. Where moisture is limited by sandy soil, low precipitation, or a short growing season, plants reduced water loss by evolving thin, needlelike evergreen leaves with a thick waxy

FIGURE 6.6 This conifer forest in Siberia is typical of the tagia, or boreal forest, that stretches around the world at high latitudes.

FIGURE 6.7 At the northern limit of the boreal forest, we find small, widely spaced black spruce intermixed with willows and heather on a wet peatland.

coating. Although not as efficient in carrying out photosynthesis during summer months as the broad, soft leaves of deciduous trees, conifer needles and scales can survive harsh winters or extended droughts and do accomplish some photosynthesis even under poor conditions. In mountain areas, particularly, fire has been an important and fairly regular factor in maintaining the coniferous forest.

The **boreal forest,** or northern coniferous forest, stretches in a broad band of mixed coniferous and deciduous trees around the world between about 45° and 60° north latitude, depending on altitude and distance from coastlines or major rivers. Among the dominant conifers are pine, hemlock, spruce, cedar, and fir (Fig. 6.6). Some common deciduous trees are birches, aspens, and maples, while mosses and lichens form much of the ground cover. In this moist, cool biome, streams and wetlands abound—especially on recently glaciated landscapes. As a result, there are many lakes, potholes, bogs, and fens. Insects that have aquatic stages in their life cycles, such as mosquitoes and biting flies, are particularly abundant, to the consternation of humans as well as wild birds and mammals.

The northernmost edge of the boreal forest is a species-poor black spruce and sphagnum moss (peat moss) woodland—often called by its Russian name, **taiga**—that forms a ragged border with the treeless arctic tundra (Fig. 6.7). The harsh climate limits both productivity and resilience of the taiga community. Cold temperatures, very wet soil during the growing season, and acids produced by fallen conifer needles and sphagnum inhibit full decay of organic matter. As a result, thick layers of semidecayed organic material, called peat, form. Boreal peat deposits are being explored as energy sources; however, the environmental disturbance caused by peat mining in this boreal community could be severe and long-lasting, perhaps even permanent.

The southern pine forest is characterized by a warm, moist climate and sandy soil and, in the past, was subjected to frequent fires. Now it is managed extensively for timber and such resinous products as turpentine and rosin. The undergrowth includes saw palmetto and various thorny bushes.

The coniferous forests of the Pacific coast represent yet another special set of environmental circumstances. Mild temperatures and abundant precipitation—up to 250 cm per year—result in luxuriant plant growth and huge trees such as the California redwood, the largest tree in the world and the largest organism of any kind known to have ever existed. Redwoods formerly grew along the Pacific coast from California to Oregon, but their distribution has been greatly reduced by logging without regard to sustainable yield or restoration.

In its wettest parts, the coastal forest becomes a **temperate rainforest,** a cool, rainy forest that often is enshrouded in fog. Condensation from the canopy (leaf drip) becomes a major form of precipitation, and annual precipitation exceeds 250 cm in some places. Mosses, lichens, and ferns cover tree branches, old stumps, and the forest floor itself. On the coast of British Columbia are some of the greatest remaining tracts of temperate rainforest in the world. There is a continual tension between those that would protect and those that wish to harvest these forests.

Temperate Forests

Forests of broad-leaved trees occur throughout the world where rainfall is plentiful. In temperate regions, the climate supports lush summer plant growth when water is plentiful but requires survival adaptations for the frozen season. A key adaptation of **deciduous**

FIGURE 6.8 The deciduous forests of eastern North America are stratified assemblages of tall trees, understory shrubs, and small, lovely, ground-covering species. A rich fauna is supported by the variety of niche opportunities that are present.

trees is the ability to produce summer leaves and then shed them at the end of the growing season. This rich and varied biome contains associations of many tree species, including oak, maple, birch, beech, elm, ash, and other hardwoods. These tall trees form a forest canopy over a diverse understory of smaller trees, shrubs, and herbaceous plants, including many annual spring flowers that grow, flower, set seed, and store carbohydrates before they are shaded by the canopy (Fig. 6.8). Where the climate is warm year-round, forests are dominated by evergreen trees such as the live oaks and cypresses of the southern United States.

Most of the dense forest that once covered Central Europe was cleared a thousand years ago. When European settlers first came to North America, dense forests of broad-leaved deciduous

trees covered most of the eastern half of what is now the United States. Much of that original deciduous forest was harvested for timber a century or more ago. Now, large areas of forest in the eastern United States have regrown and are approaching old-growth status, although with a different species composition than the original biome. Vast original deciduous forests remain in eastern Siberia, but they are being harvested at a rapid rate, perhaps the highest deforestation rate in the world. As the forests disappear, so too do Siberian tigers, bears, cranes, and a host of other unique and endangered species.

Tropical Thorn Scrub and Woodland

A Mediterranean climate is characterized by warm, dry summers and cool, moist winters. Evergreen shrubs with small, leathery, sclerophyllous (hard, waxy) leaves form dense thickets. Scrub oaks, drought-resistant pines, or other small trees often cluster in sheltered valleys. Fires burn fiercely in this fuel-rich plant assemblage and are a major factor in plant succession. Annual spring flowers often bloom profusely, especially after fires. In California, this landscape is called **chaparral,** Spanish for thicket. Some typical animals include jackrabbits, kangaroo rats, mule deer, chipmunks, lizards, and many bird species. Very similar landscapes are found along the Mediterranean coast as well as southwestern Australia, central Chile, and South Africa. Although this biome doesn't cover a very large total area, it contains a high number of unique species and is often considered a "hot-spot" for biodiversity. It also is highly desired for human habitation, often leading to conflicts with rare and endangered plant and animal species.

Areas that are drier year-round, such as the African Sahel (edge of the Sahara Desert), northern Mexico, or the American Intermountain West (or Great Basin) tend to have a more sparse, open shrubland, characterized by sagebrush, chamiso, or saltbush. In Africa, acacias and other spiny plants dominate this landscape, giving it the name **thorn scrub.** Some typical animals of this biome in America are a wide variety of snakes and lizards, rodents, birds, antelope, and mountain sheep. In Africa, this landscape is home to gazelle, rhinos, and giraffes (Fig. 6.9).

Tropical Rainforests

The humid tropical regions of South and Central America, Africa, Southeast Asia, and some of the Pacific Islands support one of the most complex and biologically rich biome types in the world (Fig. 6.10). Although there are several kinds of tropical rainforests, they share common attributes of ample rainfall and uniform temperatures. Cool **cloud forests** are found high in the mountains where fog and mist keep vegetation wet all the time. **Tropical rainforests** occur where rainfall is abundant—more than 200 cm per year—and temperatures are warm to hot year-round.

The soil of both these tropical rainforest types tends to be old, thin, acidic, and nutrient-poor, yet the number of species present can be mind-boggling. For example, the number of insect species found in the canopy of tropical rainforests has been

FIGURE 6.9 A giraffe feeds on acacias in the dry, thorn scrub of Africa.

FIGURE 6.10 Tropical rainforests, like this one in Java, support a luxuriant profusion of species and biomass. Notice the many vines and epiphytes clinging to the buttressed trunk of this giant forest tree.

estimated to be in the millions! It is estimated that one-half to two-thirds of all species of terrestrial plants and insects live in tropical forests.

The nutrient cycles of these forests also are unique. Almost all (90 percent) of the nutrients in the system are contained in the bodies of the living organisms. This is a striking contrast to temperate forests, where nutrients are held within the soil and made available for new plant growth. The luxuriant growth in tropical rainforests depends on rapid decomposition and recycling of dead organic material. Leaves and branches that fall to the forest floor decay and are incorporated almost immediately back into living biomass.

When the forest is removed for logging, agriculture, and mineral extraction, the thin soil cannot support continued cropping and cannot resist erosion from the abundant rains. And if the cleared area is too extensive, it cannot be repopulated by the rainforest community. Rapid deforestation is occurring in many tropical areas as people move into the forests to establish farms and ranches, but the land soon loses its fertility.

Tropical Seasonal Forests

Many areas in India, Southeast Asia, Australia, West Africa, the West Indies, and South America have tropical regions characterized by distinct wet and dry seasons instead of uniform heavy rainfall throughout the year, although temperatures are hot year-round. These areas have produced communities of **tropical seasonal forests:** semievergreen or partly deciduous forests tending toward open woodlands and grassy savannas dotted with scattered, drought-resistant tree species.

Tropical dry forests have typically been more attractive than wet forests for human habitation and have suffered greater degradation. Clearing a dry forest with fire is relatively easy during the dry season. Soils of dry forests often have higher nutrient levels and are more agriculturally productive than those of a rainforest. Finally, having fewer insects, parasites, and fungal diseases than a wet forest makes a dry or seasonal forest a healthier place for humans to live. Consequently, these forests are highly endangered in many places. Less than 1 percent of the dry tropical forests of the Pacific coast of Central America or the Atlantic coast of South America, for instance, remain in an undisturbed state.

AQUATIC ECOSYSTEMS

Oceans, lakes, wetlands, and other aquatic ecosystems cover more than two-thirds of the earth's surface and play a vital role in biological productivity, biogeochemical cycles, climate modification, and species diversity. We don't have space here to cover all the interesting and important aspects of limnology and oceanography, but in this section, we'll survey some aspects of aquatic biology. You'll find more on these topics in Chapters 8 and 10.

Freshwater and Saline Ecosystems

Freshwater ecosystems include rivers and streams as well as ponds, lakes, and wetlands containing water with relatively low salt concentrations. There are also some unique freshwater ecosystems that we rarely see, such as underground rivers and subterranean pools. Saltwater ecosystems cover vastly more total area and contain much greater volume of water than all freshwater bodies combined. The oceans comprise the bulk of all water in the world. In addition, some lakes such as the Great Salt Lake in Utah or the Dead Sea in Israel, as well as inland seas such as the Caspian, also contain saline (salty) water. Smaller salty and mineral-rich ponds and small lakes are fed by mineral hot springs or formed by evaporation in enclosed basins, especially in desert regions.

The amount of chlorophyll in plant cells is a good measure of photosynthetic activity. Figure 6.11 shows chlorophyll levels (as pale green) in the ocean on a June day in 2000. Where are the highest levels of primary productivity during this season. Where are the deepest green areas on land at this time?

Aquatic ecosystems are as varied as their individual sites because they are influenced not only by characteristics of local climate, soil, and resident communities but also by the adjacent terrestrial ecosystems and anything that happens uphill or upstream from them (Fig. 6.12). As with terrestrial ecosystems, the biological communities of aquatic ecosystems are largely determined by the physical characteristics of the environment, except that the surrounding medium is water instead of air.

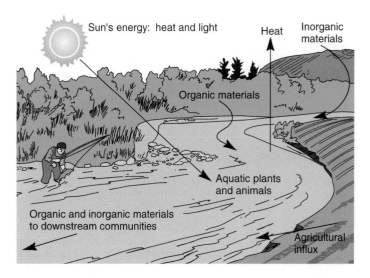

FIGURE 6.12 The character of freshwater ecosystems is greatly influenced by the immediately surrounding terrestrial ecosystems, and even by ecosystems far upstream or far uphill from a particular site.

Aquatic organisms have the same basic needs as terrestrial organisms: carbon dioxide, water, and sunlight for photosynthesis; oxygen for respiration; and food and mineral nutrients for energy, growth, and maintenance. The availability of these necessities is influenced by such site characteristics as (1) substances that are dissolved in the water, such as oxygen, nitrates, phosphates, potassium compounds, and other by-products of agriculture and industry; (2) suspended matter, such as silt and microscopic algae, that affect water clarity and, therefore, light penetration; (3) depth; (4) temperature; (5) rate of flow; (6) bottom characteristics (muddy, sandy, rocky); (7) internal convective currents; and (8) connection to, or isolation from, other aquatic ecosystems.

Vertical stratification is an important aspect of many aquatic ecosystems, especially in regard to gradients of light, temperature, nutrients, and oxygen. Organisms tend to form distinctive vertical subcommunities in response to this stratification of physical factors. The plankton subcommunity consists mainly of microscopic plants, animals, and protists (single-celled organisms such as amoebae) that float freely within the water column. Some non-planktonic organisms are specialized to live at the air-water interface (for example, insects such as water striders), and still others are able to swim freely in the open waters (such as fish).

Finally, many animal species are bottom dwellers (such as snails, burrowing worms and insect larvae, bacteria) that make up the **benthos,** or bottom subcommunity. Oxygen levels are lowest in the benthic environment. Anaerobic bacteria can live in the very low-oxygen bottom sediments. Emergent plants such as cattails and rushes are rooted in bottom sediments of the littoral (near shore) zone but spread their leaves above the water surface. They create important structural and functional links between strata of the ccosystem and often are the greatest producers of net primary productivity in the ecosystem.

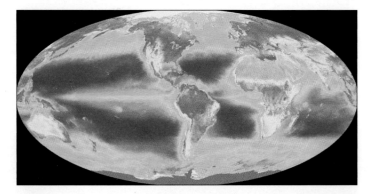

FIGURE 6.11 Satellite measurements of chlorophyll levels in the oceans and on land. Dark green to blue land areas have high biological productivity. Dark blue oceans have little chlorophyll and are biologically impoverished. Light green to yellow ocean zones are biologically rich.

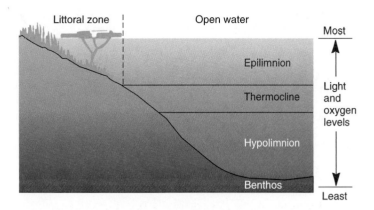

FIGURE 6.13 The layers of a deep lake are determined mainly by gradients of light, oxygen, and temperature. The epilimnion is affected by surface mixing from wind and thermal convections, while mixing between the hypolimnion and epilimnion is inhibited by a sharp temperature and density difference at the thermocline.

Deeper water bodies are characterized by the presence of a warmer, upper layer that is mixed by the wind (the epilimnion; *epi* = upon) and a colder, deep layer that is not mixed (the hypolimnion; *hypo* = below) (Fig. 6.13). The two layers are separated by a distinctive temperature transition zone called the **thermocline,** or mesolimnion.

Smaller lakes, ponds, and nearshore zones have a tendency to undergo succession, which includes changes in the biological community as a response to increases in nutrient levels. Human sources of nutrient input can radically increase the rate at which a water body "ages," as discussed in Chapter 11 (eutrophication).

Estuaries and Wetlands: Transitional Communities

Estuaries are bays or semi-enclosed bodies of brackish (salty but less so than seawater) water that form where rivers enter the ocean. Estuaries usually contain rich sediments carried downriver, forming shoals and mudflats that nurture a multitude of aquatic life. Estuaries are sheltered from the most drastic ocean action but do experience tidal ebbs and flows. Daily tides may even cause river levels to rise and fall far inland from the river mouth. The combination of physical factors in estuaries makes them very productive and of high species diversity. They are significant "nurseries" for economically important fish, crustaceans (such as crabs and shrimp), and mollusks (such as clams, cockles, and oysters).

Where the continental shelf is broad and shallow, an extensive fan-shaped sediment deposit called a **delta** may form at the river mouth. Deltas often are channelled by branches of the river, creating extensive coastal wetlands that are part of the larger estuarine zone.

Wetlands of several types form near bodies of water. **Wetlands** are ecosystems in which the land surface is saturated or covered with standing water at least part of the year and vegetation is adapted for growth under saturated conditions. There are special names for specific kinds of wetlands, but we can group them into

three major categories: swamps, marshes, and bogs and fens. Defined pragmatically, **swamps** are wetlands with trees (Fig. 6.14); **marshes** are wetlands without trees (Fig. 6.15); and **bogs** and **fens** are areas that may or may not have trees, and have waterlogged soils that tend to accumulate peat (partially decomposed plants). Swamps and marshes tend to be associated with flowing water. Fens are fed by groundwater and surface runoff, whereas bogs are fed solely by precipitation. Swamps and marshes generally have high productivity, while bogs and fens have low productivity because of their relatively low nutrients and pH.

FIGURE 6.14 Biological communities, such as the Okefenokee Swamp in Georgia, can be amazingly diverse, complex, and productive. Myriad life-forms coexist in exuberant abundance, interlinked by manifold relationships. Who lives where, and why, are central ecological questions.

FIGURE 6.15 Coastal wetlands such as this salt marsh are highly productive ecosystems and are vitally important for a wide variety of terrestrial and aquatic species.

PART ONE   Fundamental Principles of Ecology and Environmental Science

The water in marshes and swamps usually is shallow enough to allow full penetration of sunlight and seasonal warming. These mild conditions favour great photosynthetic activity, resulting in high productivity at all trophic levels. In short, life is abundant and varied. Wetlands are major breeding, nesting, and migration staging areas for waterfowl and shorebirds.

Wetlands perform major ecosystem services, the importance of which cannot be overstated. As mentioned previously, they support a great diversity of life-forms. What may be less obvious is their role in planetary water relationships. Wetlands act as traps and filters for water that moves through them. Runoff water is slowed as it passes through the shallow, plant-filled areas, reducing flooding. As a result, sediments are deposited in the wetlands instead of travelling into rivers and, eventually, the oceans. In this way, wetlands both clarify surface waters and aid in the accumulation and formation of fertile land. Furthermore, the uptake by plants of nutrients and contaminants in wetland ecosystems neutralize and detoxify substances in the water. Finally, water in wetlands seeps into the ground, helping to replenish underground water reservoirs called aquifers (see Chapter 10).

Wetlands convert naturally to terrestrial communities largely through sedimentation, eutrophication, or stream cutting and draining. Human activities have greatly accelerated these processes in many places, and wetlands are being destroyed or degraded around the world at a disturbing rate (see Chapter 22). This destruction is of great concern because it means loss of ecological services to the biosphere as well as loss of essential habitats for a myriad of species.

Shorelines and Barrier Islands

Ocean shorelines, including rocky coasts, sandy beaches, and offshore barrier islands, are particularly rich in life-forms. Rocky shorelines, in particular, support an incredible density and diversity of organisms that grow attached to any solid substrate, including each other. Sandy shorelines, on the other hand, provide homes for organisms that live among the sand grains and in burrows.

Sandy beaches are understandably popular to humans. Cities, resorts, and residences are built on beaches. Grasses and trees that hold the dunes are destroyed, destabilizing the soil system and thus increasing its susceptibility to wind and wave erosion. The constant battle to maintain such real estate is costly. Insurance is very expensive because of the hazards of natural erosion and storms. Building and maintaining sea walls and protecting and replenishing beaches are expensive, continuous processes. Protective structures often have their own effect on shore recontouring and may even increase the rate of natural shoreline loss.

Barrier islands are low, narrow, sandy islands that form offshore from a coastline. In North America they are particularly characteristic of the Atlantic and Gulf coasts (Fig. 6.16). Barrier islands protect inland shores from the onslaught of the surf, especially during severe storms. Because they are so lovely, barrier islands have been tempting targets for real estate development and about 20 percent of the barrier island surface in the United States has been developed. Unfortunately, human occupation of barrier islands often destroys the values that attract us there in the first place. Walking or driving vehicles over dune grass destroys stabilizing vegetation cover and triggers erosion. Cutting roads through dunes and building houses further destabilizes islands. A single, major storm can do massive damage, destroying houses and perhaps washing away most of an island.

Ignoring these environmental realities endangers many rare species and puts human lives and economic investment at risk. Although we now have coastal zoning and land-use regulations, we still have many environmentally damaging developments in place.

Coral reefs form in clear, warm, tropical seas and are particularly well developed in the south Pacific (Fig. 6.17). They are the accumulated calcareous skeletons of innumerable tiny colonial animals called corals. Each of the interconnected coral animals builds

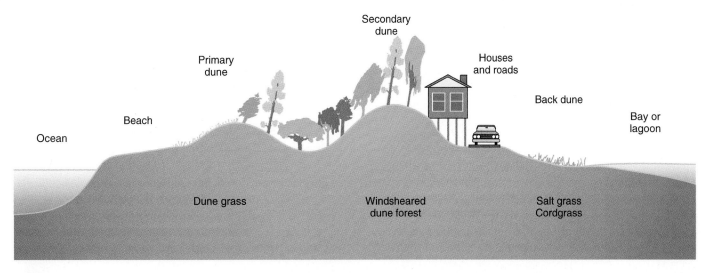

FIGURE 6.16 A cross section through a barrier island. Fragile dune grass communities on the primary dune must be protected to ensure stability of the dune complex. Roads and houses should be limited to the back dune area, both for safety and to relieve pressure on vulnerable dune vegetation. Only a few narrow footpaths should cross the dunes to the beach, where recreational use is permitted.

a calcium carbonate chamber on the surface of the accumulated secretions of previous generations of animals. Coral reefs usually form along the edges of shallow, submerged banks or shelves. The depth at which they form is limited by the depth of light penetration, due at least in part to the presence of symbiotic photosynthetic protists in their tissues. Because of the photosynthetic relationship, the presence of dead versus living coral growths can be an index of either previous ocean levels or decreased light penetration due to increased turbidity. Coral reef communities rival tropical forest communities in species diversity, numbers of individuals, brilliance of colour, and interesting forms of both plants and animals.

Reefs also are among the most endangered biological communities on the earth. Destructive fishing practices (using dynamite or cyanide to stun fish, for instance) and harvesting of coral for building or the pet trade have damaged or destroyed about three-fourths of all reefs in the world. In the next section, we will discuss some other types of human disturbance of natural habitats.

FIGURE 6.17 Coral reefs harbour some of the most diverse biological communities on the earth, rivaling tropical rainforests in species diversity, and surpassing any biome in the number of phylla (broad taxonomic groups) represented. Surprisingly, almost all the organisms you see here are animals.

HUMAN DISTURBANCE

Humans have become dominant organisms over most of the earth, damaging or disturbing more than half of the world's terrestrial ecosystems to some extent (Fig. 6.18). By some estimates, humans

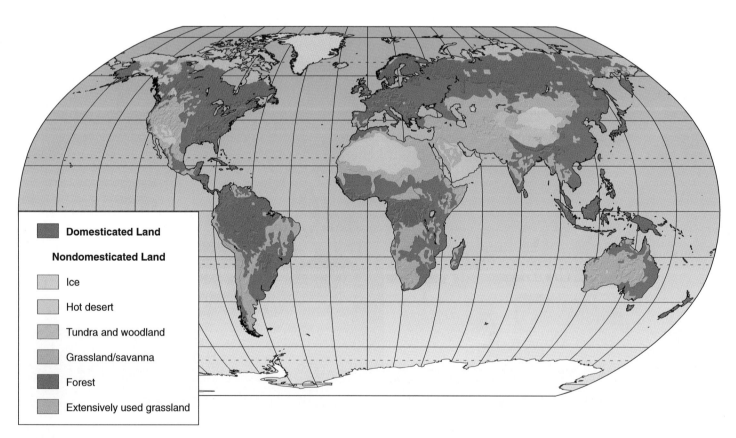

Domesticated Land

Nondomesticated Land

- Ice
- Hot desert
- Tundra and woodland
- Grassland/savanna
- Forest
- Extensively used grassland

FIGURE 6.18 Domesticated land has replaced much of the earth's original land cover.
Source: United Nations Environment Programme, *Global Environment Outlook,* 1997.

TABLE 6.1 Human Disturbance

BIOME	TOTAL AREA (10⁶ KM²)	% UNDISTURBED HABITAT	% HUMAN DOMINATED
Temperate broad-leaved forests	9.5	6.1	81.9
Chaparral and thorn scrub	6.6	6.4	67.8
Temperate grasslands	12.1	27.6	40.4
Temperate rainforests	4.2	33.0	46.1
Tropical dry forests	19.5	30.5	45.9
Mixed mountain systems	12.1	29.3	25.6
Mixed island systems	3.2	46.6	41.8
Cold deserts/semideserts	10.9	45.4	8.5
Warm deserts/semideserts	29.2	55.8	12.2
Moist tropical forests	11.8	63.2	24.9
Tropical grasslands	4.8	74.0	4.7
Temperate conifer forests	18.8	81.7	11.8
Tundra and arctic desert	20.6	99.3	0.3

Note: Where undisturbed and human dominated areas do not add up to 100 percent, the difference represents partially disturbed lands.
Source: Data from Hannah, Lee, et al., "Human Disturbance and Natural Habitat: A Biome Level Analysis of a Global Data Set," in *Biodiversity and Conservation,* 1995, Vol. 4:128–155.

preempt about 40 percent of the net terrestrial primary productivity of the biosphere either by consuming it directly, by interfering with its production or use, or by altering the species composition or physical processes of human-dominated ecosystems. Conversion of natural habitat to human uses is the largest single cause of biodiversity losses.

Researchers from the environmental group Conservation International have attempted to map the extent of human disturbance of the natural world (Fig. 6.18). The greatest impacts have been in Europe, parts of Asia, North and Central America, and islands such as Madagascar, New Zealand, Java, Sumatra, and those in the Caribbean. Data from this study are shown in Table 6.1.

Temperate broad-leaved forests are the most completely human-dominated of any major biome. The climate and soils that support such forests are especially congenial for human occupation. In eastern North America or most of Europe, for example, only remnants of the original forest still persist. Regions with a Mediterranean climate generally are highly desired for human habitation. Because these landscapes also have high levels of biodiversity, conflicts between human preferences and biological values frequently occur (Fig. 6.19).

Temperate grasslands, temperate rainforests, tropical dry forests, and many islands also have been highly disturbed by human activities. If you have travelled through the American cornbelt states such as Iowa or Illinois, you have seen how thoroughly former prairies have been converted to farmlands. Intensive cultivation of this land exposes the soil to erosion and fertility losses (see Chapter 14). Islands, because of their isolation, often have

FIGURE 6.19 "There was an environment before human beings and there will be an environment after human beings. So what's the big deal?"
© *Peter Kohlsaat/Modern Times Syndicate, 1990.*

high numbers of endemic species. Many islands, such as Madagascar, Haiti, and Java have lost more than 99 percent of their original land cover.

Tundra and Arctic deserts are the least disturbed biomes in the world. Harsh climates and unproductive soils make these areas unattractive places to live for most people. Temperate conifer forests also generally are lightly populated and large areas remain in a relatively natural state. However, recent expansion of forest harvesting in Canada and Siberia may threaten the integrity of this biome. Large expanses of tropical rainforests still remain in the Amazon and Congo basins but in other areas of the tropics such as West Africa, Madagascar, Southeast Asia, and the Indo-Malaysian peninsula and archipelago, these forests are disappearing at a rapid rate (see Chapter 17).

As mentioned earlier, wetlands have suffered severe losses in many parts of the world. About half of all original wetlands in the United States have been drained, filled, polluted, or otherwise degraded over the past 250 years. In the prairie states, small potholes and seasonally flooded marshes have been drained and converted to croplands on a wide scale. Iowa, for example, is estimated to have lost 99 percent of its presettlement wetlands. Similarly, California has lost 90 percent of the extensive marshes and deltas that once stretched across its central valley. Wooded swamps and floodplain forests in the southern United States have been widely disrupted by logging and conversion to farmland.

Similar wetland disturbances have occurred in other countries as well. In New Zealand, over 90 percent of natural wetlands have been destroyed since European settlement. In Portugal, some 70 percent of freshwater wetlands and 60 percent of estuarine habitats have been converted to agriculture and industrial areas. In Indonesia, almost all the mangrove swamps that once lined the coasts of Java have been destroyed, while in the Philippines and Thailand, more than two-thirds of coastal mangroves have been cut down for firewood or conversion to shrimp and fish ponds.

LANDSCAPE ECOLOGY

So far in this chapter, we have considered very broad ecological regions and humans solely as disturbing agents. This is a very Clemensian view (see Chapter 3) of nature as static and deterministic. Most modern scientists see nature as much more diverse, dynamic, and complex than is suggested by this perspective. Modern technology, including remote sensing, powerful computers, and geographical information systems (GIS) allow us to study temporal and spatial complexity, heterogeneity, and dynamic fluxes in ecosystems that would previously have been ignored as random noise or unwelcome complications (Fig. 6.20).

Landscape ecology is the study of reciprocal effects of spatial pattern on ecological processes. By reciprocal effects, we mean that complex spatial patterns shape, and are shaped by, the dynamic ecological processes that occur in them. In common usage a landscape is as much as you can see at one time from a high vantage point. By this definition it is smaller than a biome but generally larger than a single ecosystem. In landscape ecology, a landscape is a geographical unit with a history that shapes the features of the land and organisms in it as well as our reaction to, and interpretation of, the land.

FIGURE 6.20 Landscape ecology uses geographic information systems (GIS) to map patch size, type, and configuration as in this land cover map of part of the Upper Thames River basin in southwestern Ontario. This mapping assists planners in analyzing land-use patterns.

Landscape ecology considers humans important elements of most landscapes. Few places on the earth are devoid of human impacts. Human disturbances include hunting, timber harvesting, farming, grazing, pollution, introduction of exotic species, and alteration of atmospheric chemistry and, perhaps, global climate. Restricting our definition of "nature" to a few pristine remnants will lead us either to ignore most of the earth's surface or demand that humans remove themselves entirely from large tracts, something not likely in poor countries that need access to resources.

Patchiness and Heterogeneity

The Amazonian rainforest is a good example of the new perspective provided by landscape ecology. Usually the rainforest is depicted as a vast, homogeneous block of forest that has remained stable and unchanging for hundreds of millions of years. A detailed analysis, however, shows that it actually is a complex mosaic of many distinct habitat types and species assemblages distributed according to soil type, topography, climate, and site history. These patches are in a constant state of change as big trees fall or fires open up new ground for succession.

Landscape ecologists claim that if we look closely, all landscapes consist of similar mosaics of discrete, bounded patches with different biotic or abiotic composition. Often a predominant or continuous cover type acts as a matrix in which other patch types appear to be embedded. Like individual habitats in the rainforest,

PART ONE Fundamental Principles of Ecology and Environmental Science

these patches are dynamic and change components or functions over time. Among the causes of this patchiness are human and natural disturbances as well as underlying ecological factors such as successional processes.

Landscape heterogeneity can exist across a wide range of scale, from burned patches measuring thousands of hectares in the Interior of British Columbia to the effects of soil crumb size and insect burrows in a few square centimetres of soil. These spatial patterns exist in marine, freshwater, and wetland environments as well as terrestrial habitats, and thus the term "landscape" can apply to aquatic environments as well as dry land. A basic question in landscape ecology is whether a given phenomenon appears or applies across many different scales or is restricted to a particular scale. In other words, do patterns emerge, change, or disappear if we look at a landscape up close or from afar?

Landscape Dynamics

Time and space are of special concern in landscape ecology, and for that matter, to the rest of us as well. As humourist Garrison Keillor says, "Time exists so everything doesn't happen at once; space exists so everything doesn't happen to us." How does this apply to ecology? A physiological ecologist might want to study the effects of temperature on the rate of photosynthesis in plants. To make the study manageable, she or he might ask, "What is the photosynthetic rate in this particular *plant,* at this *exact* time, at these *specific* temperatures?" A landscape ecologist, in contrast, might ask, "What is the effect of temperature on photosynthesis in this *place,* given its unique combination of history, composition, and characteristics?" Developing a deep relationship with, and understanding of, a particular landscape is sometimes referred to as having a "sense of place."

The boundaries between habitat patches are considered especially significant by landscape ecologists. Edges can induce, inhibit, or regulate movement of materials, energy, or organisms across a landscape. Thus the dynamics between patches may be of greater overall importance than what happens within each patch. As Chapter 5 shows, wildlife managers are beginning to be very aware of edge effects as well as the size, shape, and distribution of habitat patches in fragmented landscapes.

This focus on interactions between neighbouring communities is quite different from classical ecological focus on the structure and functions of discrete communities, populations, or ecosystems. In its interest in complexity and emergent properties of systems, landscape ecology draws on theories of chaos and complexity from mathematics and physics.

There also are many similarities between landscape ecology and the equally new discipline of conservation biology. Among their shared tenets are: (1) evolutionary change is a central feature of natural systems, (2) nonequilibrium dynamics and uncertainty are more characteristic of nature than are stability and determinism, (3) heterogeneity and diversity are good and ought to be maintained, (4) the context of the surrounding landscape is important, and (5) human needs, desires, abilities, and potential must be considered in efforts to understand and protect nature. All of these points are important issues in the design of nature preserves, parks, and wildlife refuges (see Chapter 22).

SUMMARY

Major ecosystem types called biomes are characterized by similar climates, soil conditions, and biological communities. Among the major terrestrial biomes are deserts, tundra, grasslands, temperate deciduous forests, temperate coniferous forests, tropical rainforests, and tropical seasonal forests. Aquatic ecosystems include oceans and seas, rivers and lakes, estuaries, marshes, swamps, bogs, fens, and reefs. Moisture and temperature are generally the most critical determinants for terrestrial biomes. Periodic natural disturbances, such as fires, play a major role in maintaining some biomes.

Humans have disturbed, preempted, or damaged much—perhaps half or more—of all terrestrial biomes and now dominate about 40 percent of all net primary productivity on the land.

Landscape ecology is the study of reciprocal effects of spatial pattern on ecological processes. A landscape, in this sense, is a bounded geographical unit that includes both living and nonliving components. History is important in that it shapes the land and organisms as well as our reactions to, and interpretations of, the landscape. Humans are important elements of most landcapes and few places on the earth are devoid of human impacts. Restricting our definition of "nature" to a few pristine remnants will lead us either to ignore most of the earth's surface or to demand that humans remove themselves entirely from large tracts, something not very likely.

Patchiness and heterogeneity are characteristic of most landscapes. Movement of organisms and materials across boundaries between habitat patches play an important regulatory role in many ecosystems. Some landscape patterns apply across a wide range of scales, while others are restricted to a specific scale.

QUESTIONS FOR REVIEW

1. Throughout the central portion of North America is a large biome once dominated by grasses. Describe how physical conditions and other factors control this biome.

2. What is taiga and where is it found? Why might logging in taiga be more disruptive than in southern coniferous forests?

3. Why are tropical rainforests often less suited for agriculture and human occupation than tropical deciduous forests?

4. Describe four different kinds of wetlands and explain why they are important sites of biodiversity and biological productivity.

5. Which major biomes have been most heavily disturbed by human activities?

6. Define a landscape. Describe the major ecological and cultural features of the landscape in which you live.

QUESTIONS FOR CRITICAL THINKING

1. In which biome do you live? What physical and biological factors are most important in shaping your biological community? How do the present characteristics of your area differ from those 100 or 1,000 years ago?

2. Could your biome be returned to something resembling its original condition? What tools (or principles) would you use to do so?

3. In 1999 nearly 200,000 ha of coniferous forest was blown down by a windstorm in Quetico Provincial Park in Canada and the Boundary Waters Canoe Area Wilderness in the United States. The area is now a fire hazard to nearby property as well as being an aesthetic mess. Some people argue that salvage loggers should clean out the area and replant pine trees. Others argue that it was a natural forest and should be left to natural forces. What do you think?

KEY TERMS

barrier islands 123	estuaries 122
benthos 121	fens 122
biomes 114	grasslands 116
bogs 122	landscape ecology 126
boreal forest 118	marshes 122
chaparral 119	swamps 122
cloud forest 119	taiga 118
conifer 117	temperate rainforest 118
coral reefs 123	thermocline 122
deciduous 118	thorn scrub 119
delta 122	tropical rainforest 119
deserts 115	tropical seasonal forests 120
	tundra 117
	wetlands 122

Web Exercises

OCEANIC PRODUCTIVITY

Figure 6.11 is one frame from an animated map of ocean chlorophyll levels created by the NASA Earth Observatory Program. You can view the complete data set and create your own animated map by going to http://earthobservatory.nasa.gov/Observatory/Datasets/chlor.seawifs.html, or for a live link to NASA, you can go to our web page at http://www.mcgrawhill.ca/college/biosci/pae/environmentalscience.

To make an interesting map on the NASA web page, choose at least two years of data in the build an animation boxes. After the map loads on your computer, click on the play button on the left and observe the changes in colours. To step through the frames individually, use the left or right arrow buttons on your computer after the animation starts.

1. Where are the highest levels of ocean chlorophyll in January and June?

2. Why do ocean plankton grow densely in the Arctic Ocean between North America and Europe and along the west coasts of South America and Africa? (*Hint:* you might want to refer to Figure 6.11 and its associated text in your book to help answer this question.)

3. Why do most open oceans appear dark blue, which indicates a lack of chlorophyll?

4. Where do the deepest green concentrations occur on land in June and January?

Freshwater Biologist
Matt Kennedy, M.Sc., R.P.Bio.

Matt caught the bug for the outdoors during his childhood growing up first in southern Ontario and then near the Rocky Mountains in Alberta. His family was active in outside activities including hiking, skiing, and river canoeing. He was fortunate to spend many weekends, summer and winter, in and around Banff, Alberta, while going to school in Calgary. Most summers, Matt spent several weeks sailing and combing the beaches of the Pacific Ocean with his grandparents who lived on Vancouver Island, B.C. Through these experiences he developed an understanding and appreciation for many of Canada's "wild places."

Although his strengths in high school were in the sciences, Matt—mistakenly thinking that science meant life in a lab—enrolled in general arts at the University of Western Ontario (UWO) in London. After spending a year taking courses in economics, English, psychology and others, he realized his interests were in the sciences. Transferring to the science stream, he completed a four-year Bachelor of Science degree with a concentration in Ecology and Evolution. Course work included ecological subjects (zoology and botany) as well as field-based courses in the Kananaskis Valley, Alberta, and Algonquin Park in northern Ontario.

Near the end of the fourth year of his undergraduate program, Matt elected to stay on for an additional two years to complete a Master of Science degree in Zoology and Environmental Science. The degree involved conducting field research in the central Yukon Territory for two consecutive summers. The focus of the project was to develop a method to use stream-dwelling insects (commonly known as "benthic macroinvertebrates") as an indicator of the effects of alluvial placer gold mining.

Placer mining involves excavation and washing of gold-bearing streambeds to separate relatively heavy gold particles from other materials. The washing process typically results in suspension of high levels of sediment downstream of the active mining area, which can negatively affect stream biology. Matt's project involved sampling benthic communities in a series of small streams, both mined and un-mined, to determine the abundance and diversity of insects present at certain levels of mining activity. Streams where mining had been discontinued were also sampled to examine recovery of stream insect communities.

After completing his Master's degree in the mid-90s, Matt began to seek permanent employment. With the goal of incorporating his interests in both natural science and resource development, Matt took a position as a field technician for an engineering company with a specialization in the environmental sciences. While in this position, Matt gained important "hands-on" experience conducting field research in many areas of temperate and northern Canada. He also broadened his technical skills to include aspects of fisheries biology and fish habitat assessment, water and sediment quality, study design, and data analysis.

Matt's interests have primarily focused on the interaction of human activities with the natural environment. Much of Canada's economic activity depends on development of our forests, mineral deposits, and oil and gas reserves. In many cases, however, resource development has historically resulted in persistent environmental effects after the activity has ceased. Examples include hill slope erosion from clear-cut logging, acid drainage, and poor tailings management at abandoned mines, and effects from the rupture of oil pipelines near waterways. However, many of these effects related more to the use of ineffective pollution control methods, rather than negligence by the people involved in the projects.

Now nearly a decade into his professional career, Matt works for a Vancouver-based consulting company specializing in both marine and freshwater biological sciences. Now a registered professional biologist in British Columbia, Matt is involved in biological monitoring activities for pulp and paper mills in B.C. and oil sand developments in northern Alberta. He is also a project manager for baseline studies and environmental impact assessments (EIAs) for projects that are reviewed by provincial and federal environmental regulatory agencies. He continues to broaden his experiences with every new project and splits his time both in the field and in the office. Matt continues to assist resource companies to effectively design new projects and monitor their effects on the aquatic environment.

Chapter 7

Geology

Geology is intimately related to almost all the physical sciences, as history to the moral.

—Charles Lyell—

OBJECTIVES

After studying this chapter, you should be able to:

- understand some basic geologic principles, including how tectonic plate movements affect conditions for life on the earth.

- explain how the three major rock types are formed and how the rock cycle works.

- summarize economic mineralogy and strategic minerals.

- discuss the environmental effects of mining and mineral processing.

- recognize the geologic hazards of earthquakes, volcanoes, and tsunamis.

WebQuest

rocks and minerals, economic geology, mineralogy, placer mining, earthquakes, Richter scale, volcanoes, landslides, Tsunamis

Above: Steam rises as the earth's internal heat reaches the surface in Yellowstone National Park's Porcelain Basin. The park sits on the remains of an ancient—but still active—volcano.

Earthquake in India

On the morning of January 26, 2001, India's worst earthquake in half a century struck the northwestern state of Gujarat. In moments, the homes and lives of tens of thousands of people were destroyed. Measuring 7.9 on the Richter scale, the earthquake pulverized cities and towns, killing more than 7,000 people, injuring more than 35,000, and leaving nearly 100,000 homeless (Fig. 7.1). More than 100 multistory buildings collapsed, half of them in Ahmedabad, Gujarat's commercial centre. In one of the day's worst tragedies, 350 children marching in a parade to celebrate Republic Day, a national holiday, were crushed by falling buildings. In another disaster, dozens of students and teachers were caught in the flattened wreckage of a four-story high school—a school built only a year earlier. Thousands of smaller structures and houses also collapsed as the earth shook hundreds of kilometres from its epicentre near the city of Bhuj.

International donors quickly sent medical supplies, food, more than 30,000 tents, and sniffer dogs to help locate living people trapped in the wreckage. Because of damaged roads, rail lines, and bridges, though, help reached the damaged cities and towns slowly. Rescue workers struggled to find survivors using crowbars and pickaxes until help and better equipment trickled in, sometimes days later.

Why was this earthquake such a huge disaster? A combination of natural and human factors contributed to the death toll and damage. India sits in a region that is seismically active: the earth is moving slowly but continually as India pushes toward Tibet. It is this movement that is building the Himalayan Mountains, the tallest in the world. But earthquakes have always occurred periodically in India, and the human costs of this one were unsually high. As the dust cleared, accusations arose that building construction had been shoddy. Dishonest contractors had created weak structures, disregarding building codes, with the cooperation of corrupt officials. Building inspectors were accused of demanding bribes and neglecting to enforce construction standards. At the same time, cities in India have grown rapidly in recent decades, so that collapsing cities were larger and denser than in the past. Many residents noted that the older buildings in Ahmedabad stayed standing while newer ones all around them fell. India's prime minister admitted after the earthquake that the country was ill-prepared to deal with a disaster of this magnitude.

Just a month after the earthquake in Gujarat, a very different earthquake hit Seattle, Washington. This earthquake was also large (6.8 on the Richter Scale), but it was deep in the earth, so that the shaking at the surface was relatively slight. The Gujarat earthquake was closer to the surface, so that movement of the ground was more severe. In addition, Seattle has been working on earthquake preparedness for more than a decade. Building codes are strict, and hundreds of millions of dollars have been spent on reinforcing existing structures. Seattle's earthquake will cost millions in damage repair, but there was only one death.

These two earthquakes had very different natural, economic, and political situations, which contributed to very different outcomes. In this chapter we will look at the ways that earth resources and geologic forces provide both risks and opportunities in our daily lives.

A DYNAMIC PLANET

Although we think of the ground under our feet as solid and stable, the earth is a dynamic and constantly changing structure. Titanic forces stir inside the earth, causing continents to split, drift apart, and then crash into each other in slow but inexorable collisions. In this section, we will look at the structure of our planet and some of the forces that shape it.

A Layered Sphere

The **core,** or interior of the earth (Fig. 7.2), is composed of a dense, intensely hot mass of metal—mostly iron—thousands of kilometres in diameter. Solid in the centre but more fluid in the outer core, this immense mass generates the magnetic field that envelops the earth.

Surrounding the molten outer core is a hot, pliable layer of rock called the **mantle.** The mantle is much less dense than the core because it contains a high concentration of lighter elements, such as oxygen, silicon, and magnesium.

The outermost layer of the earth is the cool, lightweight, brittle **crust** of rock that floats on the mantle something like the "skin" on a bowl of warm chocolate pudding. The oceanic crust, which forms the seafloor, has a composition somewhat like that of the mantle, but it is richer in silicon. Continents are thicker, lighter regions of crust rich in calcium, sodium, potassium, and aluminum. The continents rise above both the seafloor and the ocean surface. Table 7.1 compares the composition of the whole earth (dominated by the dense core) and the crust.

Tectonic Processes and Shifting Continents

Hot enough to flow, the upper layer of the mantle has huge convection currents that break the overlying crust into a mosaic of huge blocks called **tectonic plates** (Fig. 7.3). These plates slide slowly across the earth's surface like immense icebergs, in some places breaking up into smaller pieces, in other places crashing ponderously into each other to create new, larger landmasses. Ocean basins form where continents crack and pull apart. **Magma** (molten rock) forced up through the cracks forms new oceanic crust that piles up underwater in mid-oceanic ridges. Creating the largest mountain range in the world, these ridges wind around the world for 74,000 km. Although concealed from our view, this jagged range boasts higher peaks, deeper canyons, and sheerer cliffs than any continental mountains. Slowly spreading from these fracture zones, ocean plates push against continental plates.

FIGURE 7.1 An earthquake in Gujurat, India killed thousands of people and left tens of thousands homeless.

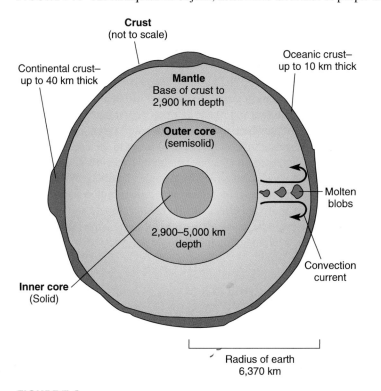

FIGURE 7.2 The layered Earth. The intensely hot, liquid or semisolid core is made mostly of molten metal. Around it is a solid but pliable mantle hot enough to bend and flow like warm taffy. Floating on top of the mantle is a thin crust of rock that breaks up into large, slowly moving tectonic plates. The crust appears 10 times thicker in this drawing than it is in reality.

Source: Carla Montgomery, Physical Geology, 3rd edition. Copyright © 1993 McGraw-Hill Company, Inc. All Rights Reserved. Reprinted by permission.

TABLE 7.1	Eight Most Common Chemical Elements (percent)		
WHOLE EARTH		**CRUST**	
Iron	33.3	Oxygen	45.2
Oxygen	29.8	Silicon	27.2
Silicon	15.6	Aluminum	8.2
Magnesium	13.9	Iron	5.8
Nickel	2.0	Calcium	5.1
Calcium	1.8	Magnesium	2.8
Aluminum	1.5	Sodium	2.3
Sodium	0.2	Potassium	1.7

Earthquakes are caused by the grinding and jerking as plates slide past each other. Mountain ranges are pushed up at the margins of colliding continental plates. The Atlantic Ocean is growing slowly as Europe and Africa drift away from the Americas. The Himalayas are still rising as the Indian subcontinent smashes into Asia. Southern California is slowly sailing north toward Alaska. In about 30 million years, Los Angeles will pass San Francisco, if either still exists by then.

When an oceanic plate collides with a continental landmass, the continental plate will ride up over the seafloor, and the oceanic plate is subducted, or pushed down into the mantle, where it melts and rises back to the surface as magma (Fig. 7.4). Deep ocean trenches mark these subduction zones, and volcanoes form where

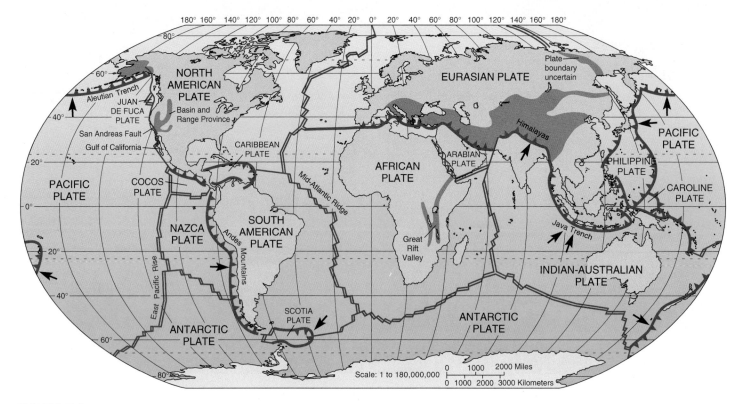

FIGURE 7.3 Map of tectonic plates. Plate boundaries are dynamic zones, characterized by earthquakes and volcanism and the formation of great rifts and mountain ranges. Arrows indicate direction of subduction where one plate is diving beneath another. These zones are sites of deep trenches in the ocean floor and high levels of seismic and volcanic activity.

Sources: U.S. Department of the Interior, U.S. Geological Survey.

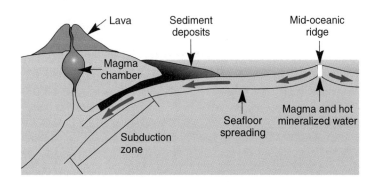

FIGURE 7.4 The rock cycle is driven by plate-tectonic movements. Old seafloor and sediment deposits are melted in subduction zones. The magma rises to erupt through volcanoes or to recrystallize at depth into igneous rocks. Weathering breaks down surface rocks and erosion deposits residue in sedimentary formations. The pressure and heat caused by tectonic movements cause metamorphism of both sedimentary and igneous rocks.

the magma erupts through vents and fissures in the overlying crust. All around the Pacific Ocean rim from Indonesia to Japan to Alaska and down the West Coast of the Americas is the so-called "ring of fire" where oceanic plates are being subducted under the continental plates. This ring is the source of more earthquakes and volcanic activity than any other place on the earth.

Over millions of years, the drifting continents can move long distances. Antarctica and Australia once were connected to Africa, for instance, somewhere near the equator, and supported luxuriant forests. Geologists suggest that several times in the earth's history most or all of the continents have gathered to form a single supercontinent, Pangaea, surrounded by a single global ocean (Fig. 7.5). The rupture of this supercontinent and redistribution of its pieces probably has profound effects on the earth's climate and may help explain the periodic mass extinctions of organisms marking the divisions between many major geologic periods.

ROCKS AND MINERALS

A **mineral** is a naturally occurring, inorganic, solid element or compound with a definite chemical composition and a regular internal crystal structure. Naturally occurring means not created by humans (or synthetic). Organic materials, such as coal, produced by living organisms or biological processes are generally not minerals. A mineral must be solid, therefore ice is a mineral but liquid water is not. The two fundamental characteristics of a mineral that distinguish it from all other minerals are its chemical composition and its crystal structure. No two minerals are identical in both respects. Once purified, metals such as iron, aluminum, or copper

FIGURE 7.5 Pangaea, the ancient supercontinent of 200 million years ago, combined all the world's continents in a single landmass.

are solid, noncrystalline elements, and thus are not minerals. The ores from which they are extracted, however, are minerals and make up an important part of economic mineralogy.

A **rock** is a solid, cohesive, aggregate of one or more minerals. Within the rock, individual mineral crystals (or grains) are mixed together and held firmly in a solid mass. The grains may be large or small, depending on how the rock was formed, but each grain retains its own unique mineral qualities. Each rock type has a characteristic mixture of minerals (and therefore of different chemical elements), grain sizes, and ways in which the grains are mixed and held together. Granite, for example, is a mixture of quartz, feldspar, and mica crystals. Different kinds of granite have distinct percentages of these minerals and particular grain sizes depending on how quickly the rock solidified. These minerals, in turn, are made up of a few elements such as silicon, oxygen, potassium, and aluminum.

Rock Types and How They Are Formed

What could be harder and more permanent than rocks? Like the continents they create, rocks are also part of a relentless cycle of formation and destruction. They are made and then torn apart, cemented together by chemical and physical forces, crushed, folded, melted, and recrystallized by dynamic processes related to those that shape the large-scale features of the crust. We call this cycle of creation, destruction, and metamorphosis the **rock cycle** (Fig. 7.6). Understanding something of how this cycle works helps explain the origin and characteristics of different types of rocks, as

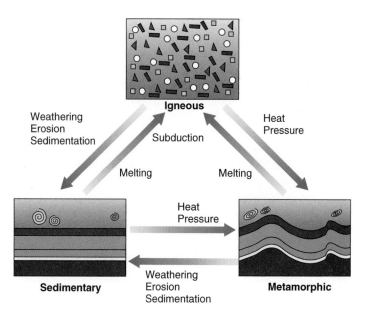

FIGURE 7.6 The rock cycle consists of processes of creation, destruction, and metamorphosis. Each of the three major rock types can be converted to either of the other types.

well as how they are shaped, worn away, transported, deposited, and altered by geologic forces.

There are three major rock classifications: igneous, sedimentary, and metamorphic. In this section, we will look at how they are made and some of their properties.

Igneous Rocks

The most common rock-type in the earth's crust is solidified from magma, welling up from the earth's interior. These rocks are classed as **igneous rocks** (from *igni,* the Latin word for fire). Magma extruded to the surface from volcanic vents cools quickly to make basalt, rhyolite, andesite, and other fine-grained rocks. Magma that cools slowly in subsurface chambers or is intruded between overlying layers makes granite, gabbro, or other coarse-grained crystalline rocks, depending on its specific chemical composition.

Weathering and Sedimentation

Most of these crystalline rocks are extremely hard and durable, but exposure to air, water, changing temperatures, and reactive chemical agents slowly breaks them down in a process called **weathering.** *Mechanical weathering* is the physical breakup of rocks into smaller particles without a change in chemical composition of the constituent minerals. You have probably seen mountain valleys scraped by glaciers or river and shoreline pebbles that are rounded from being rubbed against one another as they are tumbled by waves and currents. *Chemical weathering* is the selective removal or alteration of specific components that leads to weakening and disintegration of rock. Among the more important chemical weathering processes are oxidation (combination of oxygen with an element to form an oxide or hydroxide mineral) and hydrolysis (hydrogen atoms from water molecules combine with other chemicals to form acids). The products of these reactions are more susceptible to both mechanical weathering and to dissolving in water. For instance, when carbonic acid (formed when CO_2 and H_2O combine) percolates through porous limestone layers in the ground, it dissolves the calcium carbonate (limestone) and creates caves.

Particles of rock loosened by wind, water, ice, and other weathering forces are carried downhill, downwind, or downstream until they come to rest again in a new location. The deposition of these materials is called **sedimentation.** Waterborne particles from sediments cover ocean continental shelves and fill valleys and plains. Most of the Canadian Prairies, for instance, is covered with a layer of sedimentary material hundreds of metres thick in the form of glacierborne till (rock debris deposited by glacial ice), windborne loess (fine dust deposits), riverborne sand and gravel, and ocean deposits of sand, silt, and clay. Deposited material that remains in place long enough, or is covered with enough material to compact it, may once again become rock. Some examples of **sedimentary rock** are shale (compacted mud), sandstone (cemented sand), tuff (volcanic ash), and conglomerates (aggregates of sand and gravel).

Sedimentary rocks are formed from crystals that precipitate out of, or grow from a solution. An example is rock salt, made of the mineral halite, which is the name for ordinary table salt (sodium chloride). Salt deposits often form when a body of salt water dries up and salt crystals are left behind. Nova Scotia, Quebec, Ontario, Saskatchewan, and Alberta all have producing salt mines. The Sifto Salt Company mine in Goderich, Ontario, has been operating for more than 100 years. It is estimated that Canada has more than one million billion tonnes of salt.

Sedimentary formations often have distinctive layers that show different conditions when they were laid down. Relatively soft sedimentary rocks such as sandstone can be shaped by erosion into striking features (Fig. 7.7). Geomorphology is the study of the processes that shape the earth's surface and the structures they create.

Humans have become a major force in shaping landscapes. Geomorphologist Rodger Hooke, of the University of Maine, looking only at housing excavations, road building, and mineral production, estimates that we move somewhere around 30 to 35 gigatonnes (billion tonnes or Gt) per year worldwide. When combined with the 10 Gt each year that we add to river sediments through erosion, our earth-moving prowess is comparable to, or greater than, any other single geomorphic agent except plate tectonics.

Metamorphic Rocks

Preexisting rocks can be modified by heat, pressure, and chemical agents to create new forms called **metamorphic rock.** Deeply buried strata of igneous, sedimentary, and metamorphic rocks are subjected to great heat and pressure by deposition of overlying sediments or while they are being squeezed and folded by tectonic processes. Chemical reactions can alter both the composition and structure of the rocks as they are metamorphosed. Some common metamorphic rocks are marble (from limestone), quartzite (from sandstone), and slate (from mudstone and shale). Metamorphic

FIGURE 7.7 Soft, sedimentary rock layers can be sculpted into remarkable shapes such as Flowerpot Island in Georgian Bay, Ontario.

rocks are often the host rock for economically important minerals such as talc, graphite, and gemstones.

ECONOMIC GEOLOGY AND MINERALOGY

Economic mineralogy is the study of minerals that are valuable for manufacturing and are, therefore, an important part of domestic and international commerce. Most economic minerals are metal-bearing ores (Table 7.2). Nonmetallic economic minerals are mostly graphite, some feldspars, quartz crystals, diamonds, and many other crystals that are valued for their usefulness, beauty, and/or rarity. Metals have been so important in human affairs that major epochs of human history are commonly known by the dominant materials and the technology to use them (Stone Age, Bronze Age, Iron Age, etc.). The mining, processing, and distribution of these materials have broad and varied implications both for culture and our environment. Most economically valuable crustal resources exist everywhere in small amounts; the important thing is to find them concentrated in economically recoverable levels.

Canada is one of the world's leading mineral exporters, consistently ranking in the top five producers of 16 major minerals. The total value of mineral production in Canada, including oil and gas, amounted to $70.4 billion in 1999, some 80 percent of which was exported. This economic success, and the many social benefits it affords, has not been achieved without some social and environmental costs. Mine closures, social disruption, acid mine drainage, mine reclamation, site rehabilitation, tailings dam stability, protection of habitat, endangered species, and representative ecosystems are all issues that require ongoing attention.

TABLE 7.2	Primary Uses of Some Major Metals Consumed in Canada

METAL	USE
Aluminum	Packaging foods and beverages (38%), transportation, electronics
Chromium	High-strength steel alloys
Copper	Building construction, electric and electronic industries
Iron	Heavy machinery, steel production
Lead	Car batteries, paints
Manganese	High-strength, heat-resistant steel alloys
Nickel	Chemical industry, steel alloys
Platinum-group	Automobile catalytic converters, electronics, medical uses
Gold	Medical, aerospace, electronic uses; accumulation as monetary standard
Silver	Photography, electronics, jewellery

Metals

How has the quest for minerals and metals affected global development? We will focus first on world use of metals, earth resources that always have received a great deal of human attention. The availability of metals and the methods to extract and use them have determined technological developments, as well as economic and political power for individuals and nations. We still are strongly dependent on the unique lightness, strength, and malleability of metals.

The metals consumed in greatest quantity by world industry include iron (740 million tonnes annually), aluminum (40 million tonnes), manganese (22.4 million tonnes), copper and chromium (8 million tonnes each), and nickel (0.7 million tonnes). Most of these metals are consumed in the United States, Japan, and Europe, in that order. They are produced primarily in South America, South Africa, and the former Soviet Union. It is easy to see how these facts contribute to a worldwide mineral trade network that has become crucially important to the economic and social stability of all nations involved (Fig. 7.8). Table 7.2 shows the primary uses of these metals.

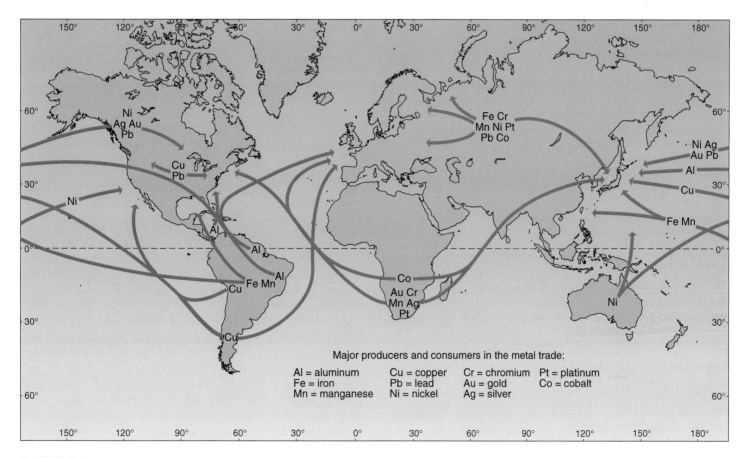

Major producers and consumers in the metal trade:

Al = aluminum   Cu = copper   Cr = chromium   Pt = platinum
Fe = iron   Pb = lead   Au = gold   Co = cobalt
Mn = manganese   Ni = nickel   Ag = silver

FIGURE 7.8   Global metal trade. Metals produced in South Africa, South America, and the former Soviet Union are shipped to markets in the United States, Europe, and Japan, creating a global economic network on which both consumers and producers depend.

The Yukon Placer Authorization

In 1898, the famous Klondike Gold Rush began an era of placer (pronounced *plasser*) gold mining in the Yukon that has continued ever since. Placer mining is essentially "panning for gold" with the help of bulldozers and other large equipment. In some areas, sediment originally deposited by rivers contains gold particles (and a few larger nuggets). Miners dredge, dig up, or dislodge the sediment with water, dump it into a large, rotating cylinder to filter out the larger rocks and stones, and then separate out the gold by gravity selection (gold is heavier than most other minerals in the sediment). All of these steps require large amounts of water, which is usually found in nearby streams and pumped to the filter system.

There are three main effects on the environment of such mining. The first is the construction of access roads and the site itself for the bulldozers and other infrastructure needed to operate a mine. Next is the obvious disruption of terrestrial ecosystems as the surface sediment and soil is scraped off for processing in the mine and extraction of the gold. Finally, the waste water from the process of gold recovery goes back into the stream ecosystem; even with sometimes large networks of settling ponds there is often a considerable increase in the suspended material in the stream, with resulting direct (toxicity) and indirect (habitat) effects on the biota in the stream.

The Canadian Fisheries Act, which has been in force since Confederation, "... prohibits the deposit of deleterious (toxic or harmful) substances into fish-frequented waters or in a place or under any conditions where it may enter fish-frequented waters." This means that it is unlawful to deposit a harmful substance either directly into a fish-bearing stream, or into a place like the top of a bank or a storm drain that leads to a fish-bearing stream. On the face of it, placer mining seems contrary to the Fisheries

Act, and in 1993, the Yukon Placer Authorization (YPA) formally recognized this contradiction. This authorization specifies the classification of streams, the allowable sediment discharges under which the placer mining industry can operate in different types of streams, and plans for inspecting and monitoring. It also provides for the temporary deferment of water quality standards on streams.

In 2002, the Minister of Fisheries and Oceans decided that the placer mining industry was not being subjected to the same rigorous standards of environmental protection as other industries. The Minister announced that the YPA would be phased out over four years and that during that time, any new mines would have to meet environmental protection standards, which the industry claimed were economically

unachievable. In the Yukon, passionate objections to this position came from all sides ... from local communities to the territorial government. Conservationists, who had lobbied hard for increased environmental protection, continued to press the Minister to go ahead with his plan. Placer miners, and the businesses and communities that depended on them, fought equally hard for concessions and retention of the YPA. Placer-mined gold production amounts to about $40 million (Cdn.) per year in the Yukon. What would you recommend to the Minister of Fisheries and Oceans in this situation? Should the same regulations apply to placer mining as other industries? Or is placer mining special because of the historical significance and the fact that it is usually a small or medium sized business, similar to family farming?

Nonmetal Mineral Resources

Nonmetal minerals are a broad class that covers resources from silicate minerals (gemstones, mica, talc, and asbestos) to sand, gravel, salts, limestone, and soils. Sand and gravel production comprise by far the greatest volume and dollar value of all nonmetal mineral resources and a far greater volume than all metal ores. Sand

and gravel are used mainly in brick and concrete construction, paving, as loose road filler, and for sandblasting. High-purity silica sand is our source of glass. These materials usually are retrieved from surface pit mines and quarries, where they have been deposited by glaciers, winds, or ancient oceans.

Limestone, like sand and gravel, is mined and quarried for concrete and crushed for road rock. It also is cut for building stone, pulverized for use as an agricultural soil additive that neutralizes acidic soil, and roasted in lime kilns and cement plants to make plaster (hydrated lime) and cement.

Evaporites (materials deposited by evaporation of chemical solutions) are mined for halite, gypsum, and potash. These are often found at or above 97 percent purity. Halite, or rock salt, is used for water softening and melting ice on winter roads in some northern areas. Refined, it is a source of table salt. Gypsum (calcium sulfate) now makes our plaster wallboard, but it has been used for plaster ever since the Egyptians plastered the walls of their frescoed tombs along the Nile River some 5,000 years ago. Potash is an evaporite composed of a variety of potassium chlorides and potassium sulphates mined in Saskatchewan. These highly soluble potassium salts have long been used as a soil fertilizer. Sulphur deposits are mined mainly for sulphuric acid production.

Strategic Metals and Minerals

World industry depends on about 80 minerals and metals, some of which exist in plentiful supplies. Three-fourths of these resources are abundant enough to meet all of our anticipated needs or have readily available substitutes. At least 18 metals, however, including tin, platinum, gold, silver, and lead, are in short supply.

Of these 80 metals and minerals, between one-half and one-third are considered "strategic" resources. **Strategic metals and minerals** are those that a country uses but cannot produce itself. As the term strategic suggests, these are materials that a government considers capable of crippling its economy or military strength if unstable global economics or politics were to cut off supplies. For this reason, wealthy industrial nations stockpile strategic resources, especially metals, in times when prices are low and a supply is available.

For less wealthy mineral- and metal-producing nations, there is another side to strategic minerals. Many less-developed countries depend on steady mineral exports for most of their foreign exchange. Zambia, for instance, relies on cobalt production for 50 percent of its national income. If a steady international market is not maintained, such producer nations could be devastated. From the small-nation producer's point of view, metal or mineral exports, like concentration on any single product or industry, are an unstable economic foundation. Often no option exists, however, if a producer is to participate in the world economy. Environmental consequences of mining may not be a high priority under these circumstances.

ENVIRONMENTAL EFFECTS OF RESOURCE EXTRACTION

Geologic resource extraction involves physical processes of mining and physical or chemical processes of separating minerals, metals, and other geologic resources from ores or other materials. An ore is a rock in which a valuable or useful metal occurs at a concentration high enough to make mining it economically

FIGURE 7.9 Some giant mining machines stand as tall as a 20-story building and can scoop up thousands of cubic metres of rock per hour.

attractive. For metals like copper, a concentration close to 1 percent is necessary to make an ore worth mining. For precious metals, like gold or platinum, 0.0001 percent may be enough to make mining and smelting worthwhile.

Mining

Extracting geologic materials is done by several different techniques depending on the accessibility of the resource and the content or concentration of the material sought. All of these methods have environmental hazards. Native metals deposited in the gravel of streambeds can be washed out hydraulically in a process called placer mining. This not only destroys streambeds but fills the water with suspended solids that smother aquatic life. Larger or deeper ore beds are extracted by strip mining or open-pit mining where overlying material is removed by large earth-moving equipment (Fig. 7.9). The resulting pits can be many kilometres across and hundreds of metres deep (Fig. 7.10). In some cases whole mountaintops or the surface of entire islands have been removed by surface mining (Case Study, p. 141). Even deeper deposits are reached by underground tunnelling, an extremely dangerous process for mine workers.

Old tunnels occasionally collapse, or subside. In coal mines, natural gas poses dangers of explosion. Uncontrollable fires, producing noxious smoke and gases, sometimes ignite coal-bearing scrap heaps stored inside or outside the mine. Surface waste deposits called tailings can cause acidic or otherwise toxic runoff when rainwater percolates through piles of stored material. Tailings from uranium mines give rise to wind scattering of radioactive dust.

Water leaking into mine shafts also dissolves metals and other toxic material. When this water is pumped out or allowed to seep into groundwater aquifers (Chapter 11), pollution occurs.

FIGURE 7.10 The world's largest open-pit mine is Bingham Canyon, near Salt Lake City, UT. More than 5 billion tonnes of copper ore and waste material have been removed since 1906 to create a hole 800 m deep and nearly 4 km wide at the top.

The process of strip mining involves stripping off the vegetation, soil, and rock layers, removing the minerals, and replacing the fill. The fill is usually replaced in long ridges, called spoil banks, because this is the easiest way to dump it cheaply and quickly. Spoil banks are very susceptible to erosion and chemical weathering. Rainfall leaches numerous chemicals in toxic concentrations from the freshly exposed earth, and the water quickly picks up a heavy sediment load. Chemical- and sediment-runoff pollution becomes a major problem in local watersheds. Problems are made worse by the fact that the steep spoil banks are very slow to revegetate. Since the spoil banks do not have natural topsoil, succession, soil development, and establishment of a natural biological community occur very slowly.

Mountaintop removal mining is a form of strip mining that has caused intense controversy in Appalachia in the southern U.S.

Since the 1980s, huge, powerful mining machinery—such as 20-story dragline excavators—have made it cheaper to remove the mountain from the coal than to dig the coal from the mountain. Hundreds of kilometres of streams have been filled or polluted by the "overburden" dumped into valleys. Entire towns and ecosystems are lost with the mountaintops.

In 1995, Natural Resources Canada released an Issues Paper entitled *Sustainable Development and Minerals and Metals*. In it, they tried to blend the goals of the Brundtland Report on Sustainable Development with the economics of mining. They recognized that mineral discovery, mine development, operation, and decommissioning were all integrated activities with important environmental and social effects. All of these had to be considered together in assessing the environmental impact of a proposed mine. It is now virtually impossible for a mining company in Canada to discover, develop, exploit, and then walk away from a mineral deposit without assessing and dealing with broader environmental consequences of the mine.

Processing

Metals are extracted from ores by heating or treatment with chemical solvents. These processes often release large quantities of toxic materials that can be even more environmentally hazardous than mining. Smelting—roasting ore to release metals—is a major source of air pollution. One of the most notorious examples of ecological devastation from smelting is a wasteland near Ducktown, Tennessee (Fig. 7.11). In the mid-1800s, mining companies began excavating the rich copper deposits in the area. To extract copper from the ore, they built huge open-air wood fires using timber from the surrounding forest. Dense clouds of sulphur dioxide released from sulphide ores poisoned the vegetation and acidified the soil over a 13,000-hectare area. Rains washed the soil off the denuded

FIGURE 7.11 A luxuriant forest once grew on this now barren hillside near Ducktown, TN. Smelter fumes killed all the vegetation more than a century ago, and erosion has washed away all the topsoil. Restoration projects are slowly bringing back ground cover and rebuilding soil.

The tiny island-nation of Nauru (pronounced NAH-roo) in the western Pacific is the smallest and most remote republic in the world. It also is a case study in humanity's ability to plunder its environment. Located on the equator some 500 km west of its nearest neighbour in the Marshall Islands, Nauru has been inhabited by Polynesian people for thousands of years. When first visited by European explorers in the eighteenth century, the island was a lush tropical paradise of swaying coconut palms and white coral beaches. Sailors called it Pleasant Island, but today the name is a bitter joke.

Compared to its former condition, Nauru is probably the most environmentally devastated nation on earth. So much land has been devoured by strip mining that residents now face the prospect of having to abandon the whole island and move elsewhere. What the miners sought was guano, a thick phosphate-rich layer of bird droppings prized by industrialized countries as fertilizer. Billions of dollars worth of this treasure have been exported, first by colonial powers and then, since independence in 1968, by the Nauruans themselves.

After a century of mining, Nauru's 7,500 residents are among the richest people in the world, but their environment has been almost totally wrecked. Eighty percent of the 21 sq km island has been stripped, leaving a bleak, barren moonscape of jagged coral pinnacles, some as tall as 25 metres. With all soil washed away, almost nothing lives in this wasteland. Travelling across it is impossible. To make things even worse, removing the vegetation has changed the climate. Heat waves rising from the sun-baked rock drive away rain clouds and the island now is plagued by constant drought.

Not only the island is ravaged. Nauruans may be among the world's most affluent people, but they are also among the most unhealthy, plagued by cardiovascular disease, diabetes, and obesity brought about by a lifestyle of idleness and imported junk food. Few islanders live past the age of 60. Since most mining is done by imported workers,

After a century of surface mining of rich phosphate deposits, four-fifths of the tiny Pacific island of Nauru is a barren moonscape of jagged coral pinnacles. Although mining has made Nauruians rich, they may soon have no place to live.

Nauruans generally lack job skills to apply elsewhere. This case study is a sad example of how easy it is for modern technology and lack of foresight to degrade both a society and its environment.

The guano deposits are expected to last only a few more years. After that, residents will be left with only a thin sliver of habitable coastline. Some efforts have been made to import new soil and to restore vegetation to the desolate interior, but attempts have been unsuccessful. It may be too late to reverse the damage. The people may be able to use some of their accumulated trust fund to buy another island, but will they find one as comfortable and beautiful as what they once had? One village leader says wistfully, "I wish Nauru could be like it was before. I remember it was so beautiful and green everywhere. We could eat coconuts and breadfruit. It makes me cry when I see what has been done. I wish we'd never discovered the phosphate."

Could Nauru's example be a warning for all of us? Humans have a long history of depleting resources and then moving on. Could we find ourselves in a similar situation someday, having exhausted our natural resources but then having an uninhabitable world?

Ethical Considerations

If the Nauruans appeal for help in finding another place to live, would we be morally obligated to assist them? Do countries that bought fertilizer from Nauru bear a special responsibility for what has happened? If we see other examples of environmental destruction occurring elsewhere, would we be ethically justified to try to intervene and stop it?

land, creating a barren moonscape where nothing could grow. Siltation of reservoirs on the Ocoee River impaired electric generation by the Tennessee Valley Authority (TVA).

Sulphur emissions from Ducktown smelters were reduced in 1907 after the U.S. Supreme Court ruled in Georgia's favour in a suit to stop interstate transport of air pollution. In the 1930s, the TVA began treating the soil and replanting trees to cut down on erosion. Recently, upwards of $250,000 per year has been spent on this effort. While the trees and other plants are still spindly and feeble, more than two-thirds of the area is considered "adequately" covered with vegetation; only 4 percent is still totally denuded. Similarly, smelting of copper-nickel ore in Sudbury, Ontario, more than a century ago caused widespread ecological destruction that is slowly being repaired following pollution control measures (see Fig. 7.17).

A recent technique called **heap-leach extraction** (Fig. 7.12) makes it possible to separate gold from extremely low-grade ores but has a high potential for water pollution. Typically, the process involves piling crushed ore in huge heaps and spraying it with a dilute alkaline-cyanide solution, which percolates through the pile to dissolve the gold. The gold-containing solution is pumped to a processing plant where gold is removed by electrolysis. A thick clay pad and plastic liner beneath the ore heap is supposed to keep the poisonous cyanide solution from contaminating surface or groundwater, but leaks are common.

The most northerly North American heap-leach gold mine is the Brewery Creek operation, near Dawson City in the Yukon. In 1999, it produced about 75,000 ounces of gold, a little less than the total declared placer gold production of about 90,000 ounces. So far, measures taken to protect the surrounding environment have worked.

In 2000, an enormous cyanide spill from a gold mining operation near Baia Mare in Romania poisoned millions of fish and threatened drinking water supplies along about 640 km of the Szamos, Tisza, and Danube Rivers in Hungary and Yugoslavia.

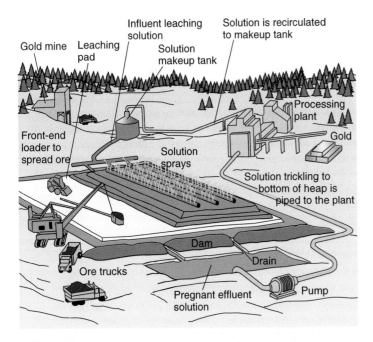

FIGURE 7.12 In a heap-leach operation, huge piles of low-grade ore are heaped on an impervious pad and sprayed continuously with a cyanide solution. As the leaching solution trickles through the crushed ore, it extracts gold and other precious metals. The "pregnant" effluent solution is then pumped to a processing plant where metals are extracted and purified. This technique is highly profitable but carries large environmental risks.
Source: George Laycock, *Audubon Magazine,* vol. 91(7), July 1989, p. 77.

CONSERVING GEOLOGIC RESOURCES

There is great potential for extending our supplies of economic minerals and reducing the effects of mining and processing through recycling. The advantages of recycling are significant: less waste to dispose of, less land lost to mining, and less consumption of money, energy, and water resources.

Recycling

Some waste products already are being exploited, especially for scarce or valuable metals. Aluminum, for instance, must be extracted from bauxite by electrolysis, an expensive, energy-intensive process. Recycling waste aluminum, such as beverage cans, on the other hand, consumes one-twentieth of the energy of extracting new aluminum. Not much data are available on the rate that consumers are recycling in Canada. We do know that Canada is a net exporter of recycled metal (mostly to the U.S.), and that "Blue Box" recycling programs for metal containers are very popular across the country. Recycling is so rapid and effective that half of all the aluminum cans now on a grocer's shelf will be made into another can within two months. The energy cost of extracting other materials is shown in Table 7.3.

Platinum, the catalyst in automobile catalytic exhaust converters, is valuable enough to be regularly retrieved and recycled from used cars (Fig. 7.13). Other metals commonly recycled are gold, silver, copper, lead, iron, and steel. The latter four are readily available in a pure and massive form, including copper pipes, lead batteries, and steel and iron auto parts. Gold and silver are valuable enough to warrant recovery, even through more difficult means. See Chapter 18 for further discussion of this topic.

TABLE 7.3 Energy Requirements in Producing Various Materials from Ore and Raw Source Materials

| | ENERGY REQUIREMENT (MJ/KG) | |
PRODUCT	NEW	FROM SCRAP
Glass	25	25
Steel	50	26
Plastics	162	n.a.
Aluminum	250	8
Titanium	400	n.a.
Copper	60	7
Paper	24	15

Source: Data from E. T. Hayes, *Implications of Materials Processing,* 1997.

FIGURE 7.13 The richest metal source we have—our mountains of scrapped cars—offers a rich, inexpensive, and ecologically beneficial resource that can be "mined" for a number of metals.

Steel and Iron Recycling: Minimills

While total global steel production has fallen in recent decades—largely because of inexpensive supplies from new and efficient Japanese steel mills—a new type of mill subsisting entirely on a readily available supply of scrap/waste steel and iron is a growing industry. Minimills, which remelt and reshape scrap iron and steel, are smaller and cheaper to operate than traditional integrated mills that perform every process from preparing raw ore to finishing iron and steel products. Minimills produce steel at between $225 and $480 per tonne, while steel from integrated mills costs $1,425 to $2,250 per tonne on average (Fig. 7.14). The energy cost is likewise lower in minimills: 5.3 million BTU/tonne of steel compared to 16.08 million BTU/tonne in integrated mill furnaces. Minimills now produce about half of all global steel production. Recycling is slowly increasing as raw materials become more scarce and wastes become more plentiful.

Substituting New Materials for Old

Mineral and metal consumption can be reduced by new materials or new technologies developed to replace traditional uses. This is a long-standing tradition, for example, bronze replaced stone technology and iron replaced bronze. More recently, the introduction of plastic pipe has decreased our consumption of copper, lead, and steel pipes. In the same way, the development of fibre-optic technology and satellite communication reduces the need for copper telephone wires.

Iron and steel have been the backbone of heavy industry, but we are now moving toward other materials. One of our primary uses for iron and steel has been machinery and vehicle parts. In automobile production, steel is being replaced by polymers (long-chain organic molecules similar to plastics), aluminum, ceramics, and new, high-technology alloys. All of these reduce vehicle weight and cost, while increasing fuel efficiency. Some of the newer alloys that combine steel with titanium, vanadium, or other metals wear much better than traditional steel. Ceramic engine parts provide heat insulation around pistons, bearings, and cylinders, keeping the rest of the engine cool and operating efficiently. Plastics and glass fibre-reinforced polymers are used in body parts and some engine components.

Electronics and communications (telephone) technology, once major consumers of copper and aluminum, now use ultra-high-purity glass cables to transmit pulses of light, instead of metal wires carrying electron pulses. Once again, this technology has been developed for its greater efficiency and lower cost, but it also affects consumption of our most basic metals.

GEOLOGIC HAZARDS

Earthquakes, volcanoes, floods, and landslides are normal earth processes, events that have made our earth what it is today. However, when they occur in proximity to human populations, their consequences can be among the worst and most feared disasters that befall us. For thousands of years people have been watching and recording these hazards, trying to understand them and learn how to avoid them.

Earthquakes

Earthquakes are sudden movements in the earth's crust that occur along faults (planes of weakness) where one rock mass slides past another one. When movement along faults occurs gradually and relatively smoothly, it is called creep or seismic slip and may be undetectable to the casual observer. When friction prevents rocks from slipping easily, stress builds up until it is finally released with a sudden jerk. The point on a fault at which the first movement occurs during an earthquake is called the epicentre.

Earthquakes, such as the one in India that opened this chapter, have always seemed mysterious, sudden, and violent, coming

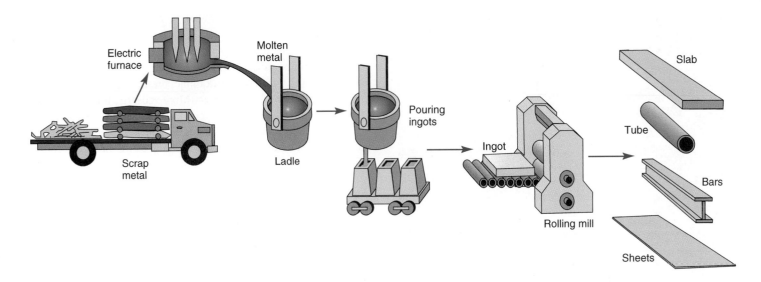

FIGURE 7.14 "Minimills" remelt and reshape scrap iron and steel. They not only extend our geologic resources by recycling discarded materials, they also conserve energy and are cheaper to operate than traditional integrated mills that depend on virgin ore.

TABLE 7.4 Worldwide Frequency and Effects of Earthquakes of Various Magnitudes

RICHTER SCALE MAGNITUDE*	DESCRIPTION	AVERAGE NUMBER PER YEAR	OBSERVABLE EFFECTS
2–2.9	Unnoticeable	300,000	Detected by instruments, but not usually felt by people.
3–3.9	Smallest felt	49,000	Hanging objects swing, vibrations like passing of light truck felt.
4–4.9	Minor earthquake	6,200	Dishes rattle, doors swing, pictures move, walls creak.
5–5.9	Damaging earthquake	800	Difficult to stand up. Windows, dishes break. Plaster cracks, loose bricks and tile fall, small slides along sand or gravel banks.
6–6.9	Destructive earthquake	120	Chimneys and towers fall, masonry walls damaged, frame houses move on foundations, small cracks in ground. Broken gas pipes start fires.
7–7.9	Major earthquake	18	General panic. Frame houses split and fall off foundations, some masonry buildings collapse, underground pipes break, large cracks in the ground.
8–8.9	Great earthquake	1 or 2	Catastrophic damage. Most masonry and frame structures destroyed. Roadways, dams, dikes collapse. Large landslides. Rails twist and bend. Underground pipes rupture.

*For every unit increase in the Richter scale, ground displacement increases by a factor of 10, while energy release increases by a factor of 30. There is no upper limit to the scale, but the largest earthquakes recorded have been 8.9.

Source: Data from B. Gutenberg in *Earth* by F. Press and R. Seiver, 1978, W. H. Freeman & Company.

without warning and leaving in their wake ruined cities and dislocated landscapes (Table 7.4). Cities such as Kobe, Japan, or Mexico City, parts of which are built on soft landfill or poorly consolidated soil, usually suffer the greatest damage from earthquakes (Fig. 7.15). Water-saturated soil can liquify when shaken. Buildings sometimes sink out of sight or fall down like a row of dominoes under these conditions.

Earthquakes frequently occur along the edges of tectonic plates, especially where one plate is subducting, or diving down, beneath another. Earthquakes also occur in the centres of continents, however. In fact, one of the largest earthquakes ever recorded in North America was one of an estimated magnitude of 8.8 that struck the area around New Madrid, Missouri, in 1812. Fortunately, few people lived there at the time and the damage was minimal. If a similar quake occurred along that fault today, the damage in Memphis, Nashville, and St. Louis would be horrendous. A calamitous earthquake was one thought to have killed 242,000 people in China in 1976.

Modern contractors in earthquake zones are attempting to prevent damage and casualties by constructing buildings that can

withstand tremors. The primary methods used are heavily reinforced structures, strategically placed weak spots in the building
that can absorb vibration from the rest of the building, and pads or
floats beneath the building on which it can shift harmlessly with
ground motion.

One of the most notorious effects of earthquakes is the
tsunami. These giant seismic sea swells (sometimes improperly
called tidal waves) can move at 1,000 km/hr, or faster, away from the
centre of an earthquake. When these swells approach the shore, they
can easily reach 15 m or more and some can be as high as 65 m.
A 1960 tsunami coming from a Chilean earthquake still caused
7-metre breakers when it reached Hawaii 15 hours later. Tsunamis
also can be caused by underwater volcanic explosions or massive
seafloor slumping. The eruption of the Indonesian volcano Krakatoa
in 1883 created a tsunami 40 m high that killed 30,000 people on
nearby islands.

While most earthquakes occur in known earthquake-prone
areas, sometimes they strike in unexpected places. In North America, major quakes have occurred in South Carolina, Missouri,
Massachusetts, Nevada, Texas, Utah, Arizona, and Washington, as
well as in California and Alaska.

FIGURE 7.15 An elevated freeway collapsed and buckled as a result
of the 1995 earthquake in Kobe, Japan.

Volcanoes

Volcanoes and undersea magma vents are the sources of most of
the earth's crust. Over hundreds of millions of years, gaseous
emissions from these sources formed the earth's earliest oceans
and atmosphere. Many of the world's fertile soils are weathered
volcanic materials. Volcanoes have also been an ever-present threat
to human populations. One of the most famous historic volcanic
eruptions was that of Mount Vesuvius in southern Italy, which
buried the cities of Herculaneum and Pompeii in A.D. 79. The
mountain had been giving signs of activity before it erupted, but
many citizens chose to stay and take a chance on survival. On
August 24, the mountain buried the two towns in ash. Thousands
were killed by the dense, hot, toxic gases that accompanied the
ash flowing down from the volcano's mouth. It continues to erupt
from time to time.

Nuees ardentes (French for "glowing clouds") are deadly,
denser-than-air mixtures of hot gases and ash like those that inundated Pompeii and Herculaneum. Temperatures in these clouds
may exceed 1,000°C, and they move at more than 100 km/hour.
Nuees ardentes destroyed the town of St. Pierre on the Caribbean
island of Martinique on May 8, 1902. Mount Pelee released a cloud
of nuees ardentes that rolled down through the town, killing somewhere between 25,000 and 40,000 people within a few minutes.
All the town's residents died except a single prisoner being held in
the town dungeon.

Mudslides are also disasters sometimes associated with volcanoes. The 1985 eruption of Nevado del Ruiz, 130 km northwest
of Bogotá, Colombia, caused mudslides that buried most of the town
of Armero and devastated the town of Chinchina. An estimated
25,000 people were killed. Heavy mudslides also accompanied the
eruption of Mount St. Helens in Washington in 1980. Sediments
mixed with melted snow and the waters of Spirit Lake at the mountain's base and flowed many kilometres from their source. Extensive damage was done to roads, bridges, and property, but because
of sufficient advance warning, there were few casualties.

Volcanic eruptions often release large volumes of ash and dust
into the air (Fig. 7.16). Mount St. Helens expelled 3 km^3 of dust and
ash, causing ash fall across much of North America. This was only
a minor eruption. An eruption in a bigger class of volcanoes was that
of Tambora, Indonesia, in 1815, which expelled 175 km^2 of dust and
ash, more than 58 times that of Mount St. Helens. These dust clouds
circled the globe and reduced sunlight and air temperatures enough
so that 1815 was known as the year without a summer.

It is not just a volcano's dust that blocks sunlight. Sulphur
emissions from volcanic eruptions combine with rain and atmospheric moisture to produce sulphuric acid (H_2SO_4). The resulting
droplets of H_2SO_4 interfere with solar radiation and can significantly
cool the world climate. In 1991, Mt. Pinatubo in the Philippines
emitted 20 million tonnes of sulphur dioxide that combined with
water to form tiny droplets of sulphuric acid. This acid aerosol
reached the stratosphere where it circled the globe for two years.
This thin haze cooled the entire earth by 1°C and postponed global
warming for several years. It also caused a 10 to 15 percent reduction in stratospheric ozone, allowing increased ultraviolet light to

FIGURE 7.16 Pyroclastic flows spill down the slopes of Mayon volcano in the Philippines in this September 23, 1984, image. Because more than 73,000 people evacuated danger zones, this eruption caused no casualties.

reach the earth's surface. One of several theories about the extinction of the dinosaurs 65 million years ago is that they were killed by acid rain and climate changes caused by massive volcanic venting in the Deccan Plateau of India.

Landslides

Landslides are mass movements of soil or rock downslope and are a major natural hazard in Canada. They have caused a substantial number of deaths, wreaked considerable damage to, and in some cases destruction of, elements of the nation's economic infrastructure (Figure 7.17).

FIGURE 7.17 This landslide, near Revelstoke, B.C., occurred in the spring of 1999. It affected the Canadian Pacific Railway (see lower part of figure).

Landslides can move rapidly or slowly and can occur in a wide variety of geologic environments, including underwater. The hazard posed by landslides can be attributable to the impact of rapidly moving debris, the failure of ground directly beneath a structure, or secondary effects such as river damming or landslide-generated waves.

SUMMARY

The earth is a complex, dynamic system. Although it seems stable and permanent to us, the crust is in constant motion. Huge blocks called tectonic plates slide over the surface of the ductile mantle. They crash into each other in ponderous slow motion, crumpling their margins into mountain ranges and causing earthquakes. Sometimes one plate will slide under another, carrying rock layers down into the mantle where they melt and flow back toward the surface to be formed into new rocks.

Rocks are classified according to composition, structure, and origin. The three basic types of rock are igneous, metamorphic, and sedimentary. These rock types can be transformed from one to another by way of the rock cycle, a continuous process of weathering, transport, burying in sediments, metamorphism, melting, and recrystallization.

During the cooling and crystallization process that forms rock from magma, minerals and metals can become concentrated enough to become economically important reserves if they are close enough to the surface to be reached by mining. Having a reliable supply of strategically important minerals and metals is vital in industrialized societies.

A few places in the world, including Canada, are especially rich in mineral deposits. Less-developed countries, most of which are in the tropics or the Southern Hemisphere, are often

PART TWO   Our Physical Environment

the largest producers of ores and raw mineral resources for the strategic materials on which the industrialized world depends.

Worldwide, only a small percentage of metals are recycled, although it is not a difficult process technically. Recycling saves energy and reduces environmental damage caused by mining and smelting. It reduces waste production and makes our metal supplies last much longer. Substitution of materials usually occurs when mineral supplies become so scarce that prices are driven up. Many of the strategic metals that we now stockpile may become obsolete when newer, more useful substitutes are found.

Both mining and processing of metals and mineral resources can have negative environmental effects. Mine drainage has polluted thousands of kilometres of streams and rivers. Fumes from smelters kill forests and spread pollution over large areas. Surface mining results in removal of natural ecosystems, soil disruption, creation of trenches and open pits, and accumulation of tailings. It is now required that strip-mined areas be recontoured, but revegetation is often difficult and limited in species composition. Smelting and chemical extraction processes also create pollution problems.

Earthquakes and volcanic events are natural geological hazards that are a result of movements of the earth's restless crust, core, and mantle. Big earthquakes are among the most calamitous natural disasters that befall people, sometimes killing hundreds of thousands in a single cataclysm.

QUESTIONS FOR REVIEW

1. Describe the layered structure of the earth.

2. Define mineral and rock.

3. What are tectonic plates and why are they important to us?

4. Why are there so many volcanoes and earthquakes along the "ring of fire" that rims the Pacific Ocean?

5. Describe the rock cycle and name the three main rock types that it produces.

6. Give some examples of strategic metals. Where are the largest supplies of these resources located?

7. Give some examples of nonmetal mineral resources and describe how they are used.

8. Describe some ways we recycle metals and other mineral resources.

9. What are some environmental hazards associated with mineral extraction?

10. Describe some of the leading geologic hazards and their effects.

QUESTIONS FOR CRITICAL THINKING

1. Look at the walls, floors, appliances, interior, and exterior of the building around you. How many earth materials were used in their construction?

2. What is the geologic history of your town or county?

3. Is your local bedrock igneous, metamorphic, or sedimentary? If you don't know, who might be able to tell you?

4. What would life be like without the global mineral trade network? Can you think of advantages as well as disadvantages?

5. Suppose a large mining company is developing ore reserves in the small, underdeveloped country where you live. How will revenues be divided fairly between the foreign company and local residents?

6. How could we minimize the destruction caused by geologic hazards? Should people be discouraged from building on floodplains or in volcanic or earthquake-prone areas?

7. How would your life be affected if Canada were to run out of strategic minerals?

8. How could our government encourage more recycling and more efficient use of minerals?

9. What can you do to protect yourself from geologic hazards?

KEY TERMS

core   132
crust   132
earthquakes   143
heap-leach extraction   142
igneous rocks   135
landslide   146
magma   132
mantle   132
metamorphic rock   136

mineral   134
rock   135
rock cycle   135
sedimentary rock   136
sedimentation   136
strategic metals and
   minerals   139
tectonic plates   132
tsunami   145
volcanoes   145
weathering   135

EARTHQUAKES IN CANADA

If you've ever thought earthquakes were something that happened in other countries and not Canada, check out Natural Resources Canada's amazing earthquake website at http://www.seismo.nrcan.gc.ca/index_e.php.

Does the number of earthquakes that have happened in Canada surprise you? How close is your home to one of the earthquake hotspots? How "hazardous" is your neighbourhood? What do you think causes the pattern you see in Canadian earthquakes?

Chapter 8

Air, Weather, and Climate

Climate is an angry beast and we are poking it with sticks.

—Wallace Broecker, Lamont-Doherty Earth Observatory—

OBJECTIVES

After studying this chapter, you should be able to:

- summarize the structure and composition of the atmosphere.
- explain how the jet streams, prevailing winds, and frontal systems determine local weather.
- describe how tornadoes and cyclonic storms form and why they are dangerous.
- comprehend how El Niño cycles change ocean surface temperatures and affect continental climate.
- understand the driving forces thought to bring about normal climatic change.
- analyze human contributions to global climate change and what effects our modifications are having on physical and biological systems.
- debate policy options for responding to the threats of global climate change.

WebQuest

atmosphere, weather, climate, tornadoes, cyclonic storms, El Niño, climate change, Kyoto Protocol

Above: Lightning flashes over the city as a summer thunderstorm approaches.

What's Happening to Our Climate?

What do skinny Canadian polar bears, Peruvian cholera epidemics, melting of Mt. Kilimanjaro's famous snows, Chinese drinking-water shortages, unusually severe Bangladeshi floods, coastal erosion in Louisiana, and disappearance of Edith's Checkerspot butterfly from parts of southern California have in common? All these phenomena are thought to be signs of human-caused global climate change, which may well be the most critical issue in environmental science today.

In February 2001, the Intergovernmental Panel on Climate Change (IPCC) released a 1,000-page report saying "with a high degree of confidence" that recent changes in the world's climate have had discernible impacts on physical and biological systems. Furthermore, the IPCC said, "most of the warming trend over the last 50 years is attributable to human activities." Composed of more than 700 scientists representing 100 countries, the IPCC reviewed results from some 3,000 scientific studies showing changes in 420 different physical and biological systems as a result of shifting global climate. This mass of evidence makes an overwhelming case that we are conducting a giant experiment to see what will happen if we increase greenhouse gas concentrations in the atmosphere. So far, the results don't look too good.

All around us evidence is accumulating that we are modifying our climate on both a local and global scale. The average world temperature is up about 1°C over the past century, and the 1990s were the warmest decade in the past millennium. Polar regions are changing even faster. The Canadian Arctic appears to be about 4°C warmer now than it was 20 years ago. Sea ice forms later in the fall and melts earlier in the spring, giving polar bears a shorter seal-hunting season. Hudson's Bay polar bears now weigh as much as 100 kg less than in the 1960s. Unusually warm waters off the coast of South America are thought to be responsible for reappearance of cholera in the 1990s after nearly a century of absence.

Glaciers are disappearing on every continent. Mt. Kilimanjaro has lost 85 percent of its ice cap since 1915. By 2015, all permanent ice on the mountaintop is expected to be gone. Alpine glaciers feed rivers such as the Indus, Ganges, Yangtze, and Yellow that supply drinking water and irrigation to more than a billion people in South and East Asia where water is already becoming a source of conflict. Ocean warming is causing severe storms and heavy monsoon rains that result in flooding in Bangladesh as well as erosion in Louisiana, where rising sea levels have inundated low-lying coastal marshes. Edith's Checkerspot is only one of many species of mammals, birds, amphibians, fish, insects, and plants that are reported to have moved their territory or migration patterns, or to have disappeared altogether as a result of changing climate.

To understand what all these observations may have to say about our future and how we should respond to the threat of global climate change, we have to know something about our atmosphere and how it produces our local weather and global climate. We also need to be acquainted with the international politics of this crucial topic. In this chapter, we'll look at how atmospheric systems work, examine the driving forces behind climate modification, and describe, in more detail, the ways that climate affects human and natural systems.

Want More?

Krupnik, I., and Jolly, D. 2002. "The Earth is Faster Now." *Indigenous Observations of Arctic Environmental Change.* Arctic Research Consortium of the United States.

COMPOSITION AND STRUCTURE OF THE ATMOSPHERE

We live at the bottom of a virtual ocean of air. Extending upward about 1,600 km, this vast, restless envelope of gases is far more turbulent and mobile than the oceans of water. Its currents and eddies are the winds.

Weather is a description of the physical conditions of the atmosphere (moisture, temperature, pressure, and wind), all of which play a vital role in shaping ecosystems. The dynamism of the atmosphere is maintained by a ceaseless flow of solar energy. Winds generated by pressure gradients push large air masses of differing temperature and moisture content around the globe. Our daily weather is created by the movement of these air masses (Fig. 8.1).

Climate is a description of the long-term pattern of weather in a particular area. Climates often undergo cyclic changes over decades, centuries, and millennia. Determining where we are in these cycles and predicting what may happen in the future is an important, but difficult, process. As human activities change the properties of the atmosphere, it becomes more difficult and more important to understand how the atmosphere works and what future weather and climate conditions may be.

Weather and climate are important, not only because they affect human activities, but because they are primary determinants of biomes and ecosystem distribution. Generalized, large-scale climate often is not as important in the life of an individual organism as is the microclimate in its specific habitat. Geographic boundaries that separate communities and ecosystems are established primarily by climatic boundaries created by temperature, moisture,

FIGURE 8.1 The atmospheric processes that purify and redistribute water, moderate temperatures, and balance the chemical composition of the air are essential in making life possible. The earth is the only planet we know with a habitable atmosphere. To a large extent, living organisms have created, and help to maintain, the atmosphere on which we all depend.

and wind-distribution patterns (Chapter 5). The movement and effects of pollutants also are strongly linked to weather and climatic conditions (Chapter 9).

Past and Present Composition

The composition of the atmosphere has changed drastically since it first formed as the earth cooled and condensed from interstellar gases and dust. Most geochemists believe that the earth's earliest atmosphere was made up mainly of hydrogen and helium and was hundreds of times more massive than it is now. Over billions of years, most of that hydrogen and helium diffused into space. At the same time, volcanic emissions have added carbon, nitrogen, oxygen, sulphur, and other elements to the atmosphere.

The current composition of the earth's atmosphere is unique in our solar system. This is the only place we know of with free oxygen and water vapour. We believe that virtually all the molecular oxygen in the air was produced by photosynthesis in blue-green bacteria, algae, and green plants. If that oxygen were not present, animals (like us) that oxidize organic compounds as an energy source could not exist. As Chapter 3 discusses, producers and consumers create a balance between carbon dioxide and oxygen levels in the atmosphere. The Gaia hypothesis, first proposed by British chemist James Lovelock, suggests that the balance between atmospheric carbon dioxide and oxygen maintained by living organisms is responsible not only for creating a unique atmospheric chemical composition but also for other environmental characteristics that make life possible. We will discuss how carbon dioxide levels regulate temperatures later in this chapter. Organisms—especially humans—continue to modify the atmosphere, but at a dramatically accelerated rate, and in ways that may be disastrous for other forms of life.

Table 8.1 presents the main components of clean, dry air. Water vapour concentrations vary from near zero to 4 percent, depending on air temperature and available moisture. Small but important concentrations of minute particles and droplets—collectively called **aerosols**—also are suspended in the air. They can have serious effects on human health and natural ecosystems as discussed later in this chapter and in Chapter 9.

Air pressure varies from day to day and from place to place as temperatures change and weather systems move across the earth's surface. This makes air pressure variations useful in forecasting local weather. At sea level the normal air pressure is generally around 1,000 mb (14.5 lbs/square inch). Air pressure drops quickly as altitude increases. On top of Mt. Everest at 8,850 m, the average air pressure is just half that at sea level. At an elevation of about 12,500 m the air pressure approaches zero.

TABLE 8.1	Present Composition of the Lower Atmosphere*	
GAS	**SYMBOL OR FORMULA**	**PERCENT BY VOLUME**
Nitrogen	N_2	78.08
Oxygen	O_2	20.94
Argon	Ar	0.934
Carbon dioxide	CO_2	0.035
Neon	Ne	0.00182
Helium	He	0.00052
Methane	CH_4	0.00015
Krypton	Kr	0.00011
Hydrogen	H_2	0.00005
Nitrous oxide	N_2O	0.00005
Xenon	Xe	0.000009

*Average composition of dry, clean air

A Layered Envelope

The atmosphere is layered into four distinct zones of contrasting temperature due to differential absorption of solar energy (Fig. 8.2). Understanding how these layers differ and what creates them helps us understand atmospheric functions.

The layer of air immediately adjacent to earth's surface is called the **troposphere.** Ranging in depth from about 12.5 km over the equator to about 8 km over the poles, this zone is where almost all weather events occur. Due to the force of gravity and the compressibility of gases, the troposphere contains about 75 percent of the total mass of the atmosphere. The troposphere's composition is relatively uniform over the entire planet because this zone is strongly stirred by winds. Air temperature drops rapidly with increasing altitude in this layer, reaching about –60°C at the top of the troposphere. A sudden transition in this temperature gradient creates a sharp boundary, the tropopause, that limits mixing between the troposphere and upper zones.

The **stratosphere** extends from the tropopause up to about 45 km. Air temperature in this zone is stable or even increases with higher altitude. Although more dilute than the troposphere, the stratosphere has a very similar composition except for two important components: water and ozone (O_3). The fractional volume of water vapour is about 1,000 times lower, and ozone is nearly 1,000 times higher than in the troposphere. Ozone is produced by lightning and solar irradiation of oxygen molecules and would not be present if photosynthetic organisms were not releasing oxygen. Ozone protects life on the earth's surface by absorbing most incoming solar ultraviolet radiation.

Recently discovered depletion of stratospheric ozone over Antarctica (and to a lesser extent over the whole planet) are of serious concern (see Chapter 9). If these trends continue, we could be exposed to increasing amounts of dangerous ultraviolet rays, resulting in higher rates of skin cancer, genetic mutations, crop failures, and disruption of important biological communities. Unlike the troposphere, the stratosphere is relatively calm. There is so little mixing in the stratosphere that volcanic ash or human-caused contaminants can remain in suspension there for many years (Chapter 7).

Above the stratosphere, the temperature diminishes again, creating the **mesosphere** or middle layer. The minimum temperature reached in this region is about –80°C. At an altitude of 80 km, another abrupt temperature change occurs. This is the beginning of the **thermosphere,** a region of highly ionized gases, extending out to about 1,600 km. Temperatures are very high in the thermosphere because molecules there are constantly bombarded by high-energy solar and cosmic radiation. There are so few molecules per unit volume, however, that if you were cruising through in a spaceship, you wouldn't notice the temperature increase.

The lower part of the thermosphere is called the **ionosphere.** This is where the aurora borealis (northern lights) appears when showers of solar or cosmic energy cause ionized gases to emit visible light. There is no sharp boundary that marks the end of the atmosphere. Pressure and density decrease gradually as one travels away from the earth until they become indistinguishable from the near vacuum of intrastellar space. The composition of the thermosphere also gradually merges with that of intrastellar space, being made up mostly of extremely dilute helium and hydrogen.

THE GREAT WEATHER ENGINE

The atmosphere is a great weather engine in which a ceaseless flow of energy from the sun causes global cycling of air and water that creates our climate and distributes energy and material through the environment.

Solar Radiation Heats the Atmosphere

The sun supplies the earth with an enormous amount of energy. Although it fluctuates from place to place and time to time, incoming solar energy at the top of the atmosphere averages about 1,330 watts per sq metre. About half of this energy is reflected or absorbed by the atmosphere, and half the earth faces away from the sun at any given time. Still, the amount reaching the earth's surface is at least 10,000 times greater than all installed electric capacity in the world.

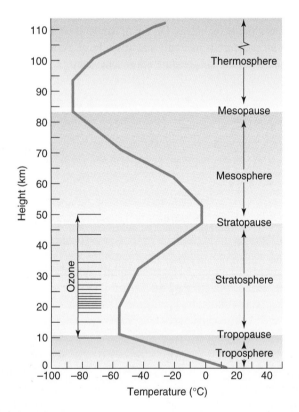

FIGURE 8.2 Temperatures change dramatically in different layers of the atmosphere. Bars in the ozone graph represent relative concentrations of the "ozone layer" with altitude.

Source: Courtesy of Dr. William Culver, St. Petersburg Junior College.

The absorption of solar energy by the atmosphere is selective. Visible light (see Fig. 3.9) passes through almost undiminished, whereas ultraviolet light is absorbed mostly by ozone in the stratosphere. Infrared radiation is absorbed mostly by carbon dioxide (CO_2) and water (H_2O) in the troposphere. Scattering of light by water droplets, ice crystals, and dust in the air also is selective. Short wavelengths (blue) are scattered more strongly than long wavelengths (red). The blue of a clear sky or clean, deep water at midday, and the spectacular reds of sunrise and sunset are the result of this differential scattering. On a cloudy day, as much as 90 percent of insolation is absorbed or reflected by clouds. Figure 8.3 shows the relative energy fluxes in the atmosphere.

Some solar energy is reflected from the earth's surfaces. **Albedo** is the term used to describe reflectivity. Fresh, clean snow can have an albedo of 90 percent, meaning that 90 percent of incident radiation falling on its surface is reflected. Dark surfaces, such as black topsoil or a dark forest canopy, absorb energy efficiently and might have an albedo of only 2 or 3 percent. The net average global albedo of the earth is about 30 percent. Clouds are responsible for most of that reflection. The earth's surface has a low average albedo (5 percent) due to the high energy absorbency of the oceans covering most of the globe.

Eventually, all the energy absorbed at the earth's surface is reradiated back into space. There is an important change in properties between incoming and outgoing radiation, however. Most of the solar energy reaching the earth is visible light, to which the atmosphere is relatively transparent; the energy reemitted by the earth is mainly infrared radiation (heat energy). These longer wavelengths are absorbed rather effectively in the lower levels of the atmosphere, trapping much of the heat close to the earth's surface. If the atmosphere were as transparent to infrared radiation as it is to visible light, the earth's surface temperature would be about 35°C colder than it is now.

This phenomenon is called the "greenhouse effect" because the atmosphere, like the glass of a greenhouse, transmits sunlight while trapping heat inside. (The analogy is not totally correct, however, because glass is much more transparent to infrared radiation than is air; greenhouses stay warm mainly because the glass blocks air movement.) Increasing atmospheric carbon dioxide due to human activities appears to be causing a global warming that could cause major climatic changes, which we will discuss later in this chapter.

Because of cycling of infrared energy between the atmosphere and the planet, the amount of energy emitted from the earth's

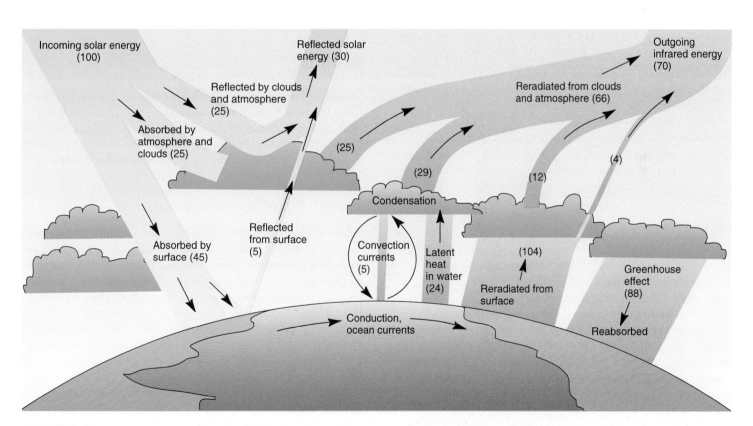

FIGURE 8.3 Energy balance between incoming and outgoing radiation. The atmosphere absorbs or reflects about half of the solar energy reaching the earth. Most of the energy reemitted from the earth's surface is long-wave, infrared energy. Most of this infrared energy is absorbed by aerosols and gases in the atmosphere and is reradiated toward the planet, keeping the surface much warmer than it would otherwise be. This is known as the greenhouse effect. The numbers shown are arbitrary units. Note that for 100 units of incoming solar energy, 100 units are reradiated to space, but more than 100 units are radiated from the earth's surface because of the greenhouse effect.

surface is about 30 percent greater than the total incoming solar radiation. The amount of energy reflected or reradiated from the top of the atmosphere must balance with the total insolation if the earth is to remain at a constant temperature.

Convection Currents and Latent Heat

Air currents, especially those carrying large amounts of water vapour, also play an important role in shaping our weather and climate. As the sun heats the earth's surface, some of that heat is transferred to adjacent air layers, causing them to expand and become less dense. This lighter air rises and is replaced by cooler, heavier air, resulting in vertical **convection currents** that stir the atmosphere and transport heat from one area to another.

Much of the solar energy absorbed by the earth is used to evapourate water. Because of the unique properties of water, it takes a significant amount of energy to change water from liquid to a vapour state, and this energy is stored in the water vapour as latent or potential energy. That latent energy subsequently is released as heat when the water condenses.

Water vapour carried into the atmosphere by rising convection currents transports large amounts of energy and plays an important role in the redistribution of heat from low to high altitudes, and from the oceans to the continental landmasses (Fig. 8.4). As warm, moist air rises, it expands (due to lower air pressures at higher altitudes) and cools. If condensation nuclei are present or if temperatures are low enough, the water will condense to form water droplets or ice crystals and precipitation will occur.

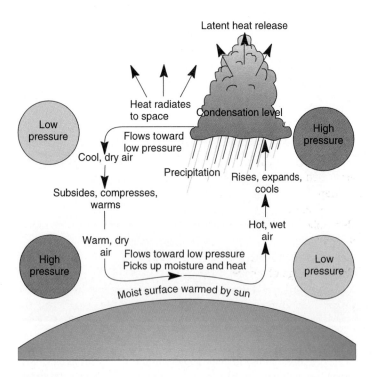

FIGURE 8.4 Convection currents and latent energy cause atmospheric circulation and redistribute heat and water around the globe.

Releasing latent heat causes the air to rise higher, cool more, and lose more water vapour. Rising, expanding air creates an area of relatively high pressure at the top of the convection column.

Air flows out of this high-pressure zone toward areas of low pressure where cool, dry air is sinking (subsiding). This subsiding air is compressed (and therefore warmed) as it approaches the earth's surface, where it piles up and creates a region of relatively high pressure at the surface. Air flows out of this region back toward the area of low surface pressure caused by rising air, thus closing the cycle. These are the driving forces of the hydrologic cycle (Chapter 10).

The convection currents just described can be as small and as localized as a narrow column of hot air rising over a sun-heated rock, or as large as the desert low-pressure cell that covers the U.S. Southwest most of the summer. The circulation patterns they create can be as mild as a gentle onshore breeze moving from the warm ocean toward the cooling shoreline as the sun goes down in the evening, or they can create monster cyclonic storms that drive hurricanes hundreds of kilometres wide across oceans and over continents. The force generated by such a storm (powered by latent heat released from condensing water vapour) can be equivalent to hundreds of megatonne-sized nuclear bombs. These circulation systems—both large and small—are the driving force behind our weather, the moment-by-moment changes in the atmosphere.

WEATHER

Even though we have been watching the skies for thousands of years, trying to forecast what the weather will be, the science of meteorology (weather studies) is still rather imprecise and uncertain. Weather forces still are so important in our lives that it makes sense to learn as much as we can about how the atmosphere makes weather.

Energy Balance in the Atmosphere

Solar energy doesn't strike the whole globe equally. At the equator, the sun is almost directly overhead all year long. Its rays are very intense because it shines through a relatively short column of air (straight down), and energy flux (flow) is high. At the poles, however, sunlight comes in at an oblique angle. The long column of air through which light must pass before it reaches the surface causes much greater energy losses from absorption and scattering. Moreover, when the light does reach the ground, it is spread over a larger area because of its angle of incidence, reducing surface heating even more.

Furthermore, the tilt of the earth's axis means that there is no sunlight at the poles during much of the winter. The equator, by contrast, has days about the same length all year long; thus, compared to the poles, the equatorial regions have an energy surplus. This energy imbalance is evened out by movement of air and water vapour in the atmosphere, and by liquid water in rivers and ocean currents. Warm, tropical air moving toward the poles and cold, polar air moving toward the equator account for about half of this energy transfer. Latent heat in water vapour (mainly from the oceans) makes up about 30 percent of the global energy redistribution. The remaining 20 percent is carried mainly by ocean currents.

Convection Cells and Prevailing Winds

As air warms at the equator, rises, and moves northward, it doesn't go straight to the pole in a single convection current. Instead, this air sinks and rises in several intermediate bands, forming circulation patterns called cells (Fig. 8.5). Nor do the returning surface flows within these cells run straight north and south. Friction, drag, and momentum cause air layers close to the earth's surface to be pulled in the direction of the earth's rotation. This deflection is called the **Coriolis effect.** In the Northern Hemisphere, the Coriolis effect deflects winds about 30° to the right of their expected path, creating clockwise or anticyclonic spiralling patterns in winds flowing out of a high-pressure centre, and cyclonic or counterclockwise winds spiralling into a low-pressure area. In the Southern Hemisphere, winds shift in the opposite direction. The earth's rotation affects only large-scale movements, however. Contrary to popular beliefs, water draining out of a bathtub in the Southern Hemisphere is controlled by the shape of the drain, and will not necessarily swirl in the opposite direction from what it would in the north.

Major zones of subsidence occur at about 30° north and south latitudes. Air flows into these regions of upper atmosphere low pressure both from the north and south. Air flowing back toward the equator from the north is turned toward the west by the Coriolis effect, creating the steady northeast "trade winds" of subtropical oceans. Their name comes from the dependable routes they provided for merchant sailing ships in earlier days. Where this dry, subsiding air falls on continents, it creates broad, subtropical desert regions (Chapter 10). Air flowing north from the horse latitudes turns eastward, giving rise to the prevailing westerlies of middle latitudes. (Notice that an eastward flowing wind is called a west wind or a westerly, due to the direction from which it originates.)

Winds directly under regions of subsiding air often are light and variable. They create the so-called horse latitudes because sailing ships bringing livestock to the New World were often becalmed here and had to throw the bodies of dead horses overboard. Rising air at the equator creates doldrums where the winds may fail for weeks at a time. Another band of variable winds at about 60° north, called the polar front, tends to block the southward flow of cold polar air. As we will see in the next section, however, all these boundaries between major air flows wander back and forth, causing great instability in our weather patterns, especially in midcontinent areas. The Southern Hemisphere has more stable wind patterns because it has more ocean and less landmass than the Northern Hemisphere.

Jet Streams

Superimposed on the major circulation patterns and prevailing surface winds are variations caused by large-scale upper air flows and shifting movements of the large air masses that they push and pull. The most massive of these rivers of air are the **jet streams,** powerful winds that circulate in shifting flows rivaling the oceanic currents in extent and effect. Generally following meandering paths from west to east, jet streams can be as much as 50 km wide and 5 km deep. The number, flowing speed, location, and size of jet streams all vary from day to day and place to place.

Wind speeds at the centre of a jet stream are often 200 km/hr and may reach twice that speed at times. Located 6 to 12 km above the earth's surface, jet streams follow discontinuities in the tropopause (the boundary between the troposphere and the stratosphere), where they are broken into large, overlapping plates that fit together like shingles on a roof. The jet streams are probably generated by strong temperature contrasts where adjacent plates overlap.

There are usually two main jet streams over the Northern Hemisphere. The subtropical jet stream generally follows a sinuous path about 30° north latitude (the southern edge of the United States), while the northern jet stream follows a more irregular path along the edge of a huge cold air mass called the circumpolar vortex (Fig. 8.6) that covers the earth's top like a cap with scalloped edges. This whole polar vortex rotates from west to east slightly faster than the planet's rotation. As it moves, the lobes, or fingers, of cold air that protrude south from the vortex sweep across Canada and the United States. The clash between cold, dry arctic air masses pushing south against warm, wet air masses moving north from the Gulf of Mexico or the Pacific Ocean causes winds, rains, and storms across the continent.

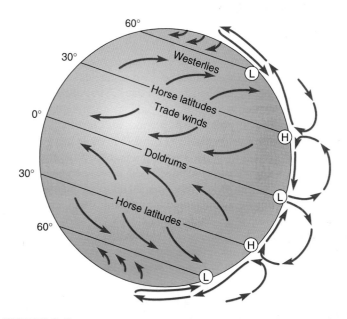

FIGURE 8.5 General circulation patterns. The actual boundaries of the circulating convection cells vary from day to day and season to season, as do the local directions of surface winds. Surface topography also complicates circulation patterns, but within these broad regions, winds usually have a predominant and predictable direction. L and H represent low- and high-pressure.

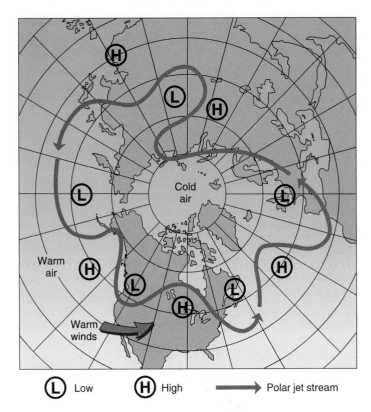

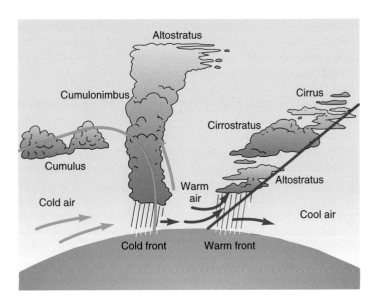

FIGURE 8.7 A cold front assumes a bulbous, "bull-nose" appearance because ground drag retards forward movement of surface air. As warm air is lifted up over the advancing cold front, it cools, producing precipitation. When warm air advances, it slides up over cooler air in front and produces a long, wedge-shaped zone of clouds and precipitation. The high cirrus clouds that mark the advancing edge of the warm air mass may be 1,000 km and 48 hours ahead of the front at ground level.

FIGURE 8.6 A typical pattern of the arctic circumpolar vortex. This large, circulating mass of cold air sends "fingers," or lobes, across North America and Eurasia, spreading storms in their path. If the vortex becomes stalled, weather patterns stabilize, causing droughts in some areas and excess rain elsewhere.

During the winter, as the Northern Hemisphere tilts away from the sun and the atmosphere cools, the polar air masses become stronger and push farther south, bringing snow and low temperatures across Canada and much of the United States. During the summer, as we tilt back toward the sun, warm air from the South pushes the polar jet stream back toward the pole.

Occasionally, the circumpolar vortex slows so that it rotates at nearly the same speed as the earth, stalling the motion of the lobes or air masses, and locking a huge ridge of hot, dry air over mid-North America for months at a time. What causes air flow to be stalled like this—or to resume normal circulation patterns—is unknown, but the amount of heat in the atmosphere surely plays a role.

Frontal Weather

The boundary between two air masses of different temperature and density is called a front. When cooler air displaces warmer air, we call the moving boundary a **cold front.** Since cold air tends to be more dense than warm air, a cold front will hug the ground and push under warmer air as it advances. As warm air is forced upward, it cools adiabatically (without loss or gain of energy), and its cargo of water vapour condenses and precipitates. Upper layers of a moving

cold air mass move faster than those in contact with the ground because of surface friction or drag, so the boundary profile assumes a curving, "bull-nose" appearance (Fig. 8.7, left). Notice that the region of cloud formation and precipitation is relatively narrow. Cold fronts generate strong convective currents and often are accompanied by violent surface winds and destructive storms. An approaching cold front generates towering clouds called thunderheads that reach into the stratosphere where the jet stream pushes the cloud tops into a characteristic anvil shape. The weather after the cold front passes is usually clear, dry, and invigorating.

If the advancing air mass is warmer than local air, a **warm front** results. Since warm air is less dense than cool air, an advancing warm front will slide up over cool, neighbouring air parcels, creating a long, wedge-shaped profile with a broad band of clouds and precipitation (Fig. 8.7, right). Gradual uplifting and cooling of air in the warm front avoids the violent updrafts and strong convection currents that accompany a cold front. A warm front will have many layers of clouds at different levels. The highest layers are often wispy cirrus (mare's tail) clouds that are composed mainly of ice crystals. They may extend 1,000 km ahead of the contact zone with the ground and appear as much as 48 hours before any precipitation. A moist warm front can bring days of drizzle and cloudy skies.

Cyclonic Storms

Few people experience a more powerful and dangerous natural force than cyclonic storms spawned by low-pressure cells over warm tropical oceans. As we discussed earlier in this chapter, low pressure is generated by rising warm air. Winds swirl into this low-pressure area, turning counterclockwise in the Northern Hemisphere due to the Coriolis effect. When rising air is laden with water vapour, the latent energy released by condensation intensifies convection currents and draws up more warm air and water vapour. As long as a temperature differential exists between air and ground and a supply of water vapour is available, the storm cell will continue to pump energy into the atmosphere.

Called **hurricanes** in the Atlantic and eastern Pacific, typhoons in the western Pacific, or cyclones in the Indian Ocean, winds near the centre of these swirling air masses can reach hundreds of kilometres per hour and cause tremendous suffering and destruction. Often hundreds of kilometres across, these giant storms can generate winds as high as 320 km/hr and push walls of water called storm surges far inland. In 1970, a killer typhoon brought torrential rains, raging winds, and a storm surge up to 5 m high that flooded thousands of square kilometres of the flat coastal area of Bangladesh, drowning more than a half million people.

Hurricane Mitch, which swept across Central America in October 1998, was the most deadly storm to strike the Western Hemisphere in 200 years. Stretching some 1,200 km across the Caribbean Ocean, it had sustained winds above 155 knots (almost 300 km/hr) and gusts well over 320 km/h (Fig. 8.8). More than two metres of rain fell in four days as the storm moved slowly across the mountains of Central America. Runoff from this one storm may have doubled the pesticide load in the Atlantic. At least 11,000 people died and more than 3 million were left homeless by the floods and mudslides triggered by the intense rain. Total damage in Honduras, Nicaragua, Guatemala, Belize, and El Salvador was estimated to be around $5 billion. This about equals the average annual cost for hurricane or flood damage for the entire United States and is about five times the annual U.S. losses from tornadoes.

FIGURE 8.8 Hurricane Mitch stretches 1,200 km from Cuba to Honduras on October 26, 1998, in this infrared spectrum image. Colours represent temperature at the top of the clouds, which correspond to wind speeds. Green = 250 km/hr, dark blue = 210 km/hr, light blue = 175 km/hr.

FIGURE 8.9 Little remained of the Evergreen Trailer Court just northeast of Edmonton, Alberta, after a twister ripped through the east side of the city in 1987.

Tornadoes, swirling funnel clouds that form over land, also are considered to be cyclonic storms although their rotation isn't generated by Coriolis forces. While never as large or powerful as hurricanes, tornadoes can be just as destructive in the limited areas where they touch down (Fig. 8.9). Some meteorologists consider smaller, less destructive phenomena such as waterspouts, dust devils, and smoke funnels over forest fires to be miniature tornadoes, but others think they have very different origins.

Tornadoes are generated on the American Great Plains by giant "supercell" frontal systems where strong, dry air cold fronts from Canada collide with warm humid air moving north from the Gulf of Mexico. Greater air temperature differences cause more powerful storms, which is why most tornadoes occur in the spring, when Arctic cold fronts penetrate far south over the warming plains. As warm air rises rapidly over dense, cold air, intense vertical convection currents generate towering thunderheads with anvil-shaped leading edges and domed tops up to 20,000 m high (Fig. 8.10). Water vapour cools and condenses as it rises, releasing latent heat and accelerating updrafts within the supercell. Penetrating into the stratosphere, the tops of these clouds can encounter jet streams, which help create even stronger convection currents.

But what sets the air mass spinning? Scientists debate this point, but one theory is that differential air speeds called shear forces—fast-moving above but slower near the ground—roll the air ahead of the advancing front in invisible horizontal tubes (or vortices) much as you might roll modelling clay between your hands. As these rolling vortex tubes are sucked into the vertical convective currents, they tilt upright and set the air column in motion, creating a spiral structure called a mesocyclone. Initially, the swirling air mass may be many kilometres wide, but cold downdrafts from the rear flank of the storm can spiral around the mesocyclone, increasing its velocity and narrowing its diameter much as a skater spins faster as she pulls in her arms. Winds at the edge of the rapidly spinning air column can reach speeds

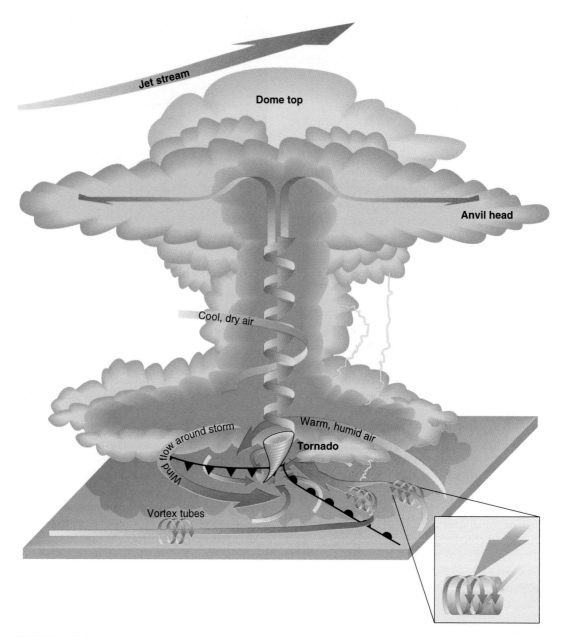

FIGURE 8.10 Tornadoes are generated by rapidly rising thermal updrafts where warm humid air is drawn up into a cold supercell thunderstorm. Rolling vortex tubes generated by differential air currents are drawn into the updraft and start it rotating. Cool, dry air spiralling down from the rear tightens the spiral and intensifies its speed. *Source:* Data from National Center for Atmospheric Research, May 20, 1996.

up to 510 km/hr that rip trees out of the ground and drive straws through telephone poles. In addition, low pressure in the centre of the funnel cloud implodes houses, sucks people out of windows, and helps hurl large objects across the countryside.

Sometimes a supercell isn't organized enough to spawn tornadoes, but strong downdrafts, called **downbursts,** can generate straight-line winds well over 160 km/hr. On July 4, 1999, a powerful straight-line wind flattened about 25 million trees in a swath 16 km by 48 km across the heart of the Boundary Waters Canoe Area in northern Minnesota. Witnesses said that as the winds hit,

the forest began rippling like a field of wheat, and then in seconds, was knocked down as if by a steamroller.

The same strong updrafts that create tornadoes can also generate hail when falling water droplets are blown back up to higher altitudes where they freeze. Each time the droplets bounce up and down, they pick up more water and become larger until finally the ice balls are heavy enough to resist being lofted by updrafts. If they remain frozen all the way to the ground, we call them hail. If you cut open large hail stones, you often can see concretion rings where successive layers of ice were added.

Seasonal Winds

A **monsoon** is a seasonal reversal of wind patterns caused by different heating and cooling rates of the ocean and continents. The most dramatic examples of monsoon weather are found in tropical or subtropical seasonal forests (see Chapter 6), where a large land area is cut off from continental air masses by mountain ranges and surrounded by a large volume of water. The Indian subcontinent is a good example (Fig. 8.11). As the summer sun heats the Indian plains, a strong low-pressure system develops. Flow of cool air from the north is blocked by the Himalayan mountains. A continuous flow of moisture-laden air from the subtropical high-pressure area over the Arabian Sea sweeps around the tip of India and up into the Ganges Plain and the Bay of Bengal. As the onshore winds are driven against the mountains and rise, the air cools and drops its load of water.

During the four months of the monsoon, which usually lasts from May until August, 10 m of rain may fall on Nepal, Bangladesh, and western India. The highest rainfall ever recorded in a season was 25 m, which fell in about five months on the southern foothills of the Himalayas in 1970. These heavy rains result in high rates of erosion and enormous floods in the flat delta of the Ganges River, but they also irrigate the farmlands that feed the second most populous country in the world. In winter, the Indian landmass is cooler than the surrounding ocean and the wind flow reverses. The northeast winds pick up moisture as they blow over the Bay of Bengal and bring winter monsoon rains to Indonesia, Australia, and the Philippines.

Africa also experiences strong seasonal monsoon winds. Hot air rising over North Africa in the summer pulls in moist air from the Atlantic Ocean along the Gulf of Guinea. The torrential rains released by this summer circulation nourish the tropical forests of Central Africa and sometimes extend as far east as the Arabian Ocean. How far north into the Sahara Desert these rain clouds penetrate determines whether crops, livestock, and people live or die in the desert margin (Fig. 8.12). In the winter, the desert cools and air flows back out over the now warmer Atlantic. These shifting winds allowed Arabian sailors, like Sinbad, to sail from Africa to India in the summer and return in the winter.

There seems to be a connection between strong monsoon winds over West Africa and the frequency of killer hurricanes in

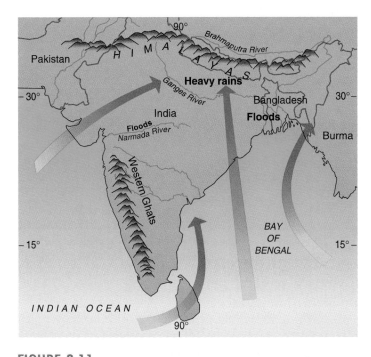

FIGURE 8.11 Summer monsoon air flows over the Indian subcontinent. Warming air rises over the plains of central India in the summer, creating a low-pressure cell that draws in warm, wet oceanic air. As this moist air rises over the Western Ghats or the Himalayas, it cools and heavy rains result. These monsoon rains flood the great rivers, bringing water for agriculture, but also causing much suffering.

FIGURE 8.12 Failure of monsoon rains brings drought, starvation, and death to both livestock and people in the Sahel desert margin of Africa. Although drought is a fact of life in Africa, many governments fail to plan for it, and human suffering is much worse than it needs to be.

the Caribbean and along the Atlantic coast of North and Central America. In years when the Atlantic surface temperatures are high, a stronger than average monsoon trough forms over Africa. This pulls in moist maritime air that brings rain (and life) to the Sahel (the margin of the Sahara). This trough gives rise to tropical depressions (low-pressure storms) that follow one another in regular waves across the Atlantic. The weak trade winds associated with "wet" years allow these storms to organize and gain strength. Evapouration from the warm surface water provides energy.

During the 1970s and 1980s when the Sahel had devastating droughts, the weather was relatively quiet in North America. Only one killer hurricane (winds over 70 km/hr) or reached the United States in two decades. By contrast, in 1995 when ocean surface temperatures were high and rain returned to the Sahel, seven hurricanes and six named tropical storms were spawned in the Atlantic. During the summer of 1996, satellite images of the Atlantic showed five active hurricanes roaring across the Caribbean at the same time. Are we witnessing a long-term climatic change or merely a temporary aberration?

Weather Modification

As author Samuel Clemens (Mark Twain) said, "Everybody talks about the weather, but nobody does anything about it." People probably always have tried to influence local weather through religious ceremonies, dancing, or sacrifices. During the drought of the 1930s in the North American Prairies, "rainmakers" fleeced desperate farmers of thousands of dollars with claims that they could bring rain.

Some recent developments appear to be effective in local weather modification, at least in some circumstances, but they are not without drawbacks and controversy. Seeding clouds with dry ice or ionized particles, such as silver iodide crystals, can initiate precipitation if water vapour is present and air temperatures are near the condensation point (Chapter 10). Dry ice also is effective at dispersing cold fog (where supercooled water droplets are present). Warm fog (air temperatures above freezing) and ice fog (ice crystals in the air) are not usually amenable to weather modification. Hail suppression by cloud seeding also can be effective, but dissipation of the clouds that generate hail diverts rain from areas that need it, as well. There are concerns that materials used in cloud seeding could cause air, ground, and water pollution.

CLIMATE

If weather is a description of physical conditions in the atmosphere (humidity, temperature, pressure, wind, and precipitation), then climate is the *pattern* of weather in a region over long time periods. The interactions of atmospheric systems are so complex that climatic conditions are never exactly the same at any given location from one time to the next. While it is possible to discern patterns of

average conditions over a season, year, decade, or century, complex fluctuations and cycles within cycles make generalizations doubtful and forecasting difficult (Fig. 8.13). We always wonder whether

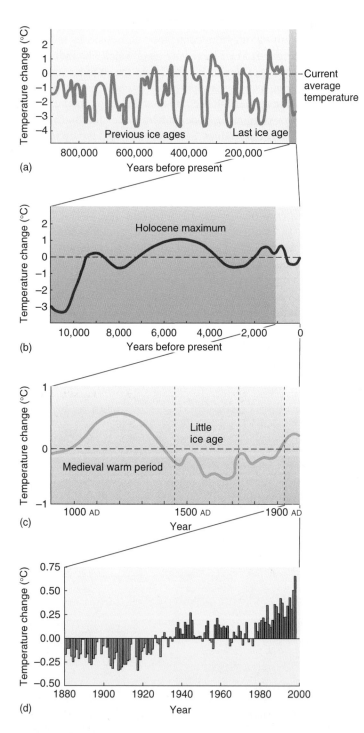

FIGURE 8.13 Global temperature variations over four time periods: (*a*) the past 900,000 years, (*b*) the past 10,000 years, (*c*) the past 1,000 years, and (*d*) the most recent 120 years. The dotted lines in each represent the mean temperature for the twentieth century. Note that both horizontal and vertical scales change in each figure.
Source: (*d*) Data from National Climate Data Center/NESDIS/NOAA, 1998.

anomalies in local weather patterns represent normal variations, a unique abnormality, or the beginnings of a shift to a new regime.

Climatic Catastrophes

Major climatic changes, such as those of the Ice Ages, have drastic effects on living organisms. When climatic change is gradual, species may have time to adapt or migrate to more suitable locations. Where climatic change is relatively abrupt, many organisms are unable to respond before conditions exceed their tolerance limits. Whole communities may be destroyed, and if the climatic change is widespread, many species may become extinct.

Perhaps the most well-studied example of this phenomenon is the great die-off that occurred about 65 million years ago at the end of the Cretaceous period. Most dinosaurs—along with 75 percent of all previously existing plant and animal species—became extinct, apparently as a result of sudden cooling of the earth's climate. Geologic evidence suggests that this catastrophe was not an isolated event. There appear to have been several great climatic changes, perhaps as many as a dozen, in which large numbers of species were exterminated (see Table 16.3, p. 344).

Driving Forces and Patterns in Climatic Changes

What causes catastrophic climatic changes? Some scientists believe that long-term climatic changes follow a purely random pattern brought about by chance interaction of unrelated events, such as asteroid impacts, cosmic radiation from exploding supernovas, massive volcanic eruptions, abrupt flooding of glacier meltwater into the ocean, and tectonic ocean spreading that changes patterns of ocean and wind circulation.

Other scientists find periodic patterns in weather cycles. One explanation is that changes in solar energy associated with 11-year sunspot cycles or 22-year solar magnetic cycles might play a role. Another theory is that a regular 18.6-year cycle of shifts in the angle at which our moon orbits the earth alters tides and atmospheric circulation in a way that affects climate. A theory that has received attention in recent years suggests that orbital variations as the earth rotates around the sun might be responsible for cyclic weather changes.

Milankovitch cycles, named after Serbian scientist Milutin Milankovitch, who first described them in the 1920s, are periodic shifts in the earth's orbit and tilt (Fig. 8.14). The earth's elliptical orbit stretches and shortens in a 100,000-year cycle, while the axis of rotation changes its angle of tilt in a 40,000-year cycle. Furthermore, over a 26,000-year period, the axis wobbles like an out-of-balance spinning top. These variations change the distribution and intensity of sunlight reaching the earth's surface and, consequently, global climate. Bands of sedimentary rock laid in the oceans (Fig. 8.15) seem to match both these Milankovitch cycles and the periodic cold spells associated with worldwide expansion of glaciers every 100,000 years or so.

A historical precedent for global climate change that had disastrous effects on humans was the "little ice age" that began in the

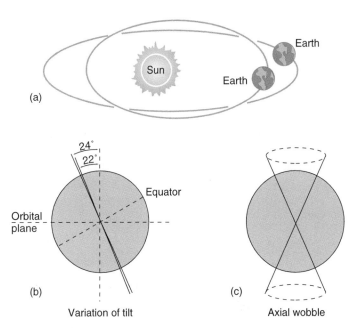

FIGURE 8.14 Milankovitch cycles, which may affect long-term climate conditions: (*a*) changes in the occupancy of the earth's orbit, (*b*) shifting tilt of the axis, and (*c*) wobble of the earth.

FIGURE 8.15 Milankovitch cycle periodicity? The light and dark bands in these seafloor sediments were laid down under different climatic regimes, each lasting 20,000 to 40,000 years. Now exposed along the French coast, these layers suggest the regularity of Milankovitch cycles.

1400s. Temperatures dropped so that crops failed repeatedly in parts of northern Europe that once were good farmland. Scandinavian settlements in Greenland founded during the warmer period around A.D. 1000 lost contact with Iceland and Europe as ice blocked shipping lanes. It became too cold to grow crops, and fish that once migrated along the coast stayed farther south. The settlers slowly died out, perhaps in battles with Inuit people who were driven south from the high Arctic by colder weather.

Evidence from ice cores drilled in the Greenland ice cap suggests that world climate may change much more rapidly than previously thought. During the last major interglacial period 135,000 to 115,000 years ago, it appears that temperatures flipped suddenly from warm to cold or vice versa over a period of years or decades rather than centuries. One possible explanation is that ocean currents acting as a conveyor belt to transport heat from the equator to the North Atlantic may have suddenly stopped or even reversed course when surges of fresh water diluted salty ocean waters. Volcanic eruptions may have played a role as well. When Mt. Pinatubo in the Philippines erupted in 1991, it ejected enough dust, aerosols, and gas into the atmosphere to cool the climate by at least 1°C. Larger volcanoes may have had greater effects.

El Niño/Southern Oscillations

What do sea surface temperatures near Indonesia, anchovy fishing off the coast of Peru, and the direction of tropical trade winds have to do with rainfall and temperatures over Canada? Strangely enough, they all may be interrelated. If the terms El Niño, La Niña, and Southern Oscillation are not yet a part of your vocabulary, perhaps they should be. They describe a connection between the ocean and atmosphere that appears to affect these factors as well as weather patterns throughout the world.

The core of this climatic system is a huge pool of warm surface water in the Pacific Ocean that sloshes slowly back and forth between Indonesia and South America like water in a giant bathtub. Most years, this pool is held in the western Pacific by steady equatorial trade winds pushing ocean surface currents westward (Fig. 8.16). These surface winds are generated by a huge low-pressure cell formed by upwelling convection currents of moist air warmed by the ocean. Towering thunderheads created by rising air bring torrential summer rains to the tropical rainforests of Northern Australia and Southeast Asia. Winds high in the troposphere carry a return flow back to the eastern Pacific where dry subsiding currents create deserts from Chile to southern California. Surface waters driven westward by the trade winds are replaced by upwelling of cold, nutrient-rich, deep waters off the west coast of South America that support dense schools of anchovies and other finfish.

Every three to five years, for reasons that we don't fully understand, the Indonesian low collapses and the mass of warm surface water surges back east across the Pacific. One theory is that the high cirrus clouds atop the cloud columns absorb enough solar radiation to cool the ocean surface and reverse trade winds and ocean surface currents so they flow eastward rather than westward. Another theory is that eastward-flowing deep currents called baroclinic waves periodically interfere with coastal upwelling, warming the sea surface off South America and eliminating the temperature gradient across the Pacific. At any rate, the shift in position of the tropical depression sets off a chain of events lasting a year or more with repercussions in weather systems across North and South America and perhaps around the world.

Peruvian fishermen were the first to notice irregular cycles of rising ocean temperatures that resulted in disappearance of the

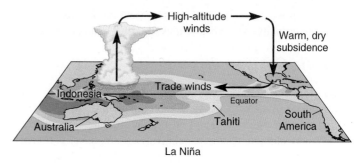

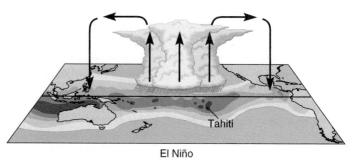

FIGURE 8.16 The El Niño/La Niña Southern Oscillation Cycle. During El Niño years, surface trade winds that normally push warm water westward toward Indonesia weaken and allow this pool of water to flow eastward, bringing storms to the Americas.

anchovy schools on which they depended. They named these events **El Niño** (Spanish for the Christ child) because they often occur around Christmas time. We have come to call the intervening years La Niña (or little girl). Together, this cycle is called the El Niño Southern Oscillation (ENSO).

Actually, sea surface temperature measurements in the central Pacific near Tahiti are the best measurement of this giant oscillation. In May 1998, toward the end of an El Niño event, surface temperatures in this region plunged 7°C in a little over a week, a dramatic event considering the enormous heat-storing capacity of ocean waters.

How does the ENSO cycle affect us? During an El Niño year, the northern jet stream—which normally is over Canada—splits and is drawn south over the United States. This pulls moist air from the Pacific and Gulf of Mexico inland, bringing intense storms and heavy rains from California across midwestern North America. The intervening La Niña years bring hot, dry weather to these same areas. An unusually long El Niño event from 1991 to 1995 broke the seven-year drought over western North America and resulted in floods of the century in the Mississippi Valley and Red River Valley. Oregon, Washington, and British Columbia, on the other hand, tend to have warm, sunny weather in El Niño years rather than their usual rain. Droughts in Australia and Indonesia during El Niño episodes cause disastrous crop failures and forest fires, including one in Borneo in 1983 that burned 3.3 million ha. An even stronger El Niño in 1997 spread health-threatening forest fire smoke over much of Southeast Asia.

Are ENSO events becoming stronger or more irregular because of global climate change? Studies of corals up to 130,000 years old suggest this is true. Furthermore, there are signs that warm ocean surface temperatures are spreading. In addition to the pool of warm water in the western Pacific associated with La Niña years, oceanographers recently discovered a similar warm region in the Indian Ocean. High sea surface temperatures spawn larger and more violent storms such as hurricanes and typhoons. On the other hand, increased cloud cover would raise the albedo while upwelling convection currents generated by these storms could pump heat into the stratosphere. This might have an overall cooling effect and act as a safety valve for global warming.

A longer-scale ocean/weather connection called the **Pacific Decadal Oscillation (PDO)** involves a very large pool of warm water that moves back and forth across the North Pacific every 30 years or so. From about 1977 to 1997, surface water temperatures in the middle and western part of the North Pacific Ocean were cooler than average, while waters off the western United States were warmer. During this time, salmon runs in northern B.C. and Alaska were bountiful, while those in southern B.C., Washington, and Oregon were greatly diminished. In 1997, however, ocean surface temperatures along the coast of western North America turned significantly cooler, perhaps marking a return to conditions that prevailed between 1947 and 1977. Under this cooler regime, northern salmon runs declined while those in the south improved somewhat. This decades-long oscillation can be seen in rainfall patterns shown by tree-ring growth.

Human-Caused Global Climate Change

As the opening story in this chapter shows, evidence is accumulating rapidly that anthropogenic (human-caused) climate modification is already beginning to occur. In its most recent report, the IPCC raised estimates for average global temperature increases over this century to 1.4 to 5.8°C. This is half again the projections made only five years earlier. The difference lies in more rapid than expected energy-consumption growth and failure of international negotiations to control greenhouse gas emissions. Although a change of a few degrees Celsius may not seem like much, remember that the difference between the average world temperature now and that during the last ice age 15,000 years ago is thought to have been only about 5°C.

Globally, the twentieth century was the warmest since thermometres were invented, and tree-ring data suggest that every year of the 1990s was among the 15 hottest of the past millennium. Historical records from North America, Europe, and Asia show that northern lakes and rivers now freeze about a week later in the fall and melt a week earlier in the spring than 150 years ago. Night temperatures generally have increased more than day temperatures as greenhouse gases prevent heat from escaping at night but don't affect incoming solar radiation during the day. While the average global temperature appears to have increased about 0.8°C over the past century, the increases at higher latitudes have been greater. Above 60°N the increase was about double that for the rest of the world. In 2000, Point Barrow, Alaska, had its first thunderstorm in recorded history. If these temperature increases continue, central Canada could be more like central Illinois in a few decades, and Georgia might be more like Guatemala.

Precipitation rates also have increased recently, especially at high latitudes during the winter. Northern regions of North America and Eurasia were about 20 percent wetter in the 1990s than in the previous 50 years (Fig. 8.17). Floods and storms that once occurred only every 100 to 500 years now may happen two or three times in a decade. It is difficult to distinguish normal weather variations from new patterns caused by human activities, but some

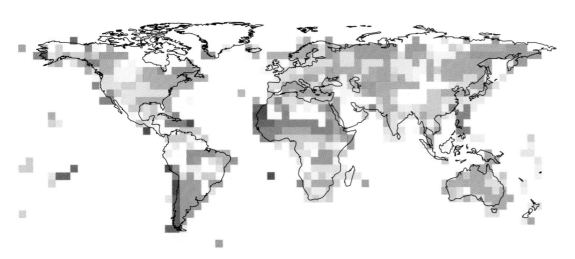

Approximate Change in Precipitation

20%   0%   20%

Decrease   Increase

FIGURE 8.17 Precipitation records for the twentieth century show more precipitation at high latitudes and decreases in equatorial zones. Green indicates more rain; brown less.
© Laurie Grace

oceanographers and climatologists suggest that the unusual frequency and length of ENSO events in recent years, together with more severe storms and climatic extremes, may signal a shift to a new climate regime in a warmer world.

In an April 2001 cover story, *Time* magazine said, "Except for nuclear war or a collision with an asteroid, no force has more potential to damage our planet's web of life than global warming." This statement shows how the issue has penetrated into the public consciousness. It may well be *the* most serious topic in environmental science. In this section, we'll look at some more evidence for, and possible consequences of, anthropogenic climate change together with policy options for reducing these impacts.

Greenhouse Gases

The first evidence for anthropogenic atmospheric changes resulted from the International Geophysical Year in 1957. As part of that study, an observatory was established on top of Mauna Loa volcano in Hawaii to study air chemistry in a remote environment assumed to be relatively pristine and unaffected by human activities. To everyone's surprise, however, CO_2 measurements showed two striking trends. The first trend is a strong annual CO_2 cycle, even in the middle of the ocean, as plants take up carbon during the summer and release it in the winter (Fig. 8.18). Because a majority of the world's land and vegetation are in the Northern Hemisphere, northern seasons dominate this signal.

The other trend in the Mauna Loa data is a steady CO_2 increase from about 315 ppm by volume in 1958 to 370 ppm in 2000. Measurements from other long-term observatories at Barrow, AK, American Somoa, and the South Pole along with 45 other monitoring sites around the world, and ships at sea confirm the Mauna Loa data.

The extra CO_2 added to the atmosphere each year is thought to come primarily from human actions. Burning fossil fuels, making cement, burning biomass, and other activities release nearly 30 billion tonnes of CO_2 every year, on average, containing some 8 billion tonnes of carbon (Fig. 8.19). About 3 billion tonnes of excess carbon is taken up by terrestrial ecosystems, and another 2 billion is absorbed by the oceans, leaving an annual atmospheric increase of about 3 billion tonnes per year in the atmosphere. The result is a 0.5 percent annual increase in atmospheric CO_2 concentrations. If current trends continue, CO_2 concentrations could reach about 500 ppm (approaching twice the preindustrial level of 280 ppm) by the end of this century. Because CO_2 is a **greenhouse gas,** one that absorbs infrared radiation and warms surface air, this change could increase mean global temperatures significantly.

Carbon dioxide is not the only anthropogenic gas causing global warming (Fig. 8.20). Methane (CH_4), chlorofluorocarbons (CFCs), nitrous oxide (N_2O), and sulphur hexafluoride also absorb

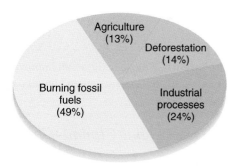

FIGURE 8.19 Contributions to global warming by different types of human activities.
Source: Data from World Resources Institute.

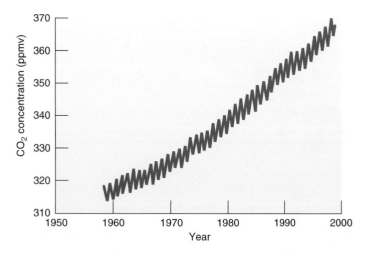

FIGURE 8.18 Carbon dioxide concentrations at Mauna Loa Observatory in Hawaii have increased dramatically since 1958. Annual fluctuations reflect differences in photosynthesis and respiration between winter and summer.
Source: Data from Dave Keeling and Tim Whorf, Scripps Institute of Oceanography.

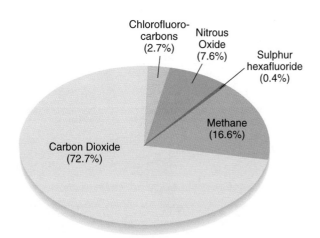

FIGURE 8.20 Relative contribution to global warming (percent of expected climate change) by human-caused releases of greenhouse gases over the next 100 years.
Source: NOAA Climate Change and Diagnostics Laboratory, 2001.

infrared radiation and warm the atmosphere. Although rarer than CO_2, some of these gases trap heat more effectively. Methane, for instance, absorbs 20 to 30 times as much infrared and is accumulating in the atmosphere about twice as fast as CO_2. Methane is released by ruminant animals, wet-rice paddies, coal mines, landfills, and pipeline leaks. CFC releases have declined in recent years as substitutes have replaced them in many uses. The CFCs already in the atmosphere will persist for many decades or centuries, however. Nitrous oxide is produced by burning organic material and soil denitrification. Sulphur hexafluoride is used in electrical insulation, magnesium production, and in medical inhalers.

Aerosol Effects

Aerosols have a tendency to reflect sunlight and cool surface air temperatures. Locally, aerosols can more than offset warming caused by greenhouse gases, but aerosols tend to be short-lived in the air and their effects are temporary. Many industrial cities have experienced a distinct cooling trend in this century. Both anthropogenic and natural aerosols can be important in our climate. The 1991 eruption of Mt. Pinatubo in the Philippines ejected enough ash and sulfate particles to cool the global climate about 1°C for nearly a year. Scientists calculate that sulfate aerosols reflect about 15 percent of the spring sunshine over India. Ironically, cleaning the air could increase warming.

Effects of Climate Change

As mentioned at the beginning of this chapter, the world climate seems already to be changing as a result of human actions. Arctic sea ice is 40 percent thinner now and the edge of the Arctic sea ice pack averages about 500 km further north than it was a century ago. Arctic wildlife, including polar bears, walruses, beluga whales, Perry caribou, and musk oxen, are declining or changing their migration and feeding patterns as a result of changing climatic conditions. In the south, the West Antarctic Ice Sheet has shrunk 25 percent over the past 25 years, including several giant icebergs—the largest about the size of Prince Edward Island—that have calved off in recent years. Here too, wildlife is suffering from loss of habitat and diminishing food supplies. Adele penguins, for instance, have declined precipitously across the Antarctic as the ice shelves on which they rest and lay their eggs have disappeared. If the entire West Antarctic Ice Sheet were to break off and melt, enough water would be released to raise sea levels five to six metres.

Alpine glaciers everywhere are retreating rapidly (Fig. 8.21). In 1972, Venezuela had six glaciers; now it has only two. The shrinkage of the Athabasca Glacier, a product of the Columbia Icefield in Jasper National Park (Alberta), is obvious from historical cairns visible from a nearby highway. Professor Mauri Pelto of the North Cascades Glacier project reports that all 107 glaciers he has been monitoring are shrinking and four have melted completely since 1984. Over the past century, glaciers in the Cascades appear to have shrunk about 30 percent. This is a serious threat to human populations in the area because glaciers provide about 760 billion litres of runoff each year. Without this resource, agriculture,

FIGURE 8.21 Alpine glaciers on every continent are retreating rapidly. Is this an indication of human-caused global climate change or merely a delayed effect of the end of the little ice age?

industry, power generation, and urban drinking water supplies will suffer. Over the next 50 years, at least half of all alpine glaciers in the world could disappear. About 1.7 billion people now live in areas where water supplies are tight. This number could increase to 5.4 billion by 2025.

Many wild plant and animal species are likely to be forced out of their current ranges as the climate warms (Fig. 8.22). Given enough time and a route for migration, these species might adapt to new conditions, but we now are forcing them to move at least 10 times the rate many achieved at the end of the last ice age. Biogeographical changes already are being observed. In Europe and North America, 57 butterfly species have either died out at the southern end of their range, or extended their northern limits, or both. The disappearance of amphibians such as the beautiful golden toads from the cloud forests of Costa Rica or western toads from Oregon's Cascade Range are thought to have been caused at least in part by changing weather patterns.

Around the world coral reefs are bleaching because of higher water temperatures. Rivalling tropical rainforests in their biological diversity and net productivity, coral reefs already are threatened by human activities such as dynamite fishing and limestone excavation. If sea temperatures continue to rise, most coral reefs will be wiped out in the next 50 years. Similarly, the Cape Floral Kingdom of South Africa, one of the most unique terrestrial ecosystems in the world, could disappear due to changing weather patterns.

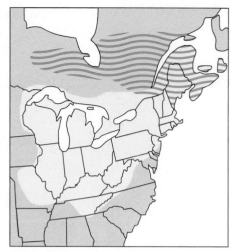

FIGURE 8.22 Change in suitable range for sugar maple, according to the climate-change model.
Source: Data from Margaret Davis and Catherine Zabriskie in *Global Warming and Biological Diversity,* ed. by Peters and Lovejoy, 1991, Yale University Press.

Winners and Losers

Local climate changes could well have severe effects on human societies, agriculture, and natural ecosystems. In many cases, both organisms and human infrastructure will not be able to move or adapt quickly enough to accommodate the rates at which climate seems to be changing. Tropical areas will probably not get much warmer than they are now, but middle and high latitudes could experience unaccustomed heat.

Does anyone win in these scenarios? Residents of northern Canada, Siberia, and Alaska probably would enjoy warmer temperatures, longer growing seasons, and longer ice-free shipping seasons from their ports. This may seem like a positive change, particularly considering that the circumpolar north is where the most significant warming has happened and will happen in the future. But warmer weather is not necessarily good news for northern people (see Want More? p.150). Stronger monsoons bringing more moisture to parts of Central Africa and South Asia could increase crop yields. But soils in some areas may not be suitable for crops no matter how agreeable the climate. Carbon dioxide is a fertilizer for plants, and higher CO_2 levels bring faster growth with less water for many plant species. Sherwood Idso of the U.S. Department of Agriculture says that increased CO_2 concentrations could trigger lush plant growth that would make crops abundant.

In studies at the Duke Forest in North Carolina, the first three years of doubled CO_2 levels produced 50 percent more photosynthesis and a 25 percent increase in net primary productivity. In the fourth year, however, growth slowed, perhaps indicating that trees are down-regulating (adapting) to the enriched atmosphere or depleting soil nutrients. Furthermore, some critics warn, even if CO_2 enrichment does produce greater harvests, protein levels in crops may decrease making them less nutritious. And while desirable plants may grow faster, weeds might flourish even

more. The greatest response in the Duke Forest experiment was poison ivy, which increased its growth rate by 70 percent with higher CO_2.

Enough water is stored in glacial ice caps in Greenland and Antarctica to raise sea levels around 100 m if they melt. About one-third of the world's population now live in areas that would be flooded if that happens. Even the 45 cm sea level rise expected by 2050 will flood much of south Florida, low-lying coastal areas of Bangladesh, Pakistan, and the Nile Delta of Egypt. Most of the world's largest urban areas are on coastlines. Wealthy cities such as New York or London can probably afford to build dikes to keep out rising seas, but poorer cities such as Jakarta, Calcutta, or Manila might simply be abandoned as residents flee to higher ground. Several small island countries such as the Maldives, the Bahamas, Tuvalu, Kiribati, and the Marshall Islands could disappear completely if sea levels rise a metre or more. The Alliance of Small Island Developing States, a coalition of about 40 countries faced with loss of some or all of their territory to rising seas, has been among the most urgent voices calling for international actions to combat climate change.

Warmer sea surface temperatures result in more water evapouration and more severe storms. In 1999, for example, the Atlantic Ocean, which usually has only one or two severe hurricanes per year, had five storms that reached category four (winds 200–275 km/hr) including Lenny on November 13–20, the latest severe storm ever recorded for the Caribbean. The ENSO events described earlier in this chapter seem to occur with greater frequency and severity now than they did in the past. Insurance companies worry that the $2 trillion in insured property along U.S. coastlines could be at risk from a combination of high seas and catastrophic storms. Some 87,000 homes in the United States within 150 m of a shoreline are in danger of coastal erosion or flooding in the next 50 years. Accountants warn that loss of land and structures to flooding and coastal erosion together with damage to fishing stocks, agriculture, and water supplies could raise worldwide insurance claims from about $40 billion in 2000 (already 10 times more than they had been 50 years earlier) to more than $300 billion in 2050. Some of this increase in insurance claims is that more people are living in more dangerous places, but extra severe storms only exacerbate this problem.

Infectious diseases are likely to increase as the insects, rodents, and ticks that carry them spread to new areas. Already we have seen diseases such as malaria, dengue fever, and encephalitis appear in parts of North America where they had never been reported before. The spread of diseases is likely to be compounded by the movement of hundreds of millions of environmental refugees and greater crowding, as people are forced out of areas made uninhabitable by rising seas and changing climates.

Permafrost is already beginning to melt in the Arctic, causing roads to buckle and houses to sink into the mud. An ominous possibility from this melting might be the release of vast stores of methane hydrate now locked in frozen ground and in sediments in the ocean floor. Together with the increased oxidation of high-latitude peat lands potentially caused by warmer and dryer conditions, release of these carbon stores could add as much CO_2 to the atmosphere as all the fossil fuels ever burned. We could trigger a disastrous positive feedback loop in which the effects of warming cause even more warming.

Another potential outcome of global climate change is that greater runoff from increased rainfall might suddenly change ocean circulation patterns that now moderate northern climates (see Fig. 10.6). Furthermore, increased ocean evapouration might intensify snowfall at high latitudes so that glaciers and snow pack would increase rather than decrease. Ironically, the increased albedo (reflectivity) of colder, snow-covered surfaces might trigger a new ice age. On the other hand, perhaps the increased energy pumped into the stratosphere by ENSO-triggered tropical storms together with greater snowfall in the Arctic and Antarctic may just balance the effects of our greenhouse gas emissions. We seem to be taking a gigantic gamble with our planet and its regulatory systems.

Climate Skeptics

While an increasing number of scientists endorse the IPCC report, some critics claim we don't know enough yet to make predictions about what may be happening to our global climate. While much of the dissenting views come from industry lobbying groups, some prominent scientists also argue that dire predictions about climate change may be overblown. John Cristy of the University of Alabama, for example, says that the upper troposphere isn't warming as fast as the earth's surface, perhaps showing that our current ideas about global climate are incorrect. Other climatologists agree that the upper troposphere is cooler than we might expect, but caution that there may be some unexplained lag between warming of different parts of the atmosphere. Another skeptic is Richard Lindzen of MIT, who maintains that current computer models don't adequately account for cloud effects. Higher evapouration rates, he believes, could produce more cumulus clouds that reflect sunlight and balance the greenhouse effect. And Vincent Gray, a climate consultant from New Zealand, contends that CO_2 concentrations measured at Mauna Loa have increased at a remarkably constant rate over the past 42 years, despite the fact that worldwide fossil fuel consumption has increased almost 50 percent during that time. Something is wrong with our models, he claims.

Patrick Michaels of the University of Virginia, who edits a publication for the Western Fuels Association, points out that weather always fluctuates unpredictably at the local scale. Severe storms always catch our attention, he asserts, but may not represent a significant pattern. Furthermore, some glaciologists report, alpine glaciers have been retreating for at least 150 years. This process apparently began long before any significant human impacts on the atmosphere. Current glacial melting, they suggest, may be just the continuing rebound from the end of the little ice age. In fact, the last 8,000 years during which human civilizations have flourished appear to have been unusually warm and stable. This is especially true of the past 150 years. Perhaps it's time for a new ice age. If so, we might be grateful for the warming provided by a little extra CO_2. Although there is no doubt that

both global and local climates can change suddenly and dramatically, critics argue, it may be too early for us to be able to draw conclusions about why this occurs, or to make predictions about what may happen next.

International Climate Negotiations

One of the centrepieces of the 1992 United Nations "Earth Summit" meeting in Rio de Janeiro was the Framework Convention on Climate Change, which set an objective of stabilizing greenhouse gas emissions to reduce the threats of global warming. At a follow-up conference in Kyoto, Japan in 1997, 160 nations agreed to roll back CO_2, methane, and nitrous oxide emissions about 5 percent below 1990 levels by 2012. Three other greenhouse gases, hydrofluorocarbons, perfluorocarbons, and sulphur hexafluoride, will also be reduced, although from what level was not decided. Known as the **Kyoto Protocol,** this treaty sets different limits for individual nations depending on their output before 1990. Poorer nations like China and India were exempted from emission limits to allow development to increase their standard of living. Wealthy countries created the problem, the poorer nations argue; and the wealthy should deal with it.

In an unusual display of international accord, 178 nations meeting in Bonn, Germany in July 2001 agreed on rules to clear the way for ratification of the Kyoto Protocol. Canada ratified the protocol in December 2002. The United States refused to sign because of concerns that developing nations won't be subjected to the same standards as industrialized countries. Other countries countered that the United States, with less than 5 percent of the world's population, now produces more than a quarter of all greenhouse gases. The United States produces about 20 tonnes of CO_2 per person per year, for instance, while Japan and most of Europe produce about half that amount, and China and India release less than one tonne per person (Fig. 8.23). The world's richest economy, other nations argue, can afford to make some genuine cuts.

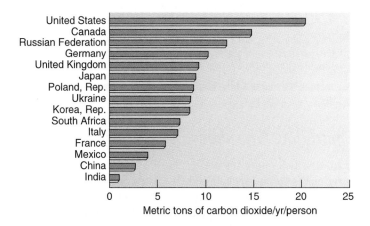

FIGURE 8.23 Annual per capita CO_2 emissions for 15 countries with highest industrial output. Although China is fourteenth in per capita emissions, it is second to the United States in total emissions due to its large population.

Source: Data from *World Resources 1998–99,* World Resources Institute.

what can you do?

Reducing Carbon Dioxide Emissions

Individuals can help reduce global warming. Although some actions may cause only a small impact, collectively they add up. Many of these options save money in the long run and have other environmental benefits such as reducing air pollution and resource consumption. Try the One Tonne Challenge (see Web Exercise) to see how well you can do!

1. Drive less, walk, bike, take public transportation, car-pool, or buy a vehicle that gets at least 12.6 km/l. Average annual CO_2 reduction: about 9 kgs for each litre of gasoline saved. Increasing your mileage by 16 km/l, for example, eliminates about 1.1 tonnes of CO_2 per year. Becoming entirely self-propelled would save, on average, about 5.45 tonnes per year.

2. Plant trees to shade your house during the summer, and paint your house a light colour if you live in a warm climate or a dark colour in a cold climate. Average annual CO_2 reduction: about 2.2 tonnes.

3. Insulate your house and seal all drafts. Average annual CO_2 reduction for the highest efficiency insulation, weatherstripping, and windows: about 2.2 tonnes.

4. Replace old appliances with new, energy-efficient models. Average annual CO_2 reduction for the most efficient refrigerator, for instance: about 1.4 tonnes.

5. Produce less waste. Buy minimally packaged goods and reusable products. Recycle. Average annual CO_2 reduction: about 0.45 metric tonnes for 25 percent less garbage.

6. Turn down your thermostat in the winter and turn it up in the summer. Average annual CO_2 reduction: about 0.23 tonnes for every 1°C change.

7. Replace standard light bulbs with long-lasting compact fluorescent ones. Average annual CO_2 reduction: about 0.23 tonnes for every bulb.

8. Wash laundry in warm or cold water, not hot. Average annual CO_2 reduction: about 0.23 tonnes for two loads per week.

9. Set your water heater thermostat no higher than 48.9°C. Average annual CO_2 reduction: about 0.23 tonnes for each 5°C temperature change.

10. Buy renewable energy from your local utility if possible. Potential annual CO_2 reduction: about 13.5 tonnes.

Controlling Greenhouse Emissions

In spite of U.S. intransigence, progress already is being made in many places toward reducing greenhouse emissions. The United Kingdom, for example, had already rolled CO_2 emissions back to 1990 levels by 2000, and vowed to reduce them 20 percent more within a decade. Britain already has started to substitute natural gas for coal, promote energy efficiency in homes and industry, and raise its already high gasoline tax. Plans are to "decarbonize" British society and to decouple GNP growth from CO_2 emissions. A revenue-neutral carbon levy is expected to lower CO_2 releases by 5 million tonnes per year by 2010, and at least 10 percent of all new energy in the United Kingdom will be from renewable sources, the government promised.

Germany, also, has reduced its CO_2 emissions at least 10 percent by switching from coal to gas and by encouraging energy efficiency throughout society. Atmospheric scientist Steve Schneider calls this a "no regrets" policy; even if we don't need to stabilize our climate, many of these steps save money, conserve resources, and have other environmental benefits. Nuclear power also is being promoted as an alternative to fossil fuels. It's true that nuclear reactions don't produce greenhouse gases, but security worries and unresolved problems of how to store wastes safely make this option unacceptable to many people.

Renewable energy sources offer a better solution to climate problems, many people believe. Chapter 13 discusses options for conserving energy and switching to renewable sources such as solar, wind, geothermal, biomass, and fuel cells. Denmark, the world's leader in wind power, now gets 20 percent of its electricity from windmills. Plans are to generate half of the nation's electricity from off-shore wind farms by 2030. Belgium, Germany, Ireland, the Netherlands, and Sweden also have announced plans for off-shore wind-energy developments. Even China reduced its CO_2 emissions 17 percent between 1997 and 1999 through greater efficiency in coal burning and industrial energy use. The U.S. energy plan, meanwhile, is to burn more coal, drill for oil in the Arctic National Wildlife Refuge, and build more nuclear plants.

In addition to reducing its output, there are options for capturing and storing CO_2. Planting trees can be effective if they're allowed to mature to old-growth status or are made into products like window frames and doors that will last for many years before they're burned or recycled. Farmland also can serve as a carbon sink if farmers change their crop mixture and practice minimum-till cultivation that keeps carbon in the soil. In 1988, the late John H. Martin suggested that phytoplankton growth in the ocean may be limited by iron deficiency. This theory was tested in 2000, when researchers spread about 3.5 tonnes of dissolved iron over 75 sq km of the South Pacific. Monitored by satellite, a ten-fold rise in chlorophyll concentration eventually spread over about 1,700 sq km and was calculated to have pulled several thousand tonnes of CO_2 out of the air. Still, some ecologists warn about such large-scale tinkering. We don't know whether the carbon fixed by phytoplankton growth will be stored in sediments or simply eaten by predators and returned immediately to the atmosphere. It's possible that dying algae might create an anoxic zone that would devastate oceanic food webs.

Another way to store CO_2 is to inject it into underground strata or deep ocean waters. Since 1996, Norway's Statoil has been pumping more than 1 million tonnes of CO_2 into an aquifer 1,000 m below the seafloor at a North Sea gas well. It is economical to do so because otherwise the company would have to pay a $50 per tonne carbon tax on its emissions. Around the world, deep, briny aquifers could store a century worth of CO_2 output at current fossil fuel consumption rates. It might also be possible to pump liquid CO_2 into deep ocean trenches, where it would form lakes contained by enormous water pressures. There are worries, however, about what this might do to deep ocean fauna and what might happen if earthquakes or landslides caused a sudden release of this CO_2 as happened in 1986 in Lake Nyos in Cameroon. In Canada, an Alberta power plant is injecting CO_2 into a coal seam too deep to be mined. The CO_2 releases natural gas, which is burned to produce electricity and more CO_2. And in Shady Point, Oklahoma, CO_2 captured from flue gas is purified and sold to carbonate beer and soft drinks. Proponents of **carbon management,** as these various projects are called, argue that it may be cheaper to clean up fossil fuel effluents than to switch to renewable energy sources.

Most attention is focused on CO_2 because it lasts in the atmosphere, on average, for about 120 years. Methane and other greenhouse gases are much more powerful infrared absorbers, but remain in the air for a much shorter time. NASA's Dr. James Hansen made a controversial suggestion, however, that the best short-term attack on warming might come by focusing not on carbon dioxide but on methane and soot, which are not mentioned in the Kyoto Protocol but are increasingly thought to contribute both to warming and to serious health problems. Reducing gas pipeline leaks would conserve this valuable resource as well as help the environment. Methane from landfills, oil wells, and coal mines that once would have been simply vented into the air is now being collected and used to generate electricity. Rice paddies are a major methane source. Changing flooding schedules and fertilization techniques can reduce anoxia (oxygen depletion) that produces marsh gas. Finally, ruminant animals (like cows, camels, and buffalo) create large amounts of digestive system gas. Modified diets can reduce flatulence significantly.

Some individual cities have announced their own plans to combat global warming. Among the first of these were Toronto, Copenhagen, and Helsinki, which pledged to reduce CO_2 emissions 20 percent from 1990 levels by 2010. Portland, Oregon, was the first U.S. city to implement a CO_2-reduction strategy. Minneapolis, Denver, and Sacramento also have similar plans. And some corporations are following suit. British Petroleum has set a goal of cutting CO_2 releases from all its facilities by 10 percent before 2010. In 2000, BP, Alcan, DuPont, and other companies joined with the Environmental Defense Fund to launch a partnership for climate action, pledging to meet or exceed Kyoto requirements. Each of us can make a contribution in this effort (see What Can You Do? p. 168).

SUMMARY

Global climate change may well be *the* most momentous issue in environmental science today. To understand why this is happening and what we can do about it, we need to know something about atmospheric processes. Weather is a description of local conditions; climate describes long-term weather patterns.

The atmosphere and living organisms have evolved together so that the present chemical composition of the air is both suitable for, and largely the result of, biological processes. Upper layers of the atmosphere play an important role in protecting the earth's surface by intercepting dangerous ultraviolet radiation from the sun. The atmosphere is relatively transparent to visible light that warms the earth's surface and is captured by photosynthetic organisms and stored as potential energy in organic chemicals.

Heat is lost from the earth's surface as infrared radiation, but fortunately for us, carbon dioxide and water vapour naturally present in the air capture the radiation and keep the atmosphere warmer than it would otherwise be. When air is warmed by conduction or radiation of heat from the earth's surface, it expands and rises, creating convection currents. These vertical updrafts carry water vapour aloft and initiate circulation patterns that redistribute energy and water from areas of surplus to areas of deficit. Pressure gradients created by this circulation drive great air masses around the globe and generate winds that determine both immediate weather and long-term climate.

The earth's rotation causes wind deflection called the Coriolis effect, which makes air masses circulate in spiralling patterns. Strong cyclonic convection currents fuelled by temperature and pressure gradients and latent energy in water vapour can create devastating storms. Tornadoes, while classified as cyclonic storms, are not set spinning by Coriolis forces. Instead, shear forces caused by differential wind speeds, together with rapidly rising warm convection currents and cold downdrafts are thought to create intensely focused spinning vortices. Although top wind speeds in tornadoes can be higher than those in hurricanes, total damage in the former is usually smaller because the area covered is smaller.

Another source of storms are the seasonal winds, or monsoons, generated by temperature differences between the ocean and a landmass. Monsoons often bring torrential rains and disastrous floods, but they also bring needed moisture to farmlands that feed a majority of the world's population. When the rains fail, as they do in drought cycles, ecosystem disruption and human suffering can be severe.

The El Niño/Southern Oscillation is a complex interaction between oceans and atmosphere that has far-reaching climatic, ecological, and social effects. ENSO cycles can affect things as widely different as forest fires in Indonesia, anchovy fishing in Peru, rainfall in the Sahara Desert, and how the corn grows in Iowa. Knowing something about how weather works can be helpful in our everyday life.

Many procedures claiming to control the weather are ineffectual, but some human actions—both deliberate and inadvertent—appear to be changing local weather and long-term climate. Many scientists warn that the gaseous pollutants we release into the atmosphere trap radiant energy and could cause a global warming trend that would drastically disrupt human activities and natural ecosystems. Understanding and protecting this complex, vital aspect of our world is clearly essential.

QUESTIONS FOR REVIEW

1. What are weather and climate? How do they differ?
2. Name and describe the four layers of air in the atmosphere.
3. What is the greenhouse effect?
4. What are the ENSO cycle and the PDO?
5. What are the jet streams? How do they influence weather patterns?
6. How do tornadoes form and why are they so destructive?
7. Describe the Coriolis effect. What causes it?
8. Summarize some of the changes thought to be caused by climate change.
9. Summarize some actions we could take, collectively and as individuals, to combat global climate change.
10. Why has the United States refused to ratify the Kyoto treaty?

QUESTIONS FOR CRITICAL THINKING

1. In most places, weather changes in highly variable, seemingly chaotic ways. How would you distinguish between meaningful patterns and random variations?
2. Should humans try to control the weather? What would be the positive effects? What would be the dangers?
3. Can we avoid great climatic changes, such as another ice age or a greenhouse effect? How can we decide what to do when the science is uncertain and the risk is great?
4. Has there been a major change recently in the weather where you live? Can you propose any reasons for such changes?

5. What forces determine the climate in your locality? Are they the same for neighbouring provinces?
6. What was the weather like when your parents were young? Do you believe the stories they tell you? Why or why not?
7. Have you ever experienced a tornado or hurricane? What was it like? What omens or warnings told you it was coming?
8. What would you do to adapt to a permanent drought? What effects would it have on your life?
9. What should we do about coastal cities threatened by rising oceans—rebuild, enclose in dikes, or just move?
10. Would you favour building nuclear power plants to reduce CO_2 emissions? Why or why not?

KEY TERMS

aerosols   151
albedo   153
carbon management   169
climate   150
cold front   156
convection currents   154
Coriolis effect   155
downbursts   158
El Niño   162
greenhouse gas   164
hurricanes   157
ionosphere   152

jet streams   155
Kyoto Protocol   167
mesosphere   152
Milankovitch cycles   161
monsoon   159
Pacific Decadal Oscillation (PDO)   163
stratosphere   152
thermosphere   152
tornadoes   157
troposphere   152
warm front   156
weather   150

Web Exercises

TRY THE ONE TONNE CHALLENGE!

If you're wondering how your present lifestyle translates into greenhouse gas emissions, and what you can do about it, take the One Tonne Challenge by visiting http://www.climatechange.gc.ca/onetonne/english/index.asp. After answering a series of questions about your home and lifestyle, strategies for reducing your contribution to climate change are presented. Determine how much effect you are now having, and how you might reduce that effect. Were there any surprises in how your lifestyle choices affected your contribution to greenhouse gases?

Chapter 9

Air Pollution

There's so much pollution in the air now that if it weren't for our lungs there'd be no place to put it all.

—Robert Orben—

OBJECTIVES

After studying this chapter, you should be able to:

- describe the major categories and sources of air pollution.
- distinguish between conventional or "criteria" pollutants and unconventional types as well as explain why each is important.
- analyze the origins and dangers of some indoor air pollutants.
- relate why atmospheric temperature inversions occur and how they affect air quality.
- evaluate the dangers of stratospheric ozone depletion and radon in indoor air.
- understand how air pollution damages human health, vegetation, and building materials.
- compare different approaches to air pollution control.
- judge how air quality around the world has improved or degraded in recent years and suggest what we might do about problem areas.

WebQuest

Criteria Air Pollutants, NO_x, sulphur dioxide, carbon monoxide

Above: Burning tropical forests to clear land for crops can create clouds of smoke thousands of kilometres across.

A Plague of Smoke

For several months during the unusually dry El Niño winter of 1997–98, a thick pall of smoke covered much of Southeast Asia. Generated primarily by thousands of forest fires on the Indonesian islands of Kalimantan (Borneo) and Sumatra, the smoke spread over eight countries and 75 million people, covering an area larger than Europe (Fig. 9.1). The air quality in Singapore and the city of Kuala Lumpur, Malaysia, just across the Strait of Malacca from Indonesia, was worse than any industrial region in the world. In towns such as Palembang, Sumatra, and Banjarmasin, Kalimantan, in the heart of the massive conflagration, the air pollution index frequently passed 800, twice the level classified in Canada and the United States as an air quality emergency, hazardous to human health.

An estimated 20 million people were treated for illnesses such as asthma, bronchitis, emphysema, eye irritation, and cardiovascular diseases, while those who couldn't afford medical care went uncounted. The number of excess deaths from this months-long episode is unknown. Unable to see through the thick haze, several boats collided in the busy Straits of Malacca, and a plane crashed on Sumatra, killing 234 passengers. Cancelled airline flights, aborted tourist plans, lost workdays, medical bills, and ruined crops are estimated to have cost countries in the afflicted area several billion dollars. Wildlife suffered as well. In addition to the loss of habitat destroyed by fires, breathing the noxious smoke was as hard on wild species as it was on people. At the Pangkalanbuun Conservation Reserve, weak and disoriented orangutans were found suffering from respiratory diseases much like those of humans.

Indonesia has the second largest expanse of tropical forest and the highest number of endemic species in the world, so fires there are of special concern. The dry season in tropical Southeast Asia has probably always been a time of burning vegetation and smoky skies. Tinder-dry forests are ignited by lightning or fires set by farmers to clear cropland. Generally burning only a hectare or two at a time, traditional shifting cultivators often help preserve plant and animal species by opening up space for early successional forest stages. Globalization and the advent of large, commercial plantations, however, have changed everything.

Although Indonesian government officials blamed the fires on small-scale farmers and indigenous people, environmental groups gathered evidence that most of the burning was caused by large agribusiness conglomerates with close ties to the government and military. Clear-cutting hardwoods for sale abroad and burning what's left to make way for huge oil-palm plantations and fast-growing pulpwood trees, these companies ignore forest protection laws. Altogether, a couple of dozen businesses owned by wealthy entrepreneurs with friends in high places are thought to be responsible for burning some 20,000 km^2 in 1997, or an area about the size of Prince Edward Island.

Conditions in Indonesia in 1997 were extreme, but forest fires and other human activities in many places around the world also cause degraded air quality. Since we all breathe air, it is important to understand how air pollution arises, why it is dangerous, and what we might do about it. Studying this chapter will help you learn about this vital topic.

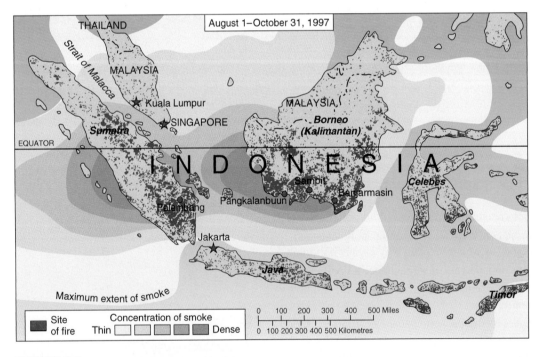

FIGURE 9.1 Thousands of fires (red dots) in Indonesia in 1997 spread a thick cloud of smoke over an area larger than Europe.
Source: Data from NOAA and NASA.

THE AIR AROUND US

How does the air taste, feel, smell, and look in your neighbourhood? Chances are that wherever you live, the air is contaminated to some degree. Smoke, haze, dust, odours, corrosive gases, noise, and toxic compounds are present nearly everywhere, even in the most remote, pristine wilderness. Air pollution is generally the most widespread and obvious kind of environmental damage. According to the Environmental Protection Agency (EPA), some 147 million tonnes of air pollution (not counting carbon dioxide or wind-blown soil) are released into the atmosphere each year in Canada and the United States by human activities (Fig. 9.2). Total worldwide emissions of these pollutants are around 2 billion tonnes per year.

Over the past 20 years, however, air quality has improved appreciably in most cities in Western Europe, North America, and Japan. Many young people might be surprised to learn that a generation ago, most North American cities were much dirtier than they are today. This is an encouraging example of improvement in environmental conditions. Our success in controlling some of the most serious air pollutants gives us hope for similar progress in other environmental problems.

While developed countries have been making progress, however, air quality in the developing world has been getting much worse. Especially in the burgeoning megacities of rapidly industrializing countries (Chapter 21), air pollution often exceeds World Health Organization standards all the time. In Lahore, Pakistan, and Xian, China, for instance, airborne dust, smoke, and dirt often are 10 times higher than levels considered safe for human health.

In this chapter, we will examine the major types and sources of air pollution. We will study how they enter and move through the atmosphere and how they are changed into new forms, concentrated or dispersed, and removed from the air by physical and chemical processes. We also will look at some of the major effects of air pollution on human health, ecosystems, and materials. Finally, we will survey some of the control methods available to reduce air pollution or mitigate its effects, and the results of air pollution control efforts on ambient air quality in Canada and the United States and elsewhere.

NATURAL SOURCES OF AIR POLLUTION

It is difficult to give a simple, comprehensive definition of pollution. The word comes from the Latin *pollutus,* which means made foul, unclean, or dirty. Some authors limit the use of the term to damaging materials that are released into the environment by human

FIGURE 9.2   Toronto, Ontario on a smoggy day (*left*) is very different from days when the air is clear (*right*).

activities. There are, however, many natural sources of air quality degradation. Some of the smoke that inflicted Southeast Asia in the opening story for this chapter came from natural fires. Volcanoes spew out ash, acid mists, hydrogen sulphide, and other toxic gases. Sea spray and decaying vegetation are major sources of reactive sulphur compounds in the air. Trees and bushes emit millions of tonnes of volatile organic compounds (terpenes and isoprenes), creating, for example, the blue haze that gave the Blue Ridge Mountains in southeastern North America their name. Pollen, spores, viruses, bacteria, and other small bits of organic material in the air cause widespread suffering from allergies and airborne infections. Storms in arid regions raise dust clouds that transport millions of tonnes of soil and can be detected half a world away. Bacterial metabolism of decaying vegetation in swamps and of cellulose in the guts of termites and ruminant animals is responsible for as much as two-thirds of the methane (natural gas) in the air.

Does it make a difference whether smoke comes from a natural forest fire or one started by humans? In many cases, the chemical compositions of pollutants from natural and human-related sources are identical, and their effects are inseparable. Sometimes, however, materials in the atmosphere are considered innocuous at naturally occurring levels, but when humans add to these levels, overloading of natural cycles or disruption of essential processes can occur. While the natural sources of suspended particulate material in the air outweigh human sources at least tenfold worldwide, in many cities more than 90 percent of the airborne particulate matter is anthropogenic (human-caused).

FIGURE 9.3 This coke oven is a point-source for a variety of primary pollutants.

HUMAN-CAUSED AIR POLLUTION

What are the major types of air pollutants and where do they come from? In this section, we will define some general categories and sources of air pollution and survey the characteristics and emission levels of Environment Canada's Criteria Air Contaminants. In Chapter 19, we will look more carefully at how Canadian government policy and legislation are affecting air quality in Canada.

Primary and Secondary Pollutants

Primary pollutants are those released directly from the source into the air in a harmful form (Fig. 9.3). **Secondary pollutants,** by contrast, are modified to a hazardous form after they enter the air or are formed by chemical reactions as components of the air mix and interact. Solar radiation often provides the energy for these reactions. Photochemical oxidants and atmospheric acids formed by these mechanisms are probably the most important secondary pollutants in terms of human health and ecosystem damage. We will discuss several important examples of such pollutants in this chapter.

Fugitive emissions are those that do not go through a smokestack. By far the most massive example of this category is dust from soil erosion, strip mining, rock crushing, and building construction (and destruction). In the United States, natural and anthropogenic sources of fugitive dust add up to some 100 million tonnes per year. The amount of CO_2 released by burning fossil fuels and biomass is nearly equal in mass to fugitive dust. Fugitive industrial emissions are also an important source of air pollution. Leaks around valves and pipe joints contribute as much as 90 percent of the hydrocarbons and volatile organic chemicals emitted from oil refineries and chemical plants.

Criteria Air Contaminants

Environment Canada has designated seven major pollutants (sulphur dioxide, carbon monoxide, nitrogen oxide, volatile organic compounds (VOCs), and three categories of particulates) for which maximum **ambient air** (air around us) levels are mandated. These seven **conventional** or **criteria air contaminants** contribute the largest volume of air-quality degradation and also are considered the most serious threat of all air pollutants to human health and welfare. Figure 9.4 shows the major sources of criteria pollutants. Table 9.1 shows an estimate of the total annual worldwide emissions of some important air pollutants both from natural and human sources. Now let's look more closely at the sources and characteristics of each of these major pollutants.

Sulphur Compounds

Natural sources of sulphur in the atmosphere include evaporation of sea spray, erosion of sulphate-containing dust from arid soils,

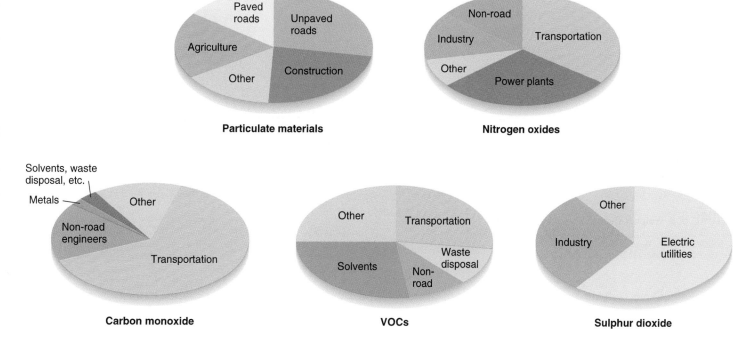

FIGURE 9.4 Anthropogenic sources of the primary "criteria" air pollutants.

Source: Data from Joyce E. Penner, "Atmospheric Chemistry and Air Quality" in W. B. Meyer and B. L. Turner (eds.), *Changes in Land Use and Land Cover: A Global Perspective*, 1994. Cambridge University Press and UNEP 1999.

TABLE 9.1 Estimated Fluxes of Pollutants and Trace Gases to the Atmosphere

| | | APPROXIMATE ANNUAL FLUX | |
| | | NATURAL | ANTHROPOGENIC |
SPECIES	SOURCES	(MILLIONS OF TONNES/YR)	
CO_2 (carbon dioxide)	Respiration, fossil fuel burning, land clearing, industrial processes	370,000	29,600*
CH_4 (methane)	Rice paddies and wetlands, gas drilling, landfills, animals, termites	155	350
CO (carbon monoxide)	Incomplete combustion, CH_4 oxidation, biomass burning, plant metabolism	1,580	930
NMHC (nonmethane hydrocarbons)	Fossil fuels, industrial uses, plant isoprenes and other biogenics	860	92
NO_x (nitrogen oxides)	Fossil fuel burning, lightning, biomass burning, soil microbes	90	140
SO_x (sulphur oxides)	Fossil fuel burning, industry, biomass burning, volcanoes, oceans	35	79
SPM (suspended particulate materials)	Biomass burning, dust, sea salt, biogenic aerosols, gas to particle conversion	583	362

*Only 27.3 percent of this amount—or 8 billion tonnes—is carbon.

Source: Data from Joyce E. Penner, "Atmospheric Chemistry and Air Quality" in W. B. Meyer and B. L. Turner (eds.), *Changes in Land Use and Land Cover: A Global Perspective*, 1994. Cambridge University Press and UNEP 1999.

fumes from volcanoes and fumaroles, and biogenic emissions of hydrogen sulphide (H_2S) and organic sulphur-containing compounds, such as dimethylsulphide, methyl mercaptan, carbon disulphide, and carbonyl sulphide. Total yearly emissions of sulphur from all sources amount to some 114 million tonnes (Fig. 9.5). Worldwide, anthropogenic sources represent about two-thirds of the total sulphur flux, but in most urban areas they contribute as much as 90 percent of the sulphur in the air. The predominant form of anthropogenic sulphur is sulphur dioxide (SO_2) from combustion of sulphur-containing fuel (coal and oil),

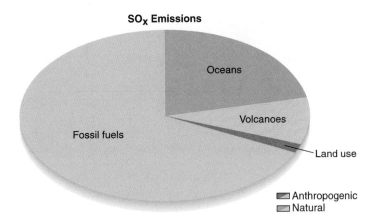

FIGURE 9.5 Sulphur fluxes into the atmosphere.
Source: Data from Joyce E. Penner, "Atmospheric Chemistry and Air Quality" in W. B. Meyer and B. L. Turner (eds.), *Changes in Land Use and Land Cover: A Global Perspective*, 1994. Cambridge University Press and UNEP 1999.

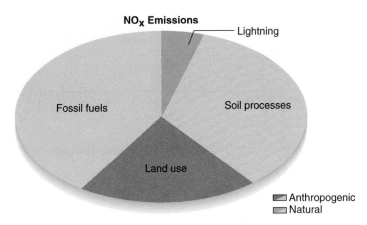

FIGURE 9.6 Worldwide sources of reactive nitrogen gases in the atmosphere.
Source: Data from Joyce E. Penner, "Atmospheric Chemistry and Air Quality" in W. B. Meyer and B. L. Turner (eds.), *Changes in Land Use and Land Cover: A Global Perspective*, 1994. Cambridge University Press and UNEP 1999.

purification of sour (sulphur-containing) natural gas or oil, and industrial processes, such as smelting of sulphide ores. China and the United States are the largest sources of anthropogenic sulphur, primarily from coal burning.

Sulphur dioxide is a colourless corrosive gas that is directly damaging to both plants and animals. Once in the atmosphere, it can be further oxidized to sulphur trioxide (SO_3), which reacts with water vapour or dissolves in water droplets to form sulphuric acid (H_2SO_4), a major component of acid rain. Very small solid particles or liquid droplets can transport the acidic sulphate ion (SO_4^{-2}) long distances through the air or deep into the lungs where it is very damaging. Sulphur dioxide and sulphate ions are probably second only to smoking as causes of air pollution-related health damage. Sulphate particles and droplets reduce visibility in the United States as much as 80 percent.

Nitrogen Compounds

Nitrogen oxides are highly reactive gases formed when nitrogen in fuel or combustion air is heated to temperatures above 650°C in the presence of oxygen, or when bacteria in soil or water oxidize nitrogen-containing compounds. The initial product, nitric oxide (NO), oxidizes further in the atmosphere to nitrogen dioxide (NO_2), a reddish brown gas that gives photochemical smog its distinctive colour. Because of their interconvertibility, the general term NO_x is used to describe these gases. Nitrogen oxides combine with water to make nitric acid (HNO_3), which is also a major component of atmospheric acidification.

The total annual emissions of reactive nitrogen compounds into the air are about 230 million tonnes worldwide (Table 9.1). Anthropogenic sources account for 60 percent of these emissions (Fig. 9.6). More than half of the NO_x generated by human activity in Canada is from air, rail, and highway transportation, while about 10 percent comes from each of oil and gas extraction, electric power generation, and perhaps surprisingly, forest fires. Nitrous oxide (N_2O) is an intermediate in soil denitrification that absorbs ultraviolet light and plays an important role in climate modification

(Chapter 8). Excess nitrogen is causing fertilization and eutrophication of inland waters and coastal seas. It also may be adversely affecting terrestrial plants both by excess fertilization and by encouraging growth of weedy species that crowd out native varieties.

Carbon Oxides

The predominant form of carbon in the air is carbon dioxide (CO_2). It is usually considered nontoxic and innocuous, but increasing atmospheric levels (about 0.5 percent per year) due to human activities appear to be causing a global climate warming that may have disastrous effects on both human and natural communities (see Chapter 8). As Table 9.1 shows, more than 90 percent of the CO_2 emitted each year is from respiration (oxidation of organic compounds by plant and animal cells). These releases are usually balanced, however, by an equal uptake by photosynthesis in green plants.

Anthropogenic (human-caused) CO_2 releases are difficult to quantify because they spread across global scales. The best current estimate from the Intergovernmental Panel on Climate Change (IPCC) is that between 7 and 8 billion tonnes of carbon (in the form of CO_2) are released each year by fossil fuel combustion and that another 1 to 2 billion tonnes are released by forest and grass fires, cement manufacturing, and other human activities. Typically, terrestrial ecosystems take up about 3 billion tonnes of this excess carbon every year, while oceanic processes take up another 2 billion tonnes. This leaves an average of at least 3 billion tonnes to accumulate in the atmosphere. The actual releases and uptakes vary greatly, however, from year to year. Some years almost all anthropogenic CO_2 is reabsorbed; in other years, almost none of it is. The ecological processes that sequester CO_2 depend strongly on temperature, nutrient availability, and other environmental factors.

Carbon monoxide (CO) is a colourless, odourless, nonirritating but highly toxic gas produced by incomplete combustion of fuel (coal, oil, charcoal, or gas), incineration of biomass or solid waste, or partially anaerobic decomposition of organic material.

PART TWO   Our Physical Environment

CO inhibits respiration in animals by binding irreversibly to hemoglobin. About 1 billion tonnes of CO are released to the atmosphere each year, half of that from human activities. In Canada, two-thirds of the CO emissions are created by internal combustion engines in transportation. Land-clearing fires and cooking fires also are major sources. About 90 percent of the CO in the air is consumed in photochemical reactions that produce ozone.

Metals and Halogens

Many toxic metals are mined and used in manufacturing processes or occur as trace elements in fuels, especially coal. These metals are released to the air in the form of metal fumes or suspended particulates by fuel combustion, ore smelting, and disposal of wastes. Worldwide lead emissions amount to about 2 million tonnes per year, or two-thirds of all metallic air pollution. Most of this lead is from leaded gasoline. Lead is a metabolic poison and a neurotoxin that binds to essential enzymes and cellular components and inactivates them. An estimated 20 percent of all inner-city children suffer some degree of mental retardation from high environmental lead levels. Leaded gasoline was phased out in Canada in 1993, and later in the United States in 1996. This has significantly reduced lead levels in the environment and in humans living near high concentrations of automobile traffic.

Mercury is another dangerous neurotoxin that is widespread in the environment. The two largest sources of atmospheric mercury appear to be coal-burning power plants and waste incinerators. Mercuric fungicides in house paint were once a major source of this deadly pollutant but now are restricted. Long-range transport of lead and mercury through the air is causing bioaccumulation in aquatic ecosystems far from the emission sources. It is now dangerous to eat fish from some once-pristine lakes and rivers because of toxic metal contamination.

Other toxic metals of concern are nickel, beryllium, cadmium, thallium, uranium, cesium, and plutonium. Some 780,000 tonnes of arsenic, a highly toxic metalloid, are released from metal smelters, coal combustion, and pesticide use each year. Halogens (fluorine, chlorine, bromine, and iodine) are highly reactive and generally toxic in their elemental form. Chlorofluorocarbons (CFCs) have been banned for most uses in industrialized countries, but about 600 million tonnes of these highly persistent chemical compounds are used annually worldwide in spray propellants, refrigeration compressors, and for foam blowing. They diffuse into the stratosphere where they release chlorine and fluorine atoms that destroy the ozone shield that protects the earth from ultraviolet radiation. We'll return to this topic later in this chapter.

Particulate Material

An **aerosol** is any system of solid particles or liquid droplets suspended in a gaseous medium. For convenience, we generally describe all atmospheric aerosols, whether solid or liquid, as **particulate material.** This includes dust, ash, soot, lint, smoke, pollen, spores, algal cells, and many other suspended materials (Fig. 9.7). Anthropogenic particulate emissions amount to about 362 million tonnes per year worldwide. Wind-blown dust, volcanic

FIGURE 9.7 Smokestacks of this Chinese cement factory are classed as primary or point pollution sources.

ash, and other natural materials may contribute considerably more suspended particulate material.

Particulates often are the most apparent form of air pollution since they reduce visibility and leave dirty deposits on windows, painted surfaces, and textiles. Respirable particles smaller than 2.5 micrometres are among the most dangerous of this group because they can be drawn into the lungs, where they damage respiratory tissues. Asbestos fibres and cigarette smoke are among the most dangerous respirable particles in urban and indoor air because they are carcinogenic.

Volatile Organic Compounds

Volatile organic compounds (VOCs) are organic chemicals that exist as gases in the air. Plants are the largest source of VOCs, releasing an estimated 350 million tonnes of isoprene (C_5H_8) and 450 million tonnes of terpenes ($C_{10}H_{15}$) each year (Fig. 9.8). About 400 million tonnes of methane (CH_4) are produced by natural wetlands and rice paddies and by bacteria in the guts of termites and ruminant animals. These volatile hydrocarbons are generally oxidized to CO and CO_2 in the atmosphere.

In addition to these natural VOCs, a large number of other synthetic organic chemicals, such as benzene, toluene, formaldehyde, vinyl chloride, phenols, chloroform, and trichloroethylene, are released into the air by human activities. About 3.5 million tonnes of these compounds are emitted each year in Canada, mainly unburned or partially burned hydrocarbons from transportation, power plants, chemical plants, and petroleum refineries. These chemicals play an important role in the formation of photochemical oxidants.

Photochemical Oxidants

Photochemical oxidants are products of secondary atmospheric reactions driven by solar energy (Fig. 9.9). One of the most

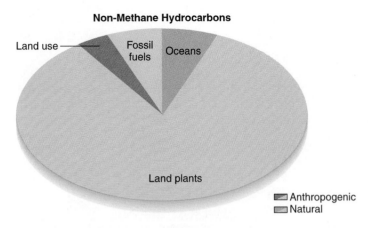

FIGURE 9.8 Sources of non-methane hydrocarbons in the atmosphere.
Source: Data from Joyce E. Penner, "Atmospheric Chemistry and Air Quality" in W. B. Meyer and B. L. Turner (eds.), *Changes in Land Use and Land Cover: A Global Perspective,* 1994. Cambridge University Press and UNEP 1999.

Atmospheric oxidant production:

1. $NO + VOC \longrightarrow NO_2$ (nitrogen dioxide)

2. $NO_2 + UV \longrightarrow NO + O$ (nitric oxide + atomic oxygen)

3. $O + O_2 \longrightarrow O_3$ (ozone)

4. $NO_2 + VOC \longrightarrow PAN$, etc. (peroxyacetyl nitrate)

Net results:

$NO + VOC + O_2 + UV \longrightarrow O_3$, PAN, and other oxidants

FIGURE 9.9 Secondary production of urban smog oxidants by photochemical reactions in the atmosphere.
Source: Data from B. J. Finlayson-Pitts and J. N. Pitts, *Atmospheric Chemistry,* 1986, John Wiley & Sons, Inc.

important of these reactions involves formation of singlet (atomic) oxygen by splitting nitrogen dioxide (NO_2). This atomic oxygen then reacts with another molecule of O_2 to make **ozone** (O_3). Ozone formed in the stratosphere provides a valuable shield for the biosphere by absorbing incoming ultraviolet radiation. In ambient air, however, O_3 is a strong oxidizing reagent and damages vegetation, building materials (such as paint, rubber, and plastics), and sensitive tissues (such as eyes and lungs). Ozone has an acrid, biting odour that is a distinctive characteristic of photochemical smog. Hydrocarbons in the air contribute to accumulation of ozone by removing NO in the formation of compounds, such as peroxyacetyl nitrate (PAN), which is another damaging photochemical oxidant.

Unconventional Pollutants

In addition to Criteria Air Contaminants there are important **unconventional** or **noncriteria air contaminants,** such as asbestos, benzene, beryllium, mercury, polychlorinated biphenyls (PCBs), and vinyl chloride. Most of these materials have no natural source in the environment (to any great extent) and are, therefore, only anthropogenic in origin.

In addition to these toxic air pollutants, some other unconventional forms of air pollution deserve mention. **Aesthetic degradation** includes any undesirable changes in the physical characteristics or chemistry of the atmosphere. Noise, odours, and light pollution are examples of atmospheric degradation that may not be life-threatening but reduce the quality of our lives. This is a very subjective category. Odours and noise (such as loud music) that are offensive to some may be attractive to others. Often the most sensitive device for odour detection is the human nose. We can smell styrene, for example, at 44 parts per billion (ppb). Trained panels of odour testers often are used to evaluate air samples. Factories that emit noxious chemicals sometimes spray "odour maskants" or perfumes into smokestacks to cover up objectionable odours.

In most urban areas, it is difficult or impossible to see stars in the sky at night because of dust in the air and stray light from buildings, outdoor advertising, and streetlights. This light pollution has become a serious problem for astronomers.

Indoor Air Pollution

We have spent a considerable amount of effort and money to control the major outdoor air pollutants, but we have only recently become aware of the dangers of indoor air pollutants. Indoor concentrations of toxic air pollutants are often higher than outdoors. Furthermore, people generally spend more time inside than out and therefore are exposed to higher doses of these pollutants.

Smoking is without doubt the most important air pollutant in terms of human health. The Surgeon General estimates that more than 400,000 people die each year in the United States from emphysema, heart attacks, strokes, lung cancer, or other diseases caused by smoking. These diseases are responsible for 20 percent of all mortality in the United States, or four times as much as infectious agents.

In some cases, indoor air in homes has concentrations of chemicals that would be illegal outside or in the workplace. Concentrations of such compounds as chloroform, benzene, carbon tetrachloride, formaldehyde, and styrene can be 70 times higher in indoor air than in outdoor air. "Green design" principles can make indoor spaces both healthier and more pleasant (In Depth, p. 181).

In the less-developed countries of Africa, Asia, and Latin America where such organic fuels as firewood, charcoal, dried dung, and agricultural wastes make up the majority of household energy, smoky, poorly ventilated heating and cooking fires represent the greatest source of indoor air pollution (Fig. 9.10). The World Health Organization (WHO) estimates that 2.5 billion people—nearly half the world's population—are adversely affected by pollution from this source. Women and small children spend long hours each day around open fires or unventilated stoves in enclosed spaces. The levels of carbon monoxide, particulates, aldehydes, and other toxic chemicals can be 100 times higher than would be

Indoor Air

How safe is the air in your home, office, or school room? As we decrease air-infiltration into buildings to conserve energy, we often trap indoor air pollutants within spaces where most of us spend the vast majority of our time. In what has come to be known as "sick building syndrome," people complain of headaches, fatigue, nausea, upper-respiratory problems, and a wide variety of allergies from workplace or home exposure to airborne toxins.

What might be making us sick? Mold spores are probably the greatest single cause of allergic reactions in indoor air. Moisture trapped in air-tight houses often accumulates in walls where moulds flourish. Air ducts provide both a good environment for growth of pathogens such as Legionnaire's disease bacteria as well as a path for their dispersal. Legionnaire's pneumonia is much more prevalent than most people realize in places like California and Australia where air-conditioning is common. Uranium-bearing rocks and sediment are widespread across North America. When uranium decays, it produces carcinogenic radon gas that can seep into buildings.

In addition, we are exposed to a variety of synthetic chemicals emitted from carpets, wall coverings, building materials, and combustion gases. You might be surprised to learn how many toxic, synthetic compounds are used to construct buildings and make furniture. Formaldehyde, for instance is a component of more than 3,000 products, including building materials such as particle board, waferboard, and urea-formaldehyde foam insulation. PVC is used in plastic plumbing pipe, floor and wall coverings, and countertops. Volatile organic solvents make up as much as half the volume of some paint. New carpets and drapes typically contain up to two dozen chemical compounds designed to kill bacteria and moulds, resist stains, bind fibres, and retain colours.

What can you do if you suspect that your living spaces are exposing you to materials that make you sick? Probably few students will be in a position anytime soon to build a new house with nontoxic materials, but there are some principles from the emerging field of "green design" that you might apply if you're house hunting, redecorating your apartment, or interviewing for a job (see Chapter 20). Low-volatile paint is now available for indoor use. Nontoxic, formaldehyde-free plywood, particle board, and insulation can be used in new construction. Nonallergenic carpets, drapes, and wall coverings are available, but some architects recommend natural wood, stone, and plaster surfaces that are easier to clean and less allergenic than any fabric.

High rates of air exchange can help rid indoor air of moisture, odours, mould spores, radon, and toxins. Does that mean energy inefficiency? Not necessarily. Air-to-air heat exchangers keep heat in during the winter and out during the summer, while still providing a healthy rate of fresh-air flow. Bathrooms and kitchens should have outdoor vents. Gas or oil furnaces should be checked for carbon monoxide production. Although many cooks prefer gas stoves because they heat quickly, they can produce toxic carbon monoxide and nitrogen oxides. Contact your local health and safety authority for further tips on how to make your home, work, or study environment healthier. No matter what your situation, there are things that each of us can do to make our indoor air cleaner and safer.

FIGURE 9.10 Smoky cooking and heating fires may cause more ill health effects than any other source of indoor air pollution except tobacco smoking. Some 2.5 billion people, mainly women and children, spend hours each day in poorly ventilated kitchens and living spaces where carbon monoxide, particulates, and cancer-causing hydrocarbons often reach dangerous levels.

legal for outdoor ambient concentrations in Canada or the United States. Designing and building cheap, efficient, nonpolluting energy sources for the developing countries would not only save shrinking forests but would make a major impact on health as well.

CLIMATE, TOPOGRAPHY, AND ATMOSPHERIC PROCESSES

Topography, climate, and physical processes in the atmosphere play an important role in transport, concentration, dispersal, and removal of many air pollutants. Wind speed, mixing between air layers, precipitation, and atmospheric chemistry all determine whether pollutants will remain in the locality where they are produced or go elsewhere. In this next section, we will survey some environmental factors that affect air pollution levels.

Inversions

Temperature inversions occur when a stable layer of warmer air overlays cooler air, reversing the normal temperature decline with increasing height and preventing convection currents from dispersing pollutants. Several mechanisms create inversions. When a cold front slides under an adjacent warmer air mass or when cool air subsides down a mountain slope to displace warmer air in the valley below, an inverted temperature gradient is established. These inversions are usually not stable, however, because winds accompanying these air exchanges tend to break up the temperature gradient fairly quickly and mix air layers.

The most stable inversion conditions are usually created by rapid nighttime cooling in a valley or basin where air movement is restricted. Los Angeles is a classic example of the conditions that create temperature inversions and photochemical smog (Fig. 9.11). The city is surrounded by mountains on three sides and the climate is dry and sunny. Extensive automobile use creates high pollution levels. Skies are generally clear at night, allowing rapid radiant heat loss, and the ground cools quickly. Surface air layers are cooled by conduction, while upper layers remain relatively warm. Density differences retard vertical mixing. During the night, cool, humid, onshore breezes slide in under the contaminated air, squeezing it up against the cap of warmer air above and concentrating the pollutants accumulated during the day.

Morning sunlight is absorbed by the concentrated aerosols and gaseous chemicals of the inversion layer. This complex mixture quickly cooks up a toxic brew of hazardous compounds. As the ground warms later in the day, convection currents break up the temperature gradient and pollutants are carried back down to the surface where more contaminants are added. Nitric oxide (NO) from automobile exhaust is oxidized to nitrogen dioxide. As nitrogen oxides are used up in reactions with unburned hydrocarbons, the ozone levels begin to rise. By early afternoon, an acrid brown haze fills the air, making eyes water and throats burn. On summer days, ozone concentrations in the Los Angeles basin can reach 0.34 ppm or more by late afternoon and the pollution index can be 300, the stage considered a health hazard.

Dust Domes and Heat Islands

Even without mountains to block winds and stabilize air layers, many large cities create an atmospheric environment quite different from the surrounding conditions. Sparse vegetation and high levels of concrete and glass in urban areas allow rainfall to run off quickly and create high rates of heat absorption during the day and radiation at night. Tall buildings create convective updrafts that sweep pollutants into the air. Temperatures in the centre of large cities are frequently 3° to 5°C higher than the surrounding countryside. Stable air masses created by this "heat island" over the city concentrate pollutants in a "dust dome." Rural areas downwind from major industrial areas often have significantly decreased visibility and increased rainfall (due to increased condensation nuclei in the dust plume) compared to neighbouring areas with cleaner air. In the late 1960s, for instance, areas downwind from Chicago and St. Louis reported up to 30 percent more rainfall than upwind regions.

Aerosols and dust in urban air seem to trigger increased cloud-to-ground lightning strikes. Houston and Lake Charles, LA, for instance, which have many petroleum refineries, have among the highest number of lightning strikes in the United States and twice as many as nearby areas with similar climate but cleaner air.

Long-Range Transport

Fine aerosols can be carried great distances by the wind. Florida researchers recently showed that as much as half of the fine reddish dust visible in Miami's air during summer months is blown across the Atlantic from the Sahara Desert. Forests in Hawaii are nourished by dust from China's Takla Makan Desert nearly 6,000 km away.

Industrial pollutants are also transported great distances by wind currents. Some of the most toxic and corrosive materials delivered by long-range transport are secondary pollutants (such as sulphuric and nitric acids or ozone), produced by the mixing and interaction of atmospheric contaminants as they travel through the air. In 1999, atmospheric scientists were surprised to find a dense haze of soot and sulphate aerosols over the Indian Ocean. Covering an area larger than the continental United States and reaching from the Bay of Bengal to the Arabian Sea, this huge cloud of pollutants is produced in South Asia and blown south by the winter monsoon. In summer it is blown back north to fall as dirty acid rain.

Increasingly sensitive monitoring equipment has begun to reveal industrial contaminants in places usually considered among

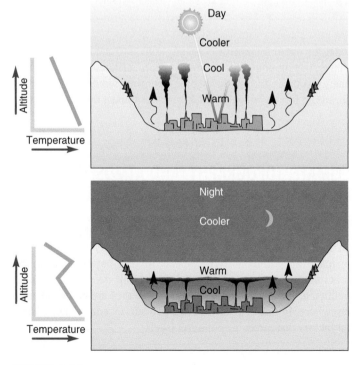

FIGURE 9.11 Atmospheric temperature inversions occur where ground level air cools more quickly than upper levels. This temperature differential prevents mixing and traps pollutants close to the ground.

PART TWO   Our Physical Environment

the cleanest in the world. Samoa, Greenland, and even Antarctica and the North Pole, all have heavy metals, pesticides, and radioactive elements in their air. Since the 1950s, pilots flying in the high Arctic have reported dense layers of reddish-brown haze clouding the arctic atmosphere. Aerosols of sulphates, soot, dust, and toxic heavy metals such as vanadium, manganese, and lead travel to the pole from the industrialized parts of Europe and Russia (Fig. 9.12). These contaminants, trapped by winds that circle the pole, concentrate at high latitudes and eventually, falling out in snow and ice, enter the food chain. The Inuit people of Broughton Island, well above the Arctic Circle, have higher levels of polychlorinated biphenyls (PCBs) in their blood than any other known population, except victims of industrial accidents. Far from any source of this industrial by-product, these people accumulate PCBs from the flesh of fish, caribou, and other animals they eat.

Stratospheric Ozone

In 1985, the British Antarctic Atmospheric Survey announced a startling and disturbing discovery: ozone levels in the stratosphere over the South Pole were dropping precipitously during September and October every year as the sun reappears at the end of the long polar winter (Fig. 9.13). This ozone depletion has been occurring at least since the 1960s but was not recognized because earlier researchers programmed their instruments to ignore changes in ozone levels that were presumed to be erroneous.

The 2000 ozone "hole" over Antarctia was the largest ever recorded, covering 44.3 million km^2 (larger than all of North America) in which all the ozone between 14 and 20 kilometres altitude was destroyed. Ominously, this phenomenon is now spreading to other parts of the world as well. About 10 percent of all stratospheric ozone worldwide was destroyed during the spring of 1998 and levels over the Arctic averaged 40 percent below normal.

Why are we worried about stratospheric ozone? At ground level, ozone is a harmful pollutant, damaging plants, building materials, and human health; in the the upper atmosphere, however, where it screens out dangerous ultraviolet (UV) rays from the sun, ozone is an irreplaceable resource. Without this shield, organisms on the earth's surface would be subjected to life-threatening radiation burns and genetic damage. A 1 percent loss of ozone results in a 2 percent increase in UV rays reaching the earth's surface and could result in about a million extra human skin cancers per year worldwide if no protective measures are taken. Thus it is urgent that we learn what is attacking the ozone layer and find ways to reverse these trends if possible.

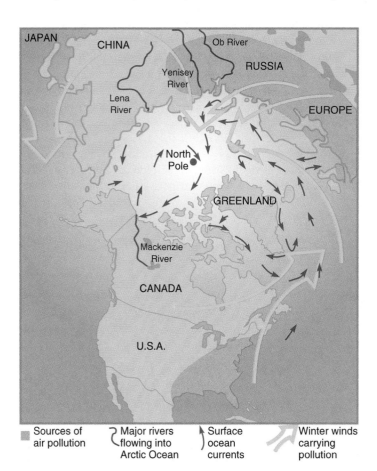

FIGURE 9.12 Air pollution from heavily industrialized regions in Europe and America are transported by circumpolar winds to the Arctic, where high levels of smog accumulate. The average transit time from Russia to Canada is only about three days.

Source: From Government of Canada, 1991. The State of Canada's Environment. Reproduced with permission of the Minister of Public Works and Government Services Canada, 1999.

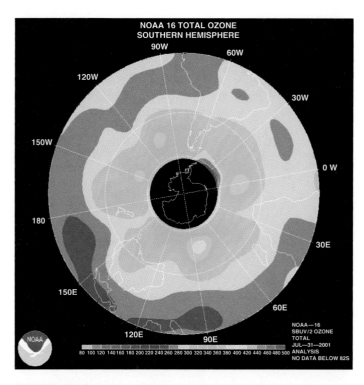

FIGURE 9.13 Ozone depletion over the South Pole is shown in this satellite image from July 31, 2001. Outside the polar vortex, green and yellow areas show elevated ozone levels. Over Antarctica (outlined in white), however, stratospheric ozone levels are reduced by 80 percent or more from normal levels. The dark circle over the South Pole is an artifact of the measuring system.

The exceptionally cold temperatures (-85 to $-90°C$) in Antarctica play a role in ozone losses. During the long, dark winter months, the strong circumpolar vortex (Chapter 8) isolates Antarctic air and allows stratospheric temperatures to drop low enough to create ice crystals at high altitudes—something that rarely happens elsewhere over the world. Ozone and chlorine-containing molecules are absorbed on the surfaces of these ice particles. When the sun returns in the spring and provides energy to liberate chlorine ions, destructive chemical reactions proceed quickly (Fig. 9.14).

Humans release a variety of chlorine-containing molecules into the atmosphere. The ones suspected of being most important in ozone losses are **chlorofluorocarbons (CFCs)** and halon gases. CFCs were invented in 1928 by scientists at General Motors who were searching for a less toxic refrigerant than ammonia. Commonly known by the trade name Freon, CFCs were regarded as wonderful compounds. They are nontoxic, nonflammable, chemically inert, cheaply produced, and useful in a wide variety of applications.

Because these molecules are so stable, however, they persist for decades or even centuries once released. When they diffuse out into the stratosphere, the intense UV irradiation releases chlorine atoms that destroy ozone. Since the chlorine atoms are not themselves consumed in these reactions, they continue to destroy ozone for years until they finally precipitate or are washed out of the air.

Until 1978, aerosol spray cans used more CFCs than any other product. Although we didn't know about the special conditions in the Antarctic at that time, it was suspected that CFCs might threaten stratospheric ozone, so laws were passed in Canada, the United States, and some European countries to ban nonessential uses. Still, CFCs continue to be used in the developing world as refrigerants, solvents, spray propellants, and foam-blowing agents. Smuggling of CFCs into Canada and the United States from Mexico and other countries where they are still legal now rivals drugs in total value of contraband.

The discovery of stratospheric ozone losses has brought about a remarkably quick international response. At a 1989 conference in Helsinki, 81 nations agreed to phase out CFC production by the end of the last century. As evidence accumulated showing that losses were larger and more widespread than previously thought, the deadline for the elimination of all CFCs (halons, carbon tetrachloride, and methyl chloroform) was moved up to 1996 and a $500 million fund was established to assist poorer countries to switch to non-CFC technologies. Fortunately, alternatives to CFCs for most uses already exist. The first substitutes are hydrochlorofluorocarbons (HCFCs), which release much less chlorine per molecule. Eventually, we hope to develop halogen-free molecules that work just as well and are no more expensive than CFCs.

There is some evidence that the CFC ban is already having an effect. CFC production in industrialized countries has fallen nearly 80 percent since 1989 (Fig. 9.15). Chlorine in the stratosphere has shown a steady decline, but unfortunately, bromine, another ozone destroyer, seems to be replacing it. Bromine is a common ingredient in weed and pest killers. This factor may

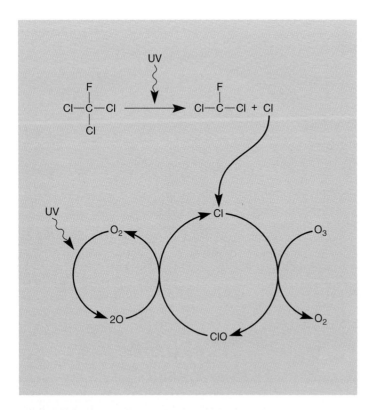

FIGURE 9.14 Destruction of stratospheric ozone by chlorine atoms. When exposed to UV radiation, CFCs release highly reactive chlorine atoms that react with ozone to produce chlorine monoxide. Singlet oxygen atoms, also produced by UV irradiation, react with ClO to start the cycle over again.

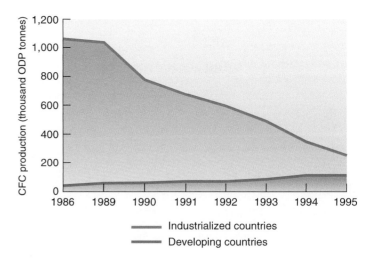

FIGURE 9.15 Production of chlorofluorocarbons in industrialized countries has fallen sharply since the Helsinki convention was passed in 1989.
Source: Data from Sebastian Oberthur, Production and Consumption of Ozone-Depleting Substances, 1986–1995 (Deutsche Gesellschaft für Technische Zusammenarbeit, Bonn Germany, 1997, p. 30).

prevent ozone recovery by 2050 as was once hoped. There may be a connection between ozone destruction and global warming. When heat is trapped in the troposphere, it allows the stratosphere to cool, triggering formation of high-altitude ice crystals. Furthermore, without ozone, which is a powerful greenhouse gas itself, the stratosphere gets even colder, setting off a negative feedback system. As so often is the case, when we disturb one environmental factor, we affect others as well.

In 1995, chemists Sherwood Rowland, Mario Molina, and Paul Crutzen shared the Nobel Prize for their work on atmospheric chemistry and stratospheric ozone. This was the first Nobel Prize for an environmental issue.

EFFECTS OF AIR POLLUTION

So far we have looked at the major types and sources of air pollutants. Now we will turn our attention to the effects of those pollutants on human health, physical materials, ecosystems, and global climate.

Human Health

Residents of the most polluted cities in North America are 15 to 17 percent more likely to die of these illnesses than those in cities with the cleanest air. Heart attacks, respiratory diseases, and lung cancer all are significantly higher in people who breathe dirty air, compared to matching groups in cleaner environments. This can mean as much as a 5- to 10-year decrease in life expectancy if you live in the worst parts of large cities, compared to a place with clean air. Of course your likelihood of suffering ill health from air pollutants depends on the intensity and duration of exposure as well as your age and prior health status. You are much more likely to be at risk if you are very young, very old, or already suffering from some respiratory or cardiovascular disease. Some people are supersensitive because of genetics or prior exposure. And those doing vigorous physical work or exercise are more likely to succumb than more sedentary folks.

Conditions are often much worse in other countries than Canada or the United States. The United Nations estimates that at least 1.3 billion people around the world live in areas where the air is dangerously polluted. In the "black triangle" region of Poland, Hungary, the Czech Republic, and Slovakia, for example, respiratory ailments, cardiovascular diseases, lung cancer, infant mortality, and miscarriages are as much as 50 percent higher than in cleaner parts of those countries. And in China, city dwellers are four to six times more likely than country folk to die of lung cancer. As mentioned earlier, the greatest air quality problem is often in poorly ventilated homes in poorer countries where smoky fires are used for cooking and heating. Billions of women and children spend hours each day in these unhealthy conditions. The World Health Organization estimates that 4 million children under age 5 die each year from acute respiratory diseases exacerbated by air pollution.

How does air pollution cause these health effects? The most common route of exposure to air pollutants is by inhalation, but direct absorption through the skin or contamination of food and water also are important pathways. Because they are strong oxidizing agents, sulphates, SO_2, NO_x, and O_3 act as irritants that damage delicate tissues in the eyes and respiratory passages. Fine suspended particulate materials (less than 2.5 µm) penetrate deep into the lungs and are both irritants and fibrotic agents. Inflammatory responses set in motion by these irritants impair lung function and trigger cardiovascular problems as the heart tries to compensate for lack of oxygen by pumping faster and harder. If the irritation is really severe, so much fluid seeps into lungs through damaged tissues that the victim actually drowns.

Carbon monoxide binds to hemoglobin and decreases the ability of red blood cells to carry oxygen. Asphyxiants such as this cause headaches, dizziness, heart stress, and can even be lethal if concentrations are high enough. Lead also binds to hemoglobin and reduces oxygen-carrying capacity at high levels. At lower levels, lead causes long-term damage to critical neurons in the brain that results in mental and physical impairment and developmental retardation.

Some important chronic health effects of air pollutants include bronchitis and emphysema.

Bronchitis is a persistent inflammation of bronchi and bronchioles (large and small airways in the lung) that causes mucus build-up, a painful cough, and involuntary muscle spasms that constrict airways. Severe bronchitis can lead to **emphysema,** an irreversible obstructive lung disease in which airways become permanently constricted and alveoli are damaged or even destroyed. Stagnant air trapped in blocked airways swells the tiny air sacs in the lung (alveoli), blocking blood circulation. As cells die from lack of oxygen and nutrients, the walls of the alveoli break down, creating large empty spaces incapable of gas exchange (Fig. 9.16). Thickened walls of the bronchioles lose elasticity and breathing becomes more difficult. Victims of emphysema make a characteristic whistling sound when they breathe. Often they need supplementary oxygen to make up for reduced respiratory capacity.

Smoking is undoubtedly the largest cause of obstructive lung disease and preventable death in the world. The World Health Organization says that tobacco kills some 3 million people each year. This makes it rank with diarrhea and AIDS as one of the world's leading killers. Because of cardiovascular stress caused by carbon monoxide in smoke and chronic bronchitis and emphysema, about twice as many people die of heart failure as die from lung cancer associated with smoking.

Plant Pathology

In the early days of industrialization, fumes from furnaces, smelters, refineries, and chemical plants often destroyed vegetation and created desolate, barren landscapes around mining and manufacturing centers. The copper-nickel smelter at Sudbury, Ontario, is a spectacular and notorious example of air pollution effects on vegetation and ecosystems. In 1886, the corporate ancestors of the International Nickel Company (INCO) began open-bed

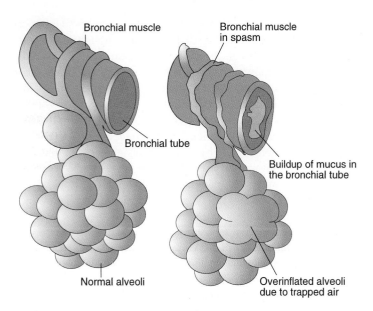

FIGURE 9.16 Bronchitis and emphysema can result in constriction of airways and permanent damage to tiny, sensitive air sacs called alveoli where oxygen diffuses into blood vessels.

FIGURE 9.17 Sulphur dioxide emissions and acid precipitation from the International Nickel Company copper smelter (*background*) killed all vegetation over a large area near Sudbury, Ontario. Even the pink granite bedrock has burned black. The installation of scrubbers has dramatically reduced sulphur emissions. The ecosystem farther away from the smelter is slowly beginning to recover.

roasting of sulphide ores at Sudbury. Sulphur dioxide and sulphuric acid released by this process caused massive destruction of the plant community within about 30 km of the smelter. Rains washed away the exposed soil, leaving a barren moonscape of blackened bedrock. Super-tall, 400 m smokestacks were installed in the 1950s and sulphur scrubbers were added 20 years later. Emissions were reduced by 90 percent and the surrounding ecosystem is beginning to recover. The area near the factory is still a grim, empty wasteland, however (Fig. 9.17). Similar destruction occurred at many other sites during the nineteenth century. Copperhill, Tennessee; Butte, Montana; and the Ruhr Valley in Germany are some well-known examples, but these areas also are showing signs of recovery since corrective measures were taken.

There are two probable ways that air pollutants damage plants. They can be directly toxic, damaging sensitive cell membranes much as irritants do in human lungs. Within a few days of exposure to toxic levels of oxidants, mottling (discolouration) occurs in leaves due to chlorosis (bleaching of chlorophyll), and then necrotic (dead) spots develop (Fig. 9.18). If injury is severe, the whole plant may be killed. Sometimes these symptoms are so distinctive that positive identification of the source of damage is possible. Often, however, the symptoms are vague and difficult to separate from diseases or insect damage.

Another mechanism of action is exhibited by chemicals, such as ethylene, that act as metabolic regulators or plant hormones and disrupt normal patterns of growth and development. Ethylene is a component of automobile exhaust and is released from petroleum refineries and chemical plants. The concentration of ethylene around highways and industrial areas is often high enough to cause injury to sensitive plants. Some scientists believe that the devastating forest destruction in Europe and North America may be partly due to volatile organic compounds.

FIGURE 9.18 Soybean leaves exposed to 0.8 parts per million sulphur dioxide for 24 hours show extensive chlorosis (chlorophyll destruction) in white areas between leaf veins.

Certain combinations of environmental factors have **synergistic effects** in which the injury caused by exposure to two factors together is more than the sum of exposure to each factor individually. For instance, when white pine seedlings are exposed to subthreshold concentrations of ozone and sulphur dioxide individually, no visible injury occurs. If the same concentrations of pollutants are given together, however, visible damage occurs. In alfalfa, however, SO_2 and O_3 together cause less damage than either one alone. These complex interactions point out the unpredictability of future effects of pollutants. Outcomes might be either more or less severe than previous experience indicates.

Pollutant levels too low to produce visible symptoms of damage may still have important effects. Field studies using open-top chambers and charcoal-filtred air show that yields in some sensitive crops, such as soybeans, may be reduced as much as 50 percent by currently existing levels of oxidants in ambient air. Some plant pathologists suggest that ozone and photochemical oxidants are responsible for as much as 90 percent of agricultural, ornamental, and forest losses from air pollution. The total costs of this damage may be as much as $10 billion per year in North America alone.

Acid Deposition

Most people in Canada and the United States became aware of problems associated with **acid precipitation** (the deposition of wet acidic solutions or dry acidic particles from the air) within the last 20 years or so, but English scientist Robert Angus Smith coined the term "acid rain" in his studies of air chemistry in Manchester, England, in the 1850s. By the 1940s, it was known that pollutants, including atmospheric acids, could be transported long distances by wind currents. This was thought to be only an academic curiosity until it was shown that precipitation of these acids can have far-reaching ecological effects.

pH and Atmospheric Acidity

We describe acidity in terms of pH (the negative logarithm of the hydrogen ion concentration in a solution). The pH scale ranges from 0 to 14, with 7, the midpoint, being neutral (Chapter 3). Values below 7 indicate progressively greater acidity, while those above 7 are progressively more alkaline. Since the scale is logarithmic, there is a tenfold difference in hydrogen ion concentration for each pH unit. For instance, pH 6 is 10 times more acidic than pH 7; likewise, pH 5 is 100 times more acidic, and pH 4 is 1,000 times more acidic than pH 7.

Normal, unpolluted rain generally has a pH of about 5.6 due to carbonic acid created by CO_2 in air. Volcanic emissions, biological decomposition, and chlorine and sulphates from ocean spray can drop the pH of rain well below 5.6, while alkaline dust can raise it above 7. In industrialized areas, anthropogenic acids in the air usually far outweigh those from natural sources. Acid rain is only one form in which acid deposition occurs. Fog, snow, mist, and dew also trap and deposit atmospheric contaminants. Furthermore, fallout of dry sulphate, nitrate, and chloride particles can account for as much as half of the acidic deposition in some areas.

Aquatic Effects

It has been known for over 30 years that acids—principally H_2SO_4 and HNO_3—generated by industrial and automobile emissions in northwestern Europe are carried by prevailing winds to Scandinavia where they are deposited in rain, snow, and dry precipitation. The thin, acidic soils and oligotrophic lakes and streams in the

mountains of southern Norway and Sweden have been severely affected by this acid deposition. Some 18,000 lakes in Sweden are now so acidic that they will no longer support game fish or other sensitive aquatic organisms.

Generally, reproduction is the most sensitive stage in fish life cycles. Eggs and fry of many species are killed when the pH drops to about 5.0. This level of acidification also can disrupt the food chain by killing aquatic plants, insects, and invertebrates on which fish depend for food. At pH levels below 5.0, adult fish die as well. Trout, salmon, and other game fish are usually the most sensitive. Carp, gar, suckers, and other less desirable fish are more resistant. There are several ways acids kill fish. Acidity alters body chemistry, destroys gills and prevents oxygen uptake, causes bone decalcification, and disrupts muscle contraction. Another dangerous effect (for us as well as fish) is that acid water leaches toxic metals, such as mercury and aluminum, out of soil and rocks.

In the early 1970s, evidence began to accumulate suggesting that air pollutants are acidifying many lakes in North America. Studies in the Adirondack Mountains of New York revealed that about half of the high-altitude lakes (above 1,000 m) are acidified and have no fish. Areas showing lake damage correlate closely with sulphate deposition (and therefore acid) in precipitation (Fig. 9.19). The positive effects of a Canada–US agreement to reduce SO_2 emissions is evident in these maps. Some 48,000 lakes in Ontario are endangered and nearly all of Quebec's surface waters, including about 1 million lakes, are believed to be highly sensitive to acid deposition.

Much of western North America has relatively alkaline bedrock and carbonate-rich soil, which counterbalance acids from the atmosphere. Recent surveys of the Rocky Mountains, the Sierra Nevadas in California, and the Cascades in Washington, however, have shown that many high mountain lakes and streams have very low buffering capacity (ability to resist pH change) and are susceptible to acidification.

Sulphates account for about two-thirds of the acid deposition in eastern North America and most of Europe, while nitrates contribute most of the remaining one-third. In urban areas, where transportation is the major source of pollution, nitric acid is equal to or slightly greater than sulphuric acids in the air. A vigorous program of pollution control has been undertaken by both Canada and the United States. Although SO_2 and NO_x emissions have decreased dramatically over the past three decades over much of Europe and eastern North America as a result of pollution control measures, rain falling in these areas remains acidic. Damage to natural ecosystems also continues to be greater than scientists expected. Apparently, alkaline dust that would once have neutralized acids in air has been depleted by years of acid rain and is now no longer effective. This may also lead to a loss of cations such as calcium, magnesium, sodium, and potassium essential to plant growth.

Forest Damage

In the early 1980s, disturbing reports appeared of rapid forest declines in both Europe and North America. One of the earliest was a detailed ecosystem inventory on Camel's Hump Mountain in

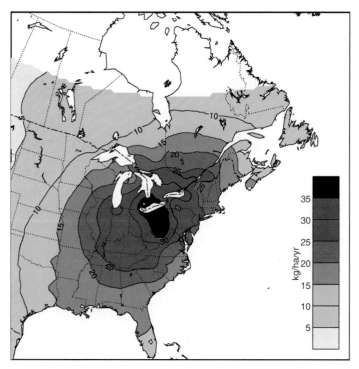

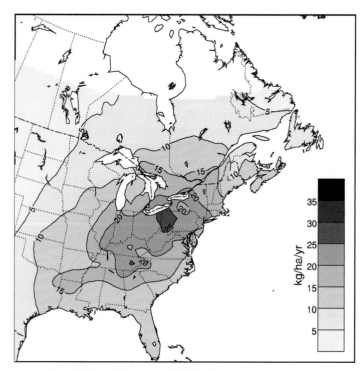

4-Year (1980–1983) Mean XSO4 Wet Deposition (kg/ha/yr)   5-Year (1996–2000) Mean XSO4 Wet Deposition (kg/ha/yr)

FIGURE 9.19 Wet sulphate deposition (1980s vs. 1990s) for eastern North America.
Source: Environment Canada, 2003. Environmental Signals: Canada's National Environmental Indicator Series.

Vermont. A 1980 survey showed that seedling production, tree density, and viability of spruce-fir forests at high elevations had declined about 50 percent in 15 years. By 1990, almost all the red spruce, once the dominant species on the upper part of the mountain, were dead or dying. A similar situation was found on Mount Mitchell in North Carolina where almost all red spruce and Fraser fir above 2,000 m are in a severe decline. Nearly all the trees are losing needles and about half of them are dead (Fig. 9.20).

Similar catastrophic declines have been reported in Europe. In Germany, for example, foresters claimed in 1985 that more than 4 million hectares (about half the total forest) was diseased or dying. Czechoslovakia, Poland, Austria, and Switzerland found similar problems. Again, high-elevation forests are most severely affected. This is very dangerous for mountain villages in the Alps that depend on forests to prevent avalanches in the winter.

Many plant pathologists now believe that dire predictions of forest death in Europe were greatly exaggerated, but air pollution and deposition of atmospheric acids are still thought to be important causes of forest destruction in many areas. In the longest-running forest-ecosystem monitoring record in North America, researchers at the Hubbard Brook Experimental Forest in New Hampshire have shown that forest soils have become depleted of natural buffering reserves of basic cations such as calcium and magnesium through years of exposure to acid rain. Replacement of these cations by hydrogen and aluminum ions seems to be one of the main causes of plant mortality.

Other mechanisms may also play a role in forest decline. Overfertilization by nitrogen compounds may make trees sensitive to early frost. Toxic metals, such as aluminum, may be solubilized by acidic groundwater. Plant pathogens and insect pests may damage trees or attack trees debilitated by air pollution. Fungi that form essential mutualistic associations (called mycorrhizae) with tree roots may be damaged by acid rain. Other air pollutants, such as sulphur dioxide, ozone, or toxic organic compounds may damage trees. Repeated harvesting cycles in commercial forests may remove nutrients and damage ecological relationships essential for healthy tree growth. Perhaps the most likely scenario is that all these environmental factors act cumulatively but in different combinations in the deteriorating health of individual trees and entire forests.

Buildings and Monuments

In cities throughout the world, some of the oldest and most glorious buildings and works of art are being destroyed by air pollution. Smoke and soot coat buildings, paintings, and textiles. Limestone and marble are destroyed by atmospheric acids at an alarming rate. The Parthenon in Athens, the Taj Mahal in Agra, the Colosseum in Rome, frescoes and statues in Florence, medieval cathedrals in Europe (Fig. 9.21), and the Parliament Buildings in Ottawa are slowly dissolving and flaking away because of acidic fumes in the air. Medieval stained glass windows in Cologne's gothic cathedral are so porous from etching by atmospheric acids

FIGURE 9.20 A Frasier fir forest on Mount Mitchell, North Carolina, killed by acid rain, insect pests, and other stressors.

FIGURE 9.21 Atmospheric acids, especially sulphuric and nitric acids, have almost completely eaten away the face of this medieval statue. Each year, the total losses from air pollution damage to buildings and materials amounts to billions of dollars.

that pigments disappear and the glass literally crumbles away. Restoration costs for this one building alone are estimated at three to four billion German marks ($1.5 to $2 billion).

On a more mundane level, air pollution also damages ordinary buildings and structures. Corroding steel in reinforced concrete weakens buildings, roads, and bridges. Paint and rubber deteriorate due to oxidization. Limestone, marble, and some kinds of sandstone flake and crumble. The Council on Environmental Quality estimates that U.S. economic losses from architectural damage caused by air pollution amount to about $4.8 billion in direct costs and $5.2 billion in property value losses each year.

AIR POLLUTION CONTROL

What can we do about air pollution? In this section we will look at some of the techniques that can be used to avoid creating pollutants or to clean up effluents before they are released.

Moving Pollution to Remote Areas

Among the earliest techniques for improving local air quality was moving pollution sources to remote locations and/or dispersing emissions with smokestacks. These approaches exemplify the attitude that "dilution is the solution to pollution." One electric utility, for example, ran newspaper and magazine ads in the early 1970s, claiming to be a "pioneer" in the use of tall smokestacks on its power plants to "disperse gaseous emissions widely in the atmosphere so that ground level concentrations would not be harmful to human health or property." The company claimed that their smoke would be "dissipated over a wide area and come down finally in harmless traces." Far from being harmless, however, those "traces" are the main source of many of our current problems. We are finding that there is no "away" to which we can throw our unwanted products. A far better solution to pollution is to prevent its release. We will now turn our attention to emission-control technology.

Particulate Removal

Filters remove particles physically by trapping them in a porous mesh of cotton cloth, spun glass fibres, or asbestos-cellulose, which allows air to pass through but holds back solids. Collection efficiency is relatively insensitive to fuel type, fly ash composition, particle size, or electrical properties. Filters are generally shaped into giant bags 10 to 15 metres long and 2 or 3 metres wide. Effluent gas is blown into the bottom of the bag and escapes through the sides much like the bag on a vacuum cleaner (Fig. 9.22a). Every few days or weeks, the bags are opened to remove the dust cake. Thousands of these bags may be lined up in a "baghouse." These filters are usually much cheaper to install and operate than electrostatic filters.

   Electrostatic precipitators (Fig. 9.22b) are the most common particulate controls in power plants. Fly ash particles pick up an electrostatic surface charge as they pass between large electrodes in the effluent stream. This causes the particle to migrate to and accumulate on a collecting plate (the oppositely charged electrode). These precipitators consume a large amount of electricity, but maintenance is relatively simple and collection efficiency can be as high as 99 percent. Performance depends on particle size and chemistry, strength of the electric field, and flue gas velocity.

   The ash collected by all of these techniques is a solid waste (often hazardous due to the heavy metals and other trace components of coal or other ash source) and must be buried in landfills or other solid waste disposal sites.

Sulphur Removal

As we have seen earlier in this chapter, sulphur oxides are among the most damaging of all air pollutants in terms of human health and ecosystem damage. It is important to reduce sulphur loading. This can be done either by using low-sulphur fuel or by removing sulphur from effluents.

Fuel Switching and Fuel Cleaning

Switching from soft coal with a high sulphur content to low-sulphur coal can greatly reduce sulphur emissions. This may eliminate jobs, however, in such areas as Appalachia in the eastern United States that are already economically depressed. Changing

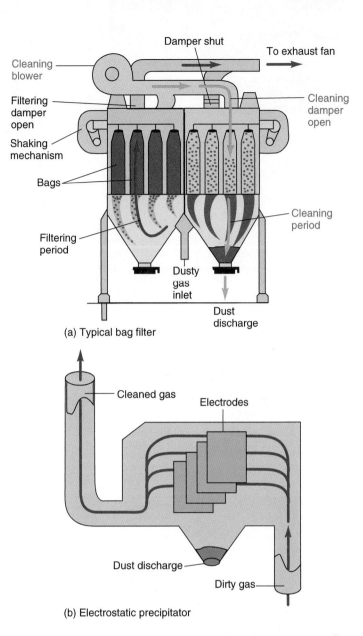

(a) Typical bag filter

(b) Electrostatic precipitator

FIGURE 9.22 Typical emission-control devices: (a) bag filter and (b) electrostatic precipitator. Note that two stages in the operational cycle are shown in (a), the filtering period on the left and the period of cleaning the filter bag on the right.

to another fuel, such as natural gas or nuclear energy, can eliminate all sulphur emissions as well as those of particulates and heavy metals. Natural gas is more expensive and more difficult to ship and store than coal, however, and many people prefer the sure dangers of coal pollution to the uncertain dangers of nuclear power (Chapter 12). Alternative energy sources, such as wind and solar power, are preferable to either fossil fuel or nuclear power, and are becoming economically competitive (Chapter 13) in many areas. In the interim, coal can be crushed, washed, and gassified to remove sulphur and metals before combustion. This improves heat content and firing properties but may replace air pollution with solid waste and water pollution problems.

Limestone Injection and Fluidized Bed Combustion

Sulphur emissions can be reduced as much as 90 percent by mixing crushed limestone with coal before it is fed into a boiler. Calcium in the limestone reacts with sulphur to make calcium sulphite ($CaSO_3$), calcium sulphate ($CaSO_4$), or gypsum ($CaSO_4 \cdot 2H_2O$). In ordinary furnaces, this procedure creates slag, which fouls burner grates and reduces combustion efficiency.

A technique called fluidized bed combustion, offers several advantages in pollution control. In this procedure, a mixture of crushed coal and limestone particles about a metre deep is spread on a perforated distribution grid in the combustion chamber (Fig. 9.23). When high-pressure air is forced through the bed, the surface of the fuel rises as much as 1 metre and resembles a boiling fluid as particles hop up and down. Oil is sprayed into the suspended mass to start the fire. During operation, fresh coal and limestone are fed continuously into the top of the bed, while ash and slag are drawn off from below. The rich air supply and constant motion in the bed make burning efficient and prevent buildup of large slag clinkers. Steam generator pipes are submerged directly into the fluidized bed, and heat exchange is more efficient than in the water walls of a conventional boiler. More than 90 percent of SO_2 is captured by the limestone particles, and NO_x formation is reduced by holding temperatures around 800°C instead of twice

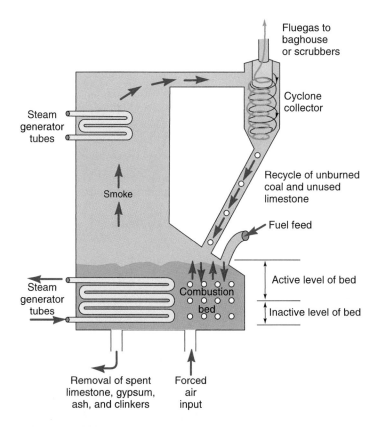

FIGURE 9.23 Fluidized bed combustion. Fuel is lifted by strong air jets from underneath the bed. Efficiency is good with a wide variety of fuels, and SO_2, NO_x, and CO emissions are much lower than with conventional burners.

that figure in other boilers. These low temperatures also preclude slag formation, which aids in maintenance. The efficient burning of this process makes it possible to use cheaper fuel, such as lignite or unwashed subbituminous coal, rather than higher priced hard coal.

Flue Gas Desulphurization

Crushed limestone, lime slurry, or alkali (sodium carbonate or bicarbonate) can be injected into a stack gas stream to remove sulphur after combustion. These processes are often called flue gas scrubbing. Spraying wet alkali solutions or limestone slurry is relatively inexpensive and effective, but maintenance can be difficult. Rock-hard plaster and ash layers coat the spray chamber and have to be chipped off regularly. Corrosive solutions of sulphates, chlorides, and fluorides erode metal surfaces. Electrostatic precipitators don't work well because of fouling and shorting of electrodes after wet scrubbing.

Dry alkali injection (spraying dry sodium bicarbonate into the flue gas) avoids many of the problems of wet scrubbing, but the expense of appropriate reagents is prohibitive in most areas. A hybrid procedure called spray drying has been tested successfully in pilot plant experiments. In this process, a slurry of pulverized limestone or slaked lime is atomized in the stack gas stream. The spray rate and droplet size are carefully controlled so that the water flash evaporates and a dry granular precipitate is produced. Passage through a baghouse filter removes both ash and sulphur very effectively.

As with coal washing, scrubbing often results in a trade-off of an air pollution problem for a solid waste disposal problem. Sulphur slag, gypsum, and other products of these processes can amount to three or four times as much volume as fly ash. A large power plant can produce millions of tonnes of waste per year.

Sulphur Recovery Processes

Instead of making a throwaway product that becomes a waste disposal problem, sulphur can be removed from effluent gases by processes that yield a usable product, such as elemental sulphur, sulphuric acid, or ammonium sulphate. Catalytic converters are used in these recovery processes to oxidize or reduce sulphur and to create chemical compounds that can be collected and sold. Markets have to be reasonably close for economic feasibility, and fly ash contamination must be reduced as much as possible.

Nitrogen Oxide Control

Undoubtedly the best way to prevent nitrogen oxide pollution is to avoid creating it. A substantial portion of the emissions associated with mining, manufacturing, and energy production could be eliminated through conservation (Chapter 12).

Staged burners, in which the flow of air and fuel are carefully controlled, can reduce nitrogen oxide formation by as much as 50 percent. This is true for both internal combustion engines and industrial boilers. Fuel is first burned at high temperatures in an oxygen-poor environment where NO_x cannot form. The residual gases then pass into an afterburner where more air is added and final combustion takes place in an air-rich, fuel-poor, low-temperature environment that also reduces NO_x formation. Stratified-charge engines and new orbital automobile engines use this principle to meet emission standards without catalytic converters.

The approach adopted by North American automakers for NO_x reductions has been to use selective catalysts to change pollutants to harmless substances. Three-way catalytic converters use platinum-palladium and rhodium catalysts to remove up to 90 percent of NO_x, hydrocarbons, and carbon monoxide at the same time (Fig. 9.24). Unfortunately, this approach doesn't work on diesel engines, power plants, smelters, and other pollution sources because of problems with back pressure, catalyst life, corrosion, and production of unwanted by-products, such as ammonium sulphate (NH_4SO_4), that foul the system.

Raprenox (*rapid removal of nitrogen oxides*) is a new technique for removing nitrogen oxides that was developed by the U.S. Department of Energy Sandia Laboratory in Livermore, California. Exhaust gases are passed through a container of common, non-poisonous cyanuric acid. When heated to 350°C, cyanuric acid releases isocyanic acid gas, which reacts with NO_x to produce CO_2, CO, H_2O, and N_2. In small-scale diesel engine tests, this system eliminated 99 percent of the NO_x. Whether it will work in full-scale applications, especially in flue gases contaminated with fly ash, remains to be seen.

Hydrocarbon Controls

Hydrocarbons and volatile organic compounds are produced by incomplete combustion of fuels or solvent evaporation from chemical factories, painting, dry cleaning, plastic manufacturing, printing, and other industrial processes that use a variety of volatile organic chemicals. Closed systems that prevent escape of fugitive gases can reduce many of these emissions. In automobiles, for instance, positive crankcase ventilation (PCV) systems collect oil that escapes from around the pistons and unburned fuel and channels it back to the engine for combustion. Modification of carburetor and fuel systems prevents evaporation of gasoline. In the

what can you do?
A GLOBAL CONCERN

Saving Energy and Reducing Pollution

- Conserve energy: carpool, bike, walk, use public transport, buy compact fluorescent bulbs, and energy-efficient appliances (see Chapter 13 for other suggestions).
- Don't use polluting two-cycle gasoline engines if cleaner four-cycle models are available for lawn mowers, boat motors, etc.
- Buy refrigerators and air conditioners designed for CFC alternatives. If you have old appliances or other CFC sources, dispose of them responsibly.
- Plant a tree and care for it (every year).
- Write to your Member of Parliament and support a transition to an energy-efficient economy.
- If green-pricing options are available in your area, buy renewable energy.
- If your home has a fireplace, install a high-efficiency, clean-burning, two-stage insert that conserves energy and reduces pollution up to 90 percent.
- Have your car tuned every 20,000 km and make sure that its anti-smog equipment is working properly. Turn off your engine when waiting longer than one minute. Start trips a little earlier and drive slower—it not only saves fuel but it's safer, too.
- Use latex-based, low-volatile paint rather than oil-based (alkyd) paint.
- Avoid spray can products. Light charcoal fires with electric starters rather than petroleum products.
- Don't top off your fuel tank when you buy gasoline; stop when the automatic mechanism turns off the pump. Don't dump gasoline or used oil on the ground or down the drain.
- Buy clothes that can be washed rather than dry-cleaned.

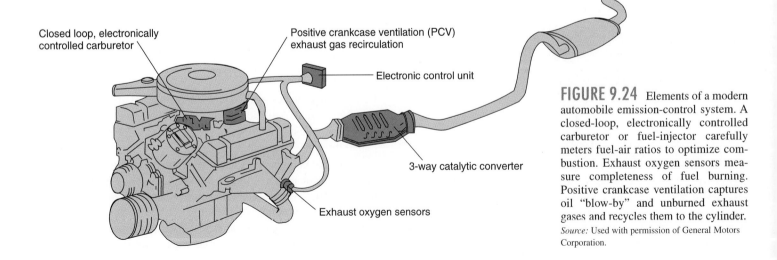

Closed loop, electronically controlled carburetor

Positive crankcase ventilation (PCV) exhaust gas recirculation

Electronic control unit

3-way catalytic converter

Exhaust oxygen sensors

FIGURE 9.24 Elements of a modern automobile emission-control system. A closed-loop, electronically controlled carburetor or fuel-injector carefully meters fuel-air ratios to optimize combustion. Exhaust oxygen sensors measure completeness of fuel burning. Positive crankcase ventilation captures oil "blow-by" and unburned exhaust gases and recycles them to the cylinder.
Source: Used with permission of General Motors Corporation.

PART TWO   Our Physical Environment

same way, controls on fugitive losses from valves, pipes, and storage tanks in industry can have a significant impact on air quality. Afterburners are often the best method for destroying volatile organic chemicals in industrial exhaust stacks. High air-fuel ratios in automobile engines and other burners minimize hydrocarbon and carbon monoxide emissions, but also cause excess nitrogen oxide production. Careful monitoring of air-fuel inputs and oxygen levels in exhaust gases can minimize all these pollutants.

CURRENT CONDITIONS AND FUTURE PROSPECTS

If we look at the changes in the concentration of air pollution emissions in urban areas of Canada over the last 20 years, together with established targets or maximum allowable levels, an interesting pattern emerges (Figure 9.25). Contaminants volatile organic compounds (VOCs) and NO_x have dropped substantially.

But ground level ozone, which is enhanced by interactions between VOCs and nitrogen oxide compounds (NO_x), has been more variable. In the last decade, the number and included geographic area of dangerously high smog days in southern Ontario have increased significantly. There are now many days in a typical summer when Algonquin Provincial Park, 250 km north Toronto, is part of a high smog warning.

With all of the enhanced technology that reduces air pollution generated by industry and personal transportation, why do we see ever-declining air quality? It is a case of numbers outrunning the efficiency and cleanliness of each individual. The number and size of industries and personal and commercial vehicles that generate the air pollution have increased dramatically in two decades. What used to be "rush hour" in the Greater Toronto Area (GTA) is a more or less continuous stream of heavy traffic on major expressways.

The outlook is not so encouraging in other parts of the world, however. The major metropolitan areas of many developing countries are growing at explosive rates to incredible sizes (Chapter 21), and environmental quality in many of them is dreadful (Fig. 9.26). The composite average annual levels of SO_2 in Tehran, Iran, for instance, are more than 150 mg/m^3, and peak levels can be up to 10 times higher. Mexico City remains notorious for bad air. Pollution levels exceed WHO health standards 350 days per year and more than half of all city children have lead levels in their blood sufficient to lower intelligence and retard development. Its 131,000 industries and 2.5 million vehicles spew out more than 5,500 tonnes of air pollutants daily. Santiago, Chile, averages 299 days per year on which suspended particulates exceed WHO standards of 90 mg/m^3.

While there are few statistics on China's pollution situation, it is known that many of China's 400,000 factories have no air pollution controls. Experts estimate that home coal burners and factories emit 10 million tonnes of soot and 15 million tonnes of sulphur dioxide annually and that emissions have increased rapidly over the past 25 years. Sheyang, an industrial city in northern China, is thought to have the world's worst continuing particulate problem with peak winter concentrations over 700 mg/m^3 (nine times North American maximum standards). Airborne particulates in Sheyang exceed WHO standards on 347 days per year. Beijing, Xian, and Guangzhou are nearly as bad. The high incidence of cancer in Shanghai is thought to be linked to air pollution.

As political walls came down across Eastern Europe and the Soviet Union at the end of the 1980s, horrifying environmental conditions in these centrally planned economies were revealed. Inept industrial managers, a rigid bureaucracy, and lack of democracy have created ecological disasters. Where governments own, operate, and regulate industry, there are few checks and balances or incentives to clean up pollution. Much of the Eastern bloc depends heavily on soft brown coal for its energy and pollution controls are absent or highly inadequate.

For many years, southern Poland, northern Czech Republic, and Slovakia were covered by a permanent cloud of smog from factories and power plants. Acid rain ate away historic buildings and damaged already inadequate infrastructures. The haze was so dark that drivers had to turn on their headlights during the day. Residents complained that washed clothes turned dirty before they could dry. Zabrze, near Katowice in southern Poland, had particulate emissions of 3,600 tonnes per square kilometre. This is more than seven times the emissions in Baltimore, Maryland, or Birmingham, Alabama, the dirtiest cities (for particulates) in the United States. Home gardening in Katowice has been banned because vegetables raised there have unsafe levels of lead and cadmium.

For miles around the infamous Romanian "black town" of Copsa Mica, the countryside is so stained by soot that it looks as if someone had poured black ink over everything. Birth defects afflict 10 percent of infants in northern Bohemia. Workers in factories there still get extra hazard pay—burial money, they call it. Life expectancy in these industrial towns is as much as 10 years less than the national average. Espenhain, in the industrial belt of the former East Germany, has one of the world's highest rates of sulphur dioxide pollution. One of every two children has lung problems, and one of every three has heart problems. Brass doorknobs and name plates have been eaten away by the acidic air in just a few months.

Not all is pessimistic, however. There have been some spectacular successes in air pollution control. Sweden and West Germany (countries affected by forest losses due to acid precipitation) cut their sulphur emissions by two-thirds between 1970 and 1985. Austria and Switzerland have gone even further. They even regulate motorcycle emissions. The Global Environmental Monitoring System (GEMS) reports declines in particulate levels in 26 of 37 cities worldwide. Sulphur dioxide and sulphate particles, which cause acid rain and respiratory disease, have declined in 20 of these cities.

Ten years ago, Cubatao, Brazil, was described as the "Valley of Death," one of the most dangerously polluted places in the world. A steel plant, a huge oil refinery, and fertilizer and chemical factories churned out thousands of tonnes of air pollutants every year. Trees died on the surrounding hills. Birth defects and respiratory diseases were alarmingly high. Since then, however, the citizens of Cubatao have made remarkable progress in cleaning up their environment. The end of military rule and restoration of

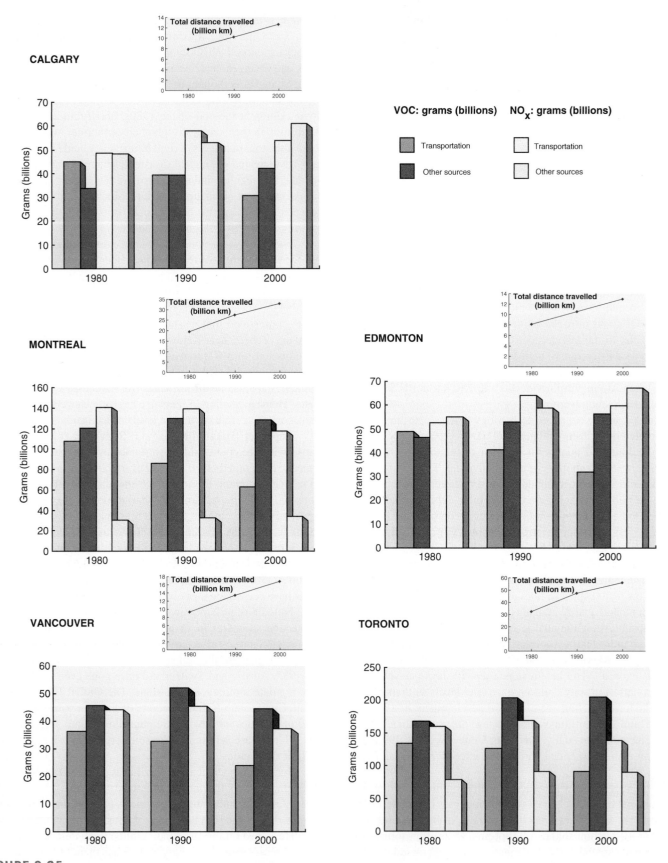

FIGURE 9.25 Air Pollution Emissions in Canada in Selected Urban Centres: 1980–2000.
Source: Energy and Environment Analysis, Arlington, Virginia.

FIGURE 9.26 Soot from factories and refineries has turned buildings, streets, and even this shepherd's sheep black in Copsa Mica, Romania.

democracy allowed residents to publicize their complaints. The environment became an important political issue. The state of São Paulo invested about $100 million, and the private sector spent twice as much to clean up most pollution sources in the valley. Particulate pollution was reduced 75 percent. Ammonia emissions were reduced 97 percent, hydrocarbons that cause ozone and smog were cut 86 percent, and sulphur dioxide production fell 84 percent. Fish are returning to the rivers, and forests are regrowing on the mountains. Progress is possible! We hope that similar success stories will be obtainable elsewhere.

SUMMARY

In this chapter, we have looked at major categories, types, and sources of air pollution. We have defined air pollution as chemical or physical changes brought about by either natural processes or human activities, resulting in air quality degradation. Air pollution has existed as long as there has been an atmosphere. Perhaps the first major human source of air pollution was fire. Burning forests, fossil fuels, and wastes continues to be the largest source of anthropogenic (human-caused) air pollution. The seven conventional large-volume pollutants are NO_x, SO_2, CO, volatile organic compounds, and three categories of particulates. The major sources of air pollution are transportation, industrial processes, stationary fuel combustion, and solid waste disposal.

We also looked at some unconventional pollutants. Indoor air pollutants, including formaldehyde, asbestos, toxic organic chemicals, radon, and tobacco smoke may pose a greater hazard to human health than all of the conventional pollutants combined. Odours, visibility losses, and noise generally are not life-threatening but serve as indicators of our treatment of the environment. Some atmospheric processes play a role in distribution, concentration, chemical modification, and elimination of pollutants. Among the most important of these processes are long-range transport of pollutants and photochemical reactions in trapped inversion layers over urban areas.

Destruction of stratospheric ozone by chlorofluorocarbons and other halogen-containing compounds is a serious global concern even though the highest levels of destruction occur over Antartica. In 2000, the area of Antarctic ozone depletion was larger than North America. Banning of CFCs has had a significant effect on reducing stratospheric chlorine levels.

QUESTIONS FOR REVIEW

1. Define primary and secondary air pollutants.
2. What are the seven Criteria Air Contaminants defined by Environment Canada? Why were they chosen?
3. What is acid deposition? What causes it?
4. What is an atmospheric inversion and how does it trap air pollutants?
5. How do electrostatic precipitators, baghouse filters, flue gas scrubbers, and catalytic converters work?
6. What is the difference between ambient standards and emission limits?
7. Which of the conventional pollutants has decreased most in the recent past and which has decreased least?

QUESTIONS FOR CRITICAL THINKING

1. Some authors limit pollution to human-caused materials. If a natural source releases the same toxic chemicals as a factory, is one pollution and the other not?

2. What might be done to improve indoor air quality? Should the government mandate such changes? What values or world views are represented by different sides of this debate?

3. Why do you suppose that air pollution is so much worse in Eastern Europe than in the West?

4. Suppose air pollution causes a billion dollars in crop losses each year but controlling the pollution would also cost a billion dollars. Should we insist on controls?

5. Utility managers once claimed that it would cost $1,000 per fish to control acid precipitation in Ontario lakes and that it would be cheaper to buy fish for anglers than to put scrubbers on power plants. Suppose that was true. Does it justify continuing pollution?

6. A worldwide ban on chlorofluorocarbons could mean that billions of people couldn't afford refrigerators and that millions might die each year from food poisoning. Is that justified by saving stratospheric ozone?

7. Is it possible to have zero emissions of pollutants? What does zero mean in this case?

8. If there are thresholds for pollution effects (at least as far as we know now), is it reasonable or wise to depend on environmental processes to disperse, assimilate, or inactivate waste products?

9. Should Canada ban all two-cycle engines (chain saws, outboard motors, lawnmowers, personal watercraft)? What evidence would you need to make a decision here?

KEY TERMS

acid precipitation   187
aerosol   179
aesthetic degradation   180
ambient air   176
bronchitis   185
carbon monoxide   178
chlorofluorocarbons
   (CFCs)   184
conventional or criteria air
   contaminants   176
dry alkali injection   191
electrostatic
   precipitators   190
emphysema   185
filters   190
fugitive emissions   176
nitrogen oxides   178
ozone   180
particulate material   179
photochemical oxidants   179
primary pollutants   176
secondary pollutants   176
sulphur dioxide   178
synergistic effects   186
temperature inversions   181
unconventional or noncriteria
   air contaminants   180
volatile organic
   compounds   179

Web Exercises

ADOPT A POLLUTANT

Go to Environment Canada's Air Pollution website at http://www.ec.gc.ca/air/air_pollution_e.html and pick an air pollutant to "adopt." Find out where it's a problem in Canada, and where it isn't. What is the source of the pollutant? Is it from a particular industry, or human activity? Does it come to us from another place? Is it becoming a more important problem or is it losing its importance? How is new technology helping or hindering progress in dealing with "your" pollutant?

Estimating Your Risk of Radon Exposure

After reading about indoor air pollution and the health risks associated with different air contaminants, you might want to look into the probabilities that you may be exposed to hazardous air. One of the most widespread indoor air health issues in North America is radon gas. The USGS provides a wealth of info about radon in indoor air on its website at   http://sedwww.cr.usgs.gov:8080/radon/georadon.html. You can use the data presented here to evaluate your own potential radon exposure as well as to learn more about the geology where you live. Read each of the sections on this page to understand what radon is, how it is formed, and how it gets into buildings. The fourth section on this page, Radon Potential, has information that will help you assess your own risk from this invisible, colourless, odourless, gas. The first illustration on the Radon Potential page is a graphic to show you how radon can move from geologic layers into your house. The second figure (131 kb Simplified geologic map of the lower 48 states of the U.S. and Puerto Rico) is too small to be very useful, but the third figure (475 kb Radioactivity map) provides very helpful information, particularly if you read the detailed analysis of uranium content of surface materials in 16 regions of the contiguous 48 states. Use this information, along with what kind of housing you live in and how it's ventilated, to evaluate your potential risk from radon in the air you breathe. If you live in an area with a high risk, read further to understand how radon is measured in indoor air and what you might do to lessen your risks.

Chapter 10

Water

If there is magic on this planet, it is in water.

—Loren Eisley—

OBJECTIVES

After studying this chapter, you should be able to:

- understand the unique properties of water and how they affect the environment at everything from micro to global scales.
- summarize how the hydrologic cycle delivers fresh water to terrestrial ecosystems and how the cycle balances over time.
- contrast the volume and residence time of water in the earth's major compartments.

WebQuest

limnology, oceanography, stream ecology, wetlands

Above: Clean water is a precious—and beautiful—resource.

Where Has the River Gone?

For more than 5,000 years, the Huang He, or Yellow River, has been the centre of Chinese civilization. The major water supply for the arid North China Plain, the river also brings fertile soil from the mountains to replenish crop fields (Fig. 10.1). But the river is known, too, as "China's sorrow" because of its frequent, catastrophic floods. More than 1,500 times in the last five millennia, the river has breached its dikes, turning the surrounding plain into a broad, shallow lake. These floods often resulted in dramatic realignments of the river channel. The last of these was in 1947, when the river shifted its mouth from the Yellow Sea, near Gansu, more than 320 km north to the Gulf of Zhili.

Water to irrigate paddy rice has long been the key to feeding China's people. With only 10 percent of its land arable, China could feed only a fraction of its population of 1.2 billion without irrigation. But with rapidly growing urban populations and increasing industrialization that compete for scarce water supplies, it isn't clear that there will be enough of this precious resource to go around. This is especially true in the arid North, where the water table beneath much of the farm land that currently supplies about 40 percent of China's food and feeds nearly 500 million people has dropped an average of 1.5 m per year for the last decade.

So much water is being diverted from the Huang He, that the river now frequently dries up completely in its lower stretches. Since 1985, the river has failed to reach the sea every year for at least a few months. In 1996, the river failed for 133 days. In 1997, a year plagued by drought in north China, there was no water in the lower Huang He for 226 days. Unless ways are found to conserve water or augment river flows, drastic changes will have to be made in Chinese agriculture, industry, and domestic water use. Some cities may have to be relocated. Farms and factories will close. Grain shortages in China could drive up world food prices with disastrous consequences for poor people everywhere.

The Huang He isn't the only river in the world that sometimes dries up completely because of human diversions and excess consumption. The mighty Colorado that roars through Arizona's Grand Canyon, is so depleted by the time it reaches the Mexican border that it seeps across the desert in a salty trickle that disappears entirely long before it reaches the Sea of Cortez. Similarly, the Amu Dar'ya and Syr Dar'ya in Kazakhstan were diverted completely in the 1970s, causing the Aral Sea to lose about two-thirds of its volume. Even the mighty Nile, the longest river in the world, and the sacred Ganges, are so overexploited that at some times of the year, hardly any of their water reaches the ocean.

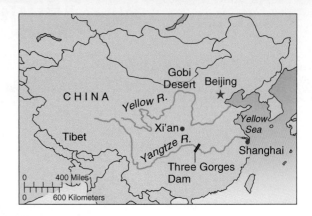

FIGURE 10.1 Flowing more than 4,800 km from the high Tibetian Plateau, across the southern Gobi Desert, to the Yellow Sea, the Huang He, or Yellow River, is the major water source for much of the arid North China Plain. Currently, agricultural, industrial, and domestic water withdrawals drain the river dry for several months each year.

The disappearance of these once mighty rivers is only a symptom of a larger dilemma. In many countries around the world, water shortages are becoming increasingly common. The United Nations warns that water supplies are likely to become one of the most pressing resource issues of this century. Already, some 2 billion people—one-third of the world's population—live in countries with moderate to high water stress. By 2025, two-thirds of all humans could be living in countries where water supplies are inadequate. Some sociologists warn that we could have water wars in the near future; others claim that we already have. In this chapter, we will look at the processes that supply fresh water and how humans access and use it. We will survey major water compartments of the environment and see how they are depleted by human activities. Finally, we will examine some ways that we can conserve and extend this vital resource.

Want More?

Dynesius, M., and Nilsson, C. 1994. "Fragmentation and Flow Regulation of River Systems in the Northern Third of the World." *Science*. 266: 753–762.

A "Water Planet"

If travellers from another solar system were to visit our lovely, cool, blue planet, they might call it Aqua rather than Terra because of its outstanding feature: the abundance of streams, rivers, lakes, and oceans of liquid water. Our planet is the only place we know where water exists as a liquid in any appreciable quantity. Water covers nearly three-fourths of the earth's surface and moves around constantly via the hydrologic cycle (discussed in this chapter) that distributes nutrients, replenishes freshwater supplies, and shapes the land. Without liquid water we would not exist. Among water's unique, almost magical qualities, are the following:

1. Water makes up 60 to 70 percent of the weight of most living organisms. It fills cells, thereby giving form and support to many tissues. Water is the medium in which all of life's chemical reactions occur, and it is an active participant in many of these reactions. Water is an excellent solvent that dissolves the nutrients cells need for life, as well as the wastes cells produce. Thus water is essential in delivering to and from cells.

2. Water is the only inorganic liquid that exists under normal conditions at temperatures suitable for life. Most substances exist as either a solid or a gas, with only a very narrow liquid temperature range. Organisms synthesize organic compounds such as oils and alcohols that remain liquid at ambient temperatures and are therefore extremely valuable to life, but the original and predominant liquid in nature is water.

3. Water molecules are cohesive, tending to stick together tenaciously. You have experienced this property if you have ever done a belly flop off a diving board. Water has the highest surface tension of any common, natural liquid. As a result, water is subject to *capillary action:* it can be drawn into small channels. Without capillary action, movement of water and nutrients into groundwater reservoirs and through living organisms might not be possible.

4. Water is unique in that it expands when it crystallizes. Most substances shrink as they change from liquid to solid. Ice floats because it is less dense than liquid water. When temperatures fall below freezing, the surface layers of lakes, rivers, and oceans cool faster and freeze before deeper water. Floating ice then insulates underlying layers, keeping most water bodies liquid (and aquatic organisms alive) throughout the winter in most places. Without this feature, lakes, rivers, and even oceans in high latitudes would freeze solid and never melt.

5. Water has a high heat of vapourization, using a great deal of heat to convert from liquid to vapour. Consequently, evaporating water is an effective way for organisms to shed excess heat. Many animals pant or sweat to moisten evaporative cooling surfaces. Why do you feel less comfortable on a hot, humid day that on a hot, dry day? Because the water vapour-laden air inhibits the rate of evaporation from

your skin, thereby impairing your ability to shed heat.

6. Water also has a high specific heat; that is, a great deal of heat is absorbed before it changes temperature. The slow response of water to temperature change helps moderate global temperatures, keeping the environment warm in winter and cool in summer. This effect is especially noticeable near the ocean, but it is important globally.

All these properties make water a unique and vitally important component of the ecological cycles that move materials and energy around and make life on earth possible.

Surface tension is demonstrated by the resistance of a water surface to penetration, as when it is walked upon by a water strider.

WATER RESOURCES

Water is a marvelous substance—flowing, rippling, swirling around obstacles in its path, seeping, dripping, trickling—constantly moving from sea to land and back again. Water can be clear, crystalline, icy green in a mountain stream or black and opaque in a cypress swamp. Water bugs skitter across the surface of a quiet lake; a stream cascades down a stairstep ledge of rock; waves roll endlessly up a sand beach, crash in a welter of foam, and recede. Rain falls in a gentle mist, refreshing plants and animals. A violent thunderstorm floods a meadow, washing away stream banks.

Water is essential for life (Chapter 3). It is the medium in which all living processes occur. Water dissolves nutrients and distributes them to cells, regulates body temperature, supports structures, and removes waste products. About 60 percent of your body is water. You could survive for weeks without food, but only a few days without water.

The earth is the only place in the universe, as far as we know, where liquid water exists in substantial quantities. Oceans, lakes, rivers, glaciers, and other bodies of liquid or solid water cover more than 70 percent of our world's surface. The total amount of water on our planet is immense—more than 1,404 million km^3

TABLE 10.1 Some Units of Water Measurement

One cubic kilometre (km^3) equals 1 billion cubic metres (m^3) or 1 trillion litres.

A modest 10 mm of rain on a hectare (100 m $\times$ 100 m) of land equals 100 cubic metres of water.

One cubic metre per second of river flow, typical for a fairly small stream, equals 1,000 litres per second.

(Table 10.1). If the earth had a perfectly smooth surface, an ocean about 3 km deep would cover everything. Fortunately for us, continents rise above the general surface level, creating dry land over about 30 percent of the planet.

The Hydrologic Cycle

The hydrologic cycle (water cycle) describes the circulation of water as it evaporates from land, water, and organisms; enters the atmosphere; condenses and is precipitated to the earth's surfaces; and moves underground by infiltration or overland by runoff into rivers, lakes, and seas (Fig. 10.2). This cycle supplies fresh water to the landmasses, maintains a habitable climate, and moderates world temperatures. Movement of water back to the sea in rivers and glaciers is a major geological force that shapes the land and redistributes material. Plants play an important role in the hydrologic cycle, absorbing groundwater and pumping it into the atmosphere by transpiration (transport plus evaporation). In tropical forests, as much as 75 percent of annual precipitation is returned to the atmosphere by plants.

Solar energy drives the hydrologic cycle by evaporating surface water. **Evaporation** is the process in which a liquid is changed to vapour (gas phase) at temperatures well below its boiling point. Water also can move between solid and gaseous states without ever becoming liquid in a process called **sublimation.** On bright, cold, windy winter days, when the air is very dry, snowbanks disappear by sublimation, even though the temperature never gets above freezing. This is the same process that causes "freezer burn" of frozen foods.

In both evaporation and sublimation, molecules of water vapour enter the atmosphere, leaving behind salts and other contaminants and thus creating purified fresh water. This is essentially distillation on a grand scale. We used to think of rainwater as a symbol of purity, a standard against which pollution could be measured. Unfortunately, increasing amounts of atmospheric pollutants are picked up by water vapour as it condenses into rain.

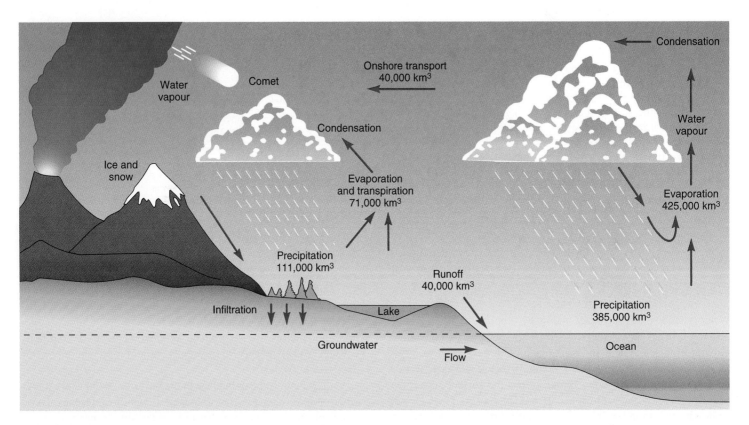

FIGURE 10.2 In the hydrologic cycle water moves constantly between aquatic, atmospheric, and terrestrial compartments driven by solar energy and gravity. The total annual runoff from land to the oceans is about 40×10^{12} m^3.

The amount of water vapour in the air is called humidity. Warm air can hold more water than cold air. When a volume of air contains as much water vapour as it can at a given temperature, we say that it has reached its **saturation point. Relative humidity** is the amount of water vapour in the air expressed as a percentage of the maximum amount (saturation point) that could be held at that particular temperature.

When the saturation concentration is exceeded, water molecules begin to aggregate in the process of **condensation.** If the temperature at which this occurs is above 0°C, tiny liquid droplets result. If the temperature is below freezing, ice forms. For a given amount of water vapour, the temperature at which condensation occurs is the **dew point.** Tiny particles, called **condensation nuclei,** float in the air and facilitate this process. Smoke, dust, sea salts, spores, and volcanic ash all provide such particles. Even apparently clear air can contain large numbers of these particles, which are generally too small to be seen by the naked eye. Sea salt is an excellent source of such nuclei, and heavy, low clouds frequently form in the humid air over the ocean.

A cloud, then, is an accumulation of condensed water vapour in droplets or ice crystals. Normally, cloud particles are small enough to remain suspended in the air, but when cloud droplets and ice crystals become large enough, gravity overcomes uplifting air currents and precipitation occurs. Some precipitation never reaches the ground. Temperatures and humidities in the clouds where snow and ice form are ideal for their preservation, but as they fall through lower, warmer, and drier air layers, reevaporation occurs. Rising air currents lift this water vapour back into the clouds, where it condenses again; thus, liquid water and ice crystals may exist for only a few minutes in this short cycle between clouds and air (Fig. 10.2).

Rainfall and Topography

Rain falls unevenly over the planet. In some places, it rains more or less constantly, while other areas get almost no precipitation of any kind. At Iquique, in the Chilean desert, for instance, no rain has fallen in recorded history. At the other end of the scale, 22 m of rain was recorded in a single year at Cherrapunji in India. Figure 10.3 shows broad patterns of precipitation around the world.

Very heavy rainfall is typical of tropical areas, especially where seasonal monsoon winds carry moisture-laden sea air onshore (Chapter 8). Mountains act as both cloud formers and rain catchers. As air sweeps up the windward side of a mountain, pressure decreases and the air cools, causing relative humidity to increase. Eventually, the saturation point is reached, and moisture

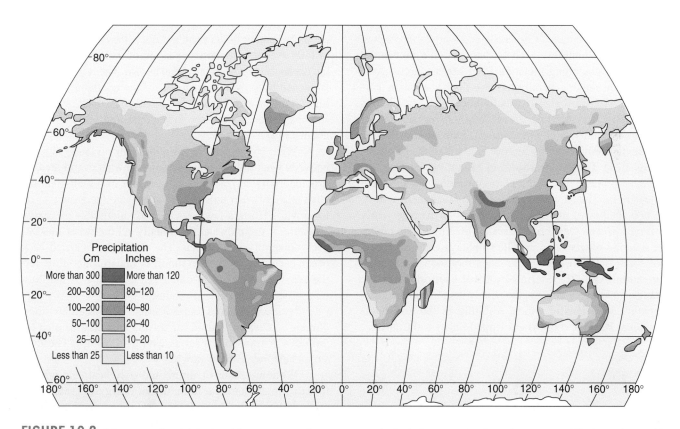

FIGURE 10.3 Mean annual precipitation. Note wet areas that support tropical rainforests occur along the equator while the major world deserts occur in zones of dry, descending air between 20 and 40 degrees North and South.

From Jerome Fellmann, et al., *Human Geography*, 4th ed. Copyright © 1995 Times Mirror Higher Education Group, Inc., Dubuque, Iowa. All Rights Reserved. Reprinted by permission.

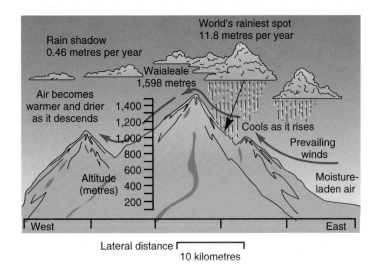

FIGURE 10.4 Rainfall on the east side of Mount Waialeale in Hawaii is more than 20 times as much as on the west side. Prevailing trade winds bring moisture-laden sea air onshore. The air cools as it rises up the flanks of the mountain and the water it carries precipitates as rain—11.8 m per year!

in the cool air condenses, forming rain drops (or snowflakes). Rain falls on the mountaintop, leaving the air drier. The cool, drier air continues over the mountain. It descends and warms once again, reducing its relative humidity even further, not only preventing rainfall there, but absorbing moisture from other sources. As a result, a mountain range generally has two distinct climatic personalities. The windward side is usually cool, wet, and cloudy, while the leeward side is warm, dry, and sunny. We call the dry area on the downwind side of a mountain its **rain shadow.**

A striking example of this contrast is found in the Hawaiian Islands (Fig. 10.4). The windward side of Mount Waialeale on the island of Kauai is one of the wettest places on earth, with an annual rainfall near 12 m. The leeward side, only a few kilometres away, is in the rain shadow of the mountain and has an average yearly rainfall of only 46 cm. On a broader scale, some mountain ranges cast rain shadows over vast areas. The Himalaya and Karakorum ranges of south Asia block moisture-laden monsoon winds from reaching central Asia. The Sierra Nevada of California and the coastal ranges of Oregon, Washington, and British Columbia intercept moisture-laden Pacific winds, resulting in the arid intermountain Great Basin of the western United States and Canada.

Desert Belts

Rising and falling air masses that result from global circulation patterns also help create deserts in two broad belts on either side of the equator around the world. Evaporation is highest near the equator where direct rays of the sun produce the greatest heat budgets. Hot air over the equator rises, cools, and drops its moisture

as rain; thus, equatorial regions are areas of high precipitation. As this cooler, drier air moves toward the poles, it condenses and sinks earthward again along the Tropics of Cancer and Capricorn (23° north and south latitude, respectively), warming as it descends.

This hot, dry air causes high evaporative losses in these subtropical regions and creates great deserts on nearly every continent: the Sahara of North Africa, the Takla Makan and Gobi of China, the Sonoran and Chihuahuan of Mexico and the United States, the Kalahari and Namib of Southwest Africa, and Australia's Great Sandy Desert. Dry air cascading down the west side of the Andes creates some of the driest deserts in the world along the coasts of Chile and Peru.

Humans and domestic animals have expanded many of these deserts by destroying forests and stripping away protective vegetation from once fertile lands, exposing the bare soil to erosion. Local weather patterns and water supplies also are adversely affected by this process of desertification. Weather and climate are discussed in Chapter 8.

Balancing the Water Budget

Everything about global hydrological processes is awesome in scale. Each year, the sun evaporates approximately 496,000 km³ of water from the earth's surface (see Fig. 10.2). More water evaporates in the tropics than at higher latitudes, and more water evaporates over the oceans than over land. Although the oceans cover about 70 percent of the earth's surface, they account for 86 percent of total evaporation. Ninety percent of the water evaporated from the ocean falls back on the ocean as rain. The remaining 10 percent is carried by prevailing winds over the continents where it combines with water evaporated from soil, plant surfaces, lakes, streams, and wetlands to provide a total continental precipitation of about 111,000 km³.

What happens to the surplus water on land—the difference between what falls as precipitation and what evaporates? Some of it is incorporated by plants and animals into biological tissues. A large share of what falls on land seeps into the ground to be stored for a while (from a few days to many thousands of years) as soil moisture or groundwater. Eventually, all the water makes its way back downhill to the oceans. The 40,000 km³ carried back to the ocean each year by surface runoff or underground flow represents the renewable supply available for human uses and sustaining freshwater-dependent ecosystems.

Evaporation and condensation help regulate the earth's climate. As warm, humid air travels from the tropics to cooler latitudes, it transports heat as well as moisture. The ocean also stores heat, releasing it slowly.

Without oceans to absorb and store heat, and wind currents to redistribute that heat in the latent energy of water vapour, the earth would probably undergo extreme temperature fluctuations like those of the moon, where it is 100°C during the day and –130°C at night. Water is able to perform this vital function because of its unique properties in heat absorption and energy of vaporization (Chapter 3).

TABLE 10.2 — Earth's Water Compartments—Estimated Volume of Water in Storage, Percent of Total, and Average Residence Time

	VOLUME (THOUSANDS OF KM^3)	% TOTAL WATER	AVERAGE RESIDENCE TIME
Total	1,403,377	100	2,800 years
Ocean	1,370,000	97.6	3,000 years to 30,000 years*
Ice and snow	29,000	2.07	1 to 16,000 years*
Groundwater down to 1 km	4,000	0.28	From days to thousands of years*
Lakes and reservoirs	125	0.009	1 to 100 years*
Saline lakes	104	0.007	10 to 1,000 years*
Soil moisture	65	0.005	2 weeks to a year
Biological moisture in plants and animals	65	0.005	1 week
Atmosphere	13	0.001	8 to 10 days
Swamps and marshes	3.6	0.003	From months to years
Rivers and streams	1.7	0.0001	10 to 30 days

*Depends on depth and other factors.
Source: Data from U.S. Geological Survey.

MAJOR WATER COMPARTMENTS

The distribution of water often is described in terms of interacting compartments in which water resides for short or long times. Table 10.2 shows the major water compartments in the world.

Oceans

Together, the oceans contain more than 97 percent of all the *liquid* water in the world. (The water of crystallization in rocks is far larger than the amount of liquid water.) Oceans are too salty for most human uses, but they contain 90 percent of the world's living biomass. While the ocean basins really form a continuous reservoir, shallows and narrows between them reduce water exchange, so they have different compositions, climatic effects, and even different surface elevations.

Oceans play a crucial role in moderating the earth's temperature (Fig. 10.5). Vast river-like currents transport warm water from the equator to higher latitudes, and cold water flows from the poles to the tropics (Fig. 10.6). The Gulf Stream, which flows northeast from the coast of North America toward northern Europe, flows at a steady rate of 10–12 km per hour and carries more than 100 times more water than all rivers on earth put together.

In tropical seas, surface waters are warmed by the sun, diluted by rainwater and runoff from the land, and aerated by wave action. In higher latitudes, surface waters are cold and much more dense. This dense water subsides or sinks to the bottom of deep ocean basins and flows toward the equator. Warm surface water of the tropics stratifies or floats on top of this cold, dense water like cream on an unstirred cup of coffee. Sharp boundaries form between different water densities, different salinities, and different temperatures, retarding mixing between these layers.

While parts of the hydrologic cycle occur on a time scale of hours or days, other parts take centuries. The average **residence time** of water in the ocean (the length of time that an individual molecule spends circulating in the ocean before it evaporates and starts through the hydrologic cycle again) is about 3,000 years. In the deepest ocean trenches, movement is almost nonexistent and water may remain undisturbed for tens of thousands of years.

Glaciers, Ice, and Snow

Of the 2.4 percent of all water that is fresh, nearly 90 percent is tied up in glaciers, ice caps, and snowfields (Fig. 10.7). Glaciers are really rivers of ice flowing downhill very slowly (Fig. 10.8). They now occur only at high altitudes or high latitudes, but as recently as 18,000 years ago about one-third of the continental landmass was covered by glacial ice sheets. Most of this ice has now melted and the largest remnant is in Antarctica. As much as 2 km thick, the Antarctic glaciers cover all but the highest mountain peaks and contain nearly 85 percent of all ice in the world.

An ice sheet that is similar in thickness but much smaller in volume covers most of Greenland. There is no landmass at the North Pole. A permanent ice pack made of floating sea ice covers much of the Arctic Ocean. Although sea ice comes from ocean water, salt is excluded in freezing so the ice is mostly fresh water. Together with the Greenland ice sheet, arctic ice makes up about 10 percent of the total ice volume. The remaining 5 percent of the world's permanent supply of ice and snow occurs mainly on high mountain peaks.

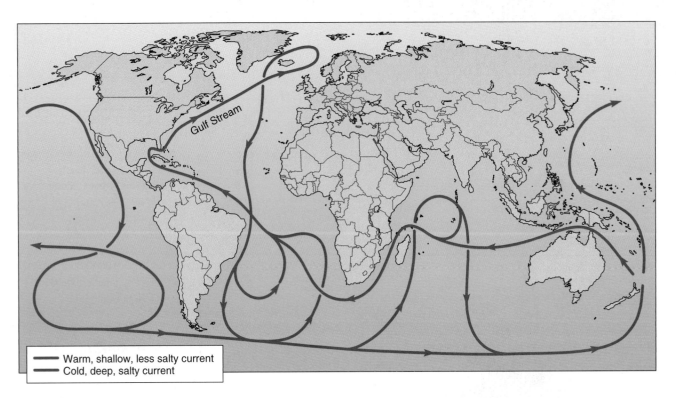

FIGURE 10.5 Ocean currents act as a global conveyor system, redistributing warm and cold water around the globe. These currents moderate our climate. For example, the Gulf Stream keeps northern Europe much warmer than northern Canada.

FIGURE 10.6 The Gulf Stream carries warm water far north, so that even this island in arctic Norway is moderately warm year-round.

Groundwater

After glaciers, the next largest reservoir of fresh water is held in the ground as **groundwater.** Precipitation that does not evaporate back into the air or run off over the surface percolates through the soil and into fractures and spaces of permeable rocks in a process called **infiltration** (Fig. 10.9). Upper soil layers that hold both air and water make up the **zone of aeration.** Moisture for plant growth comes primarily from these layers. Depending on rainfall amount, soil type, and surface topography, the zone of aeration may be very shallow or quite deep. Lower soil layers where all spaces are filled with water make up the **zone of saturation.** The top of this zone is the **water table.** The water table is not flat, but undulates according to the surface topography and subsurface structure. Nor is it stationary through the seasons, rising and falling according to precipitation and infiltration rates.

Porous layers of sand, gravel, or rock lying below the water table are called **aquifers.** Aquifers are always underlain by relatively impermeable layers of rock or clay that keep water from seeping out at the bottom.

Folding and tilting of the earth's crust by geologic processes can create shapes that generate water pressure in confined aquifers (those trapped between two impervious, confining rock layers). When a pressurized aquifer intersects the surface, or if it is penetrated by a pipe or conduit, an **artesian** well or spring results from which water gushes without being pumped.

Areas in which infiltration of water into an aquifer occurs are called **recharge zones** (Fig. 10.10). The rate at which most aquifers are refilled is very slow, however, and groundwater presently is being removed faster than it can be replenished in many areas. Urbanization, road building, and other development often block recharge zones and prevent replenishment of important aquifers. Contamination of surface water in recharge zones and seepage of pollutants into abandoned wells have polluted

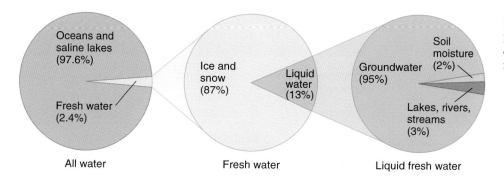

All water | Fresh water | Liquid fresh water

FIGURE 10.7 The easily accessible water in lakes, rivers, and streams represents only 3 percent of all liquid fresh water, which is 13 percent of all fresh water, which is 2.4 percent of all water.

FIGURE 10.8 Glaciers are rivers of ice sliding very slowly downhill. Together, polar ice sheets and alpine glaciers contain more than three times as much fresh water as all the lakes, ponds, streams, and rivers in the world. The dark streaks on the surface of this Alaskan glacier are dirt and rocks marking the edges of tributary glaciers that have combined to make this huge flow.

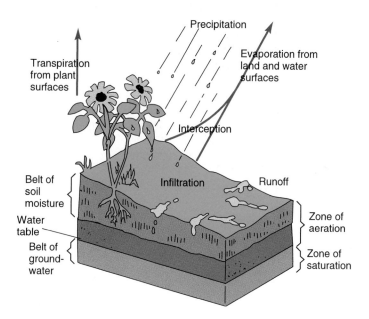

FIGURE 10.9 Precipitation that does not evaporate or run off over the surface percolates through the soil in a process called infiltration. The upper layers of soil hold droplets of moisture between air-filled spaces. Lower layers, where all spaces are filled with water, make up the zone of saturation or groundwater.

aquifers in many places, making them unfit for most uses (Chapter 11). Many cities protect aquifer recharge zones from pollution or development, both as a way to drain off rainwater and as a way to replenish the aquifer with pure water.

Some aquifers contain very large volumes of water. Even though Canada is rich in lakes and streams, there is substantially more water underground than on the surface. In the United States, where surface water is not as abundant as in Canada, there is 30 times the volume of water underground as in lakes and streams. Municipal and agricultural areas in North America are quite dependent on groundwater for their water supply. In Canada, 26 percent of the population relies on groundwater, although this varies from 100 percent in Prince Edward Island, to less than 1 percent of those

living on the Canadian Shield. The granitic bedrock of the shield is relatively impermeable to water, and groundwater is less abundant. While water can flow through limestone caverns in underground rivers, most movement in aquifers is a dispersed and almost imperceptible trickle through tiny fractures and spaces. Depending on geology, it can take anywhere from a few hours to several years for contaminants to move a few hundred metres through an aquifer.

Rivers and Streams

Precipitation that does not evaporate or infiltrate into the ground runs off over the surface, drawn by the force of gravity back toward the sea. Rivulets accumulate to form streams, and streams join to form rivers. Although the total amount of water contained at any one time in rivers and streams is small compared to the other water

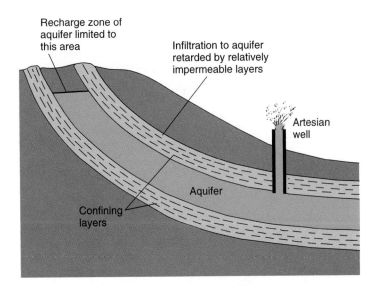

FIGURE 10.10 An aquifer is a porous, water-bearing layer of sand, gravel, or rock. This aquifer is confined between layers of rock or clay and bent by geological forces, creating hydrostatic pressure. A break in the overlying layer creates an artesian well or spring.

Labels in figure:
Recharge zone of aquifer limited to this area
Infiltration to aquifer retarded by relatively impermeable layers
Artesian well
Confining layers
Aquifer

TABLE 10.3	Major Rivers of the World	
RIVER	COUNTRIES IN RIVER BASIN	AVERAGE ANNUAL DISCHARGE AT MOUTH (CUBIC METRES/SECOND)*
Amazon	Brazil, Peru	175,000
Orinoco	Venezuela, Colombia	45,300
Congo	Congo	39,200
Yangtze	Tibet, China	28,000
Bramaputra	Tibet, India, Bangladesh	19,000
Mississippi	United States	18,400
Mekong	China, Laos, Burma, Thailand, Cambodia, Vietnam	18,300
Parana	Paraguay, Argentina	18,000
Yenisey	Russia	17,200
Lena	Russia	16,000
Irrawaddy	Burma	13,000
Ob	Russia	12,000
Ganges	Nepal, India, Bangladesh	11,600
Amur	China, Russia	11,000
St. Lawrence	Canada, United States	10,700
Mackenzie	Canada	9,600

Source: Data from World Resources Institute.

reservoirs of the world (see Table 10.2), these surface waters are vitally important to humans and most other organisms. Most rivers, if they were not constantly replenished by precipitation, meltwater from snow and ice, or seepage from groundwater, would begin to diminish in a few weeks.

The speed at which a river flows is not a very good measure of how much water it carries. Headwater streams are usually small and fast, often tumbling downhill in a continuous cascade. As the stream reaches more level terrain, it slows and generally becomes deeper and more quiet. The best measure of the volume carried by a river is its **discharge,** the amount of water that passes a fixed point in a given amount of time. This is usually expressed as litres of water per second. The 16 largest rivers in the world carry nearly half of all surface runoff on earth. The Amazon is by far the largest river in the world (Table 10.3), carrying roughly 10 times the volume of the Mississippi. Several Amazonian tributaries such as the Maderia, Rio Negro, and Ucayali would be among the world's top rivers in their own right.

Lakes and Ponds

Ponds are generally considered to be small temporary or permanent bodies of water shallow enough for rooted plants to grow over most of the bottom. Lakes are inland depressions that hold standing fresh water year-round. Maximum lake depths range from a few metres to over 1,600 m in Lake Baikal in Siberia. Surface areas vary in size from less than one-half hectare to large inland seas, such as Lake Superior or the Caspian Sea, covering hundreds of thousands of square kilometres. Both ponds and lakes are relatively temporary features on the landscape because they eventually fill with silt

or are emptied by cutting of an outlet stream through the barrier that creates them.

While lakes contain nearly 100 times as much water as all rivers and streams combined, they are still a minor component of total world water supply. Their water is much more accessible than groundwater or glaciers, however, and they are important in many ways for humans and other organisms.

Wetlands

Bogs, swamps, wet meadows, and marshes play a vital and often unappreciated role in the hydrological cycle. Their lush plant growth stabilizes soil and holds back surface runoff, allowing time for infiltration into aquifers and producing even, year-long stream flow. A substantial part of Canada's land mass (14 percent) is wetland, although wetlands in the southern strip of Canada have been drained, filled, or degraded by agricultural and urban development.

When wetlands are disturbed, their natural water-absorbing capacity is reduced and surface waters run off quickly, resulting in floods and erosion during the rainy season and dry, or nearly dry, stream beds the rest of the year. This has a disastrous effect on biological diversity and productivity, as well as on human affairs (Chapter 22).

The Atmosphere

The atmosphere is among the smallest of the major water reservoirs of the earth in terms of water volume, containing less than 0.001 percent of the total water supply. It also has the most rapid turnover rate. An individual water molecule resides in the atmosphere for about 10 days, on average. While water vapour makes up only a small amount (4 percent maximum at normal temperatures) of the total volume of the air, movement of water through the atmosphere provides the mechanism for distributing fresh water over the landmasses and replenishing terrestrial reservoirs.

SUMMARY

The hydrologic cycle constantly purifies and redistributes fresh water, providing an endlessly renewable resource. The physical processes that make this possible—evaporation, condensation, and precipitation—depend upon the unusual properties of water, especially its ability to absorb and store solar energy. More than 97 percent of all water in the world is salty ocean water. Of the 33,400 km^3 that is fresh, 99 percent is locked up in ice or snow or buried in groundwater aquifers. Lakes, rivers, and other surface freshwater bodies make up only about 0.01 percent of all the water in the world, but they provide more than half of all water for human use and habitat and nourishment for aquatic ecosystems that play a vital role in the chain of life.

QUESTIONS FOR REVIEW

1. Describe the path a molecule of water might follow through the hydrologic cycle from the ocean to land and back again.

2. Define *evaporation, sublimation, condensation, precipitation,* and *infiltration.* How do they work?

3. How do mountains affect rainfall distribution? Does this affect your part of the country?

4. What are the major water reservoirs of the world?

5. How much water is fresh (as opposed to saline) and where is it?

6. Define *aquifer.* How does water get into an aquifer?

QUESTIONS FOR CRITICAL THINKING

1. What changes might occur in the hydrologic cycle if our climate were to warm or cool significantly?

2. Why does it take so long for the deep ocean waters to circulate through the hydrologic cycle? What happens to substances that contaminate deep ocean water or deep aquifers in the ground?

3. Where would you most like to spend your vacations? Does availability of water play a role in your choice? Why?

4. How do you think strategies and techniques for protecting drinking water supplies in ground water differ from surface water source protection? What is similar about the two types of drinking water sources?

5. How does water differ from other natural liquids? How do the properties of water make the hydrologic cycle, and life, possible?

KEY TERMS

aquifer   204
artesian   204
condensation   201
condensation nuclei   201
dew point   201
discharge   206
evaporation   200
groundwater   204
infiltration   204
rain shadow   202
recharge zones   204
relative humidity   201
residence time   203
saturation point   201
sublimation   200
water table   204
zone of aeration   204
zone of saturation   204

Go to the Watershed Stewardship website at http://www.stewardshipcanada.ca/communities/watershedstewardship/home/wscnIndex.asp. Find the watershed you live in. Identify the key environmental issues in your watershed. What do you see as the major threats and opportunities to your watershed over the next 50 years?

Chapter 11

Water Quantity and Quality

Water, water, everywhere; nor any drop to drink.

—Samuel Taylor Coleridge—

OBJECTIVES

After studying this chapter, you should be able to:

- describe the important ways we use water and distinguish between withdrawal, consumption, and degradation.
- appreciate the causes and consequences of water shortages around the world and what they mean in people's lives in water-poor countries.
- understand the arguments for and against large dams.
- apply some water conservation methods in your own life.
- define *water pollution* and describe the sources and effects of some major types.
- appreciate why access to sewage treatment and clean water are important to people in developing countries.
- discuss the status of water quality in developed and developing countries.
- delve into groundwater problems and suggest ways to protect this precious resource.

- fathom the causes and consequences of ocean pollution.
- weigh the advantages and disadvantages of different human waste disposal techniques.

WebQuest

impoundments, drought, flood, International Rivers Network, RiverKeepers, WaterKeepers, StreamKeepers, eutrophication, water treatment technology

Above: Students collect invertebrates to assess the health of a stream ecosystem.

A Flood of Pigs

When Hurricane Floyd roared ashore at Cape Fear in the early morning hours of September 16, 1999, it brought much more than just strong winds to North Carolina. More than 50 cm of rain poured down in less than 24 hours on ground already saturated from previous storms. The eastern part of North Carolina is a low, coastal plain drained by an intricate web of sluggish creeks and meandering rivers. As the deluge poured down, these myriad channels filled, spilled over, and merged to create an enormous lake more than 300 km across, covering towns, farms, factories, and forests. The Great Dismal Swamp in the northeast corner of the state became truly soggy and dismal. Worst of all were the thousands of hog and poultry operations and their open manure lagoons submerged by the raging water.

In the 1990s, North Carolina became the leading turkey-producing state in the United States, and second to Iowa in pork production, with more than 45 million turkeys and 10 million pigs (but only 7 million people), mostly raised in huge, factory operations. Manure is washed out of giant confinement barns and stored in huge ponds, each typically about the size of several football fields and holding around 40,000 m^3 of liquid waste. The heart of corporate animal-farming is along the sluggish Neuse River as it winds across the flat coastal plain to its estuary on Pamlico Sound (Fig. 11.1). Even now, no flood protection measures are required for large-scale animal operations. Environmentalists have warned for years that storage of such massive amounts of manure in a storm-prone floodplain is an accident waiting to happen. A 1996 editorial in the Raleigh *News and Observer* complained: "Are we so foolish as to think the winds won't blow, the rains won't fall, and the rivers won't rise again?"

We'll probably never know the entire damage from Hurricane Floyd. Fifty-one people died in eastern North Carolina, about 8,000 homes were damaged or destroyed, and a million ha of crops were destroyed. Property losses were several billion dollars. Environmental damage is harder to measure. Millions of dead pigs, chickens, turkeys, and cattle littered the landscape or washed out to sea. Flooded junkyards, factories, and chemical plants oozed a toxic mix of gasoline, oil, and other chemicals. Twenty municipal

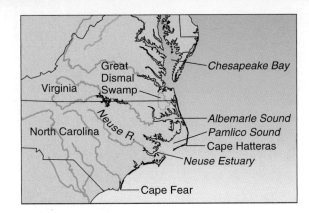

FIGURE 11.1 Heavy rains from Hurricane Floyd flooded a large area of eastern North Carolina in 1999, and washed millions of gallons of hog and turkey manure into Pamlico and Albemarle Sounds.

wastewater treatment plants were swamped. Environmentalists claim that at least 250 overflowing manure ponds disgorged around 10 million m^3 of hog and poultry waste into the flood. The Carolina Pork Council, on the other hand, says that only three of 4,000 waste lagoons overflowed. The fecal coliform bacteria now found in wells, industry claims, probably came from humans or wild Canadian geese. At the same time, North Carolina farmers are asking the federal government for $1 billion in flood relief to rebuild damaged barns and waste lagoons.

Whatever the source, decomposing sewage and dead organisms have created an anerobic "dead zone" in the Pamlico Sound, one of the largest estuary systems in North America. Nutrients also are thought to contribute to blooms of toxic microorganisms such as the dinoflagellate, which kills both finfish and shellfish, and has long-term adverse health effects on humans who eat seafood, get water on their skin, or even breathe toxic fumes from contaminated sources. When the floods finally receded, tens of thousands of dead animals were buried in shallow, watery graves, where they continue to contaminate groundwater.

Few environmental science issues are as widespread or emotionally charged as water use and pollution.

WATER AVAILABILITY AND USE

Clean, fresh water is essential for nearly every human endeavour. Perhaps more than any other environmental factor, the availability of water determines the location and activities of humans on earth (Fig. 11.2). **Renewable water supplies** are made up, in general, of surface runoff plus the infiltration into accessible freshwater aquifers. About two-thirds of the water carried in rivers and streams every year occurs in seasonal floods that are too large or violent to be stored or trapped effectively for human uses. Stable runoff is the dependable, renewable, year-round supply of surface water.

Much of this occurs, however, in sparsely inhabited regions or where technology, finances, or other factors make it difficult to use it productively. Still, the readily accessible, renewable water supplies are very large, amounting to some 1,500 km^3 per person per year worldwide.

Water-Rich and Water-Poor Countries

As you can see in Figure 10.3, South America, West Central Africa, and South and Southeast Asia all have areas of very high rainfall. Brazil and the Democratic Republic of Congo, because they have

FIGURE 11.2 Water has always been the key to survival. Who has access to this precious resource and who doesn't has long been a source of tension and conflict.

TABLE 11.1	Water-Rich Countries

COUNTRY	RENEWABLE ANNUAL SUPPLY[1] (CUBIC METRES/CAPITA)
Iceland	606,500
Surinam	452,000
Guyana	281,500
Papua New Guinea	174,000
Gabon	140,000
Solomon Islands	107,000
Canada	94,000
Norway	88,000
Panama	52,000
Brazil	31,000

[1]Renewable supplies include both surface runoff and groundwater replenishment.
Source: Data from *World Resources, 1998–99.*

TABLE 11.2	Water-Poor Countries

COUNTRY	RENEWABLE ANNUAL SUPPLY[1] (CUBIC METRES/CAPITA)
Kuwait	11
Egypt	43
United Arab Emirates	63
Malta	85
Jordan	114
Saudi Arabia	119
Singapore	172
Moldavia	225
Israel	289
Oman	393

[1]Renewable supplies include both surface runoff and groundwater replenishment.
Source: Data from *World Resources, 1998–99.*

high precipitation levels and large land areas, are among the most water-rich countries on earth. Canada and Russia, which are both very large, also have some of the highest total annual water supplies of any country. The highest per capita water supplies generally occur in countries with moist climates and low population densities (Table 11.1). Iceland, for example, has about 600 million litres per person per year. In contrast, Kuwait, where temperatures are extremely high and rain almost never falls, has less than 12,000 litres per person per year from renewable natural sources (Table 11.2). Almost all of Kuwait's water comes from imports and desalinized seawater. Egypt, in spite of the fact that the Nile River flows through it, has only about 42,000 litres of water annually per capita, or about 15,000 times less than Iceland. Singapore is an unusual case. Although it has a humid climate, this small, island nation has very little land area for its 3.5 million people, and must depend on its neighbour, Malaysia, for almost all its water.

Another important consideration is rainfall interannual variability. In some areas, such as the African Sahel region, abundant rainfall occurs some years but not others. Unless steps are taken to even out water flows, the lowest levels encountered usually limit both ecosystem functions and human activities. Some of the world's earliest civilizations, such as the Sumerians and Babylonians of Mesopotamia, the Harappans of the Indus Valley, and the early Chinese cultures, were based on communal efforts to control water, to divert floods and drain marshes during wet seasons or wet years, and to store water in reservoirs or divert it from streams so that it would be available during the dry seasons or dry years.

Drought Cycles

Rainfall is never uniform in either geographical distribution or yearly amount. Every continent has regions where rainfall is scarce because of topographic effects or wind currents. In addition, cycles of wet and dry years create temporary droughts. Water shortages have their most severe effect in semiarid zones where moisture availability is the critical factor in determining plant and animal distribution. Undisturbed ecosystems often survive extended droughts with little damage, but introduction of domestic animals and agriculture disrupts native vegetation and undermines natural adaptations to low moisture levels.

In North America, the cycle of drought seems to be about 30 years. There were severe dry years in the 1870s, 1900s, 1930s, 1950s, and 1970s. The worst of these in economic and social terms were the 1930s. Poor soil conservation practices and a series of dry years in the Great Plains combined to create the "dust bowl." Wind stripped topsoil from millions of hectares of land, and billowing dust clouds turned day into night. Thousands of families were forced to leave farms and migrate to cities. The El Niño, Southern Oscillation (ENSO) system plays an important role in droughts in North America and elsewhere. There now is a great worry that the greenhouse effect (see Chapter 8) will bring about major climatic changes and make droughts both more frequent and more severe than in the past in some places.

Types of Water Use

In contrast to energy resources, which are consumed when used, water has the potential for being reused many times. In discussing water appropriations, we need to distinguish between different kinds of uses and how they will affect the water being appropriated.

Withdrawal is the total amount of water taken from a lake, river, or aquifer for any purpose. Much of this water is employed in nondestructive ways and is returned to circulation in a form that can be used again. **Consumption** is the fraction of withdrawn water that is lost in transmission, evaporation, absorption, chemical transformation, or otherwise made unavailable for other purposes as a result of human use. **Degradation** is a change in water quality due to contamination or pollution so that it is unsuitable for other desirable services. The total quantity available may remain constant after some uses, but the quality is degraded so the water is no longer as valuable as it was.

Worldwide, humans withdraw about 10 percent of the total annual renewable supply. The remaining 90 percent is generally either uneconomical to tap (it would cost too much to store, ship, purify, or distribute), or there are ecological constraints on its use. Consumption and degradation together account for about half the water withdrawn in most industrial societies. The other half of the water we withdraw would still be valuable for further uses if we could protect it from contamination and make it available to potential consumers.

Many societies have always treated water as if there is an inexhaustible supply. It has been cheaper and more convenient for most people to dump all used water and get a new supply than to determine what is contaminated and what is not. The natural cleansing and renewing functions of the hydrologic cycle do replace the water we need if natural systems are not overloaded or damaged. Water is a renewable resource, but renewal takes time. The rate at which many of us are using water now may make it necessary to conscientiously protect, conserve, and replenish our water supply.

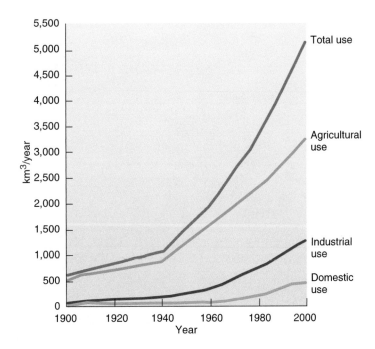

FIGURE 11.3 Growth of global water use 1900–2000. Amounts 1997–2000 are estimates.
Source: Data from L. A. Shiklomanov, "Global Water Resources" in *Nature and Resources,* vol. 26, p. 34–43, UNESCO, Paris.

Quantities of Water Used

Human water use has been increasing about twice as fast as population growth over the past century (Fig. 11.3). Water use is stabilizing in industrialized countries, but demand will increase in developing countries where supplies are available. The average amount of water withdrawn worldwide is about 646 m^3 per person per year. This overall average hides great discrepancies in the proportion of annual runoff withdrawn in different areas. As you might expect, those countries with a plentiful water supply and a small population withdraw a very small percentage of the water available to them. Canada, Brazil, and the Congo, for instance, withdraw less than 1 percent of their annual renewable supply.

By contrast, in countries such as Libya and Israel, where water is one of the most crucial environmental resources, groundwater and surface water withdrawal together amount to more than 100 percent of their renewable supply. They are essentially "mining" water—extracting groundwater faster than it is being replenished. Obviously, this is not sustainable in the long run.

The total annual renewable water supply in the United States amounts to an average of about 9,000 m^3 per person per year. About one-fifth of that amount, or some 5,000 litres per person per day is used. By comparison, the average water use in Haiti is less than 30 litres per person per day.

Use by Sector

Water use can be analyzed by identifying three major kinds of use, or sectors: domestic, industry, and agriculture. Worldwide,

PART TWO Our Physical Environment

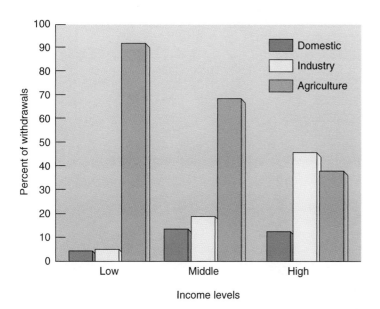

FIGURE 11.4 Water withdrawals by sector in low-, middle-, and high-income countries.
Source: Data from World Bank, 1992.

FIGURE 11.5 Rolling sprinklers allow farmers to irrigate crops on uneven terrain. Agriculture now uses 42 percent of all water withdrawn in high-income countries. In some areas, irrigation consumes 90 percent of all available water.

agriculture claims about 69 percent of total water withdrawal, ranging from 93 percent of all water used in India to only 4 percent in Kuwait, which cannot afford to spend its limited water on crops. Canada, where the fields are usually well watered by natural precipitation, uses only 12 percent of its water for agriculture. As you can see in Figure 11.4, water use by sector depends strongly on national wealth and degree of industrialization. Poorer countries with little industry and limited domestic supply systems use little of their water in these sectors.

In many developing countries, agricultural water use is notoriously inefficient and highly consumptive. Often, 60 to 70 percent of the water withdrawn for agriculture in these countries never reaches the crops for which it is intended. The most common type of irrigation is to simply flood the whole field or run water in rows between crops. As much as half is lost through evaporation or seepage from unlined irrigation canals bringing water to fields. Most of the rest runs off, evaporates, or infiltrates into the field before it can be used. While runoff can be channelled to other fields and infiltration replenishes underground aquifers, surplus water from fields is often contaminated with soil, fertilizer, pesticides, and crop residues, making it low quality. Sprinklers (Fig. 11.5) are more efficient in distributing water evenly over the field than flooding and can be used on uneven terrain, but they can lose a great deal of water to evaporation. Water-efficient drip irrigation (see Fig. 11.18) and other low-volume distribution systems can save significant amounts of water, but are used, currently, on only about 1 percent of the world's croplands.

Worldwide, industry accounts for about one-fourth of all water use, ranging from 70 percent of withdrawal in some European countries, such as Germany, to 5 percent in less industrialized countries, such as Egypt and India. Cooling water for power plants is by far the largest single industrial use of water, typically accounting for 50 to 100 percent of industrial withdrawal depending on the country. Although cooling water usually is not chemically contaminated, warm water can become thermal pollution if it is dumped directly into a stream or lake.

Industry often competes with agriculture for available water. China calculates that a given amount of water used in factories generates 60 times the value it would produce in agriculture. The future of farming looks bleak in such areas. Although Third World countries typically allocate only about 10 percent of their water withdrawal to industry, this could change rapidly as they industrialize. Water may be as important as energy in determining which countries develop and which remain underdeveloped.

FRESHWATER SHORTAGES

In 1977, the United Nations Water Conference declared that all people, regardless of their social or economic conditions, have a right to the clean drinking water and basic sanitation needed to prevent communicable diseases and provide for basic human dignity. Almost three decades later, however, an estimated 1.5 billion people lack access to adequate quantity and quality of drinking water, while nearly 3 billion, or half the world's population, do not have acceptable sanitation. As populations grow, more people move into cities, and agriculture and industry compete for increasingly scarce water supplies, water shortages are expected to become even worse in the future. A country in which consumption exceeds more than 20 percent of the available, renewable supply is considered vulnerable to **water stress.**

Globally, water supplies are abundant but unevenly distributed, and the capital needed to collect, store, purify, and distribute water is unavailable in many developing countries. Worldwide, water consumption has increased sixfold over the last century, or about twice as fast as population growth. With easily accessible water already exploited in most places, the World Bank estimates that the financial and environmental costs of tapping new supplies will be two to three times more expensive than current water projects. If present trends continue, the UN cautions, some two-thirds of the world's population will live in countries experiencing water shortages by 2025.

A Precious Resource

The World Health Organization considers 1,000 m³ of water per person per year to be the minimum level below which most countries are likely to experience chronic shortages on a scale that will impede development and harm human health. Currently, some 45 countries—most of them in Africa or the Middle East—are considered to have serious water stress, and cannot meet the minimum essential water needs of all their citizens. In some countries the problem is not so much total water supply as access to clean water. In Mali, for example, 88 percent of the population lacks clean water; in Ethiopia it is 94 percent. Rural people often have less access to clean water than do city dwellers. In the 33 worst affected countries, 60 percent of urban people can get clean water as opposed to only 20 percent of those in the country.

More than two-thirds of the world's households have to fetch water from outside the home (Fig. 11.6). For many people, this is time-consuming and heavy work; it may take up to two hours per day to gather a meagre supply. Women and children do most of this work. Improved public systems bring many benefits to poor families. In Mozambique, for example, the World Bank reports that the average time women spent carrying water decreased from two hours per day to only 25 minutes when village wells were installed. The time saved could be used to garden, tend livestock, trade in the market, care for children, or even rest!

The reasons for water shortages are many. In some cases, deficits are caused by natural forces: the rains fail; hot winds dry up reservoirs that normally would carry people through the dry season; rivers change their courses, leaving villages stranded. In other cases, shortages are human in origin: too many people compete for the resource; urbanization, overgrazing, and inappropriate agricultural practices allow water to run off before it can be captured; a lack of adequate sewage systems causes contamination of local supplies. Without money for wells, storage reservoirs, delivery pipes, and other infrastructure, people can't use the resources available to them.

Safe, clean water is available in most countries—for those who can pay the price. Water from vendors often is the only source for many in the crowded slums and shanty towns around the major cities of the developing world. Although the quality is often questionable, this water generally costs about 10 times more than a piped city supply. Naturally, sanitation levels decline when water is so expensive. A typical poor family in Lima, Peru, for instance,

FIGURE 11.6 Drawing water from a village well is a daily chore in many developing countries.

uses one-sixth as much water as a middle-class American family but pays three times as much for it. Following government recommendations that all water be boiled to prevent cholera would take up to one-third of the total income for such a poor family.

Investments in rural development have brought significant improvements over the past decade. Since 1990, nearly 800 million people—about 13 percent of the world population—have gained access to clean water (see Case Study, p. 217). The percentage of rural families with safe drinking water has risen from less than 10 percent to nearly 75 percent. Still, many people suffer from a lack of this essential resource.

Depleting Groundwater

Groundwater is the source of about 26 percent of the fresh water for agricultural and domestic use in Canada. In some provinces (such as Prince Edward Island), virtually all water used by humans is groundwater. Overuse of these supplies causes several kinds of problems, including drying of wells, natural springs, and disappearance of surface water features such as wetlands, rivers, and lakes.

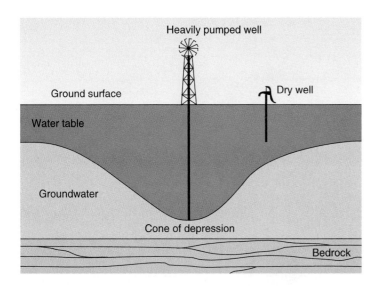

FIGURE 11.7 A cone of depression forms in the water table under a heavily pumped well. This may dry up nearby shallow wells or make pumping so expensive that it becomes impractical.

In many areas of North America, groundwater is being withdrawn from aquifers faster than natural recharge can replace it. On a local level, this causes a cone of depression in the water table, as is shown in Figure 11.7. A heavily pumped well can lower the local water table so that shallower wells go dry. On a broader scale, heavy pumping can deplete a whole aquifer. The Ogallala Aquifer in the United States, for example, underlies eight states in the arid high plains between Texas and North Dakota. As deep as 400 m in its centre, this porous bed of sand, gravel, and sandstone, once held more water than all the freshwater lakes, streams, and rivers on earth. Excessive pumping for irrigation and other uses has removed so much water that wells have dried up in many places and farms, ranches, even whole towns are being abandoned.

Many aquifers have slow recharge rates, so it will take thousands of years to refill them once they are emptied. Much of the groundwater we now are using probably was left there by the glaciers thousands of years ago. It is fossil water, in a sense. When we pump water out of a reservoir such as the Ogallala that cannot be refilled in our lifetime, we essentially are mining a nonrenewable resource. Covering aquifer recharge zones with urban development or diverting runoff that once replenished reservoirs ensures that they will not refill.

Withdrawal of large amounts of groundwater causes porous formations to collapse, resulting in **subsidence** or settling of the surface above. The U.S. Geological Survey estimates that the San Joaquin Valley in California has sunk more than 10 m in the last 50 years because of excessive groundwater pumping. Around the world, many cities are experiencing subsidence. Many are coastal cities, built on river deltas or other unconsolidated sediments. Flooding is frequently a problem as these coastal areas sink below sea level. Some inland areas also are affected by severe subsidence. Mexico City is one of the worst examples. Built on an old lake bed, it has probably been sinking since Aztec times. In recent years, rapid population growth and urbanization (Chapter 21) have caused

FIGURE 11.8 Sinkhole in Winter Park, Florida. Notice the cars and truck sliding into the hole.

groundwater overdrafts. Some areas of the city have sunk as much as 8.5 m. The Shrine of Guadalupe, the Cathedral, and many other historic monuments are sinking at odd and perilous angles.

Sinkholes form when the roof of an underground channel or cavern collapses, creating a large surface crater (Fig. 11.8). Drawing water from caverns and aquifers accelerates the process of collapse. Sinkholes can form suddenly, dropping cars, houses, and trees without warning into a gaping crater hundreds of metres across. Subsidence and sinkhole formation generally represent permanent loss of an aquifer. When caverns collapse or the pores between rock particles are crushed as water is removed, it is usually impossible to reinflate these formations and refill them with water.

Another consequence of aquifer depletion is **saltwater intrusion.** Along coastlines and in areas where saltwater deposits are left from ancient oceans, overuse of freshwater reservoirs often allows saltwater to intrude into aquifers used for domestic and agricultural purposes (Fig. 11.9).

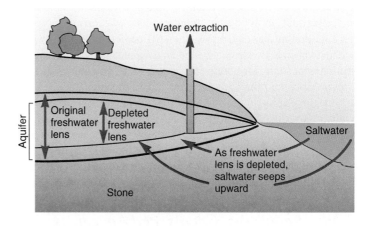

FIGURE 11.9 Saltwater intrusion into a coastal aquifer as the result of groundwater depletion. Many coastal regions of the United States are losing freshwater sources due to saltwater intrusion.

INCREASING WATER SUPPLIES

Where do present and impending freshwater shortages leave us now? On a human time scale, the amount of water on the earth is fixed, for all practical purposes, and there is little we can do to make more water. There are, however, several ways to increase local supplies.

Seeding Clouds and Towing Icebergs

In the dry prairies of North America in the 1800s and early 1900s, desperate farmers paid self-proclaimed "rainmakers" in efforts to save their withering crops. Centuries earlier, Native Americans danced and prayed to rain gods. We still pursue ways to make rain. Seeding clouds with dry ice or potassium iodide particles has been tested for many years with mixed results. Recently, researchers have been having more success using hygroscopic salts that seem to significantly increase rainfall amounts. This technique is being tested in Mexico, South Africa, and western North America. There is a concern, however, that rain induced to fall in one area decreases the precipitation somewhere else. Furthermore, there are worries about possible contamination from the salts used to seed clouds.

Another scheme that has been proposed for supplying fresh water to arid countries is to tow icebergs from the Arctic or Antarctic. Icebergs contain tremendous quantities of fresh water, but whether they can be towed without excessive melting to places where water is needed is not clear. Nor is it clear whether the energy costs to move an iceberg thousands of kilometres (if that were, in fact, possible) would be worthwhile.

Desalination

A technology that might have great potential for increasing freshwater supplies is **desalination** of ocean water or brackish saline lakes and lagoons. The most common methods of desalination are distillation (evaporation and recondensation) or reverse osmosis (forcing water under pressure through a semipermeable membrane whose tiny pores allow water to pass but exclude most salts and minerals). In 1990, the global capacity was estimated to be 13.3 million m^3 per day, or less than 0.1 percent of all freshwater withdrawals worldwide. Saudi Arabia, Kuwait, and the United Arab Emirates account for about 85 percent of all desalination, with South Africa and the United States making up most of the rest. Although desalination is still three to four times more expensive than most other sources of fresh water, it provides a welcome water supply in such places as Oman and Bahrain where there is no other access to fresh water. If a cheap, inexhaustible source of energy were available, however, the oceans could supply all the water we would ever need.

Dams, Reservoirs, Canals, and Aqueducts

People have been moving water around for thousands of years. Some of the great civilizations (Sumeria, Egypt, China, and the Inca

FIGURE 11.10 Workers drive a team of 48 mules pulling a section of conduit for the Owens Valley aqueduct to Los Angeles in 1911. Water diversions such as this have allowed Los Angeles to grow to a metropolitan area of nearly 15 million people in what originally was a semiarid desert, but the social and ecological effects have been severe in the regions from which water has been diverted.

culture of South America) were based on large-scale irrigation systems that brought river water to farm fields. In fact, some historians argue that organizing people to carry out large-scale water projects was the catalyst for the emergence of civilization. Roman aqueducts built 2,000 years ago are still in use. Those early water engineers probably never even dreamed of moving water on a scale that is being proposed and, in some cases, being accomplished now.

It is common to trap runoff with dams and storage reservoirs and transfer water from areas of excess to areas of deficit using canals, tunnels, and underground pipes. Some water transfer projects are truly titanic in scale. Los Angeles began importing water in 1913 through an aqueduct from the Owens Valley, 400 km to the north (Fig. 11.10). This project led to a statewide program known as the California Water Plan in which a system of dams, reservoirs, aqueducts, and canals transfer water from the Colorado River on the eastern border and the Sacramento and Feather rivers in the north to Los Angeles and the San Joaquin Valley in the south.

Environmental Costs

There has been much controversy about huge projects like the California Water Plan. Some of the water now being delivered to southern California is claimed by other parts of the state, neighbouring states, and even Mexico. Environmentalists claim that this water transfer upsets natural balances of streams, lakes, estuaries, and terrestrial ecosystems. Fishing enthusiasts, whitewater boaters, and others who enjoy the scenic beauty of free-running rivers mourn the loss of rivers drowned in reservoirs or dried up by diversion projects. These projects also have been criticized for using public funds to increase the value of privately held farmland and

CASE STUDY

RURAL WATER PROGRAMS IN MALAWI

Malawi is one of the poorest and least developed countries in the world. With an annual average per capita income of about $170 (US), Malawians can't afford many of the services that most of us take for granted. Having access to adequate supplies of clean, safe drinking water and basic sanitation are only a dream for many residents of Malawi. During the dry season, ponds usually turn green and smelly, or even dry up completely. Women—who do almost all domestic chores—often have to walk several kilometres to the nearest riverbed, dig down to the water table, scoop up what little water they can find, and carry it back home. It could take several hours just to get enough water for one day's drinking and cooking supply. There rarely is enough water in the dry season for bathing. Dysentery, parasitic worms, and skin diseases are considered to be normal.

But a program that provides simple tube wells and gravity-fed water systems is now bringing clean water to many people in Malawi. The gravity systems bring water to villages from year-round springs and streams in the mountains. Because the sources are located above the elevation where people live, the water is generally free of pollutants and human waste. Flowing downhill without pumps, the water passes through screens and filters to remove any particulate material, and then is distributed to public taps located in convenient spots around the villages. In villages that are too far from the mountains to be reached by piped water, shallow, hand-pumped tube wells provide access to groundwater that is much cleaner and more dependable than polluted rivers and ponds.

The water supply program was begun after the United Nations Water Conference in 1977. The first pilot project was in the Chigale region, where community members worked together to dig 10 km of trenches and lay pipes that brought fresh water to 25 taps serving 3,000 people. Each person was provided with 38 litres of good quality water per day. Because villagers provided most of the labour, the total cost of the project amounted to less than $3 (US) per person served. For the first time in their lives, many people have enough water to wash daily, even in the dry season. Women no longer need to spend so much time and energy gathering water. The community taps provide a social centre where women can meet and talk to their neighbours and friends. The success of this initial undertaking convinced residents in other villages to construct their own water systems. As of 1999, more than a million people in Malawi, mostly in rural areas, now have access to safe community water supplies.

In addition to helping individuals who now have a reliable supply of safe water, the community-building aspects of these projects are among their greatest benefits. The key to success of the project is involvement of the villagers and shared responsibility between the village residents and the government. Local communities are responsible for

With the help of a loan from international development agencies, residents of this Malawian village have a simple hand-pumped tube well that provides convenient access to safe, clean water.

constructing and maintaining the water supply system; technical advice and materials are provided by the government. Villagers generally are eager to volunteer their time and labour to ensure that their community will have a clean, lasting water supply. Many give up their ordinary wages to work on the two to three week project. When people have a sense of ownership, they are more likely to keep the system working. The Malawi rural water program shows what can be accomplished, quickly and at relatively low cost, when local communities are empowered and encouraged to organize and help themselves. How can those of us in the wealthy countries of the world help to spread this model elsewhere?

---

for encouraging agricultural development and urban growth in arid lands where other uses might be more appropriate.

Ecosystem Losses

Mono Lake, a salty desert lake in California just east of Yosemite National Park, is an example of environmental consequences of interbasin water transfers and is an important legal symbol as well (Fig. 11.11). Diversion of tributary rivers has shrunk the surface area of this lake by one-third, threatening millions of resident and migratory wading birds, ducks, and gulls that seek shelter and food there. After years of legal wrangling, the California Water Resources Control Board ruled in 1994 that Los Angeles must restrict diversions and allow the lake's surface to rise 5 km to 1,948 km above sea level. This is less than the 1941 level, but should improve conditions for birds and aquatic life.

About three-fourths of California's Trinity River is diverted to the arid Central Valley. This leaves barely enough water for a few remaining salmon. One of the final acts of the outgoing Clinton administration in 2000 was to propose reapportioning the water, but the current water users are unlikely to give up the precious resource easily. West Africa's Lake Chad, once the world's sixth-largest lake, used to support important fisheries. Water diversion for irrigation, coupled with years of drought, have reduced Lake Chad to less than one-tenth of its size four decades ago. Towns now stand on the old lake bed.

Canada has been the site of decades of protest by First Nations people over flooding of ancestral lands for hydroelectric projects. The James Bay project built by Hydro-Quebec has diverted three major rivers flowing west into Hudson Bay and created huge lakes that flooded more than 10,000 km² of forest and

FIGURE 11.11 Mono Lake in eastern California. Diversion of tributary rivers to provide water for Los Angeles has shrunk the lake's surface area by one-third, threatening migratory bird flocks that feed here. These formations were created underwater where calcium-rich springs entered the brine-laden lake.

tundra, to generate 26,000 megawatts of electrical power. In 1984, shortly after Phase I of this project was completed, 10,000 caribou drowned when trying to follow migration routes across the newly flooded land. The loss of traditional hunting and fishing sites has been culturally devastating for native Cree people. In addition, mercury leeched out of rocks in newly submerged lands has entered the food chain and many residents show signs of mercury poisoning. In a similar, but less well-known case, Manitoba Hydro has diverted the Churchill River from the west shore of Hudson Bay so that it now flows into the Nelson River and then to Lake Winnipeg. More than 125,000 km² has been inundated, much of it on permafrost, which is slowly melting and causing reservoir banks to expand continually. Here also, Cree people are being displaced against their will. As is the case in Quebec, most of the electricity generated by these dams is exported to the United States. In some places, however, dams are being removed.

Displacement of People

In China, huge dams are being built in the Three Gorges section of the Chang Jiang (Yangtze) River. More than a million people are being forced to relocate as the huge reservoir (185 m deep and 600 km long) inundates some 150 cities and villages. Because the dam is being built on an active geological fault, there is a worry that a strong earthquake could rupture the dam and drown millions of people living downstream.

Many other huge dam and water diversion projects have generated controversy and protest around the world. In Turkey, construction of a network of 22 large dams and reservoirs for the Southeast Anatolia Project on the Tigris and Euphrates Rivers is regarded as a serious threat by Turkey's downstream neighbours, Syria and Iraq, which are now at the mercy of Turkey to allow the rivers to continue to flow. In India, the Sardar Sarovar Dam on the sacred Narmada River has been the focus of decades of protest.

Many of the 1 million villagers and tribal people being displaced by this project have engaged in mass resistance and civil disobedience when police try to remove them forcibly. Some have vowed to drown rather than leave their homes. In Nepal, construction of the 240 m high Tehri Dam on the Bhagirathi River has stirred fears that a strong earthquake in this active seismic region might cause the dam to collapse and cause a catastrophic flood downstream. The Tehri Dam is only one of 17 high dams that Nepal and India plan for the Himalayan mountains.

Evaporation, Leakage, and Siltation

The main problem with dams is inefficiency. Some dams built in the western United States lose so much water through evaporation and seepage into porous rock beds that they waste more water than they make available. The evaporative loss from Lake Mead and Lake Powell on the Colorado River is about 1 km³ per year. The salts left behind by evaporation and agricultural runoff nearly double the salinity of the river and make its water unusable when it reaches Mexico. To compensate, the United States has built a $350 million desalination plant at Yuma, Arizona, to improve water quality.

As the turbulent Colorado River slows in the reservoirs created by Glen Canyon and Boulder Dams, it drops its load of suspended material. More than 10 million tonnes of silt per year collect behind these dams. Imagine a line of twenty thousand dump trucks backed up to Lake Mead and Lake Powell every day, dumping dirt into the water. Within as little as 100 years, these reservoirs could be full of silt and useless for either water storage or hydroelectric generation (Fig. 11.12).

The accumulating sediments that clog reservoirs and make dams useless also represent a loss of valuable nutrients. The Aswan High Dam in Egypt was built to supply irrigation water to make agriculture more productive. Although thousands of hectares are being irrigated, the water available is only about half that anticipated because of evaporation in Lake Nasser behind the dam and

FIGURE 11.12 This dam is now useless because its reservoir has filled with silt and sediment.

PART TWO   Our Physical Environment

seepage losses in unlined canals which deliver the water. Controlling the annual floods of the Nile also has stopped the deposition of nutrient-rich silt on which farmers depended for fertilizing their fields. This silt is being replaced with commercial fertilizer costing more than $100 million each year. Furthermore, the nutrients carried by the river once supported a rich fishery in the Mediterranean that was a valuable food source for Egypt. After the dam was installed, sardine fishing declined 97 percent. To make matters worse, growth of snail populations in the shallow permanent canals that distribute water to fields has led to an epidemic of schistosomiasis, a debilitating disease caused by parasitic flatworms in irrigation canals. In some areas, 80 percent of the residents are infected (Chapter 20).

Loss of Free-Flowing Rivers

Many water projects drown or drain free-flowing rivers, often turning them into linear, sterile irrigation canals. Conservation history records many battles between those who want to preserve wild rivers and those who would benefit from development (Fig. 11.13).

One of the first and most divisive of these battles was over the flooding of the Hetch Hetchy Valley in Yosemite National Park. In the early 1900s, San Francisco wanted to dam the Tuolumne River to produce hydroelectric power and provide water for the city water system. This project was supported by many prominent San Francisco citizens because it represented an opportunity for both clean water and municipal power. Leader of the opposition was John Muir, founder of the Sierra Club and protector of Yosemite Park. Muir said that Hetch Hetchy Valley rivalled Yosemite itself in beauty and grandeur and should be protected. After a prolonged and bitter fight, the developers won and the dam was built. Hard feelings from this controversy persisted for many years.

FIGURE 11.13 The recreational and aesthetic values of free-flowing wild rivers and wilderness lakes may be their greatest assets. Competition between *in situ* values and extractive uses can lead to bitter fights and difficult decisions.

WATER MANAGEMENT AND CONSERVATION

Watershed management and conservation are often more economical and environmentally sound ways to prevent flood damage and store water for future use than building huge dams and reservoirs.

Watershed Management

A **watershed,** or catchment, is all the land drained by a stream or river. It has long been recognized that retaining vegetation and ground cover in a watershed helps hold back rainwater and lessens downstream floods. In 1998, Chinese officials acknowledged that unregulated timber cutting upstream on the Yangtze contributed to massive floods that killed 30,000 people.

Similarly, after disastrous floods in the upper Mississippi Valley in 1993, it was suggested that, rather than allowing residential, commercial, or industrial development on floodplains, these areas should be reserved for water storage, aquifer recharge, wildlife habitat, and agriculture (Chapter 22). Sound farming and forestry practices can reduce runoff. Retaining crop residue on fields reduces flooding, and minimizing plowing and forest cutting on steep slopes protects watersheds. Wetlands conservation preserves natural water storage capacity and aquifer recharge zones. A river fed by marshes and wet meadows tends to run consistently clear and steady rather than in violent floods.

A series of small dams on tributary streams can hold back water before it becomes a great flood. Ponds formed by these dams provide useful wildlife habitat and stock-watering facilities. They also catch soil where it could be returned to the fields. Small dams can be built with simple equipment and local labour, eliminating the need for massive construction projects and huge dams.

Domestic Conservation

We could probably save as much as half of the water we now use for domestic purposes without great sacrifice or serious changes in our lifestyles. Simple steps, such as taking shorter showers, stopping leaks, and washing cars, dishes, and clothes as efficiently as possible, can go a long way toward forestalling the water shortages that many authorities predict. Isn't it better to adapt to more conservative uses now when we have a choice than to be forced to do it by scarcity in the future?

The use of conserving appliances, such as low-volume shower heads and efficient dishwashers and washing machines, can reduce water consumption greatly (What Can You Do? p. 220). You might consider whether you really need a lush green lawn that requires constant watering, feeding, and care. Planting native ground cover in a "natural lawn" or developing a rock garden or landscape in harmony with the surrounding ecosystem can be both ecologically sound and aesthetically pleasing (Fig. 11.14).

Our largest domestic water use is toilet flushing (Fig. 11.15). We dispose of relatively small volumes of waste with very large volumes of water. In many cases it is much better to treat or

Saving Water and Preventing Pollution

Each of us can conserve much of the water we use and avoid water pollution in many simple ways.

- Don't flush every time you use the toilet. Take shorter showers; don't wash your car so often.
- Don't let the faucet run while washing hands, dishes, food, or brushing your teeth. Draw a basin of water for washing and another for rinsing dishes. Don't run the dishwasher when half full.
- Dispose of used motor oil, household hazardous waste, batteries, etc., responsibly. Don't dump anything down a storm sewer that you wouldn't want to drink.
- Avoid using toxic or hazardous chemicals for simple cleaning or plumbing jobs. A plunger or plumber's snake will often unclog a drain just as well as caustic acids or lye. Hot water and soap will clean brushes more safely than organic solvents.
- If you have a lawn, or know someone who does, use water, fertilizer, and pesticides sparingly. Consider planting native plants, low-maintenance ground cover, a rock garden, or some other xeriphytic landscaping.
- Use water-conserving appliances: low-flow showers, low-flush toilets, and aerated faucets.
- Use recycled (gray) water for lawns, house plants, car washing.
- Check your toilet for leaks. A leaky toilet can waste 190 litres per day. Add a few drops of dark food colouring to the tank and wait 15 minutes. If the tank is leaking, the water in the bowl will change colour.

FIGURE 11.14 By using native plants in a natural setting, residents of Phoenix, Arizona, save water and fit into the surrounding landscape.

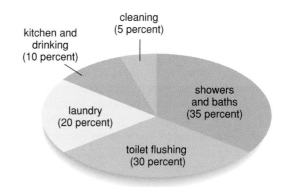

kitchen and drinking (10 percent)

cleaning (5 percent)

showers and baths (35 percent)

laundry (20 percent)

toilet flushing (30 percent)

FIGURE 11.15 Typical household water use in Canada.
Source: Environment Canada, *Primer on Fresh Water,* 5th edition, 2000.

dispose of waste at its origin before it is diluted or mixed with other materials. There are now several types of waterless or low-volume toilets. The Swedish-made Clivus Multrum (Fig. 11.16) digests both human and kitchen wastes by aerobic bacterial action, producing a rich, nonoffensive compost that can be used as garden fertilizer. There are also low-volume toilets that use recirculating oil or aqueous chemicals to carry wastes to a holding tank, from which they are periodically taken to a treatment plant. Anaerobic digesters use bacterial or chemical processes to produce usable methane gas from domestic wastes. These systems provide valuable energy and save water but are more difficult to operate than conventional toilets. Few cities are ready to mandate waterless toilets, but in 1988 a number of cities in arid areas of the United States (including Los Angeles, California; Orlando, Florida; Austin, Texas; and Phoenix, Arizona) ordered that water-saving toilets, showers, and faucets be installed in all new buildings. The motivation was twofold: to relieve overburdened sewer systems and to conserve water.

Significant amounts of water can be reclaimed and recycled. In California, water recovered from treated sewage constitutes the fastest growing water supply, growing about 30 percent per year. Despite public squeamishness, purified sewage effluent is being used for everything from agricultural irrigation to flushing toilets (Fig. 11.17). In a statewide first, San Diego is currently moving toward piping water from the local sewage plant directly into a drinking-water reservoir. Already, California uses more than 555 million m³ of recycled water annually. That's equivalent to about two-thirds of the water consumed by Los Angeles every year. Other places with growing populations and limited water supplies may soon find that it pays to follow California's example.

Recycling and Water Conservation

In many developing countries as much as 70 percent of all the agricultural water used is lost to leaks in irrigation canals, application to areas where plants don't grow, runoff, and evaporation. Better farming techniques, such as leaving crop residue on fields and ground cover on drainage ways, intercropping, use of mulches, and low-volume irrigation could reduce these water losses dramatically (Fig. 11.18).

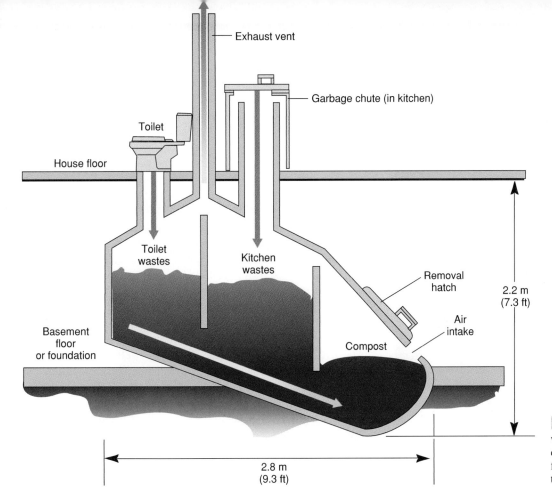

FIGURE 11.16 The Clivus Multrum waterless toilet. Wastes decompose and compost into an odourless, safe, rich fertilizer as they slowly slide down to the bottom compartment.

FIGURE 11.17 Recycled water is being used in California and Arizona for everything from agriculture, to landscaping, to industry. Some cities even use treated sewage effluent for human drinking-water supplies.

Nearly half of all industrial water use is for cooling of electric power plants and other industrial facilities. Some of this water use could be avoided by installing dry cooling systems similar to the radiator of your car. In many cases, cooling water could be reused for irrigation or other purposes in which water does not have to be drinking quality. The waste heat carried by this water could be a valuable resource if techniques were developed for using it.

WHAT IS WATER POLLUTION?

Any physical, biological, or chemical change in water quality that adversely affects living organisms or makes water unsuitable for desired uses can be considered pollution. Often, however, a change that adversely affects one organism may be advantageous to another. Nutrients that stimulate oxygen consumption by bacteria and other decomposers in a river or lake, for instance, may reduce some fish populations, but will stimulate a flourishing community of decomposers. Whether the quality of the water has suffered depends on your perspective. There are natural sources of water contamination, such as poison springs, oil seeps, and sedimentation from erosion, but in this chapter we will focus primarily on human-caused changes that affect water quality or usability.

Pollution control standards and regulations usually distinguish between point and nonpoint pollution sources. Factories, power plants, sewage treatment plants, underground coal mines,

FIGURE 11.18 Drip irrigation delivers measured amounts of water exactly where the plants need and can use it. This technique can save up to 90 percent of irrigation water usage and reduces salt buildup. It is more expensive to install and operate, however, than simple flood irrigation.

and oil wells are classified as **point sources** because they discharge pollution from specific locations, such as drain pipes, ditches, or sewer outfalls (Fig. 11.19). These sources are discrete and identifiable, so they are relatively easy to monitor and regulate. It is generally possible to divert effluent from the waste streams of these sources and treat it before it enters the environment.

In contrast, **nonpoint sources** of water pollution are scattered or diffuse, having no specific location where they discharge into a particular body of water. Nonpoint sources include runoff from farm fields and feedlots (Fig. 11.20), golf courses, lawns and gardens, construction sites, logging areas, roads, streets, and parking lots. Whereas point sources may be fairly uniform and predictable throughout the year, nonpoint sources are often highly episodic. The first heavy rainfall after a dry period may flush high concentrations of gasoline, lead, oil, and rubber residues off city streets, for instance, while subsequent runoff may have lower levels of these pollutants. Spring snowmelt carries high levels of atmospheric acid deposition into streams and lakes in some areas. The irregular timing of these events, as well as their multiple sources and scattered location, makes them much more difficult to monitor, regulate, and treat than point sources.

Perhaps the ultimate in diffuse, nonpoint pollution is **atmospheric deposition** of contaminants carried by air currents and precipitated into watersheds or directly onto surface waters as rain,

FIGURE 11.19 Sewer outfalls, industrial effluent pipes, acid draining out of abandoned mines, and other point sources of pollution are generally easy to recognize.

PART TWO Our Physical Environment

FIGURE 11.20 This bucolic scene looks peaceful and idyllic, but allowing cows to trample stream banks is a major cause of bank erosion and water pollution. Nonpoint sources such as this have become the leading unresolved cause of stream and lake pollution in Canada.

TABLE 11.3	Major Categories of Water Pollutants	
CATEGORY	**EXAMPLES**	**SOURCES**
A. CAUSES HEALTH PROBLEMS		
1. Infectious agents	Bacteria, viruses, parasites	Human and animal excreta
2. Organic chemicals	Pesticides, plastics, detergents, oil, and gasoline	Industrial, household, and farm use
3. Inorganic chemicals	Acids, caustics, salts, metals	Industrial effluents, household cleansers, surface runoff
4. Radioactive materials	Uranium, thorium, cesium, iodine, radon	Mining and processing of ores, power plants, weapons production, natural sources
B. CAUSES ECOSYSTEM DISRUPTION		
1. Sediment	Soil, silt	Land erosion
2. Plant nutrients	Nitrates, phosphates, ammonium	Agricultural and urban fertilizers, sewage, manure
3. Oxygen-demanding wastes	Animal manure and plant residues	Sewage, agricultural runoff, paper mills, food processing
4. Thermal	Heat	Power plants, industrial cooling

snow, or dry particles. The Great Lakes, for example, have been found to be accumulating industrial chemicals such as PCBs and dioxins, as well as agricultural toxins such as the insecticide toxaphene that cannot be accounted for by local sources alone. The nearest sources for many of these chemicals are sometimes thousands of kilometres away.

Amounts of these pollutants can be quite large. It is estimated that there are 600,000 kg of the herbicide atrazine in the Great Lakes, most of which is thought to have been deposited from the atmosphere. Concentration of persistent chemicals up the food chain can produce high levels in top predators. Several studies have indicated health problems among people who regularly eat fish from the Great Lakes.

Ironically, lakes also can be pollution sources as well. In the past 12 years, about 26,000 tonnes of PCBs have "disappeared" from Lake Superior. Apparently, these compounds evaporate from the lake surface and are carried by air currents to other areas where they are redeposited.

TYPES AND EFFECTS OF WATER POLLUTION

Although the types, sources, and effects of water pollutants are often interrelated, it is convenient to divide them into major categories for discussion (Table 11.3). Let's look more closely at some of the important sources and effects of each type of pollutant.

Infectious Agents

The most serious water pollutants in terms of human health worldwide are pathogenic organisms (Chapter 20). Among the most important waterborne diseases are typhoid, cholera, bacterial and amoebic dysentery, enteritis, polio, infectious hepatitis, and schistosomiasis. Malaria, yellow fever, and filariasis are transmitted by insects that have aquatic larvae. Altogether, at least 25 million deaths each year are blamed on these water-related diseases. Nearly two-thirds of the mortalities of children under 5 years old are associated with waterborne diseases.

The main source of these pathogens is from untreated or improperly treated human wastes. Animal wastes from feedlots or fields near waterways and food processing factories with inadequate waste treatment facilities also are sources of disease-causing organisms.

In developed countries, sewage treatment plants and other pollution-control techniques have reduced or eliminated most of the worst sources of pathogens in inland surface waters. Furthermore, drinking water is generally disinfected by chlorination so epidemics of waterborne diseases are rare in these countries. The United Nations estimates that 90 percent of the people in developed countries have adequate (safe) sewage disposal, and 95 percent have clean drinking water.

The situation is quite different in less-developed countries. The United Nations estimates that at least 2.5 billion people in these countries lack adequate sanitation, and that about half these people also lack access to clean drinking water. Conditions are especially bad in remote, rural areas where sewage treatment is usually primitive or nonexistent, and purified water is either unavailable or too expensive to obtain. The World Health Organization estimates that 80 percent of all sickness and disease in less-developed countries can be attributed to waterborne infectious agents and inadequate sanitation.

If everyone had pure water and satisfactory sanitation, the World Bank estimates that 200 million fewer episodes of diarrheal illness would occur each year, and 2 million childhood deaths would be avoided. Furthermore, 450 million people would be spared debilitating roundworm or fluke infections. Surely these are goals worth pursuing.

Detecting specific pathogens in water is difficult, time-consuming, and costly; thus, water quality control personnel usually analyze water for the presence of **coliform bacteria,** any of the many types that live in the colon or intestines of humans and other animals. The most common of these is *Eschericha coli* (or *E. coli*). Many strains of bacteria are normal symbionts in mammals, but some, such as *Shigella, Salmonella,* or *Lysteria* can cause fatal diseases. It is usually assumed that if any coliform bacteria are present in a water sample, infectious pathogens are present also.

To test for coliform bacteria, a water sample (or a filter through which a measured water sample has passed) is placed in a dish containing a nutrient medium that supports bacterial growth. After 24 hours at the appropriate temperature, living cells will have produced small colonies. If *any* colonies are found in drinking water samples, the water is considered unsafe and requiring disinfection. The recommended maximum coliform count for swimming water is 200 colonies per 100 ml. If the limit is exceeded, the contaminated pool, river, or lake usually is closed to swimming (Fig. 11.21).

Oxygen-Demanding Wastes

The amount of oxygen dissolved in water is a good indicator of water quality and of the kinds of life it will support. Water with an oxygen content above 6 parts per million (ppm) will support game fish and other desirable forms of aquatic life. Water with less than 2 ppm oxygen will support mainly worms, bacteria, fungi, and other detritus feeders and decomposers. Oxygen is added to water by diffusion from the air, especially when turbulence and mixing rates are high, and by photosynthesis of green plants, algae, and cyanobacteria. Oxygen is removed from water by respiration and chemical processes that consume oxygen.

The addition of certain organic materials, such as sewage, paper pulp, or food-processing wastes, to water stimulates oxygen consumption by decomposers. The impact of these materials on water quality can be expressed in terms of **biochemical oxygen demand (BOD):** a standard test of the amount of dissolved oxygen consumed by aquatic microorganisms over a five-day period. An alternative method, called the chemical oxygen demand (COD), uses a strong oxidizing agent (dichromate ion in 50 percent sulfuric acid) to completely break down all organic matter in a water sample. This method is much faster than the BOD test, but

FIGURE 11.21 Our national goal of making all surface waters in Canada "fishable and swimmable" has not been fully met, but scenes like this have been reduced by pollution control efforts.

normally gives much higher results because it oxidizes compounds not ordinarily metabolized by bacteria. A third method of assaying pollution levels is to measure **dissolved oxygen (DO) content** directly, using an oxygen electrode. The DO content of water depends on factors other than pollution (for example, temperature and aeration), but it is usually more directly related to whether aquatic organisms survive than is BOD.

The effects of oxygen-demanding wastes on rivers depends to a great extent on the volume, flow, and temperature of the river water. Aeration occurs readily in a turbulent, rapidly flowing river, which is, therefore, often able to recover quickly from oxygen-depleting processes. Downstream from a point source, such as a municipal sewage plant discharge, a characteristic decline and restoration of water quality can be detected either by measuring dissolved oxygen content or by observing the flora and fauna that live in successive sections of the river.

The oxygen decline downstream is called the **oxygen sag** (Fig. 11.22). Upstream from the pollution source, oxygen levels support normal populations of clean-water organisms. Immediately below the source of pollution, oxygen levels begin to fall as decomposers metabolize waste materials. Rough fish, such as carp, bullheads, and gar, are able to survive in this oxygen-poor environment where they eat both decomposer organisms and the waste itself. Further downstream, the water may become so oxygen-depleted that only the most resistant microorganisms and invertebrates can survive. Eventually, most of the nutrients are used up, decomposer populations are smaller, and the water becomes oxygenated once again. Depending on the volumes and flow rates of the effluent plume and the river receiving it, normal communities may not appear for several kilometres downstream.

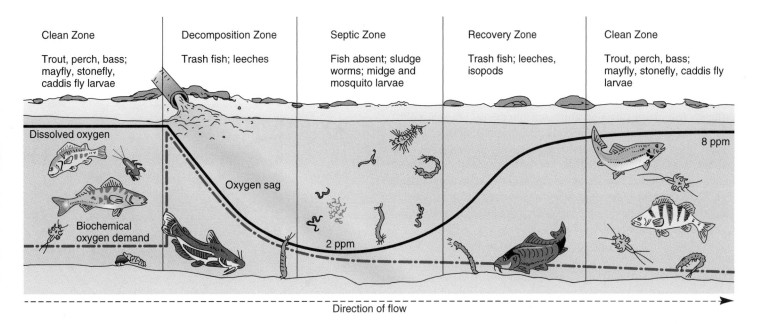

Clean Zone	Decomposition Zone	Septic Zone	Recovery Zone	Clean Zone
Trout, perch, bass; mayfly, stonefly, caddis fly larvae	Trash fish; leeches	Fish absent; sludge worms; midge and mosquito larvae	Trash fish; leeches, isopods	Trout, perch, bass; mayfly, stonefly, caddis fly larvae

Dissolved oxygen

8 ppm

Oxygen sag

Biochemical
oxygen demand

2 ppm

Direction of flow

FIGURE 11.22 Oxygen sag downstream of an organic source. A great deal of time and distance may be required for the stream and its inhabitants to recover.

Plant Nutrients and Cultural Eutrophication

Water clarity (transparency) is affected by sediments, chemicals, and the abundance of plankton organisms, and is a useful measure of water quality and water pollution. Rivers and lakes that have clear water and low biological productivity are said to be **oligotrophic** (oligo = little + trophic = nutrition). By contrast, **eutrophic** (eu + trophic = truly nourished) waters are rich in organisms and organic materials. Eutrophication, an increase in nutrient levels and biological productivity, is a normal part of successional changes (Chapter 6) in most lakes. Tributary streams bring in sediments and nutrients that stimulate plant growth. Over time, most ponds or lakes tend to fill in, eventually becoming marshes. The rate of eutrophication and succession depends on water chemistry and depth, volume of inflow, mineral content of the surrounding watershed, and the biota of the lake itself.

Human activities can greatly accelerate eutrophication. An increase in biological productivity and ecosystem succession caused by human activities is called **cultural eutrophication.** Cultural eutrophication can be brought about by increased nutrient flows, higher temperatures, more sunlight reaching the water surface, or a number of other changes. Increased productivity in an aquatic system sometimes can be beneficial. Fish and other desirable species may grow faster, providing a welcome food source. Often, however, eutrophication has undesirable results. An oligotrophic lake or river usually has aesthetic qualities and species of organisms that we value.

The high biological productivity of eutrophic systems is often seen in "blooms" of algae or thick growths of aquatic plants stimulated by elevated phosphorus or nitrogen levels (Fig. 11.23).

FIGURE 11.23 Eutrophic lake. Nutrients from agriculture and domestic sources have stimulated growth of algae and aquatic plants. This reduces water quality, alters species composition, and lowers recreational and aesthetic values of the lake.

Bacterial populations also increase, fed by larger amounts of organic matter. The water often becomes cloudy or turbid and has unpleasant tastes and odours. The deposition of silt and organic sediment caused by cultural eutrophication can accelerate the "aging" of a water body enormously over natural rates. Lakes and reservoirs that normally might exist for hundreds or thousands of years can be filled in a matter of decades.

Eutrophication also occurs in marine ecosystems, especially in near-shore waters and partially enclosed bays or estuaries. Partially enclosed seas such as the Black Sea, the Baltic, and the Mediterranean tend to be in especially critical condition. During the tourist season, the coastal population of the Mediterranean, for example, swells to 200 million people. Eighty-five percent of the effluents from large cities go untreated into the sea. Beach pollution, fish kills, and contaminated shellfish result. Extensive "dead zones" often form where rivers dump nutrients into estuaries and shallow seas. The largest in the world occurs during summer months in the Gulf of Mexico at the mouth of the Mississippi River. This hypoxic zone (less than 2 mg oxygen per litre) can cover 18,000 km^2. The hypoxic zone in Pamlico Sound following Hurricane Floyd was only about one-tenth this size.

Toxic Tides

According to the Bible, the first plague to afflict the Egyptians when they wouldn't free Moses and the Israelites was that the water in the Nile turned into blood. All the fish died and the people were unable to drink the water, a terrible calamity in a desert country. Some modern scientists believe this may be the first recorded history of a **red tide** or a bloom of deadly aquatic microorganisms called dinoflagellates. Red tides—and other colours, depending on the species involved—have become increasingly common in slow-moving rivers, brackish lagoons, estuaries, and bays, as well as near-shore ocean waters where nutrients and wastes wash down our rivers.

One of the most feared of these organisms is *Pfiesteria piscicida,* an extraordinarily poisonous dinoflagellate that only recently has been recognized as a killer of finfish and shellfish in polluted rivers and estuaries such as North Carolina's Pamlico Sound, where it recently has wiped out hundreds of thousands to millions of fish every year. Dinoflagellates are peculiar organisms with complex life cycles and many different shapes. They are unicellular, photosynthetic microorganisms distinct from both plants and animals. They typically swim with two slender, whiplike flagella, one of which vibrates in a shallow grove in the cell surface and causes the organism to rotate, while the other propels it through the water. *Pfiesteria* can change into at least two dozen distinct forms and sizes depending on water temperature, turbulence, and food supply available (Fig. 11.24). In calm, warm, brackish water, rich in sewage or other nutrients, *Pfiesteria* assumes a spherical, nontoxic, but predatory, swimming form that preys on bacteria and plankton.

Under the right conditions, a population explosion can produce a dense bloom of these cells, which can reproduce either by binary fission or sexual fusion of small gametes. If fish blunder into this profuse swarm, *Pfiesteria* quickly turn to a toxic, swimming form that attacks with soluble poisons. These toxins paralyze fish, so they can't escape, and produce skin lesions that expose the fish to infections by pathogenic bacteria and fungi. The predatory zoospores feed on substances that leak from the sores, and also eat skin and flesh directly. When the fish die, zoospores change into star-shaped or lobose (lobed) amoebae that crawl along the bottom and feed on dead carcasses. If the water is cold and calm, these

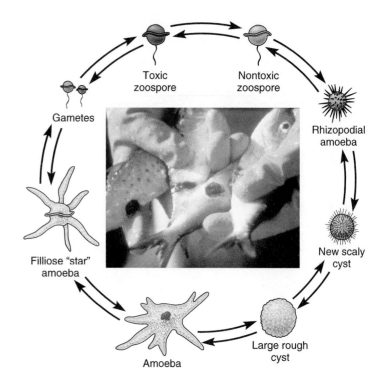

FIGURE 11.24 *Pfiesteria piscicida* life cycle. This toxic, predatory dinoflagellate has at least two dozen different sizes and cell shapes depending on nutrient supply and environmental conditions. They attack fish, causing large skin lesions that often are lethal.
Source: Adapted from J. M. Burkholder (1999), *Scientific American,* 281(2): pp. 42–49.

amoebae can attack fish directly; or if it is very turbulent, they spend their time crawling on the bottom or burrowing into the mud looking for something to eat. If the environment becomes adverse, any of these forms can turn into resistant cysts that can withstand extremely harsh conditions.

Perhaps the scariest thing about *Pfiesteria* is the recently recognized fact that it can be extremely poisonous to humans, either through eating contaminated seafood, or by breathing aerosols (airborne mist or dust) containing *Pfiesteria* cells or their secretions. The symptoms of *Pfiesteria* poisoning include headaches, blurred vision, aching joints, burning muscles, difficulty breathing, disorientation, memory loss, and long-term damage to the brain, liver, pancreas, kidneys, and immune system. Just being on a boat for a few minutes in the midst of *Pfiesteria* bloom can result in debilitating symptoms and long-term injury. This does seem like a plague, indeed, and it appears to be provoked primarily by human actions. Runoff of agricultural wastes, fertilizer, human sewage, and other nutrients into shallow, near-shore water, seems to be the main cause of this terrible affliction.

Inorganic Pollutants

Some toxic inorganic chemicals are released from rocks by weathering, are carried by runoff into lakes or rivers, or percolate into groundwater aquifers. This pattern is part of natural mineral cycles (Chapter 3). Humans often accelerate the transfer rates in these cycles thousands of times above natural background levels through the mining, processing, using, and discarding of minerals.

In many areas, toxic, inorganic chemicals introduced into water as a result of human activities have become the most serious form of water pollution. Among the chemicals of greatest concern are heavy metals, such as mercury, lead, tin, and cadmium. Supertoxic elements, such as selenium and arsenic, also have reached hazardous levels in some waters. Other inorganic materials, such as acids, salts, nitrates, and chlorine, that normally are not toxic at low concentrations may become concentrated enough to lower water quality or adversely affect biological communities.

Metals

Many metals such as mercury, lead, cadmium, and nickel are highly toxic. Levels in the parts per million range—so little that you cannot see or taste them—can be fatal. Because metals are highly persistent, they accumulate in food chains and have a cumulative effect in humans. A famous case of mercury poisoning in Japan in the 1950s was one of our first warnings of this danger.

Another mercury-poisoning disaster appears to be in process in South America. Since the mid-1980s, a gold rush has been under way in Brazil, Ecuador, and Bolivia. Forty thousand *garimperios* or prospectors have invaded the jungles along the Amazon River and its tributaries to pan for gold. They use mercury to trap the gold and separate it from sediments. Then, the mercury is boiled off with a blow torch. Miners and their families suffer nerve damage from breathing the toxic fumes. Estimates are that 130 tonnes of mercury per year are deposited in the Amazon, which will be impossible to clean up.

We have come to realize that other heavy metals released as a result of human activities also are concentrated by hydrological and biological processes so that they become hazardous to both natural ecosystems and human health. A condition known as Itai-Itai (literally ouch-ouch) disease that developed in the Japanese living near the Jintsu River was traced to cadmium poisoning from mining and smelting waste-water discharges. Bacteria forming methylated tin have been found in sediments in Chesapeake Bay, leading to worries that this toxic metal also may be causing unsuspected health effects. The use of tin compounds as antifouling agents on ship bottoms has been banned because of its toxic effects.

Lead poisoning has been known since Roman times to be dangerous to human health. Lead pipes are a serious source of drinking water pollution, especially in older homes or in areas where water is acidic and, therefore, leaches more lead from pipes. Even lead solder in pipe joints and metal containers can be hazardous. In Canada, some public health officials argue that lead is neurotoxic at any level, and the limits should be less than 10 ppb.

Mine drainage and leaching of mining wastes are serious sources of metal pollution in water. A survey of water quality in eastern Tennessee found that 43 percent of all surface streams and lakes and more than half of all groundwater used for drinking supplies was contaminated by acids and metals from mine drainage. In some cases, metal levels were 200 times higher than what is considered safe for drinking water.

Nonmetallic Salts

Desert soils often contain high concentrations of soluble salts, including toxic selenium and arsenic (In Depth, p. 229). You have probably heard of poison springs and seeps in the desert where these compounds are brought to the surface by percolating ground-water. Irrigation and drainage of desert soils mobilize these materials on a larger scale and can result in serious pollution problems, as in Kesterson Marsh in California where selenium poisoning killed thousands of migratory birds in the 1980s.

Such salts as sodium chloride (table salt) that are nontoxic at low concentrations also can be mobilized by irrigation and concentrated by evaporation, reaching levels that are toxic for plants and animals. Salt levels in the San Joaquin River in central California rose from 0.28 gm/l in 1930 to 0.45 gm/l in 1970 as a result of agricultural runoff. Salinity levels in the Colorado River and surrounding farm fields have become so high in recent years that millions of hectares of valuable croplands have had to be abandoned. The United States has built a huge desalinization plant at Yuma, Arizona, to reduce salinity in the river. In northern states, millions of tonnes of sodium chloride and calcium chloride are used to melt road ice in the winter. The corrosive damage to highways and automobiles and the toxic effects on vegetation are enormous. Leaching of road salts into surface waters may have a similarly devastating effect on aquatic ecosystems.

Acids and Bases

Acids are released as by-products of industrial processes, such as leather tanning, metal smelting and plating, petroleum distillation, and organic chemical synthesis. Coal mining is an especially important source of acid water pollution. Sulphur compounds in coal react with oxygen and water to make sulphuric acid. Thousands of kilometres of streams in the United States have been acidified by acid mine drainage, some so severely that they are essentially lifeless.

Coal and oil combustion also leads to formation of atmospheric sulphuric and nitric acids (Chapter 9), which are disseminated by long-range transport processes and deposited via precipitation (acidic rain, snow, fog, or dry deposition) in surface waters. Where soils are rich in such alkaline material as limestone, these atmospheric acids have little effect because they are neutralized. In high mountain areas or recently glaciated regions where crystalline bedrock is close to the surface and lakes are oligotrophic, however, there is little buffering capacity (ability to neutralize acids) and aquatic ecosystems can be severely disrupted. These effects were first recognized in the mountains of northern England and Scandinavia about 30 years ago.

In recent years, aquatic damage due to acid precipitation has been reported in about 200 lakes in the Adirondack Mountains of New York State and in several thousand lakes in eastern Quebec. Game fish, amphibians, and sensitive aquatic insects are generally the first to be killed by increased acid levels in the water. If acidification is severe enough, aquatic life is limited to a few resistant species of mosses and fungi. Increased acidity may result in leaching of toxic metals, especially aluminum, from soil and rocks, making water unfit for drinking or irrigation, as well.

FIGURE 11.25 The deformed beak of this young robin is thought to be due to dioxins, DDT, and other toxins in its mother's diet.

FIGURE 11.26 Sediment and industrial waste flow from this drainage canal into Lake Erie.

Organic Chemicals

Thousands of different natural and synthetic organic chemicals are used in the chemical industry to make pesticides, plastics, pharmaceuticals, pigments, and other products that we use in everyday life. Many of these chemicals are highly toxic (Chapter 20). Exposure to very low concentrations (perhaps even parts per quadrillion in the case of dioxins) can cause birth defects, genetic disorders, and cancer. Some can persist in the environment because they are resistant to degradation and toxic to organisms that ingest them. Contamination of surface waters and groundwater by these chemicals is a serious threat to human health.

The two most important sources of toxic organic chemicals in water are improper disposal of industrial and household wastes and runoff of pesticides from farm fields, forests, roadsides, golf courses, and other places where they are used in large quantities. This material washes into the nearest waterway, where it passes through ecosystems and may accumulate in high levels in certain nontarget organisms. The bioaccumulation of DDT in aquatic ecosystems was one of the first of these pathways to be understood. Dioxins, and other chlorinated hydrocarbons (hydrocarbon molecules that contain chlorine atoms) have been shown to accumulate to dangerous levels in the fat of salmon, fish-eating birds, and humans and to cause health problems similar to those resulting from toxic metal compounds (Fig. 11.25).

Sediment

Rivers have always carried sediment to the oceans, but erosion rates in many areas have been greatly accelerated by human activities. As Chapters 10 and 14 describe, some rivers carry astounding loads of sediment. Erosion and runoff from croplands contribute about 25 billion tonnes of soil, sediment, and suspended solids to world surface waters each year. Forests, grazing lands, urban construction sites, and other sources of erosion and runoff add at least 50 billion additional tonnes. This sediment fills lakes and reservoirs, obstructs shipping channels, clogs hydroelectric turbines, and makes purification of drinking water more costly. Sediments smother gravel beds in which insects take refuge and fish lay their eggs. Sunlight is blocked so that plants cannot carry out photosynthesis and oxygen levels decline. Murky, cloudy water also is less attractive for swimming, boating, fishing, and other recreational uses (Fig. 11.26).

Sediment also can be beneficial. Mud carried by rivers nourishes floodplain farm fields. Sediment deposited in the ocean at river mouths creates valuable deltas and islands. The Ganges River, for instance, builds up islands in the Bay of Bengal that are eagerly colonized by land-hungry people of Bangladesh. In Louisiana, lack of sediment in the Mississippi River (it is being trapped by dams upstream) is causing biologically rich coastal wetlands to waste away. Sediment also can be harmful. Excess sediment deposits can fill estuaries and smother aquatic life on coral reefs and shoals near shore. As with many natural environmental processes, acceleration as a result of human intervention generally diminishes the benefits and accentuates the disadvantages of the process.

Thermal Pollution and Thermal Shocks

Raising or lowering water temperatures from normal levels can adversely affect water quality and aquatic life. Water temperatures are usually much more stable than air temperatures, so aquatic organisms tend to be poorly adapted to rapid temperature changes. Lowering the temperature of tropical oceans by even one degree can be lethal to some corals and other reef species. Raising water temperatures can have similar devastating effects on sensitive organisms. Oxygen solubility in water decreases as temperatures

Arsenic in Drinking Water

When we think of water pollution, we usually visualize sewage or industrial effluents pouring out of a discharge pipe, but there are natural toxins that threaten us as well. One of these is arsenic, a common contaminate in drinking water that may be poisoning millions of people around the world. Arsenic has been known since the fourth century B.C. to be a potent poison. It has been used for centuries as a rodenticide, insecticide, and weed killer, as well as a way of assassinating enemies. Because it isn't metabolized or excreted from the body, arsenic accumulates in hair and fingernails, where it can be detected long after death. Napoleon Bonaparte was found recently to have high enough levels of arsenic in his body to suggest he was poisoned.

Perhaps the largest population to be threatened by naturally occurring groundwater contamination by arsenic is in West Bengal, India, and adjacent areas of Bangladesh. Arsenic, in the form of insoluble salts, occurs naturally in the bedrock that underlies much of this region. Under normal conditions, the groundwater stays relatively free of arsenic in a soluble form. Rapid population growth, industrialization, and intensification of agricultural irrigation, however, have put increasing stresses on the limited surface water supplies of this region (see Chapter 10). Groundwater has all but replaced other water sources for most people in West Bengal, especially in the dry season.

In the 1960s, thousands of deep tube wells were sunk throughout the region to improve water supplies. Much of this humanitarian effort was financed by loans from the World Bank in the name of human development. At first, villagers were suspicious of well water, regarding it as unnatural and possibly evil. But as surface water supplies diminished, dusty Bengali villages became more and more dependent on this new source of supposedly fresh, clean water. By the late 1980s, health workers became aware of widespread signs of chronic arsenic poisoning among villagers in both India and Bangladesh. Symptoms of chronic arsenicosis include watery and inflamed eyes, gastrointestinal cramps, gradual loss of strength, scaly skin and skin tumours, anemia, confusion, and, eventually, death.

Why is arsenic poisoning appearing now? Part of the reason is increased dependence on well water, but some villages have had wells for centuries with no problem. One theory is that excessive withdrawals now lower the water table during the dry season, exposing arsenic-bearing rocks to air, which converts normally insoluble salts to soluble oxides. When aquifers are refilled during the next rainy season, dissolved arsenic can be pumped out. Health workers estimate that the total number of potential victims in India and Bangladesh may exceed 200 million people. But with no other source of easily accessible or affordable water, few of the poorest people have much choice.

Although few places in North America have as high groundwater arsenic content as West Bengal, there are worries that millions of Americans also are exposed to dangerously high levels of this toxic element. In 1942, the U.S. Government set the acceptable level of

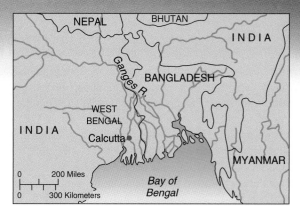

West Bengal and adjoining areas of Bangladesh have hundreds of millions of people who may be exposed to dangerous arsenic levels in well water.

arsenic in drinking water at 50 parts per billion or ppb. A 1999 study by the National Academy of Sciences found a one in 100 risk of cancer from drinking water with that level of arsenic for a lifetime. This is 10,000 times the normally accepted risk level. The U.S. limit was nearly revised to meet the World Health Organization standard of 10 ppb in 2000. The change was rejected, though. Local officials and private water supply owners complained that it would cost too much to upgrade their systems. The government has no business, they maintained, in telling us what we can or cannot drink.

Ethical Considerations

What do you think? If you choose to drink water that gives you a 1 in 100 chance of cancer, is that your business or does the government have a right (or obligation) to stop you? Is it ethical to allow customers of water districts to be exposed to a poison that they know nothing about? Would your parents or friends understand the implications of 1 in 100 risk of cancer? Do they smoke? Is it our responsibility to help the hundreds of millions of people in India and Bangladesh who now drink contaminated water? If it was our money that installed the tube wells in the first place, does that change the picture?

---

increase, so species requiring high oxygen levels are adversely affected by warming water.

Humans cause thermal pollution by altering vegetation cover and runoff patterns, as well as by discharging heated water directly into rivers and lakes. As Chapter 14 shows, nearly half the water we withdraw is used for industrial cooling. Electric power plants, metal smelters, petroleum refineries, paper mills, food-processing factories, and chemical manufacturing plants all use and release large amounts of cooling water.

The cheapest way to remove heat from an industrial facility is to draw cool water from an ocean, river, lake, or aquifer, run it through a heat-exchanger to extract excess heat, and then dump the heated water back into the original source. A **thermal plume** of heated water is often discharged into rivers and lakes, where raised temperatures can disrupt many processes in natural ecosystems and drive out sensitive organisms. To minimize these effects, power plants frequently are required to construct artificial cooling ponds or wet- or dry-cooling towers in which heat is released into the atmosphere and water is cooled before being released into natural water bodies. Wet cooling towers are cheaper to build and operate than dry systems, but lose large quantities of water to evaporation.

In some circumstances, introducing heated water into a water body is beneficial. Warming catfish-rearing ponds, for instance, can increase yields significantly. Warm water plumes from power plants often attract fish, birds, and marine mammals that find food and refuge there, especially in cold weather. This artificial environment can be a fatal trap, however. Organisms dependent on the warmth may die if they leave the plume or if the flow of warm water is interrupted by a plant shutdown. The manatee, for example, is an endangered marine mammal species that lives in Florida. Manatees are attracted to the abundant food supply and warm water in power plant thermal plumes and are enticed into spending the winter much farther north than they normally would. On several occasions, a midwinter power plant breakdown has exposed a dozen or more of these rare animals to a sudden thermal shock that they could not survive.

WATER QUALITY TODAY

Surface water pollution is often both highly visible and one of the most common threats to environmental quality. In more developed countries, reducing water pollution has been a high priority over the past few decades. Billions of dollars have been spent on control programs and considerable progress has been made. Still much remains to be done. In developed countries, poor water quality often remains a serious problem. In this section, we will look at progress as well as continuing obstacles in this important area.

Surface Waters in Canada and the United States

Like most developed countries, Canada and the United States have made encouraging progress in protecting and restoring water quality in rivers and lakes over the past 40 years. In 1948, only about one-third of North Americans were served by municipal sewage systems, and most of those systems discharged sewage without any treatment or with only primary treatment (the bigger lumps of waste are removed). Most people depended on cesspools and septic systems to dispose of domestic wastes.

Areas of Progress

Passage of the 1970 Water Act in Canada has produced striking results. Seventy percent of all Canadians in towns over 1,000 population are now served by some form of municipal sewage treatment. In Ontario, the vast majority of those systems include tertiary treatment. After 10 years of controls, phosphorus levels in the Bay of Quinte in the northeast corner of Lake Ontario have dropped nearly by half, and algal blooms that once turned waters green are less frequent and less intense than they once were. Elimination of mercury discharges from a pulp and paper mill on the Wabigoon-English River system in western Ontario has resulted in a dramatic decrease in mercury contamination that produced Minamata-like symptoms in local native people 20 years ago. Extensive flooding associated with hydropower projects has raised mercury levels in fish to dangerous levels elsewhere, however.

The 1972 Clean Water Act in the United States established a National Pollution Discharge Elimination System (NPDES), which requires an easily revoked permit for any industry, municipality or other entity dumping wastes in surface waters. The permit requires disclosure of what is being dumped and gives regulators valuable data and evidence for litigation. As a consequence, only about 10 percent of U.S. water pollution now comes from industrial or municipal point sources. One of the biggest improvements has been in sewage treatment.

Since the Clean Water Act was passed in 1972, the United States has spent more than $180 billion in public funds and perhaps ten times as much in private investments on water pollution control. Most of that effort has been aimed at point sources, especially to build or upgrade thousands of municipal sewage treatment plants. As a result, nearly everyone in urban areas is now served by municipal sewage systems and no major city discharges raw sewage into a river or lake except as overflow during heavy rainstorms.

This campaign has led to significant improvements in surface water quality in many places. Fish and aquatic insects have returned to waters that formerly were depleted of life-giving oxygen. Swimming and other water-contact sports are again permitted in rivers, lakes, and at ocean beaches that once were closed by health officials.

An encouraging example of improved water quality in North America is seen in Lake Erie. Although widely regarded as "dead" in the 1960s, the lake today is promoted as the "walleye capital of the world." Bacteria counts and algae blooms have decreased more than 90 percent since 1962. Water that once was murky brown is now clear. Interestingly, part of the improved water quality is due to immense numbers of exotic zebra mussels, which filter the lake water very efficiently. Swimming is now officially safe along 96 percent of the lake's shoreline. Nearly 40,000 nesting pairs of double-crested cormorants nest in the Great Lakes region, up from only about 100 in the 1970s. Anglers now complain that the cormorants eat too many fish. In 1998 wildlife agents found 800 cormorants shot to death in a rookery on Galloo Island at the east end of Lake Ontario.

Remaining Problems

The greatest impediments to achieving national goals in water quality in both Canada and the United States are sediment, nutrients, and pathogens, especially from nonpoint discharges of pollutants. These sources are harder to identify and to reduce or treat than are specific point sources. About three-fourths of the water pollution in the United States comes from soil erosion, fallout of air pollutants, and surface runoff from urban areas, farm fields, and feedlots. In the United States, as much as 25 percent of the 46,800,000 tonnes of fertilizer spread on farmland each year is carried away by runoff.

Cattle in feedlots produce some 129,600,000 tonnes of manure each year, and the runoff from these sites is rich in viruses, bacteria, nitrates, phosphates, and other contaminants. A single cow produces about 30 kg of manure per day, or about as much as that produced by 10 people. Some feedlots have 100,000 animals with no provision for capturing or treating runoff water. Imagine drawing your drinking water downstream from such a facility. Pets also can be a problem. It is estimated that the wastes from about a half million dogs in New York City are disposed of primarily through storm sewers, and therefore do not go through sewage treatment.

Loading of both nitrates and phosphates in surface water have decreased from point sources but have increased about fourfold since 1972 from nonpoint sources. Fossil fuel combustion has become a major source of nitrates, sulphates, arsenic, cadmium, mercury, and other toxic pollutants that find their way into water. Carried to remote areas by atmospheric transport, these combustion products now are found nearly everywhere in the world. Toxic organic compounds, such as DDT, PCBs, and dioxins, also are transported long distances by wind currents.

Surface Waters in Other Countries

Japan, Australia, and most of western Europe also have improved surface water quality in recent years. Sewage treatment in the wealthier countries of Europe generally equals or surpasses that in the United States. Sweden, for instance, serves 98 percent of its population with at least secondary sewage treatment (compared with 70 percent in the United States), and the other 2 percent have primary treatment. Poorer countries have much less to spend on sanitation. Spain serves only 18 percent of its population with even primary sewage treatment. In Ireland, it is only 11 percent, and in Greece, less than 1 percent of the people have even primary treatment. Most of the sewage, both domestic and industrial, is dumped directly into the ocean.

The fall of the "iron curtain" in 1989 revealed appalling environmental conditions in much of the former Soviet Union and its satellite states in eastern and central Europe. The countries closest geographically and socially to western Europe, the Czech Republic, Hungary, East Germany, and Poland have made massive investments and encouraging progress toward cleaning up environmental problems. Parts of Russia itself, however, along with former socialist states in the Balkans and Central Asia, remain some of the most polluted places on earth. In Russia, for example, only about half the tap water is fit to drink. In cities like St. Petersburg, even boiling and filtering isn't enough to make municipal water safe. About one-third of all Russians live in regions where air pollution levels are 10 times higher than World Health Organization safety standards. Life expectancies for Russian men have plummeted from about 72 years in 1980 to 59 years in 1999. Deaths now exceed births in Russia by about 1 million per year.

FIGURE 11.27 Two children from Copsa Micã, Romania, play in a polluted river across from the carbon factory that turns everything in town black. Eventually this filthy water will be flushed down the Danube and into the Black Sea, which has lost nearly all its commercial fisheries in only 25 years.

Only about one-quarter of all Russian children are considered healthy. In heavily industrialized cities like Magnitogorsk, a steel-manufacturing center in the Ural Mountains, 9 out of 10 children suffer from pollution-related illnesses and birth defects.

High levels of radioactivity from the Chernobyl accident in 1986 remain in many former Eastern Bloc countries. The Danube River, which originates in Bavaria and Austria, and flows through Slovakia, Poland, Hungary, Serbia, Romania, and Bulgaria before emptying into the Black Sea, illustrates some of the problems besetting the area (Fig. 11.27). Draining a landscape with a long history of unregulated mining and heavy industry, the Danube carries chrome, copper, mercury, lead, zinc, and oil to the Black Sea at 20 times the levels these materials flow into the North Sea. Just one city, Bratislava, the capital of Slovakia, dumps 73 million m^3 of industrial and municipal wastes into the river each year. With a population of about 100 million people in its catchment area, the beautiful blue Danube isn't blue anymore. Bombing in Serbia during the Kosovo War in 1999 further exacerbated the situation by releasing oil, industrial chemicals, agricultural fertilizers, pesticides, and other toxic materials into the Danube.

There are also some encouraging pollution control stories. In 1997, Minamata Bay in Japan, long synonymous with mercury poisoning, was declared officially clean again. Another important success is found in Europe, where one of its most important rivers has been cleaned up significantly through international cooperation. The Rhine, which starts in the rugged Swiss Alps and winds 1,320 km through five countries before emptying through a Dutch delta into the North Sea, has long been a major commercial artery into the heart of Europe. More than 50 million people live in its catchment basin and nearly 20 million get their drinking water from the river or its tributaries. By the 1970s, the Rhine had become so polluted that dozens of fish species disappeared and swimming was discouraged along most of its length.

Efforts to clean up this historic and economically important waterway began in the 1950s, but a disastrous fire at a chemical warehouse near Basel, Switzerland, in 1986 provided the impetus for major changes. Through a long and sometimes painful series of international conventions and compromises, land-use practices, waste disposal, urban runoff, and industrial dumping have been changed and water quality has significantly improved. Oxygen concentrations have gone up fivefold since 1970 (from less than 2 mg/l to nearly 10 mg/l or about 90 percent of saturation) in long stretches of the river. Chemical oxygen demand has fallen fivefold during this same period, and organochlorine levels have decreased as much as tenfold. Many species of fish and aquatic invertebrates have returned to the river. In 1992, for the first time in decades, mature salmon were caught in the Rhine.

The less-developed countries of South America, Africa, and Asia have even worse water quality than do the poorer countries of Europe. Sewage treatment is usually either totally lacking or woefully inadequate. In urban areas, 95 percent of all sewage is discharged untreated into rivers, lakes, or the ocean. Low technological capabilities and little money for pollution control are made even worse by burgeoning populations, rapid urbanization, and the shift of much heavy industry (especially the dirtier ones) from developed countries where pollution laws are strict to less-developed countries where regulations are more lenient.

Appalling environmental conditions often result from these combined factors (Fig. 11.28). Two-thirds of India's surface waters are contaminated sufficiently to be considered dangerous to human health. The Yamuna River in New Delhi has 7,500 coliform bacteria per 100 ml (37 times the level considered safe for swimming in North America) *before* entering the city. The coliform count increases to an incredible 24 *million* cells per 100 ml as the river leaves the city! At the same time, the river picks up some 20 million litres of industrial effluents every day from New Delhi. It's no wonder that disease rates are high and life expectancy is low in this area. Only 1 percent of India's towns and cities have any sewage treatment, and only eight cities have anything beyond primary treatment.

In Malaysia, 42 of 50 major rivers are reported to be "ecological disasters." Residues from palm oil and rubber manufacturing, along with heavy erosion from logging of tropical rainforests, have destroyed all higher forms of life in most of these rivers. In

FIGURE 11.28 Ditches in this Haitian slum serve as open sewers into which all manner of refuse and waste are dumped. The health risks of living under these conditions are severe.

the Philippines, domestic sewage makes up 60 to 70 percent of the total volume of Manila's Pasig River. Thousands of people use the river not only for bathing and washing clothes but also as their source of drinking and cooking water. China treats only 2 percent of its sewage. Of 78 monitored rivers in China, 54 are reported to be seriously polluted. Of 44 major cities in China, 41 use "contaminated" water supplies, and few do more than rudimentary treatment before it is delivered to the public.

Groundwater and Drinking Water Supplies

About 25 percent of the people in Canada, including 100 percent of those in rural areas, depend on underground aquifers for their drinking water. This vital resource is threatened in many areas by overuse and pollution and by a wide variety of industrial, agricultural, and domestic contaminants. For decades it was widely assumed that groundwater was impervious to pollution because soil would bind chemicals and cleanse water as it percolated through. Springwater or artesian well water was considered to be the definitive standard of water purity, but that is no longer true in many areas.

One of the serious sources of groundwater pollution is MTBE (methyl tertiary butyl ether), a suspected carcinogen. MTBE is a gasoline additive that has been used since the 1970s to reduce the amount of carbon monoxide and ozone in vehicle exhaust. By the time the health dangers of MTBE were confirmed in the late 1990s, aquifers across the continent had been contaminated—mainly from leaking underground storage tanks at gas stations. About 250,000 of these tanks are leaking MTBE into groundwater nationwide. In one U.S. Geological Survey (USGS) study, 27 percent of shallow urban wells tested contained MTBE. The additive is being phased out, but plumes of tainted water will continue to move through aquifers for decades to come. (Surface waters have also been contaminated, especially by two-stroke engines, such as those on personal watercraft.)

Treating MTBE-laced aquifers is expensive but not impossible. Douglas MacKay of the University of Waterloo in Ontario suggests that if oxygen could be pumped into aquifers, then naturally occurring bacteria could metabolize (digest) the compound. It could take decades or even centuries for natural bacteria to eliminate MTBE from a water supply, however. Water can also be pumped out of aquifers, reducing the flow and spread of contamination. Thus far, little funding has been invested in finding cost-effective remedies, however.

Contaminated water seeps into the ground from septic tanks, cesspools, municipal and industrial landfills and waste disposal sites, surface impoundments, agricultural fields, forests, and wells (Fig. 11.29). The most toxic of these are probably waste disposal sites. Agricultural chemicals and wastes are responsible for the largest total volume of pollutants and area affected. Because deep underground aquifers often have residence times of thousands of years, many contaminants are extremely stable once underground. It is possible, but expensive, to pump water out of aquifers, clean it, and then pump it back.

In farm country, fertilizers and pesticides commonly contaminate aquifers and wells. Herbicides such as atrazine and alachlor are widely used on corn and soybeans and show up in about half of all wells in Iowa, for example. Nitrates from fertilizers often exceed safety standards in rural drinking water. These high nitrate levels are dangerous to infants (nitrate combines with hemoglobin in the blood and results in "blue-baby" syndrome). They also are transformed into cancer-causing nitrosamines in the human gut. In Florida, 1,000 drinking water wells were shut down by state authorities because of excessive levels of toxic chemicals, mostly ethylene dibromide (EDB), a pesticide used to kill nematodes (roundworms) that damage plant roots.

Although most of the leaky, single-walled underground storage tanks once common at filling stations and factories have now been removed and replaced by more modern ones, a great deal of soil in North American cities remains contaminated by previous careless storage and disposal of petroleum products. Considering that a single litre of gasoline can make a million litres of water undrinkable, soil contamination remains a serious problem.

Direct injection of liquid wastes such as oilfield brine, effluents from chemical plants, and treated sewage down wells into deep aquifers is much less common than in the past. It is still allowed in some circumstances, however, to the dismay of some critics.

Abandoned wells represent another major source of groundwater contamination. Some domestic wells lack grouting to prevent surface contaminants from leaking into aquifers that they penetrate. When these wells are no longer in use, they often are not capped adequately, and people forget where they are. They can become direct routes for drainage of surface contaminants into aquifers.

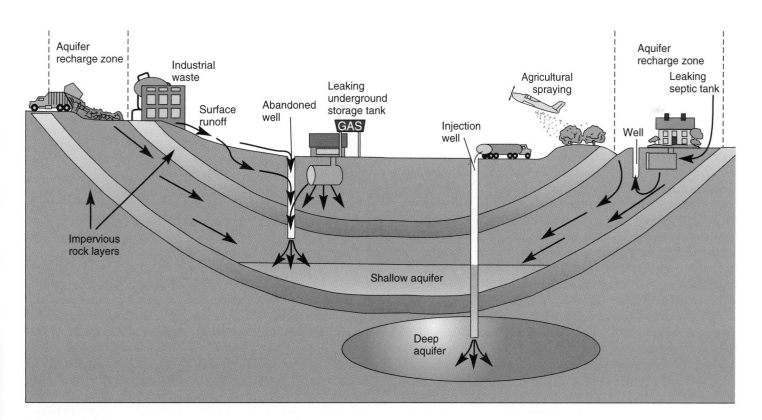

FIGURE 11.29 Sources of groundwater pollution. Septic systems, landfills, and industrial activities on aquifer recharge zones leach contaminants into aquifers. Wells provide a direct route for injection of pollutants into aquifers.

FIGURE 11.30 Beach pollution, including garbage, sewage, and contaminated runoff, is a growing problem associated with ocean pollution.

FIGURE 11.31 A deadly necklace. Marine biologists estimate that castoff nets, plastic beverage yokes, and other packing residue kill hundreds of thousands of birds, mammals, and fish each year.

Ocean Pollution

Coastal zones, especially bays, estuaries, shoals, and reefs near large cities or the mouths of major rivers, often are overwhelmed by human-caused contamination. Suffocating and sometimes poisonous blooms of algae regularly deplete ocean waters of oxygen and kill enormous numbers of fish and other marine life. High levels of toxic chemicals, heavy metals, disease-causing organisms, oil, sediment, and plastic refuse are adversely affecting some of the most attractive and productive ocean regions. The potential losses caused by this pollution amount to billions of dollars each year.

Discarded plastic flotsam and jetsam are lightweight and nonbiodegradable. They are carried thousands of kilometres on ocean currents and last for years (Fig. 11.30). Even the most remote beaches of distant islands are likely to have bits of polystyrene foam containers or polyethylene packing material that were discarded half a world away. It has been estimated that some 6 million tonnes of plastic bottles, packaging material, and other litter are tossed from ships every year into the ocean where they ensnare and choke seabirds, mammals (Fig. 11.31), and even fish. In one day, volunteers in Texas gathered more than 300 tonnes of plastic refuse from Gulf Coast beaches.

Few coastlines in the world remain uncontaminated by oil or oil products. Figure 11.32 shows locations where visible oil slicks have been reported. Oceanographers estimate that somewhere between 3 million and 6 million tonnes of oil are discharged into the world's oceans each year from both land- and sea-based operations. About half of this amount is due to maritime transport. Most of the 40 million litres of discharge is not from dramatic, headline-making accidents, such as the 1989 oil spill from the *Exxon Valdez* in Prince William Sound, Alaska, but from routine open-sea bilge pumping and tank cleaning, which are illegal but, nonetheless, are carried out once ships are beyond sight of land. Much of the rest comes from land-based municipal and industrial runoff or from atmospheric deposition of residues from refining and combustion of fuels.

The transport of huge quantities of oil creates opportunities for major oil spills through a combination of human and natural hazards. Military conflict in the Middle East and oil drilling in risky locations, such as the notoriously rough North Sea and the Arctic Ocean, make it likely that more oil spills will occur. Plans to drill for oil along the seismically active California and Alaska coasts have been controversial because of the damage that oil spills could cause to these biologically rich coastal ecosystems.

WATER POLLUTION CONTROL

Appropriate land-use practices and careful disposal of industrial, domestic, and agricultural wastes are essential for control of water pollution.

Source Reduction

The cheapest and most effective way to reduce pollution is usually to avoid producing it or releasing it to the environment in the first place. Elimination of lead from gasoline has resulted in a widespread and significant decrease in the amount of lead in surface waters in Canada and the United States. Studies have shown that as much as 90 percent less road deicing salt can be used in many areas without significantly affecting the safety of winter roads. Careful handling of oil and petroleum products can greatly reduce the amount of water pollution caused by these materials. Although we still have problems with persistent chlorinated hydrocarbons

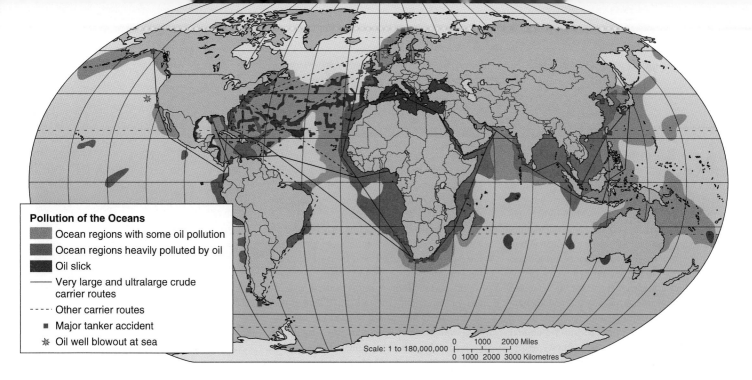

FIGURE 11.32 Location of oil pollution in the ocean. Major shipping routes of large oil tankers are very likely to be contaminated with oil.
Source: From *Student Atlas of Environmental Issues*, First Edition, edited by John L. Allen, copyright © 1997 by the McGraw-Hill Companies, Inc. Reprinted by permission of Dushkin/McGraw-Hill.

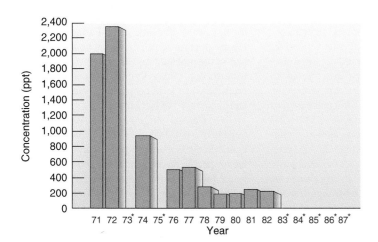

FIGURE 11.33 Dioxin concentrations in herring gull eggs on Scotch Bonnet Island in Lake Ontario have fallen sharply since 1971. Discontinued production and use of the herbicide 2,4,5-T are major reasons for this decline.
*Data not available.
Source: Data from *State of Canada's Environment,* Minister of the Environment.

spread widely in the environment, the banning of DDT and PCBs in the 1970s has resulted in significant reductions in levels in wildlife (Fig. 11.33).

Modifying agricultural practices in headwater streams in the Chesapeake Bay watershed and the Catskill Mountains of New York have had positive and cost-effective impacts on downstream water quality (Case Study, p. 236).

Industry can reduce pollution by recycling or reclaiming materials that otherwise might be discarded in the waste stream.

Both of these approaches usually have economic as well as environmental benefits. It turns out that a variety of valuable metals can be recovered from industrial wastes and reused or sold for other purposes. The company benefits by having a product to sell, and the municipal sewage treatment plant benefits by not having to deal with highly toxic materials mixed in with millions of litres of other types of wastes.

Nonpoint Sources and Land Management

Among the greatest remaining challenges in water pollution control are diffuse, nonpoint pollution sources. Unlike point sources, such as sewer outfalls or industrial discharge pipes, which represent both specific locations and relatively continuous emissions, nonpoint sources have many origins and numerous routes by which contaminants enter ground and surface waters. It is difficult to identify—let alone monitor and control—all these sources and routes. Some main causes of nonpoint pollution are:

- *Agriculture:* About 60 percent of all impaired or threatened surface waters are affected by sediment from eroded fields and overgrazed pastures; fertilizers, pesticides, and nutrients from croplands; and animal wastes from feedlots.

- *Urban runoff:* Pollutants carried by runoff from streets, parking lots, and industrial sites contain salts, oily residues, rubber, metals, and many industrial toxins. Yards, golf courses, parklands, and urban gardens often are treated with far more fertilizers and pesticides per unit area than farmlands. Excess chemicals are carried by storm runoff into waterways.

- *Construction sites:* New buildings and land development projects such as highway construction affect relatively small

CASE STUDY

WATERSHED PROTECTION IN THE CATSKILLS

New York City has long been proud of its excellent municipal drinking water. Drawn from the rugged Catskill Mountains 100 km north of the city, stored in hard-rock reservoirs, and transported through underground tunnels, the city water is outstanding for so large an urban area. Yielding 450,000 m^3 per day, and serving more than 9 million people, this is the largest surface water storage and supply complex in the world. As the metropolitan agglomeration has expanded, however, people have moved into the area around the Catskill Forest Preserve, and water quality is not as high as it was a century ago.

When the 1986 U.S. Safe Drinking Water Act mandated filtration of all public surface water systems, the city was faced with building an $8 billion water treatment plant that would cost up to $500 million per year to operate. In 1989, however, the EPA ruled that the city could avoid filtration if it could meet certain minimum standards for microbial contaminants such as bacteria, viruses, and protozoan parasites. In an attempt to avoid the enormous cost of filtration, the city proposed land-use regulations for the five counties (Green, Ulster, Sullivan, Schoharie, and Delaware) in the Catskill/Delaware watershed from which it draws most of its water.

With a population of 50,000 people, the private land within the 520 km^2 watershed is mostly devoted to forestry and small dairy farms, neither of which are highly profitable. Among the changes the city called for was elimination of storm water runoff from barnyards, feedlots, or grazing areas into watersheds. In addition, farmers would be required to reduce erosion and surface runoff from crop fields and logging operations. Property owners objected strenuously to what they regarded as onerous burdens that would cost enough to put many of them out of business. They also bristled at having the huge megalopolis impose rules on them. It looked like a long and bitter battle would be fought through the courts and the state legislature.

To avoid confrontation, a joint urban/rural task force was set up to see if a compromise could be reached, and to propose alternative solutions to protect both the water supply and the long-term viability of agriculture in the region. The task force agreed that agriculture is the "preferred land use" on private land, and that agriculture has "significant present and future environmental benefits." In addition, the task force proposed a voluntary, locally developed and administered program of "whole farm planning and best management approaches" very similar to ecosystem-based, adaptive management (see Chapter 19).

This grass-roots program, financed mainly by the city, but administered by local farmers themselves, attempts to educate landowners, and provides alternative marketing opportunities that help protect the watershed. Economic incentives are offered to encourage farmers and foresters to protect the water supply. Collecting feedlot and barnyard runoff in infiltration ponds together with solid conservation practices such as terracing, contour plowing, strip farming, leaving crop residue on fields, ground cover on waterways, and cultivation of perennial crops such as orchards and sugarbush have significantly improved watershed water quality. As of 1999, about 400 farmers—close to the 85 percent participation goal—have signed up for the program. The cost, so far, to the city has been about $50 million—or less than 1 percent of constructing a treatment plant.

In addition to saving billions of dollars, this innovative program has helped create good

Investing in soil conservation and water-quality protection on small dairy farms and agroforestry programs within the Catskill/Delaware watershed has saved New York City billions of dollars in filtration costs and has also improved community relations.

will between the city and its neighbours. It has shown that upstream cleanup, prevention, and protection are cheaper and more effective than treating water after it's dirty. Farmers have learned they can be part of the solution, not just part of the problem. And we have learned that watershed planning through cooperation is effective when local people are given a voice and encouraged to participate.

In Canada, following the Walkerton tragedy of May 2000, the Ontario government launched the Walkerton Inquiry, part of which led to specific recommendations about source water protection (http://www.ene.gov.on.ca/water.htm). We are just emerging from a time when Canadians thought they had limitless supplies of clean water. The Canadian Water Network (CWN) (http://www.cwn-rce.ca) is a collection of academic, government, and industry researchers with a common goal of ensuring safe, clean drinking water for Canadians. Check out how CWN is researching source protection in Canada!

---

areas but produce vast amounts of sediment, typically 10 to 20 times as much per unit area as farming (Fig. 11.34).

- *Land disposal:* When done carefully, land disposal of certain kinds of industrial waste, sewage sludge, and biodegradable garbage can be a good way to dispose of unwanted materials. Some poorly run land disposal sites, abandoned dumps, and leaking septic systems, however, contaminate local waters.

Generally, soil conservation methods (see Chapter 14) also help protect water quality. Applying precisely determined amounts of fertilizer, irrigation water, and pesticides saves money and reduces contaminants entering the water. Preserving wetlands that act as natural processing facilities for removing sediment and contaminants helps protect surface and groundwaters.

In urban areas, reducing materials carried away by storm runoff is helpful. Citizens can be encouraged to recycle waste oil and to minimize use of fertilizers and pesticides. Regular street sweeping greatly reduces contaminants. Runoff can be diverted away from streams and lakes. Many cities are separating storm sewers and municipal sewage lines to avoid overflow during storms.

FIGURE 11.34 Erosion on construction sites produces a great deal of sediment and is a major cause of nonpoint water pollution. Builders are generally required to install barriers to contain sediments, but these measures are often ineffectual.

Human Waste Disposal

As we have already seen, human and animal wastes usually create the most serious health-related water pollution problems. More than 500 types of disease-causing (pathogenic) bacteria, viruses, and parasites can travel from human or animal excrement through water. In this section, we will look at how to prevent the spread of these diseases.

Natural Processes

In the poorer countries of the world, most rural people simply go out into the fields and forests to relieve themselves as they have always done. Where population densities are low, natural processes eliminate wastes quickly, making this an effective method of sanitation. The high population densities of cities make this practice unworkable, however. Even major cities of many less-developed countries are often littered with human waste which has been left for rains to wash away or for pigs, dogs, flies, beetles, or other scavengers to consume. This is a major cause of disease, as well as being extremely unpleasant. Studies have shown that a significant portion of the airborne dust in Mexico City is actually dried, pulverized human feces.

Where intensive agriculture is practised—especially in wet rice paddy farming in Asia—it has long been customary to collect "night soil" (human and animal waste) to be spread on the fields as fertilizer. This waste is a valuable source of plant nutrients, but it is also a source of disease-causing pathogens in the food supply. It is the main reason that travellers in less-developed countries must be careful to surface sterilize or cook any fruits and vegetables they eat. Collecting night soil for use on farm fields was common in Europe and America until about 100 years ago when the association between pathogens and disease was recognized.

Until just over 50 years ago, most rural Canadian families and quite a few residents of towns and small cities depended on a pit toilet or "outhouse" for waste disposal. Untreated wastes tended to seep into the ground, however, and pathogens sometimes contaminated drinking water supplies. The development of septic tanks and properly constructed drain fields represented a considerable improvement in public health (Fig. 11.35). In a typical septic system, wastewater is first drained into a septic tank. Grease and oils rise to the top and solids settle to the bottom, where they are subject to bacterial decomposition. The clarified effluent from the septic tank is channeled out through a drainfield of small perforated pipes embedded in gravel just below the surface of the soil. The rate of aeration is high in this drainfield so that pathogens (most of which are anaerobic) will be killed, and soil microorganisms can metabolize any nutrients carried by the water. Excess water percolates up through the gravel and evaporates. Periodically, the solids in the septic tank are pumped out into a tank truck and taken to a treatment plant for disposal.

Where land is available and population densities are not too high, this can be an effective method of waste disposal. It is widely used in rural areas, but with urban sprawl, groundwater pollution often becomes a problem, indicating the need to shift to a municipal sewer system.

Municipal Sewage Treatment

Over more than the past 100 years, sanitary engineers have developed ingenious and effective municipal wastewater treatment systems to protect human health, ecosystem stability, and water quality. This topic is an important part of pollution control, and is a central focus of every municipal government; therefore, let's look more closely at how a typical municipal sewage treatment facility works.

Primary treatment is the first step in municipal waste treatment. It physically separates large solids from the waste stream. As raw sewage enters the treatment plant, it passes through a metal grating that removes large debris (Fig. 11.36a). A moving screen then filters out smaller items. Brief residence in a grit tank allows sand and gravel to settle. The waste stream then moves to the primary sedimentation tank where about half the suspended, organic solids settle to the bottom as sludge. Many pathogens remain in the effluent and it is not yet safe to discharge into waterways or onto the ground.

Secondary treatment consists of biological degradation of the dissolved organic compounds. The effluent from primary treatment flows into a trickling filter bed, an aeration tank, or a sewage lagoon. The trickling filter is simply a bed of stones or corrugated plastic sheets through which water drips from a system of perforated pipes or a sweeping overhead sprayer. Bacteria and other microorganisms in the bed catch organic material as it trickles past and aerobically decompose it.

Aeration tank digestion is also called the activated sludge process. Effluent from primary treatment is pumped into the tank and mixed with a bacteria-rich slurry (Fig. 11.36b). Air pumped through the mixture encourages bacterial growth and decomposition of the organic material. Water flows from the top of the tank and

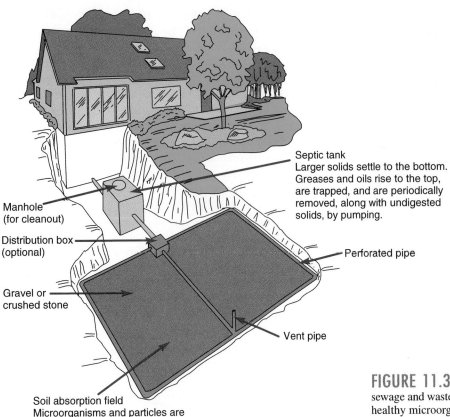

Manhole
(for cleanout)

Distribution box
(optional)

Gravel or
crushed stone

Septic tank
Larger solids settle to the bottom.
Greases and oils rise to the top,
are trapped, and are periodically
removed, along with undigested
solids, by pumping.

Perforated pipe

Vent pipe

Soil absorption field
Microorganisms and particles are
filtered out as water percolates through
the soil.

FIGURE 11.35 A domestic septic tank and drain field system for sewage and wastewater disposal. To work properly, a septic tank must have healthy microorganisms, which digest toilet paper and feces. For this reason, antimicrobial cleaners and chlorine bleach should never be allowed down the drain.

sludge is removed from the bottom. Some of the sludge is used as an inoculum for incoming primary effluent. The remainder would be valuable fertilizer if it were not contaminated by metals, toxic chemicals, and pathogenic organisms. The toxic content of most sewer sludge necessitates disposal by burial in a landfill or incineration. Sludge disposal is a major cost in most municipal sewer budgets (Fig. 11.37). In some communities this is accomplished by land farming, composting, or anaerobic digestion, but these methods don't inactivate metals and some other toxic materials.

Where space is available for sewage lagoons, the exposure to sunlight, algae, aquatic organisms, and air does the same job more slowly but with less energy costs. Effluent from secondary treatment processes is usually disinfected with chlorine, UV light, or ozone to kill harmful bacteria before it is released to a nearby waterway.

Tertiary treatment removes plant nutrients, especially nitrates and phosphates, from the secondary effluent. Although wastewater is usually free of pathogens and organic material after secondary treatment, it still contains high levels of inorganic nutrients, such as nitrates and phosphates. When discharged into surface waters, these nutrients stimulate algal blooms and eutrophication. To preserve water quality, these nutrients also must be removed. Passage through a wetland or lagoon can accomplish this. Alternatively, chemicals often are used to bind and precipitate nutrients. (see Fig. 11.36c)

In many North American cities, sanitary sewers are connected to storm sewers, which carry runoff from streets and parking lots. Storm sewers are routed to the treatment plant rather than discharged into surface waters because runoff from streets, yards, and industrial sites generally contains a variety of refuse, fertilizers, pesticides, oils, rubber, tars, lead (from gasoline), and other undesirable chemicals. During dry weather, this plan works well. Heavy storms often overload the system, however, causing bypass dumping of large volumes of raw sewage and toxic surface runoff directly into receiving waters. To prevent this overflow, cities are spending hundreds of millions of dollars to separate storm and sanitary sewers. These are huge, disruptive projects. When they are finished, surface runoff will be diverted into a river or lake and cause another pollution problem.

Low-Cost Waste Treatment

The municipal sewage systems used in developed countries are often too expensive to build and operate in the developing world where low-cost, low-tech alternatives for treating wastes are needed. One option is **effluent sewerage,** a hybrid between a traditional septic tank and a full sewer system. A tank near each dwelling collects and digests solid waste just like a septic system. Rather than using a drainfield, however, to dispose of liquids—an impossibility in crowded urban areas—effluents are pumped to a central treatment plant. The tank must be emptied once a year or so, but because only

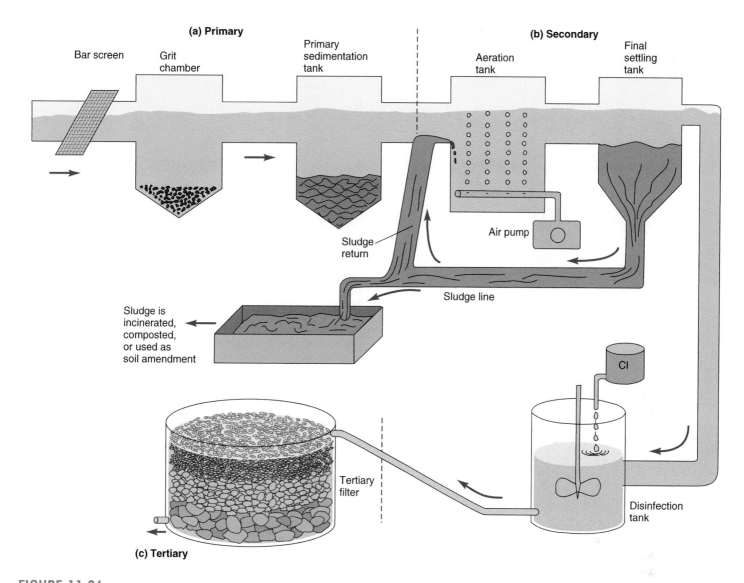

(a) Primary

Bar screen

Grit chamber

Primary sedimentation tank

(b) Secondary

Aeration tank

Final settling tank

Sludge return

Air pump

Sludge line

Sludge is incinerated, composted, or used as soil amendment

Cl

Tertiary filter

Disinfection tank

(c) Tertiary

FIGURE 11.36 (*a*) Primary sewage treatment removes only solids and suspended sediment. (*b*) Secondary treatment, through aeration of activated sludge (or biosolids), followed by sludge removal and chlorination of effluent, kills pathogens and removes most organic material. (*c*) During tertiary treatment passage through a trickling bed evaporator and/or a lagoon or marsh further removes inorganic nutrients, oxidizes any remaining organics, and reduces effluent volume.

liquids are treated by the central facility, pipes, pumps, and treatment beds can be downsized and the whole system is much cheaper to build and run than a conventional operation.

Another alternative is to use natural or artificial wetlands to dispose of wastes. Arcata, California, for instance, needed an expensive sewer plant upgrade. By transforming a 65-hectare garbage dump into a series of ponds and marshes that serve as a simple, low-cost waste treatment facility, the city saved millions of dollars and improved the environment simultaneously. Sewage is piped to holding ponds where solids settle out and are digested by bacteria and fungi. Effluent flows through marshes where it is filtered and cleansed by aquatic plants and microorganisms. The marsh is a haven for wildlife and has become a prized recreation area for the city (see Chapter 6). Eventually, the purified water flows into the bay where marine life flourishes.

Similar wetland waste treatment systems are now operating in many developing countries. Effluent from these operations can be used to irrigate crops or raise fish for human consumption if care is taken to first destroy pathogens (Fig. 11.38). Usually 20 to 30 days of exposure to sun, air, and aquatic plants is enough to make the water safe. These systems make an important contribution to human food supplies. A 2,500-hectare waste-fed aquaculture facility in Calcutta, for example, supplies about 7,000 tonnes of fish annually to local markets. The World Bank estimates that some 3 billion people will be without sanitation services by the middle of this century under a business-as-usual scenario (Fig. 11.39). With investments in innovative programs, however, sanitation could be provided to about half those people and a great deal of misery and suffering could be avoided.

FIGURE 11.37 "Well, if *you* can't use it, do you know anyone who *can* use 3,000 tons of sludge every day?"
© 2001 by Sidney Harris.

FIGURE 11.38 In India, a poplar plantation thrives on raw sewage water piped directly from nearby homes. Innovative solutions like this can make use of nutrients that would pollute water systems.
Source: FAO.

Water Remediation

Remediation means finding remedies for problems. Just as there are many sources for water contamination, there are many ways to clean it up. New developments in environmental engineering are providing promising solutions to many water pollution problems.

Containment methods confine or restrain dirty water or liquid wastes *in situ* (in place) or cap the surface with an impermeable layer to divert surface water or groundwater away from the site and to prevent further pollution. Where pollutants are buried

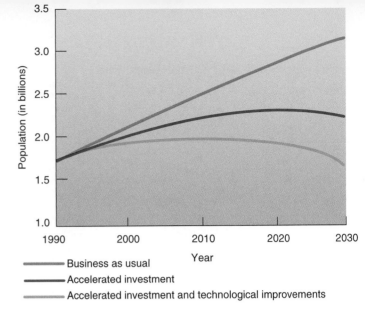

FIGURE 11.39 World population without adequate sanitation—three scenarios in the year 2030. If business as usual continues, more than 3 billion people will lack safe sanitation. Accelerated investment in sanitation services could lower this number. Higher investment, coupled with technological development, could keep the number of people without adequate sanitation from growing even though the total population increases.
Source: World Bank estimates based on research paper by Dennis Anderson and William Cavendish, "Efficiency and Substitution in Pollution Abatement: Simulation Studies in Three Sectors."

too deeply to be contained mechanically, materials sometimes can be injected to precipitate, immobilize, chelate, or solidify them. Bentonite slurries, for instance, can effectively stabilize liquids in porous substrates. Similarly, straw or other absorbent material is spread on surface spills to soak up contaminants.

Extraction techniques pump out polluted water so it can be treated. Many pollutants can be destroyed or detoxified by chemical reactions that oxidize, reduce, neutralize, hydrolyze, precipitate, or otherwise change their chemical composition. Where chemical techniques are ineffective, physical methods may work. Solvents and other volatile organic compounds, for instance, can be stripped from solution by aeration and then burned in an incinerator. Some contaminants can be removed by semipermeable membranes or resin filter beds that bind selectively to specific materials. Some of the same techniques used to stabilize liquids *in situ* can also be used *in vitro* (in a reaction vessel). Metals, for instance, can be chelated or precipitated in insoluble, inactive forms.

Often, living organisms can be used effectively and inexpensively to clean contaminated water. We call this bioremediation. Restored wetlands, for instance, along stream banks or lake margins can be very effective in filtering out sediment and removing pollutants. They generally cost far less than mechanical water treatment facilities and provide wildlife habitat as well.

Lowly duckweed, the green scum you often see covering the surface of eutrophic ponds, grows fast and can remove large amounts of organic nutrients from water. Under optimal conditions, a few square centimetres of these tiny plants can grow to cover nearly a hectare in four months. Large duckweed lagoons are being used as inexpensive, low-tech sewage treatment plants in developing countries. Where conventional wastewater purification typically costs $300 to $600 per person served, a duckweed system can cost one-tenth as much. The duckweed can be harvested and used as feed, fuel, or fertilizer. Up to 35 percent of its dry mass is protein—about twice as much as alfalfa, a popular animal feed.

Where space for open lagoons is unavailable, bioremediation can be carried out in reaction vessels. This has the advantage of controlling conditions more precisely and doesn't release organisms into the environment. Some of the most complex, holistic systems for water purification are designed by Ocean Arks International (OAI) in Falmouth, Massachusetts. Their "living machines" combine living organisms—chosen to perform specific functions—in contained environments. In a typical living machine, water flows through a series of containers, each with a distinct ecological community designed for a particular function. Wastes generated by the inhabitants of one vessel become the food for inhabitants of another. Sunlight provides the primary source of energy.

OAI has created or is in the process of building water treatment plants in a dozen states and foreign countries. Designs range from remediating toxic wastes from superfund sites to simply treating domestic wastes (Fig. 11.40). Starting with microorganisms in aerobic and anaerobic environments where different kinds of wastes are metabolized or broken down, water moves through a series of containers containing hundreds of different kinds of plants and animals, including algae, rooted aquatic plants, clams, snails, and fish, each chosen to provide a particular service. Technically,

Solar sewage wall

FIGURE 11.40 A neighbourhood sewage treatment facility designed by Jack Todd of Ocean Arks International. Water trickling through containers of aquatic plants is cleaned and purified by biological processes. Under optimal conditions, both odours and costs are minimal.
Source: Drawing is based on an original drawing by John Todd. Reprinted by permission.

the finished water is drinkable, although few people feel comfortable doing so. More often, the final effluent is used to flush toilets or for irrigation. Called ecological engineering, this novel approach can save resources and money as well as clean up our environment and serve as a valuable educational tool.

SUMMARY

Any physical, biological, or chemical change in water quality that adversely affects living organisms or makes water unsuitable for desired uses can be considered pollution. Worldwide, the most serious water pollutants, in terms of human health, are pathogenic organisms from human and animal wastes. We have traditionally taken advantage of the capacity of ecosystems to destroy these organisms, but as population density has grown, these systems have become overloaded and ineffective. Effective sewage treatment systems are needed that purify wastewater before it is released to the environment.

In industrialized nations, toxic chemical wastes have become an increasing problem. Agricultural and industrial chemicals have been released or spilled into surface waters and are seeping into groundwater supplies. The extent of this problem is probably not yet fully appreciated.

Ultimately, all water ends up in the ocean. The ocean is so large that it would seem impossible for human activities to have a significant impact on it, but pollution levels in the ocean are increasing. Major causes of ocean pollution are oil spills from tanker bilge pumping or accidents and oil well blowouts. Surface runoff and sewage outfalls discharge fertilizers, pesticides, organic nutrients, and toxic chemicals that have a variety of deleterious effects on marine ecosystems. We usually think of eutrophication (increased productivity due to nutrient addition) as a process of inland waterways, but this can occur in oceans as well.

The major water pollutants in terms of quantity are silt and sediments. Biomass production by aquatic organisms, land erosion, and refuse discharge all contribute to this problem. Addition of salts and metals from highway and farm runoff and industrial activities also damage water quality. In some areas, drainage from mines and tailings piles deliver sediment and toxic materials to rivers and lakes. Water pollution is a major source of human health problems. As much as 80 percent of all disease and some 25 million deaths each year may be attributable to water contamination.

Appropriate land-use practices and careful disposal of industrial, domestic, and agricultural wastes are essential for control of water pollution. Natural processes and living organisms have a high capacity to remove or destroy water pollutants, but these systems become overloaded and ineffective when pollution levels are too high. Municipal sewage treatment is effective in removing organic material from wastewater, but the sewage sludge is often contaminated with metals and other toxic industrial materials. Reducing the sources of these materials is often the best solution to our pollution problems.

QUESTIONS FOR REVIEW

1. What is the difference between withdrawal, consumption, and degradation of water?

2. How does water use by sector differ between rich and poor countries?

3. What is subsidence? What are its results?

4. Describe some problems associated with dam building and water diversion projects.

5. Define *water pollution*.

6. List eight major categories of water pollutants and give an example for each category.

7. Describe eight major sources of water pollution in North America. What pollution problems are associated with each source?

8. What is *Pfiesteria* and why is it dangerous?

9. What is eutrophication? What causes it?

10. What are the origins and effects of siltation?

11. Describe primary, secondary, and tertiary processes for sewage treatment. What is the quality of the effluent from each of these processes?

12. Why do combined storm and sanitary sewers cause water quality problems? Why does separating them also cause problems?

13. Describe remediation techniques and how they work.

QUESTIONS FOR CRITICAL THINKING

1. Why do we use so much water? Do we need all that we use?

2. Are there ways you could use less water in your own personal life? Would that make any difference in the long run?

3. Should we use up underground water supplies now or save them for some future time?

4. How should we compare the values of free-flowing rivers and natural ecosystems with the benefits of flood control, water diversion projects, hydroelectric power, and dammed reservoirs?

5. How precise is the estimate that 2 billion people lack access to clean water? Would it make a difference if the estimate is off by 10 percent or 50 percent?

6. How would you define *adequate* sanitation? Think of some situations in which people might have different definitions for this term.

7. Do you think that water pollution is worse now than it was in the past? What considerations go into a judgement like this? How do your personal experiences influence your opinion?

8. What additional information would you need to make a judgement about whether conditions are getting better or worse? How would you weigh different sources, types, and effects of water pollution?

9. Imagine yourself in a developing country with a severe shortage of clean water. What would you miss most if your water supply were suddenly cut by 90 percent?

10. Proponents of deep well injection of hazardous wastes argue that it will probably never be economically feasible to pump water out of aquifers more than 1 kilometre below the surface. Therefore, they say, we might as well use those aquifers for hazardous waste storage. Do you agree? Why or why not?

11. Suppose that part of the silt in a river is natural and part is human-caused. Is one pollution but the other not?

12. Suppose that you own a lake but it is very polluted. An engineer offers options for various levels of cleanup. As you increase water quality, you also increase costs greatly. How clean would you want the water to be—fishable, swimmable, drinkable—and how much would you be willing to pay to achieve your goal? Make up your own numbers. The point is to examine your priorities and values.

KEY TERMS

atmospheric deposition   222
biochemical oxygen demand
 (BOD)   224
coliform bacteria   224
consumption   212
cultural eutrophication   225

degradation   212
desalination   216
dissolved oxygen (DO)
 content   224
effluent sewerage   238
eutrophic   225
nonpoint sources   222
oligotrophic   225

oxygen sag   224
point sources   222
primary treatment   237
red tide   226
renewable water
 supplies   210
saltwater intrusion   215
secondary treatment   237

sinkholes   215
subsidence   215
tertiary treatment   238
thermal plume   229
watershed   219
water stress   213
withdrawal   212

Web Exercises

CANADA'S TOP 10 ENDANGERED RIVERS LIST

Check out National Geographic's list of Canada's Top 10 endangered rivers at http://news.nationalgeographic.com/news/2003/07/0707_030707_canadarivers.html. What do these 10 rivers have in common (if anything)? Find out every-thing you can about one of the rivers near you. What are the major stressors from human activity? What are the major natural factors that may make the river more, or less, susceptible to perturbations from human stressors?

The number one endangered river in Canada is the Petitcodiac in New Brunswick. Photos from 1954 (bottom) and 1996 (top) show the constriction of river width as sediment has been deposited.

Chapter 12

Conventional Energy

I'm not blindly opposed to progress; I'm opposed to blind progress.

—Dave Brower—

OBJECTIVES

After studying this chapter, you should be able to:

- understand the units of energy.

- know the basics of different conventional sources of energy, including wood, fossil fuels, hydroelectric, and nuclear power.

- compare the environmental benefits and costs of each type of conventional energy discussed.

WebQuest

energy reserves, fossil fuel, hydroelectrical power, nuclear energy, Atomic Energy Canada Limited, woodfuel

Above: Retreating Iraqi troops set more than 600 oil wells on fire as they left Kuwait in the 1990 Gulf War. At least 10 percent of Kuwait's oil was burned or spilled.

Oil and Wildlife in the Arctic

Beyond the jagged Brooks Range in Alaska's far northeastern corner lies one of the world's largest nature preserves, the 8-million-ha Arctic National Wildlife Refuge (ANWR). A narrow strip of treeless coastal plain in the heart of the refuge presents one of nature's grandest spectacles as well as one of the longest-running environmental battles of the last century. For a few months during the brief arctic summer, the tundra teems with wildlife. This is the calving ground of the 130,000 caribou of the Porcupine herd (Fig. 12.1). It also is important habitat for tens of thousands of snow geese, tundra swans, shorebirds, and other migratory waterfowl; a denning area for polar bears, arctic foxes, and arctic wolves; and a year-round home to about 350 shaggy musk ox. In wildlife density and diversity, it rivals Africa's Serengeti.

When the U.S. Congress established the Wildlife Refuge in 1980, a special exemption was made for about 600,000 ha of coastline between the mountains and the Beaufort Sea where geologists think sedimentary strata may contain billions of barrels of oil and trillions of cubic metres of natural gas. Called the 1002 area for the legislative provision that put it inside the wildlife refuge but reserved the right to drill for fossil fuels, this narrow strip of tundra may be the last big, on-shore, liquid petroleum field in North America. It also is critical habitat for one of the richest biological communities in the world. Can we extract the fossil fuels without driving away the wildlife and polluting the pristine landscape? Oil industry experts believe they can access resources without doing lasting environmental harm; biologists and environmentalists doubt this is so.

How much oil and gas lies beneath the tundra? No one knows for sure. Only one seismic survey has been done and a few test wells drilled. Industry geologists claim that there may be 16 billion barrels of oil under ANWR, but guessing the size and content of the formation from this limited evidence is as much an art as a science. Furthermore, the amount of oil that's economical to recover depends on market prices and shipping costs. At wholesale prices of $25 per barrel, the U.S. Geological Survey estimates that there is a 95 percent chance of producing 2 billion barrels of oil profitably, a 50–50 chance of recovering 4.5 billion barrels, and a 5 percent chance of finding 9 billion barrels. If prices drop below $10 per barrel, however, as they did in the early 1990s, the economic resource might be only a few hundred million barrels.

Energy companies are extremely interested in ANWR because any oil found there could be pumped out through existing Trans-Alaska pipeline, thus extending the life of their multibillion-dollar investment. The state of Alaska hopes that revenues from ANWR will replenish dwindling coffers as the oil supplies from nearby Prudhoe Bay wells dry up. Automobile companies, tire manufacturers, filling stations, and others who depend on our continued use of petroleum argue that domestic oil supplies are rapidly being depleted.

FIGURE 12.1 Alaska's Arctic National Wildlife Refuge is home to one of the world's largest caribou herds as well as 200 other wildlife and plant species. It may also contain North America's last big unexploited oil and gas deposit. Can we extract fossil fuels without irreparably damaging the fragile tundra ecosystem?

Oil drilling proponents point out that prospecting will occur only in winter when the ground is covered with snow and most wildlife is absent or hibernating. Once oil is located, they claim, it will take only four or five small drilling areas, each occupying no more than a few hundred hectares, to extract it. Heavy equipment would be hauled to the sites during the winter on ice roads built with water pumped from nearby lakes and rivers. Each 2-metre-thick, gravel drilling pad would hold up to 50 closely spaced wells, which would penetrate the permafrost and then spread out horizontally to reach pockets of oil up to 10 km away from the wellhead. A central processing facility would strip water and gas from the oil, which would then be pumped through elevated, insulated pipelines to join oil flowing from Prudhoe Bay (Fig. 12.2).

Opponents of this project argue that the noise, pollution, and construction activity accompanying this massive operation will drive away wildlife and leave scars on the landscape that could last for centuries. Pumping the millions of litres of water needed to build

Want More?

See information about cooperative wildlife management and the Porcupine Caribou Herd at http://www.pcmb.yk.ca/pcmb.html.

FIGURE 12.2 Nearly 1,300 km long, the Trans-Alaska Pipeline crosses three mountain ranges and more than 800 rivers or streams before reaching its terminus at Valdez.

ice roads could dry up local ponds on which wildlife depends for summer habitat. Every day six to eight aircraft—some as big as the C-130 Hercules—would fly into ANWR. The smell of up to 700 workers and the noise of numerous trucks and enormous power-generating turbines, each as large and loud as a jumbo aircraft engine, would waft out over the tundra. Pointing to the problems of other arctic oil drilling operations, where drilling crews dumped garbage, sewage, and toxic drilling waste into surface pits, environmentalists predict disaster if drilling is allowed in the refuge. Pipeline and drilling spills at Prudhoe Bay have contaminated the tundra and seeped into waterways. Scars from bulldozer tracks made 50 years ago can still be seen clearly today.

Engineers planning operations in ANWR, on the other hand, claim that old, careless ways are no longer permitted in their operations. Wastes are collected and either burned or injected into deep wells. Although some animals do seem to have been displaced at Prudhoe Bay, the central arctic herd with 27,000 caribou is five times larger now than it was in 1978 when drilling operations began there. But in ANWR the Porcupine herd has five times as many animals crowded into one-fifth the area and may be much more sensitive to disturbance than their cousins to the west.

Native people are divided on this topic. The coastal Inupiat people, many of whom work in the oil fields, support opening the refuge to oil exploration. They hope to use their share of the oil revenues to build new schools and better housing. The Gwich'in people, on the other hand, who live south of the refuge, would gain nothing from oil exploitation. They worry that migrating caribou on which they depend might decline as a result of drilling on critical calving grounds. As with so much else in this controversy, the answers you get depend on whom you ask.

Even if ANWR contains 7 billion barrels of oil, it would take at least a decade to begin to get it to market, and the peak production rate will probably be about 1 million barrels of oil per day in 2030. Flow from the 1002 area would then meet less than 4 percent of the U.S. daily oil consumption. Improving the average fuel efficiency of all cars and light trucks in America by just one kilometre per litre would save more oil than is ever likely to be recovered from ANWR, and it would do so faster and cheaper than extracting and transporting crude oil from the Arctic. Cutting our fossil fuel consumption also is vital if we are to avoid catastrophic global climate change (see Chapter 8).

WHAT IS ENERGY AND WHERE DO WE GET IT?

We discussed what energy is and how it differs from *work* (transfer of energy), *power* (rate of work), and *heat* (transfer of energy because of temperature differences) in Chapter 3. In Table 12.1, we relate some already introduced energy units to those commonly used in the collection, distribution, and consumption of energy other than food. It might strike you as unusual to consider that the same energy that makes up the "calories" in your morning doughnut, or fixed and passed on by an algal community in a lake, or that a salmon uses to swim upstream to spawn, is what runs our trains, planes, and automobiles.

A Brief Energy History

Fire was probably the first human energy technology. Charcoal from fires has been found at sites occupied by our early ancestors

TABLE 12.1 Some Energy Units
1 joule (J) = the force exerted by a current of 1 amp per second flowing through a resistance of 1 ohm
1 watt (W) = 1 joule (J) per second
1 kilowatt-hour (kWh) = 1 thousand (10^3) watts exerted for 1 hour = 36 million j = 36 thousand kj
1 megawatt (MW) = 1 million (10^6) watts
1 gigawatt (GW) = 1 billion (10^9) watts
1 petajoule (PJ) = 1 quadrillion (10^{15}) joules
1 British thermal unit (BTU) = energy to heat 1 lb of water 1°F
1 kWh = 3,412 BTU
1 standard barrel (bbl) of oil = 160 l = 5.8 million BTU = 1,700 kWh = 73,100 doughnuts
1 tonne of standard coal = 4.8 bbl oil = 6,800 kWh

1 million years ago. Muscle power provided by domestic animals has been important at least since the dawn of agriculture some 10,000 years ago. Wind and water power have been used nearly as long. The invention of the steam engine, together with diminishing supplies of wood in industrializing countries, caused a switch to coal as our major energy source in the nineteenth century. Coal, in turn, has been replaced by oil in this century due to the ease of shipping, storing, and burning liquid fuels.

World energy consumption has increased relentlessly over the past 30 years, in spite of occasional "oil crises" and political or military instability in energy producing areas such as the Middle East (Fig. 12.3). Projections by the International Atomic Energy (IEA) indicate global primary energy demand is set to increase by 1.7 percent per year from 2000 to 2030, or about two-thirds of current demand. This is a bit slower than growth over the past 30 years, which ran at 2.1 percent per year.

The IEA sees fossil fuels remaining the dominant energy source for humans, accounting for more than 90 percent of the increase in energy use to 2030. Natural gas demand will grow fastest, but oil will still be the largest individual fuel source. Though new, renewable forms of energy will grow rapidly (see Chapter 13), they are starting from a very small proportion of the total energy use and will not soon displace fossil fuels.

With these projections of growth in energy consumption, energy-related emissions of carbon dioxide will grow 1.8 percent per year from 2000 to 2030, reaching 38 billion tonnes in 2030. This is 16 billion tonnes, or 70 percent more than today.

The early 1980s saw an increased concern about conservation and development of renewable energy resources. This concern didn't last long, unfortunately, as reduced demand and increased production in the mid-1980s caused an oil glut that made prices fall almost as rapidly as they had risen a decade earlier. Through most of the 1990s, the price of a benchmark barrel of Texas light crude hovered around $15 when adjusted for inflation. Since then, production curbs instituted by the Organization of Petroleum Exporting Countries (OPEC) pushed prices above $30 per barrel, still less, in real dollars, than prices 30 years earlier.

Current Energy Sources

Fossil fuels (petroleum, natural gas, and coal) now provide about 85 percent of all commercial energy in the world (Fig. 12.4). Biomass fuels, such as wood, peat, charcoal, and manure, contribute about 6 percent of commercial energy. Other renewable sources—solar, wind, geothermal, and hydroelectricity—make up 4 to 5 percent of our commercial power (actually, hydro accounts for most of that). Although not included in these data, private systems (individual houses, businesses, industries) may capture as much renewable energy as that reported in commercial transactions. Similarly, this survey probably underrepresents the importance of biomass (fuelwood, dung) because these fuels often are gathered by the people using them, or they are sold in the informal economy and not reported internationally.

Nuclear power is roughly equal to hydroelectricity, worldwide (i.e., 4 to 5 percent of all commercial energy), but it makes up about 20 percent of all electric power in more developed countries such as Canada and the United States. We have enough nuclear fuel to produce power for a long time, and it has the benefit of not emitting greenhouse gases (Chapter 8), but as we discuss later in this chapter, safety concerns make this option unacceptable to many people.

Per Capita Consumption

Perhaps the most important facts about energy consumption are that those in the 20 richest countries consume nearly 80 percent of the natural gas, 65 percent of the oil, and 50 percent of the coal produced each year. Although we make up less than *one-fifth* of the

FIGURE 12.3 Sudden price shocks in the 1970s caused by anticipated oil shortages showed North Americans that our energy-intensive lifestyles may not continue forever.

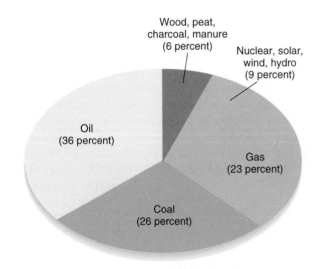

FIGURE 12.4 Worldwide commercial energy production. (Totals exceed 100 percent due to rounding.)
Source: Data from World Resources Institute, 1998–99.

PART TWO   Our Physical Environment

world's population, we use more than *one-half* of the commercial energy supply. Canada and the United States, for instance, constitute only 5 percent of the world's population but consume about one-quarter of the available energy. To be fair, however, some of that energy goes to grow crops or manufacture goods that are later shipped to developing countries.

How much energy do you use every year? Most of us don't think about it much, but maintaining the lifestyle we enjoy requires an enormous energy input. On average, each person in North America uses more than 300 GJ (equivalent to about 60 barrels of oil) per year. By contrast, in the poorest countries of the world, such as Ethiopia, Kampuchea, Nepal, and Bhutan, each person generally consumes less than one GJ per year. This means that each of us consumes, on average, almost as much energy in a single day as a person in one of these countries consumes in a year.

Clearly, having an abundant external energy supply contributes to the comfort and convenience of our lives (Fig. 12.5). Those of us in the richer countries enjoy many amenities not available to most people in the world. The linkage is not absolute, however. Several European countries, including Sweden, Denmark, and Switzerland, have higher standards of living than does Canada by almost any measure but use about half as much energy as we do. These countries have had effective energy conservation programs for many years. Japan, also, has a far lower energy consumption rate than might be expected for its industrial base and income level.

Because Japan has few energy resources of its own, it has developed very efficient energy conservation measures.

HOW ENERGY IS USED

The largest share (36.5 percent) of the energy used in Canada is consumed by industry. Mining, milling, smelting, and forging of primary metals consume about one-quarter of the industrial energy share. The chemical industry is the second largest industrial user of fossil fuels, but only half of its use is for energy generation. The remainder is raw material for plastics, fertilizers, solvents, lubricants, and hundreds of thousands of organic chemicals in commercial use. The manufacture of cement, glass, bricks, tile, paper, and processed foods also consumes large amounts of energy. Residential and commercial buildings use some 29 percent of the primary energy consumed in Canada, mostly for space heating, air conditioning, lighting, and water heating. Small motors and electronic equipment take an increasing share of residential and commercial energy.

Transportation consumes about 29 percent of all energy used in Canada each year. About 98 percent of that energy comes from petroleum products refined into liquid fuels, and the remaining 2 percent is provided by natural gas and electricity. Almost three-quarters of all transport energy is used by motor vehicles. In 1998, there were 459 passenger cars for every 1,000 people in Canada! Much of our transportation is extremely inefficient. Driving a 2,000-kg car to take one 60-kg person a few kilometres for shopping or work isn't a very wise use of resources. About 75 percent of all freight traffic in the United States is carried by trains, barges, ships, and pipelines, but because they are very efficient, they use only 12 percent of all transportation fuel.

Finally, analysis of how energy is used has to take into account waste and loss of potential energy. About *half* of all the energy in primary fuels is lost during conversion to more useful forms, while it is being shipped to the site of end use, or during its use. Electricity, for instance, is generally promoted as a clean, efficient source of energy because when it is used to run a resistance heater or an electrical appliance almost 100 percent of its energy is converted to useful work and no pollution is given off.

What happens before then, however? We often forget that huge amounts of pollution are released during mining and burning of the coal that fires power plants. Furthermore, nearly two-thirds of the energy in the coal that generated that electricity was lost in thermal conversion in the power plant. About 10 percent more is lost during transmission and stepping down to household voltages. Similarly, about 75 percent of the original energy in crude oil is lost during distillation into liquid fuels, transportation of that fuel to market, storage, marketing, and combustion in vehicles.

Natural gas is our most efficient fuel. Only 10 percent of its energy content is lost in shipping and processing since it moves by pipelines and usually needs very little refining. Ordinary gas-burning furnaces are about 75 percent efficient, and high-economy furnaces can be as much as 95 percent efficient. Because natural gas has more hydrogen per carbon atom than oil or coal, it produces

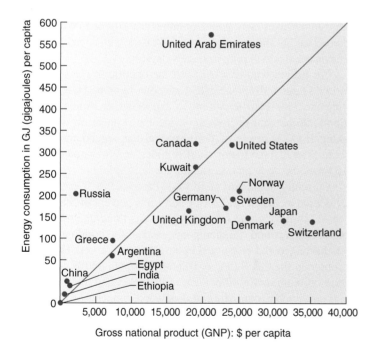

FIGURE 12.5 Per capita energy use and GNP. In general, higher energy use correlates with a higher standard of living. Denmark and Switzerland, however, use about half as much energy as we do and have a higher standard of living by most measures. Oil-rich states, such as the United Arab Emirates, use about twice as much energy as we do but are still relatively undeveloped in most areas.

Source: Data from World Resources Institute, 1998–99.

about half as much carbon dioxide—and therefore half as much contribution to global warming—per unit of energy.

WOOD AND BIOMASS

Photosynthetic organisms have been collecting and storing the sun's energy for more than 2 billion years. Plants capture about 0.1 percent of all solar energy that reaches the earth's surface. That kinetic energy is transformed, via photosynthesis, into chemical bonds in organic molecules (Chapter 3). A little more than half of the energy that plants collect is spent in such metabolic activities as pumping water and ions, mechanical movement, maintenance of cells and tissues, and reproduction; the rest is stored in biomass.

The magnitude of this resource is difficult to measure. Most experts estimate useful biomass production at 15 to 20 times the amount we currently get from all commercial energy sources. It would be ridiculous to consider consuming all green plants as fuel, but biomass has the potential to become a prime source of energy. It has many advantages over nuclear and fossil fuels because of its renewability and easy accessibility. Renewable energy resources account for about 18 percent of total world energy use, and biomass makes up three-quarters of that renewable energy supply. Biomass resources used as fuel include wood, wood chips, bark, branches, leaves, starchy roots, and other plant and animal materials.

Wood fires have been a primary source of heating and cooking for thousands of years. As recently as 1850, wood supplied 90 percent of the fuel used in Canada. Wood now provides less than 1 percent of the energy in Canada, but in many of the poorer countries of the world, wood and other biomass fuels provide up to 95 percent of all energy used. The 1,500 million cubic metres of fuelwood collected in the world each year is about half of all wood harvested.

In northern industrialized countries, wood burning has increased since 1975 in an effort to avoid rising oil, coal, and gas prices. Most of these northern areas have adequate wood supplies to meet demands at current levels, but problems associated with wood burning may limit further expansion of this use. Inefficient and incomplete burning of wood in fireplaces and stoves produces smoke laden with fine ash and soot and hazardous amounts of carbon monoxide (CO) and hydrocarbons. In valleys where inversion layers trap air pollutants, the effluent from wood fires can present a major source of air quality degradation and health risk. Polycyclic aromatic compounds produced by burning are especially worrisome because they are carcinogenic (cancer-causing).

Two billion people—about 40 percent of the total world population—depend on firewood and charcoal as their primary energy source. Of these people, three-quarters (1.5 billion) do not have an adequate, affordable supply. Most of these people are in the less-developed countries where they face a daily struggle to find enough fuel to warm their homes and cook their food. The problem is intensifying because rapidly growing populations in many developing countries create increasing demands for firewood and charcoal from a diminishing supply.

FIGURE 12.6 The firewood shortage in less-developed countries means that women and children must spend hours each day searching for fuel. Destruction of forests and removal of ground cover result in erosion and desertification.

As firewood becomes increasingly scarce, women and children, who do most of the domestic labour in many cultures, spend more and more hours searching for fuel (Fig. 12.6). In some places, it now takes eight hours, or more, just to walk to the nearest fuelwood supply and even longer to walk back with a load of sticks and branches that will only last a few days.

For people who live in cities, the opportunity to scavenge firewood is generally nonexistent and fuel must be bought from merchants. This can be ruinously expensive. In Addis Ababa, Ethiopia, 25 percent of household income is spent on wood for cooking fires. A circle of deforestation has spread more than 160 km around some major cities in India, and firewood costs up to 10 times the price paid in smaller towns.

Currently, about half of all wood harvested each year worldwide is used as fuel (Fig. 12.7). Eighty-five percent of that fuel is harvested in developing countries, whereas three-quarters of all industrial roundwood (lumber, poles, beams, and building materials) is harvested and consumed in developed countries. The poorest countries such as Ethiopia, Bhutan, Burundi, and Bangladesh depend on biomass for 90 percent of their energy. Often, the harvest is sustainable, consisting of deadwood, branches, trimmings, and shrubs. In Pakistan, for example, some 4.4 million tonnes of twigs and branches and 7.7 million tonnes of shrubs and crop residue are consumed as fuel each year with destruction of very few living trees.

In countries where fuel is scarce, however, desperate people often chop down anything that will burn. In Haiti, for instance, more than 90 percent of the once-forested land has been almost completely denuded and people cut down even valuable fruit trees to make charcoal they can sell in the marketplace. It is estimated

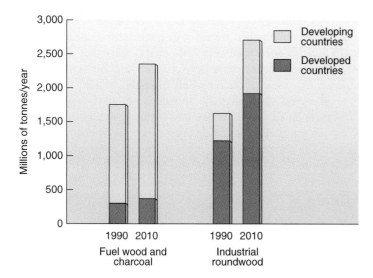

FIGURE 12.7 Worldwide, about half of all woody biomass harvested is used as fuel, mainly in developing countries. Industrial roundwood (lumber, timbers, beams) are primarily harvested and used in developed countries. A growing need for fuel in developing countries is expected to cause wood shortages in many places.
Source: Data from FAO, 1997.

that the 1,700 million tonnes of fuelwood now harvested each year globally is at least 500 million tonnes less than is needed. By 2025 the worldwide demand for fuelwood is expected to be about twice current harvest rates while supplies will not have expanded much beyond current levels. Some places will be much worse than this average. In some African countries such as Mauritania, Rwanda, and the Sudan, firewood demand already is 10 times the sustainable yield. Reforestation projects, agroforestry, community woodlots, and inexpensive, efficient, locally produced woodstoves could help alleviate expected fuelwood shortages in many places.

Where wood and other fuels are in short supply, people often dry and burn animal manure. This may seem like a logical use of waste biomass, but it can intensify food shortages in poorer countries. Not putting this manure back on the land as fertilizer reduces crop production and food supplies. In India, for example, where fuelwood supplies have been chronically short for many years, a limited manure supply must fertilize crops and provide household fuel. Cows in India produce more than 800 million tonnes of dung per year, more than half of which is dried and burned in cooking fires (Fig. 12.8). If that dung were applied to fields as fertilizer, it could boost crop production of edible grains by 20 million tonnes per year, enough to feed about 40 million people.

When cow dung is burned in open fires, more than 90 percent of the potential heat and most of the nutrients are lost. Compare that to the efficiency of using dung to produce methane gas, an excellent fuel. In the 1950s, simple, cheap methane digesters were designed for villages and homes, but they were not widely used. In China, 6 million households use biogas for cooking and lighting. Two large municipal facilities in Nanyang will soon provide fuel for more than 20,000 families. Perhaps other countries will follow China's lead.

FIGURE 12.8 In many countries, animal dung is gathered by hand, dried, and used for fuel. This girl in Mali is carrying home fuel for cooking. Using dung as fuel deprives fields of nutrients and reduces crop production.

Woodfuels consist of three main commodities: fuelwood, charcoal, and black liquor. Fuelwood and charcoal are traditional forest products derived from the forest, trees outside forests, wood-processing industries, and recycled wooden products from society. Black liquors are by-products of the pulp and paper industry.

In 1999, about 1.4 billion tonnes of fuelwood were produced worldwide, which is about 3,400 million bbl or about 5 percent of the world total energy requirement. Black liquor supplied about 500 million bbl of energy in 1997. Thus, it can be roughly estimated that woodfuels in total contribute about 3,900 million bbl annually to the world energy requirement. This amount is smaller than that of nuclear energy, which provided 4,700 million bbl in 1999, but substantially larger than the output from hydro and other renewable sources of energy.

On average, the annual per-capita consumption of woodfuels is around 5 bbl, but with considerable regional variances. People in poorer, developing countries use much more fuelwood per km^2 of forest than forested developed countries. Haiti and El Salvador use over 5,000 tonnes of firewood per km^2 of forest area, whereas Australia is less than 5 tonnes per km^2 and Canada is less than 1 tonne per km². With a similar amount of forest area (about 2.5 million km²), Canada produces far less fuelwood (2.2 million tonnes) than the United States (59 million tonnes).

COAL

Coal is fossilized plant material preserved by burial in sediments and altered by geological forces that compact and condense it into a carbon-rich fuel. Coal is found in every geologic system since the Silurian Age 400 million years ago, but graphite deposits in very old rocks suggest that coal formation may date back to Precambrian times. Most coal was laid down during the Carboniferous period (286 million to 360 million years ago) when the earth's climate was warmer and wetter than it is now. Because coal takes so long to form, it is essentially a nonrenewable resource.

Coal Resources and Reserves

World coal deposits are vast, 10 times greater than conventional oil and gas resources combined. Coal seams can be 100 m thick and can extend across tens of thousands of square kilometres that were vast swampy forests in prehistoric times. The total resource is estimated to be 10 trillion tonnes. If all this coal could be extracted, and if coal consumption continued at present levels, this would amount to several thousand years' supply. At present rates of consumption, these proven-in-place reserves—those explored and mapped but not necessarily economically recoverable—will last about 200 years.

Where are these coal deposits located? They are not evenly distributed throughout the world (Fig. 12.9). North America has one-fourth of all proven reserves located in large deposits across the central and mountain regions (Fig. 12.10). China and Russia have nearly as much. Both China and India plan to greatly increase coal consumption to raise standards of living. If they do so, CO_2 released

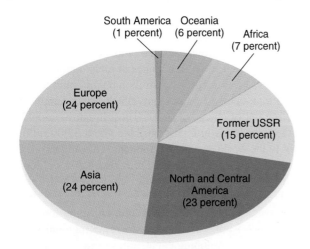

FIGURE 12.9 Proven-in-place coal reserves, by region as percent of a total 1.96 trillion tonnes.
Source: Data from World Resources Institute, 1994–95.

could exacerbate global warming. Eastern and western Europe have fairly large coal supplies in spite of their relatively small size. Africa and Latin America have very little coal despite their large size. Antarctica is thought to have large coal deposits, but they would be difficult, expensive, and ecologically damaging to mine.

It would seem that the abundance of coal deposits is a favourable situation. But do we really want to use all of the coal? What are the environmental and personal costs of coal extraction and use? In the next section, we will look at some of the disadvantages and dangers of mining and burning coal.

Mining

Coal mining is a hot, dirty, and dangerous business. Underground mines are subject to cave-ins, fires, accidents, and accumulation of poisonous or explosive gases (carbon monoxide, carbon dioxide, methane, hydrogen sulphide). Between 1870 and 1950, more than 30,000 coal miners died of accidents and injuries in Pennsylvania alone, equivalent to one man per day for 80 years. Untold thousands have died of respiratory diseases. In some mines, nearly every miner who did not die early from some other cause was eventually disabled by **black lung disease,** inflammation and fibrosis caused by accumulation of coal dust in the lungs or airways (Chapter 20). Few of these miners or their families were compensated for their illnesses by the companies for which they worked. The U.S. Department of Labor now compensates miners and their dependants under the Black Lung Benefits Program, but workers often have difficulty proving that health problems are occupationally related. Even when compensation is provided, black lung remains a wretched occupational hazard of coal mining.

Strip mining or surface mining (Fig. 12.11) is cheaper than underground mining but often makes the land unfit for any other use. Mine reclamation is now mandated in Canada and the United States, but efforts often are superficial and ineffective. Coal mining also contributes to water pollution. Sulphur and other water soluble minerals make mine drainage and runoff from coal piles and mine tailings acidic and highly toxic.

Air Pollution

Many people aren't aware that coal burning releases radioactivity and many toxic metals. Uranium, lead, cadmium, mercury, rubidium, thallium, and zinc—along with a number of other elements—are absorbed by plants and concentrated in the process of coal formation. These elements are not destroyed when the coal is burned, instead they are released as gases or concentrated in fly ash and bottom slag. You are likely to get a higher dose of radiation living next door to a coal-burning power plant than a nuclear plant under normal (nonaccident) conditions. Coal combustion is responsible for about 25 percent of all atmospheric mercury pollution in the United States.

Coal contains up to 10 percent sulphur (by weight). Unless this sulphur is removed by washing or flue-gas scrubbing, it is released during burning and oxidizes to sulphur dioxide (SO_2)

PART TWO   Our Physical Environment

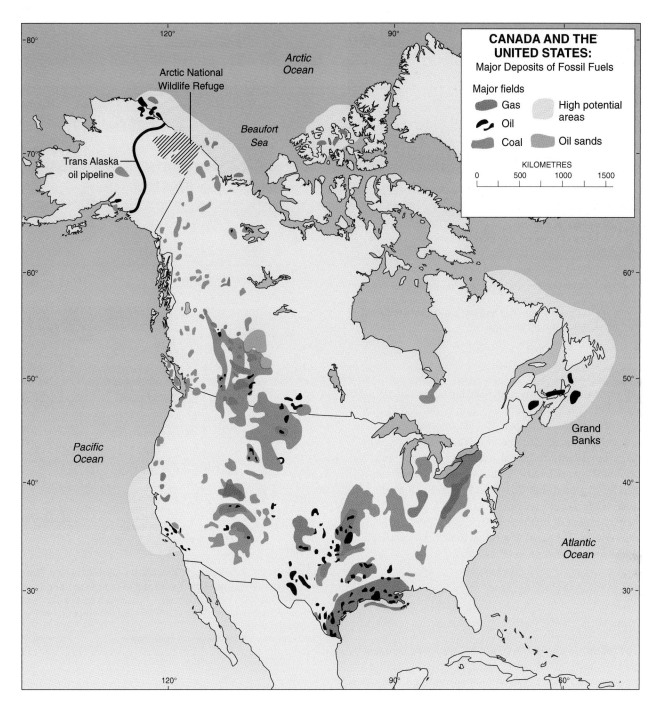

FIGURE 12.10 Canadian and U.S. fossil fuel deposits. Coloured areas show off-shore sedimentary deposits with high potential for oil or gas. The ecological risks of drilling in these areas are high; many environmentalists argue that drilling should be prohibited in these places.

From H. G. DeBlij and P. O. Muller, *Geography: Regions and Concepts,* 8th edition. Copyright © 1992 John Wiley & Sons, Inc. Reprinted by permission of John Wiley & Sons, Inc.

or sulphate (SO_4) in the atmosphere. The high temperatures and rich air mixtures ordinarily used in coal-fired burners also oxidize nitrogen compounds (mostly from the air) to nitrogen monoxide, dioxide, and trioxide. Every year the 900 million tonnes of coal burned in the United States (83 percent for electric power generation) releases 18 million tonnes of SO_2, 5 million tonnes of nitrogen oxides (NO_x), 4 million tonnes of airborne particulates, 600,000 tonnes of hydrocarbons and carbon monoxide, and close to a trillion tonnes of CO_2. This is about three-quarters of the SO_2, one-third of the NO_x, and about half of the industrial CO_2 released in the United States each year. Coal burning is the largest single source of acid rain in many areas.

Sulphur can be removed from coal before it is burned, or sulphur compounds can be removed from the flue gas after

FIGURE 12.11 A giant power shovel scoops up coal in an open pit mine in Montana. Careful recontouring and restoration can make this land productive once again, but costs are high and many old mines remain desolate and bare for decades, if not centuries.

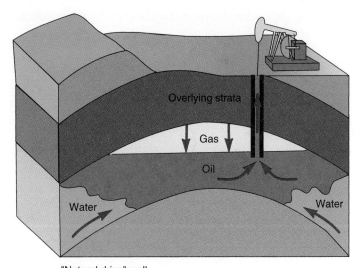

"Natural drive" well

Enhanced recovery or "stripping" well

FIGURE 12.12 Recovery process for petroleum. In a "natural drive" well, water or gas pressure forces liquid petroleum into the well. In an enhanced recovery or "stripping" well, steam and/or water mixed with chemicals are pumped down a well behind the oil pool, forcing oil up the extraction well.
Source: Data from World Resources Institute, 1998–99.

combustion. Formation of nitrogen oxides during combustion also can be minimized. Perhaps the ultimate limit to our use of coal as a fuel will be the release of carbon dioxide into the atmosphere. As we discussed in Chapter 8, carbon traps heat in the atmosphere and is a major contributor to global warming. Trapping and storing CO_2 produced by stationary facilities such as power plants is beginning to be explored, but this would be difficult to do with mobile sources such as automobiles. Coal could be used, however, to make pure hydrogen or methane for fuel cells. So-called "clean coal" technologies might make this valuable resource available to us without disastrous environmental damage.

OIL

Like coal, petroleum is derived from organic molecules created by living organisms millions of years ago and buried in sediments where high pressures and temperatures concentrated and transformed them into energy-rich compounds. Depending on its age and history, a petroleum deposit will have varying mixtures of oil, gas, and solid tarlike materials. Some very large deposits of heavy oils and tars are trapped in porous shales, sandstone, and sand deposits in the western areas of Canada and the United States.

Liquid and gaseous hydrocarbons can migrate out of the sediments in which they formed through cracks and pores in surrounding rock layers. Oil and gas deposits often accumulate under layers of shale or other impermeable sediments, especially where folding and deformation of systems create pockets that will trap upward-moving hydrocarbons (Fig. 12.12). Contrary to the image implied by its name, an oil pool is not usually a reservoir of liquid in an open cavern but rather individual droplets or a thin film of liquid permeating spaces in a porous sandstone or limestone, much like water saturating a sponge.

Pumping oil out of a reservoir is much like sucking liquid out of a sponge. The first fraction comes out easily, but removing subsequent fractions requires increasing effort. We never recover all the oil in a formation; in fact, a 30 to 40 percent yield is about average. There are ways of forcing water or steam into the oil-bearing formations to "strip" out more of the oil, but at least half the total deposit usually remains in the ground at the point at which it is uneconomical to continue pumping. Methods for squeezing more oil from a reservoir are called **secondary recovery techniques** (Fig. 12.12).

Oil Resources and Reserves

The total amount of oil in the world is estimated to be about 4 trillion barrels (600 billion tonnes), half of which is thought to be ultimately recoverable. Some 465 billion barrels of oil already have been consumed. In 1999, the proven reserves were roughly 1 trillion bbls, enough to last only 45 years at the current consumption rate of 22 billion barrels per year. It is estimated that another 800 billion barrels either remain to be discovered or are not recoverable at current prices with present technology. As oil resources become depleted and prices rise, it probably will become economical to find and bring this oil to the market unless alternative energy sources are developed. This estimate of the resource does not take into account the very large potential from unconventional liquid hydrocarbon resources, such as shale oil and tar sands, which might double the total reserve if they can be mined with acceptable social, economic, and environmental impacts.

By far the largest supply of proven-in-place oil is in Saudi Arabia, which has 250 billion barrels, about one-fourth of the total proven world reserve (Fig. 12.13). Kuwait had more than 10 percent of the proven world oil reserves before Iraq invaded in 1990. Some 600 wells were blown up and set on fire by the retreating Iraqis. Although the consequences were not as catastrophic as first feared, at least 5 billion barrels of Kuwait's oil was burned, spilled, or otherwise lost (see p. 245). Together, the Persian Gulf countries in the Middle East contain nearly two-thirds of the world's proven petroleum supplies. With our insatiable appetite for oil (some would say addiction), it is not difficult to see why this volatile region plays such an important role in world affairs.

Note that this discussion has been of *proven* reserves. If reservoirs around the Caspian Sea truly hold 200 billion barrels, they would by second only to Saudi Arabia in size (see Case Study, p. 256).

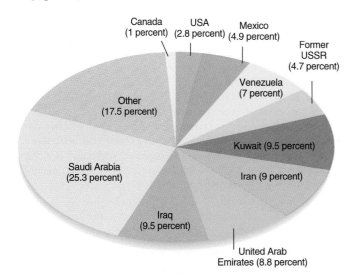

FIGURE 12.13 World proven recoverable oil reserves by country (1997). If the Caspian Sea region really has 200 billion barrels of recoverable oil, it could increase world supplies by about 25 percent, or 10 years at current rates of use.
Source: Data from World Resources Institute, 1998–99.

Although oil consumption in Europe and North America has been relatively constant in recent years, Asian demand has been growing rapidly as economies there expand and people become more affluent. China, for instance, is expanding oil use 10 percent per year. If current trends continue, about half of all oil produced by the Organization of Petroleum Exporting Countries (OPEC) will be going to Asia.

Similarly, vast deposits of oil shale occur in the western United States. Actually, **oil shale** is neither oil nor shale but a fine-grained sedimentary rock rich in solid organic material called kerogen. When heated to about 480°C, the kerogen liquefies and can be extracted from the stone. Oil shale beds up to 600 m thick occur in the Green River Formation in Colorado, Utah, and Wyoming, and lower-grade deposits are found over large areas of the eastern United States. If these deposits could be extracted at a reasonable price and with acceptable environmental impacts, they might yield the equivalent of several trillion barrels of oil.

Mining and extracting shale oil also creates many problems. It is expensive; it uses vast quantities of water, a scarce resource in the arid west; it has a high potential for air and water pollution; and it produces enormous quantities of waste. In the early 1980s, when the search for domestic oil supplies was at fever pitch, serious discussions occurred about filling whole canyons, rim to rim, with oil shale waste. One experimental mine used a nuclear explosion to break up the oil shale. All the oil shale projects dried up when oil prices fell in the mid-1980s, however.

NATURAL GAS

Natural gas is the world's third largest commercial fuel (after oil, and coal), making up 23 percent of global energy consumption. It is the most rapidly growing energy source because it is convenient, cheap, and clean burning. Because natural gas produces only half as much CO_2 as an equivalent amount of coal, substitution could help reduce global warming (Chapter 8). Natural gas is difficult to ship across oceans or to store in large quantities. North America is fortunate to have an abundant, easily available supply of gas and a pipeline network to deliver it to market. Many developing countries cannot afford such a pipeline network, and much of the natural gas produced in conjunction with oil pumping is simply burned (flared off), a terrible waste of a valuable resource (Fig. 12.15).

Natural Gas Resources and Reserves

The republics of the former Soviet Union have 41 percent of known natural gas reserves (mostly in Siberia and the Central Asian republics) and account for 36.5 percent of all production. Both eastern and western Europe buy substantial quantities of gas from these wells. Figure 12.16 shows the distribution of proven natural gas reserves in the world.

The total ultimately recoverable natural gas resources are estimated to be 230,000 million tonnes, corresponding to about 80 percent as much energy as the recoverable reserves of crude oil.

For more than the past decade, Russian troops pounded the rebellious province of Chechnya with bombs, rockets, and artillery in a war Russia could ill afford and probably never win conclusively. Why such a ferocious assault on a tiny, impoverished state at the fringes of their former empire? The answer is that Chechnya is the gateway, for Russia, to what may be one the world's richest reserves of oil and natural gas around—and under—the Caspian Sea.

It has long been known that Central Asia is rich in oil. Marco Polo wrote in the fourteenth century of "fountains from which oil springs in great abundance." An oil boom a century ago turned Baku, the capital of Azerbaijan, into a city of instant millionaires. But the Soviet Union never put much effort into developing Caspian oil, preferring to emphasize less politically challenging wells in Siberia. By the 1980s, a crumbling network of leaking pipelines, rusty drilling rigs, decaying cities, and patches of oil-soaked soil around the margins of the Caspian showed the effects of sloppy management and neglect. When the Soviet Union broke apart in 1991, a mad rush began as Western energy companies fought to be the first to exploit what may be the last really big, relatively accessible oil field in the world.

Oil deposits around the Caspian Sea are thought to hold up to 200 billion barrels, perhaps 25 percent of all the world's oil. If true, this resource would be worth about $4 trillion at today's prices, or about 30 times as much as Alaska's entire North Slope deposit. In addition, countries neighbouring the Caspian are thought to have enormous reserves of natural gas. Turkmenistan alone is thought to sit on 9 trillion m^3 of gas, making it the world's fourth-largest holder of this valuable resource. In total, the Central Asian Republics may control a quarter of the world's natural gas supply.

The biggest difficulty is how to get these resources to market from their landlocked sources. It doesn't help that the area has some of the worst weather—temperatures ranging from –40°C to +50°C—and most bellicose and unstable political climates in the world. Regional players include Russia, Chechnya, Dagestan, Abkhazia, Ingushetia, Ossetia, Kurdistan, Kazakhstan, Turkmenistan, Uzbekistan,

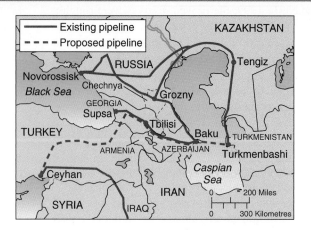

A U.S.-backed oil pipeline is proposed from Azerbaijan through Georgia, and then southwest across Turkey to the Mediterranean. Russia prefers a route across Chechnya to Nororossisk on the Black Sea.

Azerbaijan, Armenia, Georgia, Turkey, and Iran. There are more than 60 indigenous languages, and at least as many ethnic feuds in the region. Together in the 1990s, these groups had six major wars, two presidential assassination attempts, two coups, and countless guerrilla and bandit attacks.

The shortest—and probably cheapest—route for an oil or gas pipeline from the Caspian is across Iran to the Persian Gulf. The United States adamantly opposed that option, however, both to thwart Iran, and because of the risk of having additional oil passing through the vulnerable Persian Gulf. In 2001, Russia completed a pipeline running around the north side of the Caspian Sea to Novossisk on the Black Sea. This route requires tankers to pass through the highly congested straits of Bosporus and the Dardanelles. Turkey is worried about the prospects of a huge oil tanker crashing into the rocks and releasing millions of litres of toxic crude oil into this busy waterway.

Another 2,000 km pipeline is now under construction along the U.S.-preferred route from Baku, across Georgia, and then south across Turkey to the Mediterranean Sea. An extension nearly as long would cross under the Caspian and then run north to the Tengiz oil field in Kazakhstan. Costing at least $4 billion (US), the pipeline will take five years to build, and will carry about a million barrels of oil per day when finished.

Protecting this sprawling network of vulnerable pipelines in a rugged mountainous region of ancient but fierce ethnic, religious, and political hostilities is a daunting prospect. Destroying local opposition and ensuring its access to the resources of Central Asia are among the reasons that Russia has pursued the war in Chechnya with such ferocity.

Does this story suggest to you that our dependence on oil creates odd bedfellows and difficult geopolitical problems? How far will we go to ensure our access to energy resources?

Ethical Considerations

As major potential customers for Caspian oil, do we in North America have a responsibility for how the resource is tapped and shipped to market, or is that strictly for the suppliers to decide? Are the social and environmental costs of extracting fossil fuels in remote places a concern of ours? If so, how can we influence the actions of people whose religion, politics, and economics may be very different from ours? If this were the last oil in the world, would your attitude be different?

Oil Sands

2003 STATISTICS

initial volume in place	1.6 trillion barrels
remaining ultimate potential	311 billion barrels
production (marketable)	882.5 thousand barrels per day
royalties	$183 million (fiscal 2002–03)
employment (total oil, gas & oil sands)	95.4 thousand (direct upstream)
cumulative investment	$24 billion (1996–2002 CAPP)
investment	$6.7 billion (2002 CAPP)

Oil sands are deposits of bitumen, a heavy black viscous oil that must be rigorously treated to convert it into an upgraded crude oil before refineries can use it to produce gasoline and diesel fuels. Until recently, Alberta's bitumen deposits were known as **tar sands** but are now referred to as **oil sands.** Bitumen is best described as a thick, sticky form of crude oil, so heavy and viscous that it will not flow unless heated or diluted with lighter hydrocarbons. At room temperature, it is much like cold molasses.

Oil sands are substantially heavier than other crude oils. Technically speaking, bitumen is a tar-like mixture of petroleum hydrocarbons with a density greater than 960 kilograms per cubic metre; light crude oil, by comparison, has a density as low as 793 kilograms per cubic metre. Compared to conventional crude oil, bitumen requires some additional upgrading before it can be refined. It also requires dilution with lighter hydrocarbons to make it transportable by pipelines.

Bitumen makes up about 10–12 percent of the actual oil sands found in Alberta. The remainder is 80–85 percent mineral matter—including sand and clays—and 4–6 percent water. While conventional crude oil flows naturally or is pumped from the ground, oil sands must be mined or recovered *in situ* (meaning "in place"). Oil sands recovery processes include extraction and separation systems to remove the bitumen from sand and water.

Alberta's oil sands comprise one of the world's two largest sources of bitumen; the other is in Venezuela. Oil sands are found in three places in Alberta—the Athabasca, Peace River, and Cold Lake regions—and cover a total of nearly 141,000 square kilometres. They currently represent 54 percent of Alberta's total oil production, and about one-third of all the oil produced in Canada. By 2005, oil sands production is expected to represent 50 percent of Canada's total crude oil output, and 10 percent of North American production.

Mineable bitumen deposits are located near the surface and can be recovered by open-pit mining techniques. For example, the Syncrude and Suncor oil sands operations near Fort McMurray, Alberta, use the world's largest trucks and shovels to recover bitumen (Fig. 12.14). About two tonnes of oil sands must be dug up, moved, and processed to produce one barrel of oil. Roughly 75 percent of the bitumen can be recovered from sand; processed sand has to be returned to the pit and the site reclaimed.

In situ recovery is used for bitumen deposits buried too deeply—more than 75 metres—for mining to be practical. Most *in situ* bitumen and heavy oil production comes from deposits buried more than 400 metres below the surface of the earth. Cyclic steam stimulation (CSS) and steam-assisted gravity drainage (SAGD) are *in situ* recovery methods, which include thermal injection through vertical or horizontal wells, solvent injection, and CO_2 methods. Canada's largest *in situ* bitumen recovery project is at Cold Lake, Alberta, where deposits are heated by steam injection to bring bitumen to the surface, then diluted with condensate for shipping by pipelines.

Source: Text courtesy of Duncan MacDonnell and Tim Markle, Communications, Alberta Department of Energy.

FIGURE 12.14 Canada has huge reserves of tar sands such as the one being mined here in Alberta.

FIGURE 12.15 Fossil fuels provided the energy source for the Industrial Revolution and built the world in which we now live. We have used up much of the easily accessible supply of these traditional fuels, however, often through wasteful practices such as the flaring shown here.

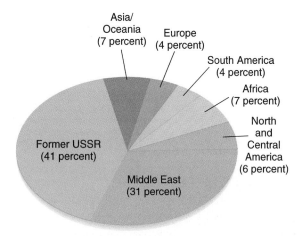

FIGURE 12.16 Percent of proven-in-place natural gas reserves, by region (1991).
Source: U.S. Congressional Research Service, 2001.

The proven world reserves of natural gas are 73,000 million tonnes. Because gas consumption rates are only about half of those for oil, current gas reserves represent roughly a 60-year supply at present usage rates.

Unconventional Gas Sources

Natural gas resources have been less extensively investigated than petroleum reserves. There may be extensive "unconventional" sources of gas in unexpected places. Prime examples are recently discovered methane hydrate deposits in arctic permafrost and beneath deep ocean sediments. **Methane hydrate** is composed of small bubbles or individual molecules of natural gas trapped in a crystalline matrix of frozen water. At least 50 oceanic deposits and a dozen land deposits are known. Altogether, they are thought to hold some 10,000 gigatonnes (10^{13} tonnes) of carbon or twice as much as the combined amount of all coal, oil, and conventional natural gas. This could be a valuable energy source but would be difficult to extract, store, and ship. If climate change causes melting of these deposits, it could trigger a catastrophic spiral of global warming because methane is 10 times as powerful a greenhouse gas as CO_2. Japan plans exploratory extraction of methane hydrate in the next few years, first on land near Prudhoe Bay, Alaska, and then in Japanese waters.

HYDROELECTRIC POWER

Falling water has been used as an energy source since ancient times. The invention of water turbines in the nineteenth century greatly increased the efficiency of hydropower dams. By 1925, falling water generated 40 percent of the world's electric power. Since then, hydroelectric production capacity has grown 15-fold, but fossil fuel use has risen so rapidly that water power is now only one-quarter of total electrical generation. Still, many countries produce most of their electricity from falling water (Fig. 12.17).

FIGURE 12.17 Hydropower dams produce clean renewable energy but can be socially and ecologically damaging.

FIGURE 12.18 A hydroelectric dam on the Magpie River in northern Ontario.

Norway, for instance, depends on hydropower for 99 percent of its electricity; Brazil, New Zealand, and Switzerland all produce at least three-quarters of their electricity with water power. Canada is the world's leading producer of hydroelectricity, running 400 power stations with a combined capacity exceeding 60,000 MW that provides most of its electricity (Fig. 12.18).

The total world potential for hydropower is estimated to be about 3 million MW. If all of this capacity were put to use, the available water supply could provide between 8 and 10 terrawatt hours (10^{12} Whr) of electrical energy. Currently, we use only about 10 percent of the potential hydropower supply. The energy derived from this source in 1994 was equivalent to about 500 million tonnes of oil, or 8 percent of the total world commercial energy consumption.

Much of the hydropower development in recent years has been in enormous dams. There is a certain efficiency of scale in giant dams, and they bring pride and prestige to the countries that build them, but they can have unwanted social and environmental effects. The largest hydroelectric dam in the world at present is the Itaipu Dam on the Parana River between Brazil and Paraguay. Designed to generate 12,600 MW of power, this dam should produce as much energy as 13 large nuclear power plants when completed. The lake that it is creating already has flooded about 1,300 km² of tropical rainforest and displaced many thousands of native people and millions of other creatures. An even larger dam that will produce twice as much electrical power is now under construction at the Three Gorges site on China's Yangtze River.

There are still other problems with big dams, besides human displacement, ecosystem destruction, and wildlife losses. Dam failure can cause catastrophic floods and thousands of deaths. Sedimentation often fills reservoirs rapidly and reduces the usefulness of the dam for either irrigation or hydropower. In China, the Sanmenxia Reservoir silted up in only four years, and the Laoying Reservoir filled with sediment before the dam was even finished. Rotting vegetation in artificial impoundments can have disastrous effects on water quality. When Lake Brokopondo in Suriname flooded a large region of uncut rainforest, underwater decomposition of the submerged vegetation produced hydrogen sulphide that killed fish and drove out villagers over a wide area. Acidified water from this reservoir ruined the turbine blades, making the dam useless for power generation.

Floating water hyacinths (rare on free-flowing rivers) have already spread over reservoir surfaces behind the Tucurui Dam on the Amazon River in Brazil, impeding navigation and fouling machinery. Herbicides sprayed to remove aquatic vegetation have contaminated water supplies. Herbicides used to remove forests before dam gates closed caused similar pollution problems. Schistosomiasis, caused by parasitic flatworms called blood flukes (Chapter 20), is transmitted to humans by snails that thrive in slow-moving, weedy tropical waters behind these dams. It is thought that 14 million Brazilians suffer from this debilitating disease.

As mentioned before, dams displace indigenous people. The Narmada Valley project in India will drown 150,000 ha of tropical forest and displace 1.5 million people, mostly tribal minorities and low-caste hill people. China's Three Gorges project will displace 1 million people. The Akosombo Dam built on the Volta River in Ghana nearly 20 years ago displaced 78,000 people from 700 towns. Few of these people ever found another place to settle, and those still living remain in refugee camps and temporary shelters.

In tropical climates, large reservoirs often suffer enormous water losses. Lake Nasser, formed by the Aswan High Dam in Egypt, loses 15 billion cubic metres each year to evaporation and seepage. Unlined canals lose another 1.5 billion cubic metres. Together, these losses represent one-half of the Nile River flow, or enough water to irrigate 2 million ha of land. The silt trapped by the Aswan Dam formerly fertilized farmland during seasonal flooding and provided nutrients that supported a rich fishery in the Delta region. Farmers now must buy expensive chemical fertilizers, and the fish catch has dropped almost to zero. As in South America, schistosomiasis is an increasingly serious problem.

If big dams—our traditional approach to hydropower—have so many problems, how can we continue to exploit the great potential of hydropower? Fortunately, there is an alternative to gigantic dams and destructive impoundment reservoirs. Small-scale, **low-head hydropower** technology can extract energy from small headwater dams that cause much less damage than larger projects. Some modern, high-efficiency turbines can even operate on **run-of-the-river flow.** Submerged directly in the stream and small enough not to impede navigation in most cases, these turbines don't require a dam or diversion structure and can generate useful power with a current of only a few kilometres per hour. They also cause minimal environmental damage and don't interfere with fish movements, including spawning migration. **Micro-hydro generators** operate on similar principles but are small enough to provide economical power for a single home. If you live close to a small stream or river that runs year-round and you have sufficient water pressure and flow, hydropower is probably a cheaper source of electricity for you than solar or wind power (Fig. 12.19).

NUCLEAR POWER

In 1953, U.S. President Dwight Eisenhower presented his "Atoms for Peace" speech to the United Nations. He announced that the United States would build nuclear-powered electrical generators to provide clean, abundant energy. He predicted that nuclear energy would fill the deficit caused by predicted shortages of oil and natural gas. It would provide power "too cheap to meter" for continued industrial expansion of both the developed and the developing world. It would be a supreme example of "beating swords into plowshares." Technology and engineering would tame the evil genie of atomic energy and use its enormous power to do useful work.

Glowing predictions about the future of nuclear energy continued into the early 1970s. Between 1970 and 1974, American utilities ordered 140 new reactors for power plants (Fig. 12.20). Some advocates predicted that by the end of the twentieth century there would be 1,500 reactors in the United States alone.

In spite of these factors, nuclear energy does continue to produce clean electricity and lighten the burden of greenhouse gases on the planet. There are 441 nuclear plants in 32 countries supplying about 21.2 percent of the world's electricity. According to the International Atomic Energy (IAEA), there were 32 new nuclear plants under construction at the end of 2002.

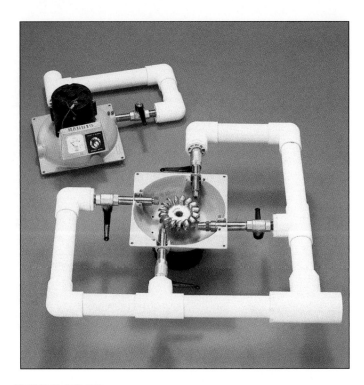

FIGURE 12.19 Solar collectors capture power only when the sun shines, but hydropower is available 24 hours a day. Small turbines such as this one can generate enough power for a single-family house with only 15 m of head and 200 l per minute flow. The turbine can have up to four nozzles to handle greater water flow and generate more power.

FIGURE 12.20 The nuclear power plant at Three Mile Island near Harrisburg, PA, was the site of a partial meltdown of the reactor core in 1979. Most of the radioactive material was kept inside the containment buildings (short, cylindrical buildings in center). The four large open cylinders are cooling towers.

How Do Nuclear Reactors Work?

The most commonly used fuel in nuclear power plants is U^{235}, a naturally occurring radioactive isotope of uranium. Ordinarily,

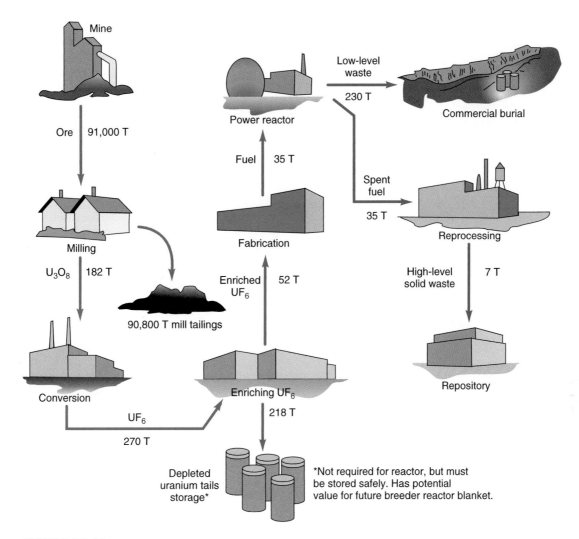

Mine

Ore 91,000 T

Milling

U_3O_8 182 T

90,800 T mill tailings

Conversion

UF_6 270 T

Enriching UF_6

218 T

Depleted
uranium tails
storage*

*Not required for reactor, but must
be stored safely. Has potential
value for future breeder reactor blanket.

Enriched UF_6 52 T

Fabrication

Fuel 35 T

Power reactor

Low-level
waste
230 T

Commercial burial

Spent
fuel
35 T

Reprocessing

High-level
solid waste 7 T

Repository

FIGURE 12.21 The nuclear fuel cycle. Quantities represent the average annual fuel requirements for a typical 1,000 MW light water reactor (T = tonnes). About 35 T or one-third of the reactor fuel is replaced every year.

U^{235} makes up only about 0.7 percent of uranium ore, too little to sustain a chain reaction in most reactors. It must be purified and concentrated by mechanical or chemical procedures (Fig. 12.21). Mining and processing uranium to create nuclear fuel is even more dirty and dangerous than coal mining. In some uranium mines 70 percent of the workers have died from lung cancer caused by high radon and dust levels. In addition, mountains of radioactive tailings and debris have been left around fuel preparation plants.

When the U^{235} concentration reaches about 3 percent, the uranium is formed into cylindrical pellets slightly thicker than a pencil and about 1.5 cm long. Although small, these pellets pack an amazing amount of energy. Each 8.5-gram pellet is equivalent to a tonne of coal or four barrels of crude oil.

The pellets are stacked in hollow metal rods approximately 4 m long. About 100 of these rods are bundled together to make a **fuel assembly.** Thousands of fuel assemblies containing 100 tonnes of uranium are bundled together in a heavy steel vessel called the reactor core. Radioactive uranium atoms are unstable—that is, when struck by a high-energy subatomic particle called a neutron, they undergo **nuclear fission** (splitting), releasing energy and more neutrons. When uranium is packed tightly in the reactor core, the neutrons released by one atom will trigger the fission of another uranium atom and the release of still more neutrons (Fig. 12.22). Thus a self-sustaining **chain reaction** is set in motion and vast amounts of energy are released.

The chain reaction is moderated (slowed) in a power plant by a neutron-absorbing cooling solution that circulates between the fuel rods. In addition, **control rods** of neutron-absorbing material, such as cadmium or boron, are inserted into spaces between fuel assemblies to shut down the fission reaction or are withdrawn to allow it to proceed. Water or some other coolant is circulated between the fuel rods to remove excess heat.

The greatest danger in one of these complex machines is a cooling system failure. If the pumps fail or pipes break during operation, the nuclear fuel quickly overheats and a "meltdown" can

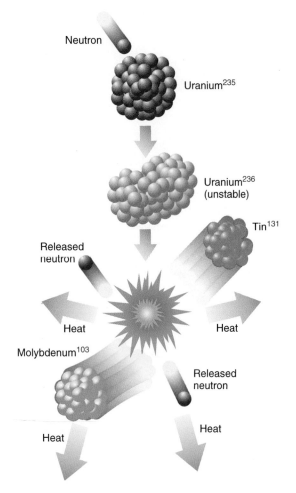

FIGURE 12.22 The process of nuclear fission is carried out in the core of a nuclear reactor. In the sequence shown here, the unstable isotope, uranium-235, absorbs a neutron and splits to form tin-131 and molybdenum-103. Two or three neutrons are released per fission event and continue the chain reaction. The total mass of the reaction product is slightly less than the starting material. The residual mass is converted to energy (mostly heat).

Source: Courtesy of Northern States Power Co., Minneapolis, MN.

result that releases deadly radioactive material. (What Do You Think? p. 264).

Canadian nuclear reactors use heavy water containing deuterium (H^2 or ^2H, the heavy, stable isotope of hydrogen) as both a cooling agent and a moderator. These *Can*adian *deu*terium (CANDU) reactors operate with natural, unconcentrated uranium (0.7 percent U^{235}) for fuel, eliminating expensive enrichment processes.

In Britain, France, and the former Soviet Union, a common reactor design uses graphite, both as a moderator and as the structural material for the reactor core. In the British MAGNOX design (named after the magnesium alloy used for its fuel rods), gaseous carbon dioxide is blown through the core to cool the fuel assemblies and carry heat to the steam generators. In the Soviet design, called RBMK (the Russian initials for a graphite-moderated, water-cooled reactor), low-pressure cooling water circulates through the core in thousands of small metal tubes.

These designs were originally thought to be very safe because graphite has high capacity for both capturing neutrons and dissipating heat. Designers claimed that these reactors could not possibly run out of control; unfortunately, they were proven wrong. The small cooling tubes are quickly blocked by steam if the cooling system fails and the graphite core burns when exposed to air. The two most disastrous reactor accidents in the world, so far, involved fires in graphite cores that allowed the nuclear fuel to melt and escape into the environment. In 1956, a fire at the Windscale Plutonium Reactor in England released roughly 100 million curies of radionuclides and contaminated hundreds of square kilometres of countryside. Similarly, burning graphite in the Chernobyl nuclear plant in Ukraine made the fire much more difficult to control than it might have been in another reactor design. The nuclear accident in 1979 at Three Mile Island near Harrisburg, Pennsylvania, released millions of times less radiation as did Chernobyl or Windscale (see Fig. 12.20).

Breeder Reactors

For more than 30 years, nuclear engineers have been proposing high-density, high-pressure, **breeder reactors** that produce fuel rather than consume it. These reactors create fissionable plutonium and thorium isotopes from the abundant, but stable, forms of uranium (Fig. 12.23). The starting material for this reaction is

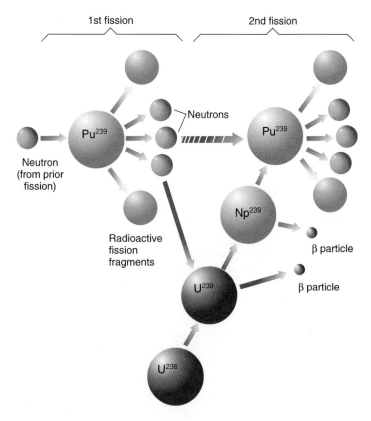

FIGURE 12.23 Reactions in a "breeder" fission process. Neutrons from a plutonium fission change U^{238} to U^{239} and then to Pu^{239} so that the reactor creates more fuel than it uses.

plutonium reclaimed from spent fuel from conventional fission reactors. After about 10 years of operation, a breeder reactor would produce enough plutonium to start another reactor. Sufficient uranium currently is stockpiled in the United States to produce electricity for 100 years at present rates of consumption, if breeder reactors can be made to work safely and dependably.

Several problems have held back the breeder reactor program. One problem is the concern about safety. The reactor core of the breeder must be at a very high density for the breeding reaction to occur. Water does not have enough heat capacity to carry away the high heat flux in the core, so liquid sodium is used as a coolant. Liquid sodium is very corrosive and difficult to handle. It burns with an intense flame if exposed to oxygen, and it explodes if it comes into contact with water. Because of its intense heat, a breeder reactor will melt down and self-destruct within a few seconds if the primary coolant is lost, as opposed to a few minutes for a normal fission reactor.

Another very serious concern about breeder reactors is that they produce excess plutonium that can be used for bombs. It is essential to have a spent-fuel reprocessing industry if breeders are used, but the existence of large amounts of weapons-grade plutonium in the world would surely be a dangerous and destabilizing development. The chances of some of that material falling into the hands of terrorists or other troublemakers are very high.

A proposed $1.7 billion breeder-demonstration project in Clinch River, Tennessee, has been on and off for 15 years. At last estimate, it would cost up to five times the original price if it is ever completed. In 1986, France put into operation a full-sized commercial breeder reactor, the SuperPhenix, near Lyons. It cost three times the original estimate to build and produces electricity at twice the cost per kilowatt of conventional nuclear power. In 1987, a large crack was discovered in the inner containment vessel of the SuperPhenix, and in 1997 it was shut down permanently.

RADIOACTIVE WASTE MANAGEMENT

One of the most difficult problems associated with nuclear power is the disposal of wastes produced during mining, fuel production, and reactor operation. How these wastes are managed may ultimately be the overriding obstacle to nuclear power.

Ocean Dumping of Radioactive Wastes

Until 1970, the United States, Britain, France, and Japan disposed of radioactive wastes in the ocean. Dwarfing all these dumps, however, are those of the former Soviet Union, which has seriously—and some fear permanently—contaminated the Arctic Ocean. Rumors of Soviet nuclear waste dumping had circulated for years, but it was not until after the collapse of the Soviet Union that the world learned the true extent of what happened. Starting in 1965, the Soviets disposed of 18 nuclear reactors—seven loaded with nuclear fuel—in the Kara Sea off the eastern coast of Novaya

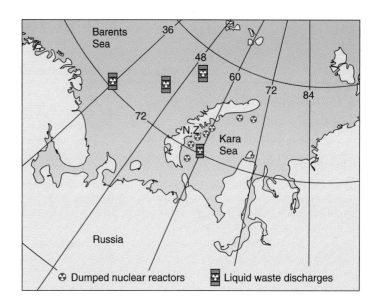

FIGURE 12.24 Sites of dumped nuclear reactors east of Novaya Zemlya. Dark blue areas indicate liquid waste discharges.

Zemlya island, and millions of litres of liquid waste in the nearby Barents Sea (Fig. 12.24). Two other reactors were sunk in the Sea of Japan.

Altogether, the former Soviet Union dumped 2.5 million curies of radioactive waste into the oceans, more than twice as much as the combined amounts that 12 other nuclear nations have reported dumping over the past 45 years. In 1993, despite protests from Japan, Russia dumped 900 tonnes of additional radioactive waste into the Sea of Japan.

Land Disposal of Nuclear Waste

In 2001, Russia began plans to store nuclear waste from other countries. Plans are to transport wastes to the Mayak in the Ural mountains. The storage site is near Chelyabinsk where an explosion at a waste facility in 1957 contaminated about 24,000 km^2. The region is now considered the most radioactive on earth, so the Russians feel it can't get much worse. They expect that storing 20,000 tonnes of nuclear waste should pay about $20 billion.

Enormous piles of mine wastes and abandoned mill tailings in all uranium-producing countries represent serious waste disposal problems. Production of 1,000 tonnes of uranium fuel typically generates 100,000 tonnes of tailings and 3.5 million litres of liquid waste. There now are approximately 200 million tonnes of radioactive waste in piles around mines and processing plants in the United States. This material is carried by the wind and washes into streams, contaminating areas far from its original source. Canada has even more radioactive mine waste on the surface than does the United States.

In addition to the leftovers from fuel production, there are about 100,000 tonnes of low-level waste (contaminated tools, clothing, building materials, etc.) and about 15,000 tonnes of high-level (very radioactive) wastes in the United States. The high-level

Chernobyl: Could It Happen Here?

In the early morning hours of April 26, 1986, residents of the Ukrainian village of Pripyat saw a spectacular and terrifying sight. A glowing fountain of molten nuclear fuel and burning graphite was spewing into the dark sky through a gaping hole in the roof of the Chernobyl Nuclear Power Plant only a few kilometres away. Although officials assured them that there was nothing to worry about in this "rapid fuel relocation," the villagers knew that something was terribly wrong. They were witnessing the worst possible nuclear power accident, a "meltdown" of the nuclear fuel and rupture of the containment facilities, releasing enormous amounts of radioactivity into the environment.

The accident was a result of a risky experiment undertaken late at night by the plant engineers in violation of a number of safety rules and operational procedures. They were testing whether the residual energy of a spinning turbine could provide enough power to run the plant in an emergency shutdown if off-site power were lost. Reactor number four had been slowed down to only 6 percent of its normal operating level. To conserve the small amount of electricity being generated, they then disconnected the emergency core-cooling pumps and other safety devices, unaware that the reactor was dangerously unstable under these conditions.

The heat level in the core began to rise, slowly at first, and then faster and faster. The operators tried to push the control rods into the core to slow the reaction, but the graphite pile had been deformed by the heat so that the rods wouldn't go in. In 4.5 seconds, the power level rose 2,000-fold, far above the rated capacity of the cooling system. Chemical explosions (probably hydrogen gas released from the expanding core) ripped open the fuel rods and cooling tubes. Cooling water flashed into steam and blew off the 1,000-tonne-concrete cap on top of the reactor. Molten uranium fuel puddled in the bottom of the reactor, creating a critical mass that accelerated the nuclear fission reactions. The metal superstructure of the containment building was ripped apart and a column of burning graphite, molten uranium, and radioactive ashes billowed 1,000 m into the air.

Panic and confusion ensued. Officials first denied that anything was wrong. The village of Pripyat was not evacuated for 36 hours. There was no public announcement for three days. The first international warning came, not from Soviet authorities, but from Swedish scientists 2,000 km away who detected unusually high levels of radioactive fallout and traced airflows back to the southern Soviet Union.

There were many acts of heroism during this emergency. Firefighters climbed to the roof of the burning reactor building to pour water into the blazing inferno. Engineers dived into the suppression pool beneath the burning core to open a drain to prevent another steam explosion. Helicopter pilots hovered over the gaping maw of the ruined building to drop lead shot, sand, clay, limestone, and boron carbide onto the burning nuclear core to smother the fire and suppress the nuclear fission reactions. More than 600,000 workers participated in putting out the fire and cleaning up contamination. Thousands who were exposed to high radiation doses already have died and many more have dim prospects for the future.

The amount of radioactive fallout varied from area to area, depending on wind patterns and rainfall. Some places had heavy doses while neighbouring regions had very little. One band of fallout spread across Yugoslavia, France, and Italy. Another crossed Germany and Scandinavia. Small amounts of radiation even reached North America. Altogether, about 7 tonnes of fuel containing 50 to 100 million curies were released, roughly 5 percent of the reactor fuel.

For several years after the accident, the Soviet government tried to suppress information and deny the consequences. The dissolution of the USSR has allowed many of these

wastes consist mainly of spent fuel rods from commercial nuclear power plants and assorted wastes from nuclear weapons production. For the past 20 years, spent fuel assemblies from commercial reactors have been stored in deep water-filled pools at the power plants. These pools were originally intended only as temporary storage until the wastes were shipped to reprocessing centres or permanent disposal sites.

David Shoesmith, a scientist in the Department of Chemistry at The University of Western Ontario, is tackling the thorny problem of nuclear waste disposal. The Canadian Nuclear Fuel Waste Program has developed a geological disposal scenario for fuel wastes resulting from the operation of nuclear generating stations. The fuel waste consists of CANDU (CANada Deuterium Uranium) fuel bundles sealed in a corrosion resistant metallic container. These containers would be placed in a disposal vault 500 to 1,000 metres deep in granite rock in the Canadian Shield (Fig. 12.25). Though a basic understanding of fuel behaviour under proposed nuclear fuel disposal conditions exists, there remains much uncertainty in the actual long-term fuel behaviour. Therefore, Shoesmith

and his lab are directing their efforts towards measuring the corrosion rates of UO_2 and used nuclear fuels in simple solutions under well-defined conditions. These experiments are directed towards developing a fuel corrosion mechanism and measuring accurate corrosion rates that can be used to develop models for the long-term performance assessment of fuel behaviour.

CHANGING FORTUNES OF NUCLEAR POWER

Although promoted originally as a new wonder of technology that could open the door to wealth and abundance, nuclear power has long been highly controversial (Fig. 12.26). For many environmental organizations, opposition to nuclear power is a perennial priority. Antinuclear groups such as the Clamshell Alliance, Northern Sun, and Greenpeace have organized mass protest rallies featuring popular actors, musicians, and celebrities who help raise funds and attract attention to their cause. Protests and civil disobedience at

Chernobyl nuclear reactor #4 after the fire and explosion (circled area).

secrets to come to light. It's difficult to separate the effects of Chernobyl from other causes of ill-health, but at least a half million people live in areas where radiation levels were high enough to be of concern. One clear effect is seen in children in Belarus where thyroid cancers have increased a hundred-fold since 1986. Childhood leukemias and some autoimmune diseases also appear to be more prevalent in highly contaminated areas.

In 2001, the last of the four Chernobyl reactors was shut down. For the present, the damaged reactor has been entombed in a giant, steel-reinforced concrete "sarcophagus." Unfortunately, this containment structure was hastily built of inferior materials, and already has begun to deteriorate. Reconstruction started in 1999, but the Ukraine is demanding billions of dollars from other countries to finance this operation.

So far, more than 250,000 people have been relocated from the contaminated area. More than 70 villages have been destroyed and millions of hectares of the richest farmland in the Commonwealth of Independent States has been abandoned. The immediate direct costs were roughly $3 billion; total costs might be 100 times that much.

Proponents of nuclear power point out that the Soviet-style reactors were much more dangerous than most of those now in Canada or the United States. Furthermore, they argue, the accident was caused by operator error and bureaucratic bungling. Something similar could never happen here, they claim. Opponents respond that we should learn from this tragedy and abandon this dangerous technology. What do you think? How likely is a similar accident in your neighbourhood? Even if the chances are very low, does the potential harm make this an unacceptable risk? How would you weigh a low probability against a terrible potential outcome?

some sites have gone on for decades, and plants that were planned—or in some cases, already built—have been abandoned or modified to burn fossil fuels because of public opposition.

On the other side, workers, consumers, utility officials, and others who stand to benefit from this new technology rally in support of nuclear power. They argue that abandoning this energy source is foolish since a great deal of money already has been invested and the risks may be lower than is commonly believed. They also point out that every MWh of energy generated at a nuclear plant eliminates green house gas emission from a fossil fuel generation.

Public opinion about nuclear power has fluctuated over the years. Before the Three-Mile Island accident in 1978, two-thirds of Americans supported nuclear power. By the time Chernobyl exploded in 1985, however, less than one-third of Americans favoured this power source. More recently, however, memories of these earlier incidents have faded. Now about half of all Americans support atomic energy, and about one-quarter say they wouldn't mind having a nuclear plant within 16 km from their home.

NUCLEAR FUSION

Fusion energy is an alternative to nuclear fission that could have virtually limitless potential. **Nuclear fusion** energy is released when two smaller atomic nuclei fuse into one larger nucleus. Nuclear fusion reactions, the energy source for the sun and for hydrogen bombs, have not yet been harnessed by humans to produce useful net energy. The fuels for these reactions are deuterium and tritium, two heavy isotopes of hydrogen.

It has been known for 40 years that if temperatures in an appropriate fuel mixture are raised to 100 million degrees Celsius and pressures of several billion atmospheres are obtained, fusion of deuterium and tritium will occur. Under these conditions, the electrons are stripped away from atoms and the forces that normally keep nuclei apart are overcome. As nuclei fuse, some of their mass is converted into energy, some of which is in the form of heat. There are two main schemes for creating these conditions: magnetic confinement and inertial confinement.

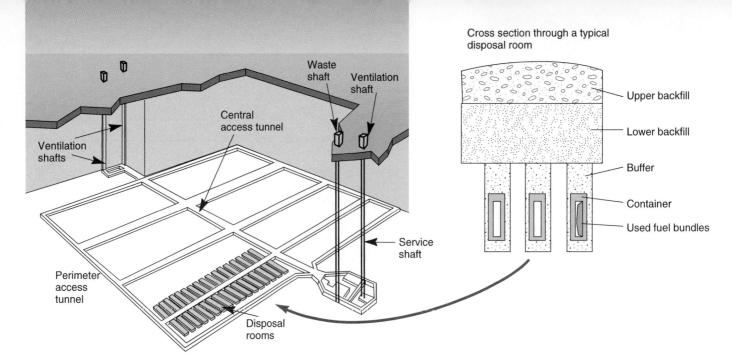

FIGURE 12.25 One proposal for the permanent disposal of used fuel is to bury the used fuel containers 500 to 1,000 m deep in the stable rock formations. Spent fuel, in the form of ceramic fuel pellets, are sealed inside the corrosion-resistant containers. Glass beads are compacted around the used fuel bundles and into the spaces between the fuel and the container shell. The containers are then buried in specially built vaults that are packed and sealed with special materials to further retard the migration of radioactive material.

Redrawn with permission of Atomic Energy of Canada Limited.

FIGURE 12.26 German protesters block a shipment of nuclear waste being sent to a permanent storage site.

Magnetic confinement involves the containment and condensation of plasma, a hot, electrically neutral gas of ions and free electrons in a powerful magnetic field inside a vacuum chamber. Compression of the plasma by the magnetic field should raise temperatures and pressures enough for fusion to occur. The most promising example of this approach, so far, has been a Russian design called *tokomak* (after the Russian initials for "torodial magnetic chamber"), in which the vacuum chamber is shaped like a large doughnut (Fig. 12.27).

Inertial confinement involves a small pellet (or a series of small pellets) bombarded from all sides at once with extremely high-intensity laser light. The sudden absorption of energy causes an implosion (an inward collapse of the material) that will increase densities by 1,000 to 2,000 times and raise temperatures above the critical minimum. So far, no lasers powerful enough to create fusion conditions have been built.

In both of these cases, high-energy neutrons escape from the reaction and are absorbed by molten lithium circulating in the walls of the reactor vessel. The lithium absorbs the neutrons and transfers heat to water via a heat exchanger, making steam that drives a turbine generator, as in any steam power plant. The advantages of fusion reactions, if they are ever feasible, include production of fewer radioactive wastes, the elimination of fissionable products that could be made into bombs, and a fuel supply that is much larger and less hazardous than uranium.

Despite 50 years of research and a $25 billion investment, fusion reactors never have reached the breakeven point at which they produce more energy than they consume. A major setback occurred in 1997, when Princeton University's Tokomak Fusion Test Reactor

PART TWO   Our Physical Environment

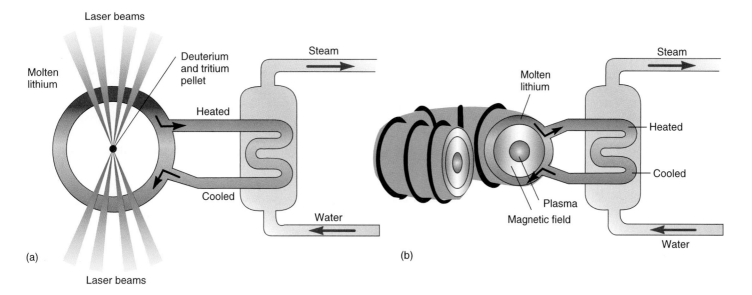

FIGURE 12.27 Nuclear fusion devices. (*a*) Inertial confinement is created by laser beams that bombard and ignite fuel pellets. Molten lithium absorbs energy from the fusion and transfers it as heat to a steam generator. (*b*) A powerful magnetic field confines the plasma and compresses it so that critical temperatures and pressures are reached. Again, molten lithium absorbs energy and transfers it to a steam generator.

was shut down. Three years earlier, this reactor had set a world's record by generating 10.7 million watts for one second, but researchers conceded that the project was still decades away from self-sustaining power generation. Proponents of fusion power urge that research be continued, maintaining that success could be just around the corner. Opponents view this technology as just another expensive wild-goose chase and predict that it will never generate enough energy to pay back the fortune spent on its development.

SUMMARY

Energy is the capacity to do work. Power is the rate of doing work. Worldwide, nearly 85 percent of all commercial energy is generated by fossil fuels, about 36 percent coming from petroleum. Next are coal, with 28 percent, and natural gas (methane), with 23 percent. Petroleum and natural gas were not used in large quantities until the beginning of this century, but supplies are already running low. Coal supplies will last several more centuries at present rates of usage, but it appears that the fossil fuel age will have been a rather short episode in the total history of humans. Nuclear power provides almost 8 percent of all commercial energy but about 20 percent of the electricity in Canada and the United States.

Energy is essential for most activities of modern society. Its use generally correlates with standard of living, but there are striking differences between energy use per capita in countries with relatively equal standards of living. Canada, for instance, consumes nearly twice as much energy per person as does Switzerland, which is higher in many categories that measure quality of life. This difference is based partly on level of industrialization and partly on policies, attitudes, and traditions in Canada that encourage extravagant or wasteful energy use.

The environmental damage caused by mining, shipping, processing, and using fossil fuels may necessitate cutting back on our use of these energy sources. Coal is an especially dirty and dangerous fuel, at least as we currently obtain and use it. Some new coal treatment methods remove contaminants, reduce emissions, and make its use more efficient (so less will be used). Coal combustion is a major source of acid precipitation that is suspected of being a significant cause of environmental damage in many areas. We now recognize that CO_2 buildup in the atmosphere has the potential to trap heat and raise the earth's temperature to catastrophic levels.

Nuclear energy offers an alternative to many of the environmental and social costs of fossil fuels, but it introduces serious problems of its own. In the 1950s, there was great hope that these problems would be overcome and that nuclear power plants would provide energy "too cheap to meter."

The greatest worry about nuclear power is the danger of accidents that release hazardous radioactive materials into the environment. Several accidents, most notably the "meltdown" at the Chernobyl plant in the Soviet Ukraine in 1986, have convinced many people that this technology is too risky to pursue.

Other major worries about nuclear power include where to put the waste products of the nuclear fuel cycle and how to ensure that it will remain safely contained for the thousands of years required for "decay" of the radioisotopes to nonhazardous levels.

Nuclear fusion occurs when atoms are forced together in order to create new, more massive atoms. If fusion actions could be created under controlled conditions in a power plant, they might provide an essentially unlimited source of energy that would avoid many of the worst problems of both fossil fuels and fission-based nuclear power. So far, however, no one has been able to sustain controlled fusion reactions that produce more energy than they consume. It remains to be seen whether this will ever be possible, and whether other unforeseen problems will arise if it does become a reality.

QUESTIONS FOR REVIEW

1. What is energy? What is power?

2. What are the major sources of commercial energy worldwide? Why are data usually presented in terms of commercial energy?

3. Where is ANWR, and why is oil drilling there controversial?

4. How does energy use in Canada compare with that in other countries?

5. How much coal, oil, and natural gas are in proven reserves worldwide? Where are those reserves located?

6. What are the most important health and environmental consequences of our use of fossil fuels?

7. Describe how a nuclear reactor works and why reactors can be dangerous.

8. Describe methods proposed for storing and disposing of nuclear wastes.

QUESTIONS FOR CRITICAL THINKING

1. We have discussed a number of different energy sources and energy technologies in this chapter. Each has advantages and disadvantages. If you were an energy policy analyst, how would you compare such different problems as the risk of a nuclear accident versus air pollution effects from burning coal?

2. If your local utility company were going to build a new power plant in your community, what kind would you prefer?

3. The nuclear industry is placing ads in popular magazines and newspapers claiming that nuclear power is environmentally friendly since it doesn't contribute to the greenhouse effect. How do you respond to that claim?

4. There are vast reserves of fossil fuels in the oil sands of Alberta, but in addition to the economic potential, development and use of this resource has environmental ramifications. Discuss the balance between socio-economic benefits and environmental problems that development of oil sands present.

5. Although we have wasted vast amounts of energy resources in the process of industrialization and development, some would say that it was a necessary investment to get to a point at which we can use energy more efficiently and sustainably. Do you agree? Might we have followed a different path?

6. Canada and the United States have about the same area of forest (2.5 million km^2) but the United States has far higher fuelwood production (59 million tonnes) than Canada (2.2 million tonnes). What are the explanations for this difference?

KEY TERMS

black lung disease   252
breeder reactor   262
chain reaction   261
control rods   261
fossil fuels   248
fuel assembly   261
inertial confinement   266
low-head hydropower   260
magnetic confinement   266
methane hydrate   258
micro-hydro generators   260
nuclear fission   261
nuclear fusion   265
oil sands   257
oil shale   255
run-of-the-river flow   260
secondary recovery
   techniques   254
tar sands   257
woodfuel   251

Web Exercises

ENERGY RESOURCES IN CANADA

A Canadian atlas of resources is available online at http://ccrs-gad1.cgdi.gc.ca/resources/EngNRAtlas.html. Look at the oil and gas reserves in Canada. Read the directions for drawing a map in the lower left frame; then, in the upper left "Topics" frame, select "Resource Development" from the drop-down menu. Click on the boxes for all energy-related layers. Note: to see where these layers are, be sure to click on the Canadian Boundaries at the bottom of the list, too. Then click on the DRAW MAP button below the list of map layers. Note that the map legend will appear below the map. Where are most of the energy sources in Canada? Are they clustered or widely distributed? Did you turn on the hydro sources as well? If not, do so now. Are they distributed in the same areas as oil and gas, or are they in different areas? Add population density to the map. Are people located where the energy is? Does this matter?

Can we continue to depend on fossil fuels or is there a more sustainable basis for our commerce and industry?

Chapter 13

New Energy Technologies

The significant problems we face cannot be solved at the same level of thinking we were at when we created them.

—Albert Einstein—

OBJECTIVES

After studying this chapter, you should be able to:

- appreciate the opportunities for energy conservation available to us.
- understand how active and passive systems capture solar energy and how photovoltaic collectors generate electricity.
- comprehend why diminishing fuelwood supplies are a crisis in less-developed countries.
- evaluate the use of dung, crop residues, energy crops, and peat as potential energy sources.
- explain how hydropower, wind, and geothermal energy contribute to our power supply.
- describe how tidal and wave energy and ocean thermal gradients can be used to generate electrical energy.

- compare and contrast different options for storing electrical energy from intermittent or remote sources.

WebQuest

energy conservation, solar energy, biomass energy, hydropower, wind energy, geothermal energy

Above: Rows of 12-story-tall, 750 kV windmills stand on the prairies.

Wind Ranching in Alberta

A steady breeze flows most of the year across the Great Plains, rippling prairie grasses in the summer and sculpting snowdrifts through the winter. When it reaches southern Alberta, it encounters row after row of tall, white columns, each holding three long, black propellers that swoosh softly as they spin. The wind turbines now standing in ordered ranks on the rolling farm and pasture land are the first stage in an exciting step toward energy sustainability. These giant windmills generate a total of 30 MW of electrical power, or enough for about 13,000 average homes.

Perhaps the most important aspect of the Suncor development is the promise it holds for the rest of the Great Plains. Millions of square kilometres, from Alberta and Saskatchewan to Texas, and from the Rocky Mountains to the tree line east of the 100th meridian, have the potential to supply considerably more energy than the total current world commercial production. As David Morris, from the Center for Local Self-Reliance says, "The Midwest could be the Saudi Arabia of safe, clean, sustainable wind power, and it would free us from dependence on foreign powers for our energy supply."

Chapter 12 detailed some of the disadvantages and limitations of fossil fuels and nuclear power; in this chapter, we will look at some options for energy conservation and renewable energy sources that might allow a real increase in comfort and standard of

Suncor has already built the Sunbridge wind farm near Gull Lake, Saskatchewan. There are 17 wind turbines here, and the 11 MW of power they generate save the equivalent of taking 5,000 vehicles off the road in greenhouse gas emissions.

living for everyone, and yet also help preserve our global environment. Even with modest investment in projects like Sunbridge and Magrath wind farms, sustainable energy could meet more than half our energy needs 50 years from now.

CONSERVATION

One of the best ways to avoid energy shortages and to relieve environmental and health effects of our current energy technologies is simply to use less. Conservation offers many benefits both to society and to the environment.

Utilization Efficiencies

Much of the energy we consume is wasted. This statement is not a simple admonishment to turn off lights and turn down furnace thermostats in winter; it is a technological challenge. Our ways of using energy are so inefficient that most potential energy in fuel is lost as waste heat, becoming a form of environmental pollution. Of the energy we do extract from primary resources, however, much is used for frankly trivial or extravagant purposes. As Chapter 12 showed, several European countries have higher standards of living than Canada, and yet use 30 to 50 percent less energy.

Many conservation techniques are relatively simple and highly cost effective. More efficient and less energy-intensive industry, transportation, and domestic practices could save large amounts of energy. Improved automobile efficiency, better mass transit, and increased railroad use for passenger and freight traffic

are simple and readily available means of conserving transportation energy. In response to the 1970s oil price shocks, automobile gas-mileage averages in the United States more than doubled from 13 mpg in 1975 to 28.8 mpg in 1988. Unfortunately, the oil glut and falling fuel prices of the late 1980s discouraged further conservation. By 2001, light trucks, vans, and sports-utility vehicles (SUVs) made up about half of all new vehicles sold in the United States, and the average mileage for all passenger vehicles had fallen to only 24 mpg.

Much more could be done. High-efficiency, low-emission automobiles are already available (What Do You Think? p. 274), and proposed rules that would make light trucks and SUVs meet the same kilometrage and emission standards as conventional cars will both clear our air and save energy. Amory B. Lovins of the Rocky Mountain Institute in Colorado estimated that raising the average fuel efficiency of the U.S. car and light truck fleet back up to the 1975 level would save more oil than the maximum expected production from Alaska's Arctic National Wildlife Refuge (see Chapter 12).

Many improvements in domestic energy efficiency have occurred in the past decade. Today's average new home uses one-half the fuel required in a house built in 1974, but much more can be done. Household energy losses can be reduced by one-half to three-fourths through better insulation, double or triple glazing of windows, thermally efficient curtains or window coverings, and by sealing cracks and loose joints. Reducing air infiltration is usually the cheapest, quickest, and most effective way of saving energy because it is the largest source of losses in a typical house. It doesn't take much skill or investment to caulk around doors, windows, foundation joints, electrical outlets, and other sources of air leakage.

According to national standards passed in 2001, all new washing machines will have to use 35 percent less water in 2007 than current models. This will add an estimated $240 to the cost of a washer, but will pay back in seven years. It also will cut water use in the United States by 40 trillion litres per year and save four times as much electricity every year as is used to light all the homes in the United States. Air conditioners will also be required to have a minimum seasonal efficiency ratio (SEER) of 12 rather than the current mimimum of 10.

For even greater savings, new houses can be built with extra thick superinsulated walls, air-to-air heat exchangers to warm incoming air, and even double-walled sections that create a "house within a house." The R-2000 program in Canada details how energy conservation can be built into homes. Special double-glazed windows that have internal reflective coatings and that are filled with an inert gas (argon or xenon) have an insulation factor of R11, the same as a standard 10-cm thick insulated wall or 10 times as efficient as a single-pane window. Superinsulated houses now being built in Sweden require 90 percent less energy for heating and cooling than the average North American home.

Orienting homes so that living spaces have passive solar gain in the winter and are shaded by trees or roof overhang in the summer also helps conserve energy. Earth-sheltered homes built into the south-facing side of a slope or protected on three sides by an earth berm are exceptionally efficient energy savers because they maintain relatively constant subsurface temperatures (Fig. 13.1). In addition to building more energy-efficient homes, there are many personal actions we can take to conserve energy (In Depth, p. 276).

Industrial energy savings are another important part of our national energy budget. More efficient electric motors and pumps, new sensors and control devices, advanced heat-recovery systems, and material recycling have reduced industrial energy requirements significantly.

Energy Conversion Efficiencies

Energy efficiency is a measure of energy produced compared to energy consumed. Table 13.1 shows the typical energy efficiencies of a variety of energy-conversion devices. Thermal-conversion machines, such as steam turbines in coal-fired or nuclear power plants, can turn no more than 40 percent of the energy in their

TABLE 13.1 Typical Net Efficiencies of Energy-Conversion Devices	
	YIELD (PERCENT)
ELECTRIC POWER PLANTS	
Hydroelectric (best case)	90
Combined-cycle steam	90
Fuel cell (hydrogen)	80
Coal-fired generator	38
Oil-burning generator	38
Nuclear generator	30
Photovoltaic generation	10
TRANSPORTATION	
Pipeline (gas)	90
Pipeline (liquid)	70
Waterway (no current)	65
Diesel-electric train	40
Diesel-engine automobile	35
Gas-engine automobile	30
Jet-engine airplane	10
SPACE HEATING	
Electric resistance	99*
High-efficiency gas furnace	90
Typical gas furnace	70
Efficient wood stove	65
Typical wood stove	40
Open fireplace	−10
LIGHTING	
Sodium vapour light	60*
Fluorescent bulb	25*
Incandescent bulb	5*
Gas flame	1

*Note that 60–70 percent of the energy in the original fuel is lost in electric power generation.
Source: U.S. Department of Energy.

FIGURE 13.1 This earth-sheltered house made of rammed dirt-filled tires and empty aluminum cans saves energy and recycles waste. South-facing windows passively collect solar energy to provide most space heating.

Hybrid Automobile Engines

It's not surprising that California was the first state to get tough on automobile emissions. Auto exhaust counts for 90 percent of the state's carbon monoxide, 77 percent of its nitrogen oxides, and 55 percent of its smog-producing hydrocarbons. In 1990, the California Air Resources Board shocked the automobile industry by ordering it to start producing emission-free vehicles or face stiff penalties. By 1998, 2 percent of all auto sales in California were supposed to be zero-emission vehicles, and 10 percent would have to be in this category by 2003. At the time this order was issued, only battery-powered electric vehicles were available. Although several manufacturers built all-electric autos—the General Motors EV1 was an example—the batteries were heavy, expensive, and required more frequent recharging than customers would accept. Even though 90 percent of all daily commutes are less than 80 km, most people want the capability to take a long road trip of several hundred kilometres without needing to stop for fuel or recharging. Although some environmentally sensitive customers bought electric vehicles, for most people they were merely an expensive novelty.

An alternative now being promoted by Toyota and Honda that appears to have much more customer appeal is the **hybrid gas-electric vehicles.** There are two basic types of hybrid electric car: series and parallel motors. In a series hybrid, a small gas or diesel engine spins a generator that recharges a small battery pack and provides electric power to motors that drive the wheels. This is essentially the design of diesel-electric train engines. The internal combustion engine is a mobile charging station. Parallel hybrids have two discrete power systems, gasoline and electric. Either one can propel the vehicle, but they work together when more power is needed for acceleration or hill climbing. The electric system operates most of the time, and the gasoline motor only runs when the batteries need recharging. Alternatively, in some designs, the gasoline engine—which can be substantially smaller than a normal car—runs most of the time and the electrical system only kicks in to give an extra boost in acceleration. In either system, kinetic energy can be recaptured during "regenerative" braking and converted back to electricity to be stored in the battery.

The first of these hybrid cars to be marketed in the United States is the two-seat Honda Insight. Its 3-cylinder, 1.0 litre gas engine is the main power source. A 7 hp electric motor helps during acceleration and hill climbing. A "continuously variable" automatic transmission cuts down on energy lost during shifting, and regenerative braking increases efficiency. With a streamlined lightweight plastic and aluminum body, the Insight is reported to use 3.1 l/100 km in highway driving and has low-enough emissions to satisfy California requirements. Cost is about $20,000. Top speed is 145 km/h.

The other entrant in the hybrid vehicle race is the Toyota Prius. Larger than the Honda, the Prius seats five adult passengers and is about the same size as the popular Toyota Camry. During most city driving, it depends only on its quiet, emission-free, battery-powered, electric motor. The 1.5 litre gas engine kicks in to help accelerate or when the batteries need recharging. Using about 4.5 l/100 km in city driving, the Prius is one of the most efficient cars on the road and can travel more than 1,000 km without refueling. Some drivers are unnerved by the noiseless electric motor. Sitting at a stoplight, it makes no sound at all. You might think it was dead, but when the light changes, you glide off silently and smoothly. Introduced in Japan in 1997, the Prius also sells in the United States for about $20,000. The Sierra Club estimates that in 160,000 km a Prius will generate 27 tonnes of CO_2, a Ford Taurus will generate 64 tonnes, while the Ford Excursion SUV will produce 134 tonnes.

In 1999, The Sierra Club awarded both the Insight and the Prius "excellence in engineering" awards, the first time this organization has ever endorsed commercial products. Ford and General Motors have announced plans to introduce "mild" hybrids that have 42 V electric systems rather than the 274 V in the Prius. These mild hybrids won't generate enough power to move the vehicle. Instead, the electricity will be used mainly to run videocassette players, dashboard computers, and other accessories. Small electric motors in mild hybrids will boost acceleration but will improve mileage only slightly.

What do you think? Would you buy a vehicle with a hybrid engine system? Would you sacrifice a little power, speed, and acceleration for a quiet, clean, efficient, environmentally friendly means of transportation?

The hybrid gas-electric Toyota Prius seats five adults, uses 4.5 l/100 km in city driving, and produces 90 percent less smog-forming exhaust gas than current North American standards.

primary fuel into electricity or mechanical power because of the need to reject waste heat. Does this mean that we can never increase the efficiency of fossil fuel use? No. Some waste heat can be recaptured and used for space heating, raising the net yield to 80 or 90 percent. In another kind of process, fuel cells convert the chemical energy of a fuel directly into electricity without an intermediate combustion cycle. Since this process is not limited by waste heat elimination, its efficiencies can approach 80 percent with such fuel as hydrogen gas or methane.

Another way to look at energy efficiency is to consider the total **net energy yield** from energy-conversion devices (Table 13.2). Net energy yield is based on the total useful energy produced during the lifetime of an entire energy system minus the energy required to make useful energy available. To make comparisons between different energy systems easier, the net energy yield is often expressed as a *ratio* between the output of useful energy and the energy costs for construction, fuel extraction, energy conversion, transmission, waste disposal, etc.

Nuclear power is a good case for net energy yield studies. We get a large amount of electricity from a small amount of fuel in a nuclear plant, but it also takes a great deal of energy to extract and process nuclear fuel and to build, operate, and eventually dismantle power plants. As Chapter 12 pointed out, we may never reach a break-even point where we get back more energy from nuclear plants than we put into them, especially considering the energy that may be required to decommission nuclear plants and guard their waste products in secure storage for thousands of years.

TABLE 13.2	Typical Net Useful Energy Yields

ENERGY SOURCE	YIELD/COST RATIO
NONRENEWABLE RESOURCES	
Coal (space or process heat)	20/1
Natural gas (as heat source)	10/1
Gasoline and fuel oil	7/1
Coal gasification (combined cycle)	5/1
Oil shale (as liquid fuel)	1/1
Nuclear (excluding waste disposal)	2/1*
RENEWABLE RESOURCES	
Hydroelectric (best case)	20/1**
Wind (electric generation)	2/1
Biomess methane	2/1
Solar electric (10 percent efficient)	1/1
Solar electric (20 percent efficient)	2/1

*Decommissioning of old nuclear plants and perpetual storage of wastes may consume more energy than nuclear power has produced.

**Hydropower yield depends on availability of water and life expectancy of dam and reservoir. Most dams produce only about 40 percent of rated capacity due to lack of water. Some dams have failed or silted up without producing any net energy yield.

Net yields and overall conversion efficiencies are not the only considerations when we compare different energy sources. The yield/cost ratio and conversion-cycle efficiency is much higher for coal burning, for instance, than for photovoltaic electrical production, making coal appear to be a better source of energy than solar radiation. Solar energy, however, is free, renewable, and nonpolluting. If we can use solar energy to get electrical energy, it doesn't matter how *efficient* the process is, as long as we get more out of it than we put in.

Negawatt Programs

Utility companies are finding it much less expensive to finance conservation projects than to build new power plants. Rather than buy megawatts of new generating capacity, power companies are investing in "negawatts" of demand avoidance. Each of us can help save energy in everyday life (What Can You Do? p. 280). Pacific Gas and Electric in California and Potomac Power and Light in Washington, D.C., both have instituted large conservation programs. They have found that conservation costs about $350 per kilowatt (kW) saved. By contrast, a new nuclear power plant costs between $3,000 and $8,000 per kW of installed capacity. New coal-burning plants with the latest air pollution-control equipment cost at least $1,000 per kW. By investing $200 million to $350 million in public education, home improvement loans, and other efficiency measures, a utility can avoid building a new power plant that would cost a billion dollars or more. Furthermore, conservation measures don't consume expensive fuel or produce pollutants.

Can application of this approach help alleviate energy shortages in other countries as well? Yes. For example, Brazil could cut its electricity consumption an estimated 30 percent with an investment of $10 billion (U.S.). It would cost $44 billion to build new power plants to produce that much electricity. South Korea has instituted a comprehensive conservation program with energy-saving building standards, efficiency labels on new household appliances, depreciation allowances, reduced tariffs on energy-conserving equipment, and loans and tax breaks for upgrading homes and businesses. It also forbids some unnecessary uses, such as air conditioning or the use of elevators between the first and third floors.

Cogeneration

One of the fastest growing sources of new energy is **cogeneration,** the simultaneous production of both electricity and steam or hot water in the same plant. By producing two kinds of useful energy in the same facility, the net energy yield from the primary fuel is increased from 30–35 percent to 80–90 percent. In 1900, half the electricity generated in the United States came from plants that also provided industrial steam or district heating. As power plants became larger, dirtier, and less acceptable as neighbours, they were forced to move away from their customers. Waste heat from the turbine generators became an unwanted pollutant to be disposed of in the environment. Furthermore, long transmission lines, which are unsightly and lose up to 20 percent of the electricity they carry, became necessary.

IN DEPTH . . . . . .

Personal Energy Efficiency

For some people, home energy efficiency means designing and building a new house that incorporates methods and materials that have been proven to conserve energy. Most of us, however, have to live in houses or apartments that already exist. How can we practise home energy conservation?

Think about the major ways we use energy. The biggest factor is space heating. Heat conservation is among the simplest, cheapest, and most effective ways to save energy in the home. Easy, surprisingly effective, and relatively inexpensive measures include weather-stripping, caulking, and adding layers of plastic to windows. Insulating walls, floors, and ceilings increases the energy conservation potential. Simply lowering your thermostat, especially at night, is a proven energy and money saver.

Water heating is the second major user of home energy, followed by large electric or gas appliances, such as stoves, refrigerators, washing machines, and dryers. Careful use of these appliances can save significant amounts of energy. Consider using less hot water, lowering the thermostat of your water heater, and buying an insulating blanket designed to wrap around old ones. Be sure that the refrigerator door doesn't stand open, and defrost it regularly (if it is not frost-free) to reduce the ice buildup that prevents the heat exchange system from working well. Make sure your oven door has a tight seal, and plan ahead when you use it; bake several things at once rather than reheating it repeatedly. Wash your clothes in cold or cool water rather than hot. Air dry your laundry, especially in the summer when sun-dried clothes are so pleasant to wear.

Some energy-efficient appliances can save substantial amounts of energy. Air conditioners and refrigerators with better condensers and heat exchangers use one-half to one-fourth as much energy as older models. New, improved furnaces can offer 95 percent efficiency, compared with conventional 70 percent efficiency in older models. The payback period may be as little as two to three years if you trade an old wasteful appliance for a newer, more efficient one. Some lightbulbs are made to save energy. High-efficiency fluorescent lights emit the same amount of light but use only one-fourth as

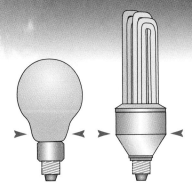

A 15 watt compact fluorescent bulb (right) produces as much light as a 60 watt incandescent bulb (left) but lasts 10 times as long.

much energy as conventional incandescent bulbs. They cost 10 times more than an ordinary bulb, but last 10 times as long. Total lifetime savings can be $30 to $50 per lamp. Almost all types of appliances, large and small, are available in efficient models and brands; we simply need to shop for them.

The easiest way to save energy is to keep an eye on energy-use habits. Turning off lights, televisions, and other devices when they are not in use is elementary. Line drying clothes and hand washing dishes are not difficult. In some cases, the most beneficial result of living a frugal, conservative life is not so much the total amount of energy you save, as the effect that this lifestyle has on you and the people around you. When you live conscientiously, you set a good example that could have a multiplying effect as it spreads through society. Making conscious ethical decisions about this one area of your life may stimulate you to make other positive decisions. You will feel better about yourself and more optimistic about the future when you make even small, symbolic gestures toward living as a good environmental citizen.

Energy Saving Potentials

	MODEL AVERAGE	BEST ON MARKET	BEST PROTOTYPE	SAVING (PERCENT)
Automobile (l/100 km)	13	4.3	2.2	83
Home (1,000 J/day)	190	68	11	94
Refrigerator (kWh/day)	4	2	1	75
Gas furnace (million J/day)	210	140	110	48
Air conditioner (kWh/day)	10	5	3	70

Source: U.S. Department of Energy.

By the 1970s, cogeneration had fallen to less than 5 percent of our power supplies, but interest in this technology is being renewed. The capacity for cogeneration more than doubled in the 1980s to about 30,000 megawatts (MW). District heating systems are being rejuvenated, and plants that burn municipal wastes are being studied. New combined-cycle coal-gasification plants or "mini-nukes" (Chapter 12) offer high efficiency and clean operation that may be compatible with urban locations. Small neighbourhood- or apartment building-sized power-generating units are being built that burn methane (from biomass digestion), natural gas, diesel fuel, or coal. The Fiat Motor Company makes a small generator for about $10,000 that produces enough electricity and heat for four or five houses.

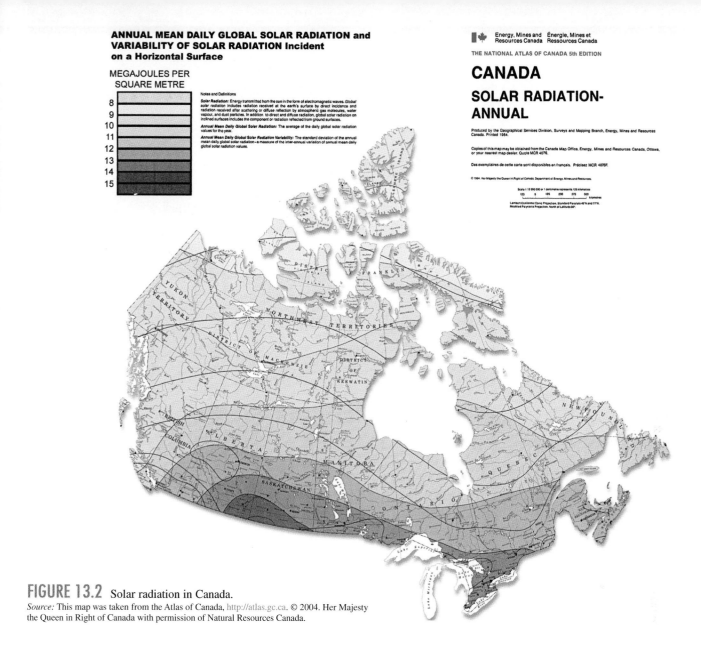

ANNUAL MEAN DAILY GLOBAL SOLAR RADIATION and VARIABILITY OF SOLAR RADIATION Incident on a Horizontal Surface

MEGAJOULES PER SQUARE METRE

8
9
10
11
12
13
14
15

Energy, Mines and Resources Canada   Énergie, Mines et Ressources Canada

THE NATIONAL ATLAS OF CANADA 5th EDITION

CANADA

SOLAR RADIATION- ANNUAL

FIGURE 13.2 Solar radiation in Canada.
Source: This map was taken from the Atlas of Canada, http://atlas.gc.ca. © 2004. Her Majesty the Queen in Right of Canada with permission of Natural Resources Canada.

TAPPING SOLAR ENERGY

The sun serves as a giant nuclear furnace in space, constantly bathing our planet with a free energy supply. Solar heat drives winds and the hydrologic cycle. All biomass, as well as fossil fuels and our food (both of which are derived from biomass), results from conversion of light energy (photons) into chemical bond energy by photosynthetic bacteria, algae, and plants.

A Vast Resource

The average amount of solar energy arriving at the top of the atmosphere is 1,330 watts per square metre (see Chapter 3). About half of this energy is absorbed or reflected by the atmosphere (more at high latitudes than at the equator), but the amount reaching the earth's surface is some 10,000 times all the commercial energy used each year. However, this tremendous infusion of energy comes in a form that, until the last century, had been too diffuse and low in intensity to be used except for environmental heating and photosynthesis.

Passive Solar Heat

Our simplest and oldest use of solar energy is **passive heat absorption,** using natural materials or absorptive structures with no moving parts to simply gather and hold heat. For thousands of years, people have built thick-walled stone and adobe dwellings that slowly collect heat during the day and gradually release that heat at night (Fig. 13.3). After cooling at night, these massive building materials maintain a comfortable daytime temperature within the house, even as they absorb external warmth.

A modern adaptation of this principle is a glass-walled "sun-space" or greenhouse on the south side of a building. Incorporating massive energy-storing materials, such as brick walls, stone floors, or barrels of heat-absorbing water into buildings also collects heat

FIGURE 13.3 Taos Pueblo in northern New Mexico uses adobe construction to keep warm at night and cool during the day.

FIGURE 13.4 A roof-top solar water heater (top dark panel) sits above eight photovoltaic electric-generating panels.

to be released slowly at night. An interior, heat-absorbing wall called a Trombe wall is an effective passive heat collector. Some Trombe walls are built of glass blocks enclosing a water-filled space or water-filled circulation tubes so heat from solar rays can be absorbed and stored, while light passes through to inside rooms.

Active Solar Heat

Active solar systems generally pump a heat-absorbing, fluid medium (air, water, or an antifreeze solution) through a relatively small collector, rather than passively collecting heat in a stationary medium like masonry. Active collectors can be located adjacent to or on top of buildings rather than being built into the structure. Because they are relatively small and structurally independent, active systems can be retrofitted to existing buildings (Fig. 13.4).

A flat black surface sealed with a double layer of glass makes a good solar collector. A fan circulates air over the hot surface and into the house through ductwork of the type used in standard forced-air heating. Alternatively, water can be pumped through the collector to pick up heat for space heating or to provide hot water. Water heating consumes 15 percent of the domestic energy budget of developed countries, so savings in this area alone can be significant. A simple flat panel with about 5 m^2 of surface can reach 95°C and can provide enough hot water for an average family of four almost anywhere in North America. In California, 650,000 homes now heat water with solar collectors. In Greece, Italy, Israel, and other countries where fuels are more expensive, up to 70 percent of domestic hot water comes from solar collectors.

Sunshine doesn't reach us all the time, of course. How can solar energy be stored for times when it is needed? There are a number of options. In a climate where sunless days are rare and seasonal variations are small, a small, insulated water tank is a good solar energy storage system. For areas where clouds block the sun for days at a time or where energy must be stored for winter use, a large, insulated bin containing a heat-storing mass, such as stone, water, or clay, provides solar energy storage (Fig. 13.5). During the summer months, a fan blows the heated air from the collector into the storage medium. In the winter, a similar fan at the opposite end of the bin blows the warm air into the house. During the summer, the storage mass is cooler than the outside air, and it

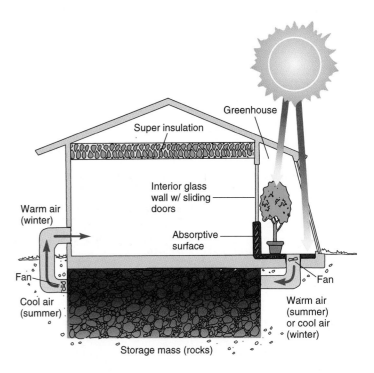

FIGURE 13.5 Underground massive heat storage unit. Heated air collected behind double- or triple-glazed windows is pumped down into a storage medium of rock, water, clay, or similar material, where it can be stored for a number of months.

helps cool the house by absorbing heat. During the winter, it is warmer and acts as a heat source by radiating stored heat. In many areas, six or seven months' worth of thermal energy can be stored in 38,000 litres of water or 40 tonnes of gravel, about the amount of water in a very small swimming pool or the gravel in two average-sized dump trucks.

Another heat storage system uses **eutectic** (phase-changing) **chemicals** to store a large amount of energy in a small volume. Heating melts these chemicals and cooling returns them to a solid state. This is an important way the hydrologic cycle stores solar energy in the natural environment, converting water from its solid phase to its liquid and gaseous phases. Most eutectic chemical systems (such as salts) do not swell when they solidify (as does

FIGURE 13.6 Parabolic mirrors focus sunlight on steam-generating tubes at this power plant in the California desert.

FIGURE 13.7 A simple box of wood, plastic, and foil can help reduce tropical deforestation, improve women's lives, and avoid health risks from smoky fires in developing countries. These inexpensive solar cookers could revolutionize energy use in developing tropical countries.

water) and undergo phase changes at higher temperatures than water and ice, making them more convenient for storing heat. Since the latent heat of crystallization is usually much greater than the specific heat required to raise temperatures, a eutectic system can store more energy in a smaller volume than can a simple, one-phase system.

Heat also can be captured in chemical reactions in the storage material. Some clays, for instance, undergo reactions that release or store heat when they absorb or release water. An insulated bin filled with a clay, such as bentonite, can be baked dry (absorbing heat) by blowing warm air into it in the summer. Sprinkling water on the clay to rewet it will release heat in winter.

HIGH-TEMPERATURE SOLAR ENERGY

Parabolic mirrors are curved reflecting surfaces that collect light and focus it into a concentrated point. There are two ways to use mirrors to collect solar energy to generate high temperatures. One technique uses long curved mirrors focused on a central tube, containing a heat-absorbing fluid (Fig. 13.6). Fluid flowing through the tubes reaches much higher temperatures than possible in a basic flat panel collector. Another high-temperature system uses thousands of smaller mirrors arranged in concentric rings around a tall central tower. The mirrors, driven by electric motors, track the sun and focus its light on a heat absorber at the top of the "power tower" where molten salt is heated to temperatures as high as 500°C, which then drives a steam-turbine electric generator.

Under optimum conditions, a 50 ha mirror array should be able to generate 100 MW of clean, renewable power. The only power tower in North America is Southern California Edison's Solar II plant in the Mojave Desert east of Los Angeles. Its 2,000 mirrors focused on a 100 m tall tower generates 10 MW or enough electricity for 5,000 homes at an operating cost far below that of nuclear power or oil. We haven't had enough experience with these facilities to know how reliable the mirrors, motors, heat absorbers, and other equipment will be over the long run.

Solar Cookers

Parabolic mirrors have been tested for home cooking in tropical countries where sunshine is plentiful and other fuels are scarce. They produce such high temperatures and intense light that they are dangerous, however. A much cheaper, simpler, and safer alternative is the solar box cooker (Fig. 13.7). An insulated box costing only a few dollars with a black interior and a glass or clear plastic lid serves as a passive solar collector. Several pots can be placed inside at the same time. Temperatures only reach about 120°C so cooking takes longer than an ordinary oven. Fuel is free, however, and the family saves hours each day usually spent hunting for firewood or dung. These solar ovens help reduce tropical forest destruction and reduce the adverse health effects of smoky cooking fires.

STORING ELECTRICAL ENERGY

Electrical energy is difficult and expensive to store. This is a problem for **photovoltaic generation** as well as other sources of electric power. Traditional lead-acid batteries are heavy and have low energy densities; that is, they can store only moderate amounts of energy per unit mass or volume. Acid from batteries is hazardous

Some Things You Can Do to Save Energy

1. Drive less: make fewer trips, use telecommunications and mail instead of going places in person.

2. Use public transportation, walk, or ride a bicycle.

3. Use stairs instead of elevators.

4. Join a car pool or drive a smaller, more efficient car; reduce speeds.

5. Insulate your house or add more insulation to the existing amount.

6. Turn thermostats down in the winter and up in the summer.

7. Weatherstrip and caulk around windows and doors.

8. Add storm windows or plastic sheets over windows.

9. Create a windbreak on the north side of your house; plant deciduous trees or vines on the south side.

10. During the winter, close windows and drapes at night; during summer days, close windows and drapes if using air conditioning.

11. Turn off lights, television sets, and computers when not in use.

12. Stop faucet leaks, especially hot water.

13. Take shorter, cooler showers; install water-saving faucets and shower-heads.

14. Recycle glass, metals, and paper; compost organic wastes.

15. Eat locally grown food in season.

16. Buy locally made, long-lasting materials.

turbine generators when extra energy is needed. Using a similar principle, pressurized air can be pumped into such reservoirs as natural caves, depleted oil and gas fields, abandoned mines, or special tanks. An Alabama power company currently uses off-peak electricity to pump air at night into a deep salt mine. By day, the air flows back to the surface through turbines, driving a generator that produces electricity. Cool night air is heated to 870°C by compression plus geothermal energy, increasing pressure and energy yield.

An even better way to use surplus electricity is in electrolytic decomposition of water to H_2 and O_2. These gases can be liquefied (like natural gas) at −252°C, making them easier to store and ship than most forms of energy. They are highly explosive, however, and must be handled with great care. They can be burned in internal combustion engines, producing mechanical energy, or they can be used to power fuel cells, which produce more electrical energy (see the following text). Liquid hydrogen cars are already being produced. They could be very attractive in smoggy cities because they produce no carbon monoxide, smog-forming hydrocarbons, carcinogenic chemicals, or soot.

Flywheels are the subject of current experimentation for energy storage. Massive, high-speed flywheels, spinning in a nearly friction-free environment, store large amounts of mechanical energy in a small area. This energy is convertible to electrical energy. It is difficult, however, to find materials strong enough to hold together when spinning at high speed. Flywheels have a disconcerting tendency to fail explosively and unexpectedly, sending shrapnel flying in all directions.

FUEL CELLS

Rather than store and transport energy, another alternative would be to generate it locally, on demand. **Fuel cells** are devices that use ongoing electrochemical reactions to produce an electric current. They are very similar to batteries except that rather than recharging them with an electrical current, you add more fuel for the chemical reaction.

Fuel cells are not new; the basic concept was recognized in 1839 by William Grove, who was studying the electrolysis of water. He suggested that rather than use electricity to break apart water and produce hydrogen and oxygen gases, it should be possible to reverse the process by joining oxygen and hydrogen to produce water and electricity. The term "fuel cell" was coined in 1889 by Ludwig Mond and Charles Langer, who built the first practical device using a platinum catalyst to produce electricity from air and coal gas. The concept languished in obscurity until the 1950s when the U.S. National Aeronautics and Space Administration (NASA) was searching for a power source for spacecraft. Research funded by NASA eventually led to development of fuel cells that now provide both electricity and drinkable water on every space shuttle flight. The characteristics that make fuel cells ideal for space exploration—small size, high efficiency, low emissions, net water production, no moving parts, and high reliability—also make them attractive for a number of other applications.

and lead from smelters or battery manufacturing is a serious health hazard for workers who handle these materials. A typical lead-acid battery array sufficient to store several days of electricity for an average home would cost about $5,000 and weigh 3 or 4 tonnes. All the components for an electric car are readily available, but, as mentioned earlier, battery technology limits how far they can go between charges.

Other types of batteries also have drawbacks. Metal-gas batteries, such as the zinc-chloride cell, use inexpensive materials and have relatively high-energy densities, but have shorter lives than other types. Sodium-sulphur batteries have considerable potential for large-scale storage. They store twice as much energy in half as much weight as lead-acid batteries. They require an operating temperature of about 300°C and are expensive to manufacture. Alkali-metal batteries have a high storage capacity but are even more expensive. Lithium batteries have very long lives and store more energy than other types, but are the most expensive. Recent advances in thin-film technology may hold promise for future energy storage but often require rare, toxic elements that are both expensive and dangerous.

Another strategy is to store energy in a form that can be turned back into electricity when needed. Pumped-hydro storage involves pumping water to an elevated reservoir at times when excess electricity is available. The water is released to flow back down through

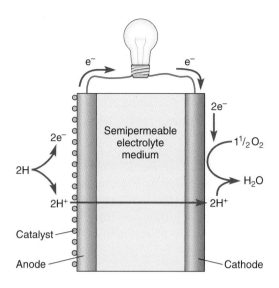

FIGURE 13.8 Fuel cell operation. Electrons are removed from hydrogen atoms at the anode to produce hydrogen ions (protons) that migrate through a semipermeable electrolyte medium to the cathode, where they reunite with electrons from an external circuit and oxygen atoms to make water. Electrons flowing through the circuit connecting the electrodes create useful electrical current.

All fuel cells consist of a positive electrode (the cathode) and a negative electrode (the anode) separated by an electrolyte, a material that allows the passage of charged atoms, called ions, but is impermeable to electrons (Fig. 13.8). In the most common systems, hydrogen or a hydrogen-containing fuel is passed over the anode while oxygen is passed over the cathode. At the anode, a reactive catalyst, such as platinum, strips an electron from each hydrogen atom, creating a positively charged hydrogen ion (a proton). The hydrogen ion can migrate through the electrolyte to the cathode, but the electron is excluded. Electrons pass through an external circuit, and the electrical current generated by their passage can be used to do useful work. At the cathode, the electrons and protons are reunited and combined with oxygen to make water.

The fuel cell provides direct-current electricity as long as it is supplied with hydrogen and oxygen. For most uses, oxygen is provided by ambient air. Hydrogen can be supplied as a pure gas, but storing hydrogen gas is difficult and dangerous because of its volume and explosive nature. Liquid hydrogen takes far less space than the gas, but must be kept below –250°C, not a trivial task for most mobile applications. The alternative is a device called a **reformer** or converter that strips hydrogen from fuels such as natural gas, methanol, ammonia, gasoline, ethanol, or even vegetable oil. Many of these fuels can be derived from sustainable biomass crops. Even methane effluents from landfills and wastewater treatment plants can be used as a fuel source. Where a fuel cell can be hooked permanently to a gas line, hydrogen can be provided by solar, wind, or geothermal facilities that use electricity to hydrolyze water.

A fuel cell run on pure oxygen and hydrogen produces no waste products except drinkable water and radiant heat. When a reformer is coupled to the fuel cell, some pollutants are released (most commonly carbon dioxide), but the levels are typically far less than conventional fossil fuel combustion in a power plant or automobile engine. Although the theoretical efficiency of electrical generation of a fuel cell can be as high as 70 percent, the actual yield is closer to 40 or 45 percent. This not much better than a very good fossil fuel power plant or a gas turbine electrical generator. On the other hand, the quiet, clean operation and variable size of fuel cells make them useful in buildings where waste heat can be captured for water heating or space heating. A new 45-story office building in New York City, for example, has two 200-kilowatt fuel cells on its fourth floor that provide both electricity and heat. This same building has photovoltaic panels on its façade, natural lighting, fresh air intakes to reduce air conditioning, and a number of other energy conservation features.

The current from a fuel cell is proportional to the size (area) of the electrodes, while the voltage is limited to about 1.23 volts per cell. A number of cells can be stacked together until the desired power level is reached. A fuel cell stack that provides almost all of the electricity needed by a typical home (along with hot water and space heating) would be about the size of a refrigerator. A 200 kilowatt unit fills a medium-size room and provides enough energy for 20 houses or a small factory. Tiny fuel cells running on methanol might soon be used in cell phones, pagers, toys, computers, videocameras, and other appliances now run by batteries. Rather than buy new batteries or spend hours recharging spent ones, you might just add an eyedropper of methanol every few weeks to keep your gadgets operating.

Fuel Cell Types

Several different electrolytes can be used in fuel cells, each with advantages and disadvantages (Table 13.3). A design, called a proton exchange membrane (PEM), was developed for use in cars, buses, and trucks in the 80s and 90s by Ballard Power Systems, Inc., a Canadian company. The membrane is a thin semipermeable layer of an organic polymer containing sulphonic acid groups that facilitate passage of hydrogen ions but block electrons and oxygen. The surface of the membrane is dusted with tiny particles of platinum catalyst. These cells have the advantage of being lightweight and operating at a relatively low temperature (80°C). The fuel efficiency of PEM systems is typically less than 40 percent. Buses equipped with PEM stacks have been demonstrated in Chicago, Miami, and Vancouver, B.C. In 1999, DaimlerChrysler unveiled their five-passenger NECAR-4. A 500 kg PEM stack that costs $30,000 (compared to about $3,000 for a conventional gasoline engine) produces performance comparable to most passenger cars, but takes up considerable interior room.

For stationary electrical generation, the most common fuel cell design uses phosphoric acid immobilized in a porous ceramic matrix as the electrolyte. Because this system operates at higher temperatures than PEM cells, less platinum is needed for the catalyst. It has a higher efficiency, 40 to 50 percent, but is heavier and larger than PEM cells. It also is less sensitive to carbon dioxide contamination than other designs. Hundreds of 200 kilovolt phosphoric acid fuel cells have been installed around the world. Some have run for decades. They supply dependable electricity in remote

TABLE 13.3 Fuel Cell Types

TYPE	PROTON EXCHANGE MEMBRANE	PHOSPHORIC ACID	MOLTEN CARBONATE	SOLID OXIDE
Electrolyte	Semipermeable organic polymer	Phosphoric acid	Liquid carbonate	Solid-oxide ceramic
Charge carrier	H^+	H^+	$CO_3^=$	$O^=$
Catalyst	Platinum	Platinum	Nickel	Perovskites (calcium titanate)
Operating temperature, °C	80	200	650	1,000
Cell material	Carbon or metal-based	Graphite-based	Stainless steel	Ceramic
Efficiency (percent)	Less than 40	40 to 50	50 to 60	More than 60
Heat cogeneration	None	Low	High	High
Status	Commercially available	Commercially available	Demonstration systems	Under development

Source: Modified from Alan C. Lloyd, *Scientific American*, July 1999, vol. 281 (1) p. 83.

locations without the spikes and sags and risk of interruption common in utility grids. The largest fuel cell ever built was an 11 MW unit in Japan, that provides enough electricity for a small town.

In 1999, the Central Park police station in New York City was equipped with a 200 kilowatt phosphoric acid fuel cell. The station is located in the middle of the park, so bringing in new electric lines would have cost $1.2 million and would have disrupted traffic and park use for months. A diesel generator was ruled out as too noisy and polluting. Solar photovoltaic panels were thought to be too obtrusive. A small, silent fuel cell provided just the right solution.

Carbonate fuel cells use an inexpensive nickel catalyst and operate at 650°C. The electrode is a very hot (thus the name molten carbonate) solution trapped in a porous ceramic. The charge carrier is carbonate ion, which is formed at the cathode where oxygen and carbon dioxide react in the presence of a nickel oxide catalyst. Migrating through the electrolyte, the carbonate ion reacts at the anode with hydrogen and carbon monoxide to release electrons. The high operating temperature of this design means that it can reform fuels internally and ionize hydrogen without expensive catalysts. Heat cogeneration is very good, but the high temperature makes these units more difficult to operate. It takes hours for carbonate fuel cells to get up to operating temperature, so they aren't suitable for short-term, quick-response uses.

The least developed of the fuel cell design is called solid oxide, because it uses a coated zirconium ceramic as an electrolyte. Oxygen ions formed by the titanium catalyst carry the charge across the electrolyte. Operating temperatures are 1,000°C. They have the highest fuel efficiency of any current design, but mass production of components has not yet been mastered, and these cells are still in the experimental stage.

METHANE AS FUEL

Methane gas is the main component of natural gas. It is produced by anaerobic decomposition (digestion by anaerobic bacteria) of any

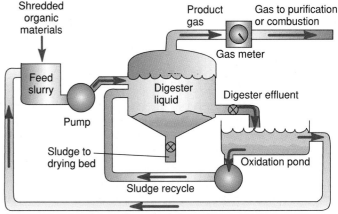

FIGURE 13.9 Continuous unit for converting organic material to methane by anaerobic fermentation. One kilogram of dry organic matter will produce 1–1.5 m³ of methane, or 2,500–3,600 million calories per tonne.
Source: *Solar Energy as a National Energy Resource,* NSF/NASA Solar Energy Panel, National Science Foundation, December 1972.

moist organic material. Many people are familiar with the fact that swamp gas is explosive. Swamps are simply large methane digesters, basins of wet plant and animal wastes sealed from the air by a layer of water. Under these conditions, organic materials are decomposed by anaerobic (oxygen-free) rather than aerobic (oxygen-using) bacteria, producing flammable gases instead of carbon dioxide. This same process may be reproduced artificially by placing organic wastes in a container and providing warmth and water (Fig. 13.9). Bacteria are ubiquitous enough to start the culture spontaneously.

Burning methane produced from manure provides more heat than burning the dung itself, and the sludge left over from bacterial digestion is a rich fertilizer, containing healthy bacteria as well as most of the nutrients originally in the dung. Whether the manure is of livestock or human origin, airtight digestion also eliminates some health hazards associated with direct use of dung, such as exposure to fecal pathogens and parasites.

How feasible is methane—from manure or from municipal sewage—as a fuel resource in developed countries? Methane is a clean fuel that burns efficiently. Any kind of organic waste material: livestock manure, kitchen and garden scraps, and even municipal garbage and sewage can be used to generate gas. In fact, municipal landfills are active sites of methane production, contributing as much as 20 percent of the annual output of methane to the atmosphere. This is a waste of a valuable resource and a threat to the environment because methane absorbs infrared radiation and contributes to the greenhouse effect (Chapter 8).

Cattle feedlots and chicken farms are a tremendous potential fuel source. In North America, collectible crop residues and feedlot wastes each year contain 58 billion gigajoules of energy, more than all the farmers use. The Haubenschild farm in central Minnesota, for instance, uses manure from 850 Holsteins to generate all the power needed for their dairy operation and still have enough excess electricity for an additional 80 homes. In January 2001, the farm saved 35 tonnes of coal, 4,500 litres of propane, and made $4,380 from electric sales.

Municipal sewage treatment plants also routinely use anaerobic digestion as a part of their treatment process, and many facilities collect the methane they produce and use it to generate heat or electricity for their operations. Although this technology is well-developed, its utilization could be much more widespread.

Alcohol from Biomass

Ethanol (grain alcohol) and methanol (wood alcohol) are produced by anaerobic digestion of plant materials with high sugar content, mainly grain and sugarcane. Ethanol can be burned directly in automobile engines adapted to use this fuel, or it can be mixed with gasoline (up to about 10 percent) to be used in any normal automobile engine. A mixture of gasoline and ethanol is often called **gasohol.** Ethanol in gasohol raises octane ratings and is a good substitute for lead antiknock agents, the major cause of lead pollution. It also helps reduce carbon monoxide emissions in automobile exhaust. Gas stations in U.S. cities that exceed U.S. EPA air quality standards have been ordered to sell so-called oxygenated fuels containing ethanol, methanol, or methyltertiarybutyl ether (MTBE). The latter, however, is a suspected carcinogen that has contaminated many lakes and aquifers in the United States (see Chapter 10).

Ethanol production could be a solution to grain surpluses and bring a higher price for grain crops than the food market offers. It also offers promise for reduced dependence on gasoline, which is refined from petroleum. Brazil has instituted an ambitious national program to substitute crop-based ethanol for imported petroleum. In 1995, the Brazilian sugar harvest produced 10 billion litres of ethanol, but falling oil prices have undercut plans to expand ethanol production to 16 billion litres per year. Both methanol and ethanol make good fuels for fuel cells.

Crop Residues, Energy Crops, and Peat

Crop residues, such as cornstalks, corncobs, or wheat straw, can be used as a fuel source, but they are expensive to gather and often are better left on the ground as soil protection (Chapter 14). The residue accumulated in food processing, however, can be a useful fuel. Hawaiian sugar growers burn bagasse, a fibrous sugarcane residue, to produce the state's second-largest energy source, surpassed only by petroleum.

Some crops are being raised specifically as an energy source. Fast-growing trees, such as *Leucaena,* and shrubs, such as alder and willow, are grown to provide wood for energy. Milkweeds, sedges, marsh grasses, cattails, and other biomass crops also might become useful energy sources, grown on land that is otherwise unsuitable for crops. Unfortunately, there may be unacceptable environmental costs from disruption of wetlands and forests that are converted to energy crop plantations. Some plant species produce hydrocarbons that can be used directly as fuel in internal combustion engines without having to be fermented or transformed to hydrogen or methane gas. Sunflower oil, for instance, can be burned directly in diesel engines. In most cases, these oils and other high-molecular-weight hydrocarbons are more valuable for other purposes, such as food, plastics, and industrial chemicals; however, they might be an attractive energy source in some circumstances.

Several other forms of biomass are actual or proposed sources of energy. People in northern climates have been digging up peat (partially decomposed plant residues) from bogs and marshes for centuries. There has been a resurgence of interest in exploiting the vast Minnesota and Canadian peat bogs, which some people regard as wastelands. After mining some or all of the peat, the land could be used for energy crops, such as cattails. Critics of these schemes point out the damaging effects of bog draining on ecosystems, watersheds, and wildlife. Burning of municipal garbage ("waste to watts") is a popular proposal; however, many people are worried about air pollution from these facilities (Chapter 18). Garbage may contain hazardous, volatile, or explosive materials that are difficult or dangerous to sort out. Effects on air quality from all forms of direct combustion of biomass are a serious concern.

ENERGY FROM THE EARTH'S FORCES

The winds, waves, tides, ocean thermal gradients, and geothermal areas are renewable energy sources. Although available only in selected locations, these sources could make valuable contributions to our total energy supply.

Wind Energy

The air surrounding the earth has been called a 20-billion-cubic-kilometre storage battery for solar energy. The World Meteorological Organization has estimated that 20 million MW of wind power could be commercially tapped worldwide, not including contributions from windmill clusters at sea. This is about 50 times the total present world nuclear generating capacity.

Wind power has advantages and disadvantages, as do other nontraditional technologies. Like solar power and hydropower, wind power taps a natural physical force. Like solar power (its ultimate source), wind power is a virtually limitless resource, is nonpolluting, and causes minimal environmental disruption. Like solar power, however, it requires expensive storage during peak production times to offset nonwindy periods.

Windmills played a crucial role in the settling of rural North America. The Great Plains had abundant water in underground aquifers, but little surface water. The strong, steady winds that blew across the prairies provided the energy to pump water that allowed agriculture to move west across the prairies. By the end of the nineteenth century, nearly every farm or ranch west of the Great Lakes had at least one windmill. Even today, some 150,000 windmills still spin productively in the Great Plains (Fig. 13.10).

As the world's conventional fuel prices rise, interest in wind energy is resurging. In the 1980s, the United States was a world leader in wind technology and California hosted 90 percent of all existing windpower generators. Some 17,000 windmills marched across windy mountain ridges at Altamont, Tehachapi,

and San Gorgino Passes. Poor management, technical flaws, and overdependence on subsidies, however, led to bankruptcy of major corporations including Kenetech, once the largest turbine producer in the United States. Now Danish, German, and Japanese wind machines are capturing the rapidly growing world market. Wind technology is now Denmark's second-largest export, employing 20,000 people and bringing in about $1 billion (U.S.) per year.

The 6,200 MW of installed wind power currently in operation worldwide demonstrate the economy of wind turbines. Theoretically up to 60 percent efficient, windmills typically produce about 35 percent efficiency under field conditions. Where conditions are favourable, wind power is already cheaper than any other new energy source, with electric prices as low as 3 cents per KWH in places with steady winds averaging at least 24 km/hr.

The World Energy Council predicts that wind could account for 200,000 to 500,000 MW of electricity by 2020, depending on how seriously politicians take global warming and how many uneconomical nuclear reactors go offline. One thousand MW meets the energy needs of about 50,000 typical North American households or is equivalent to about 6 million barrels of oil. By the middle of this century, Shell Oil suggests that half of all the world's energy could be wind and solar generated.

The standard modern wind turbine uses only two or three propeller blades. More blades on a windmill provide more torque in low-speed winds, so that the traditional Midwestern windmill, with 20 or 30 blades, was most appropriate for small-scale use in less reliable wind fields. Fewer blades operate better in high-speed winds, providing more energy for less material cost at wind speeds of 25–40 km/hr. A two-blade propeller can extract most of the available energy from a large vertical area and has less material to weaken and break in a storm. Three-bladed propellers often are preferred because they are easier to balance and spin more smoothly (Fig. 13.11).

Wouldn't wind power take up a huge land area if we were to depend on it for a major part of our energy supply? As Table 13.4 shows, the actual space taken up by towers, roads, and other

FIGURE 13.10 Millions of traditional windmills once pumped water, generated electricity, and provided power for farms and ranches across North America. Most have been displaced by internal combustion engines and rural electric power systems, but there is still potential for capturing large amounts of renewable energy from the sun, wind, and other sustainable sources.

FIGURE 13.11 A 23-metre-long, 1.6-metre-wide fiberglass blade waits to be attached to a 750 kV wind generator.

TABLE 13.4	Jobs and Land Required for Alternative Energy Sources	
TECHNOLOGY	LAND USE (m² PER GIGAWATT-HOUR FOR 30 YEARS)	JOBS (PER TERAWATT-HOUR PER YEAR)
Coal	3,642	116
Photovoltaic	3,237	175
Solar thermal	3,561	248
Wind	1,335	542

Source: Data from Lester R. Brown, et al. *Saving the Planet,* 1991, W. W. Norton & Co., Inc.

structures on a wind farm is only about one-third as much as would be consumed by a coal-fired power plant or solar thermal energy system to generate the same amount of energy over a 30-year period. Furthermore, the land under windmills is more easily used for grazing or farming than is a strip-mined coal field or land under solar panels.

When a home owner or community invests independently in wind generation, the same question arises as with solar energy: What should be done about energy storage when electricity production exceeds use? Besides the storage methods mentioned earlier, many private electricity producers believe the best use for excess electricity is to sell it to the public utility grid. When private generation is low, the public utility runs electricity through the meter and into the house or community. When the wind generator or photovoltaic systems overproduce, the electricity runs back into the grid and the meter runs backward. Ideally, the utility reimburses individuals for this electricity, for which other consumers pay the company.

Geothermal Energy

The earth's internal temperature can provide a useful source of energy in some places. High-pressure, high-temperature steam fields exist below the earth's surface. Around the edges of continental plates or where the earth's crust overlays magma (molten rock) pools close to the surface, this energy is expressed in the form of hot springs, geysers, and fumaroles (Chapter 7). Iceland, Japan, and New Zealand have high concentrations of geothermal springs and vents. Depending on the shape, heat content, and access to groundwater, these sources produce wet steam, dry steam, or hot water.

Until recently, the main use of this energy source was in baths built at hot springs. More recently, **geothermal energy** has been used in electric power production, industrial processing, space heating, agriculture, and aquaculture. In Iceland, most buildings are heated by geothermal steam. A few communities in the United States, including Boise, Idaho, use geothermal home heating. Recently, geothermal steam also has been developed for electrical generation.

California's Geysers project is the world's largest geothermal electric-generating complex, with 200 steam wells that provide some 1,300 MW of power. This system functions much like other heat-driven generators. A shaft sunk into the subsurface steam reservoir brings pressurized steam up to a turbine at the surface. The steam spins the turbine to produce electricity and then is piped off to a condensing unit and then to a cooling pond. The remaining brine (water heavily laden with dissolved minerals) is pumped back down to the reservoir. When the steam source has high mineral concentrations that would corrode delicate turbine blades, a heat exchanger is used to generate clean steam as is done in a nuclear power plant.

Among the advantages of geothermal generators are a reasonably long life span (at least several decades), no mining or transporting of fuels, and little waste disposal. The main disadvantages are the potential danger of noxious gases in the steam and noise problems from steam-pressure relief valves.

While few places have geothermal steam, the entire earth is underlain by warm rocks that can be reached by drilling down far enough into the crust. Some heating systems pump groundwater from underground aquifers or inject surface water into deep strata and then withdraw it after it has been warmed, as a way of extracting geothermal energy (Fig. 13.12). There have been experiments with pumping surplus heated water into underground reservoirs

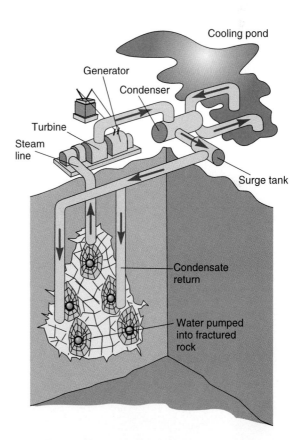

FIGURE 13.12 Geothermal steam extraction system for electricity production. Once the steam has spun the turbine, it is condensed, cooled, and pumped back into the underground steam field. The steam source for this relatively low-capital electric plant should last for several decades.
Sources: American Oil Shale Corporation and U.S. Atomic Energy Commission, *A Feasibility Study of a Plowshare of Geothermal Power Plant,* April 1971.

during the summer when the heat is not needed, and then withdrawing it in the winter as make-up water for steam-heating systems. A major concern with such systems, however, is to ensure that contaminants aren't injected into aquifers, which are important sources of drinking water.

Tidal and Wave Energy

Ocean tides and waves contain enormous amounts of energy that can be harnessed to do useful work. Tidal power exploitation is not new. The earliest recorded tide-powered mills were built in about 1100 in England. One built in Woodbridge, England, in 1170 functioned productively for 800 years. Through the seventeenth century, similar mills were built by the Dutch in the Netherlands and in New York.

The Rance River Power Station in France, in operation since 1966, was the first large tidal electric generation plant, producing 160 MW. A **tidal station** works like a hydropower dam, with its turbines spinning as the tide flows through them (Fig. 13.13). It requires a high-tide low-tide differential of several metres to spin the turbines. Unfortunately, the tidal period of 13$\frac{1}{2}$ hours causes problems in integrating the plant into the electric utility grid, as it seldom coincides with peak-use hours. Nevertheless, demand has kept the plant running for more than two decades.

The first North American tidal generator, producing 20 MW, was completed in 1984 at Annapolis Royal, Nova Scotia. A much larger project has been proposed to dam the Bay of Fundy and produce 5,000 MW of power on the Bay's 17-metre tides. The total flow at each tide through the Bay of Fundy theoretically could generate energy equivalent to the output of 250 large nuclear power plants. The environmental consequences of such a gargantuan project, however, may prevent its ever being built. The main worries are saltwater flooding of freshwater aquifers when seawater levels rise behind the dam and the flooding and destruction of rich shoals and salt flats, breeding grounds for aquatic species and a vital food source for millions of migrating shorebirds. There also would be heavy siltation, as well as scouring of the seafloor as water shoots through the dam. However, it appears that modest-sized plants like Annapolis Royal's will avoid most of these dangers. Three other medium-sized projects are under consideration for sites within the Bay of Fundy.

Ocean wave energy can easily be seen and felt on any seashore. The energy that waves expend as millions of tonnes of water are picked up and hurled against the land, over and over, day after day, can far exceed the combined energy budget for both insolation (solar energy) and wind power in localized areas. Captured and turned into useful forms, that energy could make a very substantial contribution to meeting local energy needs.

Numerous attempts have been made to use wave energy to drive electrical generators. Generally these take the form of a floating bar that moves up and down as the wave passes. When coupled to a dynamo, this mechanical energy can be converted to electricity in the same way that a waterwheel or steam turbine works. England, with a long coastline facing the stormy North Sea, plans to build an extensive system of wave-energy platforms. Unfortunately for developers of this energy source, the stormy coasts where waves are strongest are usually far from major population centers that need the power. In addition, the storms that bring this energy often destroy the equipment intended to exploit it.

An intriguing proposal has been developed for small rotating cylinders called ducks that would sit just offshore, bobbing in the surf and generating electricity. There are claims that British energy officials deliberately suppressed this low-tech approach because it might compete with conventional sources.

Ocean Thermal Electric Conversion

Temperature differentials between upper and lower layers of the ocean's water also are a potential source of renewable energy. In a closed-cycle **ocean thermal electric conversion (OTEC)** system, heat from sun-warmed upper ocean layers is used to evaporate a working fluid, such as ammonia or Freon, which has a low boiling point. The pressure of the gas produced is high enough to spin turbines to generate electricity. Cold water then is pumped from the the ocean depths to condense the gas.

As long as a temperature difference of about 20°C exists between the warm upper layers and cooling water, useful amounts of net power can, in principle, be generated with one of these systems. This differential corresponds, generally, to a depth of about 1,000 m in tropical seas. The places where this much temperature difference is likely to be found close to shore are islands that are the tops of volcanic seamounts, such as Hawaii, or the edges of continental plates along subduction zones (Chapter 7) where deep trenches lie just offshore. The west coast of Africa, the south coast

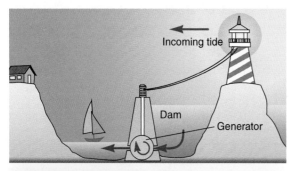

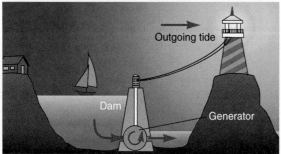

FIGURE 13.13 Tidal power station. Both incoming and outgoing tides are held back by a dam. The difference in water levels generates electricity in both directions as water runs through reversible turbogenerators.

of Java, and a number of South Pacific islands, such as Tahiti, have usable temperature differentials for OTEC power.

Although their temperature differentials aren't as great as the ocean, deep lakes can have very cold bottom water. Ithaca, NY, has recently built a system to pump cold water out of Lake Cayuga to provide natural air conditioning during the summer. Cold water discharge from a Hawaiian OTEC system has been used to cool the soil used to grow cool-weather crops such as strawberries.

SUMMARY

Exciting new technologies have been invented to use renewable energy sources. Active solar air and water heating, for instance, require less material and function more quickly than passive solar collection. Wind is now the cheapest form of new energy in many places. It has potential to supply one-third or more of our energy requirements. Parabolic mirrors can produce temperatures high enough to be used as process heat in manufacturing.

Hybrid gasoline/electric motors provide high vehicle efficiency with low pollution emissions. Even better are fuel cells, which use catalysts and semipermeable electrolytes to extract energy from fuels such as hydrogen or methanol at high efficiency and with super low emissions. Ocean thermal electric conversion, tidal and wave power stations, and geothermal steam sources can produce useful amounts of energy in some localities. One of the most promising technologies is direct electricity generation by photovoltaic cells. Since solar energy is available everywhere, photovoltaic collectors could provide clean, inexpensive, non-polluting, renewable energy, independent of central power grids or fuel-supply systems.

Biomass also may have some modern applications. In addition to direct combustion, biomass can be converted into methane or ethanol, which are clean-burning, easily storable, and transportable fuels. These alternative uses of biomass also allow nutrients to be returned to the soil and help reduce our reliance on expensive, energy-consuming artificial fertilizers.

Many of these sustainable energy sources depend on technology that is still experimental and too expensive to compete well with established energy industries. If the economies of mass production and marketing were applied, these new technologies could be made available more cheaply and more dependably. It may take special funding and other governmental incentives to make sustainable energy competitive. The subsidies for renewable energy sources have been especially meager in comparison to the billions of dollars spent on nuclear energy, large-scale hydropower, and fossil fuel extraction and utilization.

Although conventional and alternative energy sources offer many attractive possibilities, conservation often is the least expensive and easiest solution to energy shortages. Even basic conservation efforts, such as turning off lights, can save large amounts of energy when practised by many people. More major conservation methods, such as home insulation and energy-efficient appliances and transportation, can drastically reduce energy consumption and similarly reduce energy expenses. Our natural resources, our environment, and our pocketbooks all benefit from careful and efficient energy consumption.

QUESTIONS FOR REVIEW

1. What is cogeneration and how does it save energy?
2. Explain the principle of net energy yield. Give some examples.
3. What is the difference between active and passive solar energy?
4. How do photovoltaic cells generate electricity?
5. What is a fuel cell and how does it work?
6. How is methane made? Give an example of a useful methane source.
7. What are some examples of biomass fuel other than wood?
8. Describe how tidal power or ocean wave power generate electricity.

QUESTIONS FOR CRITICAL THINKING

1. What alternative energy sources are most useful in your region and climate? Why?
2. What can you do to conserve energy where you live? In personal habits? In your home, dormitory, or workplace?

3. What massive heat storage materials can you think of that could be attractively incorporated into a home?

4. Do you think building wind farms in remote places, parks, or scenic wilderness areas would be damaging or unsightly?

5. If you were the energy minister of your province, where would you invest your budget?

6. What are the advantages and disadvantages of being disconnected from central utility power?

7. Can you think of environmental consequences associated with tidal or geothermal energy? If so, how can they be mitigated?

8. You are offered a home solar energy system that costs $10,000 but saves you $1,000 a year. Will you take it at this rate? If the cost were higher and the payoff time longer, what is the threshold at which you would not buy the system?

KEY TERMS

active solar systems 278

cogeneration 275

energy efficiency 273

eutectic chemicals 278

fuel cells 280

gasohol 283

geothermal energy 285

hybrid gas-electric vehicles 274

net energy yield 275

ocean thermal electric conversion (OTEC) 286

passive heat absorption 277

photovoltaic generation 279

reformer 281

tidal station 286

Web Exercises

Use the Green Pages Search Utility at http://www.eco-web.com/ to find information about sustainable power generation companies in Canada. Are you surprised at how much we are doing? What do you think the opportunities and barriers are for such companies in Canada and internationally?

Environmental Scientist on Ice

Carl Ozyer

Growing up on the glacial sediments that formed the Scarborough Bluffs east of Toronto, little would Carl realize the significance this would have on his future. He spent many summer days climbing the bluffs, enjoying the scenic view of Lake Ontario, often wondering why other areas did not have these natural features.

Upon completion of high school he took up the position of a medical sales representative and after several successful years became unhappy with the monotony and repetition the job provided and, the lack of personal satisfaction. He decided a complete change was necessary which would enable him to gain the credentials required to work in an area he has always had an interest in, the environment. He gave up his employment and enrolled in a university program that covered a variety of environmental subjects. His enthusiasm for learning enabled him to complete his BSc in Geography specializing in Environment and Resources Management almost a year early.

During the last year of his undergraduate program it was apparent to him that he still wanted to learn more. He applied for, and was offered admittance to a BEd program, and an MSc geology graduate program. It was a tough choice to make, however, his affection for the outdoors led him to choose the geology program. That program would lead him to undertake a PhD in Quaternary Geology that would enable him to conduct fieldwork in the Arctic.

He always had a curiosity about the Arctic from stunning photos he had seen of animals such as caribou and polar bears traversing barren lands in one of the harshest environments on earth. His graduate research led him to Nunavut, where he would study the ice-movement patterns of the former Laurentide Ice Sheet during the last glaciation. To some, this may seem a simple task,

however, it required him to walk several thousand kilometres over four summers to search for clues necessary for him to reconstruct the history of the ice sheet. He also realized the ice sheet was responsible for creating and distributing glacial till, some of which is mineral rich. Since he was already studying the ice-movement history of the ice sheet, he thought he would enhance his thesis research by studying the mineral dispersal patterns of the glacial till. This required him to collect hundreds of till samples and have them analyzed for mineral content. He feels this part of his research may lead to economic opportunities for local communities in his study area.

With concern about global warming, Carl feels the history of the former Laurentide Ice Sheet, especially its deglaciation patterns may provide important insights into what can be expected should the two remaining ice sheets melt.

His arctic experience has changed the way he views the environment and the challenges posed working in such remote locations. Not only has he learned about the environment, he has learned about the people living in one of the harshest environments on earth. He feels privileged to be able to conduct research in a part of the world very few people have visited. He is driven by his fieldwork and the challenges, rewards, and sense of personal satisfaction of which the Arctic has provided him.

Although it is difficult to predict where Carl will be employed when he has completed his studies, he feels the combination of independent field research and education will provide him with employment opportunities in industry, government, or academia. The latter would be his foremost choice since it will provide him an opportunity to teach students about his experience in the Arctic.

Chapter 14

Food and Agriculture

We abuse the land because we regard it as a commodity belonging to us.
When we see land as a community to which we belong, we may begin to use it with love and respect.

—Aldo Leopold—

OBJECTIVES

After studying this chapter, you should be able to:

- describe world food supplies and some causes of chronic hunger.
- analyze some of the promises and perils of genetic engineering.
- understand the nature and causes of variability in soil.
- understand the sources and effects of land degradation, including erosion and nutrient depletion.
- explain the needs for and constraints of water, energy, and nutrients for food production.
- understand the choices available for sustainable agriculture.

WebQuest

Canada's Guide to Healthy Eating, green revolution, genetically modified organisms, soil science, no till agriculture

Above: Farmers feed the world. With proper care and stewardship, soil, water, and crops are renewable resources.

Are Shrimp Safe to Eat?

If you've bought shrimp recently at a restaurant or grocery store, chances are very good that they came from a commercial shrimp farm in a developing country such as Thailand, Ecuador, or Mexico. Once considered a luxury food, shrimp has become much more affordable in recent years, and now competes with tuna as the most popular seafood in North America. As the world's leading shrimp-consuming country, the U.S. imports around 500,000 tonnes of farm-raised shrimp every year, or about half the total world production. Although this plentiful supply of a reasonably priced, highly desirable food is a boon to diners, there are social and environmental costs associated with its production that aren't widely known.

Shrimp aquaculture or farming first became profitable about 20 years ago and has since mushroomed into a major industry in the developing world. While total catches of wild shrimp have remained relatively stable at about 2 million tonnes per year over the past two decades, farm-raised production has exploded from less than 80,000 tonnes in 1980 to more than 1 million tonnes in 2000. These shrimp are raised in shallow ponds ranging in size from a few hundred square metres to many hectares, generally constructed on or near the coastline of a tropical country. Asia has by far the largest area devoted to shrimp farming with more than 1.2 million hectares in Thailand, Indonesia, China, India, Vietnam, and Bangladesh. Ecuador, with 130,000 hectares of ponds, raises about 60 percent of all shrimp in the Western Hemisphere, and is second in the world (after Thailand) in total production.

Hailed as a "blue revolution" 25 years ago, shrimp farming and other types of aquaculture were promoted as a way to provide a nutritious, inexpensive source of protein for the growing world population as well as to reduce the pressures on already dwindling supplies of wild seafood. While much commercial fish farming has been devoted to high-price, export species such as salmon, shrimp, and oysters, cultivation of some 10 million tonnes of less expensive freshwater fish such as carp and tilapia for local consumption has, indeed, increased the protein supply available in many developing countries. Culture of saltwater species such as shrimp, however, has caused considerable damage both to wild stocks and also to ecosystems that support them.

One of the biggest problems is that flooded mangrove forests and coastal wetlands often are destroyed to build shrimp ponds. Mangrove forests are extremely important as nurseries, sheltering the young of a wide variety of ocean species. They reduce pollution by absorbing excess nutrients and sediment that would otherwise contaminate nearshore waters and threaten coral reefs. About half of all mangrove forests in the world already have been destroyed. Shrimp farms are thought to be responsible for about one-fourth of that destruction. A 2000 report in the journal *Nature* estimated that for every kilogram of shrimp raised in converted mangroves, 400 kg of wild fish production is lost. Furthermore, because shrimp farms often are stocked in very high densities, fresh seawater is flushed regularly through the ponds to wash out uneaten food, dead animals, feces, ammonia, phosphorus, and carbon dioxide. To prevent diseases among the teeming shrimp populations, most farmers also treat the ponds with antibiotics and chemicals such as formalin and calcium hypochlorite which then contaminate local surface or ground waters.

Many shrimp farms are stocked with hatchery-produced young shrimp that can be certified free of diseases. Many shrimp farmers, however, prefer to stock their ponds with juvenile shrimp caught in the wild because they are cheaper and are thought to be stronger and have a higher survival rate. Shrimp harvesters scour estuaries and tidal wetlands to collect young shrimp to sell to farmers. Their fine-mesh nets catch large numbers of unwanted "by-catch" species. Up to 150 times more fry are lost as by-catch than are used to stock shrimp farms. Furthermore, carnivorous species like salmon and shrimp often are fed high-protein fish meal made from wild ocean fish (sardines, anchovies, pilchard, and other low-value species). Because it takes roughly 5 kg of wild fish to produce 1 kg of farmed fish or shrimp, the result is a net loss of protein.

Not all aquaculture operations are environmentally harmful. With conscientious, scientific management, excess feeding can be minimized, diseases can be controlled without harmful chemical or antibiotic releases, water use can be minimized, and polluted effluent can be treated before being discharged into the environment. There isn't yet a certification process, however, so consumers can't tell whether the seafood products they buy have been obtained in an ecologically sound and sustainable manner.

So, while eating shrimp is probably safe for you, it may not be good for the environment. This example is only one of many dilemmas we face with respect to food and agriculture. In this chapter we will look at global food supplies and some of the problems associated with production and distribution of food.

Want More? Naylor, R. L., Goldburg, R. J., Primavera, J. H., Kautsky, N., Beveridge, M. C. M., Clay, J., Folke, C., Lubchenco, J., Mooney, H., and Troell, M. 2000. "Effect of Aquaculture on World Fish Supplies." *Nature* 405: 1017–1024.

NUTRITION AND FOOD SUPPLIES

Despite repeated dire predictions that runaway population growth would soon lead to terrible famines (see Chapter 4), world food supplies have more than kept up with increasing human numbers over the past three centuries. In 1950, when there were 2.5 billion people on earth, the average daily diet provided less than 2,000 calories per person. More than a billion people were considered chronically **undernourished,** meaning that they had less than the caloric intake needed for an active healthy life. In 2001, by contrast, even though the population had more than doubled to 6 billion people, the world food supply was sufficient to provide more than 2,500 calories per person per day, or enough for a healthy and productive life for everyone if it were equitably distributed.

In contrast to much of the world, the most common dietary problem in wealthy countries is now **overnutrition,** or too many calories. In North America and Europe the average daily caloric intake is about 3,500 calories. Some 1.1 billion people (16 percent of the world's population) are overweight or obese (Fig. 14.1) about the same as the number of underfed people! Excess weight raises risks of high blood pressure, heart attacks, strokes, diabetes, and other conditions. In Canada, obesity is now the greatest single health risk—greater than smoking. The challenge for these societies is to encourage more active lifestyles, more walking and biking, for example, and lower sugar and fat intake.

Asia has experienced the most rapid increase in crop production (Fig. 14.2). China and Indonesia have tripled their rice output in little more than a decade by expanding croplands and irrigation, and by using new, high-yielding rice varieties, fertilizers, and pesticides. Among the biggest questions in environmental science is whether this success can be extended throughout the world and whether we can feed another 2 or 3 billion people likely to be born in this century.

In other regions, notably sub-Saharan Africa, food production has not kept pace with rapid population growth. Thirty-five of the 40 countries in this region have had decreasing per capita food production over the past three decades. The worst declines have been in countries such as Angola, Ethiopia, Sudan, Somalia, and Uganda, where war and governmental mismanagement or corruption, combined with drought, have caused agricultural production to collapse in many areas. The collapse of the Soviet Union in 1992 has also led to a precipitous collapse in food production. While privatization of collective farms in China led to increased output, it had the opposite effect in the former USSR. Crops rot in the fields where there is no money to pay harvesters and no transportation to get commodities to the market.

Chronic Hunger and Food Security

Although the global proportion of chronically hungry people has declined over the last half of the twentieth century, the Food and Agricultural Organization (FAO) says the total number remains at about 800 million people, or about one person in five in the developing world. At least 200 million of those who don't have enough food are children. Chronic undernourishment during childhood years leads to permanently stunted growth, mental retardation, and other social and developmental disorders. Infectious diseases that are only an inconvenience for well-fed individuals become lethal threats to those who are poorly nourished. Diarrhea rarely kills a well-fed person, but a child weakened by nutritional deficiencies will be highly susceptible to this and a host of other diseases. At least 11 million children die each year from diseases exacerbated by nutritional deficiencies. And around 20 percent of the Disability-Adjusted Life Years (DALYs) suffered every year (see Chapter 20) are related to chronic nutritional deficits.

Poverty is the greatest threat to **food security,** or the ability to obtain sufficient food on a day-to-day basis. The 1.4 billion people in the world who have to live on less than $1 per day all too often can't buy the food they need and don't have access to resources to grow it for themselves. Food security occurs at multiple scales. In the poorest countries, hunger may affect nearly everyone. In other countries, although the average food availability may be good, some individual communities or families may not have enough to eat.

Where does this persistent hunger occur? Figure 14.3 shows countries with the greatest risk of food shortages. As you can see, most of sub-Saharan Africa, South and Southeast Asia, and parts of Latin America fall in this category. The largest number of hungry people in 2000, according to the FAO, was in East and Southeast Asia, where some 275 million people lacked an adequate diet. The second highest incidence of chronic undernutrition was in South Asia, where at least 250 million went hungry on a regular basis. Africa was third with about 225 million hungry people. Remarkable progress is being made in increasing food production in much of Asia, where the number of chronically hungry people is expected to drop to about 100 million by 2010.

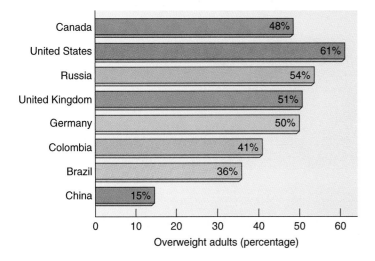

FIGURE 14.1 While much of the world is hungry, people in wealthier countries are at risk from eating too much.
Source: Worldwatch Institute 2001.

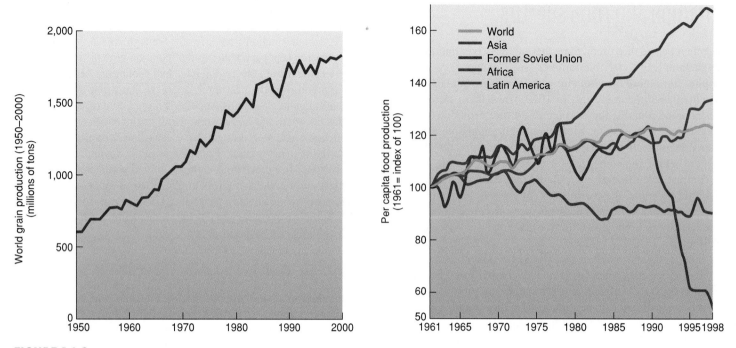

FIGURE 14.2 (*a*) Total world grain production has increased dramatically since 1950. (*b*) Relative per capita food production by region. Index is based on 1961. Notice general decline in Africa and sudden collapse of agricultural production in the former Soviet Union since 1990.
Source: (*a*) Data from USDA. (*b*) Data from Food and Agriculture Organization of the United Nations (FAO), FAOSTAT-PC on diskette (FAO, Rome, 1995).

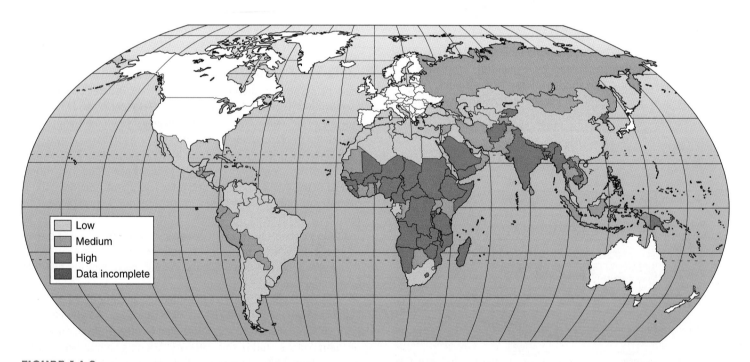

FIGURE 14.3 Countries with populations at risk of inadequate nutrition. Canada, the United States, Europe, Japan, and Australia have little risk.
Source: Food and Agriculture Organization (FAO) 2002.

Eating a Balanced Diet

What's the best way to be sure we're getting a healthy diet? Generally it isn't necessary to take synthetic dietary supplements.

Eating a balanced diet with plenty of whole grains, fruits, and vegetables should give you all the nutrients you need. For years Canadians were advised to eat daily servings of four major food groups: meat, dairy products, grains, and fruits and vegetables.

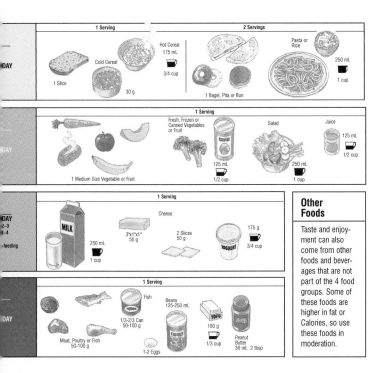

FIGURE 14.4 Healthy eating advised by Health Canada. You should eat two to four times as much bread, cereal, rice, and pasta as milk, meat, eggs, or nuts. Fats, oils, and sweets should be eaten sparingly, if at all.
Source: Health Canada.

FIGURE 14.5 Children wait for their daily ration of porridge at a feeding station in Somalia. When people are driven from their homes by hunger or war, social systems collapse, diseases spread rapidly, and the situation quickly becomes desperate.

Placing meat and dairy products first was the result of lobbying by those industries. More recently, Health Canada has revised recommendations for a balanced diet (Fig. 14.4) emphasizing grains, fruits, and vegetables, with only sparing servings of meat, dairy, fats, and sweets.

Famines

Famines are characterized by large-scale food shortages, massive starvation, social disruption, and economic chaos. Starving people eat their seed grain and slaughter their breeding stock in a desperate attempt to keep themselves and their families alive (Fig. 14.5). Even if better conditions return, they have sacrificed their productive capacity and will take a long time to recover. Famines are characterized by mass migrations as starving people travel to refugee camps in search of food and medical care. Many die on the way or fall prey to robbers.

What causes these terrible tragedies? Environmental conditions are usually the immediate trigger, but politics and economics are often equally important in preventing people from getting the food they need. Adverse weather, insect infestations, and other natural disasters cause crop failures and create food shortages. But as economist Amartya K. Sen—whose work won the Nobel Prize in 1998—points out, these factors have generally been around for a long time, and local people usually have adaptations that get them through hard times if they are allowed to follow traditional

patterns of migration and farming. Arbitrary political boundaries, however, along with wars and land seizures by the rich and powerful, block access to areas that once served as refuge during droughts, floods, and other natural disasters. Poor people can neither grow their own food nor find jobs to earn money to buy the food they need.

In 1974, for instance, a terrible famine struck Bangladesh and thousands died. The immediate cause was floods in June through August that interfered with rice planting, eliminating jobs on which many farm workers depended. Fears about impending rice shortages triggered panic buying, hoarding by speculators, and rapidly rising prices. The government began relief efforts, but it was too little, too late. By October it became apparent that rice harvests were actually higher than in previous years; prices dropped and so did mortality rates. The irony was that more food was available in 1974 than in any other year in the decade. Poor people simply couldn't afford to buy it. The United States played a role in this tragedy by cutting off aid to Bangladesh as punishment for selling jute fibres to Cuba. Withdrawal of food aid on which Bangladesh had become dependent only fuelled the panic and price gouging that occurred later.

By contrast, the state of Maharashtra in central India suffered a similar drought in 1972–73 that reduced crop yields by 50 percent. As in Bangladesh, farm labourers were thrown out of work. The Indian government moved quickly, however, to employ workers building roads, wells, and other public projects. Although wages were low, food prices didn't rise and people could afford an adequate if meagre diet. Although the amount of food available per person was less than half that in Bangladesh, few Indians starved. Since they remained in their villages and rural infrastructure was improved during the drought by public works projects, farmers recovered quickly once the rains returned.

Similarly, the world was horrified by terrible photographs of starving children in Ethiopia and the Sudan in the 1980s. More than 1 million people died in the Horn of Africa. Droughts triggered this famine, but neighbouring countries that experienced comparable weather did not have problems as severe as these unfortunate countries did. In Sudan and Ethiopia, where total food supplies dropped about 10 percent, the famine was severe. Zimbabwe and Cape Verde, by contrast, which suffered worse droughts and lost about 40 percent of their normal harvest, had no famine. In fact, social welfare programs brought about mortality *decreases* in Zimbabwe and Cape Verde during this time. Author Amartya Sen points out that armed conflict and political oppression almost always are at the root of famine. No democratic country with a relatively free press, he says, has ever had a major famine.

The aid policies of rich countries often serve more to get rid of surplus commodities and make us feel good about our generosity than to get at the root causes of starvation (Fig. 14.6). Herding people into feeding camps generally is the worst thing to do for them. The stress of getting there kills many of them, and the crowding and lack of sanitation in the camps exposes them to epidemic diseases. There are no jobs in the refugee camps, so people can't support themselves if they try. Social chaos and family breakdown expose those who are weakest to robbery and violence. Having left their land and tools behind, people can't replant crops when the weather returns to normal.

MAJOR FOOD SOURCES

Of the thousands of edible plants and animals in the world, only about a dozen types of seeds and grains, three root crops, 20 or so common fruits and vegetables, six mammals, two domestic fowl, and a few fish and other forms of marine life make up almost all of the food humans eat. Table 14.1 shows annual production of some important foods in human diets. In this section, we will highlight sources and characteristics of those foods.

Major Crops

The three crops on which humanity depends for the majority of its nutrients and calories are wheat, rice, and maize. Together, nearly 1,600 million tonnes of these three grains are grown each year. Wheat and rice are especially important since they are the staple foods for most of the 5 billion people in the developing countries of the world. These two grass species supply around 60 percent of the calories consumed directly by humans.

Potatoes, barley, oats, and rye are staples in mountainous regions and high latitudes (northern Europe, north Asia) because they grow well in cool, moist climates. Cassava, sweet potatoes, and other roots and tubers grow well in warm, wet areas and are staples in Amazonia, Africa, Melanesia, and the South Pacific. Sorghum and millet are drought resistant and are staples in the dry regions of Africa.

Fruits and vegetables—including vegetable oils—make a surprisingly large contribution to human diets. They are especially welcome because they typically contain high levels of vitamins, minerals, dietary fibre, and complex carbohydrates.

FIGURE 14.6 We rarely learn much from the media about the underlying causes of famine. Often our attention is focused on images of people from wealthy nations generously helping those who are suffering. This makes us feel good, but doesn't get to the root of the problems.

TABLE 14.1	Some Important Food Resources
CROP	**2002 YIELD (MILLION TONNES)**
Wheat	603
Rice (paddy)	593
Maize (corn)	933
Potatoes	308
Barley and oats	168
Soybeans	177
Cassava and sweet potato	315
Sugar (cane and beet)	141
Pulses (beans, peas)	54
Oil seeds	324
Vegetables and fruit	905
Meat and milk	735
Fish and seafood	140

Source: Food and Agriculture Organization (FAO), 2003.

PART THREE   Humans in the Environment

Meat, Milk, and Seafood

Beginning about 8,000 years ago with sheep and goats in what is now Iraq, humans have domesticated animals for food. Meat and milk are prized by people nearly everywhere, but their distribution is highly inequitable. Although the industrialized, more-developed countries of North America, Europe, and Japan make up only 20 percent of the world population, they consume 80 percent of all meat and milk in the world. The 80 percent of the world's people in less-developed countries raise 60 percent of the 3 billion domestic ruminants and 6 billion poultry in the world but consume only 20 percent of all animal products.

About 90 percent of the grain grown in North America is used to feed dairy and beef cattle, hogs, poultry, and other animals. It is used especially to fatten beef cattle during the last three months before they are slaughtered. As Chapter 3 shows, there is a great loss of energy with each step up the food chain. This means we could feed far more people if we ate more grain directly rather than feeding it to livestock. Every 16 kg of grain and soybeans fed to beef cattle in feedlots produce about 1 kg of edible meat. The other 15 kg are used by the animal for energy or body parts we do not eat or they are eliminated. If we were to eat the grain directly, we would get 21 times more calories and eight times more protein than we get by eating the meat it produces. Hogs and poultry are about two and four times as efficient, respectively, as cattle in converting feed to edible meat.

Fish and other seafood contribute about 94 million tonnes of high-quality food to the world's diet. This is an important protein source in many countries, making up about one-half of the animal protein and one-fourth of total dietary protein in Japan, for instance.

Unfortunately, overharvesting and habitat destruction threaten most of the world's wild fisheries. Annual catches of ocean fish rose by about 4 percent annually between 1950 and 1988. Since 1989, however, 13 of 17 major marine fisheries have declined dramatically or become commercially inviable. According to the United Nations, 70 percent of the world's edible ocean fish, crustaceans, and mollusks are declining and in urgent need of managed conservation.

An explosion of fishing technology in the 1950s and 1960s made fishers both more efficient and more lethal. Sonar, radar, remote sensing, and global positioning systems have turned fishing from an art into a science. Longlines extending 100 km and carrying 60,000 hooks, bag-shaped trawl nets large enough to engulf a dozen jumbo jets, and drift nets covering 2 million square metres, tended by floating factories that can process and freeze 500 tonnes of fish per day, make it possible to exhaust entire populations in just a few years. Between 1970 and 1990, the number and average vessel size of the world fishing fleet doubled, now representing twice the capacity needed to extract the total annual sustainable harvest. To catch $70 billion worth of fish each year, the fishing industry incurs costs totalling $124 billion. Direct and indirect subsidies make up most of the $54 billion deficit.

This intense fishing pressure destroys nontarget species as well as those sought commercially. Worldwide, one animal in four taken from the ocean is unwanted "by-catch." In 1998, for instance, high seas drift nets killed at least 30 million nontarget animals including diving sea birds and marine mammals. Trawl nets dragged across the ocean floor also destroy plants and animals that provide spawning habitat and food sources for rebuilding commercial populations. Fish farming has replaced hunting for wild animals in many cases, but serious environmental problems in this industry cause concerns.

SOIL: A RENEWABLE RESOURCE

Growing the food and fibre needed to support human life is a complex enterprise that requires knowledge from many different fields and cooperation from many different groups of people. In this section, we'll survey some of the principles of soil science and look at some of the inputs necessary for continued agricultural production.

Of all the earth's crustal resources, the one we take most for granted is soil. We are terrestrial animals and depend on soil for life, yet most of us think of it only in negative terms. English is unique in using "soil" as an interchangeable word for earth and excrement. "Dirty" has a moral connotation of corruption and impurity. Perhaps these uses of the word enhance our tendency to abuse soil without scruples; after all, it's only dirt.

The truth is that **soil** is a marvellous substance, a living resource of astonishing beauty, complexity, and frailty. It is a complex mixture of weathered mineral materials from rocks, partially decomposed organic molecules, and a host of living organisms. It can be considered an ecosystem by itself. Soil is an essential component of the biosphere, and it can be used sustainably, or even enhanced, under careful management.

There are thousands of soil types worldwide. They vary because of the influences of parent material, time, topography, climate, and organisms on soil formation. There are young soils that, because they have not weathered much, are rich in soluble nutrients. There are old soils, like the red soils of the tropics, from which rainwater has washed away most of the soluble minerals and organic matter, leaving behind clay and rust-coloured oxides.

To understand the potential for feeding the world on a sustainable basis we need to know how soil is formed, how it is being lost, and what can be done to protect and rebuild good agricultural soil. With careful husbandry, soil can be replenished and renewed indefinitely. Many farming techniques deplete soil nutrients, however, and expose the soil to the erosive forces of wind and moving water. As a result, in many places we are essentially mining this resource and using it much faster than it is being replaced.

Building good soil is a slow process. Under the best circumstances, good topsoil accumulates at a rate of about 10 tonnes per hectare per year—enough soil to make a layer about 1 mm deep when spread over a hectare. Under poor conditions, it can take thousands of years to build that much soil. Perhaps one-third to one-half of the world's current croplands are losing topsoil faster

TABLE 14.2 Soil Particle Sizes

CLASSIFICATION	SIZE
Gravel	2 to 64 mm
Sand	0.05 to 2 mm
Silt	0.002 to 0.05 mm
Clay	Less than 0.002 mm

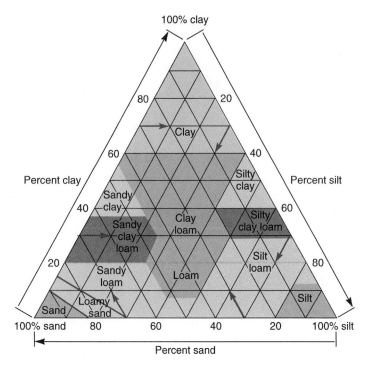

FIGURE 14.7 Soil texture is determined by the percentages of clay, silt, and sand particles in the soil. Soils with the best texture for most crops are loams, which have enough larger particles (sand) to be loose, yet enough smaller particles (silt and clay) to retain water and dissolved mineral nutrients.
Source: Data from Soil Conservation Service.

than it is being replaced. In some of the worst spots, erosion carries away about 2.5 cm of topsoil per year. With losses like that, agricultural production has already begun to fall in many areas.

Soil Composition

Most soil is about half mineral. The rest is plant and animal residue, air, water, and living organisms. The mineral particles are derived either from the underlying bedrock or from materials transported and deposited by glaciers, rivers, ocean currents, windstorms, or landslides. The weathering processes that break rocks down into soil particles are described in Chapter 17.

Particle sizes affect the characteristics of the soil (Table 14.2). The spaces between sand particles give sandy soil good drainage and usually allow it to be well aerated, but also cause it to dry out quickly when rains are infrequent. Tight packing of small particles in silty or clay soils makes them less permeable to air and water than sandy soils. Tiny capillary spaces between the particles, on the other hand, store water and mineral ions better than more porous soils. Because clay particles have a proportionately large surface area and a high ionic charge, they stick together tenaciously, giving clay its slippery plasticity, cohesiveness, and impermeability. Soils with a high clay content are called "heavy soils," in contrast to easily worked "light soils" that are composed mostly of sand or silt. Varying proportions of these mineral particles occur in each soil type (Fig. 14.7). Farmers usually consider sandy loam the best soil type for cultivating crops.

The organic content of soil can range from nearly zero for pure sand, silt, or clay, to nearly 100 percent for peat or muck, which is composed mainly of partly decomposed plant material. Much of the organic material in soil is **humus,** a sticky, brown, insoluble residue from the partially decomposed bodies of dead plants and animals. Humus is the most significant factor in the development of "structure," a description of how the soil particles clump together. Humus coats mineral particles and holds them together in loose crumbs, giving the soil a spongy texture that holds water and nutrients needed by plant roots, and maintains the spaces through which delicate root hairs grow.

Soil Organisms

Without soil organisms, the earth would be covered with sterile mineral particles far different from the rich, living soil ecosystems on which we depend for most of our food. The activity of the myriad organisms living in the soil help create structure, fertility, and tilth (condition suitable for tilling or cultivation) (Fig. 14.8).

Soil organisms usually stay close to the surface, but that thin living layer can contain thousands of species and billions of individual organisms per hectare. Algae live on the surface, while bacteria and fungi flourish in the top few centimetres of soil. A single gram of soil can contain hundreds of millions of these microscopic cells. Algae and blue-green bacteria capture sunlight and make new organic compounds. Bacteria and fungi decompose organic detritus and recycle nutrients that plants can use for additional growth. The sweet aroma of freshly turned soil is caused by actinomycetes, bacteria that grow in funguslike strands and give us the antibiotics streptomycin and tetracyclines.

Roundworms, segmented worms, mites, and tiny insects swarm by the thousands in that same gram of soil from the surface. Some of them are herbivorous, but many of them prey upon one another. Soil roundworms (nematodes) attack plant rootlets and can cause serious crop damage. A carnivorous fungus snares nematodes with tiny loops of living cells that constrict like a noose when a worm blunders into it. Burrowing animals, such as gophers, moles, insect larvae, and worms, tunnel deeper in the soil, mixing and aerating it. Plant roots also penetrate lower soil levels, drawing up soluble minerals and secreting acids that decompose

PART THREE Humans in the Environment

FIGURE 14.8 Soil ecosystems include numerous consumer organisms, as depicted here: (1) snail, (2) termite, (3) nematodes and nematode-killing constricting fungus, (4) earthworm, (5) wood roach, (6) centipede, (7) carabid (ground) beetle, (8) slug, (9) soil fungus, (10) wireworm (click beetle larva), (11) soil protozoan, (12) earthworm, (13) sow bug, (14) ants, (15) mite, (16) springtail, (17) pseudoscorpion, and (18) cicada nymph.

mineral particles. Fallen plant litter adds new organic material to the soil, returning nutrients to be recycled.

Soil Profiles

Most soils are stratified into horizontal layers called **soil horizons** that reveal much about the history and usefulness of the soil. The thickness, colour, texture, and composition of each horizon are used to classify the soil. Together these horizons make up a **soil profile** (Fig. 14.9).

The soil surface is often covered with a layer of leaf litter, crop residues, or other fresh or partially decomposed organic material. This organic layer is know as the O horizon. Below this layer is the A horizon or **topsoil,** composed of mineral particles mixed with organic material. The A horizon can range from several metres thick under virgin prairie to almost nothing in dry deserts. The O and A horizons contain most of the living organisms and organic material in the soil, and it is in these layers that most plants spread their roots to absorb water and nutrients. An eluviated (leached) Ae subhorizon often lies at the base of the A horizon. This layer is depleted of clays and soluble nutrients, which are removed by rainwater seeping down through the soil. These clays and nutrients generally accumulate in the B horizon, or **subsoil.** The B horizon may have a dense or clayey texture because of the accumulated clays. In desert regions, a dense, impermeable "hardpan" layer of accumulated minerals or salts may develop on the B horizon. This hardpan blocks plant root growth and prevents water from draining properly.

Beneath the subsoil is the parent material, or C horizon, made of weathered rock fragments with very little organic material. Weathering in the C horizon, largely accomplished by rain water mixed with organic compounds from plants, produces new soil particles and allows downward expansion of the soil profile. Parent material can be sand, solid rock, or other material. About 70 percent of the parent material in Canada was transported to its present site by glaciers, wind, or water and is not directly related to the underlying bedrock.

Soil Types

Soils are classified by how they were formed, the climate they occur in, and their structure and composition. The Canadian System of Soil Classification (CSSC) includes 9 Orders that are subdivided into 27 Great Groups, which are in turn divided into 188 Groups, within which there are hundreds of Families and Series. The *Cryosolic* order is the most common soil type in Canada, covering about 40 percent of its land mass. Cryosolic soils, underlain by permafrost within a metre or two of the surface, occur in the tundra, subarctic, and boreal forest areas of the north. Several soil types occur in forested areas, varying by both composition and interactions with the resident vegetation. The *Brunisolic* soil order is a brown soil with relatively low clay content of the B horizon, found in a forested environment in about 9 percent of the land area of Canada. *Luvisolic* soils also occur in forested areas, but have a higher clay content in the B horizon leached from the A horizon. They are found in about 8 percent of Canada's land area. *Podzolic*

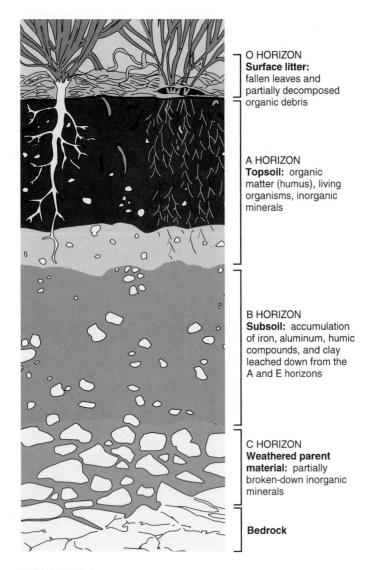

FIGURE 14.9 Soil profile showing possible soil horizons. The actual number, composition, and thickness of these layers varies in different soil types.

O HORIZON
Surface litter: fallen leaves and partially decomposed organic debris

A HORIZON
Topsoil: organic matter (humus), living organisms, inorganic minerals

B HORIZON
Subsoil: accumulation of iron, aluminum, humic compounds, and clay leached down from the A and E horizons

C HORIZON
Weathered parent material: partially broken-down inorganic minerals

Bedrock

soils underlay coniferous forests and are quite acidic. They are the substrate of 16 percent of the Canadian landscape. *Chernozemic* soils are dark in colour due to thick mats of decaying vegetation in the A horizon, and are found in the 5 percent land area of prairie and grassland regions of Canada. *Organic* soils, largely underlying wetland, bog, and fen ecosystems, occur in about 4 percent of the land area. Soils that occupy smaller areas of Canada include the frequently waterlogged *Gleysolic* (1.3 percent), the parent material *Regosolic* (1 percent), and saline *Solonetzic* (0.7 percent) orders.

WAYS WE USE AND ABUSE SOIL

Only about 11 percent of the earth's land area (14.66 million sq km out of a total of 132.4 million sq km) is currently in agricultural production. Perhaps four times as much land could potentially be converted to cropland, but much of this land serves as a refuge for cultural or biological diversity or suffers from constraints, such as steep slopes, shallow soils, poor drainage, tillage problems, low nutrient levels, metal toxicity, or excess soluble salts or acidity, that limit the types of crops that can be grown there (Fig. 14.10).

Land Resources

Table 14.3 shows the distribution of cropland by region. In parts of Canada and the United States, temperate climates, abundant water, and high soil fertility produce high crop yields that contribute to high standards of living. Other countries, although rich in land area, lack suitable soil, topography, water, or climate to sustain our levels of productivity.

If current population projections are correct, the current world average of 0.27 ha of cropland per person will decline to 0.17 ha by the year 2025. In Asia, cropland will be even more scarce—0.09 ha per person—25 years from now. If you live on a typical .10-ha suburban lot, look at your yard and imagine feeding yourself for a year on what you could produce there.

In the developed countries, 95 percent of recent agricultural growth in the twentieth century came from improved crop varieties or increased fertilization, irrigation, and pesticide use, rather than from bringing new land into production. In fact, less land is being cultivated now than 100 years ago in North America, or 600 years ago in Europe. As more effective use of labour, fertilizer, and water and improved seed varieties have increased in the more developed

FIGURE 14.10 In many areas, soil or climate constraints limit agricultural production. These farm workers in Chad weed a field that is too dry for good crop yields.

TABLE 14.3 Cropland by Region and per Capita

REGION	POPULATION (MILLIONS)	CROPLAND (10^6 HA)	CROPLAND PER CAPITA
Africa	778	190	0.24
Europe	729	316	0.43
North America	304	233	0.77
Central America	131	41	0.31
South America	332	115	0.35
Asia	3,589	622	0.17
Oceania*	29	52	1.79
World	5,892	1,569	0.27

*Oceania includes Australia, New Zealand, and many small-island nations.
Source: Data from *World Resources 1998–99*, World Resource Institute.

countries, productivity per unit of land has increased, and marginal land has been retired, mostly to forests and grazing lands. In many developing countries, land continues to be cheaper than other resources, and new land is still being brought under cultivation, mostly at the expense of forests and grazing lands. Still, at least two-thirds of recent production gains have come from new crop varieties and more intense cropping rather than expansion into new lands.

The largest increases in cropland over the last 30 years occurred in South America and Oceania where forests and grazing lands are rapidly being converted to farms. Many developing countries are reaching the limit of lands that can be exploited for agriculture without unacceptable social and environmental costs, but others still have considerable potential for opening new agricultural lands. East Asia, for instance, already uses about three-quarters of its potentially arable land. Most of its remaining land has severe restrictions for agricultural use. Further increases in crop production will probably have to come from higher yields per hectare. Latin America, by contrast, uses only about one-fifth of its potential land, and Africa uses only about one-fourth of the land that theoretically could grow crops. However, there would be serious ecological trade-offs in putting much of this land into agricultural production.

While land surveys tell us that much more land in the world *could* be cultivated, not all of that land necessarily *should* be farmed. Much of it is more valuable in its natural state. The soils over much of tropical Asia, Africa, and South America are old, weathered, and generally infertile. Most of the nutrients are in the standing plants, not in the soil. In many cases, clearing land for agriculture in the tropics has resulted in tragic losses of biodiversity and the valuable ecological services that it provides. Ultimately, much of this land is turned into useless scrub or semidesert.

On the other hand, there are large areas of rich, subtropical grassland and forest that are well watered, have good soil, and could become productive farmland without unduly reducing the world's biological diversity. Argentina, for instance, has pampas grasslands

about twice the size of Texas that closely resemble the American Midwest more than a century ago in climate and potential for agricultural growth. Some of this land could probably be farmed with relatively little ecological damage if it were done carefully.

Land Degradation

Agriculture both causes and suffers from environmental degradation. The International Soil Reference and Information Centre in the Netherlands estimates that every year 3 million ha of cropland are ruined by erosion, 4 million ha are turned into deserts, and 8 million ha are converted to nonagricultural uses such as homes, highways, shopping centres, factories, reservoirs, etc. Over the past 50 years, some 1.9 *billion* ha of agricultural land (an area greater than that now in production) have been degraded to some extent. About 300 million ha of this land is strongly degraded (deep gullies, severe nutrient depletion, crops grow poorly, restoration is difficult and expensive), while 910 million ha—about the size of China—are moderately degraded. Nearly 9 million ha of former croplands are so degraded that they no longer support any crop growth at all. The causes of this extreme degradation vary: In Ethiopia it is water erosion, in Somalia it is wind, and in Uzbekistan salt and toxic chemicals are responsible.

Definitions of degradation are based on both biological productivity and our expectations about what the land should be like. Often this is a subjective judgement and it is difficult to distinguish between human-caused deterioration and natural processes like drought. We generally consider the land degraded when the soil is impoverished or eroded, water runs off or is contaminated more than is normal, vegetation is diminished, biomass production is decreased, or wildlife diversity diminishes. On farmlands this results in lower crop yields. On ranchlands it means fewer livestock can be supported per unit area. On nature reserves it means fewer species.

The amount and degree of land degradation varies by region and country. About 20 percent of land in Africa and Asia is degraded, but most is in either the light or moderate category. In Central America and Mexico, by contrast, 25 percent of all vegetated land suffers moderate to extreme degradation. Figure 14.11 shows some areas of greatest concern for soil degradation, and Figure 14.12 shows the mechanisms for this problem. Water and wind erosion provide the motive force for the vast majority of all soil degradation, worldwide. Chemical deterioration includes nutrient depletion, salinization (salt accumulation), acidification, and pollution. Physical deterioration includes compaction by heavy machinery or trampling by cattle, waterlogging—water accumulation—from excess irrigation and poor drainage, and laterization—solidification of iron and aluminum-rich tropical soil when exposed to sun and rain.

Erosion: The Nature of the Problem

Erosion is an important natural process, resulting in the redistribution of the products of geologic weathering, and is part of both soil formation and soil loss. The world's landscapes have been sculpted by erosion. When the results are spectacular enough, we

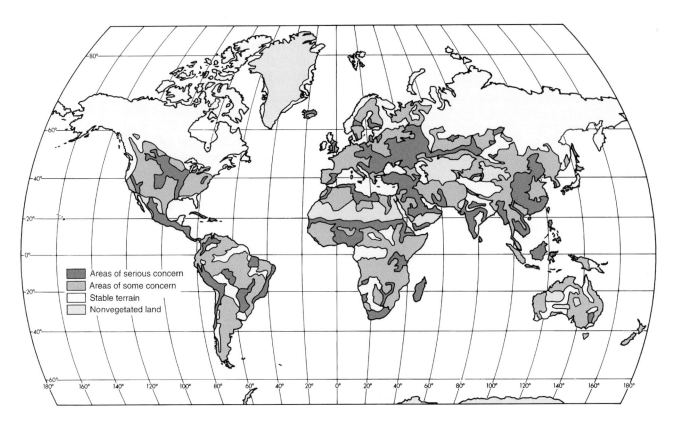

FIGURE 14.11 Areas of concern for soil degradation. In areas of serious concern, widespread moderate or localized severe degradation has already occurred, and resources are lacking or environmental conditions make rehabilitation difficult or impossible. In areas of some concern, current degradation is lighter and potential for rehabilitation is greater. How is your home area classified?

From Jerome Fellmann, et al., *Human Geography,* 4th ed., from data in *World Resources 1992–1993,* International Source Reference and Information Center.

Copyright © 1995 Times Mirror Higher Education Group, Inc., Dubuque, Iowa. All Rights Reserved. Reprinted by permission.

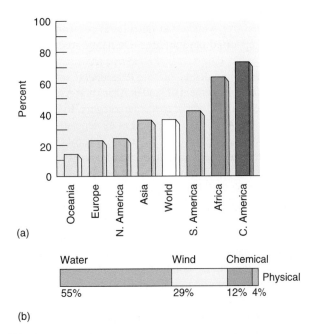

FIGURE 14.12 Degree and causes of soil degradation. (*a*) Percent of agricultural lands with human-induced soil erosion, by region. (*b*) Percent of world land area affected by different types of erosion.

Data from Glasod 2000.

enshrine them in national parks as we did with the Grand Canyon. Where erosion has worn down mountains and spread soil over the plains, or deposited rich alluvial silt in river bottoms, we gladly farm it. Erosion is a disaster only when it occurs in the wrong place at the wrong time.

In some places, erosion occurs so rapidly that anyone can see it happen. Deep gullies are created where water scours away the soil, leaving fenceposts and trees sitting on tall pedestals as the land erodes away around them. In most places, however, erosion is more subtle. It is a creeping disaster that occurs in small increments. A thin layer of topsoil is washed off fields year after year until eventually nothing is left but poor-quality subsoil that requires more and more fertilizer and water to produce any crop at all.

The net effect, worldwide, of this general, widespread top-soil erosion is a reduction in crop production equivalent to removing about 1 percent of world cropland each year. Many farmers are able to compensate for this loss by applying more fertilizer and by bringing new land into cultivation. Continuation of current erosion rates, however, could reduce agricultural production by 25 percent in Central America and Africa and 20 percent in South America by the year 2020. The total annual soil loss from croplands is thought to be 25 billion tonnes. About twice that much soil is lost from rangelands, forests, and urban construction sites each year.

In addition to reduced land fertility, this erosion results in sediment-loading of rivers and lakes, siltation of reservoirs, smothering of wetlands and coral reefs, and clogging of water intakes and waterpower turbines.

Mechanisms of Erosion

Wind and water are the main agents that move soil around (Fig. 14.12). When little rivulets of running water gather together and cut small channels in the soil, the process is called **rill erosion.** When rills enlarge to form bigger channels or ravines that are too large to be removed by normal tillage operations, we call the process **gully erosion** (Fig. 14.13). Streambank erosion refers to the washing away of soil from the banks of established streams, creeks, or rivers, often as a result of removing trees and brush along streambanks and by cattle damage to the banks.

Most soil erosion on agricultural land is rill erosion. Large amounts of soil can be transported a little bit at a time without being very noticeable. A farm field can lose 20 tonnes of soil per hectare during winter and spring runoff in rills so small that they are erased by the first spring cultivation. That represents a loss of only a few millimeters of soil over the whole surface of the field, hardly apparent to any but the most discerning eye. But it doesn't take much mathematical skill to see that if you lose soil twice as fast as it is being replaced, eventually it will run out.

Wind can equal or exceed water in erosive force, especially in a dry climate and on relatively flat land. When plant cover and surface litter are removed from the land by agriculture or grazing, wind lifts loose soil particles and sweeps them away. Windborne dust is sometimes transported from one continent to another. Scientists in Hawaii can tell when spring plowing begins in China because dust from Chinese farmland is carried by winds all the way across the Pacific Ocean. Similarly, summer dust storms in the Sahara Desert of North Africa carry about 1 billion tonnes of soil in massive airborne dust plumes over the Atlantic and Mediterranean every year. This dust creates a hazy atmosphere over islands in the Caribbean Sea, 5,000 km away and has measurable regional climatic effects. Similarly, it has been estimated that winds blowing over the Mississippi River basin have 1,000 times the soil-carrying capacity of the river itself.

Some of the highest erosion rates in the world occur in Canada and the United States. The U.S. Department of Agriculture reports that 69 million hectares of U.S. farmland and range are eroding at rates that reduce long-term productivity. Eleven tonnes per hectare is generally considered the maximum tolerable rate of soil loss because that is generally the highest rate at which soil forms under optimum conditions. Some farms lose soil at twice that rate or more.

Intensive farming practices are largely responsible for this situation. Row crops, such as corn and soybeans, leave soil exposed for much of the growing season (Fig. 14.14). Deep plowing and heavy herbicide applications create weed-free fields that look neat but are subject to erosion. Because big machines cannot easily follow contours, they often go straight up and down the hills, creating ready-made gullies for water to follow. Farmers sometimes plow through grass-lined watercourses and have pulled out windbreaks and fencerows to accommodate the large machines and to get every last square metre into production. Consequently, wind and water carry away the topsoil.

Pressed by economic conditions, many farmers have abandoned traditional crop rotation patterns and the custom of resting land as pasture or fallow every few years. Continuous monoculture cropping can increase soil loss tenfold over other farming patterns. A soil study in Iowa showed that a three-year rotation of corn, wheat, and clover lost an average of only 6 tonnes per hectare. By comparison, continuous wheat production on the same land caused nearly four times as much erosion, and continuous corn cropping resulted in seven times as much soil loss as the rotation with wheat and clover.

Erosion Hotspots

Data on soil condition and soil erosion often are incomplete, but it is evident that many places have problems as severe as, or perhaps worse than ours. China, for example, has a large area of loess (windblown silt) deposits on the North China Plain that once was covered by forest and grassland. The forests were cut down and the grasslands were converted to cropland. This plateau is now scarred by gullies 30 to 40 m deep, and the soil loss is thought to be at least 480 tonnes per hectare per year. This would be equivalent to 3 cm of topsoil per year.

One way to estimate soil loss is to measure the sediment load carried by rivers draining an area. The highest concentration of

FIGURE 14.13 Severe gullying is cutting deep trenches in this pastureland. Lost topsoil reduces productivity here while it causes siltation downstream.

FIGURE 14.14 Annual row crops leave soil bare and exposed to erosion for most of the year, especially when fields are plowed immediately after harvest, as this one always is.

sediment in any river is in the Huang (Yellow) River that originates in the loess plateau of China. Although its drainage basin is only one-fifth as big as that of the Mississippi River, the Huang carries more than four times as much soil each year.

Haiti is another country with severely degraded soil. Once covered with lush tropical forest, the land has been denuded for firewood and cropland. Erosion has been so bad that some experts now say the country has absolutely *no* topsoil left, and poor peasant farmers have difficulty raising any crops at all. Economist Lester Brown of Worldwatch Institute warns that the country may never recover from this ecodisaster.

OTHER AGRICULTURAL RESOURCES

Soil is only part of the agricultural resource picture. Agriculture is also dependent upon water, nutrients, favourable climates to grow crops, productive crop varieties, and upon the mechanical energy to tend and harvest them.

Water

All plants need water to grow. Agriculture accounts for the largest single share of global water use. Some 73 percent of all fresh water withdrawn from rivers, lakes, and groundwater supplies is used for irrigation (Chapter 10). Although estimates vary widely (as do definitions of irrigated land), about 15 percent of all cropland, worldwide, is irrigated.

Some countries are water rich and can readily afford to irrigate farmland, while other countries are water poor and must use water very carefully. The efficiency of irrigation water use is rather low in most countries. High evaporative and seepage losses from unlined and uncovered canals often mean that as much as 80 percent of water withdrawn for irrigation never reaches its intended destination. Farmers often tend to over-irrigate because water prices are relatively low and because they lack the technology to meter water and distribute just the amount needed.

Excessive use not only wastes water; it often results in **waterlogging.** Waterlogged soil is saturated with water, and plant roots die from lack of oxygen. **Salinization,** in which mineral salts accumulate in the soil, occurs particularly when soils in dry climates are irrigated with saline water. As the water evaporates, it leaves behind a salty crust on the soil surface that is lethal to most plants. Flushing with excess water can wash away this salt accumulation but the result is even more saline water for downstream users.

Worldwide, irrigation problems are a major source of land degradation and crop losses. The Worldwatch Institute reports that 60 million ha of cropland have been damaged by salinization and waterlogging. Water conservation techniques can greatly reduce problems arising from excess water use. Conservation also makes more water available for other uses or for expanded crop production where water is in short supply (see Chapter 10).

Fertilizer

In addition to water, sunshine, and carbon dioxide, plants need small amounts of inorganic nutrients for growth. The major elements required by most plants are nitrogen, potassium, phosphorus, calcium, magnesium, and sulfur. Calcium and magnesium often are limited in areas of high rainfall and must be supplied in the form of lime. Lack of nitrogen, potassium, and phosphorus even more often limits plant growth. Adding these elements in fertilizer usually stimulates growth and greatly increases crop yields. A good deal of the doubling in worldwide crop production since 1950 has come from increased inorganic fertilizer use. In 1950, the average amount of fertilizer used was 20 kg per ha. In 1990, this had increased to an average of 91 kg per ha worldwide.

Farmers may overfertilize because they are unaware of the specific nutrient content of their soils or the needs of their crops. While European farmers use more than twice as much fertilizer per hectare as do North American farmers, their yields are not proportionally higher. Phosphates and nitrates from farm fields and cattle feedlots are a major cause of aquatic ecosystem pollution. Nitrate levels in groundwater have risen to dangerous levels in many areas where intensive farming is practised. Young children are especially sensitive to the presence of nitrates. Using nitrate-contaminated water to mix infant formula can be fatal for newborns.

What are some alternative ways to fertilize crops? Manure and green manure (crops grown specifically to add nutrients to the soil) are important natural sources of soil nutrients. Nitrogen-fixing bacteria living symbiotically in root nodules of legumes are valuable for making nitrogen available as a plant nutrient (Chapter 3). Interplanting or rotating beans or some other leguminous crop with such crops as corn and wheat are traditional ways of increasing nitrogen availability.

There is considerable potential for increasing world food supply by increasing fertilizer use in low-production countries if ways can be found to apply fertilizer more effectively and reduce pollution. Africa, for instance, uses an average of only 19 kg of fertilizer per ha or about one-fourth of the world average. It has been estimated that the developing world could at least triple its crop production by raising fertilizer use to the world average. On the other hand, there is an environmental cost to increased fertilizer production, transport, and use, which needs to be factored into any potential advantages of increased fertilizer.

Energy

Farming as it is generally practised in the industrialized countries is highly energy-intensive. Fossil fuels supply almost all of this energy. Between 1920 and 1980, direct energy use on farms rose as gasoline and diesel fuels were consumed by increasing mechanization of agriculture. An even greater increase in indirect energy use, in the form of synthetic fertilizers, pesticides, and other agricultural chemicals, also occurred, especially after World War II. In Canada, although energy use by agriculture is modest compared to other industries, it is substantial and crude oil is mainly used (Fig. 14.15).

After crops leave the farm, additional energy is used in food processing, distribution, storage, and cooking. It has been estimated that the average food item in the North American diet travels 2,000 km between the farm that grew it and the person who consumes it. The energy required for this complex processing and distribution system may be five times as much as is used directly in farming. Altogether, agriculture in Canada consumes about 3 percent of the total energy we use (about a tenth of other industrial use). This results in a similar proportion of greenhouse gas emissions by agriculture relative to other sources. Most of our foods require more energy to produce, process, and get to market than they yield when we eat them.

Clearly, unless we find some new sources of energy, our present system is unsustainable. As fossil fuels become more scarce, we may need to adopt farming methods that are self-supporting. Is it possible that we may need to go back to using draft animals that can eat crops grown on the farm? Could we reintroduce natural methods of pest control, fertilization, crop drying, and irrigation? Or can we develop alternative energy sources to run our farming enterprise?

NEW CROPS AND GENETIC ENGINEERING

Although at least 3,000 species of plants have been used for food at one time or another, most of the world's food now comes from only 16 widely grown crops. Many new or unconventional varieties might be valuable human food supplies, however, especially in areas where conventional crops are limited by climate, soil, pests, or other problems. Among the plants now being investigated as potential additions to our crop roster are winged beans (Fig. 14.16), a perennial plant that grows well in hot climates where other beans will not grow. It is totally edible (pods, mature seeds, shoots, flowers, leaves, and tuberous roots), resistant to diseases, and enriches the soil. Another promising crop is tricale, a hybrid between wheat (*Triticum*) and rye (*Secale*) that grows in light, sandy, infertile soil. It is drought resistant, has nutritious seeds, and is being tested for salt tolerance for growth in saline soils or irrigation with seawater. Some traditional crop varieties grown by Native Americans, such as tepary beans, amaranth, and Sonoran panicgrass are being collected by seed conservator Gary Nabhan both as a form of cultural revival for native people and as a possible food crop for harsh environments.

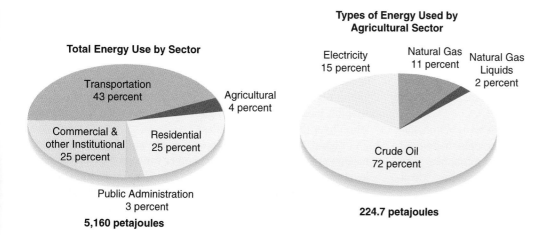

FIGURE 14.15 Energy use by Canadian agriculture in 1998.
Source: Statistics Canada, CANSIM, Matrix 7977.

FIGURE 14.16 Winged beans bear fruit year-round in tropical climates and are resistant to many diseases that prohibit growing other bean species. Whole pods can be eaten when they are green, or dried beans can be stored for later use. It is a good protein source in a vegetarian diet.

Green Revolution

So far, however, the major improvements in farm production have come from technological advances and modification of a few well-known species. Yield increases often have been spectacular. A century ago, when all maize (corn) in the United States was open pollinated, average yields were about 8 hl per hectare. In 1999, average yields from hybrid maize were around 45 hl per hectare, and under optimum conditions, 88 hl per hectare are possible. Most of this gain was accomplished by conventional plant breeding: geneticists labouriously hand-pollinating plants, moving selected genes from one variety to another.

Starting more than 50 years ago, agricultural research stations began to breed tropical wheat and rice varieties that would provide food for growing populations in developing countries. The first of the "miracle" varieties was a dwarf, high-yielding wheat (Fig. 14.17) developed by Norman Borlaug (who received a Nobel Peace Prize for his work) at a research centre in Mexico. At about the same time, the International Rice Institute in the Philippines developed dwarf rice strains with three or four times the production of varieties in use at the time. The dramatic increases obtained

FIGURE 14.17 Short-stemmed "semidwarf" wheat (*centre*) developed by Nobel prize winner, Norman Borlaug, is a high responder; that is, it can utilize high levels of fertilizer and water to produce high yields without growing so tall that the stalks fall over and are impossible to harvest. Notice how much denser the seed heads are on the semidwarf than on its normal size relative (*left*).

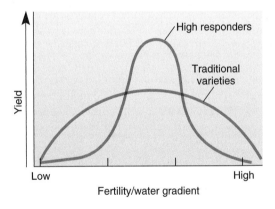

FIGURE 14.18 Green revolution miracle crops are really high responders, meaning that they have excellent yields under optimum conditions. For poor farmers who can't afford the fertilizer and water needed by high responders, traditional varieties may produce better yields.

as these new varieties spread around the world has been called the **green revolution.** It is one of the main reasons that world food supplies have more than kept pace with the growing human population over the past few decades.

Most green revolution breeds really are "high responders," meaning that they yield more than other varieties if given optimum levels of fertilizer, water, and protection from pests and diseases. Under suboptimum conditions, on the other hand, high responders may not produce as well as traditional varieties (Fig. 14.18). Poor farmers who can't afford the expensive seed, fertilizer, water, etc., which are required to become part of this movement, usually are left out of the green revolution. In fact they may be driven out of farming altogether as rising land values and falling commodity prices squeeze them from both sides.

Genetic Engineering

The most recent revolution in agriculture is the development of **genetically modified organisms (GMOs).** Sometimes called transgenic varieties, or genetically engineered crops, GMOs have DNA containing genes borrowed from entirely unrelated species (Fig. 14.19). "Golden rice," for example, contains a gene from daffodils that makes the rice produce beta carotene, a critical nutrient in many poor countries. **Genetic engineering** is even creating new animals: in 2000 a baby monkey was born containing cells of a jellyfish. The procedure of inserting bits of one DNA string into another species' DNA was only developed in the 1980s, and the nature of DNA itself was first described as recently as 1953, but GM crops have rapidly pervaded commercial agriculture, at least in Canada and the United States. In 2001, the Royal Society of Canada suggested how to regulate food biotechnology in Canada. They describe the important role of GM crops such as herbicide resistant soybeans, corn, and canola in the present Canadian agricultural context. They suggest important constraints needed to capitalize on the value of GM foods without unacceptable environmental risk.

Is this revolution a good thing? Genetic engineering can produce pest-resistant crops that, ideally, would require less insecticide or herbicide use on fields. Foods with superior nutrient content, such as golden rice, could be produced for malnourished populations, and genetic material from viruses could be inserted into potatoes or bananas, producing edible vaccines for children around the world. Crops could also be developed that would tolerate salty or waterlogged soils, improving opportunities for farming on degraded lands. On the other hand, most GM crops to date are designed to be resistant to herbicides—allowing greater use of chemicals without harming the plants themselves—or resistant to insect pests. Opponents of GM crops fear that these traits could transfer to wild plants, creating "super weeds." As in the green revolution, genetically engineered crops also are relatively expensive, and critics charge that these varieties greatly enrich the multinational corporations that develop them, while increasing farmers' dependency on the sellers of GM seeds.

Whether you are aware of it or not, you probably consume GM foods. About 70 percent of all processed foods in North American contain transgenic products, including corn (and corn sweeteners), tomatoes, oils, and potatoes.

Pest Resistance

Biotechnologists recently have created plants with genes for endogenous insecticides. *Bacillus thuringiensis* (Bt), a bacterium, makes toxins lethal to Lepidoptera (butterfly family) and Coleoptera (beetle family). The genes for some of these toxins have been transferred into crops such as maize (to protect against European cut worms), potatoes (to fight potato beetles), and cotton (for protection against boll weevils). This allows farmers to reduce insecticide spraying. In 1998, some 8 million hectares—or about 29 percent of all transgenic crops in the world—were planted to Bt-containing varieties, mostly in North America. Most European nations have not approved Bt-containing varieties. Fear of transgenic products has been a sticking point in agricultural trade between the United States and the European Union.

Entomologists worry that Bt plants churn out toxin throughout the growing season, regardless of the level of infestation, creating perfect conditions for selection of Bt resistance in pests. The

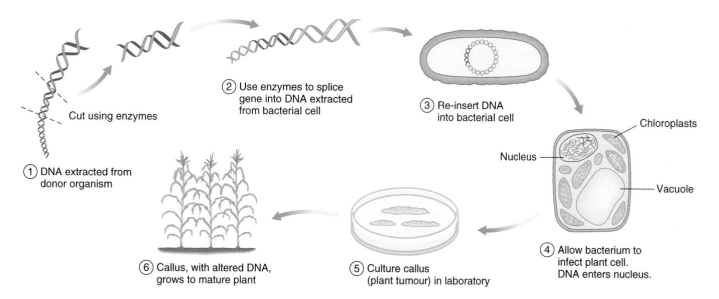

FIGURE 14.19 One method of gene transfer, using an infectious, tumor-forming bacterium such as agrobacterium. Genes with desired characteristics are cut out of donor DNA and spliced into bacterial DNA using special enzymes. The bacteria then infect plant cells and carry altered DNA into cells' nuclei. The cells multiply, forming a tumour, or callus, which can grow into a mature plant.

effectiveness of this natural pesticide—one of the few available to organic growers—is likely to be destroyed within a few years. One solution is to plant at least a part of every field in non-Bt crops that will act as a refuge for nonresistant pests. The hope is that interbreeding between these "wild-type" bugs and those exposed to Bt will dilute out recessive resistance genes. Deliberately harbouring pests and letting them munch freely on crops is something that many farmers find hard to do. In addition, devoting a significant part of their land to nonproductive crops lowers the total yield and counteracts the profitability of engineered seed.

There also is a concern about the effects on nontarget species. In laboratory tests, about half of a group of monarch butterfly caterpillars died after being fed on plants dusted with pollen from Bt corn. Under field conditions, critics point out, corn pollen can drift long distances and might contaminate the milkweeds on which monarchs normally feed. Furthermore, entomologist Karen Oberhauser notes that crop fields occupy so much of the monarchs' range that butterflies inevitably feed and reproduce within the Bt corn fields themselves. Geneticists consider this a trivial problem. They'll just create varieties in which the Bt gene is inactive in pollen, they claim. Still, others wonder if this whole enterprise is a good idea.

Weed Control

The most popular transgenic crops are not insect-resistant but those engineered to tolerate herbicides. Currently these crops occupy 20 million hectares worldwide, or about three-quarters of all genetically engineered acreage. The two main products in this category are Monsanto's "Roundup Ready" crops—so-called because they can withstand treatment with Monsanto's best-selling herbicide, Roundup (glyphosate)—and AgrEvo's "Liberty Link" crops, which resist that company's Liberty (glufosinate) herbicide. Because crops with these genes can grow in spite of high herbicide doses, farmers can spray fields heavily to exterminate weeds. This allows for conservation tillage and leaving more crop residue on fields to protect topsoil from erosion—both good ideas—but it also means using much more herbicide in higher doses than might otherwise be done.

Widespread use of these crops could create herbicide-resistant "super weeds" if genes jump from domestic species to wild relatives. This isn't much of a risk in North America where wild relatives for common crops are rare. It could be a serious problem, however, in the high biodiversity regions where our domestic crops originated, and where closely related, wild species remain common. The chances for "escaped" genes is, in fact, pretty good. In 1997, just one year after Roundup Ready canola was first planted in Canada, wild mustard plants (a canola relative) were found to have interbred and obtained the herbicide-resistant gene.

Public Opposition

Although agribusiness has been quick to adopt GM crops, the public is often more skeptical. Many people fear that GM foods are unsafe for consumers and the environment. Others object that these new, often expensive varieties tend to create unfair competition, making rich farmers richer and bankrupting poor or traditional producers. In 1999, protesters in India burned fields suspected of containing genetically engineered plants, and India's Supreme Court upheld a ban on testing transgenic crops (What Do You Think? p. 309). In 2000, Italy banned all field trials of GM crops. Objections have been especially strong in Europe, where genetically modified foods and seeds were unoffically banned from 1998 until 2001. In that year, the European parliament passed rules requiring strict testing, monitoring, and labeling of genetically engineered food products and seeds. The European parliament also banned marker genes for anti biotic resistance in plants—it is feared that antibiotic resistant crops could help spread antibiotic resistance in bacteria.

Risks to human health are not unrealistic. In 2000 it was discovered that StarLink corn, a Bt variety authorized only for livestock consumption, had been mixed into corn used in a variety of human foods, including millions of taco shells that were ultimately recalled by Kraft Foods. The danger of StarLink corn itself may be small: other Bt corn varieties have been approved for human consumption, and StarLink may ultimately be approved as well. On the other hand, this case worries producers and consumers alike because no one knows how the StarLink corn got mixed with food-grade corn. The StarLink debacle demonstrates how easy it could be for unsafe products to permeate the food industry. The main concern on the minds of GM opponents is whether other unapproved varieties, containing known allergens, could be mixed into the food stream without our knowledge.

SUSTAINABLE AGRICULTURE

How, then, shall we feed the world? Can we make agriculture compatible with sustainable ecological and social systems? Having discussed some of the problems that beset modern agriculture, we will now consider some suggested ways to overcome problems and make farming and food production just and lasting enterprises. This goal is usually termed **sustainable agriculture,** or **regenerative farming,** both of which aim to produce food and fiber on a sustainable basis and repair the damage caused by destructive practices. Some alternative methods are developed through scientific research; others are discovered in traditional cultures and practices nearly forgotten in our mechanization and industrialization of agriculture.

Soil Conservation

With careful husbandry, soil is a renewable resource that can be replenished and renewed indefinitely. Since agriculture is the area in which soil is most essential and also most often lost through erosion, agriculture offers the greatest potential for soil conservation and rebuilding. Some rice paddies in Southeast Asia, for instance, have been farmed continuously for a thousand years without any apparent loss of fertility. The rice-growing cultures that depend on

Terminator Genes

Research in biotechnology and genetic engineering is very expensive. Monsanto is reported to have spent $500 million developing Roundup Ready genes, or about as much as the entire annual USDA research budget. Naturally, they want to protect potential profits from this valuable property. Farmers who buy Monsanto seeds are required to sign a contract that stipulates what kinds of pesticides can be used on fields as well as an agreement not to save seed or allow patented crops to cross with other varieties. Seed sleuths investigate to ensure that contracts are fulfilled. By inserting unique hidden sequences in their synthetic genes, forensic molecular biologists can detect the presence of patented genetic material in fields for which royalties weren't paid. Already Monsanto has taken legal action against more than 300 farmers for replanting proprietary seeds. Farmers claim they can't prevent transgenic pollen from blowing onto their fields and introducing genes against their will. A whole new set of legal precedents is likely to be established by these suits.

A new weapon has recently been introduced in this struggle that many people regard as quite sinister. Using genetic research of a USDA scientist, a small company called Delta and Pine Land developed genetic material officially entitled "gene protection technology" but commonly known as "terminator" genes. The terminator complex includes a toxic gene from a noncrop plant stitched together with two other bits of coding that keep the killer gene dormant until late in the crop's development, when the toxin affects only the forming seeds. Thus, although the crop yield is about normal, there is no subsequent generation and no worry about farmers saving and replanting. They have to buy new seed every year. Delta was quickly purchased by Monsanto for $1 billion, or hundreds of times the small company's book value. This may have been the only time a whole company was purchased just to get a gene complex.

Engineered sterility is not uncommon; it is widely used in producing hybrid crops such as maize. What is unusual about this gene-set is that it can be moved easily from one species to another, and it can be packaged in every seed sold by the parent company. It's also unique to deliberately introduce a toxin into the part that people eat. So what's wrong with a company trying to protect its research investment? For one thing, there's a worry that the toxins might be harmful to consumers, even though toxicity tests so far show no danger. Furthermore these genes may escape. What if some of our major crops become self-sterile and can no longer reproduce? A more immediate concern is the economic effects in developing countries. While seed saving is not common on farms in most developed countries, it is customary and economically necessary in many poorer parts of the world. Melvin Oliver, the principal inventor of the terminator genes, admits that "the technology primarily targets Second and Third World markets"—in effect, guaranteeing intellectual property rights even in countries where patent protection is weak or nonexistent.

Large corporations like Monsanto argue that without patent protection, they can't afford to do the research needed to provide further advances in biotechnology. Critics charge that these companies make enough profit in developed countries to pay back their costs. Targeting less-developed countries and introducing something as potentially dangerous as the terminator gene, they claim is immoral. Some argue that the Terminator Gene would actually help ensure that GM crops would not escape from cultivation into the wild. International protests caused Monsanto to announce in 1999 that it was suspending plans to release crops with terminator genes "for the time being." Still, biotechnology research continues at a furious pace and other genetically-modified organisms are sure to be available soon. What do you think? Are those who protest this technology simply afraid of things that are new and unfamiliar, or are there legitimate reasons for concern? How can we assess risks in novel and unknown technologies such as these?

---

these fields have developed management practices that return organic material to the paddy and carefully nurture the soil's ability to sustain life.

While North American agriculture hasn't reached that level of sustainability, there is evidence that soil conservation programs are having a positive effect. In one Wisconsin study, erosion rates in one small watershed were 90 percent less in 1975–1993 than they were in the 1930s. Among the most important elements in soil conservation are land management, ground cover, climate, soil type, and tillage system.

Managing Topography

Water runs downhill. The faster it runs, the more soil it carries off the fields. Comparisons of erosion rates in Africa have shown that a 5 percent slope in a plowed field has three times the water runoff volume and eight times the soil erosion rate of a comparable field with a 1 percent slope. Water runoff can be reduced by leaving grass strips in waterways and by **contour plowing,** that is, plowing across the hill rather than up and down. Contour plowing is often combined with **strip-farming,** the planting of different kinds of crops in alternating strips along the land contours (Fig. 14.20). When one crop is harvested, the other is still present to protect the soil and keep water from running straight downhill. The ridges created by cultivation make little dams that trap water and allow it to seep into the soil rather than running off. In areas where rainfall is very heavy, tied ridges are often useful. This method involves a series of ridges running at right angles to each other, so that water runoff is blocked in all directions and is encouraged to soak into the soil.

Terracing involves shaping the land to create level shelves of earth to hold water and soil (Fig. 14.21). The edges of the terrace are planted with soil-anchoring plant species. This is an expensive procedure, requiring either much hand labour or expensive machinery, but makes it possible to farm very steep hillsides. The rice terraces in the Chico River Valley in the Philippines rise as much as 300 m above the valley floor. They are considered one of the wonders of the world.

Planting **perennial species** (plants that grow for more than two years) is the only suitable use for some lands and some soil

FIGURE 14.20 Contour plowing and strip-cropping on these farms protect the soil from erosion and help maintain fertility, as well as providing a beautiful landscape. With care and good stewardship, we can increase the carrying capacity of the land and create a sustainable environment.

FIGURE 14.21 Rice terraces on Java, Indonesia. Some rice paddies have been cultivated for hundreds or even thousands of years without any apparent loss of productivity.

types. Establishing forest, grassland, or crops such as tea, coffee, or other crops that do not have to be cultivated every year may be necessary to protect certain unstable soils on sloping sites or watercourses (low areas where water runs off after a rain).

Providing Ground Cover

Annual row crops such as corn or beans generally cause the highest erosion rates because they leave soil bare for much of the year (Table 14.4). Often, the easiest way to provide cover that protects soil from erosion is to leave crop residues on the land after harvest.

TABLE 14.4	Soil Cover and Soil Erosion	
CROPPING SYSTEM	AVERAGE ANNUAL SOIL LOSS (TONNES/HECTARE)	PERCENT RAINFALL RUNOFF
Bare soil (no crop)	41.0	30
Continuous corn	19.7	29
Continuous wheat	10.1	23
Rotation: corn, wheat, clover	2.7	14
Continuous bluegrass	0.3	12

Source: Based on 14 years of data from Missouri Experiment Station, Columbia, MO.

They not only cover the surface to break the erosive effects of wind and water, but they also reduce evaporation and soil temperature in hot climates and protect ground organisms that help aerate and rebuild soil. In some experiments, 1 tonne of crop residue per 0.4 ha increased water infiltration 99 percent, reduced runoff 99 percent, and reduced erosion 98 percent. Leaving crop residues on the field also can increase disease and pest problems, however, and may require increased use of pesticides and herbicides.

Where crop residues are not adequate to protect the soil or are inappropriate for subsequent crops or farming methods, such **cover crops** as rye, alfalfa, or clover can be planted immediately after harvest to hold and protect the soil. These cover crops can be plowed under at planting time to provide green manure. Another method is to flatten cover crops with a roller and drill seeds through the residue to provide a continuous protective cover during early stages of crop growth.

In some cases, interplanting of two different crops in the same field not only protects the soil but also is more efficient use of the land, providing double harvests. Native Americans and pioneer farmers, for instance, planted beans or pumpkins between the corn rows. The beans provided nitrogen needed by the corn, pumpkins crowded out weeds, and both crops provided foods that nutritionally balance corn. Traditional swidden (slash-and-burn) cultivators in Africa and South America often plant as many as 20 different crops together in small plots. The crops mature at different times so that there is always something to eat, and the soil is never exposed to erosion for very long.

Mulch is a general term for a protective ground cover that can include manure, wood chips, straw, seaweed, leaves, and other natural products. For some high-value crops, such as tomatoes, pineapples, and cucumbers, it is cost-effective to cover the ground with heavy paper or plastic sheets to protect the soil, save water, and prevent weed growth. Israel uses millions of square metres of plastic mulch to grow crops in the Negev Desert.

Reduced Tillage

Farmers have traditionally used a moldboard plow to till the soil, digging a deep trench and turning the topsoil upside down. In the

PART THREE Humans in the Environment

1800s, it was shown that tilling a field fully—until it was "clean"—increased crop production. It helped control weeds and pests, reducing competition; it brought fresh nutrients to the surface, providing a good seedbed; and it improved surface drainage and aerated the soil. This is still true for many crops and many soil types, but it is not always the best way to grow crops. We are finding that less plowing and cultivation often makes for better water management, preserves soil, saves energy, and increases crop yields.

There are several major **reduced tillage systems.** *Minimum till* involves reducing the number of times a farmer disturbs the soil by plowing, cultivating, etc. This often involves a disc or chisel plow rather than a traditional moldboard plow. A chisel plow is a curved chisel-like blade that doesn't turn the soil over but creates ridges on which seeds can be planted. It leaves up to 75 percent of plant debris on the surface between the rows, preventing erosion (Fig. 14.22). *Conserve-till* farming uses a coulter, a sharp disc like a pizza cutter, which slices through the soil, opening up a furrow or slot just wide enough to insert seeds. This disturbs the soil very little and leaves almost all plant debris on the surface. *No-till* planting is accomplished by drilling seeds into the ground directly through mulch and ground cover. This allows a cover crop to be interseeded with a subsequent crop.

Farmers who use these conservation tillage techniques often must depend on pesticides (insecticides, fungicides, and herbicides) to control insects and weeds. Increased use of toxic agricultural chemicals is a matter of great concern. Massive use of pesticides is not, however, a necessary corollary of soil conservation. It is possible to combat pests and diseases with integrated pest management that combines crop rotation, trap crops, natural repellents, and biological controls (Chapter 15).

Low-Input Sustainable Agriculture

Throughout Canada, the United States, and other crop-exporting countries such as Brazil and New Zealand, huge, highly mechanized farms are becoming increasingly common. Giant grain fields stretch over thousands of hectares, while large numbers of cows, hogs, or chickens live their entire life in enormous, warehouse-size confinement barns. Often owned by multinational agricultural conglomerates, these corporate farms are operated by scientific standards that might make your local hospital seem backward. Milk cows, for instance, are tracked by computer, each one automatically given an individual diet, antibiotics, and hormone injections according to their pedigree, growth rate, or milk production. Farm managers travel by helicopter to survey vast spreads. One "farm" in California covers 2,000 ha and has an annual cash-flow of around $50 million.

In contrast to this trend toward industrialized agriculture, other farmers are going back to a size and style that their grandparents might have used more than a century ago. Finding that they can't—or don't want to—compete in capital-intensive production, these farmers are making money and staying in farming by returning to small-scale, low-input farming. Rather than milking 1,000 cows in a highly mechanized operation, some farmers now raise only a couple of dozen cows and let them graze freely in pastures. Animals live outdoors in the winter (Fig. 14.23). Milk production

FIGURE 14.22 In ridge-tilling, a chisel plow is used to create ridges on which crops are planted and shallow troughs filled with crop residue. Less energy is used in plowing and cultivation, weeds are suppressed and moisture is retained by the ground cover, or crop residue, left on the field.

FIGURE 14.23 On the Minar family's 230-acre dairy farm, cows and calves spend the winter outdoors in the snow, bedding down on hay. Dave Minar is part of a growing counterculture that is seeking to keep farmers on the land and bring prosperity to rural areas.

The biggest experiment in low-input, sustainable agriculture in world history is occurring now in Cuba. The sudden collapse of the socialist bloc, upon which Cuba had been highly dependent for trade and aid, has forced an abrupt and difficult conversion from conventional agriculture to organic farming on a nationwide scale. Methods developed in Cuba could help other countries find ways to break their dependence on synthetic pesticides and fossil fuels.

Between the Cuban revolution in 1959 and the breakdown of trading relations with the Soviet Union in 1989, Cuba experienced rapid modernization, a high degree of social equity and welfare, and a strong dependence on external aid. Cuba's economy was supported during this period by the most modern agricultural system in Latin America. Farming techniques, levels of mechanization, and output often rivalled those in the United States. The main crop was sugarcane, almost all of which was grown on huge state farms and sold to the former Soviet Union at premium prices. More than half of all food eaten by Cubans came from abroad, as did most fertilizers, pesticides, fuel, and other farm inputs on which agricultural production depended.

Under the theory of comparative advantage, it seemed reasonable for Cuba to rely on international trade. With the collapse of the socialist bloc, however, Cuba's economy also fell apart. In 1990, wheat and grain imports decreased by half and other foodstuffs declined even more. At the same time, fertilizer, pesticide, and petroleum imports were down 60 to 80 percent. Farmers faced a dual challenge: how to produce twice as much food using half the normal inputs.

The crisis prompted a sudden turn to a new model of agriculture. Cuba was forced to adopt sustainable, organic farming practices based on indigenous, renewable resources. Typically, it takes three to five years for a farmer in Canada to make the change from conventional to organic farming profitable. Cuba, however, didn't have that long; it needed food immediately.

Cuba's agricultural system is based on a combination of old and new ideas. Broad community participation and use of local knowledge is essential. Scientific, adaptive management is another key. Diverse crops suitable to local microclimates, soil types, and human nutritional needs have been adopted. Natural, renewable energy sources such as wind, solar, and biomass fuels are being substituted for fossil fuels. Oxen and mules have replaced some 500,000 tractors idled by lack of fuel.

Soil management is vital for sustainable agriculture. Organic fertilizers substitute for synthetic chemicals. Livestock manure, green manure crops, composted municipal garbage, and industrial-scale cultivation of high-quality humus in earthworm farms all replenish soil fertility. In 1995 more than 100,000 tonnes of worm compost were produced and spread on fields.

Pests are suppressed by crop rotation and biological controls rather than chemical pesticides. For example, the parasitic fly controls sugarcane borers; wasps feed on the eggs of grain weevils; while the predatory ant attacks sweet potato weevils. Pest control also involves innovative use of biopesticides, such as *Bacillis thuringiensis,* that are poisonous or repellent to crop pests. Finally, integrated pest management includes careful monitoring of crops and measures to build populations of native beneficial organisms and to enhance the vigour and defences of crop species.

Worker brigades from schools and factories help provide farm labour during harvest season. In addition to state farms and rural communes, urban gardening provides a much-needed supplement to city diets. Individual gardens are encouraged, but community or institutional gardens—schools, factories, and mass organizations—also produce large amounts of food.

Although food supplies in Cuba still are limited and diets are austere, the crisis wasn't as bad as many feared. In some ways, this draconian transition is fortunate. Cuba is now on a sustainable path and is a world leader in sustainable agriculture. It could serve as a model for others who surely will face a similar transition when their supplies of fossil fuels run out.

Large-scale community gardens provide much of the fresh produce consumed by residents of a Havana housing complex.

follows natural yearly cycles rather than being controlled by synthetic hormones. The yield is lower, but so are costs.

Where 20,000 chickens living in a single immense shed—or 1,000 hogs raised in an industrial warehouse—must constantly be dosed with antibiotics to keep them healthy, animals allowed free range tend to have fewer diseases and require less care. Vast amounts of manure generated by industrial-scale factory farming are stored in football-field-sized lagoons and spread on fields. Odours from ammonia, hydrogen sulfide (rotten egg odour), and other air pollutants make life miserable for neighbours. In 1999, heavy rains from Hurricane Floyd caused hundreds of North Carolina hog manure lagoons to overflow. Around 10 billion litres of manure washed down rivers to the sea, creating a large dead zone in Palmico Sound (see Chapter 11).

One way to support local agriculture is to shop at a farmer's market. The produce is fresh, and profits go directly to the person who grows the crop. A local food co-op or owner-operated grocery store also is likely to buy from local farmers and to feature pesticide-free foods. Many co-ops and buyers' associations sign contracts directly with producers to grow the types of food they want to eat. This benefits both parties. Producers are guaranteed a local market for organic food or specialty items. Consumers can be assured of quality and can even be involved in production.

Many large supermarkets also now carry organic produce and other foods. If yours doesn't, why not ask the manager to look into it? Organic foods are becoming increasingly accepted—and profitable—as their benefits become more widely understood.

SUMMARY

Over the past 35 years, the total amount of food in the world has increased faster than the average rate of population growth, so there is now more food per person than there was in the 1970s, even though the total number of people has doubled.

There now is enough food to supply everyone in the world with more than the minimum daily food requirements, but food is inequitably distributed. The FAO estimates that 800 million people are chronically undernourished or malnourished, and at least 11 million children die each year from diseases related to malnutrition. Additional millions survive on a deficient diet, suffering from resulting stunted growth, mental retardation, and developmental disorders.

Among the essential dietary ingredients for good health are adequate calories, proteins, vitamins, and minerals. Marasmus and kwashiorkor are protein-deficiency diseases; anemia and goiter are caused by mineral deficiencies that affect millions of people worldwide.

The three major crops that are the main source of energy and nutrients for most of the world's people are rice, wheat, and maize. Some new crops or unrecognized traditional crops hold promise for increasing the nutritional status of the poorer people of the world. Scientific improvement of existing crops and modernization of agriculture (irrigation, fertilizer, and better management) are potential sources of greater agricultural production, but also raise some troubling questions about safety, equity, and the environment.

Fertile, tillable soil for growing crops is an indispensable resource for our continued existence on earth. Soil is a complex system of inorganic minerals, air, water, dead organic matter, and a myriad of different kinds of living organisms. There are hundreds of thousands of different kinds of soils, each produced by a unique history, climate, topography, bedrock, transported material, and community of living organisms.

It is estimated that 25 billion tonnes of soil are lost from croplands each year because of wind and water erosion. Perhaps twice as much is lost from rangelands and permanent pastures. This erosion causes pollution and siltation of rivers, reservoirs, estuaries, wetlands, and offshore reefs and banks. The net effect of this loss is worldwide crop reduction equivalent to losing 15 million ha, or 1 percent of the world's cropland each year.

The United States has some of the highest rates of soil erosion in the world. Soil erosion exceeds soil formation on at least 40 percent of U.S. cropland. About one-half of the topsoil that existed in North America before European settlement has been lost.

It is possible that food production could be expanded considerably, even on existing farmland, given the proper inputs of fertilizer, water, high-yield crops, and technology. This will be essential if human populations continue to grow as they have during this century. Whether it will be possible to supply agricultural inputs and expand crop production remains to be seen. Global climate change could have devastating effects on world food supplies and might necessitate conversion of forests and grasslands to feed the world's population.

Many new and alternative methods could be used in farming to reduce soil erosion, avoid dangerous chemicals, improve yields, and make agriculture just and sustainable. Returning to low-input, regenerative, "organic" farming may be more sustainable and more healthful than our current practices. Growing your own food or buying locally grown food at co-ops, farmers' markets, or through a producers' or buyers' association can provide healthy, wholesome food and also support sustainable agriculture.

QUESTIONS FOR REVIEW

1. How many people in the world are chronically undernourished and how many die each year from starvation and nutritionally related diseases?

2. How many calories does the average person need per day to maintain a healthy, active life?

3. What are proteins, vitamins, and minerals, and why do we need them in our diet?

4. What do we mean by green revolutions?

5. What is the composition of soil? What is humus? Why are soil organisms so important?

6. What are four kinds of erosion? Why is it a problem?

7. What are some possible effects of over-irrigation?

8. What is sustainable agriculture?

9. What is genetic engineering, or biotechnology, and how might they help or hurt agriculture and the environment?

QUESTIONS FOR CRITICAL THINKING

1. What worldviews might make people believe that there are already too many people to be fed or that technological progress may allow us to feed double or triple current populations? Which side of this argument do you support?

2. Suppose that a seafood company wants to start a fish farming operation in a lake near your home. What regulations or safeguards would you want to see imposed on its operation? How would you weigh the possible costs and benefits of this operation?

3. Debate the claim that famines are caused more by human actions (or inactions) than by environmental forces. What is the critical element or evidence in this debate?

4. Should farmers be forced to use environmentally sound techniques that serve farmers' best interests in the long run, regardless of short-term consequences? How could we mitigate hardships brought about by such policies?

5. What problems have accompanied the benefits of the green revolution?

6. Should we encourage (and subsidize) the family farm? What are the advantages and disadvantages (economic and ecological) of the small farm and the corporate farm?

7. Should we try to increase food production on existing farmland, or should we sacrifice other lands to increase farming areas?

8. Some rice paddies in Southeast Asia have been cultivated continuously for a thousand years or more without losing fertility. Could we, and should we, adapt these techniques to our own country?

9. Do you think that agribusinesses should be allowed to insert lethal "terminator" genes in their crop varieties? Why or why not?

KEY TERMS

contour plowing   309
cover crops   310
famines   295
food security   293
genetically modified
   organisms (GMOs)   307
genetic engineering   307
green revolution   306
gully erosion   303
humus   298
mulch   310
overnutrition   293
perennial species   309

reduced tillage systems   311
regenerative farming   308
rill erosion   303
salinization   304
soil   297
soil horizons   299
soil profile   299
strip-farming   309
subsoil   299
sustainable agriculture   308
terracing   309
topsoil   299
undernourished   293
waterlogging   304

Web Exercises

LOOKING AT SOILS

Look at Agriculture and Agri-Food Canada's Canadian Soil Information website at http://sis.agr.gc.ca/cansis/intro.html. Can you find the soil order dominant in your area? What do you think are the main reasons you have the soil you do? Climate? Geological history? Find out all you can about the dominant soil in your area. How does it determine the extent and nature of agriculture where you live? Now choose an area of Canada far removed from where you live and learn about its soils. How much do you think the difference in soils of that area from yours determines how agriculture works differently in the two places?

AGRICULTURAL PRODUCTION

Go to the U.S. National Atlas web page, http://nationalatlas.gov/natlas/natlasstart.asp. (This page displays best if you have a large or high-resolution monitor; if you don't use the scroll bars to move left and right on different parts of the page.) On the right side of the page, select several farm products that are produced in your area. (Click on the Redraw button below the map to display your choices.) How widely distributed are those crops in the United States? Can you explain the distribution of the crops you see?

Chapter 15

Pest Control

It ain't the things we know that cause all the trouble; it's the things we think we know that ain't so.

—Will Rogers—

OBJECTIVES

After studying this chapter, you should be able to:

- define the major types of pest control and describe the pests they are meant to control.
- outline the history of pest control, including the changes in pesticides that occurred in the last half of the twentieth century.
- appreciate the benefits of pest control.
- relate some of the problems of pesticide use.

WebQuest

DDT, biological control, integrated pest management, pesticide risk reduction, persistent organic pollutants (POPs)

Above: A helicopter sprays pesticides on a sugar beet field in California.

DDT and Fragile Eggshells

During the 1960s, peregrine falcons (Fig. 15.1), bald eagles, osprey, brown pelicans, shrikes, and several other predatory bird species suddenly disappeared from former territories in eastern North America. What caused this sudden decline? Studies revealed that eggs laid by these birds had thin, fragile shells that broke before hatching. Eventually, these reproductive failures were traced to residues of DDT and its degradation product, DDE, which had concentrated through food webs until reaching toxic concentrations in top trophic levels such as these bird species.

DDT (dichloro-diphenyl-trichloroethane), an inexpensive and highly effective insecticide, had been used widely to control mosquitoes, biting flies, codling moths, potato beetles, corn earworms, cotton bollworms, and a host of other costly and irritating pest species around the world. First produced commercially in 1943, more than 22 million kilograms of DDT were sprayed on fields, forests, and cities in the United States by 1950.

But as we have since discovered, there are disadvantages to widespread release of these toxic compounds in the environment. In the case of falcons, eagles, and other top predators, DDT and DDE inhibit enzymes essential for deposition of calcium carbonate in eggshells, resulting in soft, easily broken eggs. As we will discuss later in this chapter, other compounds chemically related to DDT are now thought to be disrupting endocrine hormone functions and causing reproductive losses in many species other than birds (What Do You Think? p. 325).

These effects, coupled with the discovery of a pervasive presence of various chlorinated hydrocarbons in human tissues worldwide, led to the banning of DDT in most industrialized countries in the early 1970s. Peregrine falcons, which had declined to only about 120 birds in the United States (outside of Alaska) in the mid-1970s, now number about 1,400, most of them bred in captivity and then released into the wild. Bald eagle reproduction has increased from an average of 0.46 young per nest in 1974 to 1.21

FIGURE 15.1 Peregrine falcons disappeared from the eastern United States in the 1960s as a result of excess pesticide use.

young per nest in 1994 in eastern Canada. Peregrine falcons and bald eagles had recovered enough to be removed from the endangered species list in the eastern United States in 1994.

What have we learned from this experience? Chemical pesticides offer a quick, convenient, and relatively inexpensive way to eliminate annoying or destructive organisms. At the same time, however, excessive pesticide use can kill beneficial organisms and upset the natural balance between predator and prey species. Modern pesticides undoubtedly have saved millions of human lives by killing disease-causing insects and by increasing food supplies. But we must understand what these powerful chemicals are doing and use them judiciously. In this chapter, we will study the major types of pests and pesticides along with some of the benefits and problems involved in our battle against pests.

WHAT ARE PESTS AND PESTICIDES?

A pest is something or someone that annoys us, detracts from some resource that we value, or interferes with a pursuit that we enjoy. In this chapter, we will concentrate on **biological pests,** organisms that reduce the availability, quality, or value of resources useful to humans. What's annoying or undesirable depends, of course, on your perspective. The mosquitoes that swarm in clouds over a marsh in the summer may be irritating to us, but they are an essential food source for birds and bats that feed on them. You may regard dandelions in your yard as tenacious and obnoxious weeds, but in some countries dandelions are cultivated as beautiful flowers and as a food source. Of the millions of species of organisms

only about 100 plants, animals, fungi, and microbes cause 90 percent of all crop damage worldwide.

Insects tend to be the most frequent pests, in part, because they make up at least three-quarters of all species on the earth. Most pest organisms tend to be generalists, the opportunistic species that reproduce rapidly, migrate quickly into disturbed areas, and that are pioneers in ecological succession. They compete aggressively against more specialized endemic species and can often take over a biotic community, especially where humans have disrupted natural conditions and created an opening into which they can slip. Most Canadians are familiar with dandelions, ragweed, English sparrows, starlings, European pigeons, and other "weedy" species that survive well in urban habitats. Chapter 16 describes how exotic aliens brought in by humans are crowding out native species in many places.

A **pesticide** is a chemical that kills pests. We generally think of toxic substances in this category, but chemicals that drive away pests or prevent their development are sometimes included as well. Pest control can also include activities such as killing pests by burning crop residues or draining wetlands to eliminate breeding sites. A broad-spectrum pesticide that kills a wide range of living organisms is called a **biocide** (Fig. 15.2). Fumigants, such as ethylene dibromide or dibromochloropropane, used to protect stored grain or sterilize soil fall into this category. Generally, we prefer narrower spectrum agents that attack a specific type of pest: **herbicides** kill plants; **insecticides** kill insects; **fungicides** kill fungi; acaricides kill mites, ticks, and spiders; nematicides kill nematodes (microscopic roundworms); rodenticides kill rodents; and avicides kill birds. Pesticides can also be defined by their method of dispersal (fumigation, for example) or their mode of action, such as an ovicide, which kills the eggs of pests. In a sense, the antibiotics used in medicine to fight infections are pesticides as well.

FIGURE 15.2 Synthetic chemicals can eliminate pests quickly and efficiently, but what are the long-term costs to us and to our environment?

A BRIEF HISTORY OF PEST CONTROL

Humans have probably always used chemicals to protect themselves from pests, but in the past 50 years we have entered a new era of pesticide use. How did we develop these chemical agents and how do current uses differ from previous practices?

Early Pest Controls

Using chemicals to control pests may well have been among our earliest forms of technology. People in every culture have known that salt, smoke, and insect-repelling plants can keep away bothersome organisms and preserve food. The Sumerians controlled insects and mites with sulphur 5,000 years ago. Chinese texts 2,500 years old describe mercury and arsenic compounds used to control body lice and other pests. Greeks and Romans used oil sprays, ash and sulphur ointments, lime, and other natural materials to protect themselves, their livestock, and their crops from a variety of pests.

In addition to these metals and inorganic chemicals, people have used organic compounds, biological controls, and cultural practices for a long time. Alcohol from fermentation and acids in pickling solutions prevent growth of organisms that would otherwise ruin food. Spices were valued both for their flavours and because they deterred spoilage and pest infestations. Romans burned fields and rotated crops to reduce crop diseases. The Chinese developed plant-derived insecticides and introduced predatory ants in orchards to control caterpillars 1,200 years ago. Many farmers still use ducks and geese to catch insects and control weeds (Fig. 15.3).

Synthetic Chemical Pesticides

The modern era of chemical pest control began in 1934 with the discovery of the insecticidal properties of DDT (*Dichloro-Diphenyl-Trichloroethane*) by Swiss chemist Paul Müller. DDT

FIGURE 15.3 Geese make good biological control agents. They eat weeds, grass, and insects but leave many crops alone. Their droppings enrich the soil, their down can be used to make garments and pillows, and a goose dinner makes a welcome protein source for many people.

FIGURE 15.4 Before we realized the toxicity of DDT, it was sprayed freely on people to control insects as shown here at Jones Beach, NY, in 1948.

was used to control potato beetles in Switzerland in 1939 and commercial production began in 1943. It became extremely important during World War II in areas where tropical diseases and parasites posed greater threats to soldiers than enemy bullets.

DDT seemed like a wonderful discovery. It is cheap, stable, soluble in oil, and easily spread over a wide area. It is highly toxic to insects but relatively nontoxic to mammals. The oral LD50 for DDT for humans is 113 mg/kg, compared to 50 mg/kg for nicotine and 366 mg/kg for caffeine. Where other control processes act slowly and must be started before a crop is planted, DDT can save a crop even when pests already are well established. Its high toxicity for target organisms makes DDT very effective, often producing 90 percent control with a single application. DDT seemed like the magic bullet for which science had been searching. It was sprayed on crops and houses, dusted on people and livestock, and used to combat insects all over the world (Fig. 15.4). Paul Müller received a Nobel prize in 1948 for his discovery.

As you will learn in this chapter, however, indiscriminate and excessive use of synthetic chemical pesticides has caused serious ecological damage and long-term harm to human health, and it has turned relatively innocuous species into serious pests.

PESTICIDE USES AND TYPES

Information on total world pesticide use is spotty and uncertain. The United Nations Environment Program reports sales totals rather than quantity because of difficulty in obtaining data from developing countries. In any case, it is clear that pesticide use has increased dramatically since World War II. Sales have grown from

almost nothing in 1950 to approximately $33 billion for 2.6 million tonnes in 1999. About 90 percent of all pesticides worldwide are used in agriculture or food storage and shipping. The wealthier countries of the developed world consume some three-fourths of all pesticides, but rates of use in developing countries are rising 7 to 8 percent per year compared to 2 to 4 percent annual increases in the more-developed countries.

Pesticide Use in Canada and the United States

About 20 percent of all agricultural lands in Canada was treated with herbicides in 1970 and only 2 percent was sprayed with insecticides. Fifteen years later herbicide use had more than doubled to over half of all cropland, and insecticide use had grown tenfold, especially in the prairie provinces of Alberta and parts of Saskatchewan.

The U.S. Environmental Protection Agency reports that 650,000 tonnes of pesticides were used in the United States in 1999. Germany and Italy—which have the world's next highest consumption after the United States—each use about one-fifth as much. Canada uses one-tenth as much total pesticide as does the United States but has doubled its consumption over the past decade, whereas U.S. consumption has remained relatively stable.

Pesticide Types

One way to classify pesticides is by their chemical structure. This is useful because environmental properties—such as stability, solubility, and mobility—and toxicological characteristics of members of a particular chemical group are often similar.

Inorganic pesticides include compounds of arsenic, copper, lead, and mercury. These broad-spectrum poisons are generally highly toxic and essentially indestructible, remaining in the environment forever. Seeds are sometimes coated with a mercury or arsenic powder to deter insects and rodents during storage or after planting. Handling such seeds with bare hands can be very dangerous for farmers or gardeners. They are generally neurotoxins and even a single dose can cause permanent damage.

Natural organic pesticides, or "botanicals," generally are extracted from plants. Some important examples are nicotine and nicotinoid alkaloids from tobacco; rotenone from the roots of derris and cubé plants; pyrethrum, a complex of chemicals extracted from the daisylike *Chrysanthemum cinerariaefolium* (Fig. 15.5); and turpentine, phenols, and other aromatic oils from conifers. All are toxic to insects, but nicotine is also toxic to a broad spectrum of organisms including humans. Rotenone is commonly used to kill fish. Turpentine, phenols, and other natural hydrocarbons are effective pesticides, but synthetic forms such as pentachlorophenol are more stable and more toxic than natural forms. They penetrate surfaces well and are used to prevent wood decay.

Fumigants are generally small molecules such as carbon tetrachloride, carbon disulphide, ethylene dichloride, ethylene dibromide, methylene bromide, and dibromochloropropane that

FIGURE 15.5 Harvesting pyrethrum-containing flowers in Kenya to extract natural pesticides.

FIGURE 15.6 The United Farm Workers of America estimates that 300,000 farmworkers in the United States suffer from pesticide-related illnesses each year.

gasify easily and penetrate rapidly into a variety of materials. They are used to sterilize soil and prevent decay or rodent and insect infestation of stored grain. Because these compounds are extremely dangerous for workers who apply them, use has been curtailed or banned altogether.

Chlorinated hydrocarbons, or organochlorines such as DDT, chlordane, aldrin, dieldrin, toxaphene, paradichlorobenzene (moth-balls), and lindane, are synthetic organic insecticides that inhibit nerve membrane ion transport and block nerve signal transmission. They are fast acting and highly toxic in sensitive organisms. Toxaphene, for instance, kills goldfish at 5 parts per billion (5 μg/litre), one of the highest toxicities for any compound in any organism. Chlorinated hydrocarbons may persist in soil for decades, become concentrated through food chains, and are stored in fatty tissues of a variety of organisms. The chloriphenoxy herbicides 2,4 D and 2,4,5 T have hormone-like growth-regulating properties and are selective for broad-leaved flowering plants. Because these compounds are so toxic and so long-lasting, many have been banned for most uses over much of the world.

Organophosphates, such as parathion, malathion, dichlorvos, dimethyldichlorovinylphosphate (DDVP), and tetraethylpyrophosphate (TEPP), are an outgrowth of nerve gas research during World War II. They inhibit cholinesterase, an enzyme essential for removing excess neurotransmitter from synapses in the peripheral nervous system. They are extremely toxic to mammals, birds, and fish (generally 10 to 100 times more poisonous than most chlorinated hydrocarbons). A single drop of TEPP on your skin can be lethal. Because they are quickly degraded, they are much less persistent in the environment than organochlorines, generally lasting only a few hours or a few days. These compounds are very dangerous for workers such as grape pickers who often are sent into fields too soon after they have been sprayed (Fig. 15.6).

Carbamates, or urethanes, such as carbaryl (Sevin), aldicarb (Temik), aminocarb (Zineb), carbofuran (Baygon), and Mirex share many organophosphate properties, including mode of action, toxicity, and lack of environmental persistence and low bioaccumulation. Carbamates generally are extremely toxic to bees and must be used carefully to prevent damage to these beneficial organisms.

Microbial agents and *biological controls* are living organisms or toxins derived from them used in place of pesticides. Bacteria such as *Bacillus thuringiensis* or *Bacillus popilliae* kill caterpillars or beetles by producing a toxin that ruptures the digestive tract lining when eaten. Parasitic wasps such as the tiny *Trichogramma* genus attack moth caterpillars and eggs, while lacewings and ladybugs control aphids. Viral diseases also have been used against specific pests.

PESTICIDE BENEFITS

Like all organisms, humans compete with other species for food and shelter and struggle to protect ourselves from diseases and predators. Synthetic chemical pesticides are important weapons in this fight for survival.

Disease Control

Insects and ticks serve as vectors in the transmission of a number of disease-causing pathogens and parasites. Consider malaria, for

FIGURE 15.7 Malaria, spread by the *Anopheles* mosquito, is one of the largest causes of human disease and premature death in the world. By controlling mosquitoes, pesticides save 1 million lives per year in tropical countries.

FIGURE 15.8 Elephantiasis is caused by parasitic worms (filaria) that block lymph vessels and cause fluid accumulation in various parts of the body.

example. About 500 million people suffer from this disease at any given time, and about 2 million die each year from *Plasmodium* protozoa spread to humans by *Anopheles* mosquitoes (Fig. 15.7). It is estimated that insecticidal mosquito control has prevented at least 50 million deaths from malaria over the past 50 years. Sri Lanka is a classic example of pesticide benefits. In the early 1950s, more than 2 million cases of malaria were reported in Sri Lanka each year. After DDT spraying began in 1954, new malaria cases almost completely disappeared. When DDT spraying was discontinued in 1964, however, malaria reappeared almost immediately.

Within three years the annual incidence was more than 1 million cases per year. The Sri Lankan government resumed DDT spraying in 1968 and continues limited use of this insecticide despite environmental concerns. They concluded that the reductions in medical expenses, social disruption, pain and suffering, and lost work resulting from malaria outweighed the direct costs of insecticide spraying by a thousand to one.

Some other diseases spread by biting insects include yellow fever and related viral diseases such as encephalitis, also carried by mosquitoes, and trypanosomiasis or sleeping sickness caused by protozoa transmitted by the tsetse fly. Onchocerciasis (river blindness) and filariasis (one form of which is elephantiasis, Fig. 15.8) are caused by tiny worms spread by biting flies that afflict hundreds of millions of people in tropical countries. All of these terrible diseases can be reduced by judicious use of pesticides. If you were faced with a choice of going blind before age 30 because of masses of worms accumulating in your eyeballs or a small chance of cancer due to pesticide exposures if you live to age 50 or 60, which would you choose?

Crop Protection

Although reliable data on crop losses are difficult to obtain, it is thought that plant diseases, insect and bird predation, and competition by weeds reduce crop yields worldwide by at least one-third. Postharvest losses to rodents, insects, and fungi may be as high as another 20 to 30 percent. Without modern chemical pesticides, these losses might be much higher. Although we said in Chapter 14 that there is more than enough food in the world to adequately feed everyone now living—if food were equitably distributed—this most certainly would not be true, given current intensive farming practices, if modern chemical pesticides were unavailable.

A commonly quoted estimate is that farmers save $3 to $5 for every $1 spent on pesticides. This means lower costs and generally more reliable quality for consumers. In some cases, insects and fungal diseases cause only small losses in terms of the total crop quantity, but the cosmetic damage they cause greatly reduces the economic value of crops. For example, although codling moth larvae consume very little of the apples they infest, the brown trails they leave as they crawl through the fruit and the possibility of biting into a living worm often make an unsprayed crop unsaleable. Which is the more worrisome risk for you—consuming a bug or consuming pesticides?

PESTICIDE PROBLEMS

While synthetic chemical pesticides have brought us great economic and social benefits, they also cause a number of serious problems. In this section we will examine some of the worst of those problems.

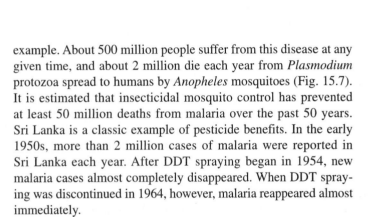

FIGURE 15.9 This machine sprays insecticide on orchard trees—and everything else in its path. Up to 90 percent of all pesticides never reach target organisms.

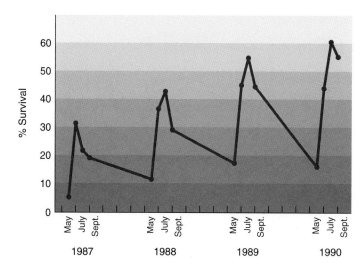

FIGURE 15.10 Survival of tobacco budworms to a constant dose of pyrethroid at different times of the year. Highest survival percentages are in July of each year, but maximum survival grows as worms become resistant.

Source: Data from S. Micinski, et al., *Resistant Pest Management Newsletter,* Vol. 3, No. 2, July 1991.

Effects on Nontarget Species

It is estimated that up to 90 percent of the pesticides we use never reach their intended targets (Fig. 15.9). Many beneficial organisms are poisoned unintentionally as a result. For instance, about 20 percent of all honeybee colonies are destroyed each year and another 15 percent are damaged by pesticide spray drift or residues on the flowers they visit. Direct losses to beekeepers amount to several million dollars per year. Losses to crops the bees would have pollinated may be 10 times higher.

In some cases, the effects of poisoning nontarget species are immediate and unmistakable. In one episode in 1972, a single application of the insecticide Azodrin to combat potato aphids on a farm in Dade County, Florida, killed 10,000 migrating robins in three days. Similarly, a 1991 derailment of a Southern Pacific tanker car on a tricky canyon bridge just north of Dunsmuir, California, dumped 75,000 litres of highly toxic metam sodium herbicide into the Sacramento River. The entire river ecosystem—including aquatic plants, insects, amphibians, and at least 100,000 trout—was completely wiped out for 45 kilometres downstream.

In other cases, the effects are more difficult to pin down, although not less serious. A study published in 1999 linked insecticide spraying on Canadian forests with dramatic declines (up to 77 percent) in Atlantic salmon. The chemical suspected in this case is 4-nonylphenol, a powerful endocrine-hormone disrupter (What Do You Think? p. 325).

Pesticide Resistance and Pest Resurgence

Pesticides almost never kill 100 percent of a target species even under the most ideal conditions. As we discussed in Chapter 5, every population contains some diversity in tolerance to adverse environmental factors. The most resistant members of a population survive pesticide treatment and produce more offspring like themselves with genes that enable them to withstand further chemical treatment (Fig. 15.10). Because most pests propagate rapidly and produce many offspring, the population quickly rebuilds with pesticide-resistant individuals. We call this phenomenon **pest resurgence** or rebound. The United Nations Environment Program reports that at least 500 insect pest species and another 250 or so weeds and plant pathogens worldwide have developed chemical resistance (Fig. 15.11). Of the 25 most serious insect pests in California, three-quarters are now resistant to one or more insecticides. This resistance means that it takes constantly increasing doses to get the same effect or that farmers who are caught on a **pesticide treadmill** must constantly try newer and more toxic chemicals in an attempt to stay ahead of the pests.

One of the most ominous developments in this race to find effective pesticides is that we increasingly find pests that are resistant to chemicals to which they have never been exposed. Apparently, genes for pesticide resistance are being transferred from one species to another by means of vectors such as viruses and plasmids (naked pieces of virus-like DNA). Often a whole cluster of genes jump between species so that multiple chemical-tolerance is inherited before a particular pest is ever exposed to any of the chemicals. Widespread use of crops genetically engineered to produce pesticides endogenously (Chapter 14) is likely to cause even more pesticide resistance.

In a way, every pesticide has a very limited useful life span before target species become resistant to it or the pesticide builds up intolerable environmental concentrations. We have made a big mistake in broadcasting pesticides recklessly and extravagantly. DDT, for instance, is such a helpful insecticide that we should have used it sparingly and carefully, so that it would still be effective against

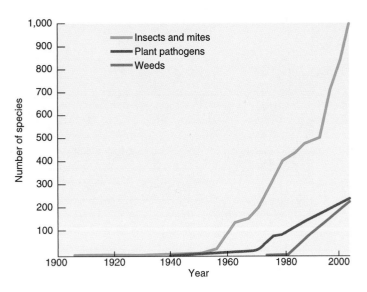

FIGURE 15.11 Many pests have developed resistance to pesticides. Because insecticides were the first class of pesticide to be used widely, selection pressures led insects to show resistance early. More recently, plant pathogens and weeds are also becoming insensitive to pesticides. *Source:* Worldwatch Institute 2003.

FIGURE 15.12 The cotton boll weevil caused terrible devastation in southern U.S. cotton fields in the 1930s and 1940s. DDT and other insecticides controlled this pest briefly, but the beetle quickly became resistant.

the worst insect pests. It has been spread so widely that 50 of the 60 malaria-carrying mosquitoes are now resistant to it and the environmental side effects outweigh its benefits. You can think of the useful life of DDT as a nonrenewable resource that we squandered.

Creation of New Pests

Often the worst side effect of broadcast spraying a pesticide is that we kill beneficial predators that previously kept a number of pests under control. Many of the agricultural pests that we want to eliminate, for instance, are herbivores such as aphids, grasshoppers, or moth larvae that eat crop plants. Under natural conditions, their populations are kept under control by predators such as wasps, lady bugs, and praying mantises. As we discussed in Chapter 3, there are generally far fewer predators in a food web than the species they prey upon. When we use a broad-spectrum pesticide, higher trophic levels are more likely to be knocked out than lower ones. This means that species that previously were insignificant can be released from natural controls and suddenly become major pests.

Consider the case of the Canete Valley in Peru. Before DDT was introduced in 1949, cotton yields were about 500 kg per ha. By 1952 yields had risen to nearly 750 kg per ha, but DDT-resistant boll weevils also had appeared (Fig. 15.12). Toxaphene replaced DDT, but within two years it also became ineffective against boll weevils, which rebounded to higher levels than ever. Even worse, *Heliothis* worms—which had not previously been a problem—began increasing rapidly. The wasps that earlier had kept both organisms in check were poisoned by the increasing pesticide doses. By 1955, cotton yields were down to 330 kg per ha, one-third less than before any pesticides were used.

Similarly, in California in 1995, cotton yields plunged by 20 percent despite a doubling of the number of insecticide applications and a 43 percent increase in pesticide costs. With the advent of chemical pest controls, farmers tend to abandon mixed crops, rotation regimes, and other traditional methods of management. This creates a greater potential for pest damage and pest resistance and thus makes it economical to use even higher levels of pesticide treatment. The application of large amounts of pesticides over wide expanses of land, together with changes in the ways the land is managed, have undermined the biological and ecological forces and interactions that previously governed population dynamics among species. A lack of understanding of the ways pesticides adversely affect beneficial organisms is one reason pest managers stick with chemical-based systems despite declining pesticide efficiency.

Persistence and Mobility in the Environment

The qualities that make DDT and other chlorinated hydrocarbons so effective—stability, high solubility, and high toxicity—also make them environmental nightmares. Because they persist for years, even decades in some cases, and move freely through air, water, and soil, they often show up far from the point of original application. Some of these compounds have been discovered far from any possible source and long after they most likely were used. Because they have an affinity for fat, many chlorinated hydrocarbons are bio-concentrated and stored in the bodies of predators— such as porpoises, whales, polar bears, trout, eagles, ospreys, and humans—that feed at the top of food webs. In a study of human pesticide uptake and storage, Canadian researchers found the levels of chlorinated hydrocarbons in the breast milk of Inuit mothers living in remote arctic villages was five times that of women from Canada's industrial region some 2,500 km to the south.

Environmental Estrogens

What might alligators in Florida, seals in the North Sea, salmon in the Great Lakes, and you have in common? All are at the top of their respective food chains and all appear to be accumulating threatening levels of toxic environmental chemicals in their body tissues. One of the most frightening possible effects of those chemicals is that they seem to be able to disrupt endocrine hormones that regulate many important bodily functions, including reproductive and immune systems. Evidence for this seems quite convincing in some wildlife populations, but whether it also is true for humans is one of the most contentious and important questions in environmental toxicology today.

One of the first examples of hormone-disrupting chemicals in the environment was a dramatic decline in alligators a decade ago in Florida's Lake Apopka. Surveys showed that 90 percent of the alligator eggs laid each year were infertile and that of the few that hatched, only about half survived more than two weeks. Male hatchlings had shrunken penises and unusually low levels of the male hormone testosterone. Female alligators, meanwhile, had highly elevated estrogen levels and abnormal ovaries. The explanation seems to be that a DDT spill in the lake in the 1980s, along with pesticide-laden runoff from adjacent farm fields, has led to high levels of DDE (a persistent breakdown product of DDT) in the reptiles' tissues and eggs. Because of a similarity in chemical structure, DDE appears to interfere with the action of androgens and estrogens, the normal sex hormones.

Researchers have begun to suspect that mysterious outbreaks of health and reproductive problems and immune system failures in other wildlife populations may have similar origins.

Humans may be affected as well. DDE and other compounds have been correlated with elevated cases of estrogen-sensitive breast and vaginal cancers in women, with developmental disorders in children, and possibly with falling sperm counts in men.

Good evidence exists from controlled laboratory experiments that rats and mice exposed *in utero* or through mother's milk to very low levels of estrogen-like compounds develop

A shocking decrease in fertility and hatchling survival of alligators in Lake Apopka, Florida, has been linked to pesticide pollution. Are humans affected by similar toxins?

physical, reproductive, and behavioural problems. We know that some of these chemicals act as synthetic hormones, others are antagonists that block normal hormone function. Furthermore, there can be striking synergy between some compounds. When endosulfan and DDT or chlordane are applied together, for example, the combination is 1,600 times more estrogenic than either chemical alone.

The question is whether these chemicals are linked to human health problems. Many of these compounds are hundreds or thousands of times less active than normal hormones, leading skeptics to doubt that they have any noticeable effects except in animals exposed to extremely high levels from a chemical spill. Furthermore, we may have protective mechanisms that are lacking in highly inbred laboratory rodents, and

we can eat a highly varied diet that includes protective factors as well as toxins.

The bottom line is that we don't know (and we may never know for sure) whether falling sperm counts, increasing cancers, birth defects, immune diseases, and behavioural disorders in humans are caused by endocrine-disrupting environmental chemicals. In 1996, the EPA ordered pesticide manufacturers to begin testing for disrupting effects. Given the continuing uncertainty about the dangers we face, what more do you think we should do? Is this threat serious enough to warrant drastic steps to reduce our risk? If you were head of Agriculture Canada or the Ministry of Agriculture and Food, how much certainty would you demand before acting to protect our environment and ourselves from this frightening potential threat?

Want More?

Semenza, J. C., Tolbert, P. E., Rubin, C. H., Guillette, L. J., Jr., and Jackson, R. J. "Reproductive Toxins and Alligator Abnormalities at Lake Apopka, Florida." *Environmental Health Perspectives.* 105 (1997): 1030–2.

Inuit people have the highest levels of these persistent pollutants of any human population except those contaminated by industrial accidents.

These compounds accumulate in polar regions by what has been called the "grasshopper effect," in which they evaporate from water and soil in warm areas and then condense and precipitate in colder regions. In a series of long-distance hops, they eventually collect in polar regions, where they accumulate in top predators. Polar bears, for instance, have been shown to have concentrations of certain chlorinated compounds 3 billion times greater than the seawater around them. In the St. Lawrence estuary, beluga (white whales), which suffer from a wide range of infectious diseases and tumours thought to be related to environmental toxins, have such high levels of chlorinated hydrocarbons that their carcasses must be treated as toxic waste.

Although DDT hasn't been used legally in Canada or the United States for more than 25 years, in 1999 researchers reported discovery of p,p'-DDE, a DDT breakdown by-product, in the amniotic fluid of 30 percent of a sample of pregnant women in Los Angeles, California. A growing baby is surrounded by amniotic fluid during its nine months in the mother's womb. The levels found in some women—up to 0.6 ppm—are sufficient to cause concern. It was once thought that these compounds wouldn't cross the placental barrier, but this apparently is not the case.

Atrazine and alochlor are the most widely used herbicides in North America. Although still a very popular herbicide for general weed control, atrazine use in Canada has declined since 1983 because of concerns about its environmental effects. In a related study, researchers measured levels of these two compounds in rain falling over much of Europe that would be illegal if it were supplied as drinking water.

Because these **persistent organic pollutants (POPs)** are so long-lasting and so dangerous, a meeting of 127 countries agreed in 2001 to a global ban on the worst of them. Use of the "dirty dozen" (Table 15.1) had been banned or severely restricted in developed countries for years. However, their production continued. Between 1994 and 1996, U.S. ports shipped more than 100,000 tonnes of DDT and POPs each year. Most of these were sent to developing countries where regulations were lax. Ironically, many of these pesticides returned to the United States on bananas and other imported crops. According to the 2001 POPs treaty, eight of the dirty dozen were banned immediately; PCBs, dioxins, and furans will be phased out; and use of DDT, still allowed for limited uses such as controlling malaria, must be pubically registered in order to permit monitoring. The POPs treaty has been hailed as a triumph for environmental health and international cooperation.

Human Health Problems

Pesticide effects on human health can be divided into two categories: (1) short-term effects, including acute poisoning and illnesses caused by relatively high doses and accidental exposures, and (2) long-term effects suspected to include cancer, birth defects, immunological problems, Parkinson's disease, and other chronic degenerative diseases.

TABLE 15.1	The "Dirty Dozen" Persistent Organic Pollutants
COMPOUND(S)	**USES**
Aldrin	Insecticide used on corn, potatoes, cotton, and for termite control
Clordane	Insecticide used on vegetables, small grains, maize, sugarcane, fruits, nuts, and cotton
Dieldrin	Insecticide used on cotton, corn, potatoes, and for termite control
DDT	Insecticide, now used primarily for disease vector control
Endrin	Insecticide used on field crops such as cotton and grains and as a rodenticide
Hexachlorobenzene (HCB)	Fungicide used for seed treatment and as an industrial chemical
Heptachlor	Insecticide used against soil insects, termites, and grasshoppers
Mirex	Insecticide used to combat fire ants, termites, mealybugs, and as a fire retardant
Toxaphene	A mixture of chemicals used as an insecticide on cotton as well as tick and mite control in livestock and fish eradication
Polychlorinated biphyenyls (PCBs)	Industrial chemicals used as insulators in electrical transformers, solvents, paper coatings, and plasticizers
Dioxins	A large family of by-products of chlorinated chemical production and incineration
Furans	A large group of by-products of chlorinated chemical production and incineration

The World Health Organization (WHO) estimates that between 3.5 and 5 million people suffer acute pesticide poisoning and at least 20,000 die each year. At least two-thirds of this illness and death results from occupational exposures in developing countries where people use pesticides without proper warnings or protective clothing (Fig. 15.13). A tragic example of occupational pesticide exposure is found among workers in the Latin American flower industry. Fueled by the year-round demand in North America for fresh vegetables, fruits, and flowers, a booming export trade has developed in countries such as Guatemala, Colombia, Chile, and Ecuador. To meet demands in North American markets for perfect flowers, table grapes and other produce, growers use high levels of pesticides, often spraying daily with fungicides, insecticides, nematicides, and herbicides. Working in warm, poorly ventilated greenhouses with little protective clothing, the workers—70 to 80 percent of whom are women—find it hard to avoid pesticide contact. Nearly two-thirds of nearly 9,000 workers surveyed in Colombia experienced blurred vision, nausea, headaches, conjunctivitis, rashes, and asthma. Although harder to document, they also reported serious chronic effects such as stillbirths, miscarriages, and neurological problems.

FIGURE 15.13 Farmworkers in developing countries often apply dangerous pesticides without any protective clothing or other safeguards. Warnings and instructions printed in English are often of little help.

Long-term health effects caused by chronic pesticide exposures are difficult to document conclusively. As we discuss in Chapter 20, isolating a specific pesticide-related disease from the myriad of other risks we encounter is complex. One well-known outcome of chronic pesticide exposure is that farmers who use 2,4-D on their crops are up to eight times as likely to contract a cancer known as non-Hodgkin's lymphoma as those who don't use herbicides.

In an ongoing study of prenatal exposure to PCBs and other persistent contaminants, researchers from Wayne State University in Detroit, Michigan, have documented significant learning and attention problems in children whose mothers ate Lake Michigan fish regularly in the years prior to pregnancy. At age 11, the most highly exposed children had difficulty paying attention, suffered from poorer short- and long-term memory, were twice as likely to be below average in reading comprehension, and were three times as likely to have low IQ scores as their age-matched peers.

Researchers in a similar study in Mexico showed striking differences in behaviour and motor skills in children exposed to pesticides compared to peers with minimal pesticide risk. Two groups of Yaqui children from northwestern Mexico were tested in this study. The children were similar in every respect except for their pesticide exposure. Children who lived in the foothills come from ranches where pesticides are used sparingly, if at all. Children from the Yaqui valley, live in a farming area where pesticide use has been heavy for the past 50 years. Samples of mother's milk and umbilical cord blood taken from valley families contained high amounts of several persistent pesticides including aldrin, endrin, dieldrin, heptachlor, and DDE. In tests designed to measure growth and development, the valley children fell far behind their foothill-dwelling peers (Fig. 15.14). Farm children also exhibited diminished memory, decreased physical stamina, higher irritability, a greater tendency to fight, and poorer eye-hand coordination.

FIGURE 15.14 Representative examples of drawings of people by 4- and 5-year-old Yaqui children relatively unexposed to pesticides (foothills) and those heavily exposed (valley).
Source: From E.A. Guillette, et al., "An Anthropological Approach to the Evaluation of Preschool Children Exposed to Pesticides in Mexico" in *Environmental Health Perspective*, 106(6):347–353, 1998. U.S. Dept. of Health and Human Services.

ALTERNATIVES TO CURRENT PESTICIDE USES

Can we avoid using toxic pesticides or lessen their environmental and human-health impacts (Fig. 15.15)? In many cases, improved management programs can cut pesticide use between 50 and 90 percent without reducing crop production or creating new diseases. Some of these techniques are relatively simple and save money while maintaining disease control and yielding crops with just as high quality and quantity as we get with current methods. In this section, we will examine behavioural changes, biological controls, and integrated pest-management systems that could substitute for current pest-control methods.

Behavioural Changes

Crop rotation (growing a different crop in a field each year in a two- to six-year cycle) keeps pest populations from building up. For instance, a soybean/corn/hay rotation is effective and economical against white-fringed weevils. Mechanical cultivation can substitute for herbicides but increases erosion. Flooding fields before planting or burning crop residues and replanting with a cover crop can suppress both weeds and insect pests. Habitat diversification, such as restoring windbreaks, hedgerows, and ground

FIGURE 15.15 Our approach to nature is to beat it into submission, but at what cost?
Ziggy © ZIGGY & FRIENDS, INC. Reprinted with permission of UNIVERSAL PRESS SYNDICATE. All rights reserved.

FIGURE 15.16 The praying mantis looks ferocious and is an effective predator against garden pests, but it is harmless to humans. They can even make interesting and useful pets.

cover on watercourses, not only prevents soil erosion but also provides perch areas and nesting space for birds and other predators that eat insect pests. Growing crops in areas where pests are absent makes good sense. Adjusting planting times can avoid pest outbreaks, while switching from huge monoculture fields to mixed polyculture (many crops grown together) makes it more difficult for pests to find the crops they like (see Case Study, p. 329). Tillage at the right time can greatly reduce pest populations. For instance, spring or fall plowing can help control overwintering corn earworms.

Biological Controls

Biological controls such as predators (wasps, ladybugs, praying mantises; Figs. 15.16 and 15.17) or pathogens (viruses, bacteria, fungi) can control many pests more cheaply and safely than broad-spectrum, synthetic chemicals. *Bacillus thuringiensis* or Bt, for example, is a naturally occurring bacterium that kills the larvae of lepidopteran (butterfly and moth) species but is harmless to mammals. A number of important insect pests such as tomato hornworm, corn rootworm, cabbage loopers, and others can be controlled by spraying bacteria on crops. Larger species are effective as well. Ducks, chickens, and geese, among other species, are used to rid fields of both insect pests and weeds. These biological organisms are self-reproducing and often have wide prey tolerance. A few mantises or ladybugs released in your garden in the spring will keep producing offspring and protect your fruits and vegetables against a multitude of pests for the whole growing season.

FIGURE 15.17 Ladybird beetles (ladybugs) prey on a variety of pests both as larvae and adults. For a few dollars you can buy several thousand of these hardy and colourful little garden protectors from organic gardening supply stores.

Herbivorous insects also have been used to control weeds (Fig. 15.18). For example, the prickly pear cactus was introduced to Australia about 150 years ago as an ornamental plant. This hardy cactus escaped from gardens and found an ideal home in the dry soils of the outback. It quickly established huge, dense stands that dominated 25 million ha of grazing land. A natural predator from South America, the cactoblastis moth, was introduced into Australia in 1935 to combat the prickly pear. Within a few years, cactoblastis larvae had eaten so much prickly pear that the cactus has become rare and is no longer economically significant.

CASE STUDY

REGENERATIVE AGRICULTURE IN IOWA

Dick and Sharon Thompson operate a diversified crop and livestock farm near Boone, Iowa. Originally, the Thompsons practised high-intensity, monocrop farming using synthetic pesticides and fertilizers just as all their neighbours did. But they felt that something was wrong. Their hogs and cattle were sick. Fertilizer, pesticide, and petroleum prices were rising faster than crop prices. They began looking for a better way to farm. Through 30 years of careful experimentation and meticulous recordkeeping, they have developed a set of alternative farming techniques they call "regenerative agriculture" because it relies on natural processes to rebuild and protect soil.

Rather than depend on synthetic chemical herbicides and pesticides to keep their fields clean of weeds and pests, the Thompsons use a variety of old and new techniques including crop rotation, cover crops, and mechanical cultivation. Instead of growing corn and beans over and over again in the same fields as most of their neighbours do, the Thompsons change crops every year so that no one weed species can become dominant and all species remain relatively easy to control. In the fall, nitrogen-fixing cover crops are planted to hold soil against wind erosion and to keep down weeds.

Before planting, animal manure is spread on fields to rebuild fertility. During the summer, cattle are pastured on fallow land, using intensive grazing techniques that discourage weed growth and spread of manure over the whole field. The soil organic content—the sentinel indicator of soil health—registers at 6 percent, which is more than twice that of their neighbours. Untouched Midwestern prairie usually has about 7 percent organic content. The capacity to store extra carbon in soil might allow farmers to bid on carbon set-aside contracts.

The high levels of organic matter and available nutrients in the Thompsons' fields, coupled with the absence of pesticides that might harm beneficial microbes and pathogens, help crops compete against weeds and insects. Weed control specialists predict that in the future more farmers will follow the Thompsons' lead and concentrate on microbial biocontrol rather than depend on conventional herbicide-dependent systems, some of which can impair soil quality and lead to carryover injury to crops.

While yields on the Thompsons' land are comparable to those of their neighbours, lower reliance on off-farm inputs—including pesticides, fertilizers, and animal drugs—keeps the Thompsons' production costs as much as 25 percent lower than their neighbours' costs. In addition to favourable financial returns, the Thompsons benefit in other ways from their innovative system. The quality of their soil is significantly better than that under conventional agriculture and is steadily improving in fertility, tilth, and health. Through their innovative work, Dick and Sharon Thompson are helping find ways to profitably produce high yields without degrading the land or the environment.

Dick and Sharon Thompson of Boone, Iowa, practise sustainable farming to produce healthy crops that are economically successful.

Genetics and bioengineering can help in our war against pests. Traditional farmers have long known to save seeds of disease-resistant crop plants or to breed livestock that tolerate pests well. Modern science can speed up this process through selection regimes or by using biotechnology to transfer genes between closely related or even totally unrelated species. (Often we don't know, however, what unintended consequences, such as super weeds or new pests,

might result from these activities.) Insect pest reproduction has sometimes been reduced by releasing sterile males. Screwworms, for example, are the flesh-eating larvae of flies that lay their eggs in scratches or skin wounds of livestock. They were a terrible problem for ranchers in Texas and Florida in the 1950s, but release of massive numbers of radiation-sterilized males disrupted reproduction and eliminated this pest in Florida. In Texas, where flies

FIGURE 15.18 Biological pest control is illustrated by the Klamath weed (*foreground*), which had invaded millions of hectares of California rangeland in the 1940s. In the background, the weeds have been eliminated by the introduction of a natural predator, the *Chrysolina* beetle. Within a few years, the weeds were entirely eradicated.
Source: Data from Natural Resources Development Council.

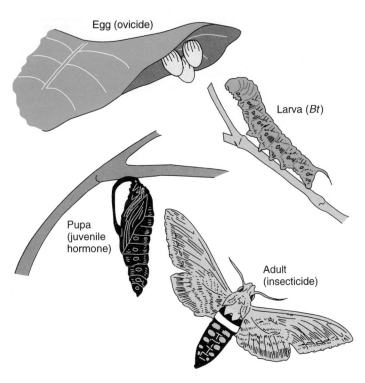

FIGURE 15.19 Different strategies can be used to control pests at various stages of their life cycles. *Bacillus thuringiensis* (Bt) kills caterpillars when they eat leaves with these bacteria on the surface. Releasing juvenile hormone in the environment prevents maturation of pupae. Predators attack at all stages.

continue to cross the border from Mexico, control has been more difficult, but continual vigilance keeps the problem manageable.

Other promising approaches are to use hormones that upset development or sex attractants to bait traps containing toxic pesticides. Many municipalities control mosquitoes with these techniques rather than aerial spraying of insecticides because of worries about effects on human health. Briquettes saturated with insect juvenile hormone are scattered in wetlands where mosquitoes breed. The presence of even minute amounts of this hormone prevent larvae from ever turning into biting adults (Fig. 15.19). Unfortunately, many beneficial insects as well as noxious ones are affected by the hormone, and birds or amphibians that eat insects may be adversely affected when their food supply is reduced. Some communities that formerly controlled mosquitoes have abandoned these programs, believing that having naturally healthy wetlands is worth getting a few bites in the summer.

Integrated Pest Management

Integrated pest management (IPM) is a flexible, ecologically based pest-control strategy that uses a combination of techniques applied at specific times, aimed at specific crops and pests. It often uses mechanical cultivation and techniques such as vacuuming bugs off crops as an alternative to chemical application (Fig. 15.20). IPM doesn't give up chemical pest controls entirely but rather tries to use the minimum amount necessary only as a last resort and avoids broad-spectrum, ecologically disruptive products. IPM relies on preventive practices that encourage growth and diversity of beneficial organisms and enhance plant defences and vigour. Careful,

FIGURE 15.20 This machine, nicknamed the "salad vac," vacuums bugs off crops as an alternative to treating them with toxic chemicals.

scientific monitoring of pest populations to determine **economic thresholds,** the point at which potential economic damage justifies pest control expenditures, and the precise time, type, and method of pesticide application is critical in IPM.

Trap crops, small areas planted a week or two earlier than the main crop, are also useful. This plot matures before the rest of the field and attracts pests away from other plants. The trap crop then is sprayed heavily with enough pesticides so that no pests are likely to escape. The trap crop is destroyed so that workers will not be exposed to the pesticide and consumers will not be at risk. The rest of the field should be mostly free of both pests and pesticides.

Some of the most dramatic IPM success stories come from the developing world. Cuba, for example (Case Study, p. 329) has turned almost entirely to organic farming. In Brazil, pesticide use on soybeans has been reduced up to 90 percent with IPM. In Costa Rica, use of IPM on banana plantations has eliminated pesticides altogether in one region. In Africa, mealybugs were destroying up to 60 percent of the cassava crop (the staple food for 200 million people) before IPM was introduced in 1982. A tiny wasp that destroys mealybug eggs was discovered and now controls this pest in over 65 million ha in 13 countries.

A successful IPM program that could serve as a model for other countries is found in Indonesia, where brown planthoppers developed resistance to virtually every insecticide and threatened the country's hard-won self-sufficiency in rice. In 1986, President Suharto banned 56 of 57 pesticides previously used in Indonesia and declared a crash program to educate farmers about IPM and the dangers of pesticide use. Researchers found that farmers were spraying their fields habitually—sometimes up to three times a week—regardless of whether fields were infested. By allowing natural predators to combat pests and spraying only when absolutely necessary with chemicals specific for planthoppers, Indonesian farmers using IPM have had higher yields than their neighbours using normal practices and they cut pesticide costs by 75 percent. In 1988, only two years after its initiation, the program was declared a success. It is being extended throughout the whole country. Since nearly half the people in the world depend on rice as their staple crop, this experiment could have important implications elsewhere (Fig. 15.21).

While IPM can be a good alternative to chemical pesticides, it also presents environmental risks in the form of exotic organisms. Wildlife biologist George Boettner of the University of

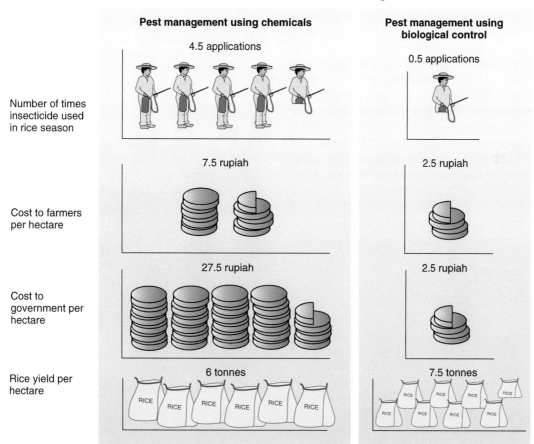

FIGURE 15.21 Indonesia has one of the world's most successful integrated pest management (IPM) programs. Switching from toxic chemicals to natural pest predators has saved money while also increasing rice production.

Source: Data from Tolba, et al., *World Environment, 1972–1992,* p. 307, Chapman & Hall, 1992 United Nations Environment Programme.

Massachusetts reported in 2000 that biological controls of gypsy moths, which attack fruit trees and ornamental plants, have also decimated populations of native North American moths. *Compsilura* flies, introduced in 1905 to control the gypsy moths, have a voracious appetite for other moth caterpillars as well. One of the largest North American moths, the Cecropia moth, with a 15 cm wingspan, was once ubiquitous in the eastern United States, but it is now rare in regions where *Compsilura* flies were released.

REDUCING PESTICIDE EXPOSURE

The Pest Control Products Act and Regulations, administered by Agriculture Agri-food Canada, regulate pesticide use. Individual Canadian provinces also set local standards. The Ontario Ministry of Agriculture and Food, for instance, introduced a comprehensive program in 1988 that set a goal of 50 percent reduction in pesticide use by the year 2002. Nonchemical pest control methods, on-farm education, and biotechnology and the development of pest-resistant crop varieties played a role in this program. Savings on chemical costs amounted to about $100 million per year.

It is wise to be aware of potential hazards from pesticides in our diet and environment, but it is important to keep these risks in perspective. While you may have heard the statement that 80 percent of all cancer is environmental, most of that risk comes from personal lifestyle choices such as smoking, unhealthy diets, and excess sun exposure. British epidemiologists Richard Doll and Richard Peto estimate that the cancer risk from commercial pesticides in food is "unimportant" compared to other health risks. Biochemist Bruce Ames agrees, claiming that 99.9 percent of all carcinogens to which we are exposed are natural. In many municipalities in Canada, citizens are advocating reduced use of

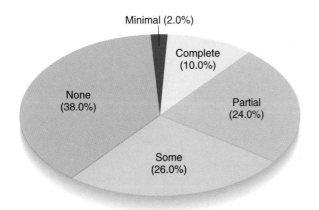

FIGURE 15.22 Level of testing for health effects of 3,350 active and inert pesticide ingredients. Complete testing includes a battery of 10 tests for both chronic and acute toxicity and genetic effects.
Source: National Academy of Sciences.

what can you do?
A GLOBAL CONCERN

Food Safety Tips

To reduce the amount of pesticide residues and other toxic chemicals in your diet, follow these simple rules.

- Wash and scrub all fresh fruits and vegetables thoroughly under running water.
- Peel fruits and vegetables when possible. Throw away the outer leaves of leafy vegetables such as lettuce and cabbage.
- Store food carefully so it doesn't get mouldy or pick up contaminants from other foods. Use food as soon as possible to ensure freshness.
- Cook or bake foods that you suspect have been treated with pesticides to break down chemical residues.
- Trim the fat from meat, chicken, and fish. Eat lower on the food chain where possible to reduce bioaccumulated chemicals (e.g., primary producers rather than secondary carnivores).
- Don't pick and eat berries or other wild foods that grow on the edges of roadsides where pesticides may have been sprayed.
- Grow your own fruits and vegetables without using pesticides or with minimal use of dangerous chemicals.
- Ask for organically grown food at your local grocery store or shop at a farmer's market or co-op where you can get such food.

Your local farmer's market is a good source of locally-grown, organic produce.

and exposure to pesticides. The Citizens Against Pesticides (http://www.caps.20m.com/index.htm) helps grassroots community groups to advocate new legislation and promote awareness of the danger of pesticides.

A Personal Plan

Even though pesticide exposure may not be the greatest threat that most of us face, why not take reasonable steps to reduce your use of and exposure to toxic materials? There are many things that each of us can do to minimize pesticides in our food and environment.

- Don't use chemicals on your yard and garden. Plant hardy ground cover and learn to appreciate that a biologically diverse lawn is healthier than a chemically sterilized one.

- Clean up spilled food and empty garbage regularly to eliminate food for ants or roaches. Keep food in tightly sealed containers. Sprinkle borax around drains and ducts where pests may be lurking. Use sticky flypaper to catch flies rather than spray dangerous chemicals inside your house or apartment.

- Get rid of aphids, scales, mites, and other houseplant pests by washing leaves and stems individually with rubbing alcohol or water, or spray plants with a dilute solution of dish soap in water.

- A saucer of stale beer in your garden will attract and drown slugs. If you have only a few worms or bugs, pick them off by hand. Purchase some predators such as ladybugs or praying mantises to protect garden plants. If you have a lot of bugs in your house or apartment, a praying mantis can make an interesting house pet.

- Drain stagnant water in or near your yard that might serve as breeding sites for mosquitoes. Pick up dead branches, fallen leaves and other litter that provide a haven for termites and other pests. Put up houses for bug-eating purple martins or bats.

- Accept slightly blemished fruits and vegetables. If you garden, trim away bad parts or give up a portion of your crop rather than saturate the environment with dangerous chemicals. Interplant insect-repellent plants such as marigolds, basil, peppermint, or garlic together with your sensitive garden plants.

- If you must use toxic chemical pesticides, use them in the smallest possible amounts and apply them only where and when it is necessary. Rather than spray chemicals, use traps or applicators that limit dispersal into the environment.

- Read organic gardening magazines that report regularly on environmental issues and alternative approaches to pest control.

Remember that being "natural" or organic does not necessarily make a chemical good, nor does being synthetic necessarily make one bad. Some natural products such as nicotine are highly toxic. Others such as mercury and lead are just as likely to bioconcentrate and persist in the environment as are synthetic compounds.

SUMMARY

Biological pests are organisms that reduce the availability, quality, or value of resources useful to humans. Pesticides are chemicals intended to kill or drive away pests. Many nonchemical pest-control approaches perform these same functions more safely and cheaply than do toxic chemicals. Of the millions of species in the world, only about 100 kinds of animals, plants, fungi, and microbes cause most crop damage. Many beneficial organisms are injured by indiscriminate pesticide use, including natural predators that serve a valuable function in keeping potential pests under control.

Humans have probably always known of ways to protect themselves from annoying creatures, but our war against pests entered a new phase with the invention of synthetic organic chemicals such as DDT. These chemicals have brought several important benefits, including increased crop production and control of disease-causing organisms. Indiscriminate and profligate

pesticide use also has caused many problems, such as killing nontarget species, creating new pests of organisms that were previously not a problem, and causing widespread pesticide resistance among pest species. Often highly persistent and mobile in the environment, many pesticides move through air, water, and soil and bioaccumulate or bioconcentrate in food chains causing serious ecological and human health problems.

A number of good alternatives offer ways to reduce our dependence on dangerous chemical pesticides. Among these are behavioural changes such as crop rotation, cover crops, mechanical cultivation, and planting mixed poly-cultures rather than vast monoculture fields. Consumers may have to learn to accept less than perfect fruits and vegetables. Biological controls such as insect predators, pathogens, or natural poisons specific for a particular pest can help reduce chemical use. Genetic breeding and biotechnology can produce pest-resistant crop and livestock strains, as well. Integrated pest management (IPM) combines all of these alternative methods together with judicious use of synthetic pesticides under precisely controlled conditions.

Regulating pesticide use is a controversial subject. Many people fear that we are exposed to far too many dangerous chemicals. Industry claims that it could not do business without these materials. Should we weaken the law and allow some carcinogens as long as the risk is "negligible"?

Many of the procedures and approaches suggested for agriculture and industry also work at home to protect us from pests and toxic chemicals alike. By using a little common sense, we can have a healthier diet, lifestyle, and environment.

QUESTIONS FOR REVIEW

1. What is a pest and what are pesticides? What is the difference between a biocide, a herbicide, an insecticide, and a fungicide?

2. How much pesticide is used worldwide? In Canada, which of the general categories of use and which specific type accounts for the greatest use? Has use been increasing or decreasing in recent years?

3. What is DDT and why was it considered a "magic bullet"? Why was it listed among the "dirty dozen" persistent organic pollutants (POPs)?

4. Describe fumigants, botanicals, chlorinated hydrocarbons, organophosphates, carbamates, and microbial pesticides.

5. Explain why pests often resurge or rebound after treatment with pesticides and how they become pesticide resistant. What is a pesticide treadmill?

6. Identify three major categories of alternatives to synthetic pesticides and describe, briefly, how each one works.

7. How did Australia fight prickly pear cactus? How did Florida eradicate screwworms?

8. List nine things you could do to reduce pesticide use in your home.

9. List eight things you could do to reduce your dietary exposure to pesticides.

QUESTIONS FOR CRITICAL THINKING

1. In retrospect, do you think Paul Müller should have received a Nobel prize for discovering the insecticidal properties of DDT?

2. If you were a public health official in a country in which malaria, filariasis, or onchocerciasis were rampant, would you spray DDT to eradicate vector organisms? Would you spray it in your own house?

3. Pesticide treadmill, "dirty dozen" pesticides, and environmental estrogens are all highly emotional terms. Why would some people choose to use or not use these terms? Can you suggest alternative terms for the same phenomena that convey different values?

4. Suppose that a developing country believes that it needs a pesticide banned in Canada to feed or protect the health of its people. Are we right to refuse to sell that pesticide?

5. How much extra would you pay for organically grown food? How would you define organic in this context?

6. If alternative pest control methods are so effective and so much safer, why aren't farmers and consumers adopting them more rapidly?

7. What would you personally consider a "negligible" risk? Would you eat grapes if you knew they had a measurable amount of some pesticide? How small would the amount have to be?

8. Why do you suppose that the Thompsons' neighbours haven't adopted regenerative agriculture?

KEY TERMS

biocide 319
biological controls 328
biological pests 318
economic thresholds 330
fungicides 319
herbicides 319
insecticides 319
integrated pest management (IPM) 330
persistent organic pollutants (POPs) 326
pesticide 319
pesticide treadmill 323
pest resurgence 323

LEARNING ABOUT LINDANE

Check out the National Roundtable on the Environment and the Economy. They have a case study on lindane, an organochlorine insecticide used for everything from treating canola seeds to getting rid of head lice. Create a balance sheet of its benefits and costs in agriculture and society.

Now check out the story on lindane in the Pesticides Action Network database at http://www.pesticideinfo.org/Detail_Chemical.jsp?Rec_Id=PC32949 and Truestar Health at http://www.truestarhealth.com/Notes/1420005.html. Is the story the same from all three sources? If not, how do you decide which one to believe?

This organic, communal garden in Havana, Cuba provides most of the fresh produce for the housing complex surrounding it.

Chapter 16

Biodiversity

The first rule of intelligent tinkering is to save all the pieces.

—Aldo Leopold—

OBJECTIVES

After studying this chapter, you should be able to:

- define *biodiversity* and *species*.
- report on the total number and relative distribution of living species on the earth.
- summarize some of the benefits we derive from biodiversity.
- describe the ways humans cause biodiversity losses.
- evaluate the effectiveness of the Species at Risk Act (SARA) and the Committee on the Status of Endangered Wildlife in Canada (COSEWIC) in protecting and recovering species at risk.
- understand how gap analysis, ecosystem management, and captive breeding can contribute to preserving biological resources.

- propose ways we could protect endangered habitats and communities through large-scale, long-range, comprehensive planning.

WebQuest

biodiversity, conservation biology, island biogeography, species at risk, extinct, extirpated, endangered, threatened, special concern

Above: A SCUBA diver explores a coral reef, one of the most diverse ecosystems in the world. Many coral reefs are threatened by "coral bleaching," a degradation of the algae that live in a symbiotic relationship with the coral that is caused by sedimentation, extreme high and low temperatures, and physical disturbance of the coral beds.

Columbia River Salmon

More than a century ago, as many as 16 million salmon and cut-throat trout migrated every year up the Columbia River system to their breeding grounds in small headwater streams and lakes over a watershed larger than Texas (Fig. 16.1). This was probably the greatest anadromous fish (spend part of life cycle in fresh water and part in saltwater) migration in the world. Some fish swam up to 2,000 km from the mouth of the Columbia on the Pacific coast to its headwaters in British Columbia or up tributary streams such as the Snake and Salmon Rivers in what is now Idaho. The fish were marvellously adapted to this remarkable journey. At least 420 separate stocks (or ecotypes) differed in the timing of their run and the subtle chemical signals they followed to find the stream where they hatched at exactly the right time for water conditions and food supplies to support their offspring. Adult salmon die after spawning and their decaying bodies nourish the ecosystem on which fry and fingerlings (young fish) will depend. The adults, each weighing as much as 45 kg, represent an enormous influx of nutrients into the small streams where they breed. Studies have shown that up to half of the nitrogen in riparian (stream side) vegetation in some streams comes from migrating fish. Without dead fish, the whole ecosystem is impoverished.

Both native people and wildlife in the Pacific Northwest depended on this prodigious bounty. For a few months every year there were more fish than anyone could eat. The runs were probably never uniform, however. For reasons that we don't fully understand, the number of fish returning from the ocean often would vary as much as 50 percent from year to year. Even now, we don't know where salmon go during the two to five years adults spend in the ocean or exactly what they eat while growing to such enormous sizes. Undoubtedly, changes in ocean temperature, circulation patterns, and food supplies caused by phenomena such as the El Niño/Southern Oscillation (Chapter 8) affect population sizes. There seems to be a 40-year cycle, for instance, when salmon runs in Alaska are high and Columbia runs are low or vice versa. Before European settlement, however, these variations didn't matter much because there were more than enough salmon for everyone.

Salmon runs in the Columbia have undergone a disastrous decline during the past 90 years (Fig. 16.2). Total numbers are down nearly 90 percent from historic highs, and only about one-tenth are now wild stocks; the rest are hatchery-reared. There are many reasons for this decrease. Overfishing by commercial operations has taken a toll. Runoff from logging, agriculture, road building, and urban areas carries warm water, sediment, and pollution that kill eggs and fry. Irrigators pump water from the river, reducing its flow. Perhaps the greatest threat to the salmon are dams that block migration. Over the last century 124 huge dams have been built on the Columbia along with 55 more on its tributaries. The river has become a chain of reservoirs. Except at its mouth, only 70 km of the Columbia runs free today. Fish ladders (stair step pools) help some adults move upstream but smolt (young fish), which depend on river currents to help them downstream, get lost in the slack water behind dams. Whole populations of smolt are now being barged downstream to get them past dams and reservoirs but many fail to survive anyway.

Hatchery rearing of salmon began on the Columbia in the 1890s. Rather than stem the decline, unfortunately, releases of hatchery fish often have exacerbated problems. Only a few genetic strains are bred in hatcheries rather than the hundreds of wild stocks. Hatchery fish have been selected for fast growth, early maturity, and aggressive behaviour. This allows them to interbreed with and outcompete native stocks. With much of their original genetic diversity gone, the fish now lack the subtle adaptations that are needed to migrate to remote streams at just the right time. At least half of the original wild salmon runs on the Columbia are now extinct and almost all that remain are in serious trouble.

In a population that fluctuates as widely as salmon, it is difficult to detect patterns until long after critical events occur. If fewer fish show up this year compared to last, is it just a natural variation or an omen that something is wrong? The last big run on the Columbia River occurred in 1924 when the commercial harvest was nearly 19 million kg. By the time fishwheels were prohibited

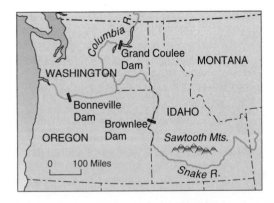

FIGURE 16.1 Up to 16 million fish per year once migrated up the Columbia River and its tributaries such as the Snake.

Want More?

Salmon Insider: The Columbia Basin Research Newsletter
http://www.cbr.washington.edu/newsletter/

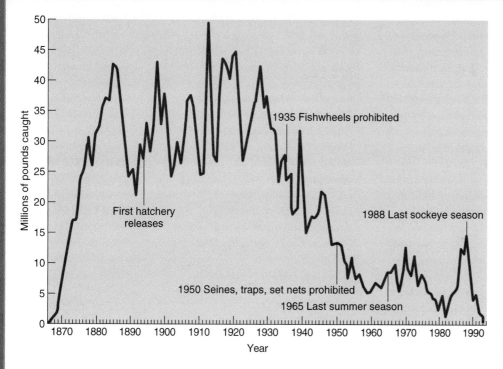

FIGURE 16.2 Commercial salmon harvest in the Columbia River 1855–1991. By the time fishwheels, seines, traps, and set nets were prohibited, the fish were already in a downward slide.

Source: Data "Columbia Fish Runs and Fisheries 1938–93" from Washington Department of Fish and Wildlife and Oregon Department of Fish and Wildlife 1994 Status Report, Olympia, WA.

in 1935, or when seines, traps, and set nets were banned in 1950, the population was already in a catastrophic decline. By 1991, when the Snake River sockeye was added to the endangered species list, only four of these fish made it to their spawning grounds in Idaho's Sawtooth Mountains. Since then, five other runs (Snake River spring and fall chinook, Umpaqua River cutthroat, and lower Columbia chum and chinook) have been added to the endangered species list while others are under consideration. Fishers and recreationists are urging the government to destroy at least four of the dams on the Snake to allow the river to run free again. Industries and residents of the Columbia River basin, who enjoy some of the lowest electric rates in the nation oppose this plan.

BIODIVERSITY AND THE SPECIES CONCEPT

From the driest desert to the dripping rainforests, from the highest mountain peaks to the deepest ocean trenches, life on earth occurs in a marvelous spectrum of sizes, colours, shapes, life cycles, and interrelationships. Think for a moment how remarkable, varied, abundant, and important the other living creatures are with whom we share this planet. How will our lives be impoverished if this biological diversity diminishes?

What Is Biodiversity?

Previous chapters of this book have described some of the fascinating varieties of organisms and complex ecological relationships that give the biosphere its unique, productive characteristics. Three kinds of **biodiversity** are essential to preserve these ecological systems: (1) *genetic diversity* is a measure of the variety of different versions of the same genes within individual species; (2) *species diversity* describes the number of different kinds of organisms within individual communities or ecosystems; and (3) *ecological diversity* assesses the richness and complexity of a biological community, including the number of niches, trophic levels, and ecological processes that capture energy, sustain food webs, and recycle materials within this system.

What Are Species?

The concept of a species occupies a central position in the concept of biodiversity, but what, exactly, do we mean by the term? In Chapter 3, we defined species as all the organisms of the same kind able to breed in nature and produce live, fertile offspring. Underlying this commonly used definition is the idea that reproductive isolation, caused by geography, physiology, or behaviour, prevents groups of similar organisms from exchanging genes, and therefore, gives them separate identities and evolutionary histories.

There are problems with species definitions based on reproductive isolation. Mating between species (hybridization) occurs in nature and may produce fertile offspring. Furthermore, we often cannot determine whether two groups that live in different places are capable of interbreeding. Sometimes, in fact, the only specimens available for study are dead. Species identification, therefore, is often based on morphological characteristics such as size, shape, colour, skeletal structure, or on either chemical or genetic traits.

Increasingly, DNA sequencing technology is also used to distinguish species. Some groups that were thought to be far apart now appear to be closely related, while others may be moved into entirely different families or even orders. Understanding molecular evolution not only helps us see how species originate, but also helps us evaluate genetic diversity.

Determining how much difference is necessary before two similar groups of organisms can be considered separate species is highly subjective and taxonomists (scientists who study classification) debate which organisms are most closely related and which should be classified as separate species. Often these debates pit "lumpers" against "splitters." Given the same group of a 100 closely related organisms, one scientist might split them into 50 or a 100 different species, while another might lump them together into only 10 or 20 species.

How Many Species Are There?

At the end of the great exploration era of the nineteenth century, some scientists confidently declared that every important kind of living thing on earth would soon be found and named. Most of those explorations focused on charismatic species such as birds and mammals. Recent studies of less conspicuous organisms such as insects and fungi suggest that millions of new species and varieties remain to be studied scientifically.

The 1.7 million species presently known (Table 16.1) probably represent only a small fraction of the total number that exist. Based on the rate of new discoveries by research expeditions—especially in the tropics—taxonomists estimate that there may be somewhere between 3 million and 50 million different species alive today. In fact, there may be 30 million species of tropical insects alone (Fig. 16.3), and by some estimates there may be 10 million species living on (or in) the ocean floor. About 70 percent of all known species are invertebrates (animals without backbones such as insects, sponges, clams, worms, etc.). This group probably makes up the vast majority of organisms yet to be discovered and may constitute 95 percent of all species. What constitutes a species in bacteria and viruses is even less certain than for other organisms, but there are large numbers of physiologically or genetically distinct varieties of these organisms.

Of all the world's species, only 10 to 15 percent live in North America and Europe. By contrast, the centres of greatest biodiversity tend to be in the tropics, especially tropical rainforests and coral reefs (Fig. 16.4). Many of the organisms in these megadiversity countries have never been studied by scientists. The Malaysian Peninsula, for instance, has at least 8,000 species of flowering plants, while Britain, with an area twice as large, has only 1,400

TABLE 16.1 Approximate Numbers of Known Living Species by Taxonomic Group	
Bacteria and cyanobacteria	4,000
Protozoa (single-celled animals)	31,000
Algae (single-celled plants)	40,000
Fungi (moulds, mushrooms)	72,000
Multicellular plants	270,000
Sponges	5,000
Jellyfish, corals, anemones	10,000
Flatworms (tapeworms, flukes)	12,000
Roundworms (nematodes, earthworms)	25,000
Clams, snails, slugs, squids, octopuses	70,000
Insects	1,025,000
Mites, ticks, spiders, crabs, shrimp, centipedes, other noninsect arthropods	110,000
Starfish, sea urchins	6,000
Fish and sharks	27,000
Amphibians	4,000
Reptiles	7,150
Birds	9,700
Mammals	4,650
Total	1,733,000

Source: Norman Myers, 2000.

species. There may be more botanists in Britain than there are species of higher plants. South America, on the other hand, has fewer than 100 botanists to study perhaps 200,000 species of plants.

HOW DO WE BENEFIT FROM BIODIVERSITY?

We benefit from other organisms in many ways, some of which we don't appreciate until a particular species or community disappears. Even seemingly obscure and insignificant organisms can play irreplaceable roles in ecological systems or be the source of genes or drugs that someday may be indispensable.

Food

All of our food comes from other organisms. Many wild plant species could make important contributions to human food supplies either as they are or as a source of genetic material to improve domestic crops. Noted tropical ecologist Norman Meyers estimates that as many as 80,000 edible wild plant species could be utilized by humans. Villagers in Indonesia, for instance, are thought to use some 4,000 native plant and animal species for food, medicine, and other valuable products. Few of these species have been explored

FIGURE 16.3 The development of flowering plants—and the insects that pollinate them—vastly increased the diversity of living things on earth.

for possible domestication or more widespread cultivation. A 1975 study by the National Academy of Science (U.S.) found that Indonesia has 250 edible fruits, only 43 of which have been cultivated widely (Fig. 16.5).

Drugs and Medicines

Living organisms provide us with many useful drugs and medicines (Table 16.2). More than half of all prescriptions contain some natural products. The United Nations Development Program estimates the value of pharmaceutical products derived from developing world plants, animals, and microbes to be more than $30 billion per year. Indigenous communities that have protected and nurtured the biodiversity on which these products are based are rarely acknowledged—much less compensated—for the resources extracted from them. Many consider this expropriation "biopiracy" and call for royalties to be paid for folk knowledge and natural assets.

Consider the success story of vinblastine and vincristine. These anticancer alkaloids are derived from the Madagascar periwinkle (Fig. 16.6). They inhibit the growth of cancer cells and are very effective in treating certain kinds of cancer. Thirty years ago, before these drugs were introduced, childhood leukemias were invariably fatal. Now the remission rate for some childhood leukemias is 99 percent. Hodgkin's disease was 98 percent fatal a few years ago, but is now only 40 percent fatal, thanks to these compounds. The total value of the periwinkle crop is roughly $15 million per year, although Madagascar gets little of those profits.

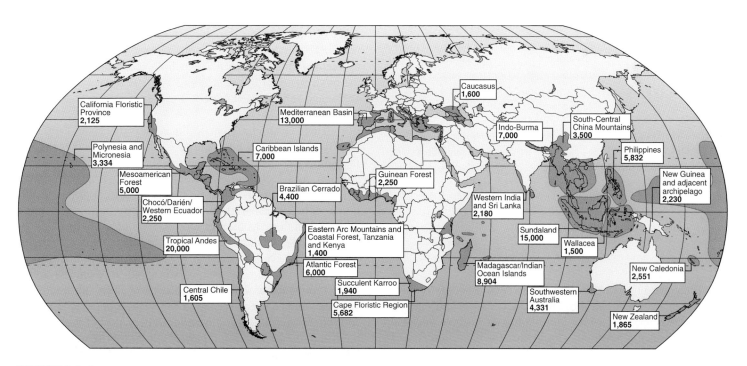

FIGURE 16.4 Biodiversity "hot spots" identified by Conservation International tend to be in tropical or Mediterranean climates and on islands, coastlines, or mountains where many habitats exist and physical barriers encourage speciation.
Source: Data from Conservation International.

FIGURE 16.5 Mangosteens from Indonesia have been called the world's best-tasting fruit, but they are practically unknown beyond the tropical countries where they grow naturally. There may be thousands of other traditional crops and wild food resources that could be equally valuable but are threatened by extinction.

FIGURE 16.6 The rosy periwinkle from Madagascar provides anti-cancer drugs that now make childhood leukemias and Hodgkin's disease highly remissible. You may recognize it as a popular flowering ornamental plant.

TABLE 16.2 Some Natural Medicinal Products

PRODUCT	SOURCE	USE
Penicillin	Fungus	Antibiotic
Bacitracin	Bacterium	Antibiotic
Tetracycline	Bacterium	Antibiotic
Erythromycin	Bacterium	Antibiotic
Digitalis	Foxglove	Heart stimulant
Quinine	Chincona bark	Malaria treatment
Diosgenin	Mexican yam	Birth-control drug
Cortisone	Mexican yam	Anti-inflammation treatment
Cytarabine	Sponge	Leukemia cure
Vinblastine, vincristine	Periwinkle plant	Anticancer drugs
Reserpine	Rauwolfia	Hypertension drug
Bee venom	Bee	Arthritis relief
Allantoin	Blowfly larva	Wound healer
Morphine	Poppy	Analgesic

Pharmaceutical companies are actively prospecting for useful products in many tropical countries. Merck, the world's largest biomedical company, is paying $1 million to the Instituto Nacional de Biodiversidad (INBIO) of Costa Rica for plant, insect, and microbe samples to be screened for medicinal applications. INBIO, a public/private collaboration, trains native people as practical "parataxonimists" to locate and catalogue all the native flora and fauna—between 500,000 and 1 million species—in Costa Rica

(Fig. 16.7). Selling data and specimens will finance scientific work and nature protection. This may be a good model both for scientific information gathering and as a way for developing countries to share in the profits from their native resources.

Ecological Benefits

Human life is inextricably linked to ecological services provided by other organisms. Soil formation, waste disposal, air and water purification, nutrient cycling, solar energy absorption, and management of biogeochemical and hydrological cycles all depend on the biodiversity of life (Chapters 3 and 20). Total value of these ecological services is at least $33 trillion per year, or more than double total world GNP.

There has been a great deal of controversy about the role of biodiversity in ecosystem stability. Mathematical models suggest that simple ecosystems can be just as stable and resilient as more complex ones. Field studies by Minnesota ecologist David Tillman have shown, however, that some diverse biological plant communities withstand environmental stress better and recover more quickly than those with fewer species.

Because we don't fully understand the complex interrelationships between organisms, we often are surprised and dismayed at the effects of removing seemingly insignificant members of biological communities. For instance, wild species provide a valuable but often unrecognized service in suppressing pests and disease-carrying organisms. It is estimated that 95 percent of the potential pests and disease-carrying organisms in the world are controlled by other species that prey upon them or compete with them in some way. Many unsuccessful efforts to control pests with synthetic chemicals (Chapter 15) have shown that biodiversity provides essential pest control services.

FIGURE 16.7 Costa Rican taxonomists study insect collections as part of an ambitious project to identify and catalogue all the species in this small, but highly diverse, tropical country. The knowledge gained may contribute toward valuable commercial products that will provide funds to help preserve biodiversity.

Aesthetic and Cultural Benefits

The diversity of life on this planet brings us many aesthetic and cultural benefits, and cultural diversity is inextricably linked to biodiversity (Chapter 1). Millions of people enjoy hunting, fishing, camping, hiking, wildlife watching, and other outdoor activities based on nature. These activities provide invigorating physical exercise. Contact with nature also can be psychologically and emotionally restorative. In some cultures, nature carries spiritual connotations, and a particular species or landscape may be inextricably linked to a sense of identity and meaning. Observing and protecting nature has religious or moral significance for many people.

Nature appreciation is economically important. The U.S. Fish and Wildlife Service estimates that Americans spend $104 billion every year on wildlife-related recreation (Fig. 16.8). This compares to $81 billion spent each year on new automobiles. Forty percent of all adults enjoy wildlife, including 39 million who hunt or fish and 76 million who watch, feed, or photograph wildlife. Ecotourism can be a good form of sustainable economic development, but we have to be careful that we don't abuse the places and cultures we visit (see Chapter 22).

For many people, the value of wildlife goes beyond the opportunity to shoot or photograph, or even see, a particular species. They argue that **existence value,** based on simply knowing that a species exists, is reason enough to protect and preserve it. We contribute to programs to save bald eagles, redwood trees, whooping cranes, whales, and a host of other rare and endangered organisms because we like to know they still exist somewhere, even if we may never have an opportunity to see them.

FIGURE 16.8 Birdwatching has become one of the top 10 outdoor activities in Canada.

WHAT THREATENS BIODIVERSITY?

Extinction, the elimination of a species, is a normal process of the natural world. Species die out and are replaced by others, often their own descendants, as part of evolutionary change. In undisturbed ecosystems, the rate of extinction appears to be about one species lost every decade. In the last century, however, human impacts on populations and ecosystems accelerated that rate, causing hundreds or perhaps even thousands of species, subspecies, and varieties to become extinct every year. If present trends continue, we may destroy *millions* of kinds of plants, animals, and microbes in the next few decades. In this section, we will look at some ways we threaten biodiversity.

Natural Causes of Extinction

Studies of the fossil record suggest that more than 99 percent of all species that ever existed are now extinct. Most of those species were gone long before humans came on the scene. Species arise through processes of mutation and natural selection and disappear the same way (Chapter 5). Often, new forms replace their own

TABLE 16.3 Mass Extinctions

HISTORIC PERIOD	TIME (BEFORE PRESENT)	EFFECTS
Ordovician	444 million	25% of all families extinct
Devonian	370 million	19% of all families extinct
Permian	250 million	54% of families, 90% of species extinct
Triassic	210 million	23% of families, 1/2 of species extinct
Cretaceous	65 million	17% of families, 1/2 of species extinct (including dinosaurs but not mammals)
Quaternary	Present	1/3 to 2/3 of all species extinct if present trends continue?

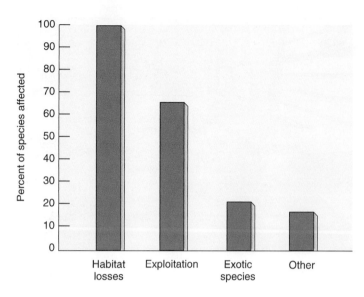

FIGURE 16.9 Threats to endangered mammals in Australasia and the Americas. For fish, exotic species represent a greater threat than shown here. For plants, trampling, grazing, collecting, flooding, and other human actions constitute serious dangers. Many species are endangered by multiple threats so that all these mammals are threatened by habitat loss but only 20 percent are affected by exotic species.
Source: Data from UNEP, 1994.

parents. The tiny Hypohippus, for instance, has been replaced by the much larger modern horse, but most of its genes probably still survive in its distant offspring.

Periodically, mass extinctions have wiped out vast numbers of species and even whole families (Table 16.3). The best studied of these events occurred at the end of the Cretaceous period when dinosaurs disappeared, along with at least 50 percent of existing genera and 15 percent of marine animal families. An even greater disaster occurred at the end of the Permian period about 250 million years ago when two-thirds of all marine species and nearly half of all plant and animal families died out over a period of about 10,000 years—a short time by geological standards. Current theories suggest that these catastrophes were caused by climate changes, perhaps triggered when large asteroids struck the earth. Many ecologists worry that global climate change caused by our release of "greenhouse" gases in the atmosphere could have similarly catastrophic effects (see Chapter 8).

Human-Caused Reductions in Biodiversity

The rate at which species are disappearing appears to have increased dramatically over the last 150 years. Between A.D. 1600 and 1850, human activities appear to have been responsible for the extermination of two or three species per decade. By some estimates, we are now losing species at thousands of times natural rates. If present trends continue, biologist Paul Ehrlich warns, somewhere between one-third to two-thirds of all current species could be extinct by the middle of this century. This would be a biodiversity loss of geologic proportions. Conservation biologists call this the sixth mass extinction, but note that this time it's not asteroids or volcanoes, but human impacts that are responsible.

Habitat Destruction

The biggest reason for the current increase in extinctions is habitat loss (Fig. 16.9). Habitat fragmentation divides populations into

isolated groups that are vulnerable to catastrophic events. Very small populations may not have enough breeding adults to be viable even under normal circumstances (Fig. 16.10). Destruction of forests, wetlands, and other biologically rich ecosystems around the world threatens to eliminate thousands or even millions of species in a human-caused mass extinction that could rival those of geologic history. By destroying habitat, we eliminate not only prominent species, but also many obscure ones of which we may not even be aware. For further discussion, see Chapters 17 and 22.

Hunting and Fishing

Overharvesting is responsible for depletion or extinction of many species. A classic example is the extermination of the American passenger pigeon. Even though it inhabited only eastern North America, 200 years ago this was the world's most abundant bird with a population of between 3 and 5 billion animals. It once accounted for about one-quarter of all birds in North America. In 1830, John James Audubon saw a single flock of birds estimated to be 16 kilometres wide, hundreds of kilometres long, and thought to contain perhaps a billion birds. In spite of this vast abundance, market hunting and habitat destruction caused the entire population to crash in only about 20 years between 1870 and 1890. The last known wild bird was shot in 1900 and the last existing passenger pigeon, a female named Martha, died in 1914 in the Cincinnati Zoo (Fig. 16.11).

Some other well-known overhunting cases include the near extermination of the great whales and the American bison (buffalo). In 1850, some 60 million bison roamed the western plains. Many were killed only for their hides or tongues, leaving millions

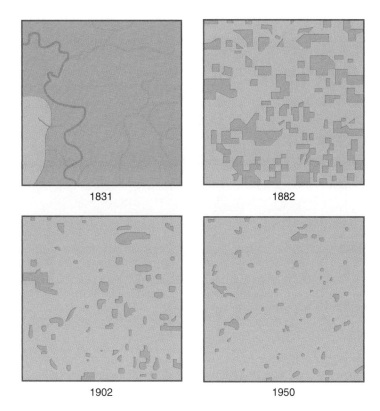

FIGURE 16.10 Decrease in wooded area of Cadiz Township in southern Wisconsin during European settlement. Shaded areas represent the amount of land in forest each year.

of carcasses to rot. Much of the bison's destruction was carried out by the U.S. Army to deprive native peoples who depended on bison for food, clothing, and shelter of these resources, thereby forcing them onto reservations. After 40 years, there were only about 150 wild bison left and another 250 in captivity.

We think that about 2.5 million great whales once inhabited the world's oceans. Their blubber was highly prized as a source of oil. At first, only the slowest whales such as the right or bowhead could be caught by human-powered dories and hand-thrown harpoons. Introduction of steamships and explosive harpoons in the nineteenth century, however, made it possible to catch and kill even the fastest whales and one species after another was driven into near extinction (Fig. 16.12).

For almost two decades, the International Whaling Commission has prohibited taking of great whales. This hunting ban seems to be having positive effects. With the exception of the North Atlantic right whales and southern blue whales, most populations appear to be recovering. There are now about 26,000 California gray whales, more than twice as many as 25 years ago and close to the prehunting number (see Chapter 1). Around 3,400 humpbacks now visit Hawaii each year compared to only one-third that many in the mid-1980s. Nearly 1 million minke whales—perhaps twice the pre-whaling number—now inhabit Arctic and Antarctic waters.

FIGURE 16.11 A stuffed passenger pigeon in a museum. The last member of this species died in the Cincinnati Zoo in 1914. Overhunting and habitat destruction caused their extinction.

Iceland, Japan, and Norway argue that whale populations have rebounded enough to support limited hunting. They kill hundreds of whales each year mostly under the guise of "scientific research," although the meat and blubber of animals taken in these programs are still sold at a handsome profit.

Fish stocks have been seriously depleted by overharvesting in many parts of the world. A huge increase in fishing fleet size and efficiency in recent years has led to a crash of many oceanic populations. Worldwide, 13 of 17 principal fishing zones are now reported to be commercially exhausted or in steep decline. At least three-quarters of all commercial oceanic species are overharvested. The collapse of the Atlantic cod fishery is symptomatic of the problem. Once present in such vast schools off the coast of New England, Nova Scotia, and Newfoundland that they could simply be dipped up in baskets, cod seemed inexhaustible. More than 300,000 tonnes per year were harvested in the 1980s, but then the

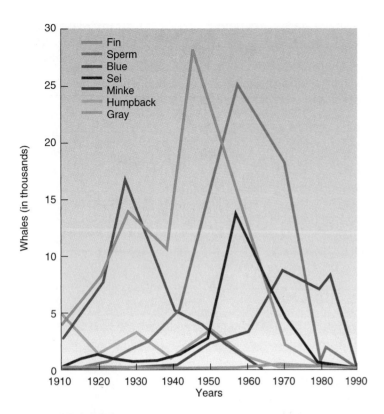

FIGURE 16.12 This world catch of whales shows a series of sharp peaks as each species in turn is hunted until commercially unprofitable. Some species such as minke, gray, and bowhead, have rebounded since commercial whaling stopped. Other species, such as blue and fin, remain rare and endangered.

Source: Data from *UNEP Environmental Data Report 1993–1994.*

FIGURE 16.13 Parts from rare and endangered species for sale on the street in China. Use of animal products in traditional medicine and prestige diets is a major threat to many species.

population crashed. In 1992, Canada banned Atlantic cod fishing, putting 20,000 people out of work and closing down a $700-million-per-year industry. Trawlers have scoured the bottom, killing fish and destroying habitat, so that some biologists doubt the population will ever recover. Other species devastated by overharvesting include top predators such as sharks and billfish as well as slow-growing, deep-water species like orange roughy. Think about what you're eating if seafood is part of your diet (What Can You Do? p. 348).

Commercial Products and Live Specimens

In addition to harvesting wild species for food, we also obtain a variety of valuable commercial products from nature. Much of this represents sustainable harvest, but some forms of commercial exploitation are highly destructive, however, and represent a serious threat to certain rare species (Fig. 16.13). Despite international bans on trade in products from endangered species, smuggling of furs, hides, horns, live specimens, and folk medicines amounts to millions of dollars each year.

Developing countries in Asia, Africa, and Latin America with the richest biodiversity in the world are the main sources of wild animals and animal products, while Europe, North America, and some of the wealthy Asian countries are the principal importers. Japan, Taiwan, and Hong Kong buy three-quarters of all cat and snake skins, for instance, while European countries buy a similar percentage of live birds. The United States imports 99 percent of all live cacti and 75 percent of all orchids sold each year.

The profits to be made in wildlife smuggling are enormous. Tiger or leopard fur coats can bring $100,000 in Japan or Europe. The population of African black rhinos dropped from approximately 100,000 in the 1960s to about 3,000 in the 1980s because of a demand for their horns. In Asia, where it is prized for its supposed medicinal properties, powdered rhino horn fetches $28,000 per kg. In Yemen, a rhino horn dagger handle can sell for up to $1,000.

Similarly, bird collectors will pay $10,000 for a rare hyacinth macaw from Brazil or up to $12,000 for a pair of golden-shouldered parakeets from Australia. An endangered albino python might bring $20,000 in Germany. The mortality rate in this live animal trade is enormous. It is generally estimated that 50 animals are caught or killed for every live animal that gets to market. Buyers, many of whom say they love animals, keep this trade going. If you collect rare plants, fish, or birds, make sure that what you buy is commercially grown, not wild-caught. If you travel abroad, don't buy products made from rare or endangered species. You not only contribute to their decline by doing so; you also expose yourself to confiscation and serious fines when you get home.

Elephants are an important example of the problems and complexities of wildlife trade (Fig. 16.14). In 1980, there were about 1.3 million African elephants; a decade later only half were left. Poachers killed more than 600,000 of these giant animals in the 1980s, mainly for their ivory tusks worth up to $100 per kilogram. An international ban on ivory trade in 1989 dried up the ivory trade and greatly reduced poaching.

In contrast to most of the continent, southern Africa has been very successful in conserving elephants. So successful, in fact, that

FIGURE 16.14 Elephant populations have been decimated throughout most of their range. The biggest threat is from poachers who kill for their valuable ivory.

FIGURE 16.15 A diver uses cyanide to stun tropical fish being caught for the aquarium trade. Many fish are killed by the method itself, while others die later during shipment. Even worse is the fact that cyanide kills the coral reef itself.

herds have had to be thinned regularly to keep them from destroying their habitat. Namibia, Botswana, and Zimbabwe were allowed to sell some of their stockpiled ivory for conservation projects and to compensate citizens whose property and livelihoods are disrupted by marauding animals. Other countries opposed lifting the ban because of the fear that any ivory sales would rekindle poaching. In fact, in 1999, more than 350 elephants were killed in Zimbabwe alone by poachers who hope to slip ivory into legal sales routes. What do you think? Should we reward those nations that have protected their elephants or continue the ban?

Plants also are threatened by overharvesting. Wild ginseng has been nearly eliminated in many areas because of the Asian demand for roots that are used as an aphrodisiac and folk medicine. Cactus "rustlers" steal cacti by the tonne from the American southwest and Mexico. With prices ranging as high as $1,000 for rare specimens, it's not surprising that many are now endangered.

The trade in wild species for pets is an enormous business. Worldwide, some 5 million live birds are sold each year for pets, mostly in Europe and North America. Currently, pet traders import (often illegally) into North America some 2 million reptiles, 1 million amphibians and mammals, 500,000 birds, and 128 million tropical fish each year. About 75 percent of all saltwater tropical aquarium fish sold come from coral reefs of the Philippines and Indonesia.

Many of these fish are caught by divers using plastic squeeze bottles of cyanide to stun their prey (Fig. 16.15). Far more fish die with this technique than are caught. Worst of all, it kills the coral animals that create the reef. A single diver can destroy all of the life on 200 square metres of reef in a day. Altogether, thousands of divers currently destroy about 50 km² of reefs each year. Net fishing would prevent this destruction, and it could be enforced if pet owners would insist on net-caught fish. More than half the world's coral reefs are potentially threatened by human activities, with up to 80 percent at risk in the most populated areas.

Predator and Pest Control

Some animal populations have been greatly reduced, or even deliberately exterminated, because they are regarded as dangerous to humans or livestock or because they compete with our use of resources. Every year, U.S. government animal control agents trap, poison, or shoot thousands of coyotes, bobcats, prairie dogs, and other species considered threats to people, domestic livestock, or crops.

This animal control effort costs about $20 million in federal and state funds each year and kills some 700,000 birds and mammals, about 100,000 of which are coyotes. Defenders of wildlife regard this program as cruel, callous, and mostly ineffective in reducing livestock losses. Protecting flocks and herds with guard dogs or herders or keeping livestock out of areas that are home range of wild species would be a better solution they believe. Ranchers and trappers, on the other hand, argue that without predator control, western livestock operations would be uneconomical.

Exotic Species Introductions

Exotic organisms—aliens introduced into habitats where they are not native—are one of the greatest threats to native biodiversity. Exotics can be thought of as biological pollution. Freed from the predators, parasites, pathogens, and competition that kept them in check in their native home, formerly mild-mannered species can turn into superaggressive "weedy" invaders in a new habitat. There are now more than 4,500 alien species in the United States. A few of those causing most trouble include:

- Kudzu vine has blanketed large areas of the southeastern United States (Fig. 16.16). Long cultivated in Japan for edible roots, medicines, and fibrous leaves and stems used for paper production, kudzu was introduced by the U.S. Soil Conservation Service in the 1930s to control erosion. Unfortunately, it succeeded too well. In the ideal conditions

Don't Eat Endangered Seafood

A good source of protein and rich with unsaturated oils, seafood is a healthy addition to your diet. But while some species of seafood are in plentiful supply, many ocean fish populations are seriously threatened by overharvesting. Which species should we avoid eating?

Shark	These top predators are slow growing and have low reproductive rates. Most commercial populations are overfished and declining rapidly.
Billfish	Swordfish, marlins, and sailfish are among the fastest and most highly adapted animals in the ocean. Like sharks, they mature late and reproduce slowly. Becoming rare in the Atlantic, their status elsewhere is unknown.
Shrimp	Plentiful in some regions but depleted in the Gulf of Mexico and the Gulf of California, shrimp harvest can result in 10 kg of unwanted species (including rare sea turtles) killed for every kilogram of shrimp caught. Shrimp farming causes serious land degradation and water pollution problems (see Chapter 14).
Orange roughy	These deep-water fish live on seamounts off New Zealand and Australia, where they take 20 years to reach maturity. They can live more than 100 years, but excess harvesting has decimated populations in only a few decades.
Groupers	This large tribe of predominantly tropical fish has been overexploited in many places. Groupers become females as they age, so heavy fishing—which takes most of the old fish—can eliminate the breeding stock and prevent reproduction.
Groundfishes	Atlantic cod, haddock, pollack, yellowtail flounder, and monkfish once were abundant but now have been driven to the brink of extinction by overfishing. The collapse of these species probably ranks as the world's greatest fishery-management disaster. Pollack, which once were considered trash fish, are now the main harvest from U.S. waters. They make up most of the artificial crab, fish sticks, fish balls, and other generic fish products we eat.
Sea scallops	While farm-raised scallops and bay scallops from shallow estuaries generally remain plentiful, deep-sea Atlantic scallops are generally overfished. Dredging for scallops takes many other species and severely disrupts habitat.
Bluefin tuna	These magnificent predators—most of which are sold as sushi or high-price steaks—are severely depleted. Most canned "white" tuna is albacore, which is becoming rare. "Chunk light" tuna is yellowfin (caught by some fishers with nets that also trap dolphins) or skipjack, which remains plentiful. Ask your grocer what "dolphin-safe" means.
Red snapper	The popularity of this Gulf Coast fish in Cajun cooking caused its depletion in only a few years. Yellowtail snapper, however, appears to be plentiful.

What, then, can you eat, if you like seafood? Farm-raised catfish, tilapia, trout, and salmon have relatively little environmental impact. Most mackerel, Pacific pollack, dolphinfish (mahimahi), squids, crabs, and crayfish (crawdads) generally are in good shape. Herring and sardines have been overfished in some areas but remain abundant elsewhere. Wild freshwater species like bass, sunfish, pike, catfish, and carp are better managed than most ocean fish, but you have to be aware of local pollution problems such as mercury accumulation. In general, avoid the top predators both because they reproduce slowly and because they are more likely to store toxins. You don't have to give up eating fish entirely, but think about the species you consume.

---

of its new home, kudzu can grow 18 to 30 m in a single season. Smothering everything in its path, it kills trees, pulls down utility lines, and causes millions of dollars in damage every year.

- Asian tiger mosquitoes are unusually aggressive species that now infest many coastal states in the United States. These species have apparently arrived on container ships carrying used tires, a notorious breeding habitat for mosquitoes. Asian tiger mosquitoes became widely recognized in 1999 because they spread West Nile virus (another species introduced with the mosquitoes), which is deadly to many wild birds and occasionally to people and livestock.

- Purple loosestrife grows in any wet soil. Originally cultivated by gardeners for its beautiful purple flower spikes, this tall wetland plant escaped into New England marshes over a century ago. Spreading rapidly across the Great Lakes, it now fills wetlands across much of southern Canada and the northern United States. Because it crowds out indigenous vegetation and has few native predators or symbionts, it tends to reduce biodiversity wherever it takes hold.

- Zebra mussels probably made their way from their home in the Caspian Sea to the Great Lakes in ballast water of transatlantic cargo ships, arriving sometime around 1985. Gluing themselves to any solid surface, zebra mussels reach enormous densities—up to 70,000 animals per square metre—covering fish spawning beds, smothering native mollusks, and clogging utility intake pipes. Found in all the Great Lakes, zebra mussels have recently been spotted in the Mississippi River and its tributaries. Public and private costs for zebra mussel removal now amount to some $400 million per year. On the good side, mussels have improved water clarity in Lake Erie at least fourfold by filtering out algae and particulates.

Purple loosestrife

Kudzu vine

Round goby

Asian long-horned beetle

Zebra mussels

FIGURE 16.16 Bioinvaders.

- The round goby, a small, pugnacious freshwater fish from the Black Sea, probably also hitched a ride in ships coming into the Great Lakes. Devouring the eggs and fry of any species that shares their territory, these tough little fish drive off native gobies and baffle native game fish with their erratic swimming and ability to hide. The round goby's single, round pelvic fin distinguishes it from native species. One possible benefit is that round gobies voraciously devour zebra mussels.

- Asian long-horned beetles are one of the most recent exotic threats in North America. Probably transported in imported logs or wooden packing crates from Asia, these wood-eating insects were first spotted in Amityville, New York, in 1996. The beetle larvae burrow into living tree trunks where they cut off sap flow between leaves and roots. Partial to oaks, maples, poplars, and other hardwood species, these shiny black bugs with distinctive white spots and enormous feelers could cause billions of dollars in damage if they spread widely. Chicago, Illinois, suburbs cut and burned 2,000 shade trees after discovering a long-horned beetle infestation in 1999.

The flow of organisms isn't just into North America; we also send exotic species to other places. The Leidy's comb jelly, for example, a jellyfish native to the western Atlantic coast, has devastated the Black Sea, now making up more than 90 percent of all biomass at certain times of the year. Similarly, the bristle worm from North America has invaded the coast of Poland and now is almost the only thing living on the bottom of some bays and lagoons. A tropical seaweed named *Caulerpa taxifolia,* originally grown for the aquarium trade, has escaped into the northern Mediterranean, where it covers the shallow seafloor with a dense, metre-deep shag carpet from Spain to Croatia. Producing more than 5,000 leafy fronds per square metre, this aggressive weed crowds out everything in its path. Rarely growing in more than scattered clumps less than 25 cm high in its native habitat, this alga was transformed by aquarium breeding into a supercompetitor that grows over everything and can withstand a wide temperature range. Getting rid of these alien species once they dominate an ecosystem is difficult if not impossible.

Diseases

Disease organisms, or pathogens, may also be considered predators. To be successful over the long term, a pathogen must establish a balance in which it is vigorous enough to reproduce, but not so lethal that it completely destroys its host. When a disease is introduced into a new environment, however, this balance may be lacking and an epidemic may sweep through the area.

The American chestnut was once the heart of many Eastern hardwood forests. In the Appalachian Mountains, at least one of every four trees was a chestnut. Often over 45 m tall, 3 m in diameter, fast growing, and able to sprout quickly from a cut stump, it was a forester's dream. Its nutritious nuts were important for birds (like the passenger pigeon), forest mammals, and humans. The wood was straight grained, light, rot-resistant and used for everything from fence posts to fine furniture and its bark was used to tan leather. In 1904, a shipment of nursery stock from China brought a fungal blight to the United States, and within 40 years, the American chestnut had all but disappeared from its native range. Efforts are now underway to transfer blight-resistant genes into the few remaining American chestnuts that weren't reached by the fungus or to find biological controls for the fungus that causes the disease.

An infection called whirling disease is decimating trout populations in many western states. It is caused by an exotic microorganism, *Myxobolus cerebralis,* which destroys cartilage in young fish causing them to swim erratically. The parasite is thought to have come into the United States in 1956 in a shipment of frozen fish.

Pollution

We have known for a long time that toxic pollutants can have disastrous effects on local populations of organisms. Pesticide-linked declines of fish-eating birds and falcons was well documented in the 1970s (Fig. 16.17). Marine mammals, alligators, fish, and other declining populations suggest complex interrelations between pollution and health (Chapter 15). Mysterious, widespread deaths of thousands of seals on both sides of the Atlantic in recent years are thought to be linked to an accumulation of persistent chlorinated hydrocarbons, such as DDT, PCBs, and dioxins, in fat, causing weakened immune systems that make animals vulnerable to infections. Similarly, mortality of Pacific sea lions, beluga whales in the St. Lawrence estuary, and striped dolphins in the Mediterranean are thought to be caused by accumulation of toxic pollutants.

Lead poisoning is another major cause of mortality for many species of wildlife. Bottom-feeding waterfowl, such as ducks, swans, and cranes, ingest spent shotgun pellets that fall into lakes and marshes. They store the pellets, instead of stones, in their gizzards and the lead slowly accumulates in their blood and other tissues. The U.S. Fish and Wildlife Service (USFWS) estimates that 3,000 tonnes of lead shot are deposited annually in wetlands and that between 2 and 3 million waterfowl die each year from lead poisoning.

Genetic Assimilation

Some rare and endangered species are threatened by **genetic assimilation** because they crossbreed with closely related species that are more numerous or more vigorous. Opportunistic plants or animals that are introduced into a habitat or displaced from their normal ranges by human actions may genetically overwhelm local populations. For example, hatchery-raised trout often are

FIGURE 16.17 Brown pelicans, and other bird species at the top of the food chain, were decimated by DDT in the 1960s. Pelicans and other species have largely recovered since DDT was banned in Canada and the United States.

Source: William P. Cunningham.

PART THREE Humans in the Environment

introduced into streams and lakes where they genetically dilute indigenous stocks. Similarly, black ducks have declined severely in Canada and the eastern United States in recent years. Hunting pressures and habitat loss are factors in this decline, but so is interbreeding with mallards forced into black duck habitat by destruction of prairie potholes in the west.

ENDANGERED SPECIES MANAGEMENT AND BIODIVERSITY PROTECTION

Over the years, we have gradually become aware of the harm we have done—and continue to do—to wildlife and biological resources. Slowly, we are adopting national legislation and international treaties to protect these irreplaceable assets. Parks, wildlife refuges, nature preserves, zoos, and restoration programs have been established to protect nature and rebuild depleted populations. There has been encouraging progress in this area, but much remains to be done. While most people favour pollution control or protection of favoured species such as whales or gorillas, surveys show that few understand what biological diversity is or why it is important.

The Endangered Species Act

Currently, the United States has 1,500 species on its endangered and threatened species lists and about 500 candidate species waiting to be considered. Canada, which generally has less diversity because of its boreal location, has designated a total of 46 endangered and 50 threatened species. The number of listed species in different taxonomic groups reflects much more about the kinds of organisms that humans consider interesting and desirable than the actual number in each group. In the United States, invertebrates make up about three-quarters of all known species but only 9 percent of those deemed worthy of protection. The International Union for Conservation of Nature and Natural Resources (IUCN) has a more comprehensive and more balanced world list (Table 16.4).

Species at Risk Act

The Canadian Species at Risk Act (SARA) was proclaimed in June 2003 after several false starts in the late 1990s. SARA was initiated as a result of Canada's 1992 commitment to protect species at risk according to the United Nations Convention on Biological Diversity. The purposes of the Act are to prevent Canadian indigenous species, subspecies, and distinct populations from becoming extirpated or extinct, to provide for the recovery of endangered or threatened species, and encourage the management of other species to prevent them from becoming at risk.

SARA is governed mainly by the federal minister of the Environment, along with the ministers of Fisheries and Oceans, and Canadian Heritage. These federal ministers, together with their provincial, territorial, and aboriginal government colleagues, administer the Act. SARA was specifically designed to:

- support and augment the Committee on the Status of Endangered Wildlife in Canada (COSEWIC), started in 1978, as an independent body of experts responsible for assessing and identifying species at risk;
- require that the best available knowledge be used to define long and short-term objectives in a recovery strategy and action plan;
- create prohibitions to protect listed threatened and endangered species and their critical habitat;
- recognize that compensation may be needed to ensure fairness following the imposition of the critical habitat prohibitions;
- be consistent with aboriginal and treaty rights and respect the authority of other federal ministers and provincial governments.

The following are the five steps in the SARA process (see Fig. 16.18):

1. Monitoring starts with an inventory of wildlife species to get an idea of the population status and trend, its ecological function, and a way of tracking information. As a result, the minister of the Environment publishes the report on the general status of wildlife species, every five years.

2. The Committee on the Status of Endangered Wildlife in Canada (COSEWIC) conducts the species assessment process. Based on the status report, a committee of experts conducts a species assessment and assigns the status of a wildlife species believed to be at some degree of risk nationally.

3. In response to an assessment and status designation, the minister issues a response statement. This document reflects the jurisdictional commitment to action and acts as a start to the national recovery process.

4. A recovery strategy outlines what is scientifically required for the successful recovery of a species at risk. This includes an identification of its critical habitat and what needs should be addressed. An action plan then identifies those specific actions needed to help in the species recovery as identified in the recovery strategy. This includes the various projects and activities with associated timelines.

5. Evaluation programs are carried out against the goals and objectives of the recovery strategy and action plan, where they are most effective. As a result, the minister must produce an annual report on the administration and implementation of the Act.

Wildlife listed by COSEWIC under SARA

The List of Wildlife Species at Risk set out in Schedule 1 of SARA includes:

17 extirpated species

- wildlife species that no longer exist in the wild in Canada, but exist elsewhere in the wild

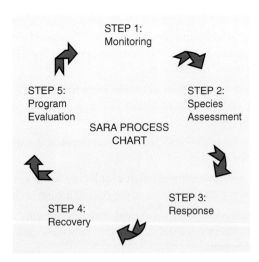

STEP 1:
Monitoring

STEP 5:
Program
Evaluation

STEP 2:
Species
Assessment

SARA PROCESS
CHART

STEP 4:
Recovery

STEP 3:
Response

FIGURE 16.18 How the Species at Risk Process works.

- 4 mammals, 2 birds, 1 amphibian, 2 reptiles, 2 fish, 1 mollusc, 3 butterflies, 2 plants

105 endangered species

- wildlife species that are facing imminent extirpation or extinction
- almost half of these are plants, mosses, or lichens; 19 of the 105 are birds

68 threatened species

- wildlife species that are likely to become endangered species if nothing is done to reverse the factors leading to their extirpation or extinction
- almost half are plants and mosses; only 7 of the 68 are birds

43 species of special concern

- wildlife species that may become threatened or endangered species because of a combination of biological characteristics and identified threats
- more than half are vertebrates (mammals, birds, fish)

It is easy to see human bias in the taxonomic distribution of species that have been identified as at risk, or even those that have been evaluated (Table 16.4). Out of those invertebrate species that have been evaluated, there is a far greater percentage that is threatened (58 percent) than vertebrate species (21 percent). Unfortunately, only about 3,400 of the over 1 million invertebrate species have been evaluated. Humans direct more of their energy and concern to larger, or more economically important species (sometimes cynically called "charismatic megafauna" by some conservation biologists). The situation has improved recently, as the value of many less obvious species as more general indicators of environmental health becomes apparent (Fig. 16.19).

Minimum Viable Populations

A critical question in all recovery programs for species at risk is the minimum population size required for long-term viability of rare and endangered species. As Chapter 4 shows, a species composed of a small number of individuals can undergo catastrophic declines due to environmental change, genetic problems, or simple random events when isolated in a limited geographic range. This phenomenon was elegantly described as **island biogeography** in the work of R. H. MacArthur and E. O. Wilson in 1967. Noticing that small islands far from a mainland have fewer terrestrial species than larger, nearer islands, MacArthur and Wilson proposed that species diversity is a balance between colonization and extinction rates (Fig. 16.20). An island far from a population source naturally has a lower rate of colonization than a nearer island because it is harder for terrestrial organisms to reach. At the same time, a large island can support more individuals of a given species and is, therefore, less likely to suffer extinction due to natural catastrophes, genetic problems, or demographic uncertainty—the chance that all the members of a single generation will be of the same sex.

Island biogeographical effects have been observed in many places. Cuba, for instance, is 100 times as large and has about 10 times as many amphibian species as its Caribbean neighbour, Monserrat. Similarly, in a study of bird species on the California Channel Islands, Jared Diamond observed that on islands with fewer than 10 breeding pairs, 39 percent of the populations went extinct over an 80-year period, while only 10 percent of populations numbering between 10 and 100 pairs went extinct in the same time (Fig. 16.21). Only one species numbering between 100 and 1,000 pairs went extinct and no species with over 1,000 pairs disappeared over this time.

Small population sizes affect species in isolated landscapes other than islands. Grizzly bears (Ursus arctos horribilis) once roamed across most of western North America. Hunting and habitat destruction reduced the number of grizzlies in the lower 48 states of the U.S. from an estimated 100,000 in 1800 to less than 1,000 animals in six separate subpopulations that now occupy less than 1 percent of the historic range. Recovery target sizes—based on estimated environmental carrying capacities—are less than 100 animals for some subpopulations. Conservation biologists predict that a completely isolated population of 100 bears cannot be maintained for more than a few generations. Even the 200 bears in Yellowstone National Park will be susceptible to genetic problems if completely isolated. Interestingly, computer models suggest that translocating only two unrelated bears into small populations every decade or so could greatly increase population viability.

For many species, loss of genetic diversity causes a variety of harmful effects that limit adaptability, reproduction, and species survival. How is diversity lost in small populations? (1) A *founder effect* occurs when a few individuals establish a new population. The limited genetic diversity from those original founders may not be enough to sustain the population. (2) A *demographic bottleneck* arises when only a few individuals survive some catastrophe. As the population replenishes itself, a limited genetic diversity similar to that in founder effect results. (3) *Genetic drift* is a reduction in gene frequency in a population due to unequal reproductive success, for example, some individuals breed more than others and their genes gradually come to dominate the population. (4) *Inbreeding*

TABLE 16.4 Numbers of Threatened Species by Major Groups of Organisms

	NUMBER OF DESCRIBED SPECIES	NUMBER OF SPECIES EVALUATED IN 2003	NUMBER OF THREATENED SPECIES IN 2000	NUMBER OF THREATENED SPECIES IN 2002	NUMBER OF THREATENED SPECIES IN 2003	NUMBER THREATENED AS % OF SPECIES THREATENED DESCRIBED AND [% EVALUATED] IN 2003
Vertebrates						
Mammals	4,842	4,789	1,130	1,137	1,130	23% [24%]
Birds	9,932	9,932	1,183	1,192	1,194	12% [12%]
Reptiles	8,134	473	296	293	293	4% [62%]
Amphibians	5,578	401	146	157	157	3% [39%]
Fishes	28,100	1,532	752	742	750	3% [49%]
Subtotal	56,586	17,127	3,507	3,521	3,524	6% [21%]
Invertebrates						
Insects	950,000	768	555	557	553	0.06% [72%]
Molluscs	70,000	2,098	938	939	967	1% [46%]
Crustaceans	40,000	461	408	409	409	1% [89%]
Others	130,200	55	27	27	30	0.02% [55%]
Subtotal	1,190,200	3,382	1,928	1,932	1,959	0.2% [58%]
Plants						
Mosses	15,000	93	80	80	80	0.5% [86%]
Ferns	13,025	180	—	—	111	1% [62%]
Gymnosperms	980	907	141	142	304	31% [34%]
Dicotyledons	199,350	7,734	5,099	5,202	5,768	3% [75%]
Monocotyledons	59,300	792	291	290	511	1% [65%]
Subtotal	287,655	9,706	5,611	5,714	6,774	2% [69%]
Others						
Lichens	10,000	2	—	—	2	0.02% [0.02%]
Subtotal	10,000	2	—	—	2	0.02% [0.02%]

Source: IUCN Red List 2003.

is mating of closely related individuals. Random, recessive, deleterious mutations—what we consider genetic diseases such as hemophilia in humans—that are usually hidden in a widely outcrossed population can be expressed when inbreeding occurs. This is why we have laws and cultural taboos prohibiting mating between siblings or first cousins in humans.

Some species seem not to be harmed by lack of genetic diversity. The northern elephant seal, for example, was reduced by overharvesting, more than a century ago, to fewer than 100 individuals. Today there are more than 150,000 of these enormous animals along the Pacific coast of Mexico and California. No marine mammal has come closer to extinction and then made such a remarkable recovery. All northern elephant seals today appear to be essentially genetically identical and yet they seem to be getting along just fine. Although interpretations are controversial, in highly

selected populations where only the most fit individuals reproduce, or in which there are few deleterious genes, inbreeding may not be such a negative factor.

Habitat Protection

Over the past decade, growing numbers of scientists, land managers, policymakers, and developers have been making the case that it is time to focus on a rational, continent-wide preservation of ecosystems that support maximum biological diversity rather than a species-by-species battle for the rarest or most popular organisms. By focusing on populations already reduced to only a few individuals, we spend most of our conservation funds on species that may be genetically doomed no matter what we do. Furthermore, by concentrating on individual species we spend millions of

FIGURE 16.19 Endangered species often serve as a barometer for the health of an entire ecosystem and as surrogate protector for a myriad of less well-known creatures.
Copyright 1990 by Herblock in the *Washington Post*.

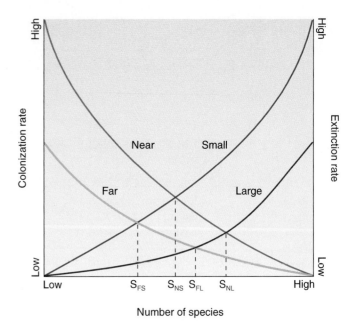

FIGURE 16.20 Predicted species richness on an island resulting from a balance between colonization (immigration) and extinction by natural causes. This island biogeography theory of MacArthur and Wilson (1967) is used to explain why large islands near a mainland (S_{NL}) tend to have more species than small, far islands (S_{FS}).
Source: Based on MacArthur and Wilson, *The Theory of Island Biogeography*, 1967, Princeton University Press.

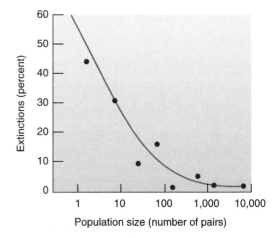

FIGURE 16.21 Extinction rates of bird species on the California Channel Islands as a function of population size over 80 years.
Source: Data from H. L. Jones and J. Diamond, "Short-term-base Studies of Turnover in Breeding Bird Populations on the California Coast Island," in *Condor,* vol. 78: 526–549, 1976.

dollars to breed plants or animals in captivity that have no natural habitat where they can be released. While flagship species such as mountain gorillas or Indian tigers are reproducing well in zoos and wild animal parks, the ecosystems that they formerly inhabited have largely disappeared.

A leader of this new form of conservation is J. Michael Scott, who was project leader of the California condor recovery program in the mid-1980s and had previously spent 10 years working on endangered species in Hawaii. In making maps of endangered species, Scott discovered that even Hawaii, where more than 50 percent of the land is federally owned, has many vegetation types completely outside of natural preserves (Fig. 16.22). The gaps between protected areas may contain more endangered species than are preserved within them.

This observation has led to an approach called **gap analysis** in which conservationists and wildlife managers look for un-protected landscapes that are rich in species. Computers and

geographical information systems (GIS) make it possible to store, manage, retrieve, and analyze vast amounts of data and create detailed, high-resolution maps relatively easily. This broad-scale, holistic approach seems likely to save more species than a piece-meal approach.

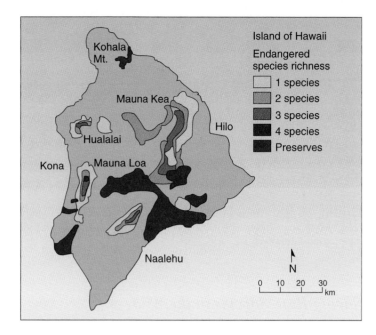

FIGURE 16.22 An example of the biodiversity maps produced by J. Michael Scott and the U.S. Fish and Wildlife Service. Notice that few of the areas of endangered species richness are protected in preserves, which were selected more for scenery or recreation than for biology.

Conservation biologist, R. E. Grumbine suggests four re management principles for protecting biodiversity in a large-scale, long-range approach:

1. Protect enough habitat for viable populations of all native species in a given region.
2. Manage at regional scales large enough to accommodate natural disturbances (fire, wind, climate change, etc.).
3. Plan over a period of centuries so that species and ecosystems may continue to evolve.
4. Allow for human use and occupancy at levels that do not result in significant ecological degradation.

International Wildlife Treaties

The 1975 Convention on International Trade in Endangered Species (CITES) was a significant step toward worldwide protection of endangered flora and fauna. It regulated trade in living specimens and products derived from listed species, but has not been foolproof. Species are smuggled out of countries where they are threatened or endangered, and documents are falsified to make it appear they have come from areas where the species are still common. Investigations and enforcement are especially difficult in developing countries where wildlife is disappearing most rapidly. Still, eliminating markets for endangered wildlife is an effective way of stopping poaching. Appendix I of CITES lists 700 species threatened with extinction by international trade.

CAPTIVE BREEDING AND SPECIES SURVIVAL PLANS

Breeding programs in zoos and botanical gardens are one way to attempt to save severely threatened species. Institutions like the Missouri Botanical Garden and the Bronx Zoo's Wildlife Conservation Society in the U.S. sponsor conservation and research programs. Botanical gardens, such as the Kew Gardens in England, and research stations, such as the International Rice Institute in the Philippines, are repositories for rare and endangered plant species that sometimes have ceased to exist in the wild. Valuable genetic traits are preserved in these collections, and in some cases, plants with unique cultural or ecological significance may be reintroduced into native habitats after being cultivated for decades or even centuries in these gardens and seed banks (Fig. 16.23).

Zoos can also help preserve wildlife, and act as repositories for genetic diversity. Until fairly recently, zoos depended on primarily wild-caught animals for most of their collections. This was a serious drain on wild populations, because up to 80 percent of the animals caught died from the trauma of capture and shipping. With better understanding of reproductive biology and better breeding facilities, most mammals in North American zoos now are produced by captive breeding programs.

Some zoos now participate in programs that reintroduce endangered species to the wild. The California condor (Fig. 16.24) is one of the best known cases of successful captive breeding. In 1986, only nine of these birds existed in their native habitat. Fearing the loss of these last condors, biologists captured them and brought them to the San Diego and Los Angeles zoos, which had begun breeding programs in the 1970s. By 2001 the population had reached 160, with several condors reintroduced to the wild. In March 2001, the first condor egg was laid in the wild since 1986, demonstrating that captive-bred condors can breed in the wild.

FIGURE 16.23 Rare plants can be preserved and studied in botanical gardens.

FIGURE 16.24 The California condor is recovering from near extinction. This is one of a few species that have received the bulk of restoration dollars.

Such breeding programs have limitations, however. Bats, whales, and many reptiles rarely reproduce in captivity and still come mainly from the wild. Furthermore, we will never be able to protect the complete spectrum of biological variety in zoos. According to one estimate, if all the space in U.S. zoos were used for captive breeding, only about 100 species of large mammals could be maintained on a long-term basis (Fig. 16.25).

These limitations lead to what is sometimes called the "Noah question:" how many species can or should we save? How much are we willing to invest to protect the slimy, smelly, crawly things? Would you favour preserving disease organisms, parasites, and vermin or should we use our limited resources to protect only beautiful, interesting, or seemingly useful organisms?

Even given adequate area and habitat conditions to perpetuate a given species, continued inbreeding of a small population in captivity can lead to the same kinds of fertility and infant survival problems described earlier for wild populations. To reduce genetic problems, zoos often exchange animals or ship individuals long distances to be bred. It sometimes turns out, however, that zoos far distant from each other unknowingly obtained their animals from the same source. Computer databases operated by the International Species Information System located at the Minnesota Zoo, now keep track of the genealogy of many species. This system can tell the complete reproductive history of every animal in every zoo in the world for some species. Comprehensive species survival plans based on this genealogy help match breeding pairs and project resource needs.

The ultimate problem with captive breeding, however, is that natural habitat may disappear while we are busy conserving the species itself. Large species such as tigers or apes are sometimes called "umbrella species." As long as they persist in their native habitat, many other species survive as well.

FIGURE 16.25 Primates endangered in the wild may be successfully preserved in zoos. Less charismatic species are rarely included in breeding programs, however.

Saving Rare Species in the Wild

Renowned zoologist George Schaller says that ultimately "zoos need to get out of their own walls and put more effort into saving the animals in the wild." An interesting application of this principle is a partnership between the Minnesota Zoo in the U.S. and the Ujung Kulon National Park in Indonesia, home to the world's few remaining Javanese rhinos. Rather than try to capture rhinos and move them to Minnesota, the zoo is helping to protect them in their native habitat by providing patrol boats, radios, housing, training, and salaries for Indonesian guards (Fig. 16.26). There are no plans to bring any rhinos to Minnesota and chances are very slight that any of us will ever see one, but we can gain satisfaction that, at least for now, a few Javanese rhinos still exist in the wild.

FIGURE 16.26 The *KM Minnesota* anchored in Tamanjaya Bay in west Java. Funds raised by the Minnesota Zoo paid for local construction of this boat, which allows wardens to patrol Ujung Kulon National Park and protect rare Javanese rhinos from poachers.

SUMMARY

In this chapter, we have briefly surveyed world biodiversity and the ways humans both benefit from and threaten it. Natural causes of wildlife destruction include evolutionary replacement and mass extinction. Among the threats from humans are overharvesting of animals and plants for food and commercial products. Millions of live wild plants and animals are collected for pets, houseplants, and medical research. Among the greatest damage we do to biodiversity are habitat destruction, the introduction of exotic species and diseases, pollution of the environment, and genetic assimilation.

The potential value of the species that may be lost if environmental destruction continues could be enormous. It is also possible that the changes we are causing could disrupt vital ecological services on which we all depend for life.

The Canadian Species at Risk Act, proclaimed in 2003, is an attempt to scientifically monitor species and develop protection and recovery strategies for those most at risk. Although it is taxonomically biased towards "charismatic megafauna," the principal behind it is protecting those species and their habitats, will also provide a benefit to the many, many species we aren't able to evaluate. The Canadian Species at Risk Act and CITES represent a new attitude toward wildlife in which we protect organisms just because they are rare and endangered. Now we are expanding our concern from individual species to protecting habitat, threatened landscapes, and entire biogeographical regions. Social, cultural, and economic factors must also be considered if we want to protect biological resources on a long-term, sustainable basis.

Zoos can be educational and entertaining while still serving important wildlife conservation and scientific functions. Modern zoos have greatly improved the living conditions for captive animals, resulting in improved breeding success; still, there are limits to the number and types of species that we could maintain under captive conditions. Zoos need to get out of their own walls and help save animals and plants in the wild.

QUESTIONS FOR REVIEW

1. What is the range of estimates of the total number of species on the earth? Why is the range so great?

2. What group of organisms has the largest number of species?

3. Define *extinction*. What is the natural rate of extinction in an undisturbed ecosystem?

4. What are rosy periwinkles and what products do we derive from them?

5. Describe some foods we obtain from wild plant species.

6. List nine seafood groups that are reduced or endangered by human activities.

7. What is the current rate of extinction and how does this compare to historic rates?

8. Compare the scope and effects of the Species at Risk Act and CITES.

9. Describe eight ways that humans directly or indirectly cause biological losses.

10. What is gap analysis and how is it related to ecosystem management and design of nature preserves?

QUESTIONS FOR CRITICAL THINKING

1. Many ecologists would like to move away from protecting individual endangered species to concentrate on protecting whole communities or ecosystems. Others fear that the public will only respond to and support glamorous "flagship" species such as gorillas, tigers, or otters. If you were designing conservation strategy, where would you put your emphasis?

2. Put yourself in the place of a fishing industry worker. If you continue to catch many species they will quickly become economically extinct if not completely exterminated. On the other hand, there are few jobs in your village and welfare will barely keep you alive. What would you do?

3. Only a few hundred grizzly bears remain in the contiguous United States, but populations are healthy in Canada and Alaska. Should millions of dollars be spent for grizzly recovery and management programs in Yellowstone National Park and adjacent wilderness areas?

4. How could people have believed a century ago that nature is so vast and fertile that human actions could never have a lasting impact on wildlife populations? Are there similar examples of denial or misjudgement occurring now?

5. Suppose you're having dinner with a friend who orders swordfish. What would you say? What are the ethical and biological arguments for or against eating endangered species?

6. In the past, mass extinction has allowed for new growth, including the evolution of our own species. Should we assume that another mass extinction would be a bad thing? Could it possibly be beneficial to us? To the world?

7. Some captive breeding programs in zoos are so successful that they often produce surplus animals that cannot be released into the wild because no native habitat remains. Plans to euthanize surplus animals raise storms of protests from animal lovers. What would you do if you were in charge of the zoo?

8. Debate with a friend or classmate the ethics of keeping animals captive in a zoo. After exploring the subject from one side, debate the issue from the opposite perspective. What do you learn from this exercise?

KEY TERMS

Web Exercises

ADOPT A SPECIES

Go to the COSEWIC website, at http://www.cosewic.gc.ca/eng/sct5/index_e.cfm and find a species that you are interested in "adopting." Pick a species that you think will not be immediately attractive to the general public. Learn everything you can about it and design a campaign to save your "adopted" species.

A coral reef is striking in its biodiversity, even to the creative eyes of a child.

Chapter 17

Land Use: Forests and Rangelands

Forests precede people; deserts succeed them.

—François-René de Chateaubriand—

OBJECTIVES

After studying this chapter, you should be able to:

- discuss the major world land uses and how human activities impact these areas.
- summarize some forest types and the products we derive from them.
- report on how and why tropical forests are being disrupted as well as how they might be better used.
- understand the major issues concerning forests in more highly developed countries such as Canada and the United States.
- understand forestry management and forestry practices.
- outline the extent, location, and state of grazing lands around the world.
- describe how overgrazing causes desertification of rangelands.

- explain why land reform and recognition of indigenous rights are essential for social justice as well as environmental protection.

WebQuest

forest management, clear-cutting, selective harvest, old-growth forests, debt-for-nature swaps

Above: Forest and rangelands, such as these near Crater Lake, Oregon, are complex ecosystems and valuable resources.

Saving an African Eden

When conservation biologist Paul Elkan first entered the Goualougo tract in remote northern Republic of Congo on a wildlife reconnaissance expedition in 1997, he was astonished at how little fear the animals showed. Generally, when wild monkeys or chimpanzees spot humans, they shriek and run away. Conditioned by years of hunting, wild animals fear humans. In the pristine forests of Goualougo, however, Elkan found that the animals followed him for hours, seemingly curious about this strange new species in their forest home. Bordered by two untamed rivers and many kilometers of dense, flooded forests, the Goualougo shows no signs of human intrusion. Even the local Bambendjelle people (called forest Pygmies by Europeans) had no memories of ever hunting there. "This is one of the last great wild places in the world," Elkan said.

The Congo basin holds about one-fourth of the world's tropical forests and is the largest stretch of lowland rainforest in the world, aside from the Amazon. Those forests are home to some of the most important wildlife populations remaining in Central Africa, including forest elephants, lowland gorillas, chimpanzees, bongo, buffalo, leopard, six species of small antelope, and eight species of monkeys. Logging and hunting are a growing threat to the forests and the wildlife they shelter. It has long been common practice for loggers to subsist on the local wildlife, or bushmeat. And logging roads provide access deep into the forest for hunters and settlers. Up to 1 million tonnes of bushmeat may be consumed in Central Africa each year.

In 1982, all the lands in the northern Republic of Congo were divided into large logging concessions. In the late 1980s, wildlife explorations revealed that the Nouabalé-Ndoki region held one of the densest assemblages of wildlife and the richest collection of primates on earth. Recognizing its ecological importance, the government of Congo upgraded the Noubalé-Ndoki forest from a forest management unit to national park in 1993. Adjacent parks in the Central African Republic and Cameroon create a contiguous area of irreplaceable lowland tropical rainforest (Fig. 17.1).

Unfortunately, when the parks were being established, the most pristine area was cut in two, and half of it, Goualougo triangle, was slated for logging. Logging rights were sold to a German-owned company, Congolaise Industrielle des Bois (C.I.B.). Biologist/explorer Mike Fay, who publicized the region's plight with his "megatransect" across the Congo basin, calls the Gualougo triangle the Inner Sanctum of this great forest. Elkan and others have confirmed the richness of this African Eden.

In July 2002, C.I.B., the Congolese government, and the Wildlife Conservation Society (WCS) announced that a 26,000 ha tract of the Goualougo triangle would be added to the Nouabalé-Ndoki National Park. This was the first time any logging company in the entire Congo Basin had voluntarily given up rights to valuable timberland. The C.I.B. estimated that it would forgo logging timber worth an estimated $40 million (U.S.). In addition, C.I.B.

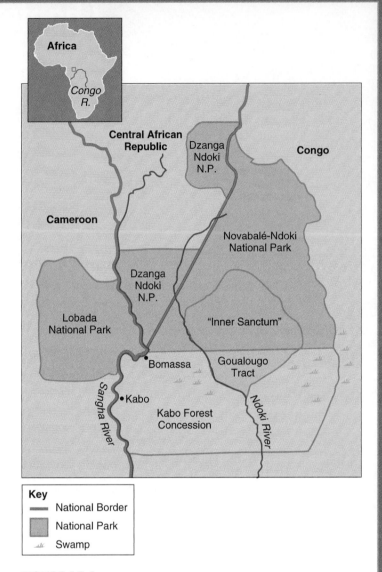

FIGURE 17.1 Adjacent parks in Cameroon, Central African Republic, and Congo protect an irreplaceable expanse of tropical rainforest.

promised to practise sustainable forest management on its remaining timber concessions and to participate in a wildlife protection program run by the WCS.

C.I.B. promises to prohibit hunting and transporting of bushmeat by company employees and promises to develop alternative protein sources (imported beef, fish, and poultry farms). The company will provide garden supplies and training to local farmers and a conservation awareness program for villagers and company employees. The forest outside the park will be zoned with community hunting areas near villages and no-hunting zones immediately surrounding the park. Protected areas will be patrolled by six government officers and 40 local ecoguards funded and managed by the WCS.

Why did the C.I.B. give up such lucrative logging concessions? Biologists played a vital role by revealing the incredible richness of the wildlife and the absence of human encroachment. Consumers who voiced concern about tropical forests were also key. More and more, consumers of paper products, furniture, and lumber are recognizing their links to tropical forests, and public pressure often leads to better policies. European environmental groups provided leadership and helped educate the public. C.I.B. gave up the Goualougo triangle as part of an effort to get its products certified as sustainably harvested, so that it could retain its European market.

Land-use policies are central to environmental quality, and they are influenced by decisions of national governments, corporations, citizen organizations, and individuals. In this chapter, we'll look at land uses, especially on forests and grasslands, and we'll consider how each of us plays a role, as individuals or as members of a community, in land-use decisions affecting our common environment.

WORLD LAND USES

The earth's total land area is about 132.4 million sq km, or about 29 percent of the surface of the globe. Figure 17.2 shows the area devoted to four major land-use categories. Much of the land that falls into the residual "other" category is tundra, marsh, desert, scrub forest, urban areas, bare rock, and ice or snow. About one-third of this land is so barren that it lacks plant cover altogether. While deserts and other unproductive lands are generally unsuitable for intensive human use, they play an important role in biogeochemical cycles and as a refuge for biological diversity. Presently, only about 4 percent of the world's land surface is formally protected in parks, wildlife refuges, and nature preserves (Chapter 22).

Notice that approximately 11 percent of the earth's landmass is now used for crops. Some agricultural experts claim that as much as half of the 7.2 billion ha of present forests and grazing lands—especially in Africa and South America—could be converted to crop production, given the proper inputs of water, fertilizer, erosion control, and mechanical preparation. Although this land could feed a vastly larger human population (perhaps 10 times the present number), sustained intensive agriculture could result in serious environmental and social problems (Chapter 14).

Rapidly increasing human populations and expanding forestry and agriculture have brought about extensive land-use changes throughout the world. Humans have affected every part of the globe, and we now dominate most areas with temperate climates and good soils. Over the past 10,000 years billions of hectares of forests, woodlands, and grasslands have been converted to cropland or permanent pasture, but overharvesting, erosion, pollution, and other forms of degradation also have turned large areas to desert or useless scrub. Biodiversity losses resulting from disruption of natural ecosystems is of great concern (Chapter 16). Air pollution-related forest declines are discussed in Chapter 9.

Cutting down forests or plowing grasslands have immediate and obvious destructive impacts on landscapes and wildlife. Given enough time, however, nature can be surprisingly resilient. New England, for example, lost most of its native forests to agriculture by the mid-nineteenth century, but today the region is largely reforested. Vermont, which had only 35 percent of its woods standing in 1850, is now 80 percent forested. Upstate New York has three times the population today than it had 150 years ago, yet it also has three times as much forest. Some rare species have been lost, but most of us would have difficulty distinguishing between second growth and primeval woods. Large mammals such as bears, moose, coyotes, and even wolves and mountain lions that had been absent for centuries now are reappearing. While this doesn't give us licence to do anything we want to natural landscapes, there is hope that damage can be repaired, as we will discuss later in this chapter.

WORLD FORESTS

Forests play a vital role in regulating climate, controlling water runoff, providing shelter and food for wildlife, and purifying the air. They produce valuable materials, such as wood and paper pulp, on which we all depend. Furthermore, forests have scenic, cultural, and historic values that deserve to be protected. In this section, we will look at forest distribution, use, and management.

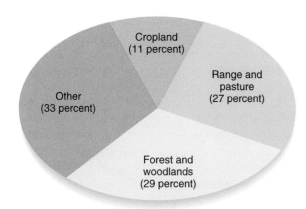

FIGURE 17.2 World land use. The "other" category includes tundra, desert, wetlands, and urban areas.
Source: Data from FAO, 1999.

Forest Distribution

Before large-scale human disturbances of the world began many thousands of years ago, forests and woodlands probably covered about 6 billion ha. Since then, about one-third of that area has been converted to cropland, pasture, settlements, or unproductive waste-lands. The 4.2 billion ha of forests and woodlands covers around 32 percent of the earth's land surface, nearly three times as much as all croplands. About four-fifths of the forest is classified as **closed canopy** (where tree crowns spread over 20 percent or more of the ground) and has potential for commercial timber harvests. The rest is **open canopy** forest or **woodland,** in which tree crowns cover less than 20 percent of the ground.

Figure 17.3 shows the distribution of forest by world region, while Figure 17.4 presents the world's main vegetation zones (see Chapter 6 for further description of these biomes). Table 17.1 describes closed-canopy forests as a percentage of original forest area by world regions. Among the forests of greatest concern are the remnants of original, primeval forests that are home to much of the world's biodiversity, endangered species, and indigenous human cultures. Sometimes called frontier forests, **old-growth forests** are those that cover a large enough area and have been undisturbed by human activities for a long enough time so that trees can live out a natural life cycle and ecological processes can occur in relatively

normal fashion. That doesn't mean that all trees need be enormous or thousands of years old. In some old-growth forests most trees may live less than a century before being killed by disease or some

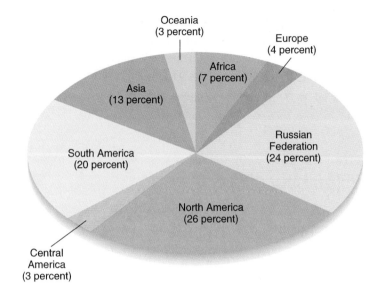

FIGURE 17.3 Current closed-canopy forests as percent of world total. *Source:* Data from *World Resources Institute 1998–99*, World Resource Institute.

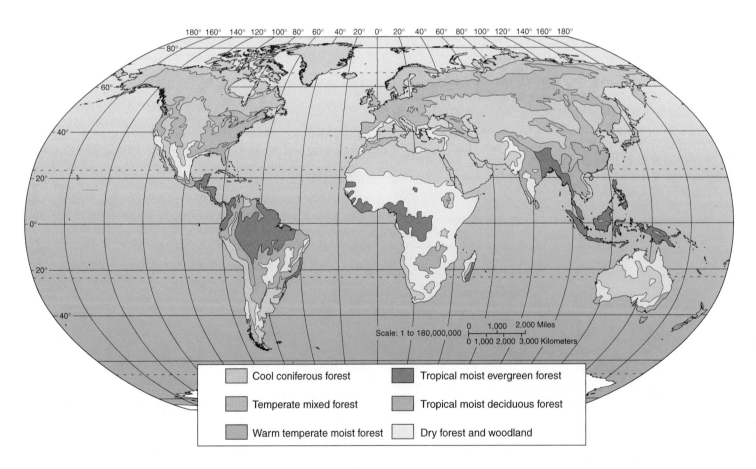

FIGURE 17.4 Main vegetation zones of the world's forests under natural conditions. Tan regions are grassland, tundra, desert, or ice.

TABLE 17.1	Current Closed-Canopy Forest Cover and Threats to Old-Growth Forests		
REGION	CURRENT FOREST AS PERCENT OF ORIGINAL	CURRENT OLD-GROWTH AS PERCENT OF ORIGINAL	PERCENT OF OLD-GROWTH THREATENED
Africa	33.9	7.8	77
Asia	28.2	5.3	60
North America	77.3	34.1	26
Central America	54.5	9.7	87
South America	69.1	45.6	54
Europe	30.5	0.1	100
Russia	68.7	29.3	19
Oceania*	64.9	22.3	76
World	53.6	21.7	39

*Oceania consists of Papua New Guinea, Australia, New Zealand, and many small-island nations.
Source: Data from *World Resources Institute 1998–99,* World Resource Institute.

natural disturbance like a fire. Nor does it mean that humans have never been present. Where human occupation entails relatively little impact, an old-growth forest may have been inhabited by people for a very long time. Even forests that have been logged or converted to cropland often can revert to old-growth characteristics if left alone long enough.

While forests still cover roughly two-thirds the area they once did worldwide, only 39 percent of those forests still retain old-growth features. The largest remaining areas of old-growth are in Russia, North America (mainly Canada), South America (mainly Brazil), and Oceania (mainly Papua New Guinea). Together these four countries account for more than three-quarters of all relatively undisturbed forests in the world. In general, remoteness rather than laws protect those forests. Although Table 17.1 describes only about one-fifth of Russian old-growth as threatened, rapid deforestation—both legal and illegal—especially in the Russian Far East, probably put a much greater area at risk. Note that all the remnants of ancient forest in Europe (mainly in Finland and Sweden) are considered threatened.

Forest Products

Wood plays a part in more activities of the modern economy than does any other commodity. There is hardly any industry that does not use wood or wood products somewhere in its manufacturing and marketing processes. Think about the amount of junk mail, newspapers, photocopies, and other paper products that each of us in developed countries handles, stores, and disposes of in a single day. Total world wood consumption is about 3.7 billion tonnes annually (Table 17.2).

Industrial timber and roundwood (unprocessed logs) are used to make lumber, plywood, veneer, particleboard, and chipboard. Together, they account for slightly less than one-half of worldwide wood consumption (about 1.57 billion tonnes per

TABLE 17.2	Annual Production and Trade of Wood Products (millions of cubic metres)			
REGION	FUEL AND CHARCOAL	INDUSTRIAL TIMBER AND ROUNDWOOD	PAPER	NET TRADE
Africa	481	59	3	–4.2
Asia	850	273	64	48.0
North and Central America	156	575	95	–25.7
South America	244	119	8	–7.6
Europe	51	268	68	13.7
Russia	57	188	5	–11.1
Oceania	9	35	3	–13.5
World	1,856	1,556	247	—

Notes: A negative number indicates net exports, while a positive number indicates net imports. Roundwood includes lumber, panels, timbers, and other industrial products.
Source: Data from *World Resources Institute 1998–99,* World Resource Institute.

year). This exceeds the use of steel and plastics combined. International trade in wood and wood products amounts to more than $100 billion each year. Developed countries produce less than half of all industrial wood but account for about 80 percent of its consumption. Less-developed countries, mainly in the tropics, produce more than half of industrial wood but use only 20 percent.

Canada, the United States, and the former Soviet Union are the largest producers of both industrial wood (lumber and panels) and paper pulp. Although old-growth, virgin forest with trees large enough to make plywood or clear furniture lumber is diminishing

FIGURE 17.5 Firewood accounts for almost half of all wood harvested worldwide and is the main energy source for one-quarter of all humans.

FIGURE 17.6 Simple, inexpensive stoves like these in Guatemala use as little as one-fourth as much wood as an open cooking fire. This simple technology can help save forests and reduce air pollution and household expenses.

everywhere, much of the industrial logging in North America and Europe occurs in managed forests where cut trees are replaced by new seedlings. In contrast, tropical hardwoods in Southeast Asia, Africa, and Latin America are being cut at an unsustainable rate, mostly from virgin forests.

Japan is by far the world's largest net importer of wood, purchasing about 46 million cubic metres per year or 96 percent of net Asian imports. Ironically, the United States is both a major exporter and importer of wood because they buy wood and paper pulp from Canada and finished wood products from Asia at the same time that they sell raw logs, rough lumber, and waste paper to Japan and other countries.

More than half of the people in the world depend on firewood or charcoal as their principal source of heating and cooking fuel (Fig. 17.5). Consequently, **fuelwood** accounts for slightly more than half of all wood harvested worldwide. Unfortunately, burgeoning populations and dwindling forests are causing wood shortages in many less-developed countries. About 1.5 billion people who depend on fuelwood as their primary energy source have less than they need. At present rates of population growth and wood consumption, the annual deficit is expected to increase from 400 million cubic metres (m^3) in 1995, to 2,600 m^3 in 2025. At that point, the demand will be twice the available fuelwood supply. The average amount of wood used for cooking and heating in 63

less-developed countries is about 1 m^3 per person per year, roughly equal to the amount of wood that each North American consumes as paper products alone.

Many people in poorer countries cook over open fires that deliver only about one-tenth of the available heat to cooking pots. Inexpensive metal stoves can double this efficiency, while locally made ceramic stoves can be four times as efficient as open fires (Fig. 17.6). These stoves can save up to 20 percent of household income for urban families.

Forest Management

Approximately 25 percent of the world's forests are managed scientifically for wood production. **Forest management** involves planning for sustainable harvests, with particular attention paid to forest regeneration (see Case Study, p. 365). Fires, insects, and diseases damage up to one-quarter of the annual growth in temperate forests. Recently, reduced forest growth and sudden die-off

Until the nineteenth century, the Menominee Nation occupied nearly half of what is now Wisconsin and northern Michigan. A woodland people, the Menominee hunted, fished, and gathered wild rice. Their name for themselves, "Mano'min ini'niwuk," means wild rice people. In 1854, besieged by smallpox, alcohol, and pressure from land-hungry European settlers, the tribe was forced onto a reservation representing less than 3 percent of their ancestral lands. Although the reservation—which lies along the Wolf River about 80 kilometres northwest of Green Bay, Wisconsin—is in the poorest county of the state, it represents a unique treasure. The forests covering 98 percent of the land make up the densest, most diverse woodlands in the Great Lakes region and the longest-running operation for sustained-yield forestry in the country.

Timber harvesting started in 1854, when 20 million board feet (1 board foot equals 2.36 dm³) were cut for lumber, planks, firewood, and fence rails. Greedy lumber barons tried to gain control of valuable white pine holdings but the tribe resisted. While the rest of the state was clear-cut, burned over, and turned into farmland, the Menominee insisted on careful, selective cutting of individual trees. By 1890, the tribe had built its own sawmill and carried out the first sustainable harvest management plan in the country. When first inventoried in 1890, the 89,000 ha reservation contained 1.3 billion board feet of lumber. Today, after 107 years in which a total of 2.25 billion board feet were harvested, the forest stock has increased to 1.7 billion board feet.

Ironically, the successful forestry operations almost brought about an end to the tribe. By 1959, the Menominee had accumulated a $10 million surplus. The Bureau of Indian affairs declared them too wealthy for continued protected status. Congress officially terminated the tribe in 1960, distributing trust funds and land allotments to individual tribal members. Forestry and mill operations were turned over to a private corporation, which immediately began liquidating reserves and racking up debts. Tribal leaders fought termination and successfully restored reservation status in 1973. Although the forest remains largely undivided, tribal enterprises still are plagued by debts incurred under privatization.

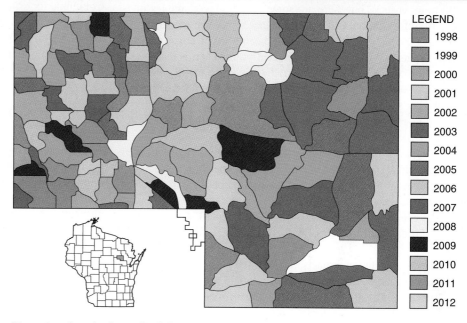

Menominee forestlands are divided into 109 compartments based on species composition, topography, stand history, and management goals. Legend shows planned year for harvest in 15-year cycle. Courtesy of Menominee Tribal Enterprises. Reprinted by permission.

Wise elders set up a simple forest management plan when they began operations more than a century ago. Rather than manage for short-term yields, as is the case for lands around them, the tribe aims for maximum quantity and quality of native species. They say, "instead of cutting the best, we cut the worst first. We're managing our resources to last forever." They have one of the few lumber operations in the country that preserves old-growth characteristics.

The forest is northern hardwood type with a mixture of sugar maple, beech, hemlock, basswood, yellow birch, white pine, jack pine, and aspen. The heart of the management plan is a continuous forest inventory to determine optimum growth, species balance, ecological health, and cutting cycles. Land is divided into 109 compartments based on 11 species combinations, topography, stand history, and management goals. Two-thirds of the forest is managed for mixed species and ages, with selective cutting on a 15-year cycle. About 20 percent is devoted to aspen and jack pine in even-age (clear-cut) stands of no more than 12 ha each. Nearly 400 ha of white pine utilize a two-step shelterwood program that mimics the natural fire-succession sequence by artificially manipulating the balance of sunlight, competition, and soil disturbance. Judicial use of herbicides, prescribed burns, selective cutting, and rock raking maintain optimum growth and regeneration of this valuable species, which has largely disappeared elsewhere in the Great Lakes forest.

Some 300 people are now employed in the tribal forestry and sawmill operations. Logging is carried out by both Indian and non-Indian private contractors. Current harvest levels are about 30 million board feet per year. Lumber from the tribal mill is certified by the Green Cross organization as "good wood" harvested in a socially and environmentally responsible manner.

As Aldo Leopold said in *Sand County Almanac,* the best definition of conservation "is written not with a pen, but with an axe. It is a matter of what a man thinks about while chopping, or while deciding what to chop." He was describing the kind of stewardship practised on the Menominee reservation. In addition to welcome economic returns, the sustainable harvesting has brought them an aesthetically pleasing forest, spiritual rejuvenation, clean water, and a sense of pride in being Menominee.

of certain tree species in industrialized countries have caused great concern. It is thought that long-range transport of air pollutants (Chapter 9) is contributing to this sudden forest death, but not all the causes and solutions are yet understood.

Most countries replant far less forest than is harvested or converted to other uses, but there are some outstanding examples of successful reforestation. China, for instance, cut down most of its forests 1,000 years ago and has suffered centuries of erosion and terrible floods as a consequence. Recently, however, a massive reforestation campaign has been started. An average of 4.5 million ha per year were replanted during the last decade. South Korea also has had very successful forest restoration programs. After losing nearly all its trees during the civil war 30 years ago, the country is now about 70 percent forested again.

In spite of being the world's largest net importer of wood, Japan has increased forests to approximately 68 percent of its land area. Strict environmental laws and constraints on the harvesting of local forests encourage imports so that Japan's forests are being preserved while it uses those of its trading partners. It is estimated that two-thirds of all tropical hardwoods cut in Asia are shipped to Japan.

Many reforestation projects involve large plantations of a single species, such as eucalyptus or hybrid poplar, in a single-use, intensive cropping called **monoculture agroforestry.** Although this produces high profits, a dense, single-species stand encourages pest and disease infestations. This type of management lends itself to mechanized clear-cut harvesting, which saves money and labour but tends to leave soil exposed to erosion. Monocultures eliminate habitat for many woodland species and often disrupt ecological processes that keep forests healthy and productive. When profits from these forest plantations go to absentee landlords or government agencies, local people have little incentive to prevent fires or keep grazing animals out of newly planted areas. In some countries, such as the Philippines, Israel, and El Salvador, government reforestation projects have been targets for destruction by antigovernment forces, with devastating environmental impacts.

Promising alternative agroforestry plans are being promoted by conservation and public service organizations such as The New Forest Fund and Oxfam. These groups encourage planting of mixed species, community woodlots including fruit and nut trees as well as fast-growing, multipurpose trees such as *Leucaena*. Millions of seedlings have been planted in hundreds of self-help projects in Asia, Africa, and Latin America. *Leucaena* is a legume, so it fixes nitrogen and improves the soil. Its nutritious leaves are good livestock fodder. It can grow up to 3 metres per year and quickly provides shade, forage for livestock, firewood, and good lumber for building. *Leucaena* can be an aggressive, weedy exotic, however, if it escapes from cultivation. As in most environmental and social solutions, a combination of species and methods are needed. Community woodlots can be planted on wasteland or along roads or slopes too steep to plow so they do not interfere with agriculture. They protect watersheds, create windbreaks, and, if composed of mixed species, also provide useful food and forest products such as fruits, nuts, mushrooms, or materials for handicrafts on a sustained-yield basis.

Crown Land

In Canada, forest management is part of the jurisdiction of the provinces and territories. Most forests are effectively owned by the province or territory (actually considered "Crown land"). In British Columbia, for example, about 95 percent of the land mass is owned by the province. Private forest companies pay stumpage fees and accept responsibilities for explicit land management plans when they log in publicly owned forests.

TROPICAL FORESTS

The richest and most diverse terrestrial ecosystems on the earth are the tropical forests. Although they now occupy less than 10 percent of the earth's land surface, these forests are thought to contain more than two-thirds of all higher plant biomass and at least one-half of all plant, animal, and microbial species in the world.

Diminishing Forests

While many temperate forests are expanding slightly due to reforestation and abandonment of marginal farmlands, tropical forests are shrinking rapidly. In 1900, an estimated 12.5 million square kilometres of tropical lands were covered with closed-canopy forest. The Food and Agriculture Organization of the United Nations estimates that about 0.8 percent of the remaining tropical forest is cleared each year (Fig. 17.7). The species extinction due to this loss of habitat is discussed in Chapter 16.

There is considerable debate about current rates of deforestation in the tropics. Scientists at Brazil's National Space Research

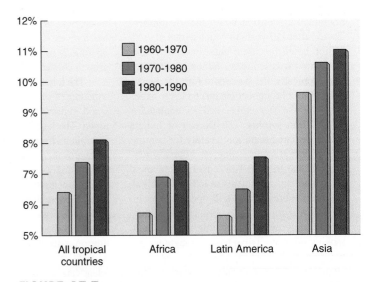

FIGURE 17.7 Estimated rate of tropical forest losses per decade 1960–1990. During this period, about one-fifth of all natural tropical forest was cleared.

Source: Data from *World Resources* 1996–97, World Resources Institute.

PART THREE Humans in the Environment

FIGURE 17.8 Cutting and burning of tropical rainforest results in wildlife destruction, habitat loss, soil erosion, rapid water runoff, and waste of forest resources. It also contributes to global climate change.

Institute (INPE) use satellite data to estimate the number of fires and the amount of forest clearing in the Amazon. In 1999, more than 31,000 fires were spotted in a single month (Fig. 17.8). Assuming that 40 percent of these fires occurred on recently cleared forest, remote sensing experts calculated that 8 million ha per year were being cut. This number has led to widespread criticism of Brazil for allowing a priceless world heritage to be destroyed. Brazilian scientists reevaluated their data and concluded that losses are far less than originally thought. Critics claim, however, that even the original estimate may be too low.

Similar uncertainty about rates of forest destruction exist in many places, partly because of divergent definitions of deforestation. Some scientists and politicians insist that it means a complete change from forest to agriculture, urban areas, or desert. Others include any area that has been logged, even if the cut is selective and regrowth will be rapid. There also are difficulties in interpreting satellite images. Savannas, open woodlands, and early successional stages following natural disturbance are hard to distinguish from logged areas. Countries also have economic and political reasons to hide or exaggerate the extent of their activities. Consequently, estimates for total tropical forest losses

range from about 5 million to more than 20 million ha per year. The FAO estimates of 14.5 million ha are generally the most widely accepted.

By most accounts, Brazil has the highest rate of deforestation in the world, but it also has by far the largest tropical forests. Indonesia and Malaysia together may be losing as much primary forest each year as Brazil even though their original amount was far lower. In 1997, forest fires on Borneo and Sumatra produced clouds of smoke that made air unbreathable in Singapore, the Philippines, and even parts of Thailand and China. Other areas have lost a greater proportion of their original forests.

In Africa, the coastal forests of Sierra Leone, Ghana, Madagascar, Cameroon, and Liberia, already have been mostly destroyed. Haiti was once 80 percent forested; today, essentially all that forest has been destroyed and the land lies barren and eroded. India, Burma, Kampuchea, Thailand, and Vietnam all have little old-growth lowland forest left. In Central America, nearly two-thirds of the original moist tropical forest has been destroyed, mostly within the last 30 years and primarily due to conversion of forest to cattle range (Fig. 17.9).

A variety of causes lead to this deforestation. In Costa Rica, in addition to cattle ranching (Fig. 17.10), banana plantations consume large areas of forest. In Brazil, a combination of land clearing for cattle ranching coupled with an invasion of immigrants from the south are responsible for forest losses. Although land-hungry farmers often are blamed for setting destructive forest fires, the culprits often are big corporations. In Indonesia and Malaysia, for example, fires are set to cover illegal logging operations or to clear land for large-scale oil palm plantations.

Swidden Agriculture

Indigenous forest people often are blamed for tropical forest destruction because they carry out shifting agriculture that requires forest clearing. Actually, this ancient farming technique can be an ecologically sound way of obtaining a sustained yield from fragile tropical soils if it is done carefully and in moderation. This practice is sometimes called "slash and burn" by people who don't realize how complex and carefully balanced this method of farming actually can be when practised sustainably. The preferred terms of **milpa** or **swidden agriculture** are taken from local names for *field*.

In this system, farmers clear a new plot of about a hectare each year. Small trees are felled and large trees are killed by girdling (cutting away a ring of bark) so that sunlight can penetrate through to the ground. After a few weeks of drying, the branches, leaf litter, and fallen trunks are burned to prepare a rich seedbed of ashes. Fast-growing crops, such as bananas and papayas, are planted immediately to control erosion and to shade root crops, such as cassava and sweet potato, which anchor the soil. Maize, rice, and up to 80 other crops are planted in a riotous profusion. Although they would not recognize the terms, these indigenous people are practising **mixed perennial polyculture.**

The diversity of the milpa plot mimics that of the jungle itself, even though the species representation is more restricted.

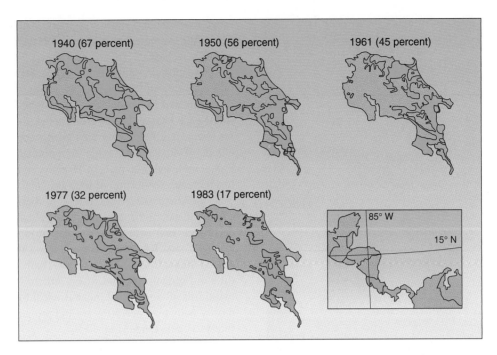

FIGURE 17.9 Loss of primary forest in Costa Rica 1940–1983. Percentages show forestland as a proportion of total land area. Since 1983 almost all remaining primary forest is in National Parks or other protected areas.

Sources: Data from Sader and Joyce, "Deforestation Rates and Trends in Costa Rica, 1940–1983," *Biotropica* 20:11–19; and T. C. Whitmore and G. T. Prance, *Biogeography and Quaternary History in Tropical America,* 1987, Clarendon Press, Oxford.

FIGURE 17.10 Cattle graze on recently cleared tropical rainforest land in Costa Rica. About two-thirds of the forest in Central America has been destroyed, mostly in the past few decades as land is converted to pasture or cropland. Unfortunately, the soil is poorly suited to grazing or farming, and these ventures usually fail in a few years.

When managed well, the soil is covered with vegetation. This variety means that crops mature in a staggered sequence and there is almost always something to eat from the plot. It also helps prevent eruptive insect infestations that would plague a monoculture crop. Annual yields from a single hectare can be as high as 6 tonnes of grain (maize or rice) and another 5 tonnes of roots, vegetable crops, nuts, and berries. This yield is comparable to the best results with intensive row cropping, and about 1,000 times as much food is produced from the same land as when it is converted to cattle pasture.

After a year or two, the forest begins to take over the garden plot again. The farmer will continue to harvest perennial crops for a while and will hunt for small animals that are attracted to the lush vegetation. Ideally, the land then will be allowed to remain covered with jungle vegetation for 10 to 15 years while nutrients accumulate before it is cleared and replanted again. In many places, population growth and displacement of farmers from other areas have forced shifting agriculturalists to reuse their traditional plots on shorter and shorter rotation. When plots are farmed every year or two, nutrients are lost faster than they can be replaced, the forest doesn't regrow as vigorously as it once did, and additional species are lost. Eventually, erosion and overuse reduce productivity so much that the land is practically useless.

Logging and Land Invasions

The other major source of forest destruction is usually the result of logging and subsequent invasion by land-hungry people from other areas. The loggers often are interested only in "creaming" the most valuable hardwoods, such as teak, mahogany, sandalwood, or

FIGURE 17.11 Logging roads open up access to landless settlers, miners, and hunters who drive away native species and indigenous people. There often is a marked difference between new migrants and indigenous swidden agriculturists.

ebony. Although only one or two trees per hectare might be taken, widespread devastation usually results. Because the canopy of tropical forests is usually strongly linked by vines and interlocking branches, felling one tree can easily bring down a dozen others. Tractors dragging out logs damage more trees, and construction of roads takes large land areas. Insects and infections invade wounded trees. Tropical trees, which usually have shallow root systems, are easily toppled by wind and erosion when they are no longer supported by their neighbours. Up to three-fourths of the canopy may be destroyed for the sake of a few logs. Obviously, the complex biological community of the layered canopy (Chapter 5) is severely disrupted by this practice.

What happens next? Bulldozed roads make it possible for large numbers of immigrants to move into the forest in search of land to farm. People with little experience or understanding of the complex rainforest ecosystem try to turn it into farms and ranches (Fig. 17.11). Too often, the result is ecological disaster. Rains wash away the topsoil and the tropical sun bakes the exposed subsoil into an impervious hardpan that is nearly useless for farming.

Degradation of rivers is another disastrous result of forest clearing. Tropical rivers carry two-thirds of all freshwater runoff in the world. In an undisturbed forest, rivers are usually clear, clean, and flow year-round because of the "sponge effect" of the thick root mat created by the trees. When the forest is disrupted and the thin forest soil is exposed, erosion quickly carries away the soil, silting river bottoms, filling reservoirs, ruining hydroelectric and irrigation projects, filling estuaries, and smothering coral reefs offshore. In Malaysia, sediment yield from an undisturbed primary forest was about 100 cubic metres per square kilometre per year. After forest clearing, the same river carried 2,500 cubic metres per square kilometre per year.

Forest Protection

What can be done to stop this destruction and encourage careful management? While much of the news is discouraging, there are some hopeful signs for tropical forest conservation. Many tropical countries have realized that forests represent a valuable resource and they are taking steps to protect them. Indonesia has announced plans to preserve 100,000 square kilometres, one-tenth of its original forest. Zaire and Brazil each plan to protect 350,000 square kilometres (about the size of Norway) in parks and forest preserves. Costa Rica is a leader in restoring—as well as conserving—tropical forest. In Costa Rica's Guanacaste National Park, ecologist Dan Janzen has led a pioneering project to reestablish dry tropical forest—one of the lesser known but most endangered ecosystems in the region.

People on the grassroots level also are working to protect and restore forests. Reforestation projects build community pride while also protecting the land (Fig. 17.12). India, for instance, has a long history of nonviolent, passive resistance to protest unfair government policies. The *satyagrahas* go back to the beginning of Indian culture and often have been associated with forest preservation. Gandhi drew on this tradition in his protests of British colonial rule in the 1930s and 1940s. During the 1970s, commercial loggers began large-scale tree felling in the Garhwal region in the state of Uttar Pradesh in northern India. Landslides and floods resulted from stripping the forest cover from the hills. The firewood on which local people depended was destroyed, and the way of life of the traditional forest culture was threatened. In a remarkable display of courage and determination, the village women wrapped their arms around the trees to protect them, sparking the *Chipko Andolan* movement (literally, movement to hug trees). They prevented logging on 12,000 square kilometres of sensitive watersheds in the Alakanada basin. Today, the *Chipko Andolan* movement has grown to more than 4,000 groups working to save India's forests.

Debt-for-Nature Swaps

Those of us in developed countries can make a contribution toward saving tropical forests as well. Financing nature protection is often a problem in developing countries where the need is

FIGURE 17.12 School children plant trees in a community reforestation project in Haiti. Grassroots efforts, such as this, benefit both local people and their environment. Many top-down development projects benefit only the ruling elite and multinational corporations.

greatest. One promising approach is called **debt-for-nature swaps.** Banks, governments, and lending institutions now hold nearly $1 trillion in loans to developing countries. There is little prospect of ever collecting much of this debt, and banks are often willing to sell bonds at a steep discount—perhaps as little as 10 cents on the dollar. Conservation organizations buy debt obligations on the secondary market at a discount and then offer to cancel the debt if the debtor country will agree to protect or restore an area of biological importance.

There have been many such swaps. Conservation International, for instance, bought $650,000 of Bolivia's debt for $100,000—an 85 percent discount. In exchange for cancelling this debt, Bolivia agreed to protect nearly 1 million ha around the Beni Biosphere Reserve in the Andean foothills. Ecuador and Costa Rica have had a different kind of debt-for-nature swap. They exchanged debt for local currency bonds that are used to fund activities of local private conservation organizations in the country. This has the dual advantage of building and supporting indigenous environmental groups while protecting the land.

Agreements have been reached with Madagascar and Zambia to swap debts for nature, and negotiations are underway with Peru, Mexico, and Tanzania. Critics charge that these swaps compromise national sovereignty and that they will do little to reduce Third World debt or change the situations that led to environmental destruction in the first place. In 2000, Conservation International made an effort to remedy this problem. By buying long-term logging concessions for 81,000 ha of tropical rainforest in Guyana,

the organization is bringing the same economic benefits that the government would get if the land were actually logged—without losing the natural resources of the forest.

TEMPERATE FORESTS

Tropical countries are not unique in harvesting forests at unsustainable rates and in an ecologically damaging fashion. Northern countries have a long history of liquidating forest resources that continues today in many places. Perhaps the largest and most destructive harvest in the world today is taking place in Eastern Russia. Siberia is larger than Amazonia and contains one-fourth of the world's timber reserves. Four million ha of Siberian taiga and deciduous forests are being felled annually—primarily by Korean loggers—to shore up Russia's faltering economy. China has announced plans for a giant hydroelectric dam on the Amur River, which forms its border with Siberia. This power source will allow China to move 100 million settlers into the region, which is home to the endangered Siberian tiger and several species of cranes. The results are likely to be similar to those we have just discussed in reference to tropical forests.

In Canada and the United States, the two main issues in timber management are (1) cutting of the last remnants of old-growth forest and (2) methods used in timber harvest.

Ancient Forests of the Pacific Northwest

Only a little more than a century ago, most of the coastal ranges of British Columbia and adjacent areas, were clothed in a lush forest of huge, ancient trees (Fig. 17.13). The moist, mild climate and rich soil of the lowland valleys nurtured magnificent stands of redwood in California and of western red cedar, Douglas fir, western hemlock, and Sitka spruce along the rest of the coast. Everyone knows that redwoods can be huge and very old, but did you know that these other species can reach 3 to 4 m in diameter, 90 m in height (as high as a 20-story building), and 1,000 or more years in age? These temperate rainforests are probably second only to tropical rainforests in terms of terrestrial biodiversity, and they accumulate more total biomass in standing vegetation per unit area than any other ecosystem on earth.

These old-growth temperate rainforests are extremely complex ecologically. Only in recent years have we begun to realize how many different species live there and how interrelated their life cycles are. Many endemic species such as the northern spotted owl (Fig. 17.14), Vaux's swift, and the marbled murrelet are so highly adapted to the unique conditions of these ancient forests that they live nowhere else.

Before loggers and settlers arrived, there were probably 12.5 million ha of virgin temperate rainforest in the Pacific Northwest. Less than 10 percent of that forest in the United States still remains, and 80 percent of what is left is scheduled to be cut down in the near future. B.C. has about 7.6 million ha of coastal temperate rainforest of which more than 800,000 ha (around 10 percent)

FIGURE 17.13 This forest in McMillan Provincial Park on Vancouver Island, B.C., is an excellent example of the old-growth temperate rain-forests of the Pacific Northwest.

is protected, but about 58,000 ha is harvested annually (less than 1 percent).

Wilderness and Wildlife Protection

Many environmentalists would like to save all remaining virgin forest in Canada as a refuge for endangered wildlife, a laboratory for scientific study, and a place for recreation and spiritual renewal. Economic pressures to harvest the valuable giant trees are considerable, however. The forest products industry employs about 150,000 people in the Pacific Northwest and adds nearly $7 billion annually to the economy. In Oregon, forest products account for one-fifth of the gross state product and many small towns depend almost entirely on logging for their economic life. On the other hand, an economic development study of Washington and Oregon suggests that recreation could provide 16 jobs for every one lost by logging. The biggest problem is how people who are currently employed in the timber industry can make a living during a transition period.

In 1989, environmentalists sued the U.S. Forest Service over plans to clear-cut most of the remaining old-growth forest, arguing that spotted owls are endangered and must be protected under the U.S. Endangered Species Act. U.S. federal courts agreed and

FIGURE 17.14 Only about 2,000 pairs of northern spotted owls remain in the old-growth forests of the Pacific Northwest. Cutting old-growth forests threatens the endangered species, but reduced logging threatens the jobs of many timber workers.

ordered some 1 million ha of ancient forest set aside to preserve the last 1,000 pairs of owls. This is about half the remaining virgin forest in Washington and Oregon. The timber industry claims that 40,000 jobs have been lost, although environmentalists dispute this number.

Environmentalists also point out that logging jobs are disappearing because of mechanization, a naturally dwindling resource base, and the shipping of raw logs to Japan. They argue that the big trees are disappearing anyway. The question is when to stop cutting—now while a few remain, or in a few years when they are all gone? The workers will have to be retrained anyway; why not sooner rather than later? In spite of complaints that logging restrictions on public land hurt business, companies with private land holdings have benefited from rising timber prices. Weyerhauser, for instance, reported a $90 million (81 percent) increase in the value of its timber holdings in 1992 as a result of the reduced cut in federal forests.

A compromise forest management plan will allow some continued cutting in the Pacific Northwest but also will protect a high percentage of prime ancient forests. In this plan, 0.9 million ha of riparian (streamside) preserves are set aside to protect salmon and other aquatic wildlife, while 2 million ha of "old-growth" reserves are established on national forest lands. The maximum annual allowable cut on public lands in the region is 1.2 billion board feet, more than environmentalists want, but far below the peak of 5 billion per year in the 1980s. Some of this logging is selective salvage and thinning in old-growth preserves.

Although current plans protect only a fraction of the original forest, local residents in many areas resent *any* restrictions on their access to resources on public or private lands. "Wise Use" groups have brought lawsuits and introduced federal, state, and local ordinances to prohibit programs that limit land use or to require repayment of losses that result from land-use regulations.

FIGURE 17.15 This huge clear-cut in Washington's Gilford Pinchot National Forest threatens species dependent on old-growth forest and exposes steep slopes to soil erosion. Restoring something like the original forest will take hundreds of years.

FIGURE 17.16 Many communities depend on timber for survival. Logging can threaten other sources of income, such as salmon fishing, however.

Harvest Methods

Most lumber and pulpwood in Canada and the United States currently is harvested by **clear-cutting,** in which every tree in a given area is cut regardless of size (Fig. 17.15). This method enables large machines to fell, trim, and skid logs rapidly, but wastes many small trees, exposes soil to erosion, and eliminates habitat for many forest species. Early successional species, such as Douglas fir, jack pine, lodgepole pine, or loblolly pine, on the other hand, flourish after clear-cutting. In a forest managed for these trees, or for raspberries, blueberries, deer, or grouse, this is a good method.

Size and shape of clear-cuts vary depending on topography and management policies. Clear-cuts can be in individual strips, alternating rows, small scattered patches, or areas as large as thousands of acres. Natural regeneration is better in small patches or strips that resemble natural forest openings than in huge clear-cuts. Small patches also are less disruptive for wildlife than are huge denuded spaces. It was once thought that good forest management required immediate removal of all dead trees and logging residue. Research has shown, however, that standing snags and coarse woody debris play important ecological roles, including soil protection, habitat for a variety of organisms, and nutrient recycling.

Other harvest practices offer variations on, or substitutes to, clear-cutting. *Coppicing* is used to encourage stump sprouts from species such as aspen, red oak, beech, or short-leaf pine and is usually accomplished by clear-cutting. In *seed tree harvesting,* some mature trees (generally two to five trees per hectare) are left standing to serve as a seed source in an otherwise clear-cut patch. *Shelterwood harvesting* involves removing mature trees in a series of two or more cuts. This encourages regeneration of wind- and sun-sensitive species such as spruce and fir. **Strip cutting** entails harvesting all the trees in a narrow corridor. For many forest types, the least disruptive harvest method is **selective cutting,** in which

only a small percentage of the mature trees are taken in each 10- or 20-year rotation. Ponderosa pines, for example, are usually selectively cut to thin stands and to improve growth of the remaining trees. A forest managed by selective cutting can retain many of the characteristics of age distribution and ground cover of a mature, old-growth forest.

The lush Douglas fir and redwood forests of the rainy Pacific Coast Range provide a vivid illustration of why clear-cutting in old-growth forests is controversial. The rich soil and mild, moist climate of the region promote rapid regeneration after clear-cutting, but many of these forests are on steep slopes where erosion is a serious problem. Hillsides are stripped of soil, which fills streams and smothers aquatic life.

Canadian First Nation people have blocked roads and brought lawsuits to protest destruction of traditional lands and subsistence ways of life. People concerned about commercial and sport fishing have joined the battle, both in British Columbia and in the United States. They argue that salmon spawning depends on the clear cold streams of the native forests. Harvesting timber often destroys this valuable resource. The income from a single year's salmon run can outweigh all the profits from timber harvesting, and the salmon return year after year, while 1,000-year-old trees will never be seen again (Fig. 17.16).

Fire Management

For more than 70 years, firefighting has been a high priority for forest managers. In the U.S., Smokey the Bear has appeared on posters, brochures, and even postage stamps to tell us of the horrors of wildfires and to warn us that "only you can prevent forest fires." To most people, a blackened, smoking, burned forest appears ruined forever. We envision raging flames devouring helpless wildlife, threatening homes, and wasting valuable timber resources (Fig. 17.17). Given such frightening images, it's no

FIGURE 17.17 By suppressing fires and allowing fuel to accumulate, we make major fires such as this more likely. The safest and most ecologically sound management policy for some forests may be to allow natural or prescribed fires, that don't threaten property or human life, to burn periodically.

wonder that the public demands fire protection and government agencies make every effort to provide it. In 1998 alone, the United States spent over $1 billion and lost 33 lives in efforts to stop forest fires.

Recent studies of the ecological role of fire in forests, however, suggest that much of our horror of fire and our attempts to suppress it may be misguided. As discussed in Chapter 5, many biological communities are fire-adapted and require periodic fires for regeneration. In the western United States, for instance, dry montane forests originally were dominated by big trees such as ponderosa pine, Douglas fir, and giant sequoias, whose thick, fire-resistant bark and lack of branches close to the ground protected them from frequent creeping ground fires. Historic accounts describe these forests as open and parklike, with little underbrush, luxuriant grass, and abundant wildlife.

Eliminating fire from these forests has allowed shrubs and small trees to fill the forest floor, crowding out grasses and forbs (herbs that are not grasses). As woody debris accumulates, the chances of a really big fire increase. Small trees act as "fire ladders" to carry flames up into the crowns of forest giants. By preventing low-intensity fires that once kept the forest open and free of fuel, we actually threaten the trees we intend to protect.

Our attempts to put fires out often cause more ecological damage than the fires themselves. Firefighters bulldoze firebreaks through sensitive landscapes such as tundra or wetlands, leaving scars that last far longer than the effects of the fire. Often the only thing that extinguishes a major fire is a change in the weather. Millions of dollars spent to dig fire lines and bomb outbreaks with chemicals and water have little effect as the fire goes where it will.

Fire fighting is also getting more expensive. Increasing numbers of isolated "end of the road" homes and cabins are being built in forested areas, and new fire-fighting equipment, including helicopters and airplanes, is expensive to operate. In the summer of 2003, more than $545 million was spent fighting almost 2,500 forest fires that burned more than 250,000 ha in British Columbia. Some critics argue that most of this money is wasted, and that the only thing that stops large fires is a change in the weather.

For 30 years the U.S. National Parks have had a policy of allowing some natural fires to burn and of setting prescribed fires in ecosystems where scientists consider periodic burns beneficial. The U.S. Forest Service has considered a similar policy for wilderness areas, but fear of public criticism has prevented much use of such a policy.

We are faced with a dilemma in many forests. After 70 years of fire suppression, fuel has now built up to a point where the next fire could be truly disastrous. The problem is how to remove excess fuel through controlled burning and thinning in a way that will return the forest to more natural conditions and yet protect property, human life, and important biological communities. We may need zoning regulations that restrict building or require fireproof construction methods and land management in highly flammable areas. Public reeducation programs to teach people the role of fire in natural systems is another important factor. Rather than seeing a burned forest as ruined, we can appreciate it as a natural stage in regeneration. This may require a very different environmental ethic than the one Smokey has been preaching and a modified view of our proper role in nature.

Sustainable Forestry and Non-Timber Forest Products

Creative solutions to forest management problems are available. In both temperate and tropical regions, scores of certification programs are being developed to identify sustainably produced wood products. One organization that is currently active in 40 countries is the Forest Stewardship Council (FSC). The FSC works to set standards for certification. SmartWood, a program of the Rainforest Alliance, is the most extensive certification program. This organization works with both tropical and temperate forest products companies. One of the promising movements in North American forestry is the development of cooperatives and networks among private landowners. In the United States alone, there are more than 9 million owners of small (about 40 ha) forest lands. Groups such as the Community Forestry Resource Center are sharing information and resources to assist in sustainable management of small working forests like these.

Logging is not the only way to make a living in a forest. Increasingly, non-timber forest products are seen as an alternative to timber production. In the United States alone, a $3 billion natural plants industry depends on healthy forests. Non-timber forest products have been around for centuries: latex (rubber), chicle (gum), nuts, and many other products have long been gathered sustainably from tropical forests (Fig. 17.18). Medicinal plants, fruits, rattan (vines), and other products can be developed from both temperate and tropical forests. The principal challenge is developing markets for these new products—something that is becoming

FIGURE 17.18 Non-timber forest products, such as natural rubber, can provide an income without destroying forests.

easier with modern communication technology and niche marketing methods.

RANGELANDS

Pasture (generally enclosed domestic meadows or managed grasslands) and **open range** (unfenced, natural prairie and open woodlands) occupy about 26 percent of the world's land surface. Much of the Prairie Provinces of Canada and the western Great Plains fall in this category. The 3.8 billion ha of permanent grazing lands in the world make up about twice the area of all agricultural crops. When you add to this about 4 billion ha of other lands (forest, desert, tundra, marsh, and thorn scrub) used seasonally for raising livestock, more than half of all land is used at least occasionally as grazing lands. More than 3 billion cattle, sheep, goats, camels, buffalo, and other domestic animals turn

Lowering Our Forest Impacts

For most urban residents, forests—especially tropical forests—seem far away and disconnected from our everyday life. There are things that each of us can do, however, to protect forests.

- Reuse and recycle paper. Make double-sided copies. Save office paper and use the back for scratch paper.

- Use e-mail. Store information in digital form rather than making hard copies of everything.

- If you build, conserve wood. Use wafer board, particle board, laminated beams or other composites rather than plywood and timbers made from old-growth trees.

- Buy products made from "good wood" or other certified sustainably-harvested wood.

- Don't buy products made from tropical hardwoods such as ebony, mahogany, rosewood, or teak unless the manufacturer can guarantee they were harvested from agroforestry plantations or sustainable-harvest programs.

- Don't patronize fast-food restaurants that purchase beef from cattle grazing on deforested rainforest land. Don't buy coffee, bananas, pineapples or other cash crops if their production contributes to forest destruction.

- Do buy Brazil nuts, cashews, mushrooms, rattan furniture, and other non-timber forest products harvested sustainably by local people from intact forests. Remember that tropical rainforest is not the only biome under attack. Contact the Taiga Rescue Network (http://www.taigarescue.org) for information about boreal forests.

- If you hike or camp in forested areas, practise minimum-impact camping. Stay on existing trails, don't build more or bigger fires than what you absolutely need. Use only downed wood for fires. Don't carve on trees or drive nails into them.

- Write to your provincial or territorial Member of Parliament and ask them to support forest protection and environmentally responsible government policies. Write to the Canadian Forest Service and voice your support for non-harvest forest values.

plants into protein-rich meat and milk that make a valuable contribution to human nutrition. Sustainable pastoralism can increase productivity while maintaining biodiversity in a grassland ecosystem.

Because grasslands and open woodlands are attractive for human occupation, they frequently are converted to cropland, urban areas, or other human-dominated landscapes. You probably have heard a great deal about tropical rainforest destruction, but worldwide the rate of grassland losses each year is three times that of tropical forest, while the area lost is nearly six times as much. Although they may appear to be uniform and monotonous to the untrained eye, native prairies can be highly productive and species-rich. More threatened American plant species occur in rangelands than any other major biome.

Range Management

By carefully monitoring the numbers of animals and the condition of the range, ranchers and pastoralists (people who live by herding animals) can adjust to variations in rainfall, seasonal plant conditions, and nutritional quality of forage to keep livestock healthy and avoid overusing any particular area. Conscientious management can actually improve the quality of the range.

Some nomadic pastoralists who follow traditional migration routes and animal management practices produce admirable yields from harsh and inhospitable regions. They can be 10 times more productive than dryland farmers in the same area and come very close to maintaining the ecological balance, diversity, and productivity of wild ecosystems on their native range. Nomadic herding requires large open areas, however, and wars, political problems, travel restrictions, incursions by agriculturalists, growing populations, and changing climatic conditions on many traditional ranges have combined to disrupt an ancient and effective way of life. The social and environmental consequences often are tragic.

Overgrazing and Land Degradation

About one-third of the world's range is severely degraded by overgrazing, making this the largest cause of soil degradation (Fig. 17.19). Among the countries with the most damage and the greatest area at risk are Pakistan, Sudan, Zambia, Somalia, Iraq, and Bolivia. Usually, the first symptom of improper range

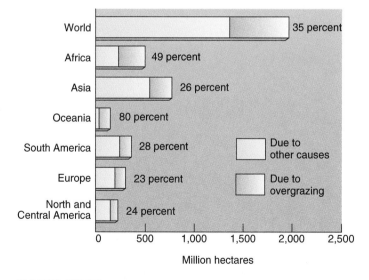

FIGURE 17.19 Rangeland soil degradation due to overgrazing and other causes. Notice that in Europe, Asia, and the Americas, farming, logging, mining, urbanization, etc., are responsible for about three-quarters of all soil degradation. In Africa and Oceania, where more grazing occurs and desert or semiarid scrub make up much of the range, grazing damage is higher.

Source: Data from *World Resources* 1994–95, World Resource Institute.

FIGURE 17.20 Sheep graze on land in Guatemala that was once tropical forest. Soils in the tropics are often thin and nutrient-poor. When forests are felled, heavy rains carry away the soil, and fields quickly degrade to barren scrubland. In some areas, land clearing has resulted in climate changes that have turned once lush forests into desert.

management is elimination of the most palatable herbs and grasses. Grazing animals tend to select species they prefer and leave the tougher, less tasty plants. When native plant species are removed from the range, weedy invaders move in. Gradually, the nutritional value of the available forage declines. As overgrazing progresses, hungry animals strip the ground bare and their hooves pulverize the soil, hastening erosion (Fig. 17.20).

The process of denuding and degrading a once-fertile land initiates a desert-producing cycle that feeds on itself and is called **desertification.** With nothing to hold back surface runoff, rain drains off quickly before it can soak into the soil to nourish plants or replenish groundwater. Springs and wells dry up. Trees and bushes not killed by browsing animals or humans scavenging for firewood or fodder for their animals die from drought. When the earth is denuded, the microclimate near the ground becomes inhospitable to seed germination. The dry barren surface reflects more of the sun's heat, changing wind patterns, driving away moisture-laden clouds, and leading to further desiccation. Because of this process, deserts have been called the footprints of civilization.

This process is ancient, but in recent years it has been accelerated by expanding populations and political conditions that force people to overuse fragile lands. Those places that are most severely affected by drought are the desert margins, where rainfall is the single most important determinant in success or failure of both natural and human systems (Fig. 17.21). In good years, herds and farms prosper and the human population grows. When drought comes, there is no reserve of food or water and starvation and suffering are widespread. Can we reverse this process? In some places, people are reclaiming deserts and repairing the effects of neglect and misuse.

As is the case in tropical forests, estimates of the extent of desertification around the world vary widely. Many arid lands are prone to highly variable rainfall with long periods of drought in which natural vegetation disappears and the land lies barren for months or years, making it difficult to distinguish between human-caused changes and normal climatic variations. A recent survey by the United Nations Environment Program found that about half of Africa, for instance, has become significantly drier over the past half century and that arid and hyperarid areas—which are most susceptible to desertification—have increased by about 50 million hectares, while humid and semiarid lands have decreased by the same amount. According to the International Soil Reference and Information Centre in the Netherlands, nearly three-quarters of all rangelands in the world show signs of either degraded vegetation or soil erosion. The highest percentage of moderate, severe, and extreme land degradation is in Mexico and Central America.

Forage Conversion by Domestic Animals

Ruminant animals, such as cows, sheep, goats, buffaloes, camels, and llamas, are especially efficient at turning plant material into protein because bacterial digestion in their multiple stomachs allows them to utilize cellulose and other complex carbohydrates that many mammals (including humans) cannot digest. As a result, they can forage on plant material from which we could otherwise extract little food value. Many grazers have very different feeding preferences and habits. Often the most effective use of rangelands is to maintain small mixed-species herds so that all vegetation types are utilized equally and none is overgrazed. Cattle and sheep, for instance, prefer grass and herbaceous plants, goats will browse on low woody shrubs, and camels can thrive on tree leaves and larger woody plants.

Worldwide, 85 percent of the forage for ruminants comes from native rangelands and pasture. In North America, however, only 15 percent of livestock feed comes from native grasslands.

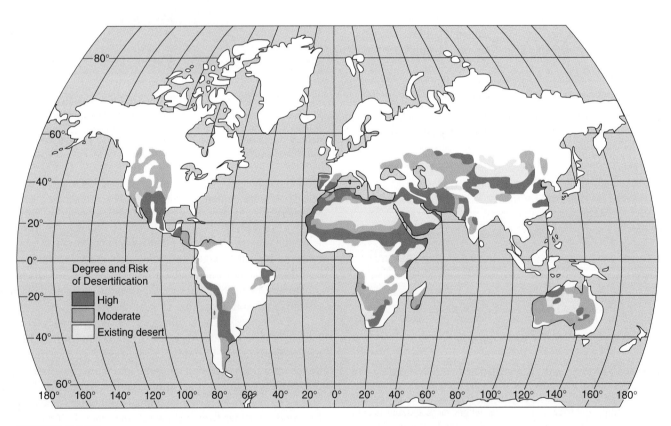

FIGURE 17.21 Degree and risk of desertification of arid lands due to human activities. Note that some existing deserts such as the Sahara in North Africa are so dry and lifeless that only a small amount of additional damage is possible.

FIGURE 17.22 Native species, such as these American bison, are a valuable part of grassland ecosystems; they also may be the best way to harvest biomass for human consumption. Native species often forage more efficiently, resist harsh climates, and are more pest- and disease-resistant than domestic livestock.

FIGURE 17.23 Landless rural peasants still work most of the land in Asia, Latin America, and Africa while they live in poverty.

The rest is made up of crops grown specifically for feed, particularly alfalfa, corn, and oats. Grain surpluses and a taste for well-marbled (fatty) meat have shifted livestock growing in the developed countries to feedlot confinement and high-quality diets. In Canada and the United States, roughly 90 percent of the total grain crop is used for livestock feed.

Harvesting Wild Animals

A few people in the world still depend on wild animals for a substantial part of their food. About one-half of the meat eaten in Botswana, for instance, is harvested from the wild. There are good reasons to turn even more to native species for a meat source. On the African savanna, researchers are finding that springbok, eland, impala, kudu, gnu, oryx, and other native animals forage more efficiently, resist harsh climates, are more pest- and disease-resistant, and fend off predators better than domestic livestock. Native species also are members of natural biological communities and demonstrate niche diversification, spreading feeding pressure over numerous plant populations in an area. A study by the U.S. National Academy of Sciences concluded that the semiarid lands of the African Sahel can support only about 20 to 28 kg of cattle per hectare but can produce nearly three times as much meat from wild ungulates (hooved mammals) in the same area. In North America, native bison and elk are raised as "novelty" meat sources; might there be a greater future role for ranching of wild animals on our own rangelands (Fig. 17.22)?

LANDOWNERSHIP AND LAND REFORM

Many of the problems discussed in this chapter have their roots in landownership and public policies concerning resource use. Some land tenure patterns have supported or even created inequality, ignorance, and environmental degradation, while other systems of land use have promoted equality, liberty, and progress. In this final section, we will look at how fair access to the land and its benefits can help bring about better conditions for humans and their environment.

Who Owns How Much?

Whatever political and economic system prevails in a given area, the largest landowners usually wield the most power and reap the most benefits, while the landless and near landless suffer at the bottom of the scale. The World Bank estimates that nearly 800 million people live in "absolute poverty ... at the very margin of existence." Three-fourths of these people are rural poor who have too little land to support themselves. Most of the other 200 million absolute poor are urban slum dwellers, many of whom migrated to the city after being forced out of rural areas (Chapter 21).

In many countries, inequitable landownership is a legacy of colonial estate systems in which landless peasants still work in virtual serfdom (Fig. 17.23). For instance, the United Nations reports that only 7 percent of the landowners in Latin America own or control 93 percent of the productive agricultural land. By far the largest share of landless poor in the world are in South Asia. In Haiti, the poorest and most environmentally degraded country in the Western Hemisphere, 1 percent of the population owns 90 percent of the land.

The 86 million landless households in India comprise about 500 million people; they are two-thirds of both the total population of India and of all the landless rural people in the world. Not surprisingly, the countries with the widest disparity in wealth are often the most troubled by social unrest and political instability. As Adam Smith said in *The Wealth of Nations* (1776), "No society can surely be flourishing and happy, of which the far greater part of the members are poor and miserable."

Land tenure is not just a question of social justice and human dignity. Political, economic, and ecological side effects of inequitable land distribution affect those of us far from the immediate location of the problem. For example, people in developing countries forced from their homes by increasing farm mechanization and cash crop production often try to make a living on marginal land that should not be cultivated. The local ecological damage caused by their struggle for land can, collectively, have a global impact. The contribution of concentrated landownership to these environmental problems, however, receives less attention than do the threatening ecological trends themselves.

Land Reform

Throughout history, some **land reform** movements seeking redistribution of landownership have been successful and others have not. Significant land reforms have occurred in many countries only after violent revolutionary movements or civil wars. An important factor in revolutions in Cuba, Mexico, China, Peru, the former Soviet Union, and Nicaragua was the breaking of feudal or colonial landownership patterns. This struggle continues in many countries. Hundreds of large and small peasant movements demand land redistribution, economic security, and political autonomy in Latin America, Africa, and Asia. Entrenched powers respond with economic reprisals, intimidation, "disappearances," and even open warfare. Millions of people have been killed in such struggles in recent centuries alone.

Consider Brazil: 340 rich landowners possess 46.8 million ha of cropland, while 9.7 million families (70 percent of all rural people) are landless or nearly landless. Studies have shown that 13 percent of the land on the biggest estates is completely idle, while another 76 percent is unimproved pasture. These inequities and the tensions they create are important factors in rainforest invasions described earlier in this chapter.

In some countries, land reform has been relatively peaceful and also successful. In Taiwan, only 33 percent of the farm families owned the land they worked in 1949, whereas nearly 60 percent of rural families farm their own land today. South Korea also has carried out sweeping land redistribution. In the 1950s, more than half of all farmers were landless, but now 90 percent of all South Korean farmers own at least part of the land they till. In general, countries that have had successful land reform also have stabilized population growth and have begun industrialization and development.

What are the ecological implications of landownership? Absentee landlords have little personal contact with the land, may not know or care about what happens to it, and often won't let sharecroppers cultivate the same land from year to year for fear they may lay claim to it. This gives tenant farmers little incentive to protect or improve the land because they can't count on reaping any long-term benefits. In fact, if tenants do improve the land or increase their yields, their rents may be raised. Also, where land is aggregated into collective farms or large estates worked by landless peasants, productivity and health of the soil tend to suffer.

Far from being a costly concession to the idea of equality, land reform often can provide a key to agricultural modernization. In many countries, the economic case for land reform rivals the social case for redistributive policies. Many studies have shown that the productivity of owner-operated farms is significantly higher than that of corporate or absentee landowner farms. Independent farmers, especially those who have only a few hectares to grow crops, tend to lavish a great deal of effort and attention on their small plots, and their yields per hectare often are twice those of larger farms.

Indigenous Lands

Indigenous (native or tribal) people make up about 10 percent of the world's population but occupy about 25 percent of the land. Many of the approximately 5,000 indigenous cultures that remain today possess ecological knowledge of their ancestral lands that is of vital importance to all of us. According to author Alan Durning, "encoded in indigenous languages, customs, and practices may be as much understanding of nature as is stored in the libraries of modern science." And indigenous people continue to play an important role as guardians of wildlife and forests. The Kuna Indians of Panama say, "Where there are forests there are indigenous people, and where there are indigenous people there are forests."

Nearly every country in the world has a sad history of decimation of indigenous cultures and exploitation or annexation of ancestral lands. Time after time, native people have been stripped of their rights through economic pressure, cunning and deceit, or outright theft (Fig. 17.24). In the past, colonial governments appropriated tribal lands under the excuse that sparsely populated or

FIGURE 17.24 George Gillette, chairman of the Fort Berthold Indian Tribe Business Council, weeps as he watches the signing of a 1948 contract to sell the tribe's best land along the Missouri River for the Garrison Dam and Reservoir Project. After signing the contract, Gillette said, "Right now, the future does not look good to us."

uncultivated land was free for the taking. Many nations still fail to recognize indigenous rights. In the Philippines, Cameroon, and Tanzania, for example, the government claims ownership of all forest lands, and indigenous people are considered squatters, even on their ancestral territories.

Our appetite for natural resources and land puts both native people and the ecosystems in which they live at risk. Decimated by plagues, violence, and corruption—and overwhelmed by the onrush of materialistic Western culture—at least half of the world's indigenous cultures are at risk of disappearing within this century. Consider the case of the 9,000 Yanomami people who live along the border between Brazil and Venezuela (Fig. 17.25). In 1991, Brazil recognized Yanomami claims to 17.6 million ha of land, but invasions by *garimpeiros* (miners) into this territory have introduced malaria, tuberculosis, flu, and respiratory diseases to which native people have no immunity. Placer mining and mercury effluents kill fish and poison the rivers. Wildlife is killed or driven off by miners, and hundreds of the Yanomami themselves have been hunted down and shot.

From the Philippines to Labrador, indigenous people are fighting for their ancestral territories. Some countries with large native populations, such as Papua New Guinea, Ecuador, Canada, and Australia, have acknowledged indigenous titles or rights to extensive areas. Often, years of history and modern development complicate these indigenous claims. One aboriginal band in Australia, for instance, claims ownership of Circular Quay, some of the most valuable land in the heart of Sidney. What would be a fair compensation after 150 years of occupation and building?

In April 1999 Nunavut (meaning "our home" in the native Inuktitut language) was created in Canada's north. Spanning nearly 2 million square kilometres or almost 20 percent of the country's landmass, this vast area of arctic tundra and glaciated mountains is managed by its 25,000 residents, 90 percent of whom are Inuit. Nunavut has some of the harshest weather anywhere in the world. Ellesmere National Park at the northern extreme of the territory

FIGURE 17.25 Native tribal people, such as these Yanomami from Brazil, are threatened by tropical forest destruction. These people have lived in harmony with nature for thousands of years. If their culture is lost, valuable knowledge about the forest will be lost as well.

typically has one week of spring, one week of summer, and one week of fall. The rest of the year is winter, and five months are completely dark. With a population spread among 25 widely scattered villages and only one short paved road in the entire territory, Nunavut faces some difficult challenges. Foremost among these is the difficulty of preserving indigenous culture, skills, and values while also becoming part of the modern world. But the right to self-rule and control over their ancestral territory has brought a new sense of energy and purpose to native people. This settlement might be a good model for resolution of native land claims elsewhere in the world.

SUMMARY

Land has traditionally been a source of wealth and power. The ways we use this limited resource shape our lives and futures. About one-third of the earth's surface is too inhospitable for agriculture, livestock, or forestry but is vital for such purposes as wilderness preservation and recreation. We grow crops on about 11 percent of the total land area. With proper preparation, we

could expand cropland in some areas. Most of the earth's land, however, is inappropriate for agriculture and is ruined by attempts to cultivate it.

Forests and woodlands cover about 32 percent of the earth's land area, providing a variety of useful products such as lumber, pulpwood, and firewood. Northern forests are growing faster in most areas of the world than they are being cut and seem in little danger of being exhausted. Tropical forests, on the other hand, are in critical danger. Irreplaceable ecosystems that are home to as many as half of all biological species are being destroyed. Quick profits encourage this exploitation, but hidden costs such as lost wildlife habitat, erosion and devaluation of exposed land, and other disastrous environmental damage will follow from small short-term gains.

A little more than one-quarter of the earth's land is used as range and pasture. Three billion grazing animals convert roughage to protein on poor land that could not otherwise be used to produce food for humans. When herds are managed properly, they actually can improve the quality of their pasture. Unfortunately, about one-third of the world's rangelands are degraded by over-grazing, with disastrous environmental consequences similar to forest destruction.

Land reform is an essential part of sound land-use management. Fair distribution of land and its benefits encourages good stewardship, increased food production, sustainable agriculture, and social justice. Inequitable land ownership, so common today in much of the world, forces the poor to use land unsuited to agriculture, while good land is monopolized by the rich. Dividing land more fairly could increase agricultural productivity and sustainability of the land because farmers who own their own land generally use it more efficiently and more carefully than do absentee landlords. Recognizing indigenous land rights is important both for preserving endangered cultures and for protecting ecological values.

QUESTIONS FOR REVIEW

1. Which type of land use occupies the greatest land area?

2. List some products that we derive from forests.

3. What are the advantages and disadvantages of monoculture forestry?

4. Describe milpa (or swidden) agriculture. Why are these techniques better for some fragile rainforest ecosystems than other types of agriculture?

5. What are some results of deforestation?

6. What are clear-cutting and below-cost timber sales, and why are they controversial?

7. What is the relationship between fair land distribution and appropriate land use?

8. Give some examples of recognition of indigenous land titles.

QUESTIONS FOR CRITICAL THINKING

1. Some forestland would be suitable as cropland. Do you think it should be converted? Why or why not?

2. Brazil needs cash to pay increasing foreign debts and to fund needed economic growth. Why shouldn't it harvest its forests and mineral resources to gain the foreign currency it wants and needs? If we want Brazil to save its forests, what can or should we do to encourage conservation?

3. Thousands of landless peasants are mining gold or cutting trees to create farms in officially protected Brazilian rainforest. What might the government do to stop these practices? Should it try to stop them?

4. What lessons do you think milpa (or swidden) agriculture has to offer large-scale commercial farming?

5. In British Columbia, the interests of environmentalists and the forestry industry often clash. How could one begin to resolve such deep conflict?

6. Aboriginal peoples often were cheated out of their land or paid ridiculously low prices for it. Present owners, however, may not have been a part of earlier land deals and may have invested a great deal to improve the land. What would be a fair and reasonable settlement of these competing land claims?

7. There is considerable uncertainty about the extent of desertification of grazing lands or destruction of tropical rainforests. Put yourself in the place of a decision maker evaluating this data. What evidence would you want to see, or how would you appraise conflicting evidence?

KEY TERMS

clear-cutting 372
closed canopy 362
debt-for-nature swaps 370
desertification 375
forest management 364
fuelwood 364
industrial timber 363
land reform 378
milpa agriculture 367
mixed perennial
   polyculture 367
monoculture
   agroforestry 366
old-growth forests 362
open canopy 362
open range 374
pasture 374
selective cutting 372
strip cutting 372
swidden agriculture 367
woodland 362

ANIMATED MAP OF WORLD FIRES

For this exercise you will need to have a Quicktime movie player loaded on your computer. If you don't have one, go to the McGraw-Hill Higher Education website for a link to the Quicktime site.

Go to the McGraw-Hill Higher Education website and find a movie called NASA_fire.mov. This movie shows all large fires (more than 4 km across) detected by a NASA satellite. Play the movie by double clicking on its name.

1. What parts of which continents have the most fires? (central Africa, Amazonia, Southeast Asia, Central America, in that order)

2. You can move through the movie slowly by moving the slider with your mouse. What are the earliest and latest dates in the animation? (January 1999–January 2000) Describe how the pattern differs between January and August, 1999 (north of the equator, in Africa in January; south, in southern Africa and Amazonia in August)

3. Look at August and September, 1999. What season is this in the areas with the biggest fires, and does the season help explain why the big fires are there? How might change of season explain the difference in fire pattern in South America from August to November? (end of winter, presumably winter is a dry season; by November a summer rainy season is arriving in much of South America)

The following movie was created using a NASA website: http://earthobservatory.nasa.gov/Observatory/datasets.html. Go to this website and look at a variety of the images to get a sense of what sort of information can be gathered and mapped using satellite images. You can also create your own animations. Just follow the directions provided.

Download the movie to your computer by clicking on its name with your right mouse button (hold control key and click if using a Mac) and selecting "save link as." Be sure to write down where you saved the movie so you can find it!

Chapter 18

Solid, Toxic, and Hazardous Waste

We have no knowledge, so we have stuff; but stuff without knowledge is never enough.

—Greg Brown—

OBJECTIVES

After studying this chapter, you should be able to:

- identify the major components of the waste stream and describe how wastes have been—and are being—disposed of in North America and around the world.
- explain the differences between dumps, sanitary landfills, and modern secure landfills.
- summarize the benefits, problems, and potential of recycling and reusing wastes.
- analyze some alternatives for reducing the waste we generate.
- understand what hazardous and toxic wastes are and how we dispose of them.
- evaluate the options for hazardous waste management.
- outline some ways we can destroy or permanently store hazardous wastes.

WebQuest

solid waste, hazardous waste, toxic waste, recycling, landfills, demanufacturing

Above: Bulldozers pack down the refuse at Fresh Kills Landfill on Staten Island, NY.

What a Long, Strange Trip It Has Been

On August 31, 1986, the cargo ship *Khian Sea* loaded 14,000 tonnes of toxic incinerator ash from Philadelphia and set off on an odyssey that symbolizes a predicament we all share: what to do with our refuse. Starting in the 1970s, Philadelphia burned most of its municipal garbage and sent the resulting incinerator ash to a landfill in New Jersey. In 1984, when New Jersey learned that the ash contained enough arsenic, cadmium, lead, mercury, dioxin, and other toxins to be classified as hazardous waste, it refused to accept any more. When six other states also rejected incinerator ash shipments, Philadelphia was in a predicament. What would they do with 180,000 tonnes of the stuff every year? The answer was to send it offshore to countries with less stringent environmental standards. A local contractor offered to transport it to the Caribbean. The *Khian Sea* was to be the first of those shipments.

When the *Khian Sea* tried to unload its cargo in the Bahamas, however, it was turned away. Over the next 14 months, the ship also was refused entry by the Dominican Republic, Honduras, Panama, Bermuda, Guinea Bissau (in West Africa), and the Netherlands Antilles. Finally in late 1987, the Haitian government issued a permit for "fertilizer" import, and the crew dumped 4,000 tonnes of ash on the beach near the city of Gonaives. Alerted by the environmental group Greenpeace that the ash wasn't really fertilizer, Haitian officials cancelled the permit and ordered everything returned to the ship, but the *Khian Sea* slipped away in the night, leaving behind a large pile of loose ash. Some of the waste has been moved inland and buried, but much of it remains on the beach, slowly being scattered by the wind and washed into the sea.

After it left Haiti, the *Khian Sea* visited Senegal, Morocco, Yugoslavia, Sri Lanka, and Singapore looking for a place to dump its toxic load. As it wandered the oceans looking for a port, the ship changed its name from *Khian Sea* to *Felicia* to *Pelacano*. Its registration was transferred from Liberia to the Bahamas to Honduras in an attempt to hide its true identity, but nobody wanted it or its contents. Like Coleridge's ancient mariner, it seemed cursed to roam the oceans forever. Two years, three names, four continents, and 11 countries later, the troublesome cargo was still on board (Fig. 18.1). Then, somewhere in the Indian Ocean between Singapore and Sri Lanka all the ash disappeared. When questioned about this, the crew had no comment except that it was all gone. Everyone assumes, of course, that once out of sight of the land, it was just dumped overboard.

If this were just an isolated incident, perhaps it wouldn't matter much. However, some 3 million tonnes of hazardous and toxic waste goes to sea every year looking for a dumping site. A 1998 report by the United Nations Human Rights Commission listed the United States as a major exporter of toxic waste. In 1989—at least in part due to the misadventures of the *Khian Sea*—33 countries met in Basel, Switzerland, and agreed to limit international shipment of toxic waste, especially from the richer countries of the world to the poorer ones. Eventually 118 countries—not including the United States—ratified the Basel Convention. In 1995, the United States announced it would ratify the Convention but reserved the right to ship "recyclable" materials to whoever will take them. Since almost everything potentially can be recycled into something, that hardly puts any limits at all on what we send offshore.

Haiti has asked Philadelphia to help pay for cleanup of the ash still sitting on the beach. Philadelphia's share would be $200,000, or about one-third of the total cleanup cost. The city of brotherly love in spite of having a $130 million budget surplus that year, claimed it couldn't afford to help out. Ultimately, the ash was removed and shipped back to the United States. At last report, the company now in charge of the problem, Waste Management, was searching for a landfill in the South to take the load. Several states have refused the ash, which sits on barges off the coast of Florida.

The *Khian Sea* saga is a notorious case, but it represents only a minute portion of the waste produced every day in North America and around the world. All of us contribute to this problem. We all generate vast amounts of unwanted stuff every year. Places to put our trash are becoming more and more scarce as the contents have become increasingly unpleasant and dangerous. We don't want it in our backyards, so it often ends up in those of the poorest and least powerful, both in Canada and around the world. In this chapter, we will look at the kinds of waste we produce, who makes them, what problems their disposal cause, as well as how we might reduce our waste production and dispose of our waste in more environmentally friendly ways.

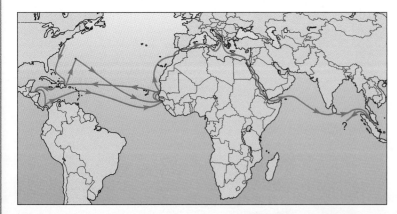

FIGURE 18.1 Route of the *Khian Sea*.

SOLID WASTE

Waste is everyone's business. We all produce wastes in nearly everything we do. According to the U.S. Environmental Protection Agency, the United States produces 11 billion tonnes of solid waste each year. About half of that amount consists of agricultural waste, such as crop residues and animal manure, which are generally recycled into the soil on the farms where they are produced. They represent a valuable resource as ground cover to reduce erosion and fertilizer to nourish new crops, but they also constitute the single largest source of nonpoint air and water pollution in the country. More than one-third of all solid wastes are mine tailings, overburden from strip mines, smelter slag, and other residues produced by mining and primary metal processing. Road and building construction debris is another major component of solid waste. Much of this material is stored in or near its source of production and isn't mixed with other kinds of wastes. Improper disposal practices, however, can result in serious and widespread pollution.

Industrial waste—other than mining and mineral production—amounts to some 400 million tonnes per year in the United States. Most of this material is recycled, converted to other forms, destroyed, or disposed of in private landfills or deep injection wells. About 60 million tonnes of industrial waste falls in a special category of hazardous and toxic waste, which we will discuss later in this chapter.

Municipal waste—a combination of household and commercial refuse—amounts to more than 200 million tonnes per year in the United States. That's approximately two-thirds of a tonne for each man, woman, and child every year—twice as much per capita as Europe or Japan, and five to ten times as much as most developing countries.

The Waste Stream

Does it surprise you to learn that you generate that much garbage? Think for a moment about how much we discard every year. There are organic materials, such as yard and garden wastes, food wastes, and sewage sludge from treatment plants; junked cars; worn out furniture; and consumer products of all types. Newspapers, magazines, advertisements, and office refuse make paper one of our major wastes (Fig. 18.2). In spite of recent progress in recycling, each Canadian disposes of more than half a tonne of solid waste per year, about 67 percent of which ends up in landfills. Wood, concrete, bricks, and glass come from construction and demolition sites, dust and rubble from landscaping and road building. All of this varied and voluminous waste has to arrive at a final resting place somewhere.

The **waste stream** is a term that describes the steady flow of varied wastes that we all produce, from domestic garbage and yard wastes to industrial, commercial, and construction refuse. Many of the materials in our waste stream would be valuable resources if they were not mixed with other garbage. Unfortunately, our collecting and dumping processes mix and crush everything together, making separation an expensive and sometimes impossible task.

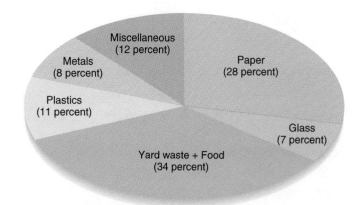

FIGURE 18.2 In spite of recent progress in recycling, each Canadian disposes of more than half a tonne of solid waste per year, about 67 percent of which ends up in landfills.
Source: Data from Environmental Protection Agency, 2000.

In a dump or incinerator, much of the value of recyclable materials is lost.

Another problem with refuse mixing is that hazardous materials in the waste stream get dispersed through thousands of tonnes of miscellaneous garbage. This mixing makes the disposal or burning of what might have been rather innocuous stuff a difficult, expensive, and risky business. Spray paint cans, pesticides, batteries (zinc, lead, or mercury), cleaning solvents, smoke detectors containing radioactive material, and plastics that produce dioxins and PCBs when burned are mixed willy-nilly with paper, table scraps, and other nontoxic materials. The best thing to do with household toxic and hazardous materials is to separate them for safe disposal or recycling, as we will see later in this chapter.

WASTE DISPOSAL METHODS

Where are our wastes going now? In this section, we will examine some historic methods of waste disposal as well as some future options. Notice that our presentation begins with the least desirable—but most commonly used—measures and proceeds to discuss some preferable options. Keep in mind as you read this that modern waste management reverses this order and stresses the "three R's" of reduction, reuse, and recycling before destruction or, finally, secure storage of wastes.

Open Dumps

For many people, the way to dispose of waste is to simply drop it someplace. Open, unregulated dumps are still the predominant method of waste disposal in most developing countries. The giant Third World megacities have enormous garbage problems. Mexico City, the largest city in the world, generates some 10,000 tonnes of trash *each day*. Until recently, most of this torrent of waste was left in giant piles, exposed to the wind and rain, as well

FIGURE 18.3 Scavengers sort through the trash at "Smoky Mountain," one of the huge metropolitan dumps in Manila, Philippines. Some 20,000 people live and work on these enormous garbage dumps. The health effects are tragic.

as rats, flies, and other vermin. Manila, in the Philippines, generates a similar amount of waste, half of which goes to a giant, constantly smoldering dump called "Smoky Mountain." More than 20,000 people live and work on this mountain of refuse, scavenging for recyclable items or edible food scraps (Fig. 18.3). In July 2000, torrential rains spawned by Typhoon "Kai Tak" caused part of the mountain to collapse, burying at least 215 people. The government would like to close these dumps, but how will the residents be housed and fed? Where else will the city put its garbage?

Most developed countries forbid open dumping, at least in metropolitan areas, but illegal dumping is still a problem. You have undoubtedly seen trash accumulating along roadsides and in vacant, weedy lots in the poorer sections of cities. Is this just a question of aesthetics? Consider the problem of waste oil and solvents. An estimated 200 million litres of waste motor oil are poured into the sewers or allowed to soak into the ground every year in

the United States. This is about five times as much as was spilled by the *Exxon Valdez* in Alaska in 1989! No one knows the volume of solvents and other chemicals disposed of by similar methods.

Increasingly, these toxic chemicals are showing up in the groundwater supplies on which nearly half the people in North America depend for drinking (Chapter 11). An alarmingly small amount of oil or other solvents can pollute large quantities of drinking or irrigation water. One litre of gasoline, for instance, can make a million litres of water undrinkable. The problem of illegal dumping is likely to become worse as acceptable sites for waste disposal become more scarce and costs for legal dumping escalate. We clearly need better enforcement of antilittering laws as well as a change in our attitudes and behaviour.

Ocean Dumping

The oceans are vast, but not so large that we can continue to treat them as carelessly as has been our habit. Every year some 25,000 tonnes of packaging, including half a million bottles, cans, and plastic containers, are dumped at sea. Beaches, even in remote regions, are littered with the nondegradable flotsam and jetsam of industrial society (see Fig. 11.30). About 150,000 tonnes of fishing gear—including more than 1,000 km of nets—are lost or discarded at sea each year. Environmental groups estimate that 50,000 northern fur seals are entangled in this refuse and drown or starve to death every year in the North Pacific alone.

Until recently, many cities in the United States dumped municipal refuse, industrial waste, sewage, and sewage sludge in the ocean. U.S. federal legislation now prohibits this dumping. New York City, the last to stop offshore sewage sludge disposal, finally ended this practice in 1992. Still, 60 million to 80 million m^3 of dredge spoil—much of it highly contaminated—are disposed of at sea. Some people claim that the deep abyssal plain is the most remote, stable, and innocuous place to dump our wastes. Others argue that we know too little about the values of these remote places or the rare species that live there to smother them with sludge and debris.

Landfills

Over the past 50 years most North American and European cities have recognized the health and environmental hazards of open dumps. Increasingly, cities have turned to **sanitary landfills,** where solid waste disposal is regulated and controlled. To decrease smells and litter and to discourage insect and rodent populations, landfill operators are required to compact the refuse and cover it every day with a layer of dirt (Fig. 18.4). This method helps control pollution, but the dirt fill also takes up as much as 20 percent of landfill space. Since 1994, all operating landfills in the United States have been required to control such hazardous substances as oil, chemical compounds, toxic metals, and contaminated rainwater that seeps through piles of waste. An impermeable clay and/or plastic lining underlies and encloses the storage area. Drainage systems are installed in and around the liner to catch drainage and to help monitor chemicals that may be leaking. Modern municipal solid-waste

Compacted waste filling trench   Original terraine

Daily 6-inch earth cover

FIGURE 18.4 In a sanitary landfill, trash and garbage are crushed and covered each day to prevent accumulation of vermin and spread of disease. A waterproof lining is now required to prevent leaching of chemicals into underground aquifers.

landfills now have many of the safeguards of hazardous waste repositories described later in this chapter.

More careful attention is now paid to the siting of new landfills. Sites located on highly permeable or faulted rock formations are passed over in favour of sites with less leaky geologic foundations. Landfills are being built away from rivers, lakes, floodplains, and aquifer recharge zones rather than near them, as was often done in the past. More care is being given to a landfill's long-term effects so that costly cleanups and rehabilitation can be avoided.

Suitable places for waste disposal are becoming scarce in many areas. Other uses compete for open space. Citizens have become more concerned and vocal about health hazards, as well as aesthetics. It is difficult to find a neighbourhood or community willing to accept a new landfill. Since 1984, when stricter financial and environmental protection requirements for landfills took effect, more than 1,200 of the 1,500 existing landfills in the United States have closed. Many major cities are running out of local landfill space. They export their trash, at enormous expense, to neighbouring communities and even other states. More than half the solid waste from New Jersey goes out of state, some of it up to 800 km away.

A positive trend in landfill management is methane recovery. Methane, or natural gas, is a natural product of decomposing garbage deep in a landfill. It is also an important "greenhouse gas." Normally methane seeps up to the landfill surface and escapes. At 300 U.S. landfills, the methane is being collected and burned. Cumulatively, these landfills could provide enough electricity for a city of a million people. Three times as many landfills could be recovering methane. Tax incentives could be developed to encourage this kind of resource recovery.

Exporting Waste

Although most industrialized nations in the world have agreed to stop shipping hazardous and toxic waste to less-developed countries, the practice still continues. In 1999, for example, 3,000 tonnes of incinerator waste from a plastics factory in Taiwan was unloaded from a ship in the middle of the night and dumped in a field near the small coastal Cambodian village of Bet Trang. The village residents thought they had been blessed with a windfall. They emptied out chunks of crumbling residue so they could use the white plastic shipping bags as bedding and roofing material. They rinsed out bags to use for rice storage, and they ripped them open with their teeth to get string to use as clotheslines and lashing for their oxcarts. Children played happily on the big pile of dusty, white material.

In the following weeks, unfortunately, the villagers discovered that rather than a treasure, they had a calamity. The first sign of trouble was when one of the dock workers who unloaded the waste died and five others were hospitalized with symptoms of nerve damage and respiratory distress. Villagers also began to complain of a variety of illnesses. The village was evacuated, and about 1,000 residents of the nearby city of Sihanoukville fled in panic. Subsequent analysis found high levels of mercury and other toxic metals in the residue. The Formosa Plastics Corp., which shipped the waste, admitted paying a $3 million bribe to Cambodian officials to permit its dumping. They said they couldn't dispose of it in Taiwan because of a threat of public protest. Following an international uproar, the plastics company agreed to go back and pick up the waste. But the villagers who handled the toxic wastes face an uncertain future. Is it safe to reinhabit their homes? Is it wise to have children? Will they suffer long-term health effect from exposure to this material?

As we will discuss later in this chapter, "garbage imperialism" also operates within richer countries as well. Poor neighbourhoods and minority populations are much more likely than richer ones to be the recipients of dumps, waste incinerators, and other locally unwanted land uses (LULUs). In recent years, attention has turned, in the United States, to Indian reservations, which are exempt from some state and federal regulations concerning waste disposal. Virtually every tribe in America has been approached with schemes to store wastes on their reservation.

Another method of disposing of toxic wastes is to "recycle" them as asphalt or concrete filler for building highways. This is considered a beneficial use, but what happens to the toxins as the roadway is slowly worn away by traffic? Similar waste products are "land farmed" or sold as fertilizer and soil amendments. There are no safety standards for fertilizer composition because it's not intended for human consumption, but much of it is used on crops that humans will eat, or will be fed to livestock that are part of our food chain. For example, Florida has about a billion m³ of phosphogypsum, a waste product of phosphate mining that producers want to market as a soil amendment. While it's true that phosphate is an essential plant nutrient, this particular product is also radioactive. In another case in Oregon, metal-rich dust and ash from steel mills is classified as hazardous waste when it leaves the mill. After being mixed with other minerals, however, it becomes fertilizer that will be spread on farm fields. Manufacturers are required to report the "active" ingredient—things like nitrogen, phosphorus, and phosphate—content of their product, but much can go unreported as "inert" matter.

Incineration and Resource Recovery

Landfilling is still the disposal method for the majority of municipal waste in Canada. Faced with growing piles of garbage and a lack of available landfills at any price, however, public officials are investigating other disposal methods. The method to which they frequently turn is burning. Another term commonly used for this technology is **energy recovery,** or waste-to-energy, because the heat derived from incinerated refuse is a useful resource. Burning garbage can produce steam used directly for heating buildings or generating electricity. Internationally, well over 1,000 waste-to-energy plants in Brazil, Japan, and western Europe generate much-needed energy while also reducing the amount that needs to be landfilled. In the United States, more than 110 waste incinerators burn 45,000 tonnes of garbage daily. Some of these are simple incinerators; others produce steam and/or electricity.

Types of Incinerators

Municipal incinerators are specially designed burning plants capable of burning thousands of tonnes of waste per day. In some plants, refuse is sorted as it comes in to remove unburnable or recyclable materials before combustion. This is called **refuse-derived fuel** because the enriched burnable fraction has a higher energy content than the raw trash. Another approach, called **mass burn,** is to dump everything smaller than sofas and refrigerators into a giant furnace and burn as much as possible (Fig. 18.5). This technique avoids the expensive and unpleasant job of sorting through the garbage for nonburnable materials, but it often causes greater problems with air pollution and corrosion of burner grates and chimneys.

In either case, residual ash and unburnable residues representing 10 to 20 percent of the original volume are usually taken to a landfill for disposal. Because the volume of burned garbage is reduced by 80 to 90 percent, disposal is a smaller task. However, the residual ash usually contains a variety of toxic components that make it an environmental hazard if not disposed of properly. Ironically, one worry about incinerators is whether enough garbage will

be available to feed them. Some communities in which recycling has been really successful have had to buy garbage from neighbours to meet contractual obligations to waste-to-energy facilities. In other places, fears that this might happen have discouraged recycling efforts.

Incinerator Cost and Safety

The cost-effectiveness of garbage incinerators is the subject of heated debates. Initial construction costs are high—usually between $100 million and $300 million for a typical municipal facility. Tipping fees at an incinerator, the fee charged to haulers for each ton of garbage dumped, are often much higher than those at a landfill. As landfill space near metropolitan areas becomes more scarce and more expensive, however, landfill rates are certain to rise. It may pay in the long run to incinerate refuse so that the lifetime of existing landfills will be extended.

Environmental safety of incinerators is another point of concern. There are high levels of dioxins, furans, lead, and cadmium in incinerator ash. These toxic materials were more concentrated in the fly ash (lighter, airborne particles capable of penetrating deep into the lungs) than in heavy bottom ash. Dioxin levels can be as high as 780 parts per billion. One part per billion of TCDD, the most toxic dioxin, is considered a health concern. All of the incinerators studied exceeded cadmium standards, and 80 percent exceeded lead standards. Proponents of incineration argue that if they are run properly and equipped with appropriate pollution-control devices, incinerators are safe to the general public. Opponents counter that neither public officials nor pollution-control equipment can be trusted to keep the air clean. They argue that recycling and source reduction efforts are better ways to deal with waste problems.

One way to reduce these dangerous emissions is to remove batteries containing heavy metals and plastics containing chlorine before wastes are burned. Bremen, West Germany, is one of several European cities now trying to control dioxin emissions by keeping all plastics out of incinerator waste. Bremen is requiring

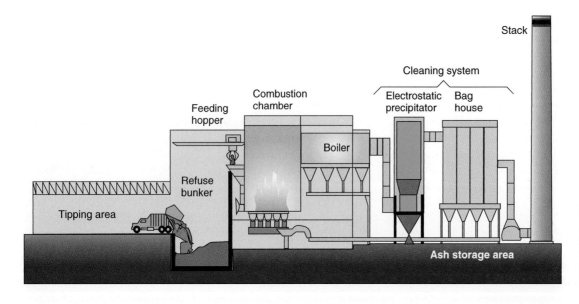

FIGURE 18.5 A diagram of a municipal "mass burn" garbage incinerator. Steam produced in the boiler can be used to generate electricity or to heat nearby buildings.

households to separate plastics from other garbage. This is expected to eliminate nearly all dioxins and other combustion by-products and prevent the expense of installing costly pollution-control equipment that otherwise would be necessary to keep the burners operating. Several cities have initiated a recycling program for the small "button" batteries used in hearing aids, watches, and calculators in an attempt to lower mercury emissions from its incinerator.

SHRINKING THE WASTE STREAM

Having less waste to discard is obviously better than struggling with disposal methods, all of which have disadvantages and drawbacks. In this section we will explore some of our options for recycling, reuse, and reduction of the wastes we produce.

Recycling

The term *recycling* has two meanings in common usage. Sometimes we say we are *recycling* when we really are *reusing* something, such as refillable beverage containers. In terms of solid waste management, however, **recycling** is the reprocessing of discarded materials into new, useful products (Fig. 18.6). Some recycling processes reuse materials for the same purposes; for instance, old aluminum cans and glass bottles are usually melted and recast into new cans and bottles. Other recycling processes turn old materials into entirely new products. Old tires, for instance, are shredded and turned into rubberized road surfacing. Newspapers become cellulose insulation, kitchen wastes become a valuable soil amendment, and steel cans become new automobiles and construction materials.

FIGURE 18.6 Trucks with multiple compartments pick up residential recyclables at curbside, greatly reducing the amount of waste that needs to be buried or burned. For many materials, however, collection costs are too high and markets are lacking for recycling to be profitable.
Source: William P. Cunningham.

There have been some dramatic successes in recycling in recent years. For instance, in the District of Lunenburg, Nova Scotia, a model recycling and composting plant was constructed in 1994 to serve 40,000 residents. The high value of aluminum scrap ($700 per tonne in 1999) has spurred a large percentage of aluminum recycling nearly everywhere. About two-thirds of all aluminum beverage cans are now recycled; up from only 15 percent 20 years ago. This recycling is so rapid that half of all the aluminum cans now on grocery shelves will be made into another can within two months. A big problem for recyclers is the wild fluctuation in market prices for commodities. Newsprint, for example, which peaked at $160 per tonne in 1995, dropped to $42 per tonne in 1999. One day, it's so valuable that people are stealing it off the curb; the next day it's literally down in the dumps. It's hard to build a recycling program when you can't count on a stable price for your product.

Another problem in recycling is contamination. Most of the 24 billion plastic soft drink bottles sold every year in the United States are made of PET (polyethylene terephthalate), which can be melted and remanufactured into carpet, fleece clothing, plastic-strapping, and nonfood packaging. However, even a smidgen of vinyl—a single PVC (polyvinyl chloride) bottle in a truckload, for example—can make PET useless. Although most bottles are now marked with a recycling number, it's hard for consumers to remember which is which. A looming worry is the prospect of single-use, plastic beer bottles. Already being test marketed, these bottles are made of PET but are amber coloured to block sunlight and have a special chemical coating to keep out oxygen, which would ruin the beer. The special colour, interior coating, and vinyl cap lining will make these bottles incompatible with regular PET, and it will probably cost more to remove them from the waste stream than the reclaimed plastic is worth. Plastic recycling already is down 50 percent from a decade ago because so many soft drink bottles are sold and consumed on the go, and never make it into recycling bins. Throw-away beer bottles may be the death knell for this industry.

Benefits of Recycling

Recycling is usually a better alternative to either dumping or burning wastes. It saves money, energy, raw materials, and land space, while also reducing pollution. Recycling also encourages individual awareness and responsibility for the refuse produced. Curbside pickup of recyclables costs around $35 per tonne, as opposed to the $80 paid to dispose of them at an average metropolitan landfill. Many recycling programs cover their own expenses with materials sales and may even bring revenue to the community.

Another benefit of recycling is that it could cut our waste volumes drastically and reduce the pressure on disposal systems. In Toronto, there is a sophisticated recycling, composting, and new technology system in place with a goal of 60 percent diversion of solid waste from landfills by 2006. So far, the city is on target to meet this goal. New York City, down to one available landfill but still producing 27,000 tonnes of garbage a day, has set a target of 50 percent waste reduction to be accomplished by recycling office paper and household and commercial waste. New York's curbside

South Africa's "National Flower"?

You've seen them in ditches, caught in fences, fluttering in trees on a windy day. Plastic shopping bags, an item of convenience world-over, are increasingly an eyesore and a nuisance. Some South Africans have begun referring to these ubiquitous bags as the country's national flower because they seem to bloom everywhere. Now the government is trying to make them disappear.

South African shops hand out about 8 billion light-weight, single-use plastic bags per year. Most may be disposed of properly (buried in landfills or burned), but many end up blowing along public streets and across the countryside. This mobile litter is not just an aesthetic nuisance: bags clog sewers and streams and threaten wildlife, as well. The principal problem is extremely fine bags, averaging 17 microns in thickness. (A micron is one-thousandth of a millimetre.) These thin bags are cheap enough to give away with groceries and other goods, but they are too fragile for reuse.

In an effort to make bags less disposable and more reusable, the South African government proposed a mandatory minimum thickness of 80 microns. Stores would be more likely to charge for these sturdier bags—and consumers would reconsider before throwing away bags they had paid for. Trade unions and plastics manufacturers insisted that existing equipment couldn't produce these bags, and they threatened job losses and factory closures. Eventually, unions and the government reached a compromise: there is now a 30 micron minimum, thick enough for reuse, and the government promises new jobs in recycling industries.

The effort to control disposable bags is just part of a wider South African effort to reduce litter and encourage recycling. Deposits have been proposed for tires, bottles, cans, and other products. Local governments are enthusiastic about these steps because litter is a chronic problem in many cities. Central Johannesburg alone is flooded with more than 200 tonnes of litter per day and spends nearly 50 million rand (nearly $7 million U.S.) a year cleaning up this debris.

South Africa is not the only country worried about this problem. Taiwan initiated a rule in 2003 that restaurants and supermarkets must charge customers for plastic bags and utensils. Australia is considering a tax on disposable bags. British supermarkets pass out some 10 billion bags each year, and the government there is considering a 9-penny-per-bag tax, which should force stores to charge for bags. Reportedly, stores support such a move. Currently, they spend £1 billion per year on bags that they give away for free. Ireland imposed such a tax, to be used for environmental cleanup, in 2002. A survey in Country Durham, England, found that 70 percent of residents favoured a system of paying for bags in order to encourage reuse. County Durham dumps more than 600 tonnes of plastic shopping bags in landfills every year, at a cost of about £20,000 ($32,000 U.S.).

By most estimates, charging customers for bags would reduce consumption by 40 to 50 percent. Germany, Norway, and others have charged shoppers for years, and as a result, reusable cloth or plastic bags are widely used.

Are disposable paper bags a better choice? They decompose or burn more readily than plastic, but they also require logging, bleaching, and waste disposal. Proponents of thicker plastic bags hope that consumers will start carrying their own shopping bags. Perhaps the best answer to the question, "paper or plastic?" is "I've brought my own, thanks."

Waste management is a growing and global problem. Often, waste production depends on our individual choices, to buy highly packaged goods, to buy unnecessary items, to reuse, recycle, or dispose of goods. Our personal choices are also complicated by cultural expectations and economic policies. In this chapter, we'll review the state of waste production, including solid waste and hazardous waste, and study efforts to reduce waste production and its environmental effects.

---

collection service, projected to be the nation's largest, should more than pay for itself simply in avoided disposal costs.

Japan probably has the most successful recycling program in the world. Half of all household and commercial wastes in Japan are recycled while the rest is about equally incinerated or landfilled. By comparison, the United States landfills more than 60 percent of all solid waste. Japanese families diligently separate wastes into as many as seven categories, each picked up on a different day (Fig. 18.7). Many large and small municipalities in Canada have successfully introduced blue and green box programs with high participation rates. Additionally, some municipalities have introduced user pay "garbage tag" systems to reduce household waste or divert more of it to recycling and composting programs (Fig. 18.8).

Recycling lowers our demands for raw resources. In the United States, 2 million trees are cut down every day to produce newsprint and paper products. Recycling the print run of a single Sunday issue of the *New York Times* would spare 75,000 trees.

Every piece of plastic we make reduces the reserves supply of petroleum and makes us more dependent on foreign oil. Recycling 1 tonne of aluminum saves 4 tonnes of bauxite (aluminum ore) and 700 kg of petroleum coke and pitch, as well as keeping 35 kg of aluminum fluoride out of the air.

Recycling also reduces energy consumption and air pollution. Plastic bottle recycling could save 50 to 60 percent of the energy needed to make new ones. Making new steel from old scrap offers up to 75 percent energy savings. Producing aluminum from scrap instead of bauxite ore cuts energy use by 95 percent, yet we still throw away more than a million tonnes of aluminum every year. If aluminum recovery were doubled worldwide, more than a million tonnes of air pollutants would be eliminated every year.

Creating a Market for Recycling

In many communities, citizens have done such a good job of collecting recyclables that a glut has developed. Mountains of some waste materials accumulate in warehouses (Fig. 18.9) because there are no markets for them. Too often, wastes that we carefully separate for recycling end up being mixed together and dumped in a landfill or incinerator, because a market for these resources is not well developed.

Our present public policies often tend to favour extraction of new raw materials. Energy, water, and raw materials are often sold to industries below their real cost to create jobs and stimulate the economy. For instance, in 1999, a pound of recycled clear PET, the material in most soft drink bottles, sold for about 40¢. By contrast, a pound of off-grade, virgin PET cost 25¢. Setting the prices of natural resources at their real cost would tend to encourage efficiency and recycling. Municipal, provincial, and federal regulations requiring government agencies to purchase a minimum amount of recycled material have helped create a market for used materials. Each of us can play a role in creating markets, as well. If we buy things made from recycled materials—or ask for them if they aren't available—we will help make it possible for recycling programs to succeed.

Composting

Pressed for landfill space, many cities have banned yard waste from municipal garbage. Rather than bury this valuable organic material, they are turning it into a useful product through **composting:** biological degradation or breakdown of organic matter under aerobic (oxygen-rich) conditions. The organic compost resulting from this process makes a nutrient-rich soil amendment that aids water retention, slows soil erosion, and improves crop yields. A home compost pile is an easy and inexpensive way to dispose of organic waste in an interesting and environmentally friendly way.

Large-scale municipal or industrial compost facilities have some of the same public relations and siting problems as incinerators and landfills. Neighbours complain of noise, odours, vermin, and increased traffic from poorly run facilities. These problems don't have to occur, but when you are dealing with huge amounts of garbage it's easy for the situation to get out of hand.

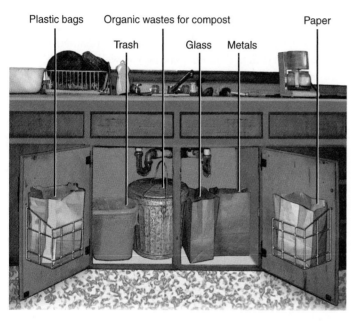

FIGURE 18.7 Source separation in the kitchen—the first step in a strong recycling program.

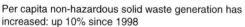
Per capita non-hazardous solid waste generation has increased: up 10% since 1998
Per capita non-hazardous solid waste disposal and recycling/reuse (kilograms per person)

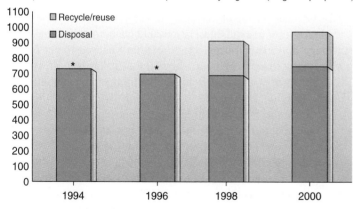

*Recycling data not available for 1994 and 1996.

Total non-hazardous solid waste disposal and recycling/reuse (million tonnes)

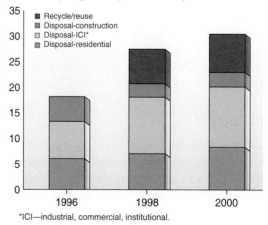

*ICI—industrial, commercial, institutional.

FIGURE 18.8 Per capita (kg per person) and total (million tonnes) non-hazardous solid waste disposal and recycling/reuse.
Source: Waste Management Industry Survey: Business and Government Sectors 1994, 1996, 1998, 2000, Statistics Canada.
Adapted by: National Indicators and Reporting Office, Environment Canada.

FIGURE 18.9 In some places, mountains of paper, plastic, and glass have accumulated in warehouses because collection programs have been more successful than material marketing efforts. Federal and provincial governments need to encourage use of recycled materials. Each of us can help by buying recycled products.

Energy from Waste

Worldwide, at least one-fifth of municipal waste is organic kitchen and garden refuse. In a landfill, much of this matter is decomposed by microorganisms generating billions of cubic metres of methane ("natural gas"), which contributes to global warming if allowed to escape into the atmosphere (Chapter 8). Many cities are drilling methane wells in their landfills to capture this valuable resource. Fuel cells (Chapter 13) are a good way to use this fuel.

This valuable organic material can be burned in an incinerator rather than being buried in landfills, but there are worries about air pollution from incineration. Organic wastes also can be decomposed in large, oxygen-free digesters to produce methane under more controlled conditions than in a landfill and with less air pollution than mass garbage burning.

Anaerobic digestion also can be done on a small scale. Millions of household methane generators provide fuel for cooking and lighting for homes in China and India (Chapter 13). In Canada, some farmers produce all the fuel they need to run their farms—both for heating and to run trucks and tractors—by generating methane from animal manure.

Demanufacturing

Demanufacturing is the disassembly and recycling of obsolete consumer products, such as TV sets, personal computers, refrigerators, washing machines, and air conditioners. Together with deconstruction of houses, it is a good way to recover valuable materials. It also can be especially suited to inner cities where there is a large supply of materials to be demanufactured and a pool of skilled and unskilled labourers who need jobs. For example, there are about 300 million TVs and personal computers in use in the United States. TVs often are discarded after only about five years, and computers, play-stations, and other electronics become obsolete even faster. Stoves, refrigerators, and other "white goods" have a much longer lifetime—typically about 12 years—but the U.S. EPA estimates that Americans dispose of 54 million of these household appliances every year. Many of these consumer products contain both valuable materials and toxins that must be kept out of the environment. Older refrigerators and air-conditioners, for example, have chlorofluorocarbons (CFCs) that destroy stratospheric ozone and cannot be released into the air. Because new production has been banned, recycled CFCs are worth about $100 per pound. For both reasons, it pays to recycle them.

Similarly, computers and other electronic equipment contain both toxic metals (mercury, lead, gallium, germanium, nickel, palladium, beryllium, selenium, arsenic) as well as valuable ones such as gold, silver, and copper. A typical personal computer, for instance, has about $6 worth of gold, $5 of copper, and $1 of silver. It's estimated that 90 percent of the cadmium, lead, and mercury contamination in our solid waste stream comes from consumer electronics, batteries, mercury lamps, and switches. Small entrepreneurial firms are emerging in many urban centres to take advantage of this valuable resource. By forming alliances with other demanufacturing firms and large manufacturers, they can both develop the skills and experience to produce quality materials and also guarantee a market.

Reuse

Even better than recycling or composting is cleaning and reusing materials in their present form, thus saving the cost and energy of remaking them into something else. We do this already with some specialized items. Auto parts are regularly sold from junkyards, especially for older car models. In some areas, stained glass windows, brass fittings, fine woodwork, and bricks salvaged from old houses bring high prices. Some communities sort and reuse a variety of materials received in their dumps (Fig. 18.10).

In many cities, glass and plastic bottles are routinely returned to beverage producers for washing and refilling. The reusable, refillable bottle is the most efficient beverage container we have. This is better for the environment than remelting and more profitable for local communities. A reusable glass container makes an average of 15 round-trips between factory and customer before it becomes so scratched and chipped that it has to be recycled. Reusable containers also favour local bottling companies and help preserve regional differences.

Since the advent of cheap, lightweight, disposable food and beverage containers, many small, local breweries, canneries, and bottling companies have been forced out of business by huge national conglomerates. These big companies can afford to ship food and beverages great distances as long as it is a one-way trip. If they had to collect their containers and reuse them, canning and bottling factories serving large regions would be uneconomical. Consequently, the national companies favour recycling rather than refilling because they prefer fewer, larger plants and don't want to be responsible for collecting and reusing containers. In some circumstances, life-cycle assessment shows that washing and

FIGURE 18.10 Reusing discarded products is a creative and efficient way to reduce wastes. This recycling centre is a valuable source of used building supplies and a money saver for the whole community.

decontaminating containers takes as much energy and produces as much air and water pollution as manufacturing new ones.

In less affluent nations, reuse of all sorts of manufactured goods is an established tradition. Where most manufactured products are expensive and labour is cheap, it pays to salvage, clean, and repair products. Cairo, Manila, Mexico City, and many other cities have large populations of poor people who make a living by scavenging. Entire ethnic populations may survive on scavenging, sorting, and reprocessing scraps from city dumps.

Producing Less Waste

What is even better than reusing materials? Generating less waste in the first place. The "What Can You Do?" box on page 394 describes some contributions you can make to reducing the volume of our waste stream. Industry also can play an important role in source reduction. The 3M Company saved over $500 million since 1975 by changing manufacturing processes, finding uses for waste products, and listening to employees' suggestions. What is waste to one division is a treasure to another.

FIGURE 18.11 How much more do we need? Where will we put what we already have?
Reprinted with special permission of King Features Syndicate.

Excess packaging of food and consumer products is one of our greatest sources of unnecessary waste. Paper, plastic, glass, and metal packaging material make up 50 percent of our domestic trash by volume. Much of that packaging is primarily for marketing and has little to do with product protection (Fig. 18.11). Manufacturers and retailers might be persuaded to reduce these wasteful practices if consumers ask for products without excess packaging. Canada's National Packaging Protocol (NPP) recommends that packaging minimize depletion of virgin resources and production of toxins in manufacturing. The preferred hierarchy is (1) no packaging, (2) minimal packaging, (3) reusable packaging, and (4) recyclable packaging.

Where disposable packaging is necessary, we still can reduce the volume of waste in our landfills by using materials that are compostable or degradable. **Photodegradable plastics** break down when exposed to ultraviolet radiation. **Biodegradable plastics** incorporate such materials as cornstarch that can be decomposed by microorganisms. Degradable plastics, such as six-pack beverage yokes, fast-food packaging, and disposable diapers, often don't decompose completely; they only break down to small particles that remain in the environment. In doing so, they can release toxic chemicals into the environment. And in modern, lined landfills they don't decompose at all. Furthermore, they make recycling less feasible and may lead people to believe that littering is okay.

Some environmental groups are beginning to think we have put too much emphasis on recycling. Many people think if they recycle aluminum cans and newspapers they are doing everything they can for the environment. Companies that make "throw away" or heavily packaged products generally favour recycling because it allows them to continue business-as-usual. While recycling is an important part of waste management, we have to remember that it is actually the third "R" in the waste hierarchy. The two preferred methods—reduction and reuse—get lost in our enthusiasm for recycling.

Reducing Waste

1. Buy foods that come with less packaging; shop at farmers' markets or co-ops, using your own containers.

2. Take your own washable refillable beverage container to meetings or convenience stores.

3. When you have a choice at the grocery store between plastic, glass, or metal containers for the same food, buy the reusable or easier-to-recycle glass or metal.

4. When buying plastic products, pay a few cents extra for environmentally degradable varieties.

5. Separate your cans, bottles, papers, and plastics for recycling.

6. Wash and reuse bottles, aluminum foil, plastic bags, etc., for your personal use.

7. Compost yard and garden wastes, leaves, and grass clippings.

8. Encourage your municipal, provincial, and federal politicians to support recycling and composting programs in your area.

Source: Minnesota Pollution Control Agency.

FIGURE 18.12 Canada ranks 24th out of 27 OECD nations in generation of hazardous waste. Industry produces about 200 kg per year per person. By comparison, the United States produces about 800 kg per person per year.

HAZARDOUS AND TOXIC WASTES

The most dangerous aspect of the waste stream we have described is that it often contains highly toxic and hazardous materials that are injurious to both human health and environmental quality. We now produce and use a vast array of flammable, explosive, caustic, acidic, and highly toxic chemical substances for industrial, agricultural, and domestic purposes (Fig. 18.12). The biggest source of these toxins are the chemical and petroleum industries (Fig. 18.13).

What Is Hazardous Waste?

A **hazardous waste** is any discarded material, liquid or solid, that contains substances known to be (1) fatal to humans or laboratory animals in low doses, (2) toxic, carcinogenic, mutagenic, or teratogenic to humans or other life-forms, (3) ignitable with a flash point less than 60°C, (4) corrosive, or (5) explosive or highly reactive (undergoes violent chemical reactions either by itself or when mixed with other materials). Notice that this definition includes both toxic and hazardous materials as defined in Chapter 22. Certain compounds are exempt from regulation as hazardous waste if they are accumulated in less than 1 kg of commercial chemicals or 100 kg of contaminated soil, water, or debris. Even larger amounts (up to 1,000 kg) are exempt when stored at an approved waste treatment facility for the purpose of being beneficially used, recycled, reclaimed, detoxified, or destroyed.

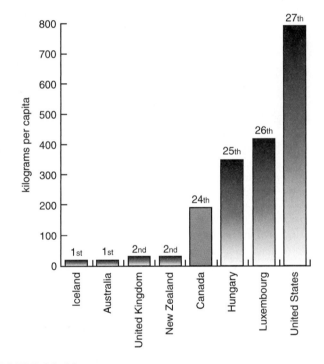

FIGURE 18.13 Kilograms of hazardous waste per capita.
Source: OECD Environmental Data 1999.

Hazardous Waste Disposal

Most hazardous waste is recycled, converted to nonhazardous forms, stored, or otherwise disposed of on-site by the generators—chemical companies, petroleum refiners, and other large industrial facilities—so that it doesn't become a public problem. Still, the hazardous waste that does enter the waste stream or the environment represents a serious environmental problem. And orphan

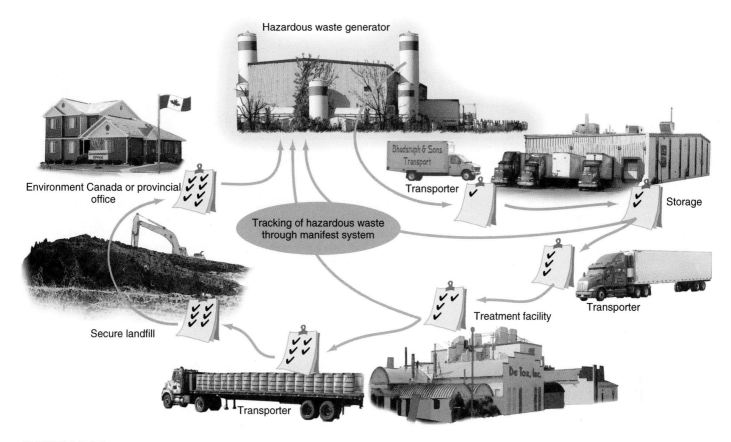

FIGURE 18.14 Toxic and hazardous wastes must be tracked from "cradle to grave" by detailed shipping manifests.

wastes left behind by abandoned industries remain a serious threat to both environmental quality and human health. For years, little attention was paid to this material. Wastes stored on private property, buried, or allowed to soak into the ground were considered of little concern to the public. An estimated 5 billion tonnes of highly poisonous chemicals were improperly disposed of in the United States between 1950 and 1975 before regulatory controls became more stringent (Fig. 18.14).

How Clean Is Clean?

Among the biggest problems in cleaning up hazardous waste sites are questions of liability and the degree of purity required. In many cities, these problems have created large areas of contaminated properties known as **brownfields** that have been abandoned or are not being used up to their potential because of real or suspected pollution. Up to one-third of all commercial and industrial sites in the urban core of many big cities fall in this category. In heavy industrial corridors the percentage typically is higher.

In 2003, Canada's National Roundtable on the Economy and the Environment produced a comprehensive report on brownfields entitled, *Cleaning up the Past, Building the Future.* The report outlined a specific plan for the reclamation of the 30,000 or so brownfields in Canada, which included both environmental improvements and social and economic revitalization of the communities where they occur. For years, no one was interested in redeveloping brownfields because of liability risks. Who would buy a property knowing that they might be forced to spend years in litigation and negotiations and be forced to pay millions of dollars for pollution they didn't create? Even if a site has been cleaned to current standards, there is a worry that additional pollution might be found in the future or that more stringent standards might be applied.

Recognizing that reusing contaminated properties can play a significant role in rebuilding old cities, creating jobs, increasing the tax base, and preventing needless destruction of open space at urban margins, programs have been established at both federal and provincial levels to encourage brownfield recycling. Adjusting purity standards according to planned uses and providing liability protection for nonresponsible parties gives developers and future purchasers confidence that they won't be unpleasantly surprised in the future with further cleanup costs. In some communities, former brownfields are being turned into "eco-industrial parks" that feature environmentally friendly businesses and bring in much needed jobs to inner city neighbourhoods.

Options for Hazardous Waste Management

What shall we do with toxic and hazardous wastes? In our homes, we can reduce waste generation and choose less toxic materials. Buy only what you need for the job at hand. Use up the last little

IN DEPTH. . . . .

Cleaning Up Toxic Waste with Plants

Getting contaminants out of soil and groundwater is one of the most widespread and persistent problems in waste cleanup. Once leaked into the ground, solvents, metals, radioactive elements, and other contaminants are dispersed and difficult to collect and treat. The main method of cleaning up contaminated soil is to dig it up, then decontaminate it or haul it away and store it in a landfill in perpetuity. At a single site, thousands of tonnes of tainted dirt and rock may require incineration or other treatment. Cleaning up contaminated groundwater usually entails pumping vast amounts of water out of the ground—hopefully extracting the contaminated water faster than it can spread through the water table or aquifer. In Canada, there are thousands of contaminated sites on factories, farms, gas stations, military facilities, sewage treatment plants, landfills, chemical warehouses, and other types of facilities. Cleaning up these sites is expected to cost billions of dollars.

Recently, a number of promising alternatives have been developed using plants, fungi, and bacteria to clean up our messes. *Phytoremediation* (remediation, or cleanup, using plants) can include a variety of strategies for absorbing, extracting, or neutralizing toxic compounds. Certain types of mustards and sunflowers can extract lead, arsenic, zinc, and other metals (*phytoextraction*). Poplar trees can absorb and break down toxic organic chemicals (*phytodegradation*). Reeds and other water-loving plants can filter water tainted with sewage, metals, or other contaminants. Natural bacteria in groundwater, when provided with plenty of oxygen, can neutralize contaminants in aquifers, minimizing or even eliminating the need to extract and treat water deep in the ground. Radioactive strontium and cesium have been extracted from soil near the Chernobyl nuclear power plant using common sunflowers.

How do the plants, bacteria, and fungi do all this? Many of the biophysical details are poorly understood, but in general, plant roots are designed to efficiently extract nutrients, water, and minerals from soil and groundwater.

The mechanisms involved may aid extraction of metallic and organic contaminants. Some plants also use toxic elements as a defense against herbivores—locoweed, for example, selectively absorbs elements such as selenium, concentrating toxic levels in its leaves. Absorption can be extremely effective. Braken fern growing in Florida was found to contain arsenic at concentrations more than 200 times higher than the soil in which it was growing.

Genetically modified plants are also being developed to process toxins. Poplars have been grown with a gene borrowed from bacteria that transform a toxic compound of mercury into a safer form. In another experiment, a gene for producing mammalian liver enzymes, which specialize in breaking down toxic organic compounds, was inserted into tobacco plants. The plants succeeded in producing the liver enzymes and breaking down toxins absorbed through their roots.

These remediation methods are not without risks. As plants take up toxins, insects could consume leaves, allowing contaminants to enter the food web. Some absorbed contaminants are volatilized, or emitted in gaseous form, through pores in plant leaves. Once toxic contaminants are absorbed into plants, the plants themselves are usually toxic and must be landfilled. But the cost of phytoremediation can be less than half the cost of landfilling or treating toxic soil, and the volume of plant material requiring secure storage ends up being a fraction of a percent of the volume of the contaminated dirt.

Cleaning up hazardous and toxic waste sites will be a big business for the foreseeable future around the world. Innovations such as phytoremediation offer promising prospects for business growth as well as for environmental health and saving taxpayers' money.

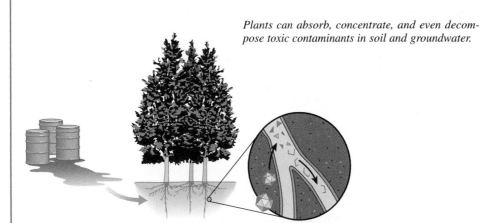

Plants can absorb, concentrate, and even decompose toxic contaminants in soil and groundwater.

bit or share leftovers with a friend or neighbour. Many common materials that you probably already have make excellent alternatives to commercial products (What Can You Do? p. 398). Dispose of unneeded materials responsibly (Fig. 18.15).

Produce Less Waste

As with other wastes, the safest and least expensive way to avoid hazardous waste problems is to avoid creating the wastes in the first place. Manufacturing processes can be modified to reduce

or eliminate waste production. In Minnesota, the 3M Company reformulated products and redesigned manufacturing processes to eliminate more than 140,000 tonnes of solid and hazardous wastes, 4 billion l of wastewater, and 80,000 tonnes of air pollution each year. They frequently found these new processes not only spared the environment but also saved money by using less energy and fewer raw materials.

Recycling and reusing materials also eliminates hazardous wastes and pollution. Many waste products of one process or

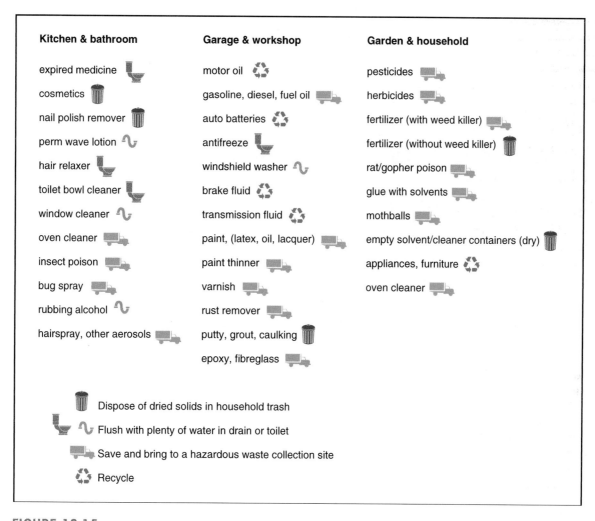

Kitchen & bathroom

expired medicine
cosmetics
nail polish remover
perm wave lotion
hair relaxer
toilet bowl cleaner
window cleaner
oven cleaner
insect poison
bug spray
rubbing alcohol
hairspray, other aerosols

Garage & workshop

motor oil
gasoline, diesel, fuel oil
auto batteries
antifreeze
windshield washer
brake fluid
transmission fluid
paint, (latex, oil, lacquer)
paint thinner
varnish
rust remover
putty, grout, caulking
epoxy, fibreglass

Garden & household

pesticides
herbicides
fertilizer (with weed killer)
fertilizer (without weed killer)
rat/gopher poison
glue with solvents
mothballs
empty solvent/cleaner containers (dry)
appliances, furniture
oven cleaner

Dispose of dried solids in household trash

Flush with plenty of water in drain or toilet

Save and bring to a hazardous waste collection site

Recycle

FIGURE 18.15 Household waste disposal guide. Solvent-containing products have the words "flammable," "combustible," or "petroleum distillates" on the labels. These materials should not be disposed of in a drain or toilet. Never mix products containing bleach with those containing ammonia. A toxic gas can form!

industry are valuable commodities in another. For example, already, about 10 percent of the wastes that would otherwise enter the waste stream in the United States are sent to surplus material exchanges where they are sold as raw materials for use by other industries. This figure could probably be raised substantially with better waste management. In Europe, at least one-third of all industrial wastes are exchanged through clearinghouses where beneficial uses are found. This represents a double savings: the generator doesn't have to pay for disposal, and the recipient pays little, if anything, for raw materials.

Convert to Less Hazardous Substances

Several processes are available to make hazardous materials less toxic. *Physical treatments* tie up or isolate substances. Charcoal or resin filters absorb toxins. Distillation separates hazardous components from aqueous solutions. Precipitation and immobilization in ceramics, glass, or cement isolate toxins from the environment so that they become essentially nonhazardous. One of the few ways to dispose of metals and radioactive substances is to fuse them in silica at high temperatures to make a stable, impermeable glass that is suitable for long-term storage.

Incineration is applicable to mixtures of wastes. A permanent solution to many problems, it is quick and relatively easy but not necessarily cheap—nor always clean—unless it is done correctly. Wastes must be heated to higher than 1,000°C for a sufficient period of time to complete destruction. The ash resulting from thorough incineration is reduced in volume up to 90 percent and often is safer to store in a landfill or other disposal site than the original wastes. Nevertheless, incineration remains a highly controversial topic (Fig. 18.16).

Several sophisticated features of modern incinerators improve their effectiveness. Liquid injection nozzles atomize liquids and mix air into the wastes so they burn thoroughly. Fluidized bed burners pump air from the bottom up through burning solid waste as it travels on a metal chain grate through the furnace. The air velocity is sufficient to keep the burning waste partially suspended. Plenty of oxygen is available, and burning is quick and complete. Afterburners add to the completeness of burning by igniting gaseous hydrocarbons not consumed in the incinerator.

FIGURE 18.16 Actor Martin Sheen joins local activists in a protest in East Liverpool, Ohio, site of the largest hazardous waste incinerator in the United States. About 1,000 people marched to the plant to pray, sing, and express their opposition. Involving celebrities draws attention to your cause. A peaceful, well-planned rally builds support and acceptance in the broader community.

Scrubbers and precipitators remove minerals, particulates, and other pollutants from the stack gases.

Chemical processing can transform materials so they become nontoxic. Included in this category are neutralization, removal of metals or halogens (chlorine, bromine, etc.), and oxidation. The Sunohio Corporation of Canton, Ohio, for instance, has developed a process called PCBx in which chlorine in such molecules as PCBs is replaced with other ions that render the compounds less toxic. A portable unit can be moved to the location of the hazardous waste, eliminating the need for shipping them.

Biological waste treatment or **bioremediation** taps the great capacity of microorganisms to absorb, accumulate, and detoxify a variety of toxic compounds. Bacteria in activated sludge basins, aquatic plants (such as water hyacinths or cattails), soil microorganisms, and other species remove toxic materials and purify effluents. Biotechnology offers exciting possibilities for finding or creating organisms to eliminate specific kinds of hazardous or toxic wastes. By using a combination of classic genetic selection techniques and high-technology gene-transfer techniques, for instance, scientists have recently been able to generate bacterial strains that are highly successful at metabolizing PCBs. There are concerns about releasing such exotic organisms into the environment, however (Chapter 16). It may be better to keep these organisms contained in enclosed reaction vessels and feed contaminated material to them under controlled conditions.

Store Permanently

Inevitably, there will be some materials that we can't destroy, make into something else, or otherwise cause to vanish. We will have to store them out of harm's way. There are differing opinions about how best to do this.

Alternatives to Hazardous Household Chemicals

Chrome cleaner: Use vinegar and nonmetallic scouring pad.

Copper cleaner: Rub with lemon juice and salt mixture.

Floor cleaner: Mop linoleum floors with 225 ml vinegar mixed with 8 l of water. Polish with club soda.

Brass polish: Use Worcestershire sauce.

Silver polish: Rub with toothpaste on a soft cloth.

Furniture polish: Rub in olive, almond, or lemon oil.

Ceramic tile cleaner: Mix 56.25 ml baking soda, 112.5 ml white vinegar, and 225 ml ammonia in 4 l warm water (good general purpose cleaner).

Drain opener: Use plunger or plumber's snake, pour boiling water down drain.

Upholstery cleaner: Clean stains with club soda.

Carpet shampoo: Mix 112.5 ml liquid detergent in .5 l hot water. Whip into stiff foam with mixer. Apply to carpet with damp sponge. Rinse with 225 ml vinegar in 4 l water. Don't soak carpet—it may mildew.

Window cleaner: Mix 75 ml ammonia, 56.25 ml white vinegar in 1 litre warm water. Spray on window. Wipe with soft cloth.

Spot remover: For butter, coffee, gravy, or chocolate stains: Sponge up or scrape off as much as possible immediately. Dab with cloth dampened with a solution of 4 ml white vinegar in 1 litre cold water.

Toilet cleaner: Pour 112.5 ml liquid chlorine bleach into toilet bowl. Let stand for 30 minutes, scrub with brush, flush.

Pest control: Spray plants with soap-and-water solution (3 tablespoons soap per litre water) for aphids, mealybugs, mites, and whiteflies. Interplant with pest repellent plants such as marigolds, coriander, thyme, yarrow, rue, and tansy. Introduce natural predators such as ladybugs or lacewings.

Indoor pests: Grind or blend 1 garlic clove and 1 onion. Add 1 tablespoon cayenne pepper and 1 litre water. Add 15 ml liquid soap.

Moths: Use cedar chips or bay leaves.

Ants: Find where they are entering house, spread cream of tartar, cinnamon, red chili pepper, or perfume to block trail.

Fleas: Vacuum area, mix brewer's yeast with pet food.

Mosquitoes: Brewer's yeast tablets taken daily repel mosquitoes.

Note: test cleaners in small, inconspicuous area before using.

Retrievable Storage. Dumping wastes in the ocean or burying them in the ground generally means that we have lost control of them. If we learn later that our disposal technique was a mistake, it is difficult, if not impossible, to go back and recover the wastes. For many supertoxic materials, the best way to store them may be in **permanent retrievable storage.** This means placing waste storage containers in a secure building, salt mine, or bedrock cavern where they can be inspected periodically and retrieved, if necessary, for repacking or for transfer if a better means of disposal is developed. This technique is more expensive than burial in a landfill because the storage area must be guarded and monitored continuously to prevent leakage, vandalism, or other dispersal of toxic materials. Remedial measures are much cheaper with this technique, however, and it may be the best system in the long run.

Secure Landfills. One of the most popular solutions for hazardous waste disposal has been landfilling. Although, as we saw earlier in this chapter, many such landfills have been environmental disasters, newer techniques make it possible to create safe, modern **secure landfills** that are acceptable for disposing of many hazardous wastes. The first line of defence in a secure landfill is a thick bottom cushion of compacted clay that surrounds the pit like a bathtub (Fig. 18.17). Moist clay is flexible and resists cracking if the ground shifts. It is impermeable to groundwater and will safely contain wastes. A layer of gravel is spread over the clay liner and perforated drain pipes are laid in a grid to collect any seepage that escapes from the stored material. A thick polyethylene liner, protected from punctures by soft padding materials, covers the gravel bed. A layer of soil or absorbent sand cushions the inner liner and the wastes are packed in drums, which then are placed into the pit, separated into small units by thick berms of soil or packing material.

When the landfill has reached its maximum capacity, a cover much like the bottom sandwich of clay, plastic, and soil—in that order—caps the site. Vegetation stabilizes the surface and improves its appearance. Sump pumps collect any liquids that filter through the landfill, either from rainwater or leaking drums. This leachate is treated and purified before being released. Monitoring wells check groundwater around the site to ensure that no toxins have escaped.

Most landfills are buried below ground level to be less conspicuous; however, in areas where the groundwater table is close to the surface, it is safer to build above-ground storage. The same protective construction techniques are used as in a buried pit. An advantage to such a facility is that leakage is easier to monitor because the bottom is at ground level.

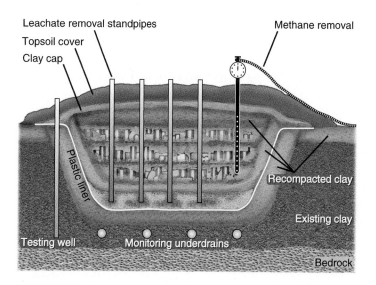

FIGURE 18.17 A secure landfill for toxic waste. A thick plastic liner and two or more layers of impervious compacted clay enclose the landfill. A gravel bed between the clay layers collects any leachate, which can then be pumped out and treated. Well samples are tested for escaping contaminants and methane is collected for combustion.

Transportation of hazardous wastes to disposal sites is of concern because of the risk of accidents. Emergency preparedness officials conclude that the greatest risk in most urban areas is not nuclear war or natural disaster but crashes involving trucks or trains carrying hazardous chemicals through densely packed urban corridors. Another worry is who will bear financial responsibility for abandoned waste sites. The material remains toxic long after the businesses that created it are gone. As is the case with nuclear wastes (Chapter 12), we may need new institutions for perpetual care of these wastes.

SUMMARY

We produce enormous volumes of waste in industrialized societies, and there is an increasing problem of how to dispose of this material in an environmentally safe manner. In this chapter, we have looked at the character of our solid and hazardous wastes. We have surveyed the ways we dispose of our wastes and the environmental problems associated with waste disposal.

Solid wastes are domestic, commercial, industrial, agricultural, and mining wastes that are primarily nontoxic. About 60 percent of our domestic and industrial wastes are deposited in landfills; most of the rest is incinerated or recycled. Old landfills were often messy and leaky, but new ones are required to have impermeable clay or plastic linings, drainage, and careful siting. Incineration can destroy organic compounds, but whether incinerators can or will be operated satisfactorily is a matter of debate. Recycling is growing nationwide, encouraged by the economic and environmental benefits it brings.

Hazardous and toxic wastes, when released into the environment, cause such health problems as birth defects, neurological disorders, reduced resistance to infection, and cancer. Environmental losses include contamination of water supplies, poisoning of the soil, and destruction of habitat. The major categories of hazardous wastes are ignitable, corrosive, reactive, explosive, and toxic. Some materials of the greatest concern are heavy metals, solvents, and synthetic organic chemicals such as halogenated hydrocarbons, organophosphates, and phenoxy herbicides.

Disposal practices for solid and hazardous wastes have often been unsatisfactory. Thousands of abandoned, often unknown waste disposal sites still leak toxic materials into the environment. Some alternative techniques for treating or disposing of hazardous wastes include not making the material in the first place, incineration, secure landfill, and physical, chemical, or biological treatment to detoxify or immobilize wastes. People are often unwilling to have transfer facilities, storage sites, disposal operations, or transportation of hazardous or toxic materials in or through their cities. Questions of safety and liability remain unanswered in solid and hazardous waste disposal.

QUESTIONS FOR REVIEW

1. What are solid wastes and hazardous wastes? What is the difference between them?

2. Describe the difference between an open dump, a sanitary landfill, and a modern, secure, hazardous waste disposal site.

3. Why are landfill sites becoming limited around most major urban centres in Canada and the United States? What steps are being taken to solve this problem?

4. Describe some concerns about waste incineration.

5. List some benefits and drawbacks of recycling wastes. What are the major types of materials recycled from municipal waste and how are they used?

6. What is composting, and how does it fit into solid waste disposal?

7. Describe some ways that we can reduce the waste stream to avoid or reduce disposal problems.

8. List 10 toxic substances in your home and how you would dispose of them.

9. What are brownfields and why do cities want to redevelop them?

10. What societal problems are associated with waste disposal? Why do people object to waste handling in their neighbourhoods?

QUESTIONS FOR CRITICAL THINKING

1. A toxic waste disposal site has been proposed for the Pine Ridge Indian Reservation in South Dakota. Many tribal members oppose this plan, but some favour it because of the jobs and income it will bring to an area with 70 percent unemployment. If local people choose immediate survival over long-term health, should we object or intervene?

2. There is often a tension between getting your personal life in order and working for larger structural changes in society. Evaluate the trade-offs between spending time and energy sorting recyclables at home compared to working in the public arena on a bill to ban excess packaging.

3. Should industry officials be held responsible for dumping chemicals that were legal when they did it but are now known to be extremely dangerous? At what point can we argue that they should have known about the hazards involved?

4. Suppose that your brother or sister has decided to buy a house next to a toxic waste dump because it costs $20,000 less than a comparable house elsewhere. What do you say to him or her?

5. Is there an overall conceptual framework or point of view in this chapter? If you were presenting a discussion of solid or hazardous waste to your class, what would be your conceptual framework?

6. Is there a fundamental difference between incinerating municipal, medical, or toxic industrial waste? Would you oppose an incinerator for one type of waste in your neighbourhood but not others? Why, or why not?

7. The Netherlands incinerates much of its toxic waste at sea by a shipborne incinerator. Would you support this as a way to dispose of our wastes as well? What are the critical considerations for or against this approach?

KEY TERMS

biodegradable plastics   393
bioremediation   398
brownfields   395
composting   391
demanufacturing   392
energy recovery   388
hazardous waste   394
mass burn   388
permanent retrievable storage   398
photodegradable plastics   393
recycling   389
refuse-derived fuel   388
sanitary landfills   386
secure landfills   399
waste stream   385

RECYCLING CHALLENGE

The Internet Consumer Recycling Guide, http://www.obviously.com/recycle/, provides a detailed list of recyclable materials and where to send them. Before you look at this website, take five minutes and write down at least 25 different types of items in the room around you. These items can be small (paper clips, hair clips, books, food packaging, etc.) or large (carpets, furniture, computers, light fixtures, plumbing, etc.). Once you have made the list, mark all those that are recyclable and note how you would recycle them.

Now look at the Recycling Guide web page, above. Look at all three "guide" pages—The World's Shortest Comprehensive Recycling Guide, the Guide to Recycling Common Materials, and Guide to Hard-to-Recycle Materials. How many additional items on your list can be recycled, according to these guides? How many could be recycled in principle but would be hard to recycle in your community? Why? What recycled products could be made by recycling the materials on your list?

How many of the items on your list might be available with recycled content? How many of them could you have avoided acquiring in the first place?

Environmental Scientist/Business Executive

Dr. Phil Whiting

Growing up in a town of 100 people in Nova Scotia, Canada, Phil has had a keen awareness of the impact of economic development on the local environment. As a young boy he observed first hand the economic benefits of the construction of a large forest products company in his hometown. But the benefits were coupled with a devastating impact on a small river running through town. His father, a noted local environmental activist, was instrumental in pressing for changes that eventually led to a major clean-up program. To Phil, he saw firsthand how industry could clean up its act, the environment could recover, and the economic benefits could still be maintained.

Phil's interest in forest products wastes enticed him to attend Acadia University in Nova Scotia so he could learn more about the "science" behind the "issues." There he gained BSc and Masters degrees in Chemistry. His thesis research topics both involved studies on the impact of forest product wastes on the environment, coupled with new methods to treat wastewater. His interest in the forest industries led him to pursue PhD studies at McGill University, where his research topic focused on developing a detailed understanding of the chemistry of lignin (a major component of wood and one of the key components in wood product wastes).

As a newly minted PhD, Phil now faced a choice. He could continue his career in the academic environment, focusing his efforts on building enhanced understanding of environmental problems. Or, he could tackle industry from the inside ... becoming a champion for environmentally responsible business practices. An industry opportunity came soon, and he took on a role as a research scientist with one of the world's biggest pulp and paper companies. His first assignment was to develop a better pulp bleaching process for a new mill. The target was to reduce the costs for the plant, reduce the amount of bleaching chemicals needed, and reduce the environmental impact of the bleaching operation. Working with a team of young scientists and engineers, in a few years a $30 million plant was built to his specifications, and his dream of an enviro-economically viable process was a reality.

This success led to a series of new opportunities, eventually leading to the role of vice president of Research & Development.

The CEO gave him his first major project— design and construct new wastewater treatment plants for all of the company's Canadian mills; and by the way, do it for half the cost that consultants had estimated. And, oh yes, make sure every mill starts up in complete adherence with all the regulations too! This exciting project led to some great results, and a key disappointment. The team met all the economic and environmental goals of the project ... except one.

As a side project, the team had designed a "zero-effluent" paper mill and convinced the company board to provide $20 million to implement these ideas, a world's first. Unfortunately, since the process used a completely new water treatment process, it was not possible to obtain the needed government permits, and the dream died.

Since then, Phil has maintained his passion for bringing environmental responsibility to the business world. This has provided him exciting opportunities to tackle a number of other interesting challenges along the same theme:

- He was vice president of R&D with the world's largest manufacturer of ultra-violet water treatment systems.

- He helped start up a new Environmental Research Centre at a leading Canadian university.

- He developed and launched a new business based on a novel home-based water treatment process.

Today, he's still enjoying playing the enviro-business game. He's maintained his passion for environmental research through his status as a volunteer university professor at two universities—for more than 20 years now. And currently he's president of a new company developing and commercializing new fuel cell and rechargeable batteries—environmentally friendly energy for the twenty-first century. Every year, North Americans throw out billions of AA and AAA batteries and his new passion is bringing to the market alkaline batteries that can be reused hundreds of times.

"It's amazing what doors are opened when you learn the science behind the issue," says Phil. "And I have no doubt, the path ahead will be just as exciting as the one behind."

Chapter 19

Environmental Law: Its Role in Guiding Governing Instruments

The best lack all conviction, while the worst are full of passionate intensity.

—W. B. Yeats—

OBJECTIVES

After studying this chapter, you should be able to:

- differentiate between statute and common law, and know the basis for each in Canadian law.
- identify federal and provincial responsibilities for resource and environmental management as defined in Canada's Constitution.
- understand the five approaches that have characterized the evolution of environmental law and policy in Canada, and remain in use today.
- consider the reasons that international treaties have not been successful.

- appreciate the importance of wicked problems, resilience and adaptive management in environmental planning.
- scrutinize collaborative, community-based planning methods.

WebQuest

Environmental Law, legislation, environmental impact assessment, economic instruments, voluntary approaches, NPRI, environmental protection

Above: The ability of ordinary citizens to petition their government and to participate in public policy formation is a hallmark of democracy.

You be the Judge

The application of the law requires one to determine what is legal from what is not legal. This should not be confused with what you or other people believe to be right and wrong, although sometimes the courts will agree with your views. If you were a court judge, how would you decide the following cases? The court decisions from these cases will be presented throughout the chapter. At the end of this chapter, you will have an appreciation for the importance of grounding any environmental law in Canada in the hands of the right level of government and the appropriate means to achieve it. The answer to these two considerations—who and how—is found in Canada's Constitution and practice of common law.

Case 1: *Palmer v. Nova Scotia Forest Industries*

In 1982, people (the plaintiffs) were living adjacent to land owned by the Government of Nova Scotia that was used to produce commercial timber. As part of a forest management plan, a multinational forest company (the defendant) obtained a licence from the Nova Scotia Department of the Environment to spray the area with phenoxy herbicides. Streams flowed from the proposed spray site through some property owners' land. They were also concerned that the spray would enter their property through the air. From the defendant's perspective, this biocide would kill undesirable species and allow the desirable species of trees to grow. The plaintiffs obtained a temporary injunction preventing the company from conducting the spray. A full trial and hearing was held beginning in May 1983 to determine if the injunction should be extended or rescinded.

The plaintiffs desired the following awards from the court:

- a permanent injunction to prevent the defendant from spraying the phenoxy herbicides 2,4-D and 2,4,5-T at the sites;
- a declaration that the plaintiffs have the right to be free of exposure to the phenoxy herbicides 2,4-D and 2,4,5-T;
- the legal costs of this action; and
- such other relief that the court thinks is just.

In order to cover legal costs, which included the cost of bringing in expert witnesses from the United States and Scandinavia, the plaintiffs raised funds through donations and a fund raising campaign. Some people put their homes up as collateral. If you were the judge, would you allow the awards the plaintiffs are seeking or not? Why?

Case 2: *Fowler v. R., and Northwest Falling Contractors Ltd. v. R.*

The federal government has responsibilities for "sea coasts and inland fisheries." This provides scope for the federal government to legislate to protect water quality on the basis of protection of

FIGURE 19.1 A helicopter spraying a forest.

fish. Prior to the passage of the Canadian Environmental Protection Act, when these cases occurred, the Fisheries Act was the primary statute for the federal government to have strong influence for water quality management. At one time, subsection 33(3) of the Fisheries Act stated:

> No person engaging in logging, lumbering, land clearing or other operations, shall put or knowingly permit to be put, any slash, stumps or other debris into any water frequented by fish or that flows into such water, or on the ice over such water, or at a place from which it is likely to be carried into either such water.

A logging company, operated by Fowler, was located in coastal British Columbia. In the case, logs were removed from the forest by dragging them across a small stream causing the deposit of debris in the stream. The stream flowed into other streams that contained fish and salmon spawning beds. The definition of "fish" in the Fisheries Act protects fish through its entire life stages and fish habitats. Fowler is charged under subsection 33(2) and is acquitted at trial. Will the federal government be successful in its appeal to the Supreme Court?

FIGURE 19.2 Logging operations and debris in a stream.

In another case (*Northwest Falling Contractors Ltd. v. R.*), 15,000 litres of diesel fuel that was stored on the appellant's property spilled into the tidal waters of British Columbia. They were charged under subsection 33(2) of the Fisheries Act, which stated:

> … no person shall deposit or permit the deposit of a deleterious substance of any type in water frequented by fish or in any place under any conditions where such deleterious substance or any other deleterious substance that results from the deposit of such deleterious substance may enter such water.

A deleterious is defined in subsequent sections of the Act. Will the federal government be successful in court?

Case 3: *Interprovincial Co-operatives Ltd. et al. v. The Queen*

The Government of Manitoba alleged that chlor-alkali plants in Saskatchewan and Ontario (specifically Interprovincial Co-Operatives and Dryden Chemicals Ltd. respectively) were discharging mercury into rivers flowing into Manitoba. They feared that fish in Manitoba were ingesting the mercury and that the fish were not suitable for human consumption. In 1975, the Manitoba Government passed The Fishermen's Assistance and Polluters' Liability Act, which removed some of the barriers to suing for compensation. Specifically, it provided for payment of compensation to 1,590 individuals. It also allowed these people the ability to assign their rights against the alleged polluters to the government, which could sue. There were two aspects of the Act that were particularly noteworthy. The first related to making it illegal to discharge any contaminant into waters in Manitoba or into any waters that flow into Manitoba. Second, defendants from other provinces could not use a defence of compliance with pollution control requirements of that province if the contaminant in question caused damage to Manitoba's fishery. The causes of action were based on the requirements of this Act as well as the causes of negligence, nuisance, and trespass. The requested award consisted of:

- damages in the amount of $2 million (the amount paid in compensation);
- an injunction to prevent further pollution; and
- a mandatory order requiring the defendant to clean up the mercury from the river beds.

The defendants believed that in passing this Act, the Manitoba Government has overstepped its legal authority (termed *ultra vires*). What do you think and why?

FIGURE 19.3 Rivers flowing into/out of Saskatchewan and Manitoba.
Source: Courtesy of Atlas Canada.

INTRODUCTION

The institutional foundations for resource and environmental management in Canada are not found in the theories contained in library books, but in the Canadian Constitution. To achieve sustainable development, we must have an understanding of the natural environment, its elements and processes, and how human interventions with these elements and processes impact both people and the environment. That is a central aspect of this book. In addition, we must also agree on what we mean by sustainable development (sustainability or whatever similar term you wish to use), and how we will measure it in order that we can determine our level of progress and success. Finally, we also must be aware of the mechanisms that governments have at

CASE STUDY

THE CONSTITUTION OF CANADA

Most countries, such as the United States, have one document that is referred to as "The Constitution." In Canada, there are 25 primary documents that comprise the Constitution, including the Constitution Act (1867), the Charter of the Hudson's Bay, the Royal Proclamation (1763), and the Quebec Act (1774). The Canadian Charter of Rights and Freedoms, which applies to all levels of government and all provincial and federal laws, was entrenched in the Constitution when it was repatriated in 1982. The Charter guarantees a set of fundamental freedoms (e.g., freedom of conscious and religion, speech, the press, peaceful assembly, and association), democratic rights, mobility rights, equality rights and legal rights. One aspect of legal rights concerns the "right to life, liberty and

security of person, and the right not to be deprived of these rights except in accordance with the principles of 'fundamental justice'." This has implications for environmental management. At a minimum, people must be fully informed of projects that might impact their quality of life, and a fair and open process of

decision-making decisions must be in place. The Constitution Act (1867), the cornerstone of the Constitution, created the federation, the provinces, the territories, the national Parliament and the provincial legislatures, and the Supreme Court of Canada. It also established the powers of each level of government.

Want More?

Estrin, D., and Swaigen, J. 1993. *Environment on Trial: A Guide to Ontario Environmental Law and Policy* (3rd Edition). Toronto: Emond Montgomery Publications.

Boyd, D. R. 2003. *Unnatural Law: Rethinking Canadian Environmental Law and Policy.* Vancouver: UBC Press.

---

their disposal to influence resource use. Indeed, it could be argued that the Constitution Act is the most important document in the practice of resource management because it guides the actual management activities of public and private resource developers and managers, and identifies what each level can do (see Case Study, above).

Webster's New Collegiate Dictionary defines law as a "rule of conduct or action prescribed or formally recognized as binding or enforced by a controlling authority." In a constitutional democracy such as Canada, that controlling authority is the elected officials that we send to federal, provincial, and territorial legislatures and local municipal councils. There are three levels to Canada's legal system: (i) the legislative branch (federal, provincial, and territorial parliaments), (ii) the executive branch (cabinets and the government bureaucracy), and (iii) the judicial branch (courts and other tribunals) of government.

David Estrin and John Swaigen, two well-known lawyers and experts in environmental law in Canada, identify common law as one of two main bodies of law in Canada. They refer to common law as **"private law"** because it is intended to resolve disputes between individuals where the outcome is of specific relevance to them and not society as a whole. When the environment became a more important issue in the 1950s, governments began to develop more focused **statute laws** or "public laws." Rather than promoting individual rights, statute law defends the right of the general public for the benefit of the public. In establishing laws, governments take away or restrict individual rights and freedoms, which are a fundamental principle of Canadian society and promoted, as we will see, in common law. This chapter will present how governments exercise control over the environment. It will also

consider three other important aspects of environment law—citizen participation in the process, self-regulation by resource development interests, and economic instruments.

Canada is a constitutional monarchy, a federal state, and a parliamentary democracy. In general, federalism is a structure and process of government that offers the benefits of political and economic union combined with local autonomy. The federation is comprised of the 10 provinces and three territories, with Nunavut being created in 1999. Canada is, in fact, one of the most highly decentralized federations in the world. One important principle that was entrenched in the 1982 Constitution Act was equalization payments, which transfer funds from rich provinces (usually Ontario and Alberta, and sometimes British Columbia) to poorer provinces. This allows all Canadians to receive roughly the same level of government services, such as hospitalization, social services, and environmental management.

As a parliamentary democracy, the federal cabinet lead by the prime minister is responsible for providing legislation to be debated and considered by Parliament. The cabinet is also responsible for the implementation of public policy through its bureaucracy, Crown agencies, and regulations. In Canada, the wishes of Parliament are the ultimate view: if a majority of the House of Commons votes "no confidence" in the government, the Cabinet must either submit its resignation to the Governor General (the representative of the British monarchy), who will then ask another political party to form a new Cabinet, or ask for a dissolution of Parliament and the holding of a general election.

An essential basis of federalism is that the level of government most appropriate to deal with an issue should be allocated relevant responsibilities. As we will see, the general approach in the

Canadian Constitution is to divide authority for what were at one time viewed as mutually exclusive areas of responsibility (an exception being agriculture, which is shared between the two levels of government) between the federal and provincial governments. At the time of Confederation, the control of natural resources was perceived to be best dealt with at a local scale than a national one. Therefore, control essentially was placed under provincial jurisdiction. Since the Constitution reflects the views of more than 100 years ago (see Case Study, p. 406), it is this mindset that primarily guides present day resource and environmental management activities. That mindset was geared to resource development based, in part, on a belief that Canada's resources were limitless. Another important aspect of environmental law is to understand what the Constitution allows any one level of government to do and to determine how a Constitution, that is in many ways well over 100 years old, can address current needs and circumstances. An ongoing challenge in environmental law has been and remains to find the "right balance" between economic development and environmental protection—the essence of sustainable development. All levels of government in Canada have been tested to meet this challenge, in part, because the relative weight given to environment, economy, and society shifts within and between generations.

Only recently has environmental law emerged as a distinct area of specialization in jurisprudence (the study of law), and we have become aware and implemented a wider range of mechanisms to influence resource use. Until about the 1950s, legal issues related to the environment were usually addressed through common law, which enables people to seek decisions from the court for damage suffered to their property. At this time, governments at all levels in Canada had little legislation (laws, regulations, policies, procedures) focused directly on "the environment." This chapter will begin by describing common law, it strengths and weaknesses. Its weaknesses, combined with changing social attitudes and needs, and increased conflicts among resource users after the 1950s prompted governments to pass laws dealing with environmental matters, although this was largely focused initially on protecting public health. The 1950s saw the introduction of statutes (laws), largely designed to establish government control over resource development and protect public health. The chapter will review the use of government control in environmental management. It will then follow the evolution of environmental law, starting with the period of government control, through to a period of public participation, and ending in a period of self-regulation and the use of economic instruments. At this time, Canada uses all five of these mechanisms (or governing instruments) to guide the management of the environment and its natural resources. The chapter concludes by discussing international treaties and the use of alternative dispute resolution mechanisms.

TYPES OF LAW IN CANADA

The term "policy" is used in many different ways to indicate both formal and informal decisions or intentions at a personal, community, national, or international level. Informal policies include decisions you make about responding to telemarketers, providing donations to charitable organizations, or your choice of university or college. Governments and businesses may have informal policies concerning dress codes in the workplace. More formal government policies related to the environment would include the regulations, programs, and procedures established under statutes such as the Canadian Environmental Assessment Act. It commits the federal government to review the desirability and feasibility of specific projects defined in the Act using specified procedures, such as public hearings. The Convention on Global Climate Change (the "Kyoto Agreement") represents the official intentions of many nations to curb greenhouse gas emissions. Some governments in Canada have supported private-public partnerships to deliver water or transportation services, which have been traditionally the exclusive domain of public agencies or crown corporations. At its core, then, a **policy** is a plan or statement of intentions—either written or stated—about a course of action or inaction intended to accomplish some end. For this chapter, **environmental policy** will refer to those official rules and regulations concerning the environment that are adopted, implemented and enforced by some government agency, regulation or contractual agreement, as well as general public opinion about environmental issues.

Common Law

In all Canadian provinces, with the exception of Quebec (which applies the Napoleonic Code from France, see Case Study, p. 408), property rights arise from English common law. In fact, England passed its common law to all its colonies, which at one time included the United States. In reading about common law in the United States, be aware that it has evolved differently than in Canada. Common law has been made and refined by judges, not by elected politicians or governments, as a result of decisions that go as far back as the Middle Ages in England. In making these early individual decisions and sharing them with other judges, a set of legal principles, based on previous decisions, emerged. These principles continue to guide common law court decisions today, although they can be modified to reflect current social contexts.

There are three important aspects to common law that have influenced the development of statute law in Canada. The first suggests that property owners are free to use (develop) their land as they see fit (subject as we will see to any restrictions applied through statute law). This supports a general value of Canadian law—people can do whatever they want with their property unless they interfere with others or until a statute law restricts this activity. Indeed, under common law, people by the simple act of owning land can consciously and negatively impact the environment, such as depleting renewable natural resources such as trees or wildlife on that property. Thus, common law has limited capacity to protect "the environment"—the focus of the law is on people's use of the environment. The only restriction to people's use of their land under common law (and the limiting of their environmental impacts) is that it may not unduly interfere with a neighbour's use and enjoyment of their property or harm a neighbour's health. It

CASE STUDY

ENVIRONMENTAL LAW IN QUEBEC

Unlike the rest of Canada, Quebec operates under a dual legal system. The Quebec Act, which was passed when it was a colony of Britain in 1774, grants the rights of citizens "to hold and enjoy their Property and Possessions together with all Customs and Usage relative thereto, and all of their civil rights." In 1774, Quebec private law was governed by the Custom of Paris but was later codified in 1866 by the Civil Code of Civil Canada, which was replaced by the Civil Code of Quebec of 1991. Civil law protects rights in different terms than common law. For instance, under "abuse of rights," "every person capable of deciding right from wrong is responsible for the damage caused by his damage fault to another, whether by positive act, imprudence, neglect or want of skill." This very general statement can be used in many environmental cases, but the fault of the part of the defendant must be proved before a plaintiff can proceed with any action. These reflect two major departures from the common law approach.

Giroux, L. 1993. "Environmental Law in Quebec" in *Environmental Law and Policy*. Hughes, E. L., Lucas, A. R., and Tilleman, W. A., II (eds.). Toronto: Emond Montgomery Publications Limited. 151–164.

was in this legal context that the landowners in New Brunswick asked the court to prevent the spraying of pesticide on the forest located adjacent to their property (Opening vignette, Case 1).

The second important aspect of common law is that it is up to the plaintiff (i.e., the land owner or tenant who might be harmed) not the proponent of the development who must prove that interference with their use and enjoyment of land is unreasonable or their health is unduly at risk. This point reinforces the first aspect of common law—people have the right to use their land—because the burden of proof lies with "harm being done" rather than "this activity does no harm." The key phrase that requires interpreting from a legal standpoint concerns the undue interference of a person's use and enjoyment of property or harm people's health. The third aspect of common law concerns agreement on the process—an impartial judge will hear the case and render a decision that is binding on all parties.

Under common law, there are five "causes of action" (rights to sue), anyone of which must be used by an individual who believes that another property owner's activities detract from their enjoyment of land or health: trespass, nuisance, riparian rights, negligence, and strict liability. Based on their ability to convince a judge of the merits of their cause, people can stop or influence other property owners from using their land as they see fit.

It is a trespass to place anything upon (or even over) someone else's property, regardless if it is harmful or not, or if it is in small or large quantities, without the landowner's consent. For instance, a biocide applied by a helicopter may drift over to an adjacent property. If the landowner can prove that the flowers that died on their property were the direct result of the spraying, they could successfully sue for damages. However, David Estrin and John Swaigen cite a similar example that killed blueberries. In this instance, they suggested that the sprayers might not be liable for the loss of the blueberry crop under trespass because the spray may have killed the bees and insects that pollinated the blueberry plants. Thus, the dead blueberry plants are, in this instance, an indirect harm, not a direct one. Common law remedies must be based on direct impacts. However, the farmer could receive damages on the basis of nuisance. Trespass has been used to fight the discharge of sawdust from lumber mills, discharges such as fluorides from aluminum plants, the application of biocide sprays, the removal of soil, the dumping of fill, or the cutting and removal of trees.

From an environmental viewpoint, nuisance is the most useful cause. While trespass deals with intentional and unintentional incursions onto property, nuisance refers to indirect invasions that are less tangible, but still interferes with the use and enjoyment of land. This cause has been particularly helpful in fighting air and water pollution, noise, vibrations, odours, soil contamination, road salt damaging plants and soil, leaking oil tanks and flooding. Determining what is an undue or unreasonable interference involves balancing many factors such as "the severity of the harm, the character of the locale, the abnormal sensitivity of the person suffering the harm or of his or her use of the property, and the utility of the defendant's conduct" (Estrin and Swaigen, 1993, 111).

Riparian rights are of particular significance to landowners and tenants who have land adjacent to water courses and lakes. Riparians are people who own or occupy land adjacent to these surface water bodies. Under common law, they have the right to the natural flow of surface water—in quantity and quality, undiminished and unpolluted—that runs beside or through their property. These rights are referred to as riparian rights. This means that upstream water users must be sensitive to the rights of all other riparian landowners and tenants. Thus, upstream diversions of water, the damming of water courses, and the discharge of substances into water must not interfere with any other riparians' rights to the water. The discharge of sawdust, pulp and paper wastes, storm water runoff, and sewage pollution are instances where riparian rights have been used in the courts. Unfortunately, groundwater is not covered by riparian rights because it is not surface water. However, groundwater contamination could be considered a nuisance. Note that the prairie provinces apply a doctrine of prior appropriation, not riparian rights, in allocating water. Different rules apply there.

Negligence refers to behaviour that falls below the standard regarded as normal or reasonable practice (also referred to the standard of care) in a community. A judgement will be made concerning the adequacy of defendant's actions to avoid probable harm to people who might suffer harm if care were not taken. Damages may be awarded if the standard of care is not met and prove that the defendant should have foreseen the damages that resulted. Strict liability arises from the concept of harm by the escape of dangerous substances. People who take these types of substances onto their property should be held accountable for any damage, even if they have taken all reasonable steps to avoid damage (i.e., not negligent). This is referred to as "absolute liability."

There are advantages to common law. It does not require a government bureaucracy to enforce, just committed people who assert their rights, and a legal system (e.g., lawyers, judges, support staff, and court rooms) to support and hear all cases. Two types of remedies are possible. Injunctions, which either prevent or clean up future damages, are one type of remedy. The second remedy concerns the monies (referred to as damages) that may be awarded to a person who has suffered injury. However, this remedy may be seen as allowing the defendant to "pay to pollute." The wide range of possible undesirable environmentally activities that are covered under the undue loss of enjoyment of property is extensive. However, there are a number of shortcomings. First, there is no true protection for the environment; it is the person's rights to enjoy the land that are protected. These ends are not necessarily synonymous. Second, there is a likely high costs and levels of effort required to prepare a case that may be successful. These costs are borne by the

landowners, although there are opportunities to receive some of the legal costs. The burden of proof lies with the plaintiff to show that the activity will or has unduly and directly caused a loss of enjoyment to their property or risk to health. This makes it an "uphill battle" for the plaintiffs. However, it is consistent with the values of Canadian society which support individual rights to develop their land, subject to restrictions imposed by common law and as we shall soon see statute law. A decision to grant an injunction or allow an activity reflects the resolution of a conflict at one point in time. The development of future scientific information and technologies, changing social norms, and standards of care cannot be anticipated and will not be reflected in any court ruling. They also reflect rulings that have limited geographic scope. Thus, the pursuit of the greater public interest might not be best achieved through the application of common law remedies.

While much of the material covered in previous chapters provides a scientific perspective on the environment, answers to questions, such as those posed in the *Herbicide Trials* (Case 1 on next page), determines to what extent they are translated into actual practice.

STATUTE LAW IN CANADA

Statute laws are those that are passed by elected members of federal, provincial, or territorial governments or municipal council or an agency that has been delegated powers to establish laws. In the words of Estrin and Swaigen (1993, 5–6), statute laws "[take] away or [restrict] people's rights and freedoms [e.g., the right to develop or apply common law remedies] by setting out conditions under which government agencies can remove or carry out these activities and the procedures must be followed, and by imposing duties, responsibilities, and obligations on people." Governments in Canada can do this because legislatures are essentially supreme, subject to the constraints imposed by the Charter of Rights and Freedoms and the other principles found in the Constitution.

Elizabeth Brubaker, executive director of Pollution Probe, believes that for centuries governments have eroded the rights of victims of pollution because they have incrementally eroded remedies and the ability to seek remedies under riparian and other property rights. For instance, in 1885 a landowner filed a lawsuit claiming that sawdust, bark, and blocks of wood detracted from his opportunity to enjoy his riverfront property on the Ottawa River. Alarmed by the potential impacts on the lumber industry, the Ontario Government passed a law that ordered courts to consider the economic importance of the lumber industry before issuing injunctions and remedies against it. This action was defended, in part, because it served the public rather the private interest. What is important for now is to realize that governments have the ability to pass laws; and that laws governing resource activities (typically focused initially on sewage disposal and pollution discharges) began to develop in the 1950s.

In addition, as managing lands on behalf of the public (Crown land), federal and provincial governments can exercise

You be the Judge: A Return to Case 1 with a Focus on Common Law

Before you read the actual ruling in Case 1, reread the opening vignette and think like a lawyer or judge. What causes of action are evident? How would you prove the use and enjoyment of land or a person's health has been impaired? Focus on the awards requested by the plaintiff—are they appropriate to raise in the context of applying common law?

The causes of action raised by the defendant were:

- private nuisance;
- trespass to land;
- the rights of riparian landowners to water undiminished in quality;
- the right of landowners to groundwater free of chemical contamination;
- a ruling based on a previous common law case that occurred in *Rylands v. Fletcher* (You would not know about this one unless you are already a legal expert!); and
- breach of the federal Fisheries Act particularly those sections dealing with the protection of fish (some of which are cited in Case 2). (We'll be finding more about this statutory aspect shortly.)

The highlights from the court written in lay terms:

- Regarding the right to be free of exposure to phenoxy herbicides, the court determined that this was beyond their power, which is limited to rights between landowners and tenants. Granting this award would affect the agricultural industry, which at the time, was the largest user of phenoxy herbicides. This is a matter of public policy to be dealt with by government and its agencies.

- A breach of the Fisheries Act may be evidence of negligence. Since negligence was not pleaded nor proven, this action failed.

- Trespass does not require proof of damage, just proof of incursion onto land. Possibility of a trespass is not equivalent to proving it is probable to occur. Nuisance requires proof that there is the significant loss of enjoyment land or health by the plaintiffs. While there was, at the time, considerable scientific uncertainty about the risk to health from dioxin, the defendant indicated the total amount and concentration of biocides it intended to use at the sites. The court noted that at the time, phenoxy herbicides had been used since the 1940s in forestry, roadside and railway rights of way, and agricultural settings, in "vast quantities and only until recently, without too much precaution." Canada has agencies, such as Health and Welfare Canada, to regulate and control the use of new chemicals, and reviews are made on a regular basis. What the plaintiffs are asking is to reverse these agencies' decisions that permit the use of these chemicals. For these reasons, as well as the commitment by the defendant to follow specified procedures to permit proper application only on the target sites, and the

inability of the plaintiff to prove a scientifically based risk to health indicates, here was no basis to support their allegations.

- The comments above make any decisions on riparian rights or groundwater rights unnecessary.

- In the *Rylands v. Fletcher* case, that decision was based on the ability of the plaintiff to show the risk of the substance and its likely escape to the plaintiff's land. This was not proven in this instance.

- The defendant is awarded costs to be paid by the plaintiffs.

This decision shows the common law in action and raises many questions including:

- Could the plaintiffs bring a legal suit against all forest operators in the province?

- Why does the court require a significant health risk as opposed to some health risk in determining private nuisance?

- How should courts treat voluntary risks to health (deciding to smoke, drive a car) from involuntary risks?

Want More?

See Wildsmith, J. 1986. "Of Herbicides and Humankind: Palmer's Common Law Lessons." *Osgoode Hall Law Journal*. 24: 161–172, and watch the movie, *Herbicide Trials* produced by the National Film Board.

considerable influence on the development of these properties. These land holding are very significant. As "owners of land," governments have **proprietary rights** that allow them to set conditions for those interests wishing to develop resources (e.g., forests, minerals, recreational opportunities), or decide to leave public lands in their natural state and forego development opportunities.

The federal government has significant land holdings in Canada's north although this is being devolved to territorial governments and to administrative structures established under land claims agreements with aboriginal peoples.

On the basis of the concepts of federalism noted earlier and the tradeoffs made by the Fathers of Confederation in 1867 to reach

You be the Judge: A Return to Case 2 with a Focus on Federal Powers over Waters Frequented by Fish and Deleterious Substances

Reread Case 2 in the vignette and consider what the federal government is trying to control? One of the most important pieces of environmental protection legislation in Canada for aquatic environments is The Fisheries Act. To conserve fish for the future, it is critical to protect the natural systems that produce the fish as well as manage a sustainable fisheries harvest. These systems are referred to as "fish habitat," and include a great variety of environments where fish live. The Fisheries Act defines "fish habitat" as: "Spawning grounds and nursery, rearing, food supply and migration areas on which fish depend directly or indirectly in order to carry out their life processes." (Subsection 34(1))

Fish habitat includes not only the water in rivers, lakes, streams and oceans, and the quality of that water; fish habitat also includes surrounding plants, life forms, and structures such as gravel beds and large woody debris that interact to make fish life possible. The federal Fisheries Act protects natural and *human-made* fish habitat, such as a drainage ditch that has become frequented by fish. Do you think they will be successful in the two cases presented? Why?

With regard, to *Fowler v. R.* the appellant was acquitted at trial. The federal government appealed and had the decision reversed at the county court. The appellant made an appeal to the Supreme Court of Canada. In a unanimous decision, it found subsection 33(3) to be *ultra vires* because of the very broad range of actions that one might be liable and the fact that the focus of the subsection was on logging activities, an area of provincial jurisdiction. In other words, the subsection did not focus enough attention on the protection of fish. There was also no evidence that supported the harm all these activities inflicted on fish.

In the *Northwest Falling Contractors* cases, the appellant filed an application of prohibition that was dismissed by the Supreme Court of British Columbia. The Court of Appeal upheld this decision on appeal. A further appeal was made to the Supreme Court of Canada. The appellant argued that the legislation was an attempt to govern over water quality generally, an area of provincial jurisdiction. There is a very broad definition of "water frequented by fish" and broad scope for defining "deleterious substance." The Court supported the federal government because subsection 33(2) is concerned with the deposit of deleterious substances in water frequented by fish, or in a place where the deleterious substance could enter such water. Deleterious substances are defined on the ability to harm fish. This approach by the federal government is clearly supported under the Constitution. The Supreme Court ruled that subsection 33(2) was *intra vires* (within the powers) of the Parliament of Canada to enact.

Want More?

See Saunders, J. O. 1988. *Inter-jurisdictional Issues in Canadian Water Management.* Calgary: Canadian Institute of Resources Law.

---

an initial agreement among the provinces to form Canada, the following division of legislative responsibilities between federal and provincial governments were agreed upon in the Constitution. These are found in sections 91, 92, and 92A of the Constitution and are referred to as the **heads of power.** The Preamble to section 91 of the Constitution, which lists federal responsibilities states "... for the Peace, Order, and Good Government of Canada, in relation to all Matters not coming within the Classes of Subjects by this Act assigned exclusively to the Legislatures of the Provinces"; the federal government shall have the following powers that are directly or indirectly related to environmental management:

- the regulation of international and interprovincial trade and commerce;
- the raising of money by any mode or system of taxation;
- to regulate navigation and fishing;
- to regulate seacoast and inland fisheries (see What Do You Think?, above);
- responsibility for Indians and lands reserved for Indians;
- to make criminal law, which has been interpreted and a power to protect public health; and
- to regulate works or undertakings that are interprovincal or international in nature (shipping, railways, pipelines, telecommunications) and works that, although they may be situated in one province, are declared by Parliament to be for the general advantage of Canada (e.g., grain elevators in the prairies, nuclear power plants throughout Canada).

The Preamble's reference to "peace, order, and good government" (denoted as POGG in some texts) has generally been interpreted as giving the federal government power to make laws to cover problems that are within provincial jurisdiction, but have become so severe, that they became "national" in either extent or significance. The control of toxic chemicals is an example of where the federal government might use POGG to establish standards and appropriate procedures to manage these substances. They could also establish these regulations in order to protect fish but these would apply only to water, and not to air and land-based discharges.

Under sections 92 and 92A of the Constitution (the latter being a resource amendment introduced in 1982), provinces may establish laws related to:

- taxation;
- management and sale of the public lands belonging to the province;
- municipal institutions;
- local works and undertakings;
- property and civil rights (see What Do You Think?, p. 413);
- generally, all matters of a merely local or private nature in the province;
- exploration for non-renewable natural resources in the province;
- development, conservation, and management of non-renewable natural resources and forestry resources in the province; and
- development, conservation, and management of sites and facilities in the province for the generation and production of electrical energy.

Agriculture is a shared responsibility between federal and provincial governments. These heads of power have prompted governments to pass statutes to or laws. A government that passes a law for which it has no jurisdiction could be judged, ultimately by the Supreme Court of Canada as being *ultra vires* extending its powers beyond its legal abilities. In response to the question: "Who manages the environment?" a correct answer would be "Everyone." A more precise answer is: "It depends." Water flowing into the United States, such as the St. John River in New Brunswick, the Great Lakes in Ontario, and the Columbia River in B.C., is considered to be in federal jurisdiction under its responsibilities for international relations. If a watercourse is used for navigation, the federal government can and has passed statutes related to this issue. A powerful tool that influences management is the federal government's spending power. By spending money itself or entering into cost-sharing programs with the provinces, the federal government can direct resource management activities. Provinces tend to resist federal intrusion into their jurisdiction and you may have read about this in newspapers related to health care and other issues. The best answer to the question: "Who manages resources" is to respond with a series of questions before answering. These would include: "Is this an international or interprovincial waterway? (see What Do You Think?, p. 413). What level of government, if any, owns the land? Is this an issue that might be considered in the national interest, or of a local and private matter? Are there activities that might be covered under the Criminal Code?"

The term "fragmented" has often been used to characterize resource management in Canada. Fragmentation begins with the division of responsibilities in the Constitution between federal and provincial governments, referred to as **vertical fragmentation** (fragmentation between levels government). It continues with the manner in which senior governments (federal and provincial) pass issue-specific statutes—energy, navigation, fisheries, and parks—**issue or interest fragmentation.** Very few statutes consider "the environment" as a whole—a notable exception being environmental impact legislation established at federal and provincial levels. Governments have established bureaucracies (agencies) to administer these acts and related programs. At the federal level, Environment Canada, the Department of Fisheries and Oceans, Natural Resources Canada, External Affairs, and Health Canada are but some of the departments that have responsibilities for resource and environmental management. The National Energy Board and Atomic Energy Canada are two of the agencies established by government that have significant responsibilities for environmental management. **Horizontal fragmentation** occurs within a single level of government. Designing effective mechanisms, such as interdepartmental committees or statutes that allow the consideration of the environment are important to ensure coordinated and effective management regimes. Effective management regimes are able to accept the realties and overcome the impediments to achieving sustainable development that are imposed by all types of fragmentation associated with resource management in Canada.

Note also that the constitutional heads of power do not directly refer to "the environment" or its specific constituents, water and air. They typically refer to resource uses and reflect the utilitarian views at the time of Confederation—the natural environment served to support economic development. Thus, the Canadian Constitution is geared towards controlling our use of resources, not the environment itself and for the sake of the environment. This orientation also promotes the potential for conflicts between competing resource users and supports the previously mentioned fragmented approach. For instance, in the court case involving forestry operations and fish in B.C., the federal government has the legislative right to create laws to protect fish. However, a province has just as legitimate a right to regulate forestry activities, some of which may negatively impact on fisheries. The province also has the right to build dams, but this would no doubt interfere with any fish in the water course and navigation. Which level of government has ultimate control over the use of a resource? The answer to this question determines whose beliefs and needs influence resource development and management in Canada. Also note that municipal governments and aboriginal communities are not mentioned in the Constitution (see Case Studies, pp. 414 and 415).

One way to examine the legal framework is to assess how the law has changed the process of management. To some extent, the following sections characterize the evolution of statute law related to the environment in Canada as one starting with government control, then shifting to a mix of government control and citizen empowerment; and finally to a period where self control and economic instruments are being used.

You be the Judge: A Return to Case 3 with a Focus on Provincial Jurisdiction based on Property and Civil Rights

Before you read the judgement, reread the Opening Vignette and consider the following questions: (1) Do you think that the Province of Manitoba is *ultra vires* because the provisions of The Fishermen's Assistance and Polluters' Liability Act go beyond to ability to legislate on issues related to property and civil rights within the province and matters of a local and private nature within the province? (2) Are the provisions of the Act not applicable because they regulate the activities of the two companies—Co-Operatives Ltd. and Dryden Chemicals Ltd.—that occur outside the province of Manitoba. If you support this argument, the Act could be applied within Manitoba.

In a 4–3 majority ruling, the Supreme Court found the statute *ultra vires*. There were three written decisions, two in the majority (by Justices Pigeon and Ritchie) that differed in the rationale for their judgement, and one in dissent (Chief Justice Laskin).

Justice Pigeon maintained that the statute was not within Manitoba's jurisdiction. He believed the scope of the case was interprovincial, not local. While the Act's focus is at addressing damage inflicted within Manitoba, the activities that caused the damage (discharge of contaminants) was conducted beyond its borders. The right to discharge the contaminant into watercourses within Ontario and Saskatchewan was legally granted by respective authorities in those provinces. Manitoba does not have the ability to remove that right. At the same time, Justice Pigeon criticized the permitting systems in Ontario and Saskatchewan when he wrote:

> It does not appear to me that a province can validly licence on its territory operations having an injurious effect outside its borders so as to afford a defence against whatever remedies are available at common law in favour of persons suffering thereby in another province.

This means that downstream fishers and other property owners, whose enjoyment of property and health was interfered with, regardless of their province of residence, should be able to seek remedies under common law. The province and the companies should not be able to hide behind the approval permits as a legitimate defence. Judge Pigeon also believed that this matter should be dealt with by the federal government.

Judge Ritchie also found that the defendants were not liable for damages under the Act. However, his argument differed from Judge Pigeon's. Ritchie believed that the Act was within the province's jurisdiction. However, since the "tests" to determine the applicability of the stature were not met, the defendants were not liable. He believed that the Act was valid based on its abilities to protect property and civil rights in Manitoba. Based on this logic, the key question now concerns if the activities undertaken in Saskatchewan and Ontario were "actionable torts" (able to go to court). Since the Ontario and Saskatchewan permits approved the companies' activities, there were no civil wrongs. The Act could not nullify civil rights granted in other provinces.

Chief Justice Laskin wrote the dissenting view. He focused attention on the purpose of the Act, which was to the damage and loss of production that occurred within the province of Manitoba. In his view, the Act was appropriately grounded in property and civil rights and did not infringe on federal jurisdiction. In considering the tests to determine the applicability of the law, Laskin believed that it was applicable because the focus was on the damage caused within the province of Manitoba, not the activities that occurred in other provinces. Thus, the tort occurs in Manitoba. The permits granted to the companies were irrelevant because they do not give them the right to injure property in other provinces. In his view, the Act did not deny rights acquired in other provinces.

This case produces considerable confusion regarding environmental law. The close 4–3 decision provides a weak basis for precedents in future cases. There is no question that the federal government has clear jurisdiction in the case and could have introduced appropriate legislation. It failed to show leadership on this issue.

Want More?

See Kennett, S. A. 1991. *Managing Interjurisdictional Waters in Canada: A Constitutional Analysis.* Calgary: Canadian Institute of Resources Law.

Government Control

There are two ways governments can control activities: **prior approval** and **penal sanctions.** Prior approval requires proponents to seek the permission of a government agency of body before an activity takes place. This is often referred to as "command and control" regulation, and is often achieved through permitting processes that proponents must obtain to undertake their activities (such as the one requested to apply biocides in Case 1). In granting a permit (also referred to as a licence or certificate of approval), a government body can establish conditions.

Provisions of Ontario's Environmental Protection Act illustrate a permitting process. It requires anyone who plans to construct or operate a potential source of pollution to obtain a

CASE STUDY

THE CONSTITUTION AND CURRENT MUNICIPAL ISSUES

Canada's municipal system of government, including that in Quebec, is almost entirely based in British law (there has been minor influence from the United States). Initially developed by individual proclamations from Parliament, by the nineteenth century, the British Parliament established a municipal system throughout that country. This allowed local governments to establish a taxing system, build streets and public markets, and enforce local regulations concerning trade and commerce. The oldest municipality in Canada is Saint John, New Brunswick, established in 1785.

The Constitution establishes two levels of government in Canada—federal and provincial. Municipal governments are mentioned in subsection 92(8) because they were already established in the three British colonies—Canada (Ontario and Quebec), New Brunswick, and Nova Scotia—that joined the confederation in 1867. This means that municipal governments are very much controlled by the provinces because it is that level of government that created them, and they could, in theory, dissolve them. Municipal governments must follow any policies that are required by the province. Sewage treatment and water supply, waste disposal, transportation, and land use planning are some of the environmental responsibilities that are undertaken by local governments.

During the federal election of 2004, the Federation of Canadian Municipalities (FCM) requested all federal parties to commit to a New Deal for Canada's communities. This New Deal concerned the need for: (1) a new intergovernmental partnership, (2) revenue sharing, (3) targeted investments, (4) capacity building for sustainable community planning, and (5) research. The FCM maintained that the principles of sustainable development were essential if the New Deal was to effectively "balance economic opportunity, social well-being and environmental conservation; that use resources efficiently; and that encourage participation in decision making and long-term planning." The comments below are focused on the need for revenue sharing and targeting investments. Chapter 21 will focus attention on the other aspects of the urban agenda.

The FCM estimated that nationwide the municipal infrastructure deficit (e.g. roads, water treatment and distribution systems, wastewater systems) was $60 billion and growing by about $2 billion annually. It also believed that the root cause of the debt reflected the outdated institutional arrangements promoted by the Constitution and inadequate fiscal resources to match the growing municipal responsibility for services and programs. Out of every tax dollar collected in Canada in 2003, the FCM estimated that $0.08 went to municipal governments, $0.42 to provincial governments, and $0.50 to the federal government. This fiscal imbalance is exacerbated by the 44 percent cut (since 1994) in the transfers payments made by senior governments to municipal governments. In addition, provincial/territorial revenues increased by 21 percent, federal revenues by 16 percent, and municipal revenues by 4 percent between 1999 and 2003. According to the FCM, relative to other levels of government in Canada, municipalities have fewer means to raise funds and primarily rely on property taxes. At this time, no municipal government in Canada levies a tax on income, sales, or gas. They are also restricted by provincial regulations to borrow money for capital expenditures and they cannot borrow to cover operating costs.

Property taxes are inadequate to support the growing responsibilities and the $60 billion infrastructure deficit. Property taxes are not responsive to periods of rapid economic growth, such as occurred in the mid and late 1990s. During this time, provincial and federal coffers grew through people paying increased taxes on their higher incomes and product/service consumption. Property taxes are not responsive to these periods of economic growth. Since all property owners, regardless of their ability to afford property taxes must pay them, they are viewed as regressive. A 2002 report on Canada by the Organization for Economic Cooperation and Development (OECD) concluded that Canadian municipal governments' high reliance on property tax lies at the root of their growing fiscal difficulties.

Among OECD federations, Canadian municipal governments are the second most dependent on property taxes (after Australia). The report also notes that Canadian cities have "relatively weak powers and resources" and should be given "some limited access to other types of taxes" to meet their increasing responsibilities.

The FCM wants to solve this problem by dividing the current monies provided by the public to governments more equitably. In partnership with provincial/territorial governments, the FCM believes that this can best be achieved through a share of sales and income taxes, in addition to a share of fuel taxes and powers to levy user fees and hotel taxes. Five priority areas were identified by the FCM: (i) infrastructure not covered by revenue sharing (including broadband), (ii) affordable housing, (iii) community and children's infrastructure, (iv) security, emergency preparedness and public health, and (v) downtown revitalization.

Sources: Sancton, A. (2000) "The Municipal Role in the Governance of Canadian Cities" and Federation of Canadian Municipalities at http://www.fcm.ca/newfcm/Java/frame.htm

"certificate of approval" for the equipment and the operational procedures that may influence the pollution potential. A homeowner may be required to install a specific type of septic tank facility in order to minimize the potential for leakage in areas close to bedrock. An industrial chemical plant might require tens or hundreds of certificates in order to ensure that the transport chemicals in and out of the plant, storage of chemicals on the property, manufacturing processes required within the plant, emissions from smokestacks, sewers or other treatment facilities minimize the potential of pollution. In this way, the Ontario Ministry of the Environment "commands and controls" the procedures and equipment used in thousands of facilities across the province. The Canadian

CASE STUDY

THE CONSTITUTION AND ABORIGINAL ISSUES

Aboriginal or First Nations people inhabited Canada long before European's "discovered" North America. For them, land is more than property to be exploited; it is home. Overcoming this cultural difference becomes more complex because many First Nations people were not conquered and there is no treaty to outline territory. In terms of the Constitution, there are several issues. One concerns the desire by First Nations to be involved at all levels of decision making in order to ensure that their concerns are addressed, and their rights respected. In this regard, First Nations expect to participate in these forums on an equal basis as other levels of government. They perceive themselves to be more than another "interest group". Instead, they are another level of government. In addition, First Nations have certain rights enshrined in the Constitution Act, 1982 that are beyond those accorded other Canadians (e.g., support of reserve lands by federal government, payment of taxes). In many instances, First Nations governments are excluded from resource and environmental management processes. When First Nations are included in decision making, they are perceived as a homogeneous group, rather than different communities. There have also been conflicts between First Nations, who have a right to access natural resources, such as trees and fish, and other primary resource users, with users who have been given access by federal, provincial, or territorial governments (e.g., fishing off the east and west coasts).

Another concern relates to the quality of life provided on First Nations' reserves. The Walkerton Inquiry identified several problems including: (i) the water and wastewater infrastructure was very inadequate; (ii) there was a shortage of trained operators; (iii) testing and inspection were inadequate; (iv) microbial contamination was frequent; and (v) distribution systems, especially on reserves, are sized to deliver about half the water available per capita to other Ontarians. Twenty-two reserves across Ontario were found to have problems with their drinking water supply. These findings reflect the state of living on reserves across the country. The federal Department of Indian and Northern Affairs (2004) found that 75 percent of the 740 water treatment systems and 70 percent of the 462 waste water treatment systems on reserve posed a medium to high risk to drinking water and wastewater quality. Many reserves have been under boil water orders for years and many members of these communities rely on bottled water for drinking and have to boil water for everything from bathing to washing the dishes.

Want More?

See the Assembly of First Nations at: http://www.afn.ca/Assembly_of_First_Nations.htm and Indian and Northern Affairs at: http://www.ainc-inac.gc.ca

---

Environmental Protection Act is a second example or government control through an approval process. This **federal statute** requires manufacturers and importers of new chemicals that may be toxic to inform the government and allows the federal government to demand information on the testing of new and existing chemicals. The federal government regulates the manufacture, sale, release, and disposal of chemical deemed toxic. Since provincial governments may also regulate these same chemicals, the Canadian Environmental Protection Act requires the federal government to consult with the provinces to ensure that inappropriate duplications of effort are avoided.

Effectively implementing permitting systems, monitoring the activities of successful applicants, and prosecuting those who fail to comply requires a significantly large, highly trained, and well-funded bureaucracy to administer. Plans submitted by applicants must be reviewed carefully, fairly, and in a timely manner. Inspections must take place on a random, yet regular basis. Investigations must take place when problems are suspected. Failure to comply with the conditions established in a permit can lead to warnings, directions to correct problems provided, prosecution and potential conviction. Past problems with the implementation of command and control approaches have pertained to the high cost, the inconsistent judgements made by officials across a jurisdiction (Canada, province, municipality), and the long time periods required for approval.

Environmental assessment is another way in which governments approve projects prior to their initiation. The United States National Environmental Policy Act (NEPA) passed in 1969 but signed by President Nixon in 1970 was the first of its kind because it required the federal department in the United States to consider the environmental aspects of these activities. NEPA forms the cornerstone of both U.S. environmental policy and law. It does three important things: (1) it authorizes the Council on Environmental Quality (CEQ), the oversight board for general environmental conditions; (2) it directs federal agencies to take environmental consequences into account in decision making; and (3) it requires an **Environmental Impact Statement (EIS)** be prepared for every major federal project having a significant impact on the quality of the human environment. When NEPA was being debated, there were suggestions that it should be considered a constitutional amendment guaranteeing the right of a clean environment to everyone. A similar debate has occurred in Canada in the context of the Charter of Rights and Freedoms. Difficulties in defining what "clean" means along with worries about how we could achieve this ambitious goal, limited NEPA to a statute with more limited, but still important, powers than a constitutional amendment. What do you think? Should everyone have an inherent right to a clean environment? How would you define "clean"?

In theory, the environmental impact assessment (EIA) process requires the gathering and analyzing of data about a proposed program or project and assessing the merits in the context of its economic, social, and environmental impacts on the natural and human environment. It can be a powerful tool in the environmental arsenal. In addition to requiring sound scientific and socioeconomic data, the EIA process requires more open planning by both public agencies and businesses. An EIA can bring to light adverse aspects of a project that otherwise may have remained hidden. It can provide valuable information about a proposal to opponents who cannot afford to do their own research. It promotes the consideration of alternatives to the project being proposed.

There are several threshold considerations that determine whether a project requires an EIS (Fig. 19.4). Typically, government-sponsored or government projects are subject to an EIA process, and the legislation is specific to a single jurisdiction. For instance, the CEAA is binding only on federal departments, agencies and federally sponsored projects. It does not apply to any provincial agencies. The private sector may be involved if specific types of projects (e.g., toxic chemical plant or other activities that have a significant impact on the environment), which are identified in the legislation require that one be completed. To do a complete EIA for a project is usually time consuming and costly. The final document is often hundreds of pages long making it somewhat incomprehensible to much of the public, and it can take many months to complete. General contents of an EIA includes: (1) purpose and need for the project, (2) alternatives to the proposed action (including no action), and (3) a statement of positive and negative environmental impacts of the proposed activities. In addition, an EIA should make clear the relationship between short-term resources and long-term productivity, identify cumulative effects, as well as any irreversible commitment of resources resulting from project implementation.

At the federal level, the Canadian Environmental Assessment Act is the key statute in triggering the need for an environmental impact assessment. Introduced to Parliament in 1992 and passed in 1995, it has four main objectives:

- to ensure that environmental effects of projects receive careful consideration before responsible authorities take action;
- to encourage responsible authorities to take actions that promote sustainable development, thereby achieving or maintaining a healthy environment and a healthy economy;
- to ensure that projects to be carried out in Canada or on federal lands do not cause significant adverse environmental effects outside the jurisdiction in which projects are carried out; and
- to ensure that there is opportunity for public participation in the environmental assessment process.

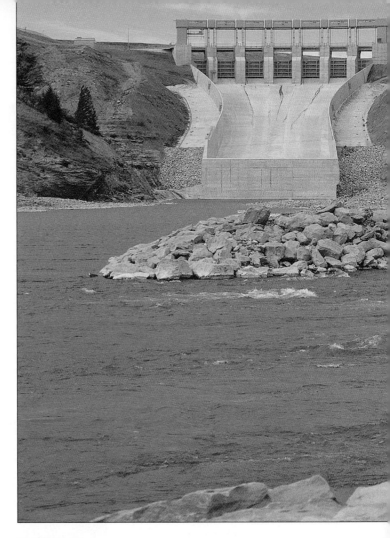

FIGURE 19.4 Every major federal and provincial government has passed legal provisions requiring environmental assessments on projects that have a significant impact on the environment. This is the Oldman Dam in Picher Creek, Alberta.

The CEAA requires an environmental assessment where a federal authority supports a private or public sector project in one or more of four ways:

- by being the proponent of the project,
- by providing money for the project,
- by providing land for the project, or
- by issuing some form of regulatory approval for the project.

Unlike most environmental assessment acts, there is no threshold clearly defined. Instead, there is a list of types of projects, called the "comprehensive study list." If a project falls into one of these project types, and if the CEAA otherwise requires the project to undergo a federal environmental assessment, then the assessment must be more rigorous and systematic than if the project were not on the comprehensive study list. The decision to do an environmental impact assessment is made by the federal authority promoting the project. This situation can lead to a potential conflict of interest for the federal proponent.

In 1999, the CEAA was reviewed as stipulated in section 72. Amendments to address weaknesses were proposed in 2002. The review found that the current environmental impact assessment process contained several strong points in its objectives and principles, the basic structure of the process, the aspects to be addressed, and the role of the Canadian Environmental Assessment Agency, which administers the CEAA. However, it suggested improvement in the following three areas:

- making the process more predictable, consistent and timely;

- ensuring all environmental impact assessment documents were of high quality by: (i) increasing compliance with the CEAA; (ii) strengthening the role of monitoring, and (iii) placing greater emphasis on studying cumulative effects; and

- enhancing opportunities for public participation.

In Canada, all provinces and the federal government have passed environmental assessment processes. According to environmental lawyer and former executive director of the Sierra Legal Defence Fund David Boyd, "the environmental assessment process prescribed by Canadian laws has been described as totalitarian, a boondoggle, a hoax, a paper tiger, a Trojan horse, and a nasty game. A seemingly straightforward planning process has evolved into a lightening rod for criticism, litigation, and debate about the nature of progress in the twenty-first century." At the same time, the Canadian Environmental Assessment Agency claims that the Canadian Environmental Assessment Act (CEAA) has "helped achieve sustainable development through the promotion of sound economic development while reducing adverse effects on our environment." Clearly, there are strengths to be built upon and weaknesses to address.

Penal sanctions refer to a failure to meet standards specified in a permit or as defined in government regulations. A violation of environmental legislation or regulations is usually treated as a criminal offence. This would be in addition to any administrative penalties (e.g., suspension, termination of permit) for non-compliance with the conditions of a permit, which are applied infrequently. Criminal proceedings for environmental violations are initiated usually by federal or provincial authorities. Private citizens also have a common law right to initiate criminal proceedings, but these may be taken over by provincial authorities if it is deemed in the public interest. Applying penal sanctions has been problematic in Canada. In the 1990s, cutbacks in federal and provincial budgets exacerbated this problem. When prosecutions have occurred they have focused on obvious violations. These shortcomings are captured in Figure 19.5, which demonstrates a relatively low level of prosecutions and convictions for offences as reflected in the enforcement activities conducted under the Canadian Environmental Protection Act for the period 1991/92 to 2001/02. This would typify most efforts in this area for all governments in Canada.

A more serious form of penal sanction is to use criminal law to protect the environment. This provision has been incorporated into a few statutes, such as CEPA, and the Pest Control Products Act. In addition to the usual set of penalties established in environmental statutes—fines and/or terms of imprisonment, as well as powers of a court to make various orders against an offender—

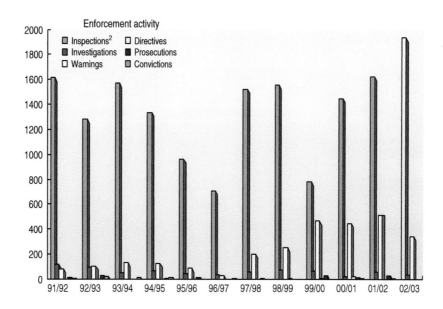

FIGURE 19.5 Canadian Environmental Protection Act's Enforcement Activities 1991/92 to 2002/03.

Notes:
1. This date is based upon a fiscal year.
2. This number represents the number of on-site inspections (field/site inspections). It does not include off-site inspections (administrative verifications).
Source: Environment Canada. Canadian Environmental Protection Act annual reports.

there are provisions for imprisonment. For example, CEPA 1999 provides for maximum fines of up to $1 million or imprisonment for up to five years, or both. Any criminal proceedings would be dealt with by the criminal courts. In addition, CEPA allows people who have suffered a loss as a result of a violation of the Act to collect financial damages from the offender.

The two government control approaches, prior approval and penal sanctions, require complete and costly administrative structures to ensure that successful applicants comply with permit requirements. Rather than taking this route, government officials have often negotiated individual agreements with specific plant operators, rather than using prosecution and sanctions. This detracts from the strength of the system and erodes public confidence. This lack of public confidence in public agencies prompted, in part, the need for citizen empowerment in the process.

Citizen Empowerment

In the 1970s, public involvement or participation in decision making became more widespread. In Canada, there are three ways to get the public informed and involved: participation in the decision-making processes, sharing information, and **whistle-blower legislation.**

Citizen Participation

Figure 19.6 shows the process of decision making used in applying the federal government's Canadian Environmental Assessment Act. Passed in 1992, replacing the Federal Environmental Review Process, the CEAA requires that a public registry be established in order that there is access to a wide range of EIA documents on every project covered by the Act. The extent of participation depends on the type of assessment conducted. At the screening stage, when the federal proponent is determining if a project should be subject to the CEAA, public involvement is discretionary. Public involvement is required for comprehensive studies, and the public receives notice about a project and has the opportunity to make written comments. If a review panel is held, the public has an opportunity to participate in the public hearing process by presenting evidence and questioning the proponent's

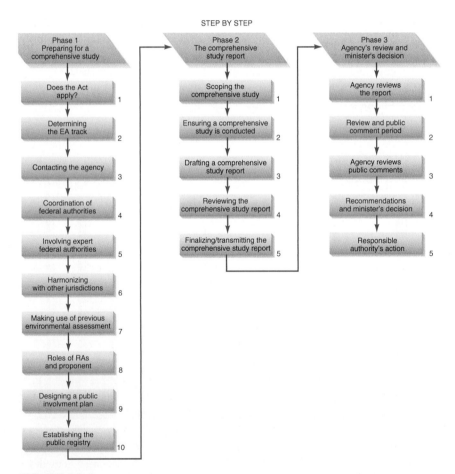

FIGURE 19.6 The CEAA Process for Comprehensive Environmental Assessments.

Note the opportunities for public involvement in this decision-making process. Consult the Canadian Environmental Assessment Agency's website for more information and consider what would you do to improve the level of public participation, or do you think it is adequate as is?
Source: http://www.ceaa.gc.ca/013/0001/0003/comps_e.htm.

Using NPRI data, Pollution Probe provides people the opportunity to find out the levels of pollution emitted in their own neighbourhoods. Have your postal code ready and go to this site to find out what is happening on the chemical release front in your own back yard!

http://www.pollutionwatch.org/home.jsp

Want More?

Go to http://www.cec.org/home/index.cfm?varlan=english for information on "Taking Stock."

---

experts. Participation is supported in CEAA through the establishment of intervener funding, which totalled about $840,000 (about 0.5 percent of the total amount of funds spent by the federal government on environmental impact assessment between 1992 and 1997). These provisions and this level of funding severely limit effective public participation. Public participation should be supported during the screening phase because it is at this point that the key decision concerning the applicability of the CEAA are made. In 1998, Canada's Commissioner on Sustainable Development noted that important projects were being excluded from the Act. This shortcoming could be corrected through earlier opportunities for public participation.

Other federal, provincial, and territorial environmental statutes enshrine public participation in legislation. For instance, the Canadian Environmental Protection Act requires the minister to publish the list of proposed regulation in the Canada Gazette. Within 60 days of the publication of a regulation, any person may file a notice of objection requesting a board of review. The Yukon Environment Act also applies this approach, which should add to the effectiveness of the prior approval approach to environmental management. The CEPA also encourages the federal minister of the Environment to consult with any interested person and to work jointly with the representatives of different sectors interested in the protection of the environment. Some provinces have also adopted this prior publication approach.

Although there is no "right to a clean environment" in the Canadian Constitution, the Commission on Environmental Cooperation notes that some legislation contains provisions that support the concept of an "environmental bill of rights." These include Ontario's Environmental Bill of Rights Act, Yukon Environment Act, and the Northwest Territories' Environmental Rights Act. More limited support of the concept can be found in Quebec's Environmental Quality Act, Alberta's Environmental Protection and Enhancement Act, and Nova Scotia's Environmental Act. The Commission on Environmental Cooperation states that although these and other statutes have different provisions, they contain many of the same central concepts and introduce many of the same types of actions, such as:

- a citizen's right to reasonable environmental quality;

- a declaration of the public trust doctrine, i.e., that natural resources are held in trust by the government, such that a citizen can bring an action for damages against the government in cases of environmental degradation;

- a right to protect the environment, that is, a right for a citizen to bring an action or application against another individual in order to protect the environment, though remedies and defences may vary; and,

- a legal remedy for violation of an environmental Act or standard, that is, a specific provision that a person may bring an action to protect the environment to the extent provided for by the law.

Given the array of other pieces of legislation, why do you believe that it is necessary to have an "environmental bill of rights" enshrined in specific laws? Why would/would you not support its inclusion in the Constitution Act?

Information Sharing

The saying "information is power," describes one underlying truth in environmental management. It is often proponents of development and government bodies who have the funds to conduct studies and have access to information that others do not. To address this problem and provide a balanced "playing field," some laws provide for access to information. At the federal level, the Access to Information Act provides an opportunity for the public to request specific pieces of information on any subject not deemed to be "classified" from all government agencies. Most provinces have an equivalent law. Moreover, in many provinces, citizens complaining about unfair treatment by government agencies or officials can consult the provincial office of the Ombudsman. The Ombudsman is an independent and impartial investigator whom a person can consult without having to go through the expense and process of a judicial review. In 1995, the federal government created a Commissioner of the Environment and Sustainable Development who reports directly to the Auditor General of Canada. The public can contact the Commissioner on any matter related to the environment who will contact the Minister responsible for the topic of interest. The Minister must respond to the Commissioner. The Commissioner also provides annual reports.

Since 1994, when Ontario's Environmental Bill of Rights was passed, residents have been notified about proposed legislation, policies, regulations, and other legal instruments that could have a significant effect on the environment. The government must consider the public's input before it makes a final decision. These proposals are posted on a website that anyone can access and then provide comments to the government. The effectiveness of this mechanism is questionable. In 1998, two proposals to export 600,000,000 litres of water per year from the Great Lakes by tanker to Asia were approved by the Ministry of the Environment. Even though the application was posted on the Web page (home page at http://www. ene.gov.on.ca/envision/env_reg/ebr/english/), no one likely noticed it and no one opposed it until after it was approved, at which time, the strong public outcry forced the government to withdrawal its approval. In addition, many government agencies provide data and reports on the Web. These would include water use data, weather forecasts, scientific reports, policies, and media releases.

The Commission on Environmental Cooperation concludes that private industry is also providing information though five mechanisms. First, industry must promptly report any spill or unusual pollution incident to government authorities. Second, businesses that operate under a "certificate of approval" must report its discharges to government agencies on a regular basis. Third, many businesses that operated under a "certificate of approval" may be required to report information on ambient environmental conditions in the vicinity of the business. These data help to determine whether and how the discharge of pollutants is actually affecting the environment. Fourth, particular high-risk activities, such as the transport of toxic material, are subject to special reporting requirements (e.g., reporting systems that track each shipment from origin to destination). In addition, businesses that import or manufacture a new chemical substance to Canada must provide toxicological testing data to the federal government, as required by Canadian Environmental Protection Act. Fifth, the law requires some companies to participate in a special data gathering programs, such as the National Pollutant Release Inventory (NPRI). The NPRI contains information collected from businesses on emissions of certain pollutants into the environment (see Case Study, 419). These data have been analyzed and published by the Commission on Environmental Cooperation (its *"Taking Stock"* series), which are widely reported in the news media (Figs. 19.7 and 19.8).

Whistle-Blower Protection

Why don't people working for government agencies or the private sector that are conducting their operations in an environmentally inappropriate and possibly illegal way report them to the relevant authorities? The answer, without the advantage of "whistle-blower protection," they would have been crazy to do so because the whistle-blower would likely feel the wrath of their bosses and fellow

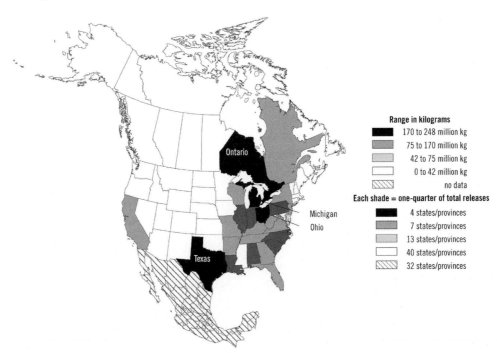

FIGURE 19.7 Largest Sources of Total Reported Amounts of Releases and Transfers in North America, 2001: States and Provinces.
Source: "Taking Stock 2001: North American Pollutant Releases and Transfers," Commission for Environmental Cooperation of North America 2004.

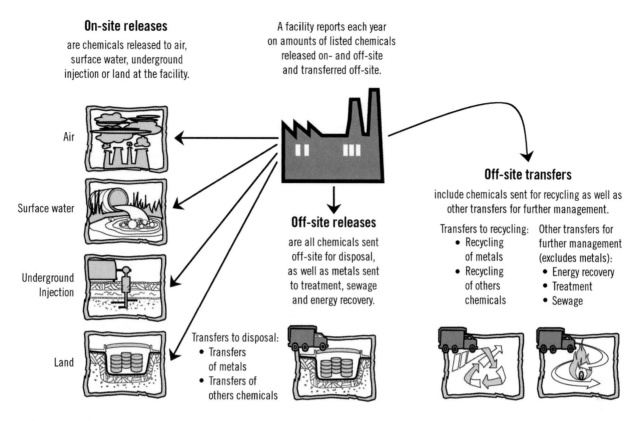

On-site releases
are chemicals released to air, surface water, underground injection or land at the facility.

A facility reports each year on amounts of listed chemicals released on- and off-site and transferred off-site.

Air

Surface water

Underground Injection

Land

Off-site releases
are all chemicals sent off-site for disposal, as well as metals sent to treatment, sewage and energy recovery.

Transfers to disposal:
• Transfers of metals
• Transfers of others chemicals

Off-site transfers
include chemicals sent for recycling as well as other transfers for further management.

Transfers to recycling:
• Recycling of metals
• Recycling of others chemicals

Other transfers for further management (excludes metals):
• Energy recovery
• Treatment
• Sewage

FIGURE 19.8 PRTR Releases and Transfers in North America

Source: "Taking Stock 2001: North American Pollutant Releases and Transfers," Commission for Environmental Cooperation of North America 2004.

employees. Reprisals by people at their work might include being harassed by your boss, ostracized by co-workers, fired, demoted, sued, or otherwise made miserable. If a whistle-blower chooses to fight any of this through the courts, they usually face a long, uphill battle at their own expense.

In the United States, both federal and state governments have established laws that prevent whistle-blowers working for government or the private sector to be harassed or fired. Some of these laws also provide financial compensation, recognizing that laws alone cannot fully protect whistle-blowers from vengeful employers. Ontario's Environmental Bill of Rights (EBR) provides protection for employees who may be harassed, disciplined, or fired as a result of their participation in activities under the Act. Any whistle-blower who reports activities related to any of the 20 statutes covered by the EBR is protected. Any person can file a written complaint with the Ontario Labour Relation Board alleging that an employer has taken "reprisals" against an employee on "prohibited ground." Prohibited ground refers to reprisals taken because the whistle-blower, in good faith, did or may:

• participate in the decision-making process in connection with any activity covered by the Act;

• apply for a review of a policy, statute, regulation, or instrument under Part IV of the EBR;

• apply for an investigation under the EBR;

• seek the enforcement of a prescribed statute, regulation, or instrument;

• give information to an appropriate authority for the purposes of an investigation, review or hearing related to a prescribed policy, statute, regulation, or instrument, or;

• give evidence in a proceeding under the EBR or a statute prescribed under it.

Voluntary Controls

In the 1980s, government and industry, particularly those in North America, Japan, and Europe, became motivated to support voluntary initiatives. Four reasons prompted government support. First, the importance of the private sector in environmental management was realized. It should be perceived as part of the solution as well as part of the problem. The previous text supports the view that the private sector is part of the problem. Changing the business culture could promote sustainable development. Second, there was a realization that although government control, citizen empowerment, and information sharing had produced some successes, there was a perception that other forms of responses were required to achieve further improvements. While discharge and emission limits can be established through government control, other methods might focus attention on the cleaner methods of production. Third, **government voluntary programs** (Table 19.1) can allow an

TABLE 19.1 Characteristics and Types of Voluntary Programs

PARTICIPANTS IN VOLUNTARY PROGRAM	GENERAL CHARACTERISTICS	GENERAL TYPES OF PROGRAMS
Industry	Industry initiates and has exclusive management responsibilities for all aspects pf implementation. Government agencies may recognize the voluntary initiative. These initiatives may be company-based or industry wide.	**Individual company initiatives.** Companies establish their own environmental goals and programs such as, corporate environmental management systems, corporate environmental policy, guidelines, principles or codes of conduct, corporate environmental programs, and corporate environmental reports. **Collective industry initiatives.** These are generally industry association initiatives in order to promote better industry environmental performance (e.g., Global Environmental Management Initiative, Public Environmental Reporting Initiative). General principles for environmental protection, ideally providing some form of follow-up to ensure implementation of principles, are established. **Cross-sector initiatives.** Provide broad, general guidelines or qualitative objectives (e.g., continuous improvement, best practice) for an often diffuse and heterogeneous membership. They may provide essential leadership for industry sector branches but in being unable or unwilling to set specific environmental goals or targets, their real effectiveness is usually difficult to measure. **Industry-specific initiatives.** Offer the potential to translate qualitative objectives into quantifiable goals and targets for specific industries. Usually pursued only by large companies.
Government	Programs are developed and run by government. Environmental need(s) and goals for industry are established, with possible involvement of industry. Likely have specific targets and time frames. They may be industry-specific or cross-sector (e.g., energy efficiency, reduction of the harm from toxic substances). They are not legally binding or enforceable, and generally carry no implicit threat of regulation if objectives are not met.	**Challenge programs.** Participants are challenged by government or their industry to achieve specific environmental targets or performance goals. Intent is to motivate, facilitate and encourage cooperation and information sharing among competitors. **Technology upgrade programs.** Participants encouraged to undertake specific category of technology improvement such as installing energy-efficient lighting. **Ecolabelling programs.** Products (or services) are labelled to designate that they are environmentally friendly, energy efficient, cost effective, etc. Consumers can make informed purchases. **Regulatory relief programs.** In exchange for superior environmental performance and stakeholder accountability, companies/industry sectors are offered simplified environmental permitting, exemption from existing environmental regulations or taxes (in Denmark participating companies are reimbursed CO_2 tax), or the promise of a "stable" regulatory climate during the period of agreement. **Award or prize programs.** Rewards (e.g., public recognition, financial awards) are provided to stimulate good performance. May be offered at local to national levels. **R&D/innovation programs.** These programs are most useful when there is a need to encourage industry to develop better technologies that require large/long-term investments.
Joint Government and Industry Initiatives	Jointly developed by government and industry, usually involving some form of negotiations and the sharing of management responsibilities such as monitoring and evaluation.	**Covenants.** "Voluntary agreements" concluded between different levels of government and representatives of industry with the status of binding contracts in civil law. Covenants are being used within industry as implementation instruments in areas likely covered by government control. Covenants serve to establish a specific concrete implementation program that supports existing laws and regulations. **Voluntary agreements.** Generally consist of an agreement between industry and public authorities concerning environmental goals to be met by industry within a specified time frame. The most common distinctions made are between those voluntary agreements, which are non-binding, "gentlemen's agreements" and those which are contractual, and contain specific control measures (e.g., monitoring and reporting requirements). **Environmental agreements** or **negotiated agreements.** Alternative terms used by those who point out that agreements are by definition voluntary or that an agreement that involves sanctions (e.g., implicit threat of government control) is not, by definition, voluntary.
Third Party Initiatives	Developed and run by non-governmental organizations	**ISO 14000.** An environmental management standard that is gradually becoming an international market requirement, involving non-governmental standards associations in numerous countries. **Social investment groups.** Apply environmental and social principles and criteria to guide investment management. Hope to change corporate behaviour (e.g., Coalition for Environmentally Responsible Societies Principles in the USA; UNEP Statement by Financial Institutions on the Environment and Sustainable Development; Statement of Environmental Commitment by the Insurance Industry). **Labour agreement.** Trade unions and employers have also come to agreements about access to workplace information and how it will be used and disseminated.

Source: With information available from UNEP Division of Technology, Industry and Economics. http://www.unepie.org/media/review/vol21no1-2/vol21no1-2.htm#initiatives.

environmental agency to address environmental problems or allow the development of solutions in areas in which its statutory authority is non-existent or weak. Fourth, from an ideological perspective, there was a desire to reduce the level of government regulation in the market (business).

From industry's perspective, three factors prompted their interest in voluntary programs. First, they wished to remain competitive and this might be achieved if they could influence public policy, including environmental policy. From their vantage point, this is likely the prime motivation. Second, they wished to undertake environmental management in as cost-efficient manner as possible. Third, voluntary programs can be used to advantage by environmentally progressive and financially healthy companies because if they adopt an **individual company initiative** (Table 19.1), it can signal that new regulations should and/or could be developed for the entire industry. This could remove weaker competitors from the market.

Four general types of voluntary initiatives, based on who initiates the activity, can be identified: (i) industry initiatives, (ii) government initiatives, (iii) joint government/industry initiatives, and (iv) third-party initiatives. A summary of their essential characteristics is contained in Table 19.1.

Lyon (2003) summarizes the activities at the World Resources Institute's Sixth Annual Sustainable Enterprise Summit held in 2003. It was attended by leaders of business, government, academia, and non-governmental organizations, all of who shared an interest in how business can bring about environmental improvement. On a positive note, a third-party initiative lead by the Natural Resources Defence Council (NRDC) resulted in toxic emission reductions at a Dow Chemical plant at Midland, Michigan, by 37 percent and yielded savings of $5.6 million annually. Despite this high level of success, the NRDC was unable to get any other industries involved in the program.

According to Lyon (2003), preempting regulatory threats is one way industry can "influence" the policy process. For instance, shortly after the Bhopal, India, disaster that saw the release of toxic chemicals into the town killing thousands of people, the Chemical Manufacturers Association (CMA) surveyed Americans about their support for the chemical industry. The results indicated, "Americans' trust of the chemical industry was in the same league with their trust of tobacco and nuclear power firms." In order to increase public confidence in the industry, the CMA created the Responsible Care Program. Lyon (2003) reported that all companies wishing to belong to the CMA (which has now changed its name to the American Chemistry Council—perhaps part of its public relations strategy) had to commit themselves to the four principles of the Responsible Care Program: (i) a formal company commitment to the program's abiding principles; (ii) adoption of the prescribed Codes of Practice; (iii) creation of a National Community Advisory Panel; and (iv) the use of certain environmental performance indicators. The United Kingdom, Canada, Australia, and the United States are some of the countries actively participating in the program. Lyon cites a secondary source to quote Fred Webber, past president of the CMA who said, "In my opinion, Responsible Care is more than a good initiative—

it's the industry's franchise to operate." The timing, intent, and high level of industry participation in programs such as Responsible Care can "convince politicians and the public that the most urgent problems in a particular area have already been solved, and the effort required to address any remaining concerns would be better spent elsewhere" (Lyon, 2003). In this way, the industry benefits because it has significant control over the nature of the response rather than the solution being legislated through traditional government control mechanisms. If the industry's efforts are insufficient, the government could use regulatory controls to force more effective solutions. In the case of the Responsible Care Program, it was successful at increasing the public's confidence for the chemical industry and its environmental performance. The jury is still out on whether the long-term public interest is well served by voluntary initiatives.

Voluntary programs, such as the Accelerated Reduction/ Elimination of Toxics (ARET) and the National Packaging Protocol (NaPP), are part of Canadian efforts to management the environment. Voluntary initiatives are central to Canada's commitment to the Kyoto Protocol. As noted earlier in this chapter, the Canadian target for the Kyoto Protocol is to reduce greenhouse gas emissions by 6 percent below their 1990 levels by 2012. However, while the Kyoto Protocol establishes Canada's emission reduction commitment, how this reduction is to be achieved is not specified. Canada's *Climate Change Plan* answers this question in identifying a strategy based on the use of a mix of instruments government grants, tax measures and financial incentives, regulatory and market-based measures. The plan targets specific industry sectors— Transportation, Housing and Commercial/Institutional Buildings, Large Industrial Emitters and Small and Medium-Sized Enterprises—to participate in the program. It is intended that each sector will establish **covenants** (Table 19.1) that establish the targeted emission reduction levels that would take into account "the competitiveness issues in each sector and could address a number of elements, such as emissions intense to the undertakings,

See Thomas Lyon's article at: http://www.cato.org/pubs/ regulation/regv26n3/v26n3-10.pdf and UNEPs Division of Technology, Industry and Economics at http://www.unepie.org/media/review/ vol21no1-2/vol21no1-2.htm

See http://www.climatechange. gc.ca/plan_for_canada/

technological investments and any other initiatives to reduce emissions, as well as partnership activities." While the plan recognizes that some Canadian companies have already taken action to reduce their GHG emissions, largely through a mix of **industry initiatives** (Table 19.1) it is unclear if these industries will be given credit for those reductions. Instead, it suggests that these companies will not be disadvantaged and that the government will continue to work with industry to design a system consistent with that principle. Thus, if industries were trying to be preemptive, they may not achieve their ends. On the other hand, if increasing their competitive advantage motivated them, these companies could end up benefiting from the plan.

Like the experience worldwide, Canada's record with voluntary initiatives is mixed at best. In its *1999 Economic Outlook of Canada,* the OECD (2000) stated, "a strategy based on voluntary agreements *alone* cannot be expected to correct completely for the external costs of pollution" (emphasis added). They must be used in conjunction with government control programs, information sharing initiatives, such as the NPRI which serve to monitor performance, and a greater degree of citizen involvement in government and decision making. In addition, the OECD identified the need to increase the use of economic instruments. The next section turns attention to this approach to environmental management.

Economic Instruments

In its *1999 Economic Outlook of* Canada report, the OECD (2000) concluded the following:

> Policies and regulatory mechanisms aimed at curbing air, soil and water pollution have been successful in some areas, notably the Great Lakes. However, there is scope for improvement. The polluter-pays principle is not systematically applied, and reliance on voluntary agreements has not been sufficient to achieve environmental objectives (for instance, in the case of the management of toxic substances). No-cost opportunities for curbing pollution are rare, and, a strategy based on voluntary agreements alone cannot be expected to correct completely for the external costs of pollution. Hence there is a need to increase the use of economic instruments (for instance, charges on toxic emissions and waste, and disposal fees for products containing toxic substances) to reinforce the polluter-pays principle.

Economic instruments are defined as the "use market-based signals to motivate desired types of decision making. They either provide financial rewards for desired behaviour or impose costs for undesirable behaviour" (Stratos, Inc., 2003). They identify four types of instruments: (i) property rights, (ii) fee-based measures, (iii) liability and assurance regimes, and (iv) tradable permits (Table 19.2).

Economic instruments are created through legislation and should be viewed as an emerging approach that complements rather than replaces traditional government control approaches. It is important that governments adhere to the powers provided to them

TABLE 19.2 Types of Economic Instruments
• **Property rights:** ownership rights, use rights, development rights and transferable development rights can all be used to promote responsible resource management.
• **Fee-based measures:** fees, charges, taxes, deposit-refunds and revenue-neutral rebates all impose payments of specified amounts, thereby creating an explicit cost associated with environmentally damaging activities and an easily quantifiable incentive for reducing the activity.
• **Liability and assurance regimes:** liability rules and various types of bonds can provide strong incentives to avoid environmental impacts and to cleanup and restore environmental damage.
• **Tradeable permits:** provide mechanisms for minimizing the social and private costs of meeting a cap on emissions.

Source: Santos, Inc., 2003. http://www.smartregulation.gc.ca/en/06/01/su-11.asp

in the Constitution Act (Case Study, p. 425). Estrin and Swaigen (1993) maintain that economic instruments are "a less direct form of regulation ... and may be employed by policy makers to affect the costs and benefits of different behavioural options." The intent of the "behaviour options" is to affect the way companied produce or individuals buy those goods and services that are less harmful to the environment. This is achieved by using the power of "the market" (price) to influence corporate and human behaviour.

The idea of economic instruments has been with us for some time. The OECD advocated the "polluter-pay" principle (a fee-based measure (Table 19.2)) in 1972. The intent is to have the costs of waste management transferred to companies. In the past, companies have been able to avoid these costs because they discharged wastes (often with the government approval through a certificate of approval or compliance) into water, air and landfill sites. This approach transfers the costs to the next user of the resource. For instance, at one time people drank water from Canadian streams without a big concern about getting sick. It was truly a "free resource." If an industry discharges wastes into a river that serves as the water supply for a downstream community, that community and not the industry must pay to millions of dollars to construct and operate a water treatment plant. This impact is known as an **externality**—the discharger does not pay to have the discharge fully treated and the cost is paid by others. This might be classified as a **spatial fragmentation,** which is in addition to the previously mentioned vertical, horizontal, and issue forms of fragmentation.

The advantages of economic instruments are as follows:

- they are more efficient than government control approaches;
- a 25 percent reduction in the cost of compliance could be achieved. In the U.S., $200 billion is spent annually on pollution control;
- industries that apply continuous improvement programs would be rewarded;

CASE STUDY

THE CONSTITUTION ACT AND ECONOMIC INSTRUMENTS

Governments must stay within the powers granted to them under the Constitution or their programs may be ruled *ultra vires* by the judicial system.

Rolfe and Nolan (1993) concluded that relative to the provinces, the federal government had relatively limited power to establish economic instruments for environmental protection. However, its power is still significant. It may use them in the context of global, national or regional problems, which need a coordinated national approach. For instance, it

can enter the greenhouse gas emission trading system that will be established under the Kyoto Protocol. Other caution would have to be used in considering other forms of economic instruments, such as deposit refund systems for persistent toxics, carbon taxes, tradable permit systems for discharges into fish bearing waters, discharge taxes, and tradeable permit systems for the phase out of health threatening substances, to ensure they are constitutionally valid. Provinces have wide ranging abilities in this area. Rolfe and Nolan suggested that they

can "establish discharge permit trading systems for any discharges within the province; establish revenue generating discharge fee systems; place charges on products that lead to emissions; impose a system of discharge fees or tradeable emission permits for vehicles; and establish a system of deposits and refunds for products or substances that are sold or produced within the province."

Source: Rolfe, C. and L. Nowlan (1993) available at http://www.wcel.org/celpub/4994.html.

- the development of new technologies would be stimulated and assist in Canada's status international trade;
- they can address the shortcomings of government control approaches; and
- they can correct for the "externality problem" (Santos, Inc., 2003).

The OECD (2000) concluded that although the nation "uses economic instruments such as user fees, tax incentives and 'sin' taxes to promote a range of social and economic policies, Canada lags most of the rest of the OECD in its use of economic instruments in support of environmentally sustainable development." Tradeable permits have not been widely used and there has been little effort devoted to changing consumption patterns through pricing, and almost no experience with economic instruments that encourage the technological improvement of products. This finding is supported by Santos, Inc. (2003), which reviewed the Canadian experience with economic instruments. The federal government uses tradeable permits to manage some fisheries and to address ozone-depleting substances, and has supported pilot greenhouse gas trading projects. Tax incentives are used to promote ethanol, renewable energy and gifts of ecologically significant land. They noted, "Canada generates only about 5 percent of its tax revenues from environmentally related taxes. Over three-quarters of these revenues are from the gasoline excise tax, which arguably is designed to generate revenue rather than to reduce fuel consumption. This is in contrast to over 8 percent for the U.K., over 10 percent in Norway and over 12 percent in the case of Netherlands and Korea."

Ontario has introduced a trading program to reduce emissions from large coal and oil-fired electricity generators. In order to address water shortages on the South Saskatchewan River, Alberta is establishing a transferable water rights regime. Winnipeg may soon be the most advanced jurisdiction in the use of economic instruments in Canada. Under its proposed "New Deal,"

homeowners and businesses would see a reduction in their property taxes, which would be offset by suite of taxes including a sales tax, a fuel tax (to fund transit expenditures), taxes on hotels, gas, electricity (not used for heating), and garbage disposal fees. These taxes are more extensive than the pay-for-use approaches that some municipalities have applied to water and sewage services and garbage collection.

The custom of underpricing the use of resources, such as water, can contribute to environmental problems. Canadians are among the highest per capita users of water in the world (Fig. 19.9). Part of this reflects the myth of an abundant supply of water resources. High use is also promoted by a very low price relative to other countries (Fig. 19.10). It costs money to support the pumps, pipes, water and sewage treatment, water testing billing systems, and other required components that supports our water use.

Want More?

See Stratos, Inc. 2003 at http://www.smartregulation.gc.ca/en/06/01/su-11.asp

Rolfe, C. and Nowlan, L. 1993. *Economic Instruments and the Environment: Selected Legal Issues.* Ann Hillyer. ed. Vancouver: West Coast Environmental Law Research Foundation and Environment Canada. Available at http://www.wcel.org/wcelpub/4994.html.

More information on Winnipeg's New Deal can be found at http://www.city.winnipeg.mb.ca/interhom/mayors percent5Foffice/newdeal/.

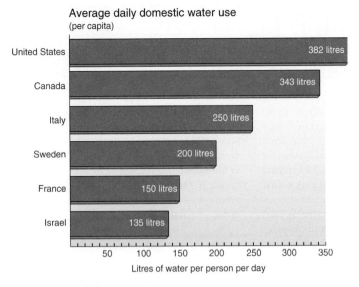

Average daily domestic water use
(per capita)

FIGURE 19.9 Water use in Canada and other countries.

Source: Environment Canada's freshwater website (www.ec.gc.ca), 2004. Reproduced with permission of the Minister of Public Works and Government Services 2004.

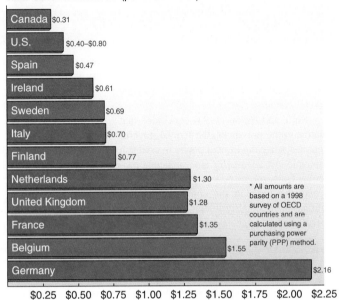

Typical municipal water prices in Canada and other countries (per cubic metre)

* All amounts are based on a 1998 survey of OECD countries and are calculated using a purchasing power parity (PPP) method.

FIGURE 19.10 Typical municipal water prices in Canada and other countries.

Relative to other developed nations, Canadians pay a low price for their water. This contributes to overuse and abuse of the water resource base.

Source: Environment Canada's freshwater website (www.ec.gc.ca), 2004. Reproduced with permission of the Minister of Public Works and Government Services 2004.

In 1996, the National Round Table on the Environment and Economy estimated that Canada's water and wastewater infrastructure required between $38 billion and $49 billion to meet needs at that time. There had been a clear and substantial under-investment. In looking at meeting future needs until 2016, it suggested that $70 billion to $90 billion was required. Paying realistic charges for water and wastewater services is one potential solution to meeting this need. Raising money through increased taxes, while providing funds, does little to encourage a reduction in water use. Privatization (partial or whole) is another controversial alternative. Increased water rates can provide for needed infrastructure funds and provide incentives to conserve. While there is little debate about the need for increased levels of investment, the willingness of people to pay increased water rates is not clear. It is also unclear what the economic impacts of increased water rates on the industrial and agricultural sectors would have on the Canadian economy, and the cost consumers would pay for goods ranging from milk and other beverages to cars and clothes made in Canada.

Some critics of economic instruments that are applied to pollution permits suggest that they are in essence "licences to pollute." The same comment can be made with regard to certificates of approval that are based on standards that fail to protect the environment and human health. There is also some resistance from moving from a system that has treated many resources, particularly water and air, as free goods to one based on "payments for use." The effective implementation of economic instruments requires the support of government control, citizen empowerment, information sharing and voluntary approaches. According to Santos, Inc. (2003), other barriers to the implementation of economic instruments include lack of awareness, resistance to what are perceived to be new taxes, concerns over maintaining and improving the competitiveness of industry and agriculture, the lack of strong advocates for their use, and a general lack of experience in dealing

with them. Overcoming these barriers is critical if Canada is to apply economic instruments in an effective, efficient and equitable manner.

INTERNATIONAL TREATIES AND CONVENTIONS

As recognition of the interconnections in our global environment has advanced, the willingness of nations to enter into protective covenants and treaties has grown concomitantly. Table 19.3 lists some major international treaties and conventions, while Figure 19.11 shows the number of participating parties in them. Note that the earliest of these conventions has no nations as participants; they were negotiated entirely by panels of experts. Not only the number of parties taking part in these negotiations has grown, but the rate at which parties are signing on and the speed at which agreements take force also have increased rapidly. The Convention on International Trade in Endangered Species (CITES), for example, was not enforced until 14 years after ratification, but the Convention on Biological Diversity was enforceable after just one year, and had 160 signatories only four years after introduction. Over the past 25 years, more than 170 treaties and conventions have been negotiated to protect our global environment. Designed to regulate activities ranging from intercontinental shipping of hazardous waste, to deforestation, overfishing, trade in endangered species, global warming, and wetland protection, these agreements

TABLE 19.3 Some Important International Treaties Related to the Environment

CBD: Convention on Biological Diversity 1992 (1993)

CITES: Convention on International Trade on Endangered Species of Wild Fauna and Flora 1973 (1987)

CMS: Convention on the Conservation of Migratory Species of Wild Animals 1979 (1983)

Basel: Basel Convention on the Transboundary Movements of Hazardous Wastes and their Disposal 1989 (1992)

Ozone: Vienna Convention for the Protection of the Ozone Layer and Montreal Protocol on Substances that Deplete the Ozone Layer 1985 (1988)

UNFCCC: United Nations Framework Convention on Climate Change 1992 (1994)

CCD: United Nations Convention to Combat Desertification in those Countries Experiencing Serious Drought and/or Desertification, Particularly in Africa 1994 (1996)

Ramsar: Convention on Wetlands of International Importance especially as Waterfowl Habitat 1971 (1975)

Heritage: Convention Concerning the Protection of the World Cultural and Natural Heritage 1972 (1975)

UNCLOS: United Nations Convention on the Law of the Sea 1982 (1994)

theoretically cover almost every aspect of human impacts on the environment.

Unfortunately, many of these environmental treaties constitute little more than vague, good intentions. In spite of the fact that we often call them laws, there is no body that can legislate or enforce international environmental protection. The United Nations and a variety of regional organizations bring stakeholders together to negotiate solutions to a variety of problems but the agreed-upon solutions generally rely on moral persuasion and public embarrassment for compliance. Most nations are unwilling to give up sovereignty. There is an international court, but it has no enforce-

ment power. Nevertheless, there are creative ways to strengthen international environmental protection.

One of the principal problems with most international agreements is the tradition that they must be by unanimous consent. A single recalcitrant nation effectively has veto power over the wishes of the vast majority. For instance, more than 100 countries at the U.N. Conference on Environment and Development (UNCED), held in Rio de Janeiro in 1992, agreed to restrictions on the release of greenhouse gases. At the insistence of U.S. negotiators, however, the climate convention was reworded so that it only *urged*—but did not require—nations to stabilize their emissions.

As a way of avoiding this problem, some treaties incorporate innovative voting mechanisms. When a consensus cannot be reached, they allow a qualified majority to add stronger measures in the form of amendments that do not need ratification. All members are legally bound to the whole document unless they expressly object. This approach was used in the Montreal Protocol, passed in 1987 to halt the destruction of stratospheric ozone by chlorofluorocarbons (CFCs). The agreement allowed a vote of two-thirds of the 140 participating nations to amend the protocol. Although initially the protocol called for only a 50 percent reduction in CFC production, subsequent research showed that ozone was being depleted faster than previously thought. The protocol was strengthened by amendment to an outright ban on CFC production in spite of the objection of a few countries.

Where strong accords with meaningful sanctions cannot be passed, sometimes the pressure of world opinion generated by revealing the sources of pollution can be effective. NGOs and others can use this information to expose violators. For example, the environmental group Greenpeace discovered monitoring data in 1990 showing that Britain was disposing of coal ash in the North Sea. Although not explicitly forbidden by the Oslo Convention on ocean dumping, this evidence proved to be an embarrassment, and the practice was halted.

Trade sanctions can be an effective tool to compel compliance with international treaties. The Montreal Protocol, for example, bound signatory nations not to purchase CFCs or products made using them from countries that refused to ratify the treaty.

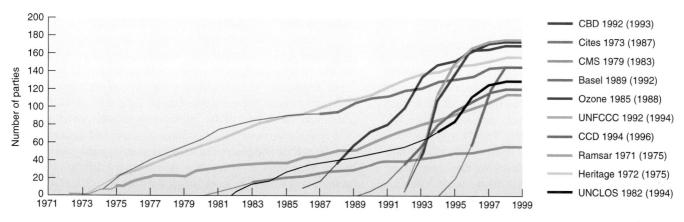

FIGURE 19.11 Additions of participating parties to some major international environmental treaties. The thick portion of each line shows when the agreement went into effect (date in parentheses). See Table 19.3 for complete treaty names.
Source: United Nations Environment Programme from Global Environment Outlook—2000.

Because many products employed CFCs in their manufacture, this stipulation proved to be very effective. On the other hand, trade agreements also can work against environmental protection. The World Trade Organization was established to make international trade more fair and to encourage development. It has been used, however, to subvert national environmental laws. In a ruling in 1998, the WTO forbid the United States from restricting imports of shrimp from Thailand, Malaysia, India, and Pakistan that were caught with nets that trap endangered sea turtles. Similarly, under provisions of the North American Free Trade Agreement, the Ethyl Corporation of the United States sued Canada for $250 million in compensation for banning the manganese-based gasoline additive MMT, which is suspected to be a neurotoxin. Environmentalists are concerned about a powerful international treaty currently being negotiated by the Organization of Economic Cooperation and Development, a group of 29 of the wealthiest countries in the world. The Multilateral Agreement on Investments, or MAI, may be more threatening to social justice and environmental protection than either the WTO or NAFTA.

DISPUTE RESOLUTION AND PLANNING

The adversarial approach of our current legal system often fails to find good solutions for many complex environmental problems. Identifying an enemy and punishing him for transgressions seems more important to us than finding win/win compromises. Gridlocks occur in which conflicts between adversaries breed mutual suspicion and decision paralysis. The result is continuing ecosystem deterioration, economic stagnation, and growing incivility and confrontation. The complexity of many environmental problems arises from the fact that they are not purely ecological, economic, or social, but a combination of all three. They require an understanding of the interrelations between nature and people. Are there ways to break these logjams and find creative solutions? In this section we will look at some new developments in mediation, dispute resolution, and alternative procedures for environmental decision making.

Wicked Problems and Adaptive Management

Rational choice theories of planning and decision making assume that if we just collect more data, buy faster computers to crunch numbers, build more complex models, and spend more money, any problem should be resolvable. More information, it's assumed, will lead automatically to better management. This "bigger hammer" approach may be effective in problems that are difficult but relatively straightforward. Increasingly, however, we have come to recognize that many of the most important problems we face don't fit this pattern. Questions like what ecosystem health means, or how clean is clean, don't have simple right or wrong answers. They depend on your worldview and how you define these terms. Different people come to different conclusions even if they share the same information.

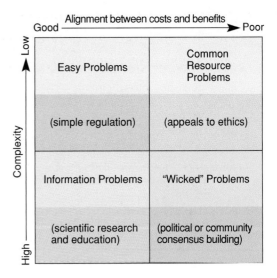

FIGURE 19.12 The difficulty of environmental decision making increases as problems become more complex and the congruence between those who bear costs and enjoy benefits decreases.
From K. Sexton, et. al., *Better Environment Decisions*, 1999. Copyright © 1999 Island Press. Reprinted by permission of Alexander Hoyt & Associates.

Environmental scientists describe problems with no simple right or wrong answers as being **wicked problems,** not in the sense of having malicious intent, but rather as obstinate or intractable. These problems often are nested within other sets of interlocking issues. The definition of both the problem and its solutions differ for various stakeholders. There are no value-free, objective answers for these dilemmas, only choices that are better or worse depending on your viewpoint. Wicked problems are important and have serious consequences, but also are complex and have a poor match between who bears the costs and who bears the benefits on any proposed solution (Fig. 19.12). They usually can't be solved by simple rules and regulations, more scientific research, or appeals to ethics. Often the best solution comes from community-based planning and consensus building. Inherent uncertainty gives these questions no clear end point. You cannot know when all possible solutions have been explored.

Recent advancement in understanding how ecological systems work gives us some insight into many wicked problems. Like biological organisms, social problems often change and evolve over time. Their history unfolds in complex ways, depending on chance interactions and unpredictable events. Like ecological systems, there may never be a stable equilibrium in many environmental issues. Each involves an assemblage of issues and actors that are unique in time and place. They can't be standardized. There are no good precedents from previous experience. Their solutions are unique, and what may work today, may not be applicable tomorrow. How can we learn to cope with such uncertainty?

One promising approach to solving wicked environmental problems comes from the work of ecologists C. S. Holling and Lance Gunderson, and planners Steven Light and Kai Lee, among others. Starting with the observation that human understanding of

TABLE 19.4 Institutional Conditions for Adaptive Management

1. There is a mandate to take action in the face of uncertainty.
2. Decision makers are aware they are experimenting anyway.
3. Decision makers care about improving outcomes over biological time scales.
4. Preservation of pristine environments is no longer an option, and human intervention cannot produce desired outcomes predictably.
5. Resources are sufficient to measure ecosystem-scale behaviour.
6. Theory, models, and field methods are available to estimate and infer ecosystem-scale behaviour.
7. Hypotheses can be formulated.
8. Organizational culture encourages learning from experience.
9. There is sufficient stability to measure long-term outcomes; institutional patience is essential.

From Kai N. Lee, *Compass and Gyroscope*, 1993. Copyright © 1993 Island Press. Reprinted by permission of Alexander Hoyt & Associates.

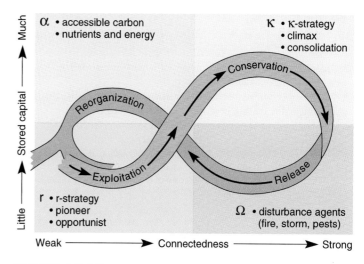

FIGURE 19.13 The creative-destruction cycle. Resilience, or the ability to reorganize and recover from disturbance, is the most important characteristic of both natural and human systems.
From Buzz Holling, *Barriers & Bridges*, edited by Gunderson, Holling, and Light, 1995. Columbia University Press. Reprinted by permission of the authors.

nature is imperfect, this group believes that all human interactions with nature should be experimental. They suggest that environmental policies should incorporate **adaptive management,** or "learning by doing," designed from the outset to test clearly formulated hypotheses about the ecological, social, and economic impacts of the actions being undertaken. Rather than assume that what seemed the best initial policy option will always remain so, we need to carefully monitor how conditions are changing and what effects we are having on both target and nontarget elements of the system. If our policy succeeds, the hypothesis is affirmed. But if the policy fails, an adaptive design still permits learning, so that future decisions can proceed from a better base of understanding. The goal of adaptive management and experimental design is to enable us to live with the unexpected. They aim to yield understanding as much as to produce answers or solutions (Table 19.4). This approach to natural resources is similar to—but more explicitly experimental than—ecosystem management (see Chapter 2).

Resilience in Ecosystem and Institutions

The great economist Joseph Schumpeter described "waves of creative destruction" that transform economic systems. Another insight from Holling and his collaborators is that similar cycles of destructive creation operate in both ecological systems and in policy institutions (Fig. 19.13). This is a familiar process that occurs in secondary succession (Chapter 5). The release phase of the cycle occurs when factors such as fires, storms, or pests disturb a biological community, mobilizing nutrients and making space available for new growth. During the reorganization phase, pioneer and opportunist species colonize the new habitat. These species grow rapidly on the accessible carbon, nutrient, and energy sources dur-

ing the exploitation stage. As the community matures, both the stored capital and connectedness increase until the ecosystem reaches a stage at which the system is poised for some new disturbance that starts the cycle again.

The most important characteristic of natural systems is their **resilience,** or ability to recover from disturbance. This doesn't imply that the ecosystem always returns to the exact condition it was in before the disturbance. It may have a new assemblage of species, or different set of physical conditions, but if it is resilient, the system has the ability to reorganize itself in creative and constructive ways. "Environmental quality is not achieved by attempting to eliminate change or surprises," Holling observed. The goal, instead, is resilience in the face of surprise. Surprise can be counted on. Resilience comes from adaptation to stress, from survival of the fittest in a turbulent environment.

In studying a variety of natural resource management regimes, Holling and others observed that human institutions also follow a similar pattern. In studying a variety of natural resource management issues ranging from restoration of the Florida everglades, to control of spruce budworm in New England forests, to cattle grazing in South Africa, to protection of salmon in the Pacific Northwest, they observed that every attempt to manage ecological variables one factor at a time inexorably leads to less resilient ecosystems, more rigid management systems, and more dependent societies. Initial success sets the conditions for eventual collapse. Take the example of forest fire suppression. For 70 years, the U.S. Forest Service has had a very effective policy of putting out all forest fires. The result has been that flammable debris has built up in the forests so that major conflagrations are now inevitable. During this time, however, people have felt safe moving to the borders of the forests and now there is a large population with a huge investment in property that needs to be protected from fire. Furthermore, a big bureaucracy has built up whose

TABLE 19.5	Planning for Resilience

1. Interdisciplinary, integrated modes of inquiry are needed for adaptive management of wicked problems.
2. We must recognize that these problems are fundamentally nonlinear and that we need nonlinear approaches to them.
3. Interactions between slow ecological processes such as global climate change or soil erosion in the American cornbelt are difficult to study together with the fast processes that bring creative destruction such as potential collapse of Antarctic ice sheets or appearance of a dead zone in the Gulf of Mexico, but we need to look for connections.
4. The spatial and temporal scales of our concerns are widening. We now need to consider global connections and problems in our planning.
5. Both ecological and social systems are evolutionary and are not amenable to simple solutions based on knowledge of small parts of the whole or on assumptions of constancy or stability of fundamental relationships.
6. We need adaptive management policies that focus on building resilience and the capacity of renewal in both ecosystems and human institutions.

Source: L. Gunderson, C. Holling, and S. Light, *Barriers and Bridges to the Renewal of Ecosystems and Institutions*, 1995, Columbia University Press.

raison d'etre is to fight fires. It takes more and more money to forestall a calamity that becomes increasingly likely because of our efforts to prevent it.

What happens in each of these cases is that our goal to control variability in ecological systems leads us to a narrow purpose and to focus exclusively on solving a single problem. But elements of the system change gradually as a consequence of our management success in ways that we did not anticipate. As more homogenous ecosystems develop over a landscape scale, resilience decreases, and it becomes more likely that the system will flip suddenly into a new regime. What can we do to avoid this trap? Table 19.5 suggests some important lessons for ecological managers.

The Precautionary Principle

One response to the uncertainty of wicked problems and chaotic, nonlinear, discontinuous systems is to plan a margin of safety for error or surprises. Drawing on studies of ecological systems, many conservation biologists advocate a **precautionary principle** that says that when an activity raises threats of harm to human health or the environment, precautionary measures should be taken even if some cause and effect relationships are not fully established scientifically. At a meeting at the Wingspread Center in 1998, an international group of scientists, government officials, lawyers, and grassroots environmental activists agreed on four basic tenets of precautionary action:

- People have a duty to take anticipatory steps to prevent harm. If you have a reasonable suspicion that something bad might be going to happen, you have an obligation to try to stop it.
- The burden of proof of carelessness of a new technology, process, activity, or chemical lies with the proponents, not with the general public.
- Before using a new technology, process, or chemical, or starting a new activity, people have an obligation to examine a full range of alternatives, including the alternative of not using it.
- Decisions applying the precautionary principle must be open, informed, and democratic, and must include the affected parties.

In 2001, the Supreme Court of Canada supported a by-law introduced by the town of Hudson, Quebec, which prohibited the use of pesticides for non-essential uses. Biocide companies challenged the ruling arguing that the regulation and control of biocides was the responsibility of federal and provincial governments, not local. (Elements of this argument should sound familiar, Recall "You be the Judge, Case 1, p. 404.) The Supreme Court supported Hudson's by-law concluding that all levels of government in Canada have vital roles to play in environmental management. Rather than diving for the lowest common denominator, the Supreme Court took the view "local governments should be empowered to exceed, but not to lower national norms." Many other municipalities are expected to follow Hudson's lead.

This decision also supports the application of the Precautionary Principle. The Supreme Court decision legitimizes the application of this principle as well as an enhanced role for municipalities in environmental management.

Arbitration and Mediation

Another set of alternatives to the adversarial nature of litigation and administrative challenges is the growing field of dispute resolution. Increasingly used to avoid the time, expense, and winner-take-all confrontation inherent in tort law, these techniques encourage compromise and workable solutions with which everyone can live.

Arbitration is a formal process of dispute resolution somewhat like a trial. There are stringent rules of evidence, cross-examination of witnesses, and the process results in a legally binding decision. The arbitrator takes a more active role than a judge, however, and is not as constrained by precedent. The arbitrator is more interested in resolving the dispute rather than strict application or interpretation of the law. Arbitration is usually an attractive prospect if you don't think you could win a formal lawsuit, but why would anyone agree to arbitration if they think they can win the whole enchilada in court? They might take this route just to avoid disagreeable surprises. Juries can be fickle. Furthermore, they might want to get an unpleasant process over with sooner rather than later. In addition, arbitration often is written into contracts so that the disputants have no choice over the matter.

There are disadvantages to arbitration. It doesn't create a legally binding precedent, something that often is the main motivation for a lawsuit. There is less opportunity to appeal if you don't like the decision you get. There also is less protection from self-incrimination, false witnesses, or evidence you didn't expect. You don't generate nearly as much publicity because the proceedings and record are not public. For some litigants, the publicity generated by a trial is more valuable than the settlement itself. Finally, you are less likely to win the whole thing. Some sort of compromise is the most likely outcome.

Mediation is a process in which disputants are encouraged to sit down and talk to see if they can come up with a solution by themselves (Fig. 19.14). The mediator makes no final decision but is simply a facilitator of communication. This process is especially useful in complex issues where there are multiple stakeholders with different interests, as is often the case in environmental controversies. For example, a mediation was attempted to work out policy and management disputes about the Boundary Waters Canoe Area Wilderness in northern Minnesota. Local property owners, resort operators, anglers who wanted to use motorboats on wilderness lakes, wildlife protection groups, wilderness advocates, and the U.S. Forest Service all were represented around the table. Each group had its own agenda, although they all quickly coalesced into pro-motor and anti-motor factions. Interestingly, although we usually think of environmentalists as being relatively disadvantaged in these issues, in this case, the local folks felt they were unfairly outgunned by the high-powered lawyers who represented some of the pro-wilderness and wildlife groups. Although the participants agreed on many points and it seemed as if the mediation was about to come up with a workable compromise, at the last minute everything fell apart and the groups refused to agree about anything. Each side accused the other of intransigence and bad will, and the issues they disagreed upon will likely end up in court.

This example illustrates both the promises and perils of mediation. It can be quicker and cheaper than court battles. It can lead to compromise and understanding that will lead to further cooperation, and that will solve problems faster than endless appeals. And it may find creative solutions that satisfy multiple parties and interests. On the other hand no one can be forced to mediate or to do so in good faith. Rancorous participants can tie up the process in long, pointless arguments that only make others more angry. Ultimately, there can be a tyranny of the minority. A single person can veto an agreement that everyone else wants. Furthermore, mediation represses or denies certain irreconcilable structural conflicts, giving the impression of equality between disputants when none really exists. Unequal negotiating skills of the participants can lead to unfair outcomes and even more rancor and paranoia than before the mediation was attempted. As is the case in arbitration, mediation doesn't generate the publicity and complete victory that some groups may desire.

Collaborative Approaches to Community-Based Planning

Over the past several decades, natural resource managers have come to recognize the value of holistic, adaptive, multiuse, multivalue approach to planning. Involving all stakeholders and interest groups early in the planning process can help avoid the "train wrecks" in which adversaries become entrenched in non-negotiable positions. Working with local communities can tap into traditional knowledge and gain acceptance for management plans that finally emerge from policy planning. This approach is especially important in nonlinear, nonequilibrium systems and wicked problems. Among the more important reasons to use collaborative approaches are:

- The way wicked problems are formulated depends on your worldview. Incorporating a variety of perspectives early in the process is more likely to lead to the development of acceptable solutions in the end.

- People have more commitment to plans they have helped develop. The first stage is therefore to identify those involved and to engage them in the process.

- There is truth in the old adage that "two heads are better than one." Involving multiple stakeholders and multiple sources of information enriches the process.

- Community-based planning provides access to situation-specific information and experience that can often only be obtained by active involvement of local residents.

- Participation is an important management tool. Project-threatening resistance on the part of certain stakeholders can be minimized by inviting active cooperation of all stakeholders throughout the planning process.

- The knowledge and understanding needed by those who will carry out subsequent phases of a project can only be gained through active participation.

A good example of community-based planning can be seen in the Atlantic Coastal Action Programme (ACAP) in eastern Canada.

FIGURE 19.14 Mediation encourages stakeholders to discuss issues and try to find a workable compromise.

The Quincy Library Group

Northern California has had intense debates about forest management policy for many years. On one hand, many small rural communities are almost totally dependent on the forest industry for economic survival. On the other hand, as native, old-growth forests become increasingly scarce, many environmental groups call for a sharp decrease in logging on public lands, and campaign to preserve as much as possible of what wilderness is left in its original state. The Forest Service is caught in the middle of this debate, with a mandate for both environmental protection and economic production. How can it reconcile these competing demands?

An experiment in community-based environmental planning in the small Sierra Nevada town of Quincy, California, has recently generated a great deal of interest and controversy. Some people praise it as an exciting model for cooperative management, while others deplore it as a fraud and a hoax designed to circumvent existing environmental controls and give away precious natural resources to local industry.

Several wicked problems had led the citizens of Quincy to feel that they were facing a crisis that demanded some drastically new approaches. Decades of fire suppression had left the forest surrounding the town choked with dead trees and woody debris. In the dry climate of the northern Sierras, a catastrophic fire seemed inevitable. Calls for protection of roadless areas and old-growth-associated species such as the marbled murlet and the northern spotted owl, worried the forest products industry that it might not have a continuing supply of wood. With a single-industry economy, and timber harvest down as much as 80 percent from historic highs, Quincy, like many of its neighbouring communities, felt like an endangered species itself. Attempts at

dialogue among the various factions in town usually ended up in shouting matches.

One day in the early 1990s, three Quincy residents with very different backgrounds—a local timber industry employee, an ardent environmentalist, and a county supervisor—got together to talk about their common concerns. Agreeing to meet in the only building in town where they weren't allowed to shout at each other, the Quincy Library Group was born. As other residents joined the conversation, a consensus began to emerge about how the forest could be managed, environmental quality could be protected, and the local economy could survive.

The Quincy Library Plan calls for new management plans on 1 million ha of land on the Plumas, Lassen, and Tahoe National Forests. "Fuel breaks" created by thinning out dense sections of forest would allow harvesting of up to 24,000 ha of public land each year. In addition, another 24,000 ha of forest would be harvested to support the local forest industry. Roadless areas, riparian zones, and habitat for endangered species such as the spotted owl would be protected. Hailed as a breakthrough in innovative, cooperative planning, the Library Group's proposal was introduced in Congress by California Senator Dianne Feinstein and Representative Vic Fazio, as an experimental, adaptive management strategy. Each year of the "pilot project" the Forest Service must report on the economic, social, and environmental effects of its actions. After five years of environmental monitoring and science-based assessment, the whole plan will be reexamined.

Failing to muster enough support in Congress to pass on its own, the Quincy Library Plan was attached as a rider on the 1999 Omnibus Appropriation Bill, which was signed into law by President Clinton in 1998. Environmental groups denounced both the method

of passage as well as the content of the plan, claiming that it will double the timber harvest in the affected forest region, and subvert existing environmental laws. They claim it is a corporate welfare handout to the Sierra Pacific Industries, and will open the door to privatizing national forests. Interestingly, a split occurred between some national environmental groups, which opposed excess local control, and their northern California chapters, which defended local knowledge and autonomy.

What do you think? Is community-based planning a recognition of the wisdom and practical experience of local residents, or simply a way to give special favors to local industry? Is this courageous innovation, or simply sleeping with the enemy? Is the fact that participants come to understand and like each other a healthy development or the beginning of a sell-out? Would it be better to maintain an adversarial stance in this case, or seek compromise?

Located in the northern Sierra Nevada mountains, the small town of Quincy, California, is the site of a controversial experiment in community-based resource planning.

The purpose of this project is to develop blueprints for the restoration and maintenance of environmentally degraded harbors and estuaries in ways that are both biologically and socially sustainable. Officially established under Canada's Green Plan and supported by Environment Canada, this program created 13 community groups, some rural and some urban, with membership in each dominated by local residents. Federal and provincial government agencies are represented primarily as nonvoting observers and resource people.

Each community group is provided with core funding for full-time staff who operate an office in the community and facilitate meetings.

Four of the 13 ACAP sites are in the Bay of Fundy, an important and unique estuary lying between New Brunswick and Nova Scotia. Approximately 270 km long, and with an area of more than 12,000 sq km, the bay, together with the nearby Georges Bank and the Gulf of Maine form one of the richest fisheries in the world. With the world's highest recorded tidal range (up to 16 m at maximum

FIGURE 19.15 The Bay of Fundy has the greatest tidal range in the world. It is the site of innovative community-based environmental planning process.

spring tide), the bay sustains a great variety of fishery and wildlife resources, and provides habitat for a number of rare or endangered species. Now home to more than 1 million people, the coastal region is an important agricultural, lumbering, and paper-producing region (Fig. 19.15).

Since European settlement began in 1604, the Bay of Fundy region has experienced great changes in population growth, resource use, and human-induced ecosystem change. More than 80 percent of the saltmarshes present in 1604 have been eliminated or degraded. Pollution and sediment damage harbours and biological communities. Overfishing and introduction of exotic species have resulted in endemic species declines. The collapse of cod, halibut, and haddock fishing has had devastating economic effects on the regional economy and the livelihoods of local residents. Aquaculture is now a more valuable activity than all wild fisheries.

To cope with these complex, intertwined social and biological problems, ACAP is bringing together different stakeholders from around the bay to create comprehensive plans for ecological, economic, and social sustainability. Through citizen monitoring and adaptive management, the community builds social capital (knowledge, cooperative spirit, trust, optimism, working relations), develops a sense of ownership in the planning process, and eliminates some of the fears and sectorial rivalry that often divides local groups, outsiders, and government agents.

On the other hand, giving a greater voice and increased power to local communities isn't always seen as a positive step by those from the outside (What Do You Think?, p. 432).

Green Plans

Several national governments have undertaken integrated environmental planning that incorporates community round-tables for vision development. Canada, New Zealand, Sweden, and Denmark all have so-called **green plans** or comprehensive, long-range national environmental strategies. The best of these plans weave together complex systems, such as water, air, soil, and energy, and mesh them with human factors such as economics, health, and carrying capacity. Perhaps the most thorough and well-thought-out green plan in the world is that of the Netherlands.

Developed in the 1980s through a complex process involving the public, industry, and government, the 400-page Dutch plan contains 223 policy changes aimed at reducing pollution and establishing economic stability. Three important mechanisms have been adopted for achieving these goals: integrated life-cycle management, energy conservation, and improved product quality. These measures should make consumer goods last longer and be more easily recycled or safely disposed of when no longer needed. For example, auto manufacturers are now required to design cars so they can be repaired or recycled rather than being discarded.

Among the guiding principles of the Dutch green plan are: (1) the "stand-still" principle that says environmental quality will not deteriorate, (2) abatement at the source rather than cleaning up afterward, (3) the "polluter pays" principle that says users of a resource pay for negative effects of that use, (4) prevention of unnecessary pollution, (5) application of the best practicable means

FIGURE 19.16 Under the Dutch Green Plan, 250,000 ha of drained agricultural land are being restored to wetland and 40,000 ha are being replanted as woodland.

for pollution control, (6) carefully controlled waste disposal, and (7) motivating people to behave responsibly.

The Netherlands have invested billions of guilders in implementing this comprehensive plan. Some striking successes already have been accomplished. Between 1980 and 1990, emissions of sulphur dioxide, nitrogen oxides, ammonia, and volatile organic compounds were reduced 30 percent. By 1995, pesticide use had been reduced 25 percent from 1988 levels, and chlorofluorocarbon use had been virtually eliminated. By 1998, industrial wastewater discharge into the Rhine River was 70 percent less than a decade earlier. Some 250,000 ha of former wetlands that had been drained for agriculture are being restored as nature

preserves and 40,000 ha of forest are being replanted. This is remarkably generous and foresighted in such a small, densely populated country, but the Dutch have come to realize they cannot live without nature (Fig. 19.16).

Not all goals have been met so far. Planned reductions in CO_2 emissions failed to materialize when cheap fuel prices encouraged fuel-inefficient cars. Currently a carbon tax is being considered. A sudden population increase caused by immigration from developing countries and Eastern Europe also complicates plan implementation, but the basic framework of the Dutch plan has much to recommend it, nevertheless. Other countries would be more sustainable and less environmentally destructive if they were to adopt a similar plan.

SUMMARY

Fragmentation is one word that characterizes resource and environmental management efforts in Canada. Four types of fragmentation were noted—vertical, horizontal, issue and spatial. The Brundtland Report challenges all nations and its citizens to overcome the institutional arrangements that promote a fragmented approach. Instead, decision makers should be exposed to the full consequences of their decisions. In the words of the Brundtland

Commission, "the real world of interlocking economic and economic systems will not change, the policies and institutions must." Time will tell if Canada and the other nations of the world are up to the challenge of reforming constitutional, administrative, policy and financial arrangements to meet the challenge of sustainable development.

QUESTIONS FOR REVIEW

1. What is the difference between common and statute law?
2. What is the significance of the Fisheries Act?
3. What are the four general approaches to voluntary initiatives?
4. Why have some international environmental treaties and conventions been effective while most have not? Describe two such treaties.

5. What are wicked problems? Why are they difficult?
6. What is resilience? Why is it important?
7. Describe adaptive management.
8. What is collaborative, community-based planning?
9. What is unique about the Dutch green plan?

QUESTIONS FOR CRITICAL REVIEW

1. What are three key constitutional obstacles that must be overcome if Canada is to achieve a sustainable development?
2. Assess the ability of common law to provide for effective resource and environmental management.
3. Discuss how "certificates of approval" or permitting systems and economic instruments can be considered as "permissions to pollute."

4. Try creating a list of arguments for and against an international body with power to enforce global environmental laws. Can you see a way to create a body that could satisfy both reasons for and against this power?
5. Think of a familiar example of a wicked environmental problem. What are the most important elements that make it wicked? What institutional changes could we implement to make this issue less wicked?

6. The Holling diagram for the creative-destruction cycle (Fig. 19.13) is described in terms of ecological change. Try applying this model to cycles of change in human institutions. Describe the actors, conditions, and forcing factors in each of the four quadrants of the model.

7. Take a current wicked environmental problem. If you were an environmental leader trying to resolve this problem, would you choose litigation, arbitration, or mediation? What are your reasons for favouring or rejecting each one?

8. Based on the information you have now, do you favour or oppose the Quincy Library Plan? What additional information would you need to make a good decision about this case?

KEY TERMS

adaptive management   429
arbitration   430
covenants   423
Environmental Impact
   Statement (EIS)   415
environmental policy   407
externality   424
federal statute   415
government voluntary
   programs   421

green plans   433
heads of power   411
horizontal
   fragmentation   412
individual company
   initiatives   423
industry initiatives   424
issue or interest
   fragmentation   412
mediation   431
penal sanctions   413
policy   407

precautionary
   principle   430
prior approval   413
private law   406
proprietary rights   410
resilience   429
spatial fragmentation   424
statute laws   406
vertical fragmentation   412
whistle-blower
   legislation   418
wicked problems   428

Environmental Health

I'm so worried about bein' so full of doubt about everything, anyway.

—Eric Idle—

OBJECTIVES

After studying this chapter, you should be able to:

- define *health* and *disease* in terms of some major environmental factors that affect humans.
- identify some major infectious organisms and hazardous agents that cause environmental diseases.
- identify examples of emergent human diseases.
- distinguish between toxic and hazardous chemicals and between chronic and acute exposures and responses.
- compare the relative toxicity of some natural and synthetic compounds as well as report on how such ratings are determined and what they mean.
- evaluate the major environmental risks we face and how risk assessment and risk acceptability are determined.

WebQuest

bioaccumulation, biomagnification, LD50, bioassay, ecosystem health, severe acute respiratory syndrome

Above: Hong Kong residents protect themselves from Severe Acute Respiratory Syndrome (SARS), a highly infectious, pneumonia like emergent disease.

The Cough Heard Round the World

Early in 2003, news began to trickle out of China that a very infectious "atypical pneumonia" was spreading rapidly in hospitals around Guangzhou (formerly known as Canton) in Guangdong Province. Symptoms included fever, chills, headaches, muscle pains, and a dry cough, but in many patients, especially the elderly, the disease would quickly turn into a deadly pneumonia. In February of 2003, a doctor from Guangzhou, who had contracted this disease from his patients, travelled to Hong Kong. There he passed the infection to other travellers, who carried what is now known as **Severe Acute Respiratory Syndrome** (SARS) to Beijing, Canada, Taiwan, Singapore, and Vietnam.

Within six months, SARS spread to 31 countries around the world, where more than 8,500 probable cases and 812 deaths were reported. Fear of SARS travelled even faster and farther than the disease itself. Rumours multiplied across the Internet as conferences and sporting events were cancelled, factories closed, and tourism to China fell by as much as 85 percent after the epidemic was revealed. The economic impact of SARS is estimated to be at least $30 billion (U.S.) in 2003 alone. By July 2003, the World Health Organization (WHO) declared the outbreak contained, but suggested that the world remain vigilant for further infections.

The rapid transmission of this disease shows how interconnected we all are. A virus can travel in just a few days anywhere a plane flies. One flight attendant is thought to have been the source of infection for 160 people in seven countries. SARS also points to the need for better communication and identification of new diseases. Because the Chinese government hid the extent and

FIGURE 20.2 Often cited for his love of Toronto, lead singer of the Rolling Stones Mick Jagger helped Toronto overcome the shadow of SARS by performing a free concert in the city. The Concert for SARS Relief was held in July 2003 and included 15 other acts. More than 500,000 people filled Downsview Park to see the concert, seen by many as a light at the end of the dark tunnel that SARS had put Canada in.

severity of the disease for several months, patients with this highly infectious disease mingled with the general hospital population, spreading the illness. Medical professionals, not knowing how serious the infection was, treated patients without wearing protective clothing. Major hospitals in Beijing, Taipei, Hong Kong, and Singapore were closed because so many staff members were sick, making treatment of the contagion even more difficult.

While globalization helped spread SARS, it also helped in rapid recognition and treatment of the disease. It took centuries to discover the cause of cholera. Identifying the virus that causes AIDS took two years. But within weeks after the WHO issued its first warning about SARS, an electron microscopist at the Centers for Disease Control in Atlanta found a new coronavirus in cell cultures infected with tissue from SARS patients. Less than a month later, labs from Vancouver to Singapore were sequencing its RNA. In May 2003, Scientists at Hong Kong University announced they had found coronavirus nearly identical to those from SARS patients in civets, badgers, and raccoon dogs being sold in Guangdong meat markets.

Wild species had been suspected as a source of the SARS virus since some of the first infections occurred among chefs and animal merchants. Exotic animals are regarded as delicacies in southern China, where they are featured at banquets and dinners at expensive restaurants. In April of 2003, Chinese police raided animal markets and hotels and restaurants, seizing 838,500 animals including many rare and endangered species. The emergence of SARS may reduce animal smuggling, but the existence of a reservoir of this virus in the wild may mean that it will be impossible to completely eradicate the disease.

FIGURE 20.1 In spring 2003, Canada was in the midst of what could be described as its worst disease epidemic in modern times. The World Health Organization advised against travel to Toronto and Vancouver and health care workers who treated the sick were at high risk of contracting SARS themselves. Masks were the first response to the feared airborne virus that caused SARS, especially among health care workers in Vancouver and Toronto.

SARS deaths, so far, are relatively insignificant compared to the 3 million people who die from AIDS or the 1 million who die from malaria every year, but we tend to fear new risks with unknown causes, while ignoring more routine but perhaps more dangerous risks that we believe we can control. Still, the emergence of SARS reminds us how susceptible we remain to infectious diseases in an interconnected world. Dr. Jong-wook Lee, the newly elected head of the WHO said, "SARS is the first new disease of the twenty-first century, but it will not be the last." The U.S. Institute of Medicine warns that gaps in our defences against biological assault from both terrorists and natural sources makes us vulnerable to other deadly epidemics. In this chapter, we'll look at some principles of environmental health to help you understand some of the risks we face and what we might do about them.

TYPES OF ENVIRONMENTAL HEALTH HAZARDS

What is health? The World Health Organization defines **health** as a state of complete physical, mental, and social well-being, not merely the absence of disease or infirmity. By that definition, we all are ill to some extent. Likewise, we all can improve our health to live happier, longer, more productive, and more satisfying lives if we pay attention to what we do.

What is a disease? A **disease** is a deleterious change in the body's condition in response to an environmental factor that could be nutritional, chemical, biological, or psychological. Diet and nutrition, infectious agents, toxic chemicals, physical factors, and psychological stress all play roles in **morbidity** (illness) and mortality (death). To understand how these factors affect us, let's look at some of the major categories of environmental health hazards.

The World Health Organization (http://www.who.int/en/) monitors and researches human health and disease issues around the world. In Canada, the Health, Environment, and Economy program of the Canadian Round Table (http://www.nrtee-trnee.ca/eng/programs/ArchivedPrograms/Health/health_e.htm) considers links between human health and our environment.

Infectious Organisms

For most of human history, the greatest health threats have been pathogenic (disease-causing) organisms and accidents or violence. Although cardiovascular diseases (heart attacks and strokes), cancer, injuries (intentional and unintentional), and other ills of modern life now have become the leading killers almost everywhere in the world, infectious diseases still kill more than 13 million people every year, or about 24 percent of all disease-related deaths (Table 20.1). A majority of these deaths are in the poorer countries of the world where better nutrition, improved sanitation, and inexpensive vaccinations could save millions of lives each year. How does that affect us? Our world is increasingly interconnected. New, extremely virulent diseases like Ebola as well as drug-resistant forms of more familiar diseases such as tuberculosis and cholera can arise in remote areas and then spread rapidly throughout the world.

Before you learn more about some of the terrible illnesses that afflict people, it's worth pointing out that in many ways most people are healthier than ever before. One evidence of this is the

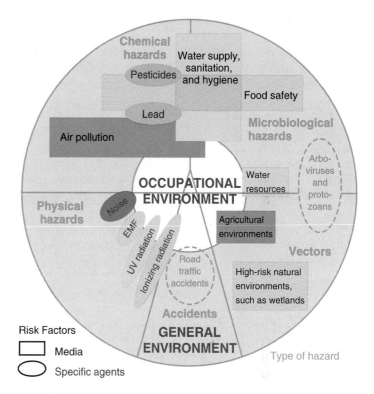

FIGURE 20.3 Major sources of environmental health risks. *Source:* WHO, 2002.

fact that people are living longer almost everywhere. As Chapter 4 points out, life expectancies have increased around the world from about 40 years to 65 years on average over the last century. Although developing countries lag behind the wealthier, industrialized nations, progress in education and health care is being made among some of the world's poorest people.

Cardiovascular disease (heart attacks, strokes) are by far the largest killers in the world now (Table 20.1). Malignant tumours (cancer) and benign tumours have become the second biggest cause of mortality. Of the broad categories of communicable diseases, infectious lung diseases such as tuberculosis cause the most deaths each year. Accidents and trauma (including traffic deaths, poisons, falls, fires, drowning, suicide, wars, and homicide) kill more than 5 million people each year, while perinatal conditions including infections and other problems associated with birth take some

TABLE 20.1	Leading Causes of Death Worldwide

CAUSE	MILLIONS OF DEATHS PER YEAR
Cardiovascular disease	16.9
Cancers and tumours	7.2
Infectious and parasitic diseases	
Acute respiratory diseases*	5.7
HIV/AIDS	3.0
Diarrheal diseases	2.2
Childhood diseases**	1.5
Malaria	1.1
Accidents and trauma	5.1
Chronic respiratory disease	3.5
Maternal and perinatal	2.8
Digestive and nutritional	2.3
Neuropsychiatric disorders	0.9
Diabetes	0.8
Congenital abnormalities	0.7
Other	2.2
Total	55.9

*Includes TB, influenza, etc.
**Includes pertussis, polio, measles, and tetanus.
Source: World Health Organization, 2001.

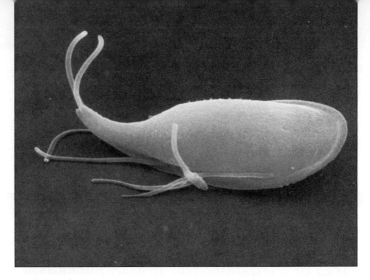

FIGURE 20.4 *Giardia,* a parasitic intestinal protozoan, is resported to be the largest single cause of diarrhea in Canada and the United States. It is spread from human feces through food and water. Even pristine wilderness areas have *Giardia* outbreak due to careless campers.

2.8 million lives. Some 4 *billion* new cases of diarrhea each year are responsible for about 2.2 million deaths, mostly in children under 5 years old. Bacteria, viruses, and protozoa can cause diarrhea (Fig. 20.4). Many children who suffer severe diarrhea early in life subsequently show signs of permanent mental and developmental retardation.

Although virtually unknown 20 years ago, acquired immune deficiency syndrome (AIDS) has now become the largest single cause of communicable deaths in the world, killing an estimated 3 million people in 2000. At least 36 million people—70 percent of them in Africa—are now living with the human immune-deficiency virus (HIV), which causes AIDS. The disease is continuing to spread rapidly with at least 5.3 million new infections occuring each year. In Eastern Europe and Central Asia, for example, the number of recorded cases of HIV infections jumped from 420,000 to 700,000 in just one year. Some of this increase may be better screening, but rapid increases in drug addiction and unsafe sex practices in the region are primary risk factors.

Surely the most tragic epidemic of AIDS has occurred in sub-Saharan Africa, where in countries such as Botswana, Zimbabwe, and Zambia, up to 36 percent of the adult population is now HIV positive. A whole generation of young, productive adults has died, and some villages consist only of the very old or very young. No one is left to support families or keep the economy and society functioning. There may be 10 million African children living with

AIDS or who are orphans whose parents have died from the disease. Heterosexual sex is the main cause of HIV transmission in Africa, and 55 percent of all HIV-positive adults there are women. One sign of hope is that several western firms have offered to sell anti-AIDS drugs in Africa for a fraction of the cost of those same drugs in the United States or Canada. This may make it more feasible to treat AIDS sufferers and to prevent transmission from mothers to newborn children.

A cluster of vaccine-preventable diseases including measles, polio, hepatitis B, and tetanus claim some 1.5 million lives each year, mostly among children in developing countries (Fig. 20.5). Considerable progress is being made in eliminating many of these diseases. Smallpox, for instance, was eradicated in 1977. A massive inoculation campaign in 1999 and 2000 immunized more than 450 million children against polio. An encouraging step in the crusade to eliminate many infectious diseases is the creation of edible vaccines by transferring antigenic genes into crop plants by Dr. Charles Arntzen and his colleagues at the Boyce Thompson Institute. These transgenic crops could be grown very cheaply and would avoid the need for refrigeration, needles, and inoculations. Unfortunately, total U.S. expenditures for research on all infectious diseases amount to only $75 million per year. To put that in perspective, the United States spends $225 million per year to support military bands and about $3 billion per year on beer.

Morbidity and Quality of Life

Death rates don't tell us everything about the burden of disease. Obviously many people suffer from illness who don't die but whose quality of life is severely diminished. When people are sick, work doesn't get done, crops don't get planted or harvested, meals aren't cooked, children can't study and learn. The 2 billion people who suffer from various worms, flukes, and other internal parasites at any given time rarely die from their affliction but the

FIGURE 20.5 About 1.5 million children die every year from easily preventable diseases. This billboard in Guatemala encourages parents to have their children vaccinated against polio, diphtheria, TB, tetanus, pertussis (whooping cough), and scarlet fever (1 to r).

TABLE 20.2 Leading Causes of Global Disease Burden

RANK	1990	RANK	2020
1	Pneumonia	1	Heart disease
2	Diarrhea	2	Depression
3	Perinatal conditions	3	Traffic accidents
4	Depression	4	Stroke
5	Heart disease	5	Chronic lung disease
6	Stroke	6	Pneumonia
7	Tuberculosis	7	Tuberculosis
8	Measles	8	War
9	Traffic accidents	9	Diarrhea
10	Birth defects	10	HIV/AIDS
11	Chronic lung disease	11	Perinatal conditions
12	Malaria	12	Violence
13	Falls	13	Birth defects
14	Iron anemia	14	Self-inflicted injuries
15	Malnutrition	15	Respiratory cancer

Source: World Health Organization, 2002.

suffering and debilitation is real, nonetheless. Statistics on the total economic and social consequences of diseases are difficult to obtain, but the annual toll is clearly very high (Table 20.2).

Over the years health agencies have attempted to develop new ways to measure disability or loss of quality life along with mortality. One of the newest—and still controversial—measures is the **Disability-Adjusted Life Year** or **DALY.** DALYs combine premature deaths and loss of a healthy life resulting from illness or disability. This is an attempt to evaluate the total burden of disease, not just how many people die. Clearly, many more years of expected life are lost when a child dies of neonatal tetanus than when a 70-year-old dies of pneumonia. Similarly, a child permanently paralyzed by polio will have many more years of suffering and lost potential than will a senior citizen who has a stroke. According to the World Health Organization World Health Report 2000, communicable diseases are responsible for nearly half of all the 1.4 billion DALYs lost each year, and children under age 15 account for at least half of that total.

The heaviest burden of illness is borne by the poorest people who can afford neither a healthy environment nor adequate health care. About 90 percent of all DALY losses occur in the developing world where less than one-tenth of all health care dollars is spent. Sub-Saharan Africa, for example, has about twice as many people as Canada plus the United States, but suffers 100 times the DALY losses associated with maternal and perinatal conditions. Similarly, Africa has more than 1,000 times the DALY losses from childhood infectious diseases than North America. The 1.3 billion people worldwide who live on less than $1 per day generally lack access to adequate housing, sanitation, and safe drinking water, all of which increase their exposure to pathogens such as those responsible for the roughly 4 billion cases of diarrhea each year. Malnutrition exacerbates many diseases. Well-fed children rarely die from diarrhea or tuberculosis. Better garbage and human waste disposal,

clean water, inexpensive childhood inoculations, and a number of other interventions could feasibly prevent somewhere between 10 and 40 percent of this disease burden.

Somewhere around 2 billion people at any given time suffer from worms, flukes, protozoa, and other internal parasites. While victims rarely die from these infections, their quality of life clearly suffers. People whose nutrition is marginal at best can ill afford to donate blood to uninvited guests. The combination of malnutrition and anemia from blood loss can retard normal childhood growth and development. They also can impair the immune system, leaving afflicted children and adults less able to battle common diseases such as measles, diarrhea, respiratory infections, tuberculosis, and malaria.

Progress is being made, however, in eradicating some dreadful parasitic diseases. Onchocerciasis (river blindness) is caused by tiny nematodes (round worms) transmitted by the bite of black flies. Masses of dead worms accumulate in the eyeball, destroying vision. This dreadful disease affects some 18 million people and permanently blinds about 500,000 every year. In some African villages, nearly every adult over 30 is blind from onchocerciasis (Fig. 20.6). In a multinational effort, governments of affected countries working together with the United Nations and the World Bank have begun to control the blackfly by insecticide sprays. In addition, the U.S. pharmaceutical manufacturer Merck & Co. has committed to providing ivermectin—a drug that safely and effectively kills the larvae in the body—free of charge as long as river blindness exists. Similarly, SmithKline Beecham has pledged to freely distribute the drug albendazole to about 1 billion people in an effort to eliminate the grossly disfiguring disease known as

FIGURE 20.6 A child leads a blind adult in West Africa. River blindness, caused by tiny nematodes that are transmitted by biting flies, affects 18 million people. In some African villages, nearly every adult over age 30 is blind due to this disease.

elephantiasis (see Fig. 15.8) that now afflicts about 120 million people. This effort will cost the company about $500 million over the next two decades. Drancunculiasis or guinea worm, another dreadful tropical disease, is being eliminated through community health education and provision of safe drinking water.

Emergent Diseases and Environmental Change

Although many diseases such as measles, pneumonia, and pertussis (whooping cough) have probably inflicted humans for millenia, at least 30 new infectious diseases have appeared in the past two decades while many well-known diseases have reappeared in more virulent, drug-resistant forms. Frightening maladies such as Ebola and Marburg fever have made sudden and highly lethal outbreaks in many places around the world (Fig. 20.7). An **emergent disease** is one never known before or one that has been absent for at least 20 years.

Probably the largest loss of life from an individual disease in a single year was the great influenza epidemic of 1918. Somewhere between 30 and 40 million people succumbed to this virus in less than 12 months. This was more than twice the total number killed in all the battles of World War I, which was occurring at the time. Crowded, unsanitary troop ships carrying American soldiers to Europe started the epidemic. War refugees, soldiers from other nations returning home, and a variety of other travellers quickly spread the virus around the globe. Flu is especially contagious, spreading either by direct contact with an infected person or by breathing airborne particles released by coughing or sneezing. Most flu strains are zoonotic (transmitted from an animal host

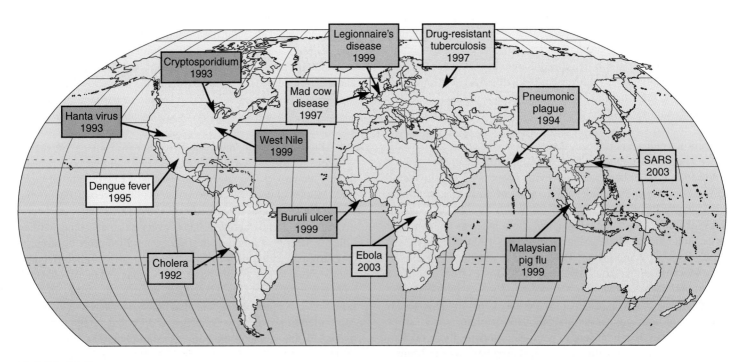

FIGURE 20.7 Some recent outbreaks of highly lethal, infectious diseases. Why are supercontagious organisms emerging in so many different places?
Source: U.S. Centers for Disease Control and Prevention.

to humans). Pigs, birds, monkeys, and rodents often serve as reservoirs from which viruses can jump to humans. Although new flu strains seem to appear nearly every year, we're fortunate that no epidemic has been as deadly as that of 1918, but we never know what might be waiting for the right moment to strike.

The story of SARS that introduced this chapter is a good example of an emergent disease. Although coronaviruses have long been known to cause a variety of diseases—some lethal—in animals, and two members of this virus family cause about 30 percent of all human colds, the particularly virulent form that appears to have jumped from wild animals to humans in southern China had been previously unknown to science. Figure 20.7 shows some recent outbreaks of emergent diseases around the world. Ebola, which can kill 90 percent of those infected, is thought to have spread to humans in central Africa when infected monkeys or chimpanzees were eaten.

Some health experts warn that the West Nile virus may be more deadly for Canada and the U.S. than SARS. The West Nile virus belongs to a family of mosquito-transmitted viruses that cause encephalitis, or brain inflammation. Although recognized in Africa in 1937, the West Nile virus was absent from North America until 1999, when a bird or mosquito from the Middle East apparently introduced it. In only two years, the disease spread rapidly from New York, where it was first reported, through the eastern U.S. and up to Canada. The virus infects 230 species of animals, including 130 bird species. By September 2002, the National Microbiology Laboratory in Winnipeg had confirmed a West Nile virus infection in one Ontario citizen. It was the first confirmed case in Ontario although it was already known that the virus was circulating through the province of Ontario—possibly related to cross-border travel into the U.S. See Figure 20.8.

The largest recent death toll from an emergent disease is HIV/AIDS. Although virtually unknown 15 years ago, acquired immune deficiency syndrome has now become the fifth leading cause of contagious deaths. The WHO estimates that 60 million people are now infected with HIV and that 3 million die every year from AIDS complications. Although two-thirds of all current HIV infections are now in sub-Saharan Africa, the disease is spreading rapidly in South and East Asia. Over the next 20 years, there could be an additional 65 million AIDS deaths.

In Botswana, health officials estimate about 40 percent of all adults are HIV-positive and that two-thirds of all current 15-year-olds will die of AIDS before age 50. Botswana's average life expectancy is now 36.1 years. Worldwide, more than 14 million children have lost one or both parents to AIDS. The economic costs of treating patients and lost productivity from premature deaths are estimated to be at least $35 billion (U.S.) per year.

A number of factors contribute currently to the appearance and spread of contagious diseases. With a population of 6 billion, human densities are much higher, enabling germs to spread further and faster than ever before. Expanding populations push into remote areas, hunting animals that once would have never seen humans, converting lands to agriculture, and being exposed to exotic pathogens and parasites. We have caused environmental change on a massive scale: cutting forests, polluting air and water, creating unhealthy urban

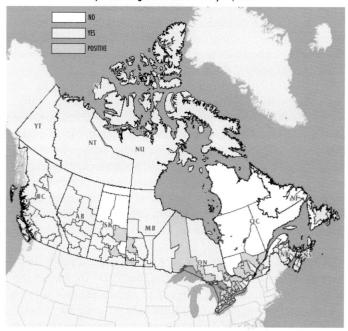

Dead Birds Submitted for West Nile Virus Diagnosis by Health Region Canada as of July 14, 2004

FIGURE 20.8 Birds submitted for diagnosis (by Health Region).
Source: Health Canada Population and Public Health Branch.

surroundings, causing local—and perhaps global—climate change. Nutrients and pesticides running off farm fields cause blooms of toxic algae in lakes and near-shore ocean bodies. Elimination of predators and habitat changes favour disease-carrying organisms such as mice, rats, cockroaches, and mosquitoes.

Dengue or breakbone fever is spread by mosquitoes. In the 1960s about 30,000 cases were reported each year. The WHO believes that some 20 million new infections now occur annually and that at least 2.5 billion people live in areas where mosquitoes carry this disease. An important factor in the spread of many diseases is the speed and frequency of modern travel. Millions of people go every day from one place to another by airplane, boat, train, or automobile. Very few places on earth are more than 24 hours by jet plane from any other place.

In 2001, a woman flying from the Congo arrived in Canada delirious with a high fever. She didn't, in fact, have Ebola, but Canadian officials were very nervous for a few days. What if she had been carrying the virus? How many people might she have infected in her 18-hour flight?

Humans aren't the only species to suffer from emergent and resurgent diseases. In February 2001, an outbreak of foot and mouth disease was reported on a farm near London, England. Within a month, it had infected pigs, sheep, and cows on more than 700 British farms. The virus, which is carried on human clothes, shoes, and hair, also can travel up to 60 km by air. All livestock within 3 km of affected farms were destroyed, but the infection quickly spread to Ireland, the Netherlands, and France. Altogether, the four countries ordered more than half a million animals slaughtered, burned, and buried in an effort to contain the disease. Faced with the prospect of killing up to 30 million head of livestock, Britain has turned to a massive inoculation program. British detectives traced the outbreak to pig swill made from garbage containing illegally imported meat.

Before foot and mouth hit, the European livestock industry was already reeling from an epidemic of bovine spongiform encephalopathy (BSE), better known as mad cow disease. Caused by a mysterious, nonliving but infectious protein called a prion, this disease kills nerve cells and leaves large, spongy holes in the brain. Prions have been known for many years to infect sheep and cause a disease called scrapie. The disease was probably transmitted to cattle by feeding them the ground-up carcasses of infected sheep. It was first detected in British cows in the 1980s and then spread to beef cattle in France, Germany, Portugal, Spain, and Switzerland. It is suspected as well in about a dozen other countries. Over the past decade, at least 90 people—mostly in Britain—have died from Creuztfeldt-Jacob Disease (CJD), the human form of this illness, which is spread by eating contaminated meat. Prions are extremely hardy; it takes at least 1,000°C to destroy them. In an effort to control the disease, nearly 5 million European cows and sheep have been slaughtered.

The North American cattle market, which is highly integrated between Canada and the United States, was thrown into chaos when in 2003 a BSE-infected cow was discovered on a farm in Alberta. The border was closed to Canadian cattle and large scale consumer concern about BSE and CJD resulted in very difficult times for Canadian beef farmers.

A BSE-like illness, called chronic wasting disease, has infected wild and domestic elk and deer herds in Saskatchewan, Colorado, Wyoming, and South Dakota. At least two hunters who died of Creuztfeld-Jacob Disease may have been exposed to chronic wasting disease, although it's impossible to prove the connection. A recent study at the University of Pittsburgh suggested that about 5 percent of the people who were thought to have died from Alzheimer's may have had CJD. If this is true of the general population, e.g., some 200,000 of the supposed Alzheimer's patients in the United States might really have CJD, we may be in the midst of a major prion epidemic.

Antibiotic and Pesticide Resistance

Malaria, the most deadly of all insect-borne diseases, is an example of the return of a disease that once was thought to be nearly vanquished. Malaria now claims about 3 million lives every year—90 percent in Africa and most of them children. With the advent of modern medicines and pesticides, malaria had nearly been wiped out in many places but recently has come roaring back. The protozoan parasite that causes the disease is now resistant to most antibiotics, while the mosquitoes that transmit it have developed resistance to many insecticides. Spraying of DDT in India and Sri Lanka reduced malaria from millions of infections per year to only a few thousand in the 1950s and 1960s. Now South Asia is back to

its pre-DDT level of some 2.5 million new cases of malaria every year. Other places that never had malaria or dengue now have them as a result of climate change and habitat alteration. Gulf-coast states in America, for example, are now home to the *Anopheles* mosquito that carries these diseases.

Why have vectors such as mosquitoes and pathogens such as the malaria parasite become resistant to pesticides and antibiotics? Part of the answer is natural selection and the ability of many organisms to evolve rapidly. Another factor is the human tendency to use control measures carelessly. When we discovered that DDT and other insecticides could control mosquito populations, we spread them indiscriminately without much thought to ecological considerations. In the same way, antimalarial medicines such as chloroquine were given to millions of people, whether they showed symptoms or not. This was a perfect recipe for natural selection. Many organisms were exposed only minimally to control measures. This allowed those with natural resistance to outcompete others and spread their genes through the population (Fig. 20. 9). After repeated cycles of exposure and selection, many microorganisms and their vectors are insensitive to almost all our weapons against them.

There are many examples of drug resistance in pathogens. Tuberculosis, once the foremost cause of death in the world, had nearly been eliminated—at least from the developed world—by the end of the twentieth century. Drug-resistant varieties of TB are now spreading rapidly, however. One of the places these strains arise is in Russia, where crowded prisons with poor sanitation, little medical care, gross overcrowding, and inadequate nutrition serve as a breeding ground for this deadly disease. Inmates who are treated with antibiotics rarely get a complete dose. Those with TB aren't segregated from healthy inmates. Patients with active TB are released from prison and sent home to spread the disease further. And migrants fleeing from Russian economic and social chaos carry the disease to other countries.

Another frightening development is the appearance of drug-resistant strains of common bacteria, such as *Streptococcus pyogenes* (the cause of strep throat) and *Staphylococcus aureus,* the most common form of hospital-acquired infections. Especially virulent forms of these bacteria occasionally appear. Toxic-shock syndrome, in which staphylococcus toxins spread through the body, sometimes brings death in a matter of hours. "Flesh-eating bacteria," generally a strain of streptococcus, creates gruesome wounds as it rapidly destroys living flesh. Hard-to-control strains of both strep and staph are increasingly common. According to the Centers for Disease Control, half of hospital-contracted staph infections in 1997 were resistant to most antibiotics, up from just two percent in 1974.

Toxic Chemicals

Dangerous chemical agents are divided into two broad categories: hazardous and toxic. **Hazardous** means dangerous. This category includes flammables, explosives, irritants, sensitizers, acids, and caustics. Many chemicals that are hazardous in high concentrations are relatively harmless when dilute. **Toxins** are poisonous. This means they react with specific cellular components to kill cells. Because of this specificity, they often are harmful even in dilute

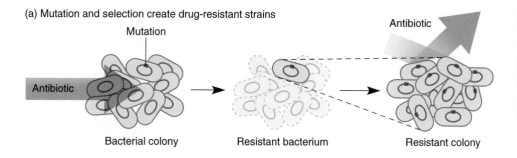

(a) Mutation and selection create drug-resistant strains

Mutation

Antibiotic

Antibiotic

Bacterial colony   Resistant bacterium   Resistant colony

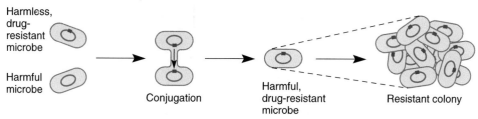

(b) Conjugation transfers drug resistance from one strain to another

Harmless, drug-resistant microbe

Harmful microbe

Conjugation

Harmful, drug-resistant microbe

Resistant colony

FIGURE 20.9 How microbes acquire antibiotic resistance. Random mutations make a few cells resistant. When challenged by antibiotics, only those cells survive to give rise to a resistant colony. Sexual reproduction (conjugation) or plasmid transfer moves genes from one strain or species to another.
Source: From Dr. Steven Morse, Rockefeller University in Time, September 12, 1994. Copyright © Time, Inc.

TABLE 20.3	Top 20 Toxic and Hazardous Substances

1. Arsenic	10. Chloroform
2. Lead	11. DDT
3. Mercury	12. Aroclor1254
4. Vinyl chloride	13. Aroclor 1260
5. Benzene	14. Trichloroethylene
6. Cadmium	15. Dibenz(a,h)anthacene
7. Benzo(a)pyrene	16. Dieldrin
8. Polycyclic aromatic hydrocarbons	17. Chromium, Hexavalent
	18. Chlordane
9. Benzo(b)fluoranthene	19. Hexachlorobutadiene

Source: U.S. Environmental Protection Agency, 2003.

concentrations. Toxins can be either general poisons that kill many kinds of cells, or they can be extremely specific in their target and mode of action. Ricin, for instance, is a protein found in castor beans and one of the most toxic organic compounds known. Three hundred picograms (trillionths of a gram) injected intravenously is enough to kill an average mouse. A single molecule can kill a cell. This is about 200 times more lethal than any dioxin, which often are claimed to be the most toxic substances known. Table 20.3 shows some of the toxic chemicals and elements of greatest concern to the EPA.

Allergens are substances that activate the immune system. Some allergens act directly as **antigens;** that is, they are recognized as foreign by white blood cells and stimulate the production of specific antibodies. Other allergens act indirectly by binding to other materials and changing their structure or chemistry so they become antigenic and cause an immune response.

Formaldehyde is a good example of a widely used synthetic chemical that is a powerful sensitizer. It is both directly and indirectly allergenic. Some people who are exposed to formaldehyde in plastics, wood products, insulation, glue, fabrics, and a variety of other products become hypersensitive not only to formaldehyde itself but also to many other materials in their environment, sometimes called the "sick building" syndrome. These individuals may have to go to great lengths to protect themselves from these allergenic substances.

Immune system depressants are pollutants that seem to suppress the immune system rather than activate it. Little is known about how this occurs or which chemicals are responsible. Immune system failure is thought to have played a role, however, in widespread deaths of seals in the North Atlantic and of dolphins in the Mediterranean. These dead animals generally contain high levels of pesticide residues, polychlorinated biphenyls (PCBs), and other contaminants, that may disrupt normal endocrine hormone functions (see What Do You Think? Chapter 15) and make them susceptible to a variety of opportunistic infections. Similarly, some humans with "sick house" syndrome or other environmental illnesses seem to have defective immune responses. Demonstrating a clear cause-and-effect relationship in these cases usually is difficult, however. Exactly which pollutants affect which people often remains unclear.

Neurotoxins are a special class of metabolic poisons that specifically attack nerve cells (neurons). The nervous system is so important in regulating body activities that disruption of its activities is especially fast-acting and devastating. Different types of neurotoxins act in different ways. Heavy metals such as lead and mercury kill nerve cells and cause permanent neurological damage. Anesthetics (ether, chloroform, halothane, etc.) and chlorinated hydrocarbons (DDT, Dieldrin, Aldrin) disrupt nerve cell membranes necessary for nerve action. Organophosphates (Malathion, Parathion) and carbamates (carbaryl, zeneb, maneb) inhibit acetylcholinesterase, an enzyme that regulates signal transmission

Tips for Staying Healthy

- Eat a balanced diet with plenty of fresh fruits, vegetables, legumes, and whole grains. Wash fruits and vegetables carefully, they may well have come from a country where pesticide and sanitation laws are lax.

- Use unsaturated oils such as olive or canola rather than hydrogenated or semisolid fats such as margarine.

- Cook meats and other foods at temperatures high enough to kill pathogens; clean utensils and cutting surfaces; store food properly.

- Wash your hands frequently. You transfer more germs from hand to mouth than any other means of transmission.

- When you have a cold or flu, don't demand antibiotics from your doctor—they aren't effective against viruses.

- If you're taking antibiotics, continue for the entire time prescribed—quitting as soon as you feel well is an ideal way to select for anti-biotic-resistant germs.

- Practise safe sex.

- Don't smoke and avoid smoky places.

- If you drink, do so in moderation. Never drive when your reflexes or judgement are impaired.

- Exercise regularly: walk, swim, jog, dance, garden. Do something you enjoy that burns calories and maintains flexibility.

- Get enough sleep. Practise meditation, prayer, or some other form of stress reduction.

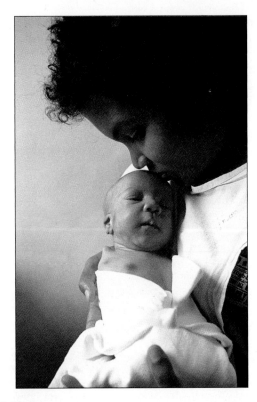

FIGURE 20.10 Development of this baby's arms and legs was blocked when its mother took the sedative thalidomide early in her pregnancy. Although the drug has been banned in Europe and North America for 20 years, it is still used to treat leprosy in some countries. Unfortunately, some of this potent teratogen is used by pregnant women who are unaware of its tragic side effects.

between nerve cells and the tissues or organs they innervate (for example, muscle). Most neurotoxins are both extremely toxic and fast-acting.

Mutagens are agents, such as chemicals and radiation, that damage or alter genetic material (DNA) in cells. This can lead to birth defects if the damage occurs during embryonic or fetal growth. Later in life, genetic damage may trigger neoplastic (tumour) growth. When damage occurs in reproductive cells, the results can be passed on to future generations. Cells have repair mechanisms to detect and restore damaged genetic material, but some changes may be hidden, and the repair process itself can be flawed. It is generally accepted that there is no "safe" threshold for exposure to mutagens. Any exposure has some possibility of causing damage.

Teratogens are chemicals or other factors that specifically cause abnormalities during embryonic growth and development. Some compounds that are not otherwise harmful can cause tragic problems in these sensitive stages of life. One of the most well-known examples of teratogenesis is that of the widely used sedative thalidomide. In the 1960s, thalidomide (marketed under the trade name Cantergan) was the most widely used sleeping pill in Europe. It seemed to have no unwanted effects and was sold without prescription. When used by pregnant women, however, it caused abnormal fetal development resulting in phocomelia (meaning seal-like limbs), in which there is a hand or foot, but no arm or leg (Fig. 20.10). There is evidence that taking a single thalidomide pill in the first weeks of pregnancy is sufficient to cause these tragic birth defects. Altogether, at least 12,000 children were affected before this drug was withdrawn from the market. Fortunately, thalidomide was not approved for sale in the United States because the Food and Drug Administration was not satisfied with the laboratory tests of its safety.

Ironically, thalidomide has positive as well as negative features. The drug has been found to be effective in treating leprosy and is being tested against AIDS, cancer, retinal degeneration, and tissue rejection in organ transplants. Tragically, these beneficial applications continue to have a dark side. In Brazil, where thalidomide has been used widely to treat leprosy, some doctors failed to warn patients about the dangers of becoming pregnant while on the drug. Other people, hearing about miraculous cures with thalidomide, obtained it from unlicensed laboratories without knowing about its side effects. In 1994, more than 50 cases of thalidomide-related birth defects were reported in Brazil.

Perhaps the most prevalent teratogen in the world is alcohol. Drinking during pregnancy can lead to **fetal alcohol syndrome**—a cluster of symptoms including craniofacial abnormalities, developmental delays, behavioural problems, and mental defects that last throughout a child's life. Even one alcoholic drink a day during pregnancy has been associated with decreased birth weight.

Carcinogens are substances that cause **cancer,** invasive, out-of-control cell growth that results in malignant tumors. Cancer rates rose in most industrialized countries during the twentieth century, and cancer is now the second leading cause of death in the United States, killing 539,533 people in 1999. According to the American Cancer Society, 1 in 2 males and 1 in 3 females in the United States will have some form of cancer in their lifetime. Some authors blame this cancer increase on toxic synthetic chemicals in their environment and diet (Fig. 20.11). Others argue that it is attributable mainly to lifestyle (smoking, sunbathing, alcohol) or simply living longer.

If the number of deaths from cancer is adjusted for age, the only major types that have become more prevalent in recent years are prostate cancer in men and lung cancer in women (Fig. 20.12). In fact, smoking has become the leading cause of death for women in the United States. Every year, 165,000 women die from smoking-related illnesses. This is 39 percent of all female deaths, or 27,000 more than breast cancer, the next biggest killer for women. With lung cancer among women up six-fold since 1950, the slogan, "You've come a long way, baby!" takes on a more sinister meaning. The average loss of life associated with smoking is 14 years, a penalty not readily appreciated by young women just starting the habit.

Natural and Synthetic Toxins

There has been so much bad news lately about the dangers of industrial chemicals that some people assume that all human-made compounds are poisonous while all natural materials must be benign and innocuous. In fact, many natural chemicals are very dangerous while many synthetic ones are relatively harmless. Since most plants and many animal species can't escape from predators or defend themselves by fighting back, many of them have evolved a kind of chemical warfare, secreting or storing in their tissues a vast arsenal of irritants, toxins, metabolic disrupters, and other chemicals that discourage competitors and predators. Some of the chemical defences employed by organisms are very sophisticated and specific. Both plants and animals make chemicals similar to—or even identical with—neurotransmitters, hormones, or regulatory molecules of predators or potential enemies. Our cells don't distinguish whether these chemicals are natural or synthetic.

You may have heard the argument that we evolved together with natural toxins whereas synthetic chemicals are so new that we haven't had time to adapt to them. But, in fact, natural chemicals such as arsenic and cyanide have been around longer than we have but are still toxic to us. Organic compounds like solanine, caffeic acid, and oxalic acid are common in many of the foods we eat but we haven't developed an immunity to them.

Toxicologist Bruce Ames claims that there are 10,000 times as many natural pesticides in our diets as synthetic ones. He argues that our fear of synthetic chemicals may divert our attention from more important issues. He finds natural compounds in crops such as potatoes, tomatoes, coffee, celery, and mushrooms, for instance, that are more carcinogenic than some commercial products. Ironically, plants attacked by insects may synthesize natural toxins that are more dangerous than the residues left from protective treatment with synthetic pesticides. Similarly, treatment of crops with fungicides (many of which are carcinogenic) may prevent growth of molds that are even more carcinogenic. Simply because food is raised "organically" may not necessarily make it safer than food raised by current commercial practices.

Other environmental health specialists, however, argue that we should not underrate the dangers of toxic synthetic chemicals. Natural chemicals in our diet are significantly different when mixed with fibre and a multitude of other substances than they may appear as pure chemicals in laboratory tests. Broccoli, for example, clearly contains carcinogens, but reduces the risk of cancer when added to the diet of laboratory animals because it also contains anticancer factors.

Diet

Diet also has an important effect on health. For instance, there is a strong correlation between cardiovascular disease and the amount of salt and animal fat in one's diet.

Fruits, vegetables, whole grains, complex carbohydrates, and dietary fibre (plant cell walls) often have beneficial health effects. Certain dietary components, such as pectins; vitamins A, C, and E; substances produced in cruciferous vegetables (cabbage, broccoli, cauliflower, brussels sprouts); and selenium, which we get from plants, seem to have anticancer effects.

Eating too much food is a significant dietary health factor in developed countries. Almost 50 percent of Canadian adults are now considered obese, and the worldwide total of obese or overweight people is estimated to be more than 1 billion. Perhaps more disturbing is the increasing number of young people in Canada who are overweight. About 50 percent of children from ages 7 to 13 are overweight or obese, compared to about 20 percent in 1981. Obesity increases the chance of type 2 diabetes and heart disease.

MOVEMENT, DISTRIBUTION, AND FATE OF TOXINS

There are many sources of toxic and hazardous chemicals in the environment and many factors related to each chemical itself, its route or method of exposure, and its persistence in the environment, as well as characteristics of the target organism (Table 20.4), that determine the danger of the chemical. We can think of an ecosystem as a set of interacting compartments between which a chemical moves, based on its molecular size, solubility, stability, and reactivity (Fig. 20.13). The routes used by chemicals to enter our bodies also play important roles in determining toxicity (Fig. 20.14). In this section, we will consider some of these characteristics and how they affect environmental health.

Solubility

Solubility is one of the most important characteristics in determining how, where, and when a toxic material will move through

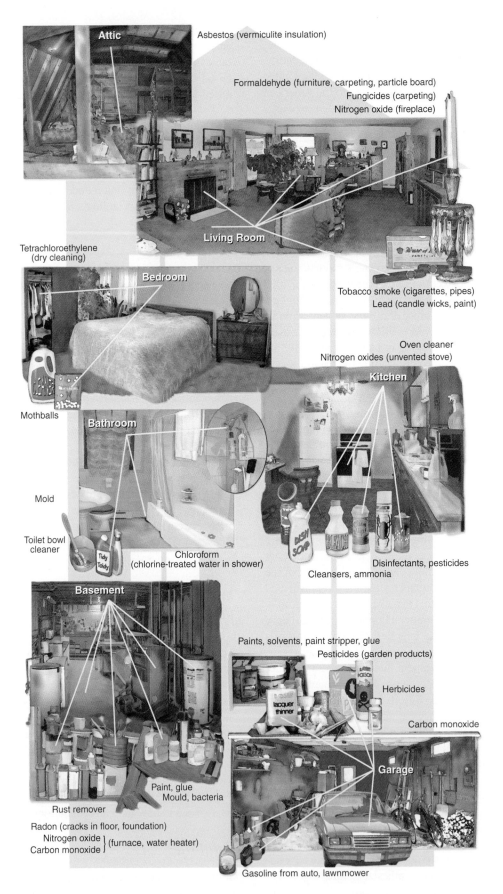

Asbestos (vermiculite insulation)

Formaldehyde (furniture, carpeting, particle board)
Fungicides (carpeting)
Nitrogen oxide (fireplace)

Attic

Living Room

Tetrachloroethylene
(dry cleaning)

Bedroom

Tobacco smoke (cigarettes, pipes)
Lead (candle wicks, paint)

Oven cleaner
Nitrogen oxides (unvented stove)

Kitchen

Mothballs

Bathroom

Mold

Toilet bowl
cleaner

Chloroform
(chlorine-treated water in shower)

Disinfectants, pesticides

Cleansers, ammonia

Basement

Paints, solvents, paint stripper, glue

Pesticides (garden products)

Herbicides

Carbon monoxide

Garage

Paint, glue
Mould, bacteria

Rust remover

Radon (cracks in floor, foundation)
Nitrogen oxide } (furnace, water heater)
Carbon monoxide

Gasoline from auto, lawnmower

FIGURE 20.11 Some sources of toxic and hazardous substances in a typical home.

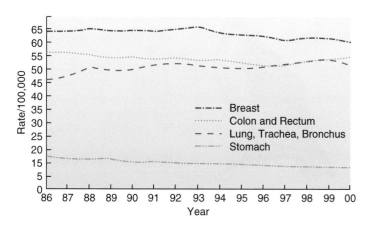

FIGURE 20.12 Age-adjusted cancer incidence over time in Canada between 1986 and 2000. Better treatment means more people survive. When adjusted for an aging population, mortality for most major cancers has been stable or falling.
Source: Health Canada Population and Public Health Branch.

TABLE 20.4	Factors in Environmental Toxicity

FACTORS RELATED TO THE TOXIC AGENT

1. Chemical composition and reactivity
2. Physical characteristics (such as solubility, state)
3. Presence of impurities or contaminants
4. Stability and storage characteristics of toxic agent
5. Availability of vehicle (such as solvent) to carry agent
6. Movement of agent through environment and into cells

FACTORS RELATED TO EXPOSURE

1. Dose (concentration and volume of exposure)
2. Route, rate, and site of exposure
3. Duration and frequency of exposure
4. Time of exposure (time of day, season, year)

FACTORS RELATED TO ORGANISM

1. Resistance to uptake, storage, or cell permeability of agent
2. Ability to metabolize, inactivate, sequester, or eliminate agent
3. Tendency to activate or alter nontoxic substances so they become toxic
4. Concurrent infections or physical or chemical stress
5. Species and genetic characteristics of organism
6. Nutritional status of subject
7. Age, sex, body weight, immunological status, and maturity

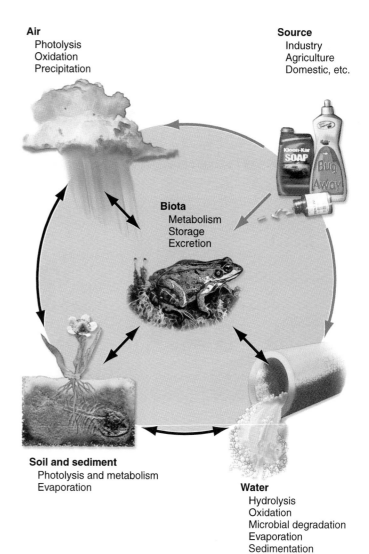

FIGURE 20.13 Movement and fate of chemicals in the environment. Processes that modify, remove, or sequester compounds are shown in parentheses. Toxins also move directly from a source to soil and sediment.

the environment or through the body to its site of action (What Do You Think? p. 451). Chemicals can be divided into two major groups: those that dissolve more readily in water and those that dissolve more readily in oil. Water-soluble compounds move rapidly and widely through the environment because water is ubiquitous. They also tend to have ready access to most cells in the body because aqueous solutions bathe all our cells. Molecules that are oil- or fat-soluble (usually organic molecules) generally need a carrier to move through the environment, into, and within, the body. Once inside the body, however, oil-soluble toxins penetrate readily into tissues and cells because the membranes that enclose cells are themselves made of similar oil-soluble chemicals. Once they get inside cells, oil-soluble materials are likely to be accumulated and stored in lipid deposits where they may be protected from metabolic breakdown and persist for many years.

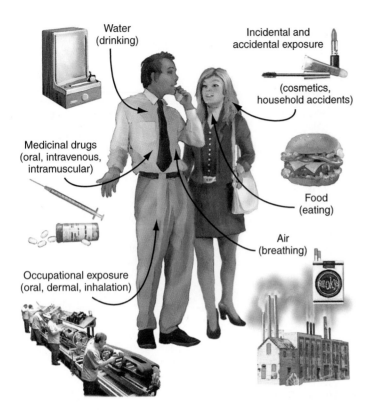

FIGURE 20.14 Routes of exposure to toxic and hazardous environmental factors.

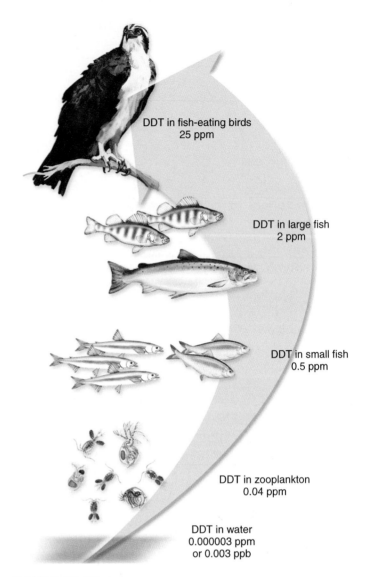

DDT in fish-eating birds
25 ppm

DDT in large fish
2 ppm

DDT in small fish
0.5 ppm

DDT in zooplankton
0.04 ppm

DDT in water
0.000003 ppm
or 0.003 ppb

FIGURE 20.15 Bioaccumulation and biomagnification. Organisms lower on the food chain take up and store toxins from the environment. They are eaten by larger predators, who are eaten, in turn, by even larger predators. The highest members of the food chain can accumulate very high levels of the toxin.

Bioaccumulation and Biomagnification

Cells have mechanisms for **bioaccumulation,** the selective absorption and storage of a great variety of molecules. This allows them to accumulate nutrients and essential minerals, but at the same time, they also may absorb and store harmful substances through these same mechanisms. Toxins that are rather dilute in the environment can reach dangerous levels inside cells and tissues through this process of bioaccumulation.

The effects of toxins also are magnified in the environment through food webs. **Biomagnification** occurs when the toxic burden of a large number of organisms at a lower trophic level is accumulated and concentrated by a predator in a higher trophic level. Phytoplankton and bacteria in aquatic ecosystems, for instance, take up heavy metals or toxic organic molecules from water or sediments (Fig. 20.15). Their predators—zooplankton and small fish—collect and retain the toxins from many prey organisms, building up higher concentrations of toxins. The top carnivores in the food chain—game fish, fish-eating birds, and humans—can accumulate such high toxin levels that they suffer adverse health effects (Chapter 11). One of the first known examples of bioaccumulation and biomagnification was DDT, which accumulated through food chains so that by the 1960s it was shown to be interfering with reproduction of peregrine falcons, brown pelicans, and other predatory birds at the top of their food chains (Chapter 15).

Persistence

Some chemical compounds are very unstable and degrade rapidly under most environmental conditions so that their concentrations decline quickly after release. Some of the modern herbicides, for instance, quickly lose their toxicity. Other substances are more persistent and last for long times. Some of the most useful chemicals such as chlorofluorocarbons, PVC plastics, chlorinated hydrocarbon pesticides, and asbestos are valuable because they are resistant to degradation. This stability also causes problems because these materials persist in the environment and have unexpected effects far from the sites of their original use. DDT, for instance, is a useful pesticide because it breaks down very slowly and doesn't have to be reapplied very often. Its toxic effects may spread to

Soft Vinyl Toys and Medical Supplies

Are soft vinyl baby toys and medical products safe or should they be banned? Rival groups on opposite sides of this question have recently issued numerous reports denouncing each other and attacking the motives, methods, integrity, and conclusions of their opponents. How can we know whom to believe or what to do?

At stake in this controversy is a wide range of useful products ranging from medical IV bags, catheters, disposable gloves, and surgical equipment, to squeezable children's toys, baby bottles, rattles, and teething rings. What these things have in common is that they are made of polyvinyl chloride (PVC) softened with phthalate (pronounced "thalate") plasticizers. There are many phthalates, but the two most common are di (2-ethyl-hexyl) phthalate (DEHP)—found generally in medical devices—and diisononyl phthalate (DINP), which is used primarily in baby toys. These compounds can make up as much as half the weight of some soft vinyls. Because plasticizers don't bind chemically to the polymer, they can diffuse out of the final product under the right conditions.

Both DEHP and DINP are known to be toxic at high doses to laboratory animals, having been linked to a variety of illnesses including reproductive abnormalities as well as kidney and liver damage and possibly some cancers. In addition, these phthalates are known to disrupt important endocrine hormone functions in laboratory animals. The question is whether high enough levels of plasticizers are likely to leech out of PVC products to be dangerous.

According to the U.S. Consumer Product Safety Commission, there is little to fear. "Based on scientific studies currently available," they reported, "the staff concludes that few, if any, children are at risk from liver or other organ toxicity from the release of DINP from these products. This is because the amount they might ingest does not reach a level that would be harmful." They also suggested, however, that we should "continue to work on better estimates of the amount of phthalate released when products are mouthed by children."

Not everyone agrees that phthalates are safe. A group of American health workers and environmentalists calling themselves Health Care Without Harm (HCWH) claim soft PVC in medical products and children's toys represents an unacceptable and unnecessary risk. They argue that the "precautionary principle" says that we shouldn't be exposed to toxic chemicals in consumer products and that the burden of proof of harmlessness lies with manufacturers, not the general public. In 1999, the European Union banned all sales of soft PVC toys.

"Nonsense," says former Surgeon General C. Everett Koop. "This is just the latest phony chemical scare. . . The ceaseless obsession with ousting the frequently nonexistent bogeymen from our chemical cornucopia does quite a lot to strengthen the ranks of consumer groups but very little to actually improve the health and quality of our lives. And while it provides television newsmagazines with a well-worn story line, it ultimately diverts our attention from real opportunities to enhance life and longevity. . . Highly inflammatory scare campaigns and dubious news stories have unnecessarily frightened families into believing their baby's teething rings and vinyl toys were conduits of cancer-causing chemicals. A distinguished panel of 17 scientists and physicians, under my chairmanship, concluded that the chemicals DINP and DEHP used to make toys and medical devices soft and flexible, are safe and pose no harm to adults or children."

HCWH, on the other hand, led by the environmental group Greenpeace, charges that Koop's "blue ribbon" panel of experts, which was funded primarily by the chemical industry, rejected important scientific studies and ignored safe, cost-competitive substitutes for every current soft PVC product. HCWH claims that the chlorine industry is trying to bog down the debate in a "dueling risk assessment strategy perfected long ago by the tobacco industry." Although they acknowledge being considered "on the fringe" by some mainstream environmental groups, HCWH points out that a number of products such as asbestos, tobacco, DDT, and tetraethyl lead were promoted by industry as perfectly safe but subsequently turned out to be quite dangerous.

Consumer pressure and fears of lawsuits have lead many manufacturers to quickly pull soft PVC products—especially baby toys and teethers—off store shelves. Most soft plastics in stores today are polyethylene or polypropylene, which do not contain plasticizers. What do you think of this debate? Are manufacturers caving in to mass hysteria, or should they have known better than to expose us to soft PVC in the first place? How would you evaluate the claims of the various sides of this argument? Can you think of reasons that one side or the other might be trying to mislead you? What additional information would you need to make an informed judgement about the safety of soft PVC?

Ethical Considerations

Is this a moral issue or simply a question of scientific facts? Does the fact that babies and sick people are exposed to these products make them more unethical, or is this just an appeal to our emotions? The effects of chronic, low-level exposure to phthalates won't be known for many years. Does this make the argument for a precautionary approach stronger?

unintended victims, however, and it may be stored for long periods of time in organisms that lack mechanisms to destroy it.

Chemical Interactions

Some materials produce *antagonistic* reactions. That is, they interfere with the effects or stimulate the breakdown of other chemicals. For instance, vitamins E and A can reduce the response to some carcinogens. Other materials are *additive* when they occur together in exposures. Rats exposed to both lead and arsenic show twice the toxicity of only one of these elements. Perhaps the greatest concern is *synergistic* effects. Synergism is an interaction in which one substance exacerbates the effects of another. For example, occupational asbestos exposure increases lung cancer rates 20-fold. Smoking increases lung cancer rates by the same amount. Asbestos workers who also smoke, however, have a 400-fold increase in cancer rates. How many other toxic chemicals are we exposed to that are below threshold limits individually but combine to give toxic results?

MECHANISMS FOR MINIMIZING TOXIC EFFECTS

A fundamental concept in toxicology is that every material can be poisonous under some conditions, but most chemicals have some safe level or threshold below which their effects are undetectable or insignificant. Each of us consumes lethal doses of many chemicals over the course of a lifetime. One hundred cups of strong coffee, for instance, contain a lethal dose of caffeine. Similarly, 100 aspirin tablets, or 10 kilograms of spinach or rhubarb, or a litre of alcohol would be deadly if consumed all at once. Taken in small doses, however, most toxins can be broken down or excreted before they do much harm. Furthermore, damage they cause can be repaired. Sometimes, however, mechanisms that protect us from one type of toxin or at one stage in the life cycle become deleterious with another substance or in another stage of development. Let's look at how these processes help protect us from harmful substances as well as how they can go awry.

Metabolic Degradation and Excretion

Most organisms have enzymes that process waste products and environmental poisons to reduce their toxicity. In mammals, most of these enzymes are located in the liver, the primary site of detoxification of both natural wastes and introduced poisons. Sometimes, however, these reactions work to our disadvantage. Compounds, such as benzopyrene, for example, that are not toxic in their original form are processed by these same enzymes into cancer-causing carcinogens. Why would we have a system that makes a chemical more dangerous? Evolution and natural selection are expressed through reproductive success or failure. Defense mechanisms that protect us from toxins and hazards early in life are "selected for" by evolution. Factors or conditions that affect postreproductive ages (like cancer or premature senility) usually don't affect reproductive success or exert "selective pressure."

We also reduce the effects of waste products and environmental toxins by eliminating them from our body through excretion. Volatile molecules, such as carbon dioxide, hydrogen cyanide, and ketones are excreted via breathing. Some excess salts and other substances are excreted in sweat. Primarily, however, excretion is a function of the kidneys, which can eliminate significant amounts of soluble materials through urine formation. Accumulation of toxins in the urine can damage this vital system, however, and the kidneys and bladder often are subjected to harmful levels of toxic compounds. In the same way, the stomach, intestine, and colon often suffer damage from materials concentrated in the digestive system and may be afflicted by diseases and tumours.

Repair Mechanisms

In the same way that individual cells have enzymes to repair damage to DNA and protein at the molecular level, tissues and organs that are exposed regularly to physical wear-and-tear or to toxic or hazardous materials often have mechanisms for damage repair. Our skin and the epithelial linings of the gastrointestinal tract, blood vessels, lungs, and urogenital system have high cellular reproduction rates to replace injured cells. With each reproduction cycle, however, there is a chance that some cells will lose normal growth controls and run amok, creating a tumour. Thus any agent, such as smoking or drinking, that irritates tissues is likely to be carcinogenic. And tissues with high cell-replacement rates are among the most likely to develop cancers.

MEASURING TOXICITY

In 1540, the German scientist Paracelsus said "the dose makes the poison," by which he meant that almost everything is toxic at some level. This remains the most basic principle of toxicology. Sodium chloride (table salt), for instance, is essential for human life in small doses. If you were forced to eat a kilogram of salt all at once, however, it would make you very sick. A similar amount injected into your bloodstream would be lethal. How a material is delivered—at what rate, through which route of entry, and in what medium—plays a vitally important role in determining toxicity.

This does not mean that all toxins are identical, however. Some are so poisonous that a single drop on your skin can kill you. Others require massive amounts injected directly into the blood to be lethal. Measuring and comparing the toxicity of various materials is difficult because not only do species differ in sensitivity, but individuals within a species respond differently to a given exposure. In this section, we will look at methods of toxicity testing and at how results are analyzed and reported.

Animal Testing

The most commonly used and widely accepted toxicity test is to expose a population of laboratory animals to measured doses of a specific substance under controlled conditions. This procedure is expensive, time-consuming, and often painful and debilitating to the animals being tested. It commonly takes hundreds—or even thousands—of animals, several years of hard work, and hundreds of thousands of dollars to thoroughly test the effects of a toxin at very low doses. More humane toxicity tests using computer simulation of model reactions, cell cultures, and other substitutes for whole living animals are being developed. However, conventional large-scale animal testing is the method in which we have the most confidence and on which most public policies about pollution and environmental or occupational health hazards are based.

In addition to humanitarian concerns, there are several problems in laboratory animal testing that trouble both toxicologists and policymakers. One problem is differences in sensitivity to a

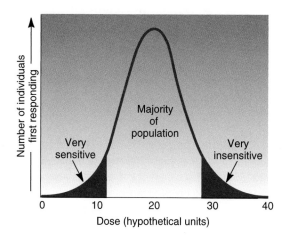

FIGURE 20.16 Probable variations in sensitivity to a toxin within a population. Some members of a population may be very sensitive to a given toxin, while others are much less sensitive. The majority of the population falls somewhere between the two extremes.

toxin of the members of a specific population. Figure 20.16 shows a typical dose/response curve for exposure to a hypothetical toxin. Some individuals are very sensitive to the toxin, while others are insensitive. Most, however, fall in a middle category forming a bell-shaped curve. The question for regulators and politicians is whether we should set pollution levels that will protect everyone, including the most sensitive people, or only aim to protect the average person. It might cost billions of extra dollars to protect a very small number of individuals at the extreme end of the curve. Is that a good use of resources?

Dose/response curves are not always symmetrical, making it difficult to compare toxicity of unlike chemicals or different species of organisms. A convenient way to describe toxicity of a chemical is to determine the dose to which 50 percent of the test population is sensitive. In the case of a lethal dose (LD), this is called the **LD50** (Fig. 20.17).

Unrelated species can react very differently to the same toxin, not only because body sizes vary but also because of differences in physiology and metabolism. Even closely related species can have very dissimilar reactions to a particular toxin. Hamsters, for instance, are nearly 5,000 times less sensitive to some dioxins than are guinea pigs. Of 226 chemicals found to be carcinogenic in either rats or mice, 95 caused cancer in one species but not the other. These variations make it difficult to estimate the risks for humans since we don't consider it ethical to perform controlled experiments in which we deliberately expose people to toxins.

Toxicity Ratings

It is useful to group materials according to their relative toxicity. A moderate toxin takes about one gram per kilogram of body weight for an average human to make a lethal dose. Very toxic materials take about one-tenth that amount, while extremely toxic substances take one-hundredth as much (only a few drops) to kill most people. Supertoxic chemicals are extremely potent; for some,

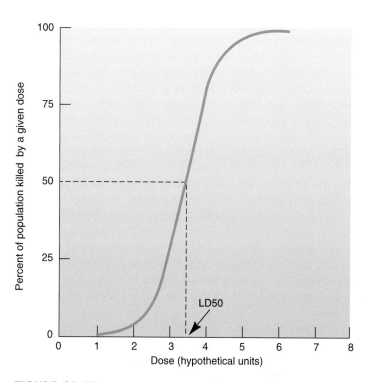

FIGURE 20.17 Cumulative population response to increasing doses of a toxin. The LD50 is the dose that is lethal to half the population.

a few micrograms (millionths of a gram—an amount invisible to the naked eye) make a lethal dose. These materials are not all synthetic. One of the most toxic chemicals known, for instance, is ricin, a protein found in castor bean seeds. It is so toxic that 0.3 billionths of a gram given intravenously will generally kill a mouse. If aspirin were this toxic, a single tablet, divided evenly, could kill 1 million people.

Many carcinogens, mutagens, and teratogens are dangerous at levels far below their direct toxic effect because abnormal cell growth exerts a kind of biological amplification. A single cell, perhaps altered by a single molecular event, can multiply into millions of tumour cells or an entire organism. Just as there are different levels of direct toxicity, however, there are different degrees of carcinogenicity, mutagenicity, and teratogenicity. Methanesulphonic acid, for instance, is highly carcinogenic, while the sweetener saccharin is a suspected carcinogen whose effects may be vanishingly small.

Acute versus Chronic Doses and Effects

Most of the toxic effects that we have discussed so far have been **acute effects.** That is, they are caused by a single exposure to the toxin and result in an immediate health crisis of some sort. Often, if the individual experiencing an acute reaction survives this immediate crisis, the effects are reversible. **Chronic effects,** on the other hand, are long lasting, perhaps even permanent. A chronic effect can result from a single dose of a very toxic substance, or it can be the result of a continuous or repeated sublethal exposure.

We also describe long-lasting *exposures* as chronic, although their effects may or may not persist after the toxin is removed. It usually is difficult to assess the specific health risks of chronic exposures because other factors, such as aging or normal diseases, act simultaneously with the factor under study. It often requires very large populations of experimental animals to obtain statistically significant results for low-level chronic exposures. Toxicologists talk about "megarat" experiments in which it might take a million rats to determine the health risks of some supertoxic chemicals at very low doses. Such an experiment would be terribly expensive for even a single chemical, let alone for the thousands of chemicals and factors suspected of being dangerous.

An alternative to enormous studies involving millions of animals is to give massive amounts—usually the maximum tolerable dose—of a toxin being studied to a smaller number of individuals and then to extrapolate what the effects of lower doses might have been. This is a controversial approach because it is not clear that responses to toxins are linear or uniform across a wide range of doses.

Figure 20.18 shows three possible results from low doses of a toxin. Curve (*a*) shows a baseline level of response in the population, even at zero dose of the toxin. This suggests that some other factor in the environment also causes this response. Curve (*b*) shows a straight-line relationship from the highest doses to zero exposure. Many carcinogens and mutagens show this kind of response. Any exposure to such agents, no matter how small, carries some risks. Curve (*c*) shows a threshold for the response where some minimal dose is necessary before any effect can be observed. This generally suggests the presence of some defence mechanism that prevents the toxin from reaching its target in an active form or repairs the damage that it causes. Low levels of exposure to the toxin in question may have no deleterious effects, and it might not be necessary to try to keep exposures to zero.

Which, if any, environmental health hazards have thresholds is an important but difficult question. The 1958 Delaney Clause to the U.S. Food and Drug Act forbids the addition of *any* amount of known carcinogens to food and drugs, based on the assumption that any exposure to these substances represents unacceptable risks. This standard was replaced in 1996 by a *no reasonable harm* requirement, defined as less than one cancer for every million people exposed over a lifetime. This change was supported by a report from the National Academy of Sciences concluding that synthetic chemicals in our diet are unlikely to represent an appreciable cancer risk. We will discuss risk analysis in the next section.

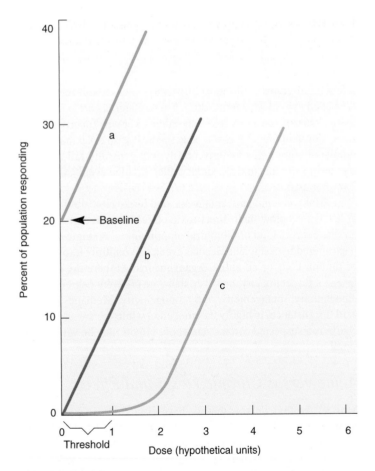

FIGURE 20.18 Three possible dose-response curves at low doses. (*a*) Some individuals respond, even at zero dose, indicating that some other factor must be involved. (*b*) Response is linear down to the lowest possible dose. (*c*) Threshold must be passed before any response is seen.

Detection Limits

You may have seen or heard dire warnings about toxic materials detected in samples of air, water, or food. A typical headline announced recently that 23 pesticides were found in 16 food samples. What does that mean? The implication seems to be that any amount of dangerous materials is unacceptable and that counting the numbers of compounds detected is a reliable way to establish danger. We have seen, however, that the dose makes the poison. It matters not only what is there, but how much, where it is located, how accessible it is, and who is exposed. At some level, the mere presence of a substance is insignificant.

Toxins and pollutants may seem to be more widespread now than in the past, and this is surely a valid perception for many substances. The daily reports we hear of new materials found in new places, however, are also due, in part, to our more sensitive measuring techniques. Twenty-five years ago, parts per million were generally the limits of detection for most chemicals. Anything below that amount was often reported as zero or absent rather than more accurately as undetected. Fifteen years ago, new machines and techniques were developed to measure parts per billion. Suddenly, chemicals were found where none had been suspected. Now we can detect parts per trillion or even parts per quadrillion in some cases. Increasingly sophisticated measuring capabilities may lead us to believe that toxic materials have become more prevalent. In fact, our environment may be no more dangerous; we are just better at finding trace amounts.

RISK ASSESSMENT AND ACCEPTANCE

Even if we know with some certainty how toxic a specific chemical is in laboratory tests, it still is difficult to determine **risk** (the probability of harm times the probability of exposure) if that chemical is released into the environment. As you already have seen, many factors complicate the movement and fate of chemicals both around us and within our bodies. Furthermore, public perception of relative dangers from environmental hazards can be skewed so that some risks seem much more important than others.

Assessing Risks

A number of factors influence how we perceive relative risks associated with different situations.

- People with social, political, or economic interests—including environmentalists—tend to downplay certain risks and emphasize others that suit their own agendas. We do this individually as well, building up the dangers of things that don't benefit us while diminishing or ignoring the negative aspects of activities we enjoy or profit from.

- Most people have difficulty understanding and believing probabilities. We feel that there must be patterns and connections in events, even though statistical theory says otherwise. If the coin turned up heads last time, we feel certain that it will turn up tails next time. In the same way, it is difficult to understand the meaning of a 1-in-10,000 risk of being poisoned by a chemical.

- Our personal experiences often are misleading. When we have not personally experienced a bad outcome, we feel it is more rare and unlikely to occur than it actually may be. Furthermore, the anxieties generated by life's gambles make us want to deny uncertainty and to misjudge many risks.

- We have an exaggerated view of our own abilities to control our fate. We generally consider ourselves above-average drivers, safer than most when using appliances or power tools, and less likely than others to suffer medical problems, such as heart attacks. People often feel they can avoid hazards because they are wiser or luckier than others.

- News media give us a biased perspective on the frequency of certain kinds of health hazards, overreporting some accidents or diseases while downplaying or underreporting others. Sensational, gory, or especially frightful causes of death like murders, plane crashes, fires, or terrible accidents occupy a disproportionate amount of attention in the public media. Heart diseases, cancer, and stroke kill nearly 15 times as many people in the United States as do accidents and 75 times as many people as do homicides, but the emphasis placed by the media on accidents and homicides is nearly inversely proportional to their relative frequency compared to either cardiovascular disease or cancer. This gives us an inaccurate picture of the real risks to which we are exposed.

- We tend to have an irrational fear or distrust of certain technologies or activities that leads us to overestimate their dangers. Nuclear power, for instance, is viewed as very risky, while coal-burning power plants seem to be familiar and relatively benign; in fact, coal mining, shipping, and combustion cause an estimated 10,000 deaths each year in the United States, compared to none known so far for nuclear power generation. An old, familiar technology seems safer and more acceptable than does a new, unknown one.

Accepting Risks

How much risk is acceptable? How much is it worth to minimize and avoid exposure to certain risks? Most people will tolerate a higher probability of occurrence of an event if the harm caused by that event is low. Conversely, harm of greater severity is acceptable only at low levels of frequency. A 1-in-10,000 chance of being killed might be of more concern to you than a 1-in-100 chance of being injured. For most people, a 1-in-100,000 chance of dying from some event or some factor is a threshold for changing what we do. That is, if the chance of death is less than 1 in 100,000, we are not likely to be worried enough to change our ways. If the risk is greater, we will probably do something about it. The U.S. Environmental Protection Agency generally assumes that a risk of 1 in 1 million is acceptable for most environmental hazards. Critics of this policy ask, acceptable to whom?

For activities that we enjoy or find profitable, we are often willing to accept far greater risks than this general threshold. Conversely, for risks that benefit someone else we demand far higher protection. For instance, your chance of dying in a motor vehicle accident in any given year is about 1 in 5,000, but that doesn't deter many people from riding in automobiles. Your chances of dying from lung cancer if you smoke one pack of cigarettes per day is about 1 in 1,000. By comparison, the risk from drinking water with the U.S. EPA limit of trichloroethylene is about 2 in 1 billion. Strangely, many people demand water with zero levels of trichloroethylene, while continuing to smoke cigarettes.

Table 20.5 lists some activities estimated to increase your chances of dying in any given year by 1 in 1 million. These are statistical averages, of course, and there clearly are differences in where one lives or how one rides a bicycle that affect the danger level of these activities. Still, it is interesting how we readily accept some risks while shunning others.

Our perception of relative risks is strongly affected by whether risks are known or unknown, whether we feel in control of the outcome, and how dreadful the results are. Risks that are unknown or unpredictable and results that are particularly gruesome or disgusting seem far worse than those that are familiar and socially acceptable.

Studies of public risk perception show that most people react more to emotion than statistics. We go to great lengths to avoid some dangers while gladly accepting others. Factors that are involuntary, unfamiliar, undetectable to those exposed, catastrophic, or that have delayed effects or are a threat to future generations are

TABLE 20.5 Activities Estimated to Increase your Chances of Dying in Any Given Year by 1 in 1 Million	
ACTIVITY	RESULTING DEATH RISK
Smoking 1.4 cigarettes	Cancer, heart disease
Drinking 0.5 litre of wine	Cirrhosis of the liver
Spending 1 hour in a coal mine	Black lung disease
Living 2 days in New York or Boston	Air pollution
Travelling 6 minutes by canoe	Accident
Travelling 16 km by bicycle	Accident
Travelling 240 km by car	Accident
Flying 1,600 km by jet	Accident
Flying 9,600 km by jet	Cancer caused by cosmic radiation
Living 2 months in Denver	Cancer caused by cosmic radiation
Living 2 months in a stone or brick building	Cancer caused by natural radioactivity
One chest X ray	Cancer caused by radiation
Living 2 months with a cigarette smoker	Cancer, heart disease
Eating 40 tablespoons of peanut butter	Cancer from aflatoxin
Living 5 years at the site boundary of a typical nuclear power plant	Cancer caused by radiation from routine leaks
Living 50 years 8 kilometres from a nuclear power plant	Cancer caused by accidental radiation release
Eating 100 charcoal-broiled steaks	Cancer from benzopyrene

Source: From William Allman, "Staying Alive in the Twentieth Century," *Science 85,* 5(6): 31, October 1985. Used by permission of the author.

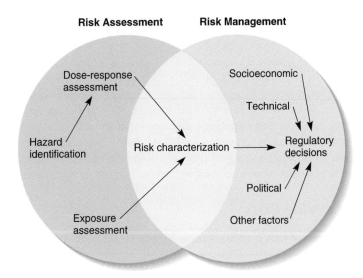

FIGURE 20.19 Risk assessment organizes and analyzes data to determine relative risk. Risk management sets priorities and evaluates relevant factors to make regulatory decisions.
Source: Data from D. E. Patton, "USEPA's Framework for Ecological Risk Assessment" in *Human Ecological Risk Assessment,* Vol. 1 No. 4.

especially feared while those that are voluntary, familiar, detectable, or immediate cause less anxiety. Even though the actual number of deaths from automobile accidents, smoking, or alcohol, for instance, are thousands of times greater than those from pesticides, nuclear energy, or genetic engineering, the latter preoccupy us far more than the former.

ESTABLISHING PUBLIC POLICY

Risk management combines principles of environmental health and toxicology together with regulatory decisions based on socio-economic, technical, and political considerations (Fig. 20.19). The biggest problem in making regulatory decisions is that we are usually dealing with many sources of harm to which we are exposed, often without being aware of them. It is difficult to separate the effects of all these different hazards and to evaluate their risks accurately, especially when the exposures are near the threshold of

measurement and response. In spite of often vague and contradictory data, public policymakers must make decisions.

The case of the sweetener saccharin is a good example of the complexities and uncertainties of risk assessment in public health. Studies in the 1970s at the Canadian Health Protection Branch and the University of Wisconsin suggested a link between saccharin and bladder cancer in male rats. Critics of these studies pointed out that humans would have to drink 800 cans of diet soda *per day* to get a saccharin dose equivalent to that given to the rats. Furthermore, they argued this response may be unique to male rats. In 2000, the U.S. Department of Health concluded a study that found no association between saccharin and cancer in humans. All warnings were removed from saccharin-containing products. Still, some groups like the U.S. Center for Science in the Public Interest consider this sweetener dangerous and urge us to avoid it if possible.

In setting standards for environmental toxins, we need to consider (1) combined effects of exposure to many different sources of damage, (2) different sensitivities of members of the population, and (3) effects of chronic as well as acute exposures. Some people argue that pollution levels should be set at the highest amount that does *not* cause measurable effects. Others demand that pollution be reduced to zero if possible, or as low as is technologically feasible. It may not be reasonable to demand that we be protected from every potentially harmful contaminant in our environment, no matter how small the risk. As we have seen, our bodies have mechanisms that enable us to avoid or repair many kinds of damage so that most of us can withstand some minimal level of exposure without harm (Fig. 20.20).

On the other hand, each challenge to our cells by toxic substances represents stress on our bodies. Although each individual stress may not be life-threatening, the cumulative effects of all the environmental stresses, both natural and human-caused, to which

FIGURE 20.20 "Do you want to stop reading those ingredients while we're trying to eat?"
Reprinted with permission of the *Star-Tribune*, Minneapolis-St. Paul.

we are exposed may seriously shorten or restrict our lives. Furthermore, some individuals in any population are more susceptible to those stresses than others. Should we set pollution standards so that no one is adversely affected, even the most sensitive individuals, or should the acceptable level of risk be based on the average member of the population?

Finally, policy decisions about hazardous and toxic materials also need to be based on information about how such materials affect the plants, animals, and other organisms that define and maintain our environment. In some cases, pollution can harm or destroy whole ecosystems with devastating effects on the life-supporting cycles on which we depend. In other cases, only the most sensitive species are threatened. Table 20.6 shows the U.S. Environmental Protection Agency's assessment of relative risks to human welfare. This ranking reflects a concern that our exclusive focus on reducing pollution to protect human health has neglected risks to natural ecological systems. While there have been many benefits from a case-by-case approach in which we evaluate the health risks of individual chemicals, we have often missed broader ecological problems that may be of greater ultimate importance.

SUMMARY

Health is a state of physical, mental, and social well-being, not merely the absence of disease or infirmity. The cause or development of nearly every human disease is at least partly related to environmental factors. For most people in the world, the greatest health threat in the environment is still, as it always has been, from pathogenic organisms. Bacteria, viruses, protozoa, parasitic worms, and other infectious agents probably kill more people each year than any other cause of death. Highly lethal emergent diseases such as SARS, West Nile virus, Ebola, and AIDS along with newly drug-resistant forms of old diseases are an increasing worry everywhere in the world.

Stress, diet, and lifestyle also are important health factors. Our social or cultural environment may be as important as our physical environment in determining the state of our health. People in some areas in the world live exceptionally long and healthful lives. We might be able to learn from them how to do so as well.

Estimating the potential health risk from exposure to specific environmental factors is difficult because information on the precise dose, length and method of exposure, and possible interactions between the chemical in question and other potential toxins to which the population may have been exposed is often lacking. In addition, individuals have different levels of sensitivity and response to a particular toxin and are further affected by general health condition, age, and sex.

The distribution and fate of materials in the environment depend on their physical characteristics and the processes that transport, alter, destroy, or immobilize them. Uptake of toxins into organisms can result in accumulation in tissues and transfer from one organism to another.

Estimates of health risks for large, diverse populations exposed to very low doses of extremely toxic materials are inexact because of biological variation, experimental error, and the necessity of extrapolating from results with small numbers of laboratory animals. In the end, we are left with unanswered questions. Which are the most dangerous environmental factors that we face? How can we evaluate the hazards of all the natural and synthetic chemicals that now exist? What risks are acceptable? We have not yet solved these problems or answered all the questions raised in this chapter, but it is important that these issues be discussed and considered seriously.

QUESTIONS FOR REVIEW

1. What is SARS? How is it thought to have started?

2. What is the difference between toxic and hazardous? Give some examples of materials in each category.

3. What are some of the most serious infectious diseases in the world? How are they transmitted?

4. How do stress, diet, and lifestyle affect environmental health? What diseases are most clearly related to these factors?

5. How do the physical and chemical characteristics of materials affect their movement, persistence, distribution, and fate in the environment?

6. Define LD50. Why is it more accurate than simply reporting toxic dose?

7. What is the difference between acute and chronic toxicity?

8. Define *carcinogenic, mutagenic, teratogenic,* and *neurotoxic.*

9. Why are soft PVC toys considered dangerous to children?

10. How do organisms reduce or avoid the damaging effects of environmental hazards?

11. What are the relative risks of smoking, driving a car, and drinking water with the maximum permissible levels of trichloroethylene? Are these relatively equal risks?

QUESTIONS FOR CRITICAL THINKING

1. How would you feel or act if your child were dying of diarrhea? Why do we spend more money on heart or cancer research than childhood illnesses?

2. Some people seem to have a poison paranoia about synthetic chemicals. Why do we tend to assume that natural chemicals are benign while industrial chemicals are evil?

3. Analyze the claim that we are exposed to thousands of times more natural carcinogens in our diet than industrial ones. Is this a good reason to ignore pollution?

4. What are the premises in the discussion of assessing risk? Could conflicting conclusions be drawn from the facts presented in this section? What is your perception of risk from your environment?

5. Table 20.5 equates activities such as smoking 1.4 cigarettes, having one chest X ray, and riding 16 kilometres on a bicycle. Do you agree with this assessment? Do some items on this list require further clarification?

6. Should pollution levels be set to protect the average person in the population or the most sensitive? Why not have zero exposure to all hazards?

7. What level of risk is acceptable to you? Are there some things for which you would accept more risk than others?

KEY TERMS

acute effects   453
allergens   445
antigens   445
bioaccumulation   450
biomagnification   450
cancer   447
carcinogens   447
chronic effects   453
Disability-Adjusted Life Year (DALY)   441
disease   439

emergent disease   442
fetal alcohol syndrome   446
hazardous   444
health   439
LD50   453
morbidity   439
mutagens   446
neurotoxins   445
risk   455
Severe Acute Respiratory Syndrome   438
teratogens   446
toxins   444

Web Exercises

LEARNING ABOUT DISEASES

The World Health Organization has a wealth of information about mortality, disability burdens, life expectancies, demographics, and other topics related to environmental health. Visit our website at http://www.mcgrawhill.ca/college/environmentalscience to find live links for information about disease and health, or go directly to the websites listed below. You can find massive amounts of data in table form by country, disease, and death rates at www.who.int/whosis/. Go to the information about specific diseases at www.who.int/health-topics/ idindex.hum. Look up the current status of ebola and leischmaniasis, and compare them to a disease that occurs closer to where you live. What organisms cause these diseases? What is their current distribution and prevalance? What environmental and social factors contribute to their spread? What treatment (if any) exists for these diseases, and how might they be prevented? See links on this web page for the newsletter *Action Against Infection* for up-to-date information about efforts to stop the spread of contagious diseases.

Chapter 21

Urbanization and Sustainable Cities

The problems that overwhelm us today are precisely those we failed to solve decades ago.

—Mostafa K. Tolba—

OBJECTIVES

After studying this chapter, you should be able to:

- distinguish between a rural village, a city, and a megacity.
- recognize the push and pull factors that lead to urban growth.
- report on the growth rate of giant metropolitan urban areas such as Mexico City, as well as the problems this growth engenders.
- understand the causes and consequences of urban sprawl.
- explain the principles of smart growth, garden cities, and conservation designs.
- evaluate the ways that cities can be ecologically, socially, and economically sustainable.

WebQuest

urban planning, megacities, sustainable cities

Above: Cities like Hong Kong are powerful engines for cultural, economic, and political change. They also consume resources and create environmental problems.

Canada's Urban Centres in Transition

Urbanization has been an essential part of Canada's development of a strong and stable economy and society. Cities play an important role in fostering scientific, technological, and artistic innovation, and are centres of culture and education. As measured by the gross domestic product (GDP), Canada's urban centres contribute to the wealth of the nation, as follows:

- Halifax contributes 47 percent of Nova Scotia's GDP
- Montreal Region accounts for 49 percent of Quebec's GDP
- Greater Toronto Area accounts for 44 percent of Ontario's GDP
- Winnipeg accounts for 67 percent of Manitoba's GDP
- Calgary and Edmonton account for 64 percent of Alberta's GDP
- Vancouver accounts for 53 percent of B.C.'s GDP

While there are many attributes of urban centres in Canada, there are a number of issues that remain to be addressed.

In Canada's major cities—Halifax, Montreal, Toronto, Winnipeg, Edmonton, and Vancouver—most of their infrastructures, including the roads, water treatment, and distribution systems, wastewater treatment systems, flood protection works (i.e., dams and flood walls), were built before the 1960s. Since that time, not only has there been increased use as population numbers grew and deterioration in their structural integrity over time (often seen as pot holes in roads and leaks in water pipes), but also health and safety standards established by senior governments have increased (e.g., more stringent water quality standards). In some municipalities, the cast iron pipes from the 1880s are still being used for water distribution. Air and water quality is also declining in many cities.

Traffic gridlock, poverty, and lack of funding are becoming more problematic in many areas. To compound the problem of meeting the need to refurbish and expand their infrastructure, the federal government reduced transfer payments to provinces, which in turn slashed funding to municipalities in the 1990s. Since the 1960s, peoples' expectations for a high-quality and safe living environment have increased. During the 1990s, provincial governments downloaded the responsibility for delivering some social programs (e.g., ambulance, affordable housing) to local governments.

Accommodating those who are adjusting to living in a new country or setting (such as recent immigrants and refugees who often move to urban centres), promoting cultural diversity, meeting the needs of an aging population, effectively managing urban growth, and dealing with a widening gap between rich and poor are some of the emerging social issues local governments in Canada are now facing. Thus, there are increased demands and expectations placed on municipalities by citizens and reduced levels of funding to municipalities from senior governments. Municipalities have limited means to raise funds, such as through property taxes and user fees. Property taxes ignore a person's

FIGURE 21.1 Montreal and the surrounding region accounts for 49 percent of Quebec's gross domestic product. As a large urban centre, Montreal can boast art, technology, culture, and science, while at the same time having difficulty maintaining its infrastructure or people. This is the difficulty most urban centres across the world face today.

ability to pay and reflect only the assessed value of their home. Thus, there is a funding squeeze and dilemma.

The urban challenges facing Canada cannot be ignored. In 2001, 80 percent of Canadians lived in urban areas with 51 percent of this concentrated in four centres: the Golden Horseshoe in Ontario, the Montreal region, the Vancouver and lower mainland, and the Calgary–Edmonton corridor. Between 1996 and 2001, virtually all of Canada's population growth was concentrated in its four largest urban areas. These grew by 7.6 percent while the rest of the country grew by 0.5 percent during this period. Population projections suggest that by 2020 between 85 percent and 90 percent of Canada's 36 million people will live in urban areas.

As the Canadian economy shifts from a manufacturing and resource-based economy to a knowledge-based economy,

Want More?

Tyler, M. E. 2000. "The Ecological Restructuring of Urban Form." In *Canadian Cities in Transition: The Twenty-first Century.* T. Bunting and P. Filion eds. Toronto: Oxford University Press.

Jenks, M., Burton, E., and Williams, K. eds. 1996. *"The Compact City" A Sustainable Urban Form?* New York: E&FN Spon.

sources of competitive advantage tend to be best developed in localized geographic clusters. Universities, colleges, and research facilities tend to be located in urban centres. Canada's ability to compete in the global economy does and will increasingly rely on its urban areas. Although the federal government provided $55 billion in direct and indirect services and programs in urban areas in 2001–2002, this amount, combined with provincial transfers and local revenues, is inadequate to meet current needs.

Given the problems facing municipalities and their importance to Canadian society, the prime minister established a Caucus Task Force on Urban Issues in 2001. It believes that balancing economic competitiveness, social harmony, and a sustainable environment will support a high quality of life for Canadians. Urban issues were an important concern in the platforms of the major federal parties in the 2004 election. We await the delivery on this commitment.

URBANIZATION

Since their earliest origin, **cities** have been centres of education, religion, culture, commerce, record keeping, communication, and political power. As cradles of civilization, cities have influenced culture and society far beyond their proportion of the total population (Fig. 21.2). Until recently, however, only a small percentage of the world's people lived in urban areas and even the greatest cities of antiquity (Athens, Rome, and Constantinople) were small by modern standards. The vast majority of humanity has always lived in rural areas where farming, fishing, hunting, timber harvesting, animal herding, mining, or other natural resource-based occupations provided the means of individual and family support.

Since the beginning of the Industrial Revolution some 300 years ago, however, cities have grown rapidly in size and power. In every developing country, the transition from an agrarian society to an industrial one has been accompanied by **urbanization,** an increasing concentration of people in cities and a transformation of land use and society to a metropolitan pattern of organization. Industrialization and urbanization bring many benefits, especially

to the top members of society, but they also cause many problems, as we will detail in this chapter.

Half of the people in the world now live in urban areas. The other half is increasingly dependent on urban centres for their economic survival. In addition, telephones, Internet communication, and satellite links are often provided through urban centres to rural areas and/or other countries. Demographers predict by 2035, the population of the world's cities will double to well over 5 billion, with virtually this entire growth taking place in developing countries. This trend, combined with other factors such as globalization and democratization, has reinforced the importance of cities in sustainable development. While cities have played a leadership role in national economies and societies, many cities, particularly in the developing world, have very high rates of poverty, and many residents live in unsanitary and degraded environments (Fig. 21.3).

It was in this context that the concept of sustainable cities was promoted in Agenda 21, which was formulated at the 1992 Earth Summit. In this chapter, we look at what defines a city, how cities came into existence, why people live in cities, and what the environmental conditions cities have been, are now, and might be in the future.

FIGURE 21.2 Since their earliest origins, cities have been centres of education, religion, commerce, politics, and culture. Unfortunately, they have also been sources of pollution, crowding, disease, and misery.

FIGURE 21.3 Half of the world's population now lives in urban areas. Despite the economic success that can come from cities and urban centres, most have very high rates of poverty, in which residents live in unsanitary, broken-down environments.

Some of the most severe urban problems in the world are found in the giant megacities of the developing countries. Far more lives may be threatened by the desperate environmental conditions in these cities than by any other issue we have studied. As described in the opening section, Canada's cities also face a number of challenges. We will look at a few of those problems and possible solutions that could improve the urban environment abroad and in Canada. Human ingenuity gave rise to the development of cities. Harnessing and refocusing that ingenuity is key to solving our current set of problems.

What is an Urban Area?

Just what defines an urban area or city? Definitions differ. In the 2001 Census, Statistics Canada defined an urban area as one with a population of at least 1,000 and no fewer than 400 persons per square kilometre. The definition of "urban" changes over time within a country such as Canada and differs between countries. For instance, in 1931 all incorporated cities, towns, and **villages** in Canada, regardless of population size and density, were defined as urban. In the United States, any incorporated city, regardless of size, is considered to be a "city," and it defines any "city" with more than 2,500 residents as "urban." A village of more than 200 people is considered "urban" in Sweden and Denmark. Thus, students and researchers must be careful when comparing urban populations over time within a country and between countries.

Another way of thinking about where people live is to consider community *functions*. In Canada, all land outside an urban area is defined as rural. In **rural areas,** more residents typically are more reliant on agriculture or other ways of harvesting natural resources for their livelihood. In an **urban area,** by contrast, the majority of the people are not directly dependent on natural resource-based operations. However, an emerging problem on the fringe of many urban areas is to meet the urban-oriented demands (e.g., waste disposal, roads, and amenities) of people who reside in rural areas but work in urban areas. Other areas that have been traditional seasonal tourist areas have seen the construction of permanent residences or mega-homes (e.g., Muskoka lakes in Ontario and ocean-front properties in the Maritimes), which have changed the demand for services and the traditional character of rural areas.

Statistics Canada provides for a functional view of major population centres by defining census metropolitan areas (CMAs), which are areas comprising a core population of at least 100,000 people combined with adjacent urban and rural areas that have a high degree of economic and social integration with that urban area. In 2001, about 67 percent of Canada's population lived in one of the largest 27 CMAs. There are four CMAs that have populations of more than 1 million: Toronto (4.3 million), Montreal (3.3 million), Vancouver (1.8 million), and Ottawa-Hull (1 million). Collectively, these four CMAs contain about 33 percent of Canada's total population. Another five CMAs have populations between 500,000 and 1 million: Edmonton, Calgary, Quebec City, Hamilton, and Winnipeg. Most of the CMAs are located relatively close to the Canada-U.S. border. Similar to Canada's definition

FIGURE 21.4 A traditional hill-tribe village in northern Thailand is closely tied to the land through culture, economics, and family relationships. While the timeless pattern of life here gives a great sense of identity, it can also be stifling and repressive.

of CMAs, the U.N. defines "urban agglomerations" as areas that incorporate the population within a city or town and the adjacent suburban fringe.

Particularly in Third World contexts, small communities or villages are small collections of rural households linked by culture, custom, family ties, and close associations with the land and water resources (Fig. 21.4). By contrast, cities all over the world comprise a differentiated community with a population and resource base large enough to allow residents to specialize in arts, crafts, services, or professions rather than natural resource-based occupations. While the rural village often has a sense of security and connection, it can also be stifling. A city offers more freedom to experiment, to be upwardly mobile, and to break from restrictive traditions, but it can be harsh and impersonal (Fig. 21.5).

Beyond about 10 million inhabitants, an urban area is considered a supercity or **megacity.** Megacities in many parts of the world have grown to enormous size. Chongqing, China, having annexed a large part of Sichuan province and about 30 million people, claims to be the biggest city in the world. In the United States, urban areas between Boston and Washington, DC, have merged into a nearly continuous megacity (sometimes called Bos-Wash) containing about 35 million people. The Tokyo-Yokohama-Osaka-Kobe corridor contains nearly 50 million people.

Want More?

Knox, P. L. 1994. *Urbanization: An Introduction to Urban Geography.* Englewood Cliffs, New Jersey: Prentice Hall.

FIGURE 21.5 A city is a differentiated community with a large enough population and resource base to allow specialization in arts, crafts, services, and professions. Although there are many disadvantages in living so closely together, there are also many advantages. Cities are growing rapidly, and most of the world's population will live in urban areas if present trends continue.

In Ontario, the area referred to as the Golden Horseshoe extends around the western end of Lake Ontario from Niagara Falls through to Oshawa and contains about 50 percent of Ontario's total population. Because these agglomerations have expanded beyond what we normally think of as a city, some geographers prefer to think of urbanized **core regions** that dominate the social, political, and economic life of most countries. The forces of globalization are and will continue to influence urban development in the future. Geographer Paul Knox notes that the emergence of transnational organizations, the development of faster forms of transportation, telecommunications and information technologies, and geopolitical changes that have promoted capitalist development are modifying the traditional role of cities. Rather than serving national needs and wants, cities are now integrated into the international economy and geopolitical life. Outcomes from this pressure for change can be seen in the centralization of important corporate functions to a few "world cities," such as New York, London, and Tokyo, and in the developed world, a shift away from a manufacturing-based economy to an information economy. Many companies have moved or threatened to move manufacturing operations from Canada to other countries that offer incentives and/or cheaper labour. Cities in Canada are very much connected through economic activity, financial markets, and the flow of information to other cities in the world.

World Urbanization

At the turn of the twentieth century, almost two-thirds of Canadians lived in rural areas: small towns and villages or individual farms (Fig. 21.6). In 2001, 79.4 percent of Canadians lived in cities with a population of 10,000 or more, up from 78.5 percent in 1996. Between 1996 and 2001, the rural or small town popula-

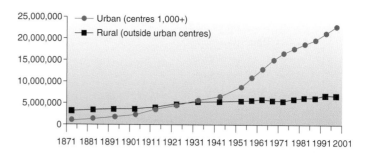

FIGURE 21.6 Urbanization in Canada—1871 to 1996.
Source: Statistics Canada, Census of populations 1871–1996.

tion shrank by 0.4 percent. United Nations Secretary-General Kofi Annan believes the world has entered the "urban millennium." In 2000, about 3 billion people, or 50 percent of the world's population, lived in urban areas; by 2030 more than 60 percent of the world's population is projected to live in cities. Many countries are seeing an increase in the size of urban populations and decrease in rural ones.

In 1850, only about 2 percent of the world's population lived in cities. By 2000, 47 percent of the people in the world were urban dwellers. Only Africa and South Asia remain predominately rural, but people there are swarming into cities in ever-increasing numbers. About 75 percent of the people in Europe, North America, and Latin America already live in cities (Table 21.1). Some urban experts predict that by 2010 the whole world will be urbanized to the levels now seen in developed countries.

As Figure 21.7 shows, 90 percent of the population growth over the next 25 years is expected to occur in less-developed countries of the world. Most of that growth will be in already overcrowded cities and least affluent countries such as Mexico, China, and Brazil in the world. The combined population of these cities is projected by the Population Reference Bureau to jump from

TABLE 21.1	Urban Share of Total Population (percent)		
REGION	1975	1995	2025*
Africa	25	34	54
Asia	25	35	55
Europe	67	74	83
Russian Federation	66	76	86
North and Central America	57	68	79
South America	64	78	88
Oceania	72	70	75
World	38	45	61

*Estimated.

Source: Data from *World Resources 1996–97.*

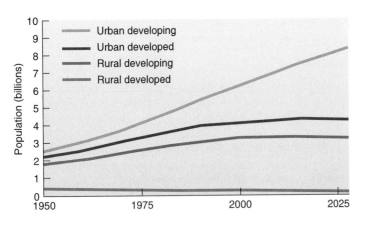

FIGURE 21.7 Urban and rural growth in developed and developing countries.
Source: Data from United Nations (U.N.) Population Division, *World Urbanization Prospects (The 1996 Revision),* on diskette (U.N. New York, 1996).

6 billion (2000) to more than 8 billion by the year 2025 (Fig. 21.8). Although these countries recently have become more affluent, their capacity to address urban growth in the short and long term is very uncertain. Meanwhile, rural populations in these countries are expected to remain constant or even decline somewhat as people migrate into the cities.

Recent urban growth has been particularly dramatic in the largest cities, especially those of the developing world. In 1900, 13 cities had populations of more than 1 million; except for Tokyo and Peking, all were in Europe or North America (Table 21.2). By 1995, there were 235 metropolitan areas of more than 1 million but only three of the largest cities, Tokyo, New York, and Los Angeles, were in developed countries. None of the 13 largest urban centres in the world has ever been nor is projected to be located in Canada. The demographic trend in developing-world cities will be a unique aspect for the millennium. The U.N. estimates that more than 2.4 billion people, nearly 90 percent of the total population growth expected between 1995 and 2030, are expected to live in cities. The scale of this growth is unparalleled.

Can cities function in a sustainable manner with 20 or 25 million people? Can they supply the public services necessary to sustain a civilized society? Adding 750,000 new people annually to Mexico City amounts to building a new city the size of Edmonton's CMA (783,100) or the combined populations of Saint John, Halifax, St. John's, Chicoutimi-Jonquière, and Trois-Rivières (combined population of 781,600). This growth is occurring, as is the case in the most rapidly growing urban centres in the world, in a country that has a less developed and diversified economy than Canada's, less stable government, and a higher foreign debt load. Their capacity to deal with the challenges of urbanization will be severely tested. The remainder of this chapter provides some thoughts on these questions and a sense of the current challenges facing cities the world over and in Canada.

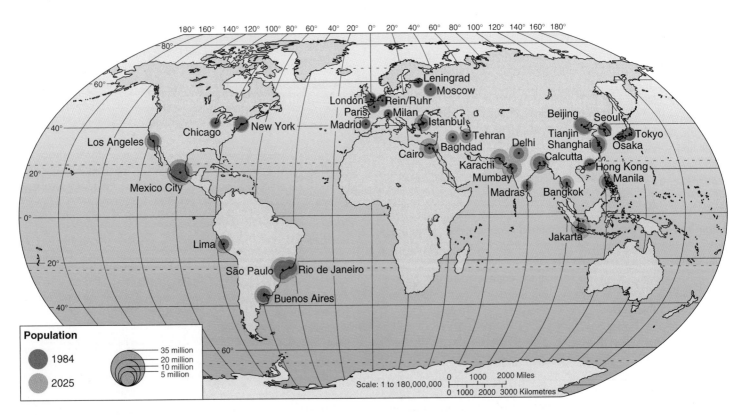

FIGURE 21.8 By 2025, at least 400 cities will have populations of 1 million or more, and 93 supercities will have populations above 5 million. Three-fourths of the world's largest cities will be in developing countries that already have trouble housing, feeding, and employing their people.

1900		1995		2015	
London, England	6.6	Tokyo, Japan	26.9	Tokyo, Japan	29.0
New York, USA	4.2	Mexico City, Mexico	16.6	Mumbay, India**	26.2
Paris, France	3.3	São Paulo, Brazil	16.5	Mexico City, Mexico	25.1
Berlin, Germany	2.4	New York, USA	16.3	Lagos, Nigeria	24.6
Chicago, USA	1.7	Mumbay, India**	15.1	São Paulo, Brazil	20.3
Vienna, Austria	1.6	Shanghai, China	13.6	Shanghai, China	18.0
Tokyo, Japan	1.5	Los Angeles, USA	12.4	New York, USA	17.6
St. Petersburg, Russia	1.4	Calcutta, India	11.9	Calcutta, India	17.3
Philadelphia, USA	1.4	Seoul, S. Korea	11.6	Beijing, China	16.0
Manchester, England	1.3	Beijing, China	11.3	Los Angeles, USA	14.2
Birmingham, England	1.2	Osaka, Japan	10.6	Buenos Aires, Argentina	13.9
Moscow, Russia	1.1	Lagos, Nigeria	10.3	Rio de Janeiro, Brazil	11.9
Peking, China*	1.1	Rio de Janeiro, Brazil	10.2	Osaka, Japan	10.6

*Now spelled Beijing.

**Formerly known as Bombay.

Source: Data from T. Chandler, *Three Thousand Years of Urban Growth,* 1974, Academic Press; and United Nations Populations Division, 1998.

What is Urban Sustainability?

According to Mark Roseland, who is a geography professor at Simon Fraser University and associate director of its Community and Economic Development Centre, the origins of sustainable cities arose in the 1970s and can be benchmarked to the founding of Urban Ecology in Berkeley (1975), a non-profit organization to "rebuild cities in balance with nature." That organization published *Eco-city Berkeley* (1987), a book dedicated to how that city could be ecologically rebuilt, and the journal *Urban Ecologist.* Five international Eco-City Conferences have been held in: Berkeley (1990), Adelaide Australia (1992), Yoff, Senegal (1996), Curitiba, Brazil (2000), and Shenzhen, China (2002).

There is no single accepted definition of urban sustainability. However, its essential elements that are reflected in the quotes below suggest the integration of economic, social (broadly defined), and environmental considerations into decision-making structures that directly and indirectly influence the urban quality of life over the long term. Some definitions include:

- "Urban sustainability can only be assured with a human ecological understanding of the complex interactions among environmental, economic, political, and social/cultural factors and with careful planning and management grounded in ecological principles." (http://www.societyforhuman ecology.org/China/background.htm)

- "Striving for harmony in the development of a civil society, economy, environment, culture and political institutions." (Judith Maxwell, chair, Economic Council of Canada, 1985–1992)

- "The enhanced well-being of cities or urban regions, including integrated economic, ecological, and social components, which maintain the quality of life for future generations." (National Roundtable on the Environment and Economy)

Much of literature suggests a decentralized administrative structure and local participation in planning are important in order that communities can define sustainability from their perspective. It calls for social equity, between current and future generations, between residents in urban and their surrounding regions' hinterland, and through improved governance and adequate housing.

Implementing the concept of urban sustainability will require making difficult choices, such as decisions about production and consumption patterns in urban areas. Some of the hard choices we must come to terms with include balancing regional versus urban demands, development versus non-development, the abilities of ecosystems to support human and non-human needs, and the nature of what we define as a quality of life.

Governments must play a fundamental role in being active agents of positive change. New administrative arrangements, at national, provincial, regional and local levels must be carefully considered. Some of these alternatives include entering into public/ private partnerships, establishing competitive but regulated land markets, providing housing finance and building materials, restructuring production regulations, improving urban and regional planning, and reforming tax systems that support rather than detract from urban sustainability.

Urban sustainability issues are highly interdependent. Urban transport congestion and air pollution reflect the geography, land use patterns, and structure of cities. In the context of improving environmental quality in cities, NTREE suggested that Canada's efforts should focus on urban form, urban transportation systems, and energy use in buildings. In considering urban form, there are two competing perspectives. On the one hand, there is a view that compact cities—large, dense, and concentrated urban centres—best support the principles of sustainable development. This approach offers the advantage of focusing on promoting energy intensive patterns of activity, which would assist in dealing with the reduction of greenhouse gases. Disadvantages of the compact city include a rejection of the current and dominant form of living—suburban and semi-rural, less green space in urban areas, and a centralization of decision making. On the other hand, others suggest that green cities, which have a more open structure and a landscape mosaic comprised of buildings, agricultural fields, and other open spaces support the principles of sustainable development.

Mary Ellen Tyler of the University of Calgary's Faculty of Environmental Design suggests four approaches to measuring urban sustainability. These are:

- The **ecological footprint** championed by people such as Rees and Wackernagel. It measures the inputs required to sustain human activities and the waste this use generates. Vancouver covers an area of 4,000 km^2 and has a regional population of 1.8 million people. However, its ecological footprint is about 19 times its geographic extent (see Chapter 1).

- The **urban metabolism/life support system** approach emphasizes the types and quantities of resources a city consumes and releases, and determines the extent and significance of this resource use and disposal pattern on ecological functions, i.e., air, water, soil, and energy.

- **Urban sustainability/sustainable cities,** which has its roots in the Brundtland Commission (see Chapter 1). It emphasizes the need to merge and give equal consideration to social, economic, and environmental values. It reflects a process of open decision making, and the outcomes and impacts arising from that process.

- **Urban environment reporting** such as Statistics Canada's "Stress-Response Environmental Statistical System" (STRESS), which organized information on human activity by focusing attention on environmental media (i.e., air, land, and water) and specific activities (e.g., agriculture, forestry, and fishing), with thematic "stresses" imposed on the environment and the environment's response to these stresses.

These academic debates surrounding the concept and measurement of sustainable cities highlight that there are no simple and universal solutions to these problems, but we do have the ability to discuss and reshape the urban world we live in. One of the problems we face in considering solutions is to understand the processes that contribute to the problem. The following discussion describes world urbanization, the causes of urban growth, and key problems in the developing and developed world.

CAUSES OF URBAN GROWTH

Urban populations grow in two ways: by natural increase (more births than deaths) and by immigration. Natural increase is fueled by improved food supplies, better sanitation, and advances in medical care that reduce death rates and cause populations to grow both within cities and in the rural areas around them (Chapter 4). In Latin America and East Asia, natural increase is responsible for two-thirds of urban population growth. In Africa and West Asia, immigration is the largest source of urban growth. Immigration to cities can be caused both by **push factors** that force people out of the country and by **pull factors** that draw them into the city.

Immigration Push Factors

People migrate to cities for many reasons. In some areas, the countryside is overpopulated and simply can't support more people. The "surplus" population is forced to migrate to cities in search of jobs, food, and housing. Not all rural-to-urban shifts are caused by overcrowding in the country, however. In some places, economic forces or political, racial, or religious conflicts drive people out of their homes. The countryside may actually be depopulated by such demographic shifts. The United Nations estimated that in 1992 at least 10 million people fled their native country and that another 30 or 40 million were internal refugees within their own country, displaced by political, economic, or social instability. Many of these refugees end up in the already overcrowded megacities of the developing world.

Land tenure patterns and changes in agriculture also play a role in pushing people into cities. The same pattern of agricultural mechanization that made farm labour largely obsolete in the United States early in the last century is spreading now to developing countries. Furthermore, where land ownership is concentrated in the hands of a wealthy elite, subsistence farmers are often forced off the land so it can be converted to grazing lands or monoculture cash crops. Speculators and absentee landlords let good farmland sit idle that otherwise might house and feed rural families.

Immigration Pull Factors

Even in the largest and most hectic cities, many people are there by choice, attracted by the excitement, vitality, and opportunity to meet others like themselves. Cities offer jobs, housing, entertainment, and freedom from the constraints of village traditions. Possibilities exist in the city for upward social mobility, prestige, and power not ordinarily available in the country. Cities support specialization in arts, crafts, and professions for which markets don't exist elsewhere.

Modern communications also draw people to cities by broadcasting images of luxury and opportunity. An estimated 90 percent of the people in Egypt, for instance, have access to a television set. The immediacy of television makes city life seem more familiar and attainable than ever before. We generally assume that beggars and homeless people on the streets of teeming Third World cities have no other choice of where to live, but many of these people want to be in the city. In spite of what appears to be dismal conditions, living in the city may be preferable to what the country had to offer.

Government Policies

Government policies often favour urban over rural areas in ways that both push and pull people into the cities. Developing countries commonly spend most of their budgets on improving urban areas (especially around the capital city where leaders live), even though only a small percentage of the population lives there or benefits directly from the investment. This gives the major cities a virtual monopoly on new jobs, housing, education, and opportunities, all of which bring in rural people searching for a better life. In Peru, for example, Lima accounts for 20 percent of the country's population, but has 50 percent of the national wealth, 60 percent of the manufacturing, 65 percent of the retail trade, 73 percent of the industrial wages, and 90 percent of all banking in the country. Similar statistics pertain to São Paulo, Mexico City, Manila, Cairo, Lagos, Bogotá, and a host of other cities.

Governments often manipulate exchange rates and food prices for the benefit of more politically powerful urban populations but at the expense of rural people. Importing lower-priced food pleases city residents, but local farmers then find it uneconomical to grow crops. As a result, an increased number of people leave rural areas to become part of a large urban workforce, keeping wages down and industrial production high. Zambia, for instance, sets maize prices below the cost of local production to discourage farming and to maintain a large pool of workers for the mines. Keeping the currency exchange rate high stimulates export trade but makes it difficult for small farmers to buy the fuels, machinery, fertilizers, and seeds that they need. This depresses rural employment and rural income while stimulating the urban economy. The effect is to transfer wealth from the country to the city.

CURRENT URBAN PROBLEMS

Large cities in both developed and developing countries face similar challenges in accommodating the needs and by-products of dense populations. The problems are most intense, however, in rapidly growing cities of developing nations.

Depending on what we do over the next few years, urbanization holds the promise of an unfettered and high quality of life in the future as well as the threat of tremendous human suffering. At a global level, the Habitat II Conference, the second United Nations Conference on Human Settlements held in Istanbul, Turkey, in 1996, is an important milestone. Referred to as "the City Summit," it followed the first one held in Vancouver in 1976. In Istanbul, the intent was to consider the implementation of that portion of Agenda 21 that stressed the need to improve the social, environmental, and economic quality of human settlements and living and working environments for all people, in particular, the urban

and rural poor. The discussions at Habitat II were partly based on the United Nation's publication *An Urbanizing World: Global Report on Human Settlements 1996*. That document details the staggering problems facing us and why. It also suggested we have the power, knowledge, and resources to do something about it.

The Developing World

As Figure 21.7 shows, 90 percent of the human population growth in this century is expected to occur in the developing world, mainly in Africa, Asia, and South America. Almost all of that growth will occur in cities—especially the largest cities—which already have trouble supplying food, water, housing, jobs, and basic services for their residents. The unplanned and uncontrollable growth of those cities causes tragic urban environmental problems. Consider as you study urban conditions what responsibilities we in the richer countries have to help others who are less fortunate and how we might do so.

Traffic and Congestion

A first-time visitor to a supercity—particularly in a less-developed country—is often overwhelmed by the immense crush of pedestrians and vehicles of all sorts that clog the streets. The noise, congestion, and confusion of traffic make it seem suicidal to venture onto the street. Jakarta, for instance, is one of the most densely populated cities in the world (Fig. 21.9). Traffic is chaotic almost all the time. People commonly spend three or four hours each way commuting to work from outlying areas. Bangkok also has monumental traffic problems. The average resident spends the equivalent of 44 days a year sitting in traffic jams. About 20 percent of all fuel is consumed by vehicles standing still. Hours of work lost each year are worth at least $3 billion.

FIGURE 21.9 Motorized rickshaws, motor scooters, bicycles, street vendors, and pedestrians all vie for space on the crowded streets of Jakarta. But in spite of the difficulties of living here, people work hard and have hope for the future.

Air Pollution

The dense traffic (commonly old, poorly maintained vehicles), smoky factories, and use of wood or coal fires for cooking and heating often create a thick pall of air pollution in the world's supercities. Lenient pollution laws, corrupt officials, inadequate testing equipment, ignorance about the sources and effects of pollution, and lack of funds to correct dangerous situations usually exacerbate the problem. What is its human toll? An estimated 60 percent of Calcutta's residents are thought to suffer from respiratory diseases linked to air pollution. Lung cancer mortality in Shanghai is reported to be four to seven times higher than rates in the countryside. Mexico City, which sits in a high mountain bowl with abundant sunshine, little rain, high traffic levels, and frequent air stagnation, has one of the highest levels of photochemical smog (Chapter 9) in the world.

Sewer Systems and Water Pollution

Few cities in developing countries can afford to build modern waste treatment systems for their rapidly growing populations. The World Bank estimates that only 35 percent of urban residents in developing countries have satisfactory sanitation services (Fig. 21.10). The situation is especially desperate in Latin America, where only 2 percent of urban sewage receives any treatment. In Egypt, Cairo's sewer system was built about 50 years ago to serve a population of

FIGURE 21.10 This tidal canal in Jakarta serves as an open sewer. By some estimates, about half of the 10 million residents of this city have no access to modern sanitation systems.

2 million people. It is now being overwhelmed by more than 10 million people. Less than one-tenth of India's 3,000 towns and cities have even partial sewage systems and water treatment facilities. Some 150 million of India's urban residents lack access to sanitary sewer systems. In Colombia, the Bogotá River, 200 km downstream from Bogotá's 5 million residents, still has an average fecal bacteria count of 7.3 million cells per litre, more than 700,000 times the safe drinking level and 3,500 times higher than the limit for swimming.

Some 400 million people, or about one-third of the population, in developing world cities do not have safe drinking water, according to the World Bank. Although city dwellers are somewhat more likely than rural people to have clean water, this still represents a large problem. Where people have to buy water from merchants, it often costs 100 times as much as piped city water and may not be any safer to drink. Many rivers and streams in Third World countries are little more than open sewers, and yet they are all that poor people have for washing clothes, bathing, cooking, and—in the worst cases—for drinking. Diarrhea, dysentery, typhoid, and cholera are widespread diseases in these countries, and infant mortality is tragically high (Chapter 11).

Housing

The United Nations estimates that at least 1 billion people—20 percent of the world's population—live in crowded, unsanitary slums of the central cities and in the vast shantytowns and squatter settlements that ring the outskirts of most Third World cities. Around 100 million people have no home at all. In Mumbay, India, for example, it is thought that half a million people sleep on the streets, sidewalks, and traffic circles because they can find no other place to live (Fig. 21.11). In São Paulo, perhaps 1 million "street kids" who have run away from home or been abandoned by their parents live however and wherever they can. This is surely a symptom of a tragic failure of social systems.

Slums are generally legal but inadequate multifamily tenements or rooming houses, either custom built to rent to poor people or converted from some other use. The chals of Mumbay, India, for example, are high-rise tenements built in the 1950s to house immigrant workers. Never very safe or sturdy, these dingy, airless buildings are already crumbling and often collapse without warning. Eighty-four percent of the families in these tenements live in a single room; half of those families consist of six or more people. Typically, they have less than 2 square metres of floor space per person and only one or two beds for the whole family. They may share kitchen and bathroom facilities down the hall with 50 to 75 other people. Even more crowded are the rooming houses for mill workers where up to 25 men sleep in a single room only 7 metres square. Because of this crowding, household accidents are a common cause of injuries and deaths in Third World cities, especially to children. Charcoal braziers or kerosene stoves used in crowded homes are a routine source of fires and injuries. With no place to store dangerous objects beyond the reach of children, accidental poisonings and other mishaps are a constant hazard.

Shantytowns are settlements created when people move onto undeveloped lands and build their own houses. Shacks are

FIGURE 21.11 In Mumbay, as many as half a million people sleep on the streets because they have no other place to live. Ten times as many live in crowded, dangerous slums and shantytowns throughout the city.

FIGURE 21.12 Shantytowns called *favelas* perch on hillsides above Rio de Janeiro.

built of corrugated metal, discarded packing crates, brush, plastic sheets, or whatever building materials people can scavenge. Some shantytowns are simply illegal subdivisions where the landowner rents land without city approval. Others are spontaneous or popular settlements or **squatter towns** where people occupy land without the owner's permission. Sometimes this occupation involves thousands of people who move onto unused land in a highly organized, overnight land invasion, building huts and laying out streets, markets, and schools before authorities can root them out. In other cases, shantytowns just gradually "happen."

Called *barriads, barrios, favelas,* or *turgios* in Latin America, *bidonvillas* in Africa, or *bustees* in India, shantytowns surround every megacity in the developing world (Fig. 21.12). They are not an exclusive feature of poor countries, however. Some 250,000 immigrants and impoverished citizens live in the *colonias* along the southern Rio Grande in Texas. Only 2 percent have access to adequate sanitation. Many live in conditions as awful as you would see in any Third World city.

About three-quarters of the residents of Addis Ababa, Ethiopia, or Luanda, Angola, live in squalid refugee camps. Two-thirds of the population of Calcutta live in unplanned squatter

settlements and nearly half of the 20 million people in Mexico City live in uncontrolled, unauthorized shantytowns. Many governments try to clean out illegal settlements by bulldozing the huts and sending riot police to drive out the settlers, but the people either move back in or relocate to another shantytown.

These popular but unauthorized settlements usually lack sewers, clean water supplies, electricity, and roads. Often the land on which they are built was not previously used because it is unsafe or unsuitable for habitation. In Bhopal, India, and Mexico City, for example, squatter settlements were built next to deadly industrial sites. In Rio de Janeiro, La Paz (Bolivia), Guatemala City, and Caracas (Venezuela), they are perched on landslide-prone hills. In Lima (Peru), Khartoum (Sudan), and Nouakchott (Mauritania), shantytowns have spread onto sandy deserts. In Manila, thousands of people live in huts built on towering mounds of garbage and burning industrial waste in city dumps (see Fig. 18.3).

As desperate and inhumane as conditions are in these slums and shantytowns, many people do more than merely survive there. They keep themselves clean, raise families, educate their children, find jobs, and save a little money to send home to their parents. They learn to live in a dangerous, confusing, and rapidly changing world and have hope for the future. The people have parties; they sing and laugh and cry. They are amazingly adaptable and resilient. In many ways, their lives are no worse than those in the early industrial cities of Europe and America more than a century ago. Perhaps continuing development will bring better conditions to cities of the Third World as it has for many in the First World.

The Developed World

For the most part, the rapid growth of central cities that accompanied industrialization in nineteenth and early-twentieth century Europe and North America has now slowed or even reversed. For instance, London, England, once the most populous city in the world, has lost nearly 2 million people, dropping from its high of 8.6 million in 1939 to 6.7 now. While the greater metropolitan area surrounding London has been expanding to about 10 million inhabitants, the city itself is now only the 12th largest city in the world.

Many of the worst environmental problems of the more developed countries have been substantially reduced in recent years. Improved sanitation and medical care have reduced or totally eliminated many communicable diseases that once afflicted urban residents, although outbreaks of sickness and death in North Battleford, Saskatchewan, and Walkerton Ontario, highlight the need for water quality monitoring, reliable testing, and source area protection.

Air and water quality have improved since the beginning of the century as heavy industries such as steel smelting and chemical manufacturing have moved to developing countries. However, the closing of these types of industrial sites have sometimes left a legacy of pollution, as is the case with the Sydney Tar Ponds (see Case Study, p. 473). In consumer and information economies, workers no longer need to be concentrated in central cities. They can live and work in dispersed sites, including their homes! Automobiles now make it possible for the working class to enjoy amenities, such as single-family homes, yards, and access to recreational facilities that were once available only to the rich.

Urbanization Issues in Canada

Like the rest of the world, Canada's economy has shifted from a rural-based one to an urban one. Canada is now one of the most highly urbanized nations in the world, with nearly 80 percent of its population living in large urban centres and this is expected to increase (Table 21.3). Canadian cities have emerged as the main engines of the national economy. Dealing with the transition from a rural to urban-based economy has been challenging. The vitality and fabric of many rural areas have declined as people have left, and businesses and services (e.g., stores, farm suppliers, post offices, and schools) have closed. In urban areas, the challenge to meet current and future environmental, social, and economic issues is significant.

Accommodating this growth without exacerbating urban sprawl and placing increased demands on a generally fragile road and water infrastructure and inadequate public transit system will be a challenge. Designing and building office buildings,

TABLE 21.3	Projected Population by 2020 in Canada's Eight Largest CMAs	
Vancouver	3,000,000
Calgary	1,200,000
Edmonton	1,100,000
Winnipeg	700,000
Toronto	6,600,000
Montreal	3,800,000
Ottawa-Gatineau	1,300,000
Halifax	500,000

Source: Statistics Canada.

transit systems, water and sewage treatment plants, electrical generating and distribution systems, and increasing the efficiency of water and energy use are requirements if Canadian cities are to be considered environmentally sustainable. This will also require a shift from a fossil fuel-based society to one that increasingly relies on renewable forms of energy, such as ethanol and bio-diesel vehicles, hydrogen fuel cells, wind and solar power, and district energy systems.

Immigration has been a major factor behind the increase in the urban population in Canada and will continue to be so in the future. Canadian policy has generally encouraged immigrants to come to Canada and many of these people have located in the urban centres of Toronto, Vancouver, and Montreal. Much of this growth occurred after World War II and continues today. Statistics Canada reports that 445,000 immigrants settled in the Toronto area and 180,000 around Vancouver between 1996 and 2001. Edmonton and Calgary owe their recent growth to migrants from different parts of Canada seeking economic opportunities in those cities. If Canada is to remain committed to the principles of a multicultural society, all levels of government in Canada must be prepared to strengthen and support cultural diversity within its cities. Closing the opportunity gap between aboriginal and non-aboriginal people must also be reduced.

In the 1990s, provinces downloaded responsibilities for housing, welfare, and urban transit without any additional funding. The municipal property tax base, the primary source for local governments to generate revenue, has been inadequate to meet the additional responsibilities associated with these services. In 2001, then Prime Minister Chrétien announced the formation of a caucus Task Force on Urban Issues. Its creation highlights and illustrates the idea of a New Deal for Canadian Cities and Communities, which has become an important theme on the national policy scene. According to the Canadian Policy Research Network, this idea has been driven by two factors. First, there is growing recognition that the "prosperity and well-being of all Canadians depends greatly on the 'place quality' of our cities and communities. " Second, "cross-national studies now show that Canadian urban centres, not long

CASE STUDY

THE SYDNEY TAR PONDS

In the late 1880s, coal mining became a mainstay of the Sydney, Nova Scotia, economy. Later, the construction of a steel plant and coke plant meant further prosperity for the region. The coke ovens baked the locally mined coal at high temperatures in order to remove undesirable by-products to produce a purer and hotter-burning fuel source for the steel mill's open-hearth blast furnaces. Sydney Steel Co. was the cornerstone of this activity and it provided employment for almost 100 years. Changing economic circumstances and technological developments started to spell the end of the prosperity in the 1960s. Although federal and provincial governments provided financial support, the coke ovens closed in 1988 and the steel plant ceased operations in 2001, having played a pivotal role in the history, economy, and culture of the surrounding community.

While it was operating, the smoke that came out of the smoke stacks and the water that discharged from water pipes translated to industrial activity and local jobs. However, with little government environmental regulations, little money was spent to reduce the amount of water and air emissions that were by-products from the coke ovens and steel furnaces. The coke oven is at the centre of the tar ponds disaster. During the coking process, undesirable tar and gases were separated off from the desired coke, with the water being discharged into a creek before spilling into Sydney Harbour. This area is known as Muggah Creek. This waste is now considered toxic and contains benzene, kerosene, arsenic, lead, naphthalene, and other toxics.

After 80 years, water discharged from the coke ovens into Muggah Creek has left an environmental disaster covering an area of 95 ha, the largest such site in Canada, about 700,000 tonnes of chemical waste and raw sewage discharge, and 40,000 tonnes that are contaminated with polychlorinated biphenyls (PCBs). Hence the reference to "Tar Ponds." Thirty-six sewage outfalls continue to pour 13 million litres of

FIGURE 21.13 This aerial view of Sydney, Nova Scotia, shows part of the "tar ponds," possibly one of North America's worst environmental disasters. The water of Muggah Creek flows by the old coking operations before spilling into the tidal estuary. The ponds contain approximately 700,00 tonnes of toxic sludge.

untreated waste daily into the tar ponds. If water levels are too high in the south pond, sewers do not drain, causing backups in adjacent buildings.

In the 1980s, government studies found high levels of polycyclic aromatic hydrocarbons (PAHs) at the site, some of which are linked to cancer in humans, as well as polychlorinated biphenyls (PCBs), other chemicals, and metals. The site is associated with many negative health effects. Cape Breton has the highest cancer rate and lowest life expectancy in Canada. The soil located near the toxic site is contaminated with arsenic and other hazardous chemicals. Residents living in close proximity to the Tar Ponds have reported "orange goo" seeping into their cellars and basements. This indicates the groundwater has also been contaminated. Lobsters and other aquatic plant and animal life have been affected, bringing about the closing of the lobster.

Federal and provincial governments responded to this problem by agreeing to clean up the site. In a 1986 agreement, the governments committed $55 million to excavate the sludge and eliminate PAH contamination by burning it at very high temperatures in a specially designed incinerator. Unforeseen

problems related to higher than expected PCB levels, combined with difficulties moving the large volumes of PCB-laden sludge, contributed to termination of the project in 1994. In 1996, the provincial government proposal to contain the area was rejected by the public.

The community began to become more assertive in defining acceptable methods of disposable and formed a Joint Action Group (JAG). Comprised of local residents, business people, local representatives, and youth, it made a series of recommendations to the government on acceptable measures. In 1998, an agreement formalizing JAG's relationship with the three levels of government was signed. This was followed by a commitment by governments to provide $62 million to carry out the studies, design work, and other essential preparations. In May 2004, federal and provincial governments allocated $400 million to the project. It is expected to take 10 years to complete. The community remains fully involved in the remediation process as it proceeds.

See video clip at: http://www.courseworld.com/bio/biogr/tarpond.html.

---

ago recognized for their high quality, are falling behind the international standard as investments from an earlier generation run their course." The remainder of this chapter examines urban issues in the developed world and in Canada by examining economic, social, and environmental challenges and opportunities.

Economic Challenges and Opportunities

In promoting and supporting growth, cities must provide infrastructure and services. However, the capacity of Canadian municipalities to adequately service growth is constrained by its limited taxing powers, such as the property tax, which fails to consider a

CASE STUDY

ONTARIO'S HIGHWAY 407: PRIVATIZING HIGHWAY SERVICES

Rather than being like most highways, which are government-operated without any direct fees for use, Highway 407 is not government-operated and is a pay-for-use highway. Established in 1999, the highway is operated by 407/ETR International Incorporated through a 99-year lease arrangement that was sold by the Ontario provincial government for approximately $3.1 billion. 407/ETR International Incorporated is owned by a consortium comprised of the Canadian subsidiary of Cintra Concesiones de Infraestructuras be Transporte, which is co-owned by Spain's Grupo Ferrivial and Australian-headquartered Macquarie Infrastructure Group, and Canadian-based engineering firm SNC-Lavalin.

The lease agreement is the key mechanism that guides operation of the highway and establishes the responsibilities of the 407/ETR International Incorporated and the provincial government. It traverses the area north of Toronto from Hamilton/Burlington in the west to Pickering in the east. On an average workday, over 320,000 trips are taken on Highway 407 (up 40 percent from 1999), which relieves congestion on other already over-congested highways, including the 401 and 427.

One innovative aspect of the highway is the use of transponders that toll drivers automatically. For those drivers who do not have transponders, a picture of their licence plate is taken and they are billed through the mail. Thus, there is no need to slow down for toll booths. It is perceived to be the first financially successful privately-owned toll road. Despite the use of this technology and its strong financial performance, there has been controversy.

In 2004, a major difference of opinion arose between the province of Ontario and the 407/ETR International Incorporated. In February, the government served notice that it considered the consortium in default of the contract because the consortium failed to seek government approval for toll increases. Since 1999, rates have increased by more than 200 percent for some peak driving hours. In early 2004, the tolls were $0.131 per kilometre for cars and light trucks during off-peak times and $0.1395 per kilometre for cars and light trucks during peak times. If the vehicle does not have a transponder, there is an additional charge of $3.35 per trip. In July 2004, an arbitrator ruled in favour of 407 International Inc., thus government approval is not required.

Believing that it is unwise for a private consortium to have unfettered rights in delivering this essential service to the public, the provincial government has indicated it will appeal this ruling and renegotiate the lease agreement. One outcome could see the provincial government reacquiring the highway.

---

person's or business's ability to pay. If Canada's cities are to develop the new knowledge-based economy (e.g., bio-medical, fibre-optics, media, information technology), urban areas must offer the infrastructure and services to attract people and at the same time make it affordable for businesses, and people of different income levels.

Water and sewage treatments plants, highways, public transit, airports, and power grids are critical in supporting Canadians' quality of life. While it will likely be necessary to change the way in which Canadians use all of these services in order to be economically and environmentally sustainable, the Federation of Canadian Municipalities estimates a $44 billion shortfall to maintain, improve, and (when appropriate) expand the municipal infrastructure. Since municipalities are constrained by the limitations of the relatively small property tax base, some have privatized water systems. Some provincial governments have also privatized some of their services, such as highway systems (e.g., Highway 407 in Ontario (see Case Study, above)) or entered public private partnerships to finance these services.

The efficient and effective transport of goods and people are central to Canada's economy. Gridlock is seriously impeding the competitiveness of Canada's urban regions. The prime minister's Caucus Task Force on Urban Issues reported that traffic congestion in the Greater Toronto Area cost $2 billion annually through lost productivity. Montreal, Ottawa-Gatineau, downtown Calgary, and the Lion's Gate Bridge in Vancouver are well-known points of congestion in the current urban landscape. While the immediate response is to construct more roads, congestion likely reflects inadequate investments in public transport.

Canada is the only G8 country without a national urban transit investment program. Some progress is being made. In York Region (Ontario), a strong emphasis has been placed on public transit in its Transportation Master Plan. There will be additional comments on transportation issues when we discuss urban sprawl in a later section of this chapter.

Social Challenges and Opportunities

Maxwell noted that after World War II, one significant difference between Canadian and American cities was the lack of spatial segregation and the relatively minor inner city problems. Since 1980, there has been a concentration of poverty in specific parts of many of Canadian cities. The polarization of income and the spatial segregation between rich and poor has led to the **"ghettoization,"** in which a diversity of our most vulnerable and poverty-stricken people—visible minorities, aboriginals, single-parent families, and disabled people—cluster to poor areas. Higher levels of drug and alcohol abuse, crime and marginalization are associated with these areas. Poverty often prevents people from fully accessing needed services and adequate housing. In many cities, 1 in 5 people have very low incomes (Table 21.4). Groups most often at risk include the working poor and their children, single-parent families, seniors, urban aboriginals, people with disabilities, immigrants,

In Vancouver's Downtown Eastside, federal, provincial, and local governments are working to solve an escalating rate of disease, crime, poverty, and homelessness. Vancouver's Downtown Eastside was once a vibrant commercial and entertainment district in the economic heart of the city. However, this changed between the 1960s and 1990s as economic decline, business closures, a large open market in illegal drugs, poverty, and homelessness undermined the vitality of the community. In 1997, the region's health authority declared that rising HIV infection rates among intravenous drug users constituted a public health crisis. Government leaders jointly initi-ated the Vancouver Agreement to ensure a coordinated, effective response.

In March 2000, the federal, provincial, and local governments signed the "Vancouver Agreement" that committed them to work cooperatively in order to:

- improve economic development by increasing legitimate business activity;
- improve social well-being of residents by: (i) increasing the employment rate of community residents; (ii) ensuring there is no net loss of low-income housing; and (iii) decreasing the number of children (especially aboriginal children) living in the community who are removed from their homes;
- improve the health of local residents by: (i) reducing preventable deaths, and (ii) reducing levels of communicable disease;
- improve community safety by: (i) reducing open drug use; and (ii) reducing theft, property crime and violent crime.

Similar initiatives have also been undertaken in Winnipeg and Edmonton. The Winnipeg agreement was completed on August 31, 2001.

TABLE 21.4 Persons with Low Income in Selected Urban Regions (1998)

Montreal	29.0 percent
Winnipeg	21.0 percent
Ottawa-Gatineau	20.9 percent
Quebec City	20.3 percent
All urban regions with populations over 500,000	19.4 percent
Halifax	17.9 percent
Edmonton	17.8 percent
Calgary	16.7 percent
Vancouver	16.1 percent
London	15.5 percent
Toronto	14.8 percent
Hamilton-Burlington	12.7 percent

Source: Prime Minister's Caucus Task Force on Urban Issues.

refugees, and women. Child and family poverty is increasing. If we can improve the development and learning of children by reducing poverty, long-term personal and social development could be greatly enhanced.

There are a number of other important social challenges in cities. Nearly every major urban centre in Canada has a shortage of affordable housing. More than 50 percent of aboriginal people live in cities, not reserves. Many are attracted to urban areas by a perception of opportunity, such as employment, education, the good life. However, the prime minister's Task Force maintained that many aboriginal people are ill equipped to adjust to urban living and with a lack of affordable housing, some turn to crime and substance abuse. Any Canadian urban strategy must meet their needs and circumstances. Urban centres are the final destination of about 85 percent of the 220,000 immigrants who come to Canada every year. Supporting the need for housing, employment, and child support strains the capacity of local governments.

In the 1990s, governments tightened the requirements for people to qualify for social assistance, and even once on assistance it became easier to lose it. Maxwell cites the example of a person making $25,000 per year (a little more than $10 per hour on a full-time wage) and offered a raise. They may face increased levels of taxation or lose social service support because the income and social benefit they receive will be taxed. Thus, they essentially are penalized for accepting the raise—a real dilemma for the working poor. Poverty has at least two costs to society. The first is the lack of meaningful and innovative contributions people in these circumstances could make to the Canadian society and economy. Second, the high cost of health and social services, and policing increases demands on an already strained system. Alcohol and substance abuse are also problems requiring innovative solutions and strong political commitment. Initiatives, such as the Vancouver Agreement (see Case Study, above) may provide one approach to redress these problems.

Unfortunately, the prime minister's Caucus Task Force had few targeted comments on what to do about urban social issues such as spatially segregated poverty and inadequate child care. Instead, it focused on transportation, housing, and infrastructure.

Environmental Challenges and Opportunities

Safe water, clean air, dealing with brownfields (contaminated sites), and promoting healthy well-planned communities are four key environmental challenges.

Safe Water

Although Canada is perceived by many people as a country of abundant and clean water resources, there is growing concern about the issue of "safe water" after unfortunate experiences in Walkerton, Ontario, and North Battleford, Saskatchewan. Water and waste water treatment varies across the country and some municipalities and aboriginal communities in particular use outdated purification and/or treatment methods.

Clean Air

Like water, clean air is essential in sustaining healthy communities. Despite improved fuel efficiency from passengers cars (SUVs and light trucks being the exception) and billions being spent to support public transit, NTREE noted most key indicators suggest negative trends: the use of cars is on the rise and urban rider ship is down. Concentrations of ground level ozone, a by-product of fossil fuel emissions, are on the rise (see Chapter 9).

The release of greenhouse gases and the implications for climate change is one of the most important and controversial issues confronting the global community. Although Canada and other nations (although not the United States) have ratified the Kyoto Protocol, which establishes national targets for greenhouse emissions, its implementation is unclear.

Indeed, during the 2004 federal election the Conservative Party of Canada suggested Canada should withdraw from the Protocol and put the effort into solving other air quality problems. Key to reducing the emission of greenhouse gases is to reduce energy use and adopt the strategies discussed in Chapter 13. These include reducing energy for heating and cooling our buildings and reducing our transport requirements and emissions.

Urban Sprawl

While the move to the suburbs and rural areas has brought many benefits to the average citizen, it has also caused numerous urban problems. Cities that were once compact now spread over the landscape, consuming open space, and making inefficient use of available resources. This pattern of growth is known as **sprawl.** While there is no universally accepted definition of the term, sprawl generally includes the characteristics outlined in Table 21.5. In many cities in Europe and North America, the number of people living within the political boundaries of major cities declined, but populations in adjacent centres has increased, requiring roads, water works, and public transit services.

In many North American metropolitan areas, the bulk of new housing is in large, tract developments that leapfrog beyond the edge of a city in search of inexpensive rural land. The U.S. Department of Housing and Urban Development estimates that urban sprawl consumes some 200,000 ha of farmland each year. Canadians and North Americans continue to move and live in the suburbs. Since 1981, Statistics Canada indicates that urban population densities have declined, suggesting that Canadians are living on larger

TABLE 21.5	Characteristics of Urban Sprawl

1. Unlimited outward extension.
2. Low-density residential and commercial development.
3. Leapfrog development that consumes farmland and natural areas.
4. Fragmentation of power among many small units of government.
5. Dominance of highways and private automobiles.
6. No centralized planning or control of land uses.
7. Widespread strip malls and "big-box" shopping centres.
8. Great fiscal disparities among localities.
9. Reliance on deteriorating older neighbourhoods for low-income housing.
10. Decaying city centres as new development occurs in previously rural areas.

Source: Excerpt from speech by Anthony Downs at the CTS Transportation Research Conference, as appeared on Website by Planners Web, Burlington, VT, 2001.

FIGURE 21.14 Sprawl consumes natural areas, farmland, and open space. It wastes resources, results in traffic congestion, and requires large investments to replace schools, roads, parks, and shopping that is abandoned in the inner city.

lots. This trend makes it more expensive to service with water, roads, police and fire protection, and public transit. It also means that more land, some of which may have agricultural, environmental, or amenity values are removed from the landscape (Fig. 21.14).

The amount of urbanized land has increased from 16,000 km^2 to 28,000 km^2 between 1971 and 1996. (See, for example, Figure 21.15.) According to the David Suzuki Foundation, the rate at which land is being urbanized in Canada is sometimes double the rate of population increase. Between 1996 and 2001, this meant land conversion rates of some 15 percent in the four largest urban areas in Canada. The difference between the rates of population growth and the land they occupy is one way of measuring urban sprawl.

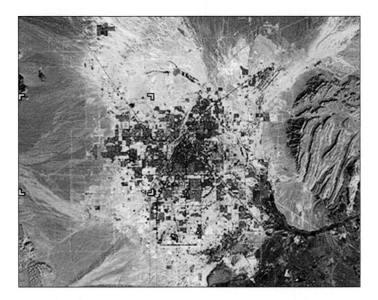

FIGURE 21.15 Satellite images of Las Vegas, NV, in 1972 (*a*) and 1992 (*b*) show how the metropolitan area has grown over two decades.
Source: U.S. Geological Survey.

Since cities are often located in fertile river valleys or shorelines, much of that land could be valuable in producing crops. However, since planning authority is divided among many small jurisdictions, metropolitan areas have no way to regulate growth or provide for rational, efficient use. Small towns and townships generally welcome this growth because it profits local landowners and business people in the short term. However, as the demand for urban-like services from the rural-urban dweller increase, the cost for new and upgraded roads, sewers, water mains, power lines, schools, shopping centres, and other extra infrastructure, becomes expensive to support this low-density development. The Ontario and Quebec governments forced the merger of the smaller communities that formed Metropolitan Toronto and Montreal, in part, to provide for more systematic and regional planning. Initially, some people and groups fiercely resisted these decisions. Promoting changes to well-established political and administrative arrangements are difficult; much like implementing sustainable urban development.

Several factors promote development outside of urban cores. In its report titled, *Driven to Action: Stopping Sprawl in Your Community,* the Suzuki Foundation suggests that personal automobiles are perceived to be a convenient means of transportation. Governments also subsidize the infrastructure required to support automobiles and for water, sewage, and electric and gas utilities. These provide a bias that favours low-density development. Landowners, builders, real estate agents, and others who profit have a vested interest in maintaining the *status quo*. Since residents will require regular car purchases and maintenance, the auto and auto-repair sectors benefit from urban sprawl.

Another dimension of the urban sprawl problem in Canada concerns the manner in which the tax system and cost-sharing programs work against sustainability. For instance, the property tax system, which is based on current value assessment, reflects market value and not the costs of serving properties. NTREE noted that the property tax system is also biased against multi-unit residential development, which allows for increased population densities, because a higher rate of taxation is applied to these buildings than to single detached buildings.

Development charges are another means for municipalities to obtain income. The intent is to ensure new development costs (e.g., roads, transit, sewers, schools, fire, police) are covered. However, since they are typically structured on a citywide, average cost per dwelling basis, a single detached home with a 25-metre frontage pays the same as someone with a 10-metre frontage. The Town of Markham (Ontario) has established a varying development charge for regions across the town, although within the regions the development costs are averaged. Reversing the indirect manner in which these types of tax and fee structures would go a long way to supporting sustainable cities.

Since many Canadians and Americans live far from where they work, shop, and recreate, they consider it essential to own their own vehicle. The Suzuki Foundation cited a study that estimated the average vehicle ownership cost (2002) in Canada was $6262.43 for the year, or about $17.16 per day. In addition, operating costs were $0.1225 cents per kilometre. The average U.S. driver spends about 443 hours per year sitting behind a steering wheel. This means that for most people, the equivalent of one full 8-hour day per week is spent sitting in an automobile. Sprawling suburbs and a relative lack of physical activity contribute to the near-epidemic levels of obesity for all age groups in North America. A study in the *American Journal of Public Health* concluded that people living in the most sprawling communities are likely to weigh 3 kg more than those living in the most compact communities. Chapter 9 discussed the implications from exhaust emissions from automobiles.

Building the roads, parking lots, filling stations and other facilities needed for an automobile-centred society takes a vast amount of space and resources. In some North American metropolitan areas, it is estimated that one-third of all land is devoted to the automobile. To make it easier for suburban residents to get from their homes to their jobs and shopping, we provide an amazing network of highways. At a cost of several trillion dollars to build, the intense highway systems in North America were designed allow us to drive at high speeds from source to destination without ever having to stop. As more and more drivers clog the highways, however, the reality is far different. For example, in Los Angeles, which has the worst traffic congestion in the United States, the average speed in 1982 was 93 km/hr, and the average driver spent less than four hours in traffic jams. In 2000, the average speed in Los Angeles was only 57.3 km/hr, and the average driver spent 82 hours per year waiting in traffic. Motorists in Toronto, Vancouver, Montreal, and other urban areas can sympathize with these statistics. Rather than ease congestion and save fuel, more highways can exacerbate the problem (Fig. 21.16).

Sprawl also creates problems in central cities from which residents and businesses have fled. With a reduced tax base and fewer civic leaders living or working in downtown areas, the city is unable to maintain it infrastructure. This exit is accelerated due globalization, which often sees the centralization of business in large but fewer centres in Canada or abroad. Streets, parks, schools, and public buildings fall into disrepair at the same time that these types of facilities are being built in the suburbs. The poor who are left behind often cannot find jobs where they live and have no way to commute to more distant jobs. The low-density development in suburbs is radically and economically exclusionary because it provides no affordable housing and makes it impractical to design a viable public transit system.

Finally, sprawl fosters uniformity and alienation from local history and the natural environment. Housing developments often are based on only a few standard housing designs, while shopping centres and strip malls everywhere feature the same national and multinational chains. You could drive off the highway in the outskirts of almost any big city in Canada or the United States and see exactly the same brands of fast-food restaurants, motels, stores, gas stations, and big-box shopping centres.

Smart Growth

Are there alternatives to unplanned sprawl and wasteful resource use? One option proposed by many urban planners is **smart growth** that makes efficient and effective use of land resources and existing infrastructure by encouraging in-fill development that avoids costly duplication of services and inefficient land use. Smart growth aims to provide a mix of land uses to create a variety of affordable housing choices and opportunities. It also attempts to provide a variety of transportation choices including pedestrian friendly neighbourhoods. This approach to planning also seeks to maintain a unique sense of place by respecting local cultural and natural features.

FIGURE 21.16 Building new highways to reduce congestion is like trying to diet by loosening your belt.

Smart growth, which is also called new urbanism by some people, is an alternative to the traditional planning approach that produces sprawl (Table 21.6).

In short, according to the Conservation Council of Ontario, smart growth aims to:

- protect agricultural land resources in order to provide food for our communities;
- protect and enhance natural areas and greenspace;
- create livable communities;
- improve the quality of our environment;
- conserve energy and other resources; and
- promote a sustainable economy.

Some Canadian examples in British Columbia and Toronto, Ontario, of smart growth may be seen at: www.smartgrowth.bc.ca.index.cfm and www.city.toronto.on.ca/sustainability.

TABLE 21.6 Comparing Smart Growth with Sprawl Growth Approaches

	SMART GROWTH	SPRAWL
Density	Higher density, clustered activities	Lower density, dispersed activities
Growth pattern	Infill (brownfield) development	Urban periphery (greenfield) development
Land use mix	Mixed	Single use, segregated
Scale	Human scale. Smaller buildings, blocks and roads. Attention to detail, since people experience the landscape up close, as pedestrians.	Large scale. Larger buildings, blocks, wide roads. Less detail since people experience the landscape at a distance, as motorists.
Public services (shops, schools, parks)	Local, distributed, smaller. Accommodates walking access.	Regional, consolidated, larger. Requires automobile access.
Transport	Multi-modal transportation and land use patterns that support walking, cycling and public transit.	Automobile-oriented transportation and land use patterns, poorly suited for walking, cycling and public transit.
Connectivity	Highly connected roads, sidewalks and paths allowing more direct travel by motorized and non-motorized modes.	Hierarchal road network with many unconnected roads and walkways and barriers to non-motorized travel.
Street design	Streets designed to accommodate a variety of activities. Traffic calming.	Streets designed to maximize motor vehicle traffic volume and speed.
Planning process	Planned and coordinated between jurisdictions and stakeholders.	Unplanned, with little coordination between jurisdictions and stakeholders.
Public space	Emphasis on the public realm (streetscapes, pedestrian areas, public parks, public facilities)	Emphasis on the private realm (yards, shopping malls, gated communities, private clubs)

Source: Victoria Transport Policy Institute.

By making land-use planning open and democratic, smart growth makes urban expansion fair, predictable, and cost-effective. All stakeholders are encouraged to participate in creating a vision for the city and to collaborate rather than confront each other. Goals are established for staged and managed growth in urban transition areas with compact development patterns. This approach is not opposed to growth. It recognizes that the goal is not to block growth but to channel it to areas where it can be sustained over the long term. It strives to enhance access to equitable public and private resources for everyone and to promote the safety, livability, and revitalization of existing urban and rural communities.

As cities grow and transportation and communications enable communities to interact more, the need for regional planning becomes both more possible and more pressing. Community and business leaders need to make decisions based on a clear understanding of regional growth needs and how infrastructure can be built most efficiently and for the greatest good.

Smart Growth, Brownfields, and Contaminated Sites

Smart growth protects environmental quality. It attempts to reduce traffic and to conserve farmlands, wetlands, and open space. This may mean restricting land use, which is strongly opposed by some people. However, it also means finding economically sound ways to reuse polluted industrial land, "brownfields" within cities. A **brownfield** is an area of land that has been abandoned because of known or suspected soil contamination.

These are often located in downtown areas near old railway and industrial sites, or underground oil and gas storage containers. With about 3,000 brownfield sites across Canada, there is a potential health and environmental risk as well as a potential opportunity for many communities. If these sites could be cleaned up they could provide land for affordable or other forms of housing, parks and recreational facilities, and economic development (see Case Study, p. 480).

The prime minister's Caucus Task Force on Urban Issues identified three major impediments to brownfield development. First, there are uncertain and unfair liability regimes. Second, there is a lack of financing for brownfield remediation and redevelopment. Third, there is a lack of awareness about the importance and benefits of reinvesting in brownfields.

What are the potential benefits of smart growth in Canada? A Canada Mortgage and Housing Corporation study found that using such alternative standards lowers public and private capital costs of infrastructure by 16 percent, or approximately $5,300 (in 1995 Canadian dollars) per unit. Infrastructure replacement, and operating and maintenance costs are cheaper by almost 9 percent or almost $11,000 per unit over a 75-year period. In another study completed in 1995, it was concluded that more compact and efficient urbanization in the Greater Toronto Area would save (in 1995 dollars) about $10 billion to $16 billion in infrastructure costs and about $2.5 billion to $4 billion in operating and maintenance costs over 25 years.

BROWNFIELD REDEVELOPMENT OF THE ANGUS SHOPS, MONTREAL

Between 1904 and 1992, the Angus Shops of Montreal was used to maintain and repair railway rolling stock and build new railway equipment. During World War II, more than 12,000 men and women worked in the more than 60 buildings that comprised the 500 ha site and built war armaments. Like the Sydney Tar Ponds, it belched smoke into the adjacent neighbourhoods, a sign of industrial activity of the day. With the opening of the St. Lawrence Seaway in 1959 and the increase in the use of transport trucks to move goods, the facility closed in 1992. Unfortunately the soil was contaminated with heavy metals, petroleum hydrocarbons, and polycyclic aromatic hydrocarbons (PAHs) that were discarded on the site during assembly and repair activities.

Deserted and unproductive, the Canadian Pacific Railways spent $1 million annually for taxes and maintenance. The cleanup required $12 million, including $3 million from a $180-million fund established by the Quebec government. After the cleanup was completed, the residential development required $204 million, the commercial development $20 million, and the industrial development $250 million. Seven hundred homes of varying sizes and styles, as well as commercial and industrial buildings have been built. The heritage values of some buildings have been preserved and enhanced. Nine parks and other green spaces of different sizes have been established. A linear park at the west end of the site is connected to an existing bicycle path network

FIGURE 21.17 The remaining original Angus shops: left, the former fire station, centre, the locoshop, and right, the former general office.

NTREE estimates the benefits of this development as:

- $12 million has been invested to clean up the environment ($8.64 million invested by private parties);

- $391.6 million invested to date by private parties to build up a residential neighbourhood, supermarket, light industry, and a biotechnology centre;

- property taxes (of 2002) had increased to $2.19 million annually;

- transformation of a vast non-productive former industrial site into a new dynamic neighbourhood, integrating commercial development and a light industrial park; and

- new social and economic development of the area.

Other brownfield development sites in Canada include:

- Moncton Shops Project, Moncton, New Brunswick
- Voisey's Bay Project, Argentia, Newfoundland
- ICI, Shawinigan, Quebec
- Centre de la Petite Enfance (Familigarde), Ville La Salle, Quebec
- Barton and Crooks Streets, Hamilton, Ontario
- Spencer Creek Village, Hamilton (Dundas) Ontario
- West Harbour Lands, Cobourg, Ontario
- Courtald's Fibres Project, Cornwall, Ontario
- Finishing Mill Lofts, Cornwall, Ontario
- False Creek, Vancouver, British Columbia

If external costs (e.g., air emissions, health care, traffic policing) are considered as well as these capital and maintenance costs, then the annual savings to be achieved by increasing development efficiency are in the range of $700 million to $1 billion annually.

One of the best examples of successful urban land-use planning in the United States is Portland, Oregon, which has rigorously enforced a boundary on its outward expansion, requiring, instead, that development be focused on in-filling unused space within the city limits. Because of its many urban amenities, Portland is considered one of the best cities in America. Between 1970 and 1990, the Portland population grew by 50 percent but its total land area grew only 2 percent. During this time, Portland property taxes decreased 29 percent and vehicle miles traveled increased only 2 percent. By contrast, Atlanta, which had similar population growth, experienced an explosion of urban sprawl that increased its land area three-fold, drove up property taxes 22 percent, and increased traffic miles by 17 percent. A result of this expanding traffic and increasing congestion was that Atlanta's air pollution increased by 5 percent, while Portland's, which has one of the best public transit systems in the nation, decreased by 86 percent. Portland shares many of the same goals as Chattanooga, described at the beginning of this chapter and in Table 21.7.

TABLE 21.7 Goals for Smart Growth

1. Create a positive self-image for the community.
2. Make the downtown vital and livable.
3. Alleviate substandard housing.
4. Solve problems with air, water, toxic waste, and noise pollution.
5. Improve communication between groups.
6. Improve community member access to the arts.

Source: Vision 2000, Chattanooga, TN.

New Urbanist Movement

The new urbanist movement is different from smart growth because it places more emphasis on the cultural and heritage values of the community, rather than resource use efficiency. Rather than abandon the cultural history and infrastructure investment in existing cities, a group of architects and urban planners is attempting to redesign metropolitan areas to make them more appealing, efficient, and livable. European cities such as Stockholm, Sweden; Helsinki, Finland; Leichester, England: and Neerlands, the Netherlands have a long history of innovative urban planning. In the United States, Andres Duany, Elizabeth Plater-Zyberk, Peter Calthorpe, and Sym Van Der Ryn have been leaders in this movement. Sometimes called a neo-traditionalist approach, these designers attempt to recapture some of the best features of small towns and the best cites of the past. They are designing urban neighbourhoods that integrate houses, offices, shops, and civic buildings. Ideally, no house should be more than a five-minute walk from a neighbourhood centre with a convenience store, a coffee shop, a bus stop, and other amenities. A mix of apartments, townhouses, and detached houses in a variety of price ranges insures that neighbourhoods will include a diversity of ages and income levels. Some design principles of this movement include:

- Limit city size or organize them in modules of 30,000 to 50,000 people, large enough to be a complete city but small enough to be a community. A greenbelt of agricultural and recreational land around the city limits growth while promoting efficient land use. By careful planning and cooperation with neighbouring regions, a city of 50,000 people can have real urban amenities such as museums, performing arts centres, schools, hospitals, etc.

- Determine in advance where development will take place. This protects property values and prevents chaotic development in which the lowest uses drive out the better ones. It also recognizes historical and cultural values, agricultural resources, and such ecological factors as impact on wetlands, soil types, groundwater replenishment and protection, and preservation of aesthetically and ecologically valuable sites.

- Turn shopping malls into real city centres that invite people to stroll, meet friends, or listen to a debate or a street

FIGURE 21.18 Many cities have redesigned core shopping areas to be more "user friendly." Pedestrian shopping malls, such as this one in Boulder, Colorado, create space for entertainment, dining, and chance encounters and provide opportunities for building a sense of community.

musician (Fig. 21.18). If there aren't 100 places for an impromptu celebration, a place isn't a real city. Another test of a city is a vital nightlife. Design city spaces with sidewalk cafes, pocket parks, courtyards, balconies, and porticoes that shelter pedestrians, bring people together, and add life and security to the street. Restaurants, theatres, shopping areas, and public entertainment that draw people to the streets generate a sense of spontaneity, excitement, energy, and fun.

- Locate everyday shopping and services so people can meet daily needs with greater convenience, less stress, less automobile dependency, and less use of time and energy. This might be accomplished by encouraging small-scale commercial development in or close to residential areas. Perhaps we should once again have "mom and pop" stores on street corners or in homes.

- Increase jobs in the community by locating offices, light industry, and commercial centres in or near suburbs, or by enabling work at home via computer terminals. These alternatives save commuting time and energy and provide local jobs. There are

CHAPTER 21 Urbanization and Sustainable Cities

www.mcgrawhill.ca/college/cunningham 481

FIGURE 21.19 Bicycles provide low-cost, nonpolluting, energy-efficient urban transportation. Guangzhou (Canton), China, a city of 4 million people, is said to have 3 million bicycles.

also concerns, however, about work-at-home employees being exploited in low-paying "sweatshop" conditions by unscrupulous employers. Some safeguards may be needed.

- Encourage walking or the use of small, low-speed, energy-efficient vehicles (microcars, motorized tricycles, bicycles, etc.) for many local trips now performed by full-size automobiles. Creating special traffic lanes, reducing the number or size of parking spaces, or closing shopping streets to big cars might encourage such alternatives (Fig. 21.19).

- Promote more diverse, flexible housing as alternatives to conventional, detached single-family houses. "In-fill" building between existing houses saves energy, reduces land costs, and might help provide a variety of living arrangements. Allowing owners to turn unused rooms into rental units provides space for those who can't afford a house and brings income to retired people who don't need a whole house themselves. Allowing single-parent families or groups of unrelated adults to share housing and to use facilities cooperatively also provides alternatives to those not living in a traditional nuclear family. One of the great "discoveries" of urban planning is that mixing various types of housing—individual homes, townhouses, and high-rise apartments—can be attractive if buildings are aesthetically arranged in relation to one another.

- Create housing "superblocks" that use space more efficiently and foster a sense of security and community. Widen peripheral arterial streets and provide pedestrian overpasses so traffic flows smoothly around residential areas; then reduce interior streets within blocks to narrow access lanes with speed bumps and barriers to through traffic so children can play more safely. The land released from streets can be used for gardens, linear parks, playgrounds, and other public areas that will foster community spirit and encourage people to get out and walk. Cars can be parked in remote lots or parking ramps, especially where people have access to public transit and can walk to work or shopping.

FIGURE 21.20 Communal gardens in Seattle, Washington, use once-vacant land to grow food and give local residents a chance to experience nature in a highly urban setting.

- Make cities more self-sustainable by growing food locally, recycling wastes and water, using renewable energy sources, reducing noise and pollution, and creating a cleaner, safer environment (Fig. 21.20). Reclaimed inner-city space or a greenbelt of agricultural and forestland around the city provides food and open space as well as such valuable ecological services as purifying air, supplying clean water, and protecting wildlife habitat and recreation land.

- Rooftop gardens and natural plantings can absorb up to 70 percent of rain water. They provide habitat for birds and insects, and ameliorate climate. They save energy and provide contact with nature for building residents. Many German cities now require that at least half of all new development must be covered with vegetation. The least expensive way to do this is with green roofs on buildings and parking structures.

- Invite public participation in decision making as was done in Chattanooga. Emphasize local history, culture, and environment to create a sense of community and identity. Create local networks in which residents take responsibility for crime prevention, fire protection, and home care of children, the elderly, sick, and disabled. Coordinate regional planning through metropolitan boards that cooperate with but do not supplant local governments.

Ecologists and conservation biologists have, in the past, tended to study pristine nature, untouched—as much as possible—by human influences. In recent years, however, we have come to recognize that cities are ecological systems too. If ecology is the relationship among organisms and their environment, what could be more ecological than studying how humans impact other species and our environment? And for a vast majority of us, that means studying urban ecosystems where we are the keystone species. Rather than merely seeing humans as disturbing factors in these ecosystems, some scientists are welcoming people as participants in research and project-based service learning.

One of the first things that scientists are discovering is that cities are not homogenous; like other ecosystems, they have patches at varying scales that change over time. Using computer models and sophisticated mapping techniques, ecologists can explore how patches shift around in space and time.

Environmental justice concerns are at the forefront of many urban ecology research projects. Where are toxic and hazardous materials generated, stored, and released in the city? How do they move around and where do they accumulate? In Detroit, for example, a group of students worked with experts to map data on more than 5,000 children with elevated blood lead levels. Not surprisingly, they found a correlation between low income, old housing, incidence of poisoning, and concentration of special-education students. Sometimes public awareness of a problem is one of the best outcomes of this research. Another large group of students in the Detroit area used geographic information systems together with chemical and biological analysis to prepare a detailed study of water quality and ecosystem health in the Rouge River. Participatory planning coupled with citizen science enable some cities to identify indicators of urban sustainability (Table 21.8).

Garden Cities and New Towns

The twentieth century saw numerous experiments in building **new towns** for society at large that try to combine the best features of

---

TABLE 21.8 Urban Sustainability Indicators

Children in poverty
Violent crime
Access to health care
Air and water quality, litter
Vacant or deteriorating housing
Participation in neighbourhood organizations
Money earned and spent in neighbourhood
Access to public transportation
Shopping and services within walking distance
Quality of schools
Cultural and recreational opportunities

Source: MN Citizen's Environment Action Committee, 1999.

---

the rural village and the modern city. One of the most influential of all urban planners was Ebenezer Howard (1850–1929), who not only wrote about ideal urban environments but also built real cities to test his theories. In his *Garden Cities of Tomorrow,* written in 1898, Howard proposed that the congestion of London could be relieved by moving whole neighbourhoods to **garden cities** separated from the central city by a greenbelt of forests and fields.

In the early 1900s, Howard worked with architect Raymond Unwin to build Letchworth and Welwyn Garden just outside of London. Interurban rail transportation provided access to these cities. Houses were clustered in "superblocks" surrounded by parks, gardens, and sports grounds. Streets were curved. Safe and convenient walking paths and overpasses protected pedestrians from traffic. Businesses and industries were screened from housing areas by vegetation. Each city was limited to about 30,000 people to facilitate social interaction. Housing and jobs were designed to create a mix of different kinds of people and to integrate work, social activities, and civic life. Trees and natural amenities were carefully preserved and the towns were laid out to maximize social interactions and healthful living. Care was taken to meet residents' psychological needs for security, identity, and stimulation.

Letchworth and Welwyn Garden each have 35 to 50 people per hectare. This is a true urban density, about the same as New York City in the early 1800s and five times as many people as most suburbs today. By planning the ultimate size in advance and choosing the optimum locations for housing, shopping centres, industry, transportation, and recreation, Howard believed he could create a hospitable and satisfying urban setting while protecting open space and the natural environment. He intended to create parklike surroundings that would preserve small-town values and encourage community spirit in neighbourhoods.

Letchworth and Welwyn Garden were the first of 32 new towns established in Great Britain, which now house about 1 million people. The Scandinavian countries have been especially successful in building garden cities. The former Soviet Union also built about 2,000 new towns. Some are satellites of existing cities, and others are entirely new communities far removed from existing urban areas.

Planned communities also have been built in the United States following the theories of Ebenezer Howard, but most plans have been based on personal automobiles rather than public transit. In the 1920s, Lewis Mumford, Clarence Stein, and Henry Wright drew up plans that led to the establishment of Radburn, New Jersey, and Chatham Village (near Pittsburgh). Reston, Virginia, and Columbia, Maryland, both were founded in the early 1960s and are widely regarded as the most successful attempts to build new towns of their era. Another movement to build new towns according to Howard's principles has sprung up in the 1990s. Towns such as Seaside in northern Florida and Kentlands outside Washington, D.C., cluster houses to save open space and create a sense of community. Commercial centres are located within a few minutes walk of most houses, and streets are designed to encourage pedestrians and to provide places to gather and visit.

Designing for Open Space

Traditional suburban development typically divides land into a checkerboard layout of nearly identical 1 to 5 ha parcels with no designated open space (Fig. 21.21, *top*). The result is a sterile landscape consisting entirely of house lots and streets. This style of development, which is permitted—or even required—by local zoning and ordinances, consumes agricultural land and fragments wildlife habitat. Many of the characteristics that people move to the country to find—space, opportunities for outdoor recreation, access to wild nature, a rural ambience—are destroyed by dividing every hectacre into lots that are "too large to mow but too small to plow."

An interesting alternative known as **conservation development,** cluster housing, or open space zoning preserves at least half of a subdivision as natural areas, farmland, or other forms of open space. Among the leaders in this design movement are landscape architects Ian McHarg, Frederick Steiner, and Randall Arendt. They have shown that people who move to the country don't necessarily want to own a vast tract of land or to live kilometres from the nearest neighbour; what they most desire is long views across an interesting landscape, an opportunity to see wildlife, and access to walking paths through woods or across wildflower meadows.

By carefully clustering houses on smaller lots, a conservation subdivision can provide the same number of buildable lots as a conventional subdivision and still preserve 50 to 70 percent of the land as open space (Fig. 21.21, *bottom*). This not only reduces development costs (less distance to build roads, lay telephone lines, sewers, power cables, etc.) but also helps foster a greater sense of community among new residents. Walking paths and recreation areas get people out of their houses to meet their neighbours. Home owners have smaller lots to care for and yet everyone has an attractive vista and a feeling of spaciousness.

Some good examples of this approach are Farmview near Yardley, Pennsylvania, and Hawksnest in Delafield Township, Wisconsin. In Farmview, 332 homes are clustered in six small villages set in a 160 ha rural landscape, more than half of which is dedicated as permanent farmland. House lots and villages were strategically placed to maximize views, helping the development to lead its county in sales for upscale developments. Hawksnest is situated in dairy-farming country outside of Waukesha, Wisconsin. Seventy homes are situated amid 70 ha of meadows, ponds, and woodlands. Restored prairies, neighbourhood recreational facilities, and connections to a national scenic trail have proved to be valuable marketing assets for this subdivision.

And urban habitat can make a significant contribution toward saving biodiversity. In a ground-breaking series of habitat conservation plans triggered by the need to protect the endangered California gnatcatcher, some 85,000 ha of coastal scrub near San Diego was protected as open space within the rapidly expanding urban area. This is an area larger than Yosemite Valley, and will benefit many other species as well as humans.

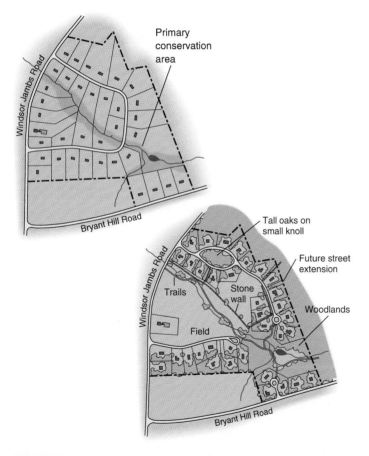

FIGURE 21.21 Conventional subdivision (*top*) and an open space plan (*bottom*). Although both plans provide 36 home sites, the conventional development allows for no open space. Cluster housing on smaller lots in the open space design preserves at least half the area as woods, prairie, wetlands, farms, or other conservation lands, while providing residents with more attractive vistas and recreational opportunities than a checkerboard development.

SUSTAINABLE URBAN DEVELOPMENT IN CANADA

As the previous text suggests, local governments face tremendous and varied challenges. The range of options to address these problems range from constitutional reforms that would support more autonomy for cities to changing and enhancing financial arrangements. In a speech to the Canadian Federation of Municipalities in Winnipeg in the summer of 2003, Paul Martin expressed support for cities getting a share of the federal tax on gasoline. When he became leader of the Liberal Party in November 2003, he stated that he intended to make "good on the promise of a New Deal for our municipalities." While there has been no effective action to implement this specific commitment as of July 2004, the Liberal government before the election of 2004 exempted municipalities from paying the GST, which is expected to save them $465 million Cdn per year. Some provinces have taken action to provide additional sources of funding to cities. For instance, Calgary and Edmonton get a share of the provincial gas tax: $0.05 from every litre sold within their respective boundaries. In the Greater Vancouver area, the regional transportation authority gets $0.115 per litre from provincial gas taxes. Other sources being discussed across the country include additional fees for garbage collection, and new municipal taxes on liquor and hotel rooms. These proposals are controversial—some people will see these options as tax grabs.

Money alone will not solve urban problems in Canada. Other key elements of Canada's urban strategy will also include:

- *coordination and integration* of all government programs related to immigration, employment, health care, crime prevention, infrastructure, transit/transportation and the environment;
- *collaboration and consultation* among all levels of government, the private sector, non-governmental organizations and the public;
- clearly defined objectives that advance urban revitalization and sustainability; and
- implementation on *best practices,* which is supported by *research*.

Maxwell maintains that translating these principles into practice will require all participants to build on the attributes and refute the negative aspects of past decision-making behaviours, and consider issues in three new dimensions that are characterized by thinking:

- regionally, to see the full scope of the economic, environmental and infrastructure needs and possibilities, not just those within their own jurisdiction;
- inclusively, to see that people from different walks of life and socioeconomic status are all part of the solution; and
- from the bottom up, to ensure that needs are defined locally based on neighbourhood and family needs.

The prime minister's Caucus Task Force on Urban Issues suggested priority areas for national programs were related to three areas: (i) affordable housing, (ii) transit/transportation, and (iii) sustainable infrastructure.

There will be no universal and simple solutions to the challenges confronting Canada's settlements. Given its geographic and historic diversity, what is the balance between the needs of large Canadian metropolitan areas and the surrounding countryside? Given the Canadian Constitutional setting, what is the appropriate role of the federal, provincial, territorial, and local governments and Aboriginal peoples? What financial arrangements and decision-making mechanisms will best support Canadian cities to become sustainable?

In the final analysis, three major themes will permeate the sustainable urban management strategies in Canada and abroad: (i) demonstrated leadership by governments, particularly national governments, to get their own house in order, including ensuring their buildings and transport fleets are energy (and water) efficient; (ii) fiscal reform aimed at ensuring effective, efficient and adequate forms of funding are available; (iii) multi-level government with possible involvement from the private sector on encouraging investments in public transit and urban infrastructure; and (iv) ensuring investments that increase energy efficiency, including shifting the form of energy from fossil fuels to renewable forms of energy, are made.

In this way, a better balance between the demands made by cities on the natural environment and the natural environment's ability to support these demands can be achieved in a socially, fiscally and economically responsible manner.

SUSTAINABLE URBAN DEVELOPMENT IN THE THIRD WORLD

What can be done to improve conditions in Third World cities? Curitiba, Brazil, is an outstanding example of what can be done, even in relatively undeveloped countries, to improve transportation, protect central cities, and create a sense of civic pride (see Case Study, p. 486). Other cities have far to go, however, before they reach this standard. Among the immediate needs are housing, clean water, sanitation, food, education, health care, and basic transportation for their residents. The World Bank estimates that interventions to improve living conditions in urban households in the developing world could average the annual loss of almost 80 million "disability-free" years of life. This is about twice the feasible benefit estimated from all other environmental programs studied by the World Bank.

Some countries, recognizing the need to use vacant urban land, are redistributing unproductive land or closing their eyes to illegal land invasions. Indonesia, Peru, Tanzania, Zambia, and Pakistan have learned that squatter settlements make a valuable contribution to meeting national housing needs. Squatters' rights are being upheld in some cases, and such services as water, sewers, schools, and electricity are being provided to the settlements (Fig. 21.22). Some countries intervene directly in land distribution and land prices. Tunisia, for instance, has a "rolling land bank" to buy and sell land. This strong and effective program controls urban land prices and reduces speculation and unproductive land ownership.

FIGURE 21.22 In this *colonia* on the outskirts of Mexico City, residents work with the government to bring in electricity, water, and sewers to shantytowns and squatter settlements. Like many Third World countries, Mexico recognizes that helping people help themselves is the best way to improve urban living.

CASE STUDY

CURITIBA: AN ENVIRONMENTAL SHOWCASE

Curitiba, a Brazilian city of about 2 million people located on the Atlantic coast about 650 km southwest of Rio de Janeiro, has acquired a worldwide reputation for its innovative urban planning and environmental protection policies. Tree-lined streets, clean air, smoothly flowing traffic, and absence of litter and garbage make this one of Latin America's most livable cities.

The architect of this remarkable program is mayor Jaime Lerner, who began in 1962 as a student leader protesting the proposed destruction of Curitiba's historic downtown centre. Elected mayor nine years later, he has worked for more than two decades to preserve the city and make it a more beautiful and habitable place. Now Governor of the State of Parana, Lerner's success has thrust the city into the international spotlight as a shining example of what conservation and good citizenship can do to improve urban environments, even in Third World cities.

The heart of Curitiba's environmental plan is education for both children and adults. Signs posted along roadways proclaim "50 kg of paper equals one tree" and "recycle; it pays." School children study ecology along with Portuguese and math. With the assistance of children who encourage their parents, the city has successfully instituted a complex recycling plan that requires careful separation of different kinds of materials. The city calculates that 1,200 trees per day are saved by paper recycled in this program. "Imagine if the whole of Brazil did this with an urban population 60 times greater than Curitiba," Lerner exclaims. "We could save 26 million trees per year!" More than 70 percent of the residents now recycle. Food and bus coupons are given out in exchange for recyclables, providing an alternative to welfare while also protecting the environment.

Another area in which Curitiba is setting an example is transportation. Faced with a population that tripled in two decades, bringing increasing levels of traffic congestion and air pollution, Curitiba had a transportation dilemma. The choices were either to bulldoze freeways through the historic heart of the city or to institute mass transit. The city chose mass transit.

Special articulated buses each carry up to 300 passengers on dedicated transit lanes. You pay your fare before entering the platform waiting area. When the bus pulls in, multiple doors open flush with the platform so that many people can enter or exit at once (even in wheelchairs) and each stop takes only seconds. A feeder network of smaller buses and vans collects riders from neighbourhoods, so that everyone has convenient, frequent, and inexpensive transit service. Now more than three-quarters of the city's population leave their automobiles at home every day and take special express buses to work. The system is so successful that ridership has increased from 25,000 passengers a day 20 years ago to more than 1.5 million per day now, or 70 percent of all trips in the city. The result is not only less congestion and pollution but major energy savings.

Other measures to clear the air and reduce congestion include a limit on building height and construction of an industrial park outside city boundaries. Maximum use is made of all materials and buildings. Worn-out buses become city training centres, an old military fort is a cultural centre, and a gunpowder depot is now a theatre. Water and energy conservation are practised widely. Even litter—an ubiquitous component of most Brazilian cities—is absent in Curitiba. More than one million trees have been planted and the green area ratio of 52 square metres of park per capita is higher than most American cities.

People are so imbued with city pride that they keep their surroundings spotless. To further beautify their city, civic volunteers have planted 1.5 million trees—more than any other place in Brazil.

Although many residents initially were skeptical of this environmental plan, now a remarkable 99 percent of the city's inhabitants would not want to live anywhere else. The World Bank uses Curitiba as an example of what can be done through civic leadership and public participation to clean up the urban environment. Some people claim that Curitiba, with its cool climate and high percentage of European immigrants, may be a special case among Brazilian cities. Lerner claims that Curitiba has no special features except concern, creativity, and communal efforts to care for its environment. Could you start a similar program in your hometown?

Many planners argue that social justice and sustainable economic development are answers to the urban problems we have discussed in this chapter. If people have the opportunity and money to buy better housing, adequate food, clean water, sanitation, and other things they need for a decent life, they will do so. Democracy, security, and improved economic conditions help in slowing population growth and reducing rural-to-city movement. An even more important measure of progress may be institution of a social welfare safety net guaranteeing that old or sick people will not be abandoned and alone.

Some countries have accomplished these goals even without industrialization and high incomes. Sri Lanka, for instance, has lessened the disparity between the core and periphery of the country. Giving all people equal access to food, shelter, education, and health care eliminates many incentives for interregional migration.

Both population growth and city growth have been stabilized, even though the per capita income is only $800 per year. China has done something similar on a per capita income of around $300 per year.

Whether sustained, environmentally sound economic development is possible for a majority of the world's population remains one of the most important and most difficult questions in environmental science. The unequal relationship between the richer "Northern" countries and their impoverished "Southern" neighbours is a major part of this dilemma. Some people argue that the best hope for developing countries may be to "delink" themselves from the established international economic systems and develop direct south-south trade based on local self-sufficiency, regional cooperation, barter, and other forms of nontraditional exchange that are not biased in favour of the richer countries.

SUMMARY

A rural area is one in which a majority of residents are supported by methods of harvesting natural resources. An urban area is one in which a majority of residents are supported by manufacturing, commerce, or services. A village is a rural community. A city is an urban community with sufficient size and complexity to support economic specialization and to require a higher level of organization and opportunity than is found in a village.

Urbanization in developed countries including Canada over the past 200 years has caused a dramatic demographic change. A similar shift is now occurring in most parts of the world. Only Africa and South Asia remain predominantly rural, but cities are growing rapidly there as well. In 2000, for the first time in history, more than half the world's people lived in urban areas. Most future urban growth in the next century will be in the supercities of the Third World. A century ago only 13 cities had populations above 1 million; now there are 235 such cities. By 2050 that number will probably double again, and three-fourths of those cities will be in the Third World.

Cities grow by natural increase (births) and migration. People move into the city because they are "pushed" out of rural areas or because they are "pulled" in by the advantages and opportunities of the city. Huge, rapidly growing cities in the developing world often have appalling environmental conditions. Among the worst problems faced in these cities are traffic congestion, air pollution, inadequate or nonexistent sewers and waste disposal systems, water pollution, and housing shortages. Millions of people live in slums and shantytowns where conditions are frightful, yet these people raise families, educate their children, learn new jobs and new ways of living, and have hope for the future.

The problems of developed world cities tend to be associated with urban sprawl around the outskirts and decay and blight in the core. Unlimited expansion into rural areas, leapfrog development, and lack of coordinated land-use planning lead to loss of farmlands and open space, traffic congestion, air and water pollution, and numbingly uniform housing tracts and shopping centres. Sprawl also requires local government to spend millions of dollars to replace roads, sewers, water lines, schools, parks, power grids, and other infrastructure being abandoned in the inner city. Still, there are ways that we can improve cities in both the developed and the developing world to make them healthier, safer, and more environmentally sound, socially just, and culturally fulfilling than they are now. Smart growth, garden cities, new traditionalist urban movements, and conservation development are among the ideas advanced for improving our cities. Curitiba, Brazil, is an encouraging example of how these principles can be applied in the developing world.

QUESTIONS FOR REVIEW

1. What is the difference between a city and a village and between rural and urban?

2. How many people now live in cities, and how many live in rural areas worldwide?

3. What changes in urbanization are predicted to occur in the next 50 years, and where will that change occur?

4. Identify the 13 largest cities in the world. Has the list changed in the past 50 years? Why?

5. When did Canada pass the point at which more people live in the city than the country? When will the rest of the world reach this point?

6. Describe the current conditions in a typical megacity of the developing world. What forces contribute to its growth?

7. Describe the difference between slums and shantytowns.

8. How has transportation affected the development of cities? What have been the benefits and disadvantages of freeways?

9. Describe some ways in which Canadian cities and suburbs could be redesigned to be more ecologically sound, socially just, and culturally amenable.

QUESTIONS FOR CRITICAL THINKING

1. Picture yourself living in a rural village or a Third World city. What aspects of life there would you enjoy? What would be the most difficult for you to accept?

2. Why would people move to one of the megacities of the developing world if conditions are so difficult there?

3. A city could be considered an ecosystem. Using what you learned in Chapters 2 and 3, describe the structure and function of a city in ecological terms.

4. Look at the major urban area(s) in your province. Why were they built where they are? Are those features now a benefit or drawback?

5. Who benefits from urban sprawl and who suffers? Why is this process so powerful and persistent?

6. Weigh the costs and benefits of automobiles in modern Canadian life. Is there a way to have the freedom and convenience of a private automobile without its negative aspects?

7. A number of proposals are presented in this chapter for urban redesign. Which of them would be appropriate or useful for your community? Try drawing up the ideal plan for your neighbourhood.

Web Exercise

YOUR DREAM CITY

It is almost overwhelming to think of being the mayor of a large city like Toronto or Vancouver and making it work in an environmentally sustainable way. Even urban planning at a small scale, such as a town or a village, can be challenging. In any town or city, there always seems to be challenges in balancing urban development with environmental concerns of the city's inhabitants, human and otherwise. You can try out your skills as a city planner by going to the Maxis website and playing SimCity Classic at http://simcity.ea.com/play/simcity_classic.php. See if you can make your city work in an environmentally friendly way!

Chapter 22

Preserving Our Natural Environment

Never doubt that a small group of thoughtful, committed people can change the world; indeed it is the only thing that can.

—Margaret Mead—

OBJECTIVES

After studying this chapter, you should be able to:

- understand the origins and current problems of national parks.
- explain the need for and problems with wildlife refuges and protected wilderness areas.
- understand the factors that are important to consider in the design of nature reserves.
- evaluate the tension between conservation and economic development and how the Man and Biosphere (MAB) program and ecotourism projects address this tension.
- demonstrate why wetlands are valuable ecologically and culturally and why they are currently threatened worldwide.
- report on current management policies and problems concerning floodplains, coastlines, and barrier islands.

WebQuest

protected areas, biosphere reserves, Man and Biosphere (MAB), conservation biology, ecotourism

Above: Kluane National Park in the southwestern Yukon, part of a UN World Heritage site and a popular place for wilderness hiking.

Ecotourism on the Roof of the World

Rising dramatically from the steamy southern jungles of the Ganges River Valley to the icy peaks of the Himalayan mountains on the Tibetan border, Nepal is one of the most scenic countries in the world. Tourists savour the exotic culture of Katmandu or Namche Bazar or hike through lush mountain forests of rhododendron and pine. Offering spectacular scenery, friendly people, and low prices, this charismatic country has become a premiere destination for adventure travellers.

With an annual per capita income of only $170, Nepal is among the poorest countries in the world. The phenomenal increase in visitors over the past 20 years has brought much-needed income but also has caused severe environmental degradation. Forests along popular trekking trails have been decimated to provide firewood for cooking and heating of water for the numerous wealthy outsiders, while tonnes of garbage and discarded gear litter popular campsites.

One of the most popular Nepalese trekking routes is a three- to four-week circuit of the Annapurna Range in the centre of the Himalayan Range (Fig. 22.1). Crossing rushing rivers on swaying suspension bridges, passing between the 8,167-m Dhaulagiri and the 8,091-m Annapurna I (the seventh and tenth highest mountains in the world, respectively), this ancient pilgrim trail follows the Kali Gandaki Valley to holy shrines at Muktinath. Surmounting the 5,416-m Thorung La pass north of Annapurna, hikers follow the Marsyangdi valley back to the regional centre at Pokhara. First opened to foreigners in 1977, this trail now attracts over 45,000 visitors each year.

Most Nepalese benefit very little from tourists who congest their villages, consume resources, and snap photographs incessantly, but the Annapurna region is different. An innovative project was launched in 1985 to alleviate the destructive impact of masses of trekkers and to maximize the income-generating potential of ecotourism. The Annapurna Conservation Area Project (ACAP) is a 2,590-km² biosphere reserve that serves as an encouraging model for conservation and development in the Third World.

Far different from Western ideals of parks composed of empty, virgin land, the ACAP is home to more than 100,000 people who continue to use resources in traditional ways. The area is divided into five different zones: intensive farming lands around the periphery, protected forest and seasonal grazing areas in the foothills, special management zones along tourist routes, protected regions with high biological or cultural richness, and wilderness areas in the high peaks.

FIGURE 22.1 The Annapurna Conservation Area Project directs money from visitors into development and environmental programs directed by, and of benefit to, local residents. This may be a model for preserving nature in other developing countries.

Recognizing that there can be no meaningful conservation without the active involvement of local people, fees paid by visitors to ACAP go directly to residents to manage the preserve. About $500,000 per year finances a variety of conservation, education, and development projects. More than 700 local entrepreneurs have been trained in lodge management, hygiene, and marketing. Forest guards have been hired, latrines built, trails repaired, and schools and clinics built for local people. Trekkers now are required to use kerosene rather than wood. Local tree nurseries provide stock for reforestation projects. Solar panels and water turbines provide renewable energy for both tourists and residents. The area is cleaner, healthier, and more enjoyable for everyone.

This unique and successful experiment gives us a different view of the meaning and purpose of parks and nature preserves than the ideal of pristine nature conveyed by most Canadian national parks. It raises some interesting questions about competing needs of human and nonhuman residents and how they might be balanced sustainably. It also provides a model of how protected areas might be designed and managed in other developing countries.

In this chapter, we will study the history of parks, preserves, and wildlife refuges around the world. We will examine other success stories as well as problems facing efforts to protect nature. Because of their great ecological importance, we will pay special attention to wetlands, floodplains, and coastal regions and efforts to protect them.

PARKS AND NATURE PRESERVES

Throughout the world, much of the most biologically productive land is in private hands. If we hope to preserve a meaningful sample of natural biodiversity and ecological functions, private lands will have to play a major role. Furthermore, the world is rapidly urbanizing. Already more than half of all humans live or work in urban areas, so if we are to have contact with nature on a regular basis, urban open space will be vitally important. In many cultures, people have set aside special places as parks, preserves, or sacred groves. We can learn much about environmental values and attitudes by studying those places. Furthermore, they serve as benchmarks to gauge what we have lost and what we might aspire to recover in nature.

Park Origins and History

Since ancient times, sacred groves have been set aside for religious purposes and hunting preserves or pleasuring grounds for royalty. As such, they have been reserved primarily for elite members of society. The imperial retreat of the Han emperors of China, for example, built in the second century B.C. near their capital Ch'ang-an, is the earliest landscaped park of which we have detailed description. Large enough to encompass mountains, forests, and marshes as well as palaces and formal gardens, the park reflected Taoist beliefs about the ideal landscape and our place in the cosmos. Great towers and mountaintop pavilions served as retreats from which the emperor could contemplate nature in tranquility. Not all was peace and serenity, however; the emperor and his entourage also enjoyed hunting herds of wildlife maintained for their enjoyment. Although much of the park appeared natural, it required just as much engineering and earth moving to achieve and maintain this wild appearance as in the formal geometric gardens of Renaissance Europe.

Natural landscaping became popular in England during the eighteenth century under the leadership of architects such as Lancelot Brown (known as Capability Brown because he saw a capability for improving nature everywhere). This new organic design rejected the straight lines and rigid symmetry of earlier gardens, opting for sweeping vistas over rolling hills, meadows, forests, and natural-looking ponds and marshes. The illusion of wild nature was carefully contrived, however. Clumps of trees were sculpted to create vistas, and miniature buildings were strategically placed to emphasize the receding perspective. Brown built moats and fences hidden in ditches to control access to private property without interrupting the vista, a concept rediscovered by modern zoos.

Perhaps the first public parks open to ordinary citizens were the grand esplanades and the tree-sheltered agora that served as a gathering place in the planned Greek city. Central Park in New York City is an important successor to both this democratic ideal and the naturalistic principles of romantic landscaping. Promoted in 1844 by newspaper editor William Cullen Bryant as a "pleasuring ground in the open air for the benefit of all," the park was to provide healthful open space and contact with nature for the crowded masses of the city (Fig. 22.2). A worldwide competition for design of the park was won by Frederick Law Olmstead, who became the father of landscape architecture in the United States.

Olmstead left New York in 1864 to become the original commissioner of Yosemite Park in California, the first area set aside to protect wild nature in the United States. Yosemite was authorized by President Abraham Lincoln in the midst of the Civil War to protect its resources from the unbridled exploitation common in frontier areas. Because there was no mechanism for running a park at the national level at that time, Yosemite was deeded to the State of

FIGURE 22.2 Central Park in New York City was one of the first large urban parks designed to provide nature experience and healthful recreation for common people.

California. It was transferred back to the federal government as a national park in 1890.

In 1872, President Ulysses S. Grant signed an act designating about 800,000 ha of land in the Wyoming, Montana, and Idaho territories as Yellowstone National Park, the first *national* park in the world. Although the initial interest of both the founders and visitors to Yellowstone was the spectacular "curiosities" and natural "wonders" of the geysers, hot springs, and canyons (Fig. 22.3), the park was large enough to encompass and preserve real wilderness. Because the territories had no means to manage the area, Yellowstone was made a national park and guarded by the army until the National Park Service was founded in 1916.

As it became apparent that wild nature and places of scenic beauty and cultural importance were rapidly disappearing with the closing of the North American frontier, the drive to set aside more national parks accelerated. Canada's Banff National Park was established in 1885. In the United States, Mount Rainier was authorized in 1899, Crater Lake in 1902, Mesa Verde in 1906, Grand Canyon in 1908, Glacier Park in 1910, and Rocky Mountain National Park in 1915. In 1911, the Canadian Parks Service was established as the first institution to be charged with the mandate of managing national parks and protected wilderness areas.

North American Parks

Parks serve a variety of purposes. They can teach us about our past and provide sanctuaries where nature is allowed to evolve in its own way. They are havens not only for wildlife but also for the human spirit. Canada and the United States have greater total amounts of land dedicated to protected areas than any country except Denmark (which protects vast areas of Greenland's ice and snow) and Australia (which has designated great expanses of outback as aboriginal lands and parks). Although Mexico's parks are newer and less extensive than its wealthy neighbours', they contain

FIGURE 22.3 Yellowstone National Park, established in 1872, is regarded as the first national park in the world. Although focused initially on the spectacle of natural curiousities and wonders, it has come to be appreciated for its beauty and wilderness values.

far more biological and cultural diversity than do parks in either Canada or the United States.

Existing Systems

Canada has a total of 1,471 parks and protected areas occupying about 150,000 sq km. Among this group are national parks, provincial parks, outdoor recreation parks, and historic parks. They range in size from vast wilderness expanses such as Wood Buffalo National Park in northern Manitoba or Ellesmere Island National Park Reserve in the Northwest Territories, to tiny pockets of cultural or natural history occupying only a few hectares. Kluane National Park in the Yukon, the new Tatshenshini-Alsek Wilderness in British Columbia, the adjoining Wrangell-St. Elias National Park, and Glacier Bay National Park in Alaska together encompass an area of about 10 million ha, roughly the size of Belgium or 10 times as big as Yellowstone National Park. While many ecological reserves in Canada enforce strictly controlled access, other protected areas encourage intensive recreation, allow hunting, logging, or mining, and permit environmental manipulation for management purposes.

The U.S. national park system has grown to more than 280,000 sq km in 376 parks, monuments, historic sites, and recreation areas. Each year about 300 million visitors enjoy this system. The most heavily visited units are the urban recreation areas, parkways, and historic sites. The jewels of the park system, however, and what most people imagine when they think of a national park, are the great wilderness parks of the West. Passage of the Alaska Lands Act of 1980 nearly doubled the national park system. State and local parks occupy only about one-sixteenth as much area as national parks yet have about twice as many visitors.

Park Problems

Originally, the great wilderness parks of Canada and the United States were seen as fortresses protected from development or exploitation by legal boundaries and diligent park rangers. Most were buffered from human impacts by their remote location and the wild lands surrounding them. Today, the situation has changed. Many parks have become islands of nature surrounded and threatened by destructive land uses and burgeoning human populations that crowd park boundaries. Forests are clear-cut right up to the edges of some parks, while mine drainage contaminates streams and groundwater. Garish tourist traps clustered at park entrances detract from the beauty and serenity that most visitors seek.

Threats to park values come from within as well. The four contiguous mountain parks of the Canadian Rockies (Banff, Jasper, Yoho, and Kootenay) receive about 10 million visitors each year. Four towns are located within their boundaries and pressures are mounting to expand hotels, shops, downhill ski facilities, convention centres, and condominiums. In 1999 a proposal was rejected for an open-pit coal mine adjacent to Jasper National Park. On Cape Cod National Seashore and in the new California Desert Park, dune buggies, dirt bikes, and off-road vehicles (ORV) run over fragile sand dunes, disturbing vegetation and wildlife and destroying the aesthetic experience of those who come to enjoy nature (Fig. 22.4). In Florida's Everglades National Park, water flow through the "river of grass" has been disrupted and polluted by encroaching farms and urban areas. Wading-bird populations have declined by 90 percent, down from 2.5 million in the 1930s to 250,000 now.

Other parks experience similar difficulties. Yosemite National Park exemplifies these problems. On popular three-day weekends, as many as 25,000 visitors crowd into the 18-square-kilometre valley floor, which is less than 1 percent of the total park area (Fig. 22.5). They fill the valley with noise and smoke, trample fragile meadows and riverbanks, and spend hours in traffic jams. You can buy a pizza, play video games, do your laundry, play golf or tennis, and shop for curios and souvenirs in Yosemite Valley today, but you are less and less likely to experience the solitude and peace of nature extolled by John Muir. Park rangers have become traffic cops and crowd-control specialists rather than naturalists. A general management plan for Yosemite directed that both Park Service and Curry Company—the concessionaire for tourist services—headquarters be removed from the valley along with 370 buildings, employee housing, 17 percent of existing guest rooms, and all automobiles. Little progress has been made toward meeting these goals, however.

FIGURE 22.4 Off-road machines, such as motorcycles, dune buggies, and four-wheel drive vehicles can cause extensive and long-lasting damage to sensitive ecosystems. Tracks can persist for decades in deserts and wetlands where recovery is slow.

FIGURE 22.5 Visitors crowd popular trails in Yosemite National Park on summer weekends, making quiet and solitude impossible to find.

Air pollution is a serious threat to many parks. The haze over the Blue Ridge Parkway is no longer blue but gray-brown because of air pollution carried in by long-range transport. Sulphate concentrations in Shenandoah and Great Smoky Mountains National Parks are five times human health standards, and ozone levels in Acadia National Park in Maine exceed primary air quality standards by as much as 50 percent on some summer days. Visitors to the Grand Canyon once could see mountains 160 km away; now the air is so smoggy you can't see from one rim to the other during one-third of the year. The main culprits are power plants in Utah and Arizona that supply electricity to Los Angeles, Phoenix, and other urban areas. Acid rain threatens sensitive lakes in the high mountains of the West, as well as in eastern Canada and New England. Photochemical smog is damaging the giant redwoods in California's Sequoia National Park and contributing to forest declines in the Adirondacks.

Mining and oil interests continue to push for permission to dig and drill in the parks, especially on the 1.2 million hectares of private inholdings (private lands) in the parks. These forces were successful in excluding mineral lands in the Misty Fjords and Cape Kruzenstern National Monuments in Alaska. Uranium mines at the edge of the Grand Canyon threaten to contaminate the Colorado River and the park's water supply with radioactive contaminants.

Some 80,000 ha of **inholdings**—land already in private ownership when parks were established—are in danger of conversion to incompatible uses. Conservation groups urge Congress to appropriate the hundreds of millions of dollars it would take to buy this property to prevent it from being turned into luxury housing and condominiums.

Wildlife

Wildlife is at the centre of many arguments regarding whether the purpose of the parks is to preserve nature or to provide entertainment for visitors (Fig. 22.6). In the early days of the parks, "bad" animals (such as wolves and mountain lions) were killed so that populations of "good" animals (such as deer and elk) would be high. Rangers cut trees to improve views, put out salt blocks to lure animals to good viewing points close to roads, and otherwise manipulated nature to provide a more enjoyable experience for the guests.

Critics of this policy claim that favouring some species over others has unbalanced ecosystems and created a sad illusion of a natural system. They claim that excessively large elk populations in Yellowstone and Grand Teton National Parks, for instance, have degraded the range so badly that other species such as mice and ground squirrels are being crowded out. Park rangers tried hiring professional hunters to reduce the elk herd, but a storm of protest was raised. Sportsmen want to be able to hunt the elk themselves, animal lovers don't want them to be killed at all, and wilderness advocates don't like the precedent of hunting in national parks. The Park Service has retreated to a policy of "natural regulation," intended to let nature take its course. When elk starve to death, as thousands do in a hard winter, however, many people are appalled.

In Yellowstone National Park, unusually cold temperatures and deep snow during the winter of 1996–97 drove bison out of the high country and into surrounding national forest lands in search of food. More than 2,000 animals—two-thirds of the entire park herd either starved to death or were gunned down by hunters and game wardens as they crossed park boundaries. Ranchers claimed that bison carry brucellosis and could infect domestic cattle, although there is no record of this ever happening. Native American tribes and animal rights activists protested this slaughter as cruel and

FIGURE 22.6 Wild animals have always been one of the main attractions in national parks. Many people lose all common sense when interacting with big, dangerous animals. This is not a petting zoo.

TABLE 22.1	IUCN Categories of Protected Areas

CATEGORY	DEGREE OF HUMAN IMPACT OR INTERVENTION
1. Ecological reserves and wilderness areas	Little or none
2. National parks	Low
3. Natural monuments and archeological sites	Low to medium
4. Habitat and wildlife management areas	Medium
5. Cultural or scenic landscapes, recreation areas	Medium to high

Source: Data from World Conservation Union, 1990.

unnecessary. Part of the problem is that a herd of 3,000 bison is probably more than the carrying capacity of the land within the park, which has become an island of habitat encircled by lands claimed for other purposes by humans. Another reason for bison migration is that heavy snowmobile traffic packs down trails that invite animals to move through deep snow. In 1998, the U.S. Fish and Wildlife Service spent nearly $500,000 to build fences and holding corrals to prevent bison from leaving the park. Ironically, the fees paid by grazing rights in the national forests from which bison were excluded amount to about $15,000 per year. Proposals to reduce elk and bison populations by reintroducing predators such as wolves and mountain lions have been highly controversial (see What Do You Think? p. 497).

New Directions

What else is being done to enhance the visitor's experience of a national park while maintaining the natural ecosystem as much as possible? Several parks have removed facilities that conflict with natural values. In Yellowstone, a big laundry and a cabin ghetto next to Old Faithful have been torn down. Shabby hotels and filling stations that once stood at Norris Junction and Yancy's Hole are gone. The golf course, slaughterhouse, and tent camps at Mammoth Hot Springs also have been removed. In Yosemite, Grand Canyon, and Denali, tourists park their cars and take shuttle buses into the park to reduce congestion and pollution. A poorly

performing concessionaire was denied renewal of a contract for Yosemite, something that had never before happened.

There are proposals that a number of parks be closed to cars, and some areas might be closed to tourists altogether to protect wildlife and fragile ecosystems. The International Union for the Conservation of Nature and Natural Resources (IUCN) divides protected areas into five categories with increasing levels of protection and decreasing human impacts (Table 22.1). Many of our parks have tried to meet all these goals simultaneously but may have to select those of highest value. Most parks limit the number of overnight visitors. The time may come when park permits will need to be reserved years in advance and visits to certain parks will be limited to once in a lifetime! How would you feel about such a policy? Would you rather visit a pristine, uncrowded park only once or a less perfect place whenever you wanted?

One of the biggest problems with managing parks and nature preserves is that boundaries usually are based on political rather than ecological considerations. Airsheds, watersheds, and animal territories or migration routes often extend far beyond official boundaries and yet profoundly affect communities that we are attempting to preserve. Yellowstone and Grand Teton Parks in northwestern Wyoming are examples of this concept. Although about 1 million ha in total size, these parks probably cannot preserve viable populations of large predators such as grizzly bears. Management policies in the surrounding national forests and private lands seriously affect conditions in the park. The natural **biogeographical area** (an entire self-contained ecosystem and its associated land, water, air, and wildlife resources) must be managed as a unit if we are to preserve all its values.

World Parks and Preserves

The idea of setting aside nature preserves has spread rapidly over the past 60 years as people around the world have become aware

TABLE 22.2	World Nature Preserves by Area and Type	
BIOME	NUMBER OF AREAS	PERCENT OF PROTECTED AREA
Tropical dry forests	907	21
Tropical humid forests	355	12
Temperate deciduous forests	682	6
Temperate coniferous forests	114	7
Deserts	215	17
Tundra	31	26
Grasslands	116	3
Mountain regions	318	7
Islands	74	<1
Lakes and Wetlands	10	<1

Source: Norman Meyers, 1998.

FIGURE 22.7 Hikers explore Ellesmere National Park in Canada's Nunuvat Territory. Although rich in scenery and solitude, this park has very little biodiversity.

of the growing scarcity of wildlife and wild places. Still, in many cultures wilderness and wild lands are regarded as useless wastelands that should be put to some productive use if possible. So far, more than 530 million ha (nearly 4 percent of the earth's land) is designated as parks, wildlife refuges, and nature preserves worldwide (Table 22.2).

The largest number of protected areas is in tropical dry forests and savannas—principally in Africa—and temperate deciduous forests—mainly in North America and Europe—but many of these preserves are too small to maintain significant biological populations over the long term. Vast stretches of arctic tundra in a few parks and preserves in Alaska, Canada, Greenland, and Scandinavia make up about 26 percent of all protected areas (Fig. 22.7). Deserts and tropical humid forests also are well represented, but grasslands, aquatic ecosystems, and islands are badly underrepresented.

According to the United Nations Environment Program, North and Central America have the largest fraction—33 percent of all protected land or nearly 10 percent of their land area—designated for protection of any continent. The former Soviet Union, with 17 percent of the world's land, has only 3 percent of the officially protected area. The rapid destruction of Siberian forests (see Chapter 17) and terrible pollution problems in Russia and its former allies raises serious concerns about these natural areas. The IUCN has identified an additional 3,000 areas, totalling about 3 billion hectares worthy for national park or wildlife refuge status. The most significant of these areas are designated world **biosphere reserves** or world heritage sites. Currently, about 300 of these special refuges have been designated in 75 countries.

There are 12 UN Biosphere Reserves in Canada, including Long Point in southern Ontario, Clayoquot Sound on the west coast of British Columbia, and Redberry Lake in Saskatchewan. Among the individual countries with the most admirable plans to

protect natural resources are Costa Rica, Tanzania, Rwanda, Botswana, Benin, Senegal, Central Africa Republic, Zimbabwe, Butan, and Switzerland, each of which has designated 10 percent or more of its land as ecological protectorates. Brazil has even more ambitious plans, calling for some 231,600 sq km, or 18 percent of the country to be protected in nature preserves. So far, however, many of these areas are parks in name only. Lacking guards, visitor centres, administrative personnel, or even boundary fences, they are open to vandals and thieves to loot as they will.

Protecting Natural Heritage

Even parks with systems in place for protection and management are not always safe from exploitation or changes in political priorities. Many problems threaten natural resources and environmental quality in the parks. In Greece, the Pindos National Park is threatened by plans to build a hydroelectric dam in the centre of the park. Furthermore, excessive stock grazing and forestry exploitation in the peripheral zone are causing erosion and loss of wildlife habitat. In Colombia, the Paramillo National Park also is threatened by dam building. Oil exploration along the border of the Yasuni National Park in Ecuador pollutes water supplies, while miners and loggers in Peru have invaded portions of Huascaran National Park. In Palau, coral reefs identified as a potential biosphere reserve are damaged by dynamiting, while on some beaches in Indonesia every egg laid by endangered sea turtles is taken by egg hunters. These are just a few of the many problems in parks around the world. Often countries with the most important biomes lack funds, trained personnel, and experience to manage some of the areas under their control (Fig. 22.8).

The IUCN has developed a **world conservation strategy** for natural resources that includes the following three objectives: (1) to

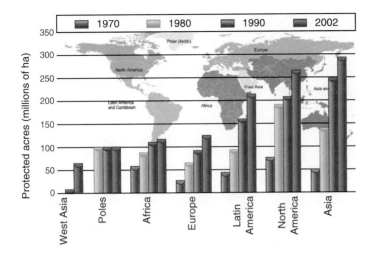

FIGURE 22.8 Change in amount of protected areas by region.
Source: UN Environment Program, 2003.

TABLE 22.3	IUCN Ecological Plan of Action

1. Launch a consciousness-raising exercise to bring the issue of biological resources to the attention of policymakers and the public at large.

2. Design national conservation strategies that take explicit account of the values at stake.

3. Expand our network of parks and preserves to establish a comprehensive system of protected areas.

4. Undertake a program of training in the fields relevant to biological diversity to improve the scientific skills and technological grasp of those charged with its management.

5. Work through conventions and treaties to express the interest of the community of nations in the collective heritage of biological diversity.

6. Establish a set of economic incentives to make species conservation a competitive form of land use.

Source: Data from International Union for Conservation and National Resources.

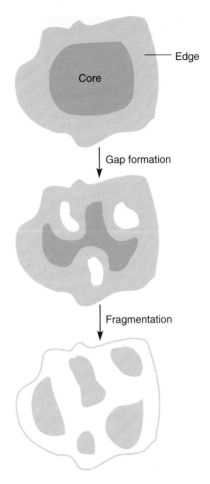

FIGURE 22.9 A patch or "island" of habitat becomes reduced and fragmented as vegetation gaps, or other human disturbances, expand until only small fragments remain. Note that core regions disappear early in this process. Finally, only edges are left in a matrix of disturbed habitat.

Size and Design of Nature Preserves

What is the optimum size and shape of a wildlife preserve? For many years, conservation biologists have disputed whether it is better to have a *single large or several small* reserves (the SLOSS debate). Ideally, a reserve should be large enough to support viable populations of endangered species, keep ecosystems intact, and isolate critical core areas from damaging external forces (see Chapter 16). But as gaps are opened in habitat by human disturbance (Fig. 22.9), and eventually areas are fragmented into isolated islands, edge effects may eliminate core characteristics everywhere.

To satisfy the conflicting needs and desires of humans and nature, we may need a spectrum of preserves with decreasing levels of interference and management ranging from: (1) *recreation areas,* designed primarily for human entertainment, aesthetics, and enjoyment; (2) *historic areas,* intended to preserve a landscape as we imagine it looked in some previous time—such as pre-settlement or pioneer days; (3) *conservation reserves,* set aside to maintain essential ecological functions, preserve biodiversity, or

maintain essential ecological processes and life-support systems (such as soil regeneration and protection, recycling of nutrients, and cleansing of waters) on which human survival and development depend; (2) to preserve genetic diversity, which is the foundation of breeding programs necessary for protection and improvement of cultivated plants and domesticated animals; (3) to ensure that any utilization of species and ecosystems is sustainable.

These goals are further elaborated in the ecological plan of action adopted by the IUCN and shown in Table 22.3. A promising approach for financing these objectives is debt-for-nature swaps (Chapter 17).

Yellowstone Wolves

On a bright moonlit night, a chilling chorus of howls drifts across the Lamar River Valley at the northern edge of Yellowstone National Park in Wyoming. Elk and deer shift nervously at this signal that a new predator—the gray or timber wolf—now roams their territory. Part of a group of about 120 gray wolves in the park, this pack is among the first of their species to inhabit Yellowstone in nearly 70 years. Hunted to extinction throughout the western United States, gray wolves have been making a comeback in the northern Rocky Mountains since the U.S. Fish and Wildlife Service transplanted 66 Canadian wolves into Yellowstone and central Idaho in 1995 and 1996. This reintroduction program is part of a long-discussed effort to restore ecosystems deprived of natural predators. The project has become mired in controversy, however, and the fate of the wolves—and their prey—now depends on decisions of judges far from Yellowstone wilderness.

Few animals arouse as much animosity and admiration as does the wolf. Children's bedtime tales from many cultures teach us to beware of the big, bad wolf. Ancestor to all our pet dogs, the wolf embodies some of our deepest fears and strongest impulses. It is a potent symbol of much that we both love and fear in wild nature. Less than a century ago, an estimated 100,000 wolves roamed the western United States. As settlers moved west, however, wolves were poisoned, shot, trapped, clubbed, or killed wherever and whenever they could be found.

Ecosystems in many predator-less places show clear signs of overpopulation by prey species such as elk and deer. The Yellowstone Park elk herd had grown to some 25,000 animals, probably four or five times the habitat's carrying capacity. Vegetation was overgrazed and populations of smaller animals such as ground squirrels were declining. Several decades ago, ecologists recommended that keystone predators such as wolves be reintroduced to control prey populations. These proposals bring howls of angry protest from local ranchers, who regard wolves as sinister killers that threaten children, pets, livestock and their whole way of life. "Shoot, shovel, and shut-up" is the preferred wolf management of many Westerners. It

took more than 20 years to get approval for wolf reintroduction. One of the key compromises was that the transplanted animals were classified as a "nonessential, experimental population," which meant that any animals that threatened livestock could be shot.

Surprisingly, once back in the park, wolves became established very quickly. Taking advantage of the abundant food supply, they tripled their population in just three years. The effects on the ecosystem were immediate and striking. Biodiversity increased noticeably. Fewer elk, deer, and moose meant more food for squirrels, gophers, voles, and mice. Abundant small prey, in turn, led to increased numbers of eagles, hawks, fox, pine martens, and weasels. Large animal carcasses left by the wolves provided a feast for scavengers such as bears, ravens, and magpies. Coyotes, which had become common in the wolf's absence, suffered a 50 percent decline when wolves hunted them down and killed them. This helped small mammals that once were coyote prey. Plants such as grasses, forbs, willows, and aspen flourished in the absence of grazing and browsing pressure. Rangers and naturalists were delighted that the ecosystem was balanced once again. And tourists were thrilled to catch a glimpse of a wolf or to hear them howl.

Most wolf supporters believe that better livestock management and payments to ranchers for any authenticated losses can take care of any problems wolves might cause. Not everyone was happy, however, with the reintroduction program. The American Farm Bureau sued to force removal of the transplanted wolves, claiming they were a different subspecies from the former (extinct) residents. A group of environmental organizations, including the Audubon Society and the Sierra Club, also filed a suit asking that any "native" wolves that migrate into the park on their own not be part of the experimental population. A U.S. federal judge joined the two lawsuits. Most environmental groups then withdrew, but a few remained litigants because they opposed killing wolves under any circumstances. In 1998, the U.S. district court ruled that the nonessential experimental designation violates the U.S. Endangered Species Act. The judge ordered the transplanted wolves removed from Yellowstone, which really means they will be shot, because there is no other place

to which they could be moved and not be an experimental population. This decision was overturned, however, by the Tenth Circuit Court of Appeals.

Wolf reintroduction raises some important questions about the purposes of parks. It also reveals different attitudes toward nature. Which of the ethical perspectives described in Chapter 2 do you see represented here? Why do you think some animal rights groups would regard the suffering of individual wolves, who may be shot because they threaten livestock, as more important than the interests of the species as a whole? What role should science play in this dispute? Suppose you were a mediator charged with bringing all the stakeholders together to try to design an acceptable wolf management program in Yellowstone. What strategies would you use and where would you start?

Are wolves beautiful, thrilling symbols of wild nature or ruthless killers? Reintroduction of these top predators into Yellowstone National Park has enthusiastic support from environmental groups but passionate opposition from local ranchers and hunters.

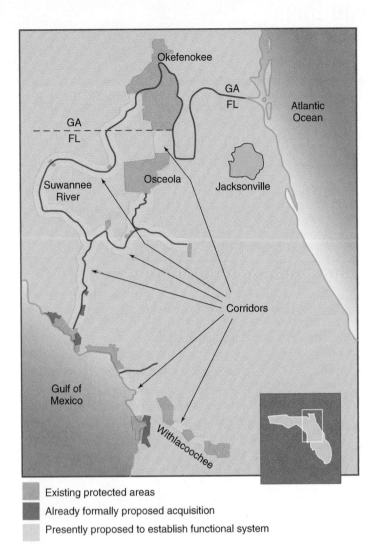

Existing protected areas

Already formally proposed acquisition

Presently proposed to establish functional system

FIGURE 22.10 Corridors serve as routes of migration, linking isolated populations of plants and animals in scattered nature preserves. Although individual preserves may be too small to sustain viable populations, connecting them through river valleys and coastal corridors can facilitate interbreeding and provide an escape route if local conditions become unfavourable.

Source: R. F. Noss and L. D. Harris, "Nodes, Networks and MUMs: Preserving Diversity at All Levels," in *Environmental Management,* vol. 10: 299–309, 1986.

protect a particular species or group of organisms; (4) *pristine research areas,* to serve as a baseline of undisturbed nature; and (5) *inviolable preserves,* for sensitive species from which all human entrance is strictly prohibited.

For some species with small territories, several small isolated refuges can support viable populations and provide insurance against a disease or other calamity that might wipe out a single population. But small preserves can't support species such as elephants or tigers that need large amounts of space. Given human needs and pressures, however, big preserves aren't always possible. Establishing **corridors** of natural habitat to allow movement of species from one area to another (Fig. 22.10) can help maintain genetic exchange and prevent the high extinction rates often characteristic of isolated and fragmented areas.

FIGURE 22.11 How small can a nature preserve be? In an ambitious research project, scientists in the Brazilian rainforest are carefully tracking wildlife in plots of various sizes, either connected to existing forests or surrounded by clear-cuts. As you might expect, the largest and most highly specialized species are the first to disappear.

An interesting experiment funded by the World Wildlife Fund and the Smithsonian Institution is being carried out in the Brazilian rainforest to determine the effects of shape and size on biological reserves. Some 23 test sites, ranging in size from one hectare to 10,000 hectares have been established. Some areas are surrounded by clear-cuts and newly created pastures (Fig. 22.11), while others remain connected to the surrounding forest. Selected species are regularly inventoried to monitor their dynamics after a disturbance. As was expected, some species disappear very quickly, especially from small areas. Sun-loving species flourish in the newly created forest edges, but deep-forest, shade-loving species move out, particularly when size or shape reduces the distance from the edge to the centre below a certain minimum. This demonstrates the importance of surrounding some reserves with buffer zones that maintain the balance of edge and shade species.

Conservation and Economic Development

Many of the most seriously threatened species and ecosystems of the world are in the developing countries, especially in the tropics. This situation concerns us all because these countries are the guardians of biological resources that may be vital to all of us. Unfortunately, where political and economic systems fail to provide people with land, jobs, and food, disenfranchised citizens turn to legally protected lands, plants, and animals for their needs. Immediate human survival always takes precedence over long-term environmental goals. Clearly the struggle to save species and unique ecosystems cannot be divorced from the broader

FIGURE 22.12 Ecotourism can be a sustainable resource use. If local communities share in the revenue, it gives them an incentive to value and protect biodiversity and natural beauty.

TABLE 22.4	Ecotourism Suggestions

1. *Pretrip preparation.* Learn about the history, geography, ecology, and culture of the area you will visit. Understand the do's and don'ts that will keep you from violating local customs and sensibilities.

2. *Environmental impact.* Stay on designated trails and camp in established sites, if available. Take only photographs and memories and leave only goodwill wherever you go.

3. *Resource impact.* Minimize your use of scarce fuels, food, and water resources. Do you know where your wastes and garbage go?

4. *Cultural impact.* Respect the privacy and dignity of those you meet and try to understand how you would feel in their place. Don't take photos without asking first. Be considerate of religious and cultural sites and practices. Be as aware of cultural pollution as you are of *environmental pollution.*

5. *Wildlife impact.* Don't harass wildlife or disturb plant life. Modern cameras make it possible to get good photos from a respectful, safe distance. Don't buy products such as ivory, tortoise shell, animal skins, or feathers from endangered species.

6. *Environmental benefit.* Is your trip strictly for pleasure or will it contribute to protecting the local environment? Can you combine ecotourism with work on cleanup campaigns or delivery of educational materials or equipment to local schools or nature clubs?

7. *Advocacy and education.* Get involved in letter writing, lobbying, or educational campaigns to help protect the lands and cultures you have visited. Give talks at schools or to local clubs after you get home to inform your friends and neighbours about what you have learned.

struggle to achieve a new world order in which the basic needs of all are met.

The tropics are suffering the greatest destruction and species loss in the world, especially in humid forests and coastal ecosystems. People in some of the affected countries are beginning to realize that the biological richness of their environment may be their most valuable resource and that its preservation is vital for sustainable development. Ecotourism can be more beneficial to many of these countries over the long term than extractive industries such as logging and mining (Fig. 22.12 and Case Study, p. 502). Table 22.4 suggests some ways to ensure that tourism is ecologically and socially beneficial.

In many cases, sustainable production of food, fiber, medicines, and water in rural areas depends on ecosystem services derived from adjacent conservation reserves. Tourism associated with wildlife watching and outdoor recreation can be a welcome source of income for underdeveloped countries. If local people share in the benefits of saving wildlife, they probably will cooperate and the programs will be successful. To reformulate Thoreau's famous dictum, "In broadly shared economic progress is preservation of the wild."

Indigenous Communities and Biosphere Reserves

Areas chosen for nature preservation are often traditional lands of indigenous people who cannot simply be ordered out. Finding ways to integrate human needs with those of wildlife is essential for local acceptance of conservation goals in many countries. In 1986, UNESCO initiated its **Man and Biosphere (MAB) program** that encourages division of protected areas into zones with different purposes. Critical ecosystem functions and endangered wildlife are protected in a central core region where limited scientific study is the only human access allowed. Ecotourism and research facilities are located in a relatively pristine buffer zone around the core, while sustainable resource harvesting and permanent habitation are allowed in multiple-use peripheral regions (Fig. 22.13).

Mexico's 545,000-hectare Sian Ka'an Reserve on the Caribbean coast is a good example of a MAB reserve. The core area includes 528,000 ha of coral reef and adjacent bays, marshes, and lowland tropical forest. More than 335 bird species have been observed within the reserve, along with endangered manatees, five types of jungle cats, spider and howler monkeys, and four species of increasingly rare sea turtles. Approximately 25,000 people live in communities in peripheral regions around the reserve, and the

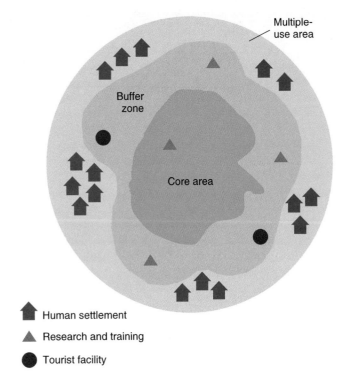

Buffer
zone

Core area

▲ Human settlement

▲ Research and training

● Tourist facility

FIGURE 22.13 A model biosphere reserve. Traditional parks and wildlife refuges have well-defined boundaries to keep wildlife in and people out. Biosphere reserves, by contrast, recognize the need for people to have access to resources. Critical ecosystem is preserved in the core. Research and tourism are allowed in the buffer zone, while sustainable resource harvesting and permanent habitations are situated in the multiple-use area around the perimeter.

resort developments of Cancun are located just to the north. In addition to tourism, the economic base of the area includes lobster fishing, small-scale farming, and coconut cultivation.

The Amigos de Sian Ka'an, a local community organization, played a central role in establishing the reserve and is working to protect the resource base while it improves living standards for local people. New intensive farming techniques and sustainable harvesting of forest products enable people to make a living without destroying the resource base. Better lobster harvesting techniques developed at the reserve have improved the catch without depleting native stocks. Local people now see the preserve as a benefit rather than an imposition from outside. Unfortunately, the government has very limited funds to develop or patrol the reserve.

An even grander plan called *Passeo Pantera* (path of the panther) envisions a thousand-mile-long series of reserves and protected areas interconnected by natural corridors and managed buffer zones that would preserve both wildlife and native cultures along the entire Caribbean coast of Central America from the Yucatan to Panama. Whether the politically unstable and financially troubled countries involved could get together to establish such a plan is questionable. Competing goals and objectives of conservation groups, international development banks, and powerful neighbours such as Canada and the United States also complicate the picture, but the idea is a noble one whose time perhaps has come.

In some cases, human occupancy can help maintain natural values. The Keolado National Park in India is such a case. After years of struggle, native villagers were able to show that grazing by domestic water buffalo is essential to maintaining wetlands and protecting biodiversity.

PROTECTED AREAS

Although indigenous people had lived in the Americas for thousands of years before the first Europeans arrived, introduced diseases killed up to 90 percent of the existing population so that the continent appeared a vast, empty wilderness to early explorers. As historian Frederick Jackson Turner pointed out in a series of articles and speeches around the turn of the century, a belief that wilderness was not only a source of wealth but also the origin of strength, self-reliance, wisdom, and character is deeply embedded in our culture. The frontier was seen as a place for continuous generation of democracy, social progress, economic growth, and national energy. A number of authors, including Henry Thoreau, Aldo Leopold, Sigurd Olson, Edward Abbey, and Wallace Stegner, have written about the physical, mental, and social benefits for modern people of rediscovering solitude and challenge in wilderness (Fig. 22.14).

A protected area is one measure of societal response aimed at conserving ecosystems. It is generally agreed that the greater the area protected from development, the better the chance for a healthy sustaining ecosystem. Protected areas include parks, wildlife and forest reserves, **wilderness,** and other conservation areas designated through federal, provincial, and territorial legislation (Fig. 22.15).

While there are over 3,500 of these protected areas in Canada, only about 800 of these are larger than 1,000 hectares, and they capture over 98 percent of the total area protected in Canada. While there are more protected areas in southern Canada, these are generally small compared with the few, but very large, protected areas in Canada's North.

Federal, provincial, and territorial governments have collectively designated about 9 percent of Canada as "protected." This protection ranges to over 90 percent for some ecoregions to none in other ecoregions. About two-thirds of the land occupied by Canada's ecoregions has some protection, leaving about one third with virtually no protection.

Species recovery projects are another means to protect biodiversity in ecosystems. Canada has also introduced tax measures that favour donations of lands for conservation purposes. As well, efforts are being made to conserve and restore selected ecosystems such as wetlands. Sustaining native species and their habitats is also a key component of sustainable development strategies, particularly for resource activities such as agriculture, forestry, and fisheries that heavily modify or harvest Canada's ecosystems.

FIGURE 22.14 Glacier National Park preserves a piece of relatively untouched natural ecosystem. Areas such as this serve as a refuge for endangered wildlife, a place for outdoor recreation, a laboratory for scientific research, and a source of wonder, awe, and inspiration.

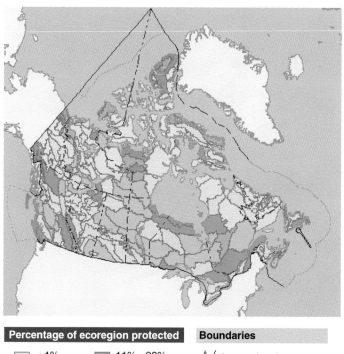

Percentage of ecoregion protected

< 1%	11% - 23%
1% - 5%	23% - 47%
5% - 11%	47% - 100%

Other Features

National Park Area
Water area
Regions outside Canada

Boundaries

International
Provincial / Territorial
Canada / Kalaallit Nunaat dividing line
EEZ (200 mile)

FIGURE 22.15 Proportion of ecoregions in Canada that are protected.
Source: Natural Resources Canada.

A prolonged battle has been waged over these *de facto* wilderness areas, pitting environmental groups who want more wilderness against loggers, miners, ranchers, and others who want less wilderness. The arguments for saving wilderness are that it provides (1) a refuge for endangered wildlife, (2) an opportunity for solitude and primitive recreation, (3) a baseline for ecological research, and (4) an area where we have chosen simply to leave things in their natural state. The arguments against more wilderness are that timber, energy resources, and critical minerals contained on these lands are essential for economic development.

To people who live in remote areas, jobs, personal freedom, and local control of resources seem more important than abstract values of wilderness. They often see themselves as an embattled minority trying to protect an endangered, traditional way of life against a wealthy elite who want to lock up huge areas for recreation or aesthetic purposes.

For many people, especially those in developing countries, the idea of pristine wilderness untouched by humans is regarded as neither very important nor very interesting. In most places, all land is occupied fully—if sparsely—by indigenous people. To them the area is home no matter how empty it may look to outsiders. From this perspective, preserving biological diversity, scenic beauty, and other natural resources may be a good idea, but excluding humans and human features from the land does not necessarily make it more valuable. In fact, saving cultural heritage, working landscapes, and historical evidence of early human occupation can often be among the most important reasons to protect an area.

Refuge Management

Although **wildlife refuges** were originally intended to be sanctuaries in which wildlife would be protected from hunting and other

CASE STUDY

ZIMBABWE'S "CAMPFIRE" PROGRAM

Stretching across Zimbabwe from Victoria Falls to the Mozambique border is a broad swath of acacia thornbrush known as the Zambezi River Valley. Home to some of the world's greatest numbers and diversity of big game animals, this area is a magnet for tourists who come to hunt, photograph, or just watch elephants, cape buffalo, antelope, lions, leopards, and other charismatic megafauna. Tourism is a lucrative business in Africa. Every year, tourists spend more than $250 million on wildlife-related activities, but little of that wealth makes its way to local residents. Foreign tour operators pocket the vast proportion of the profits, while native people get little or nothing.

Unfortunately, both the survival of the vast herds of wildlife and the tourist business based upon them are threatened as habitat is converted to farmland to support growing human populations. In response to declining wildlife numbers and threats to the lucrative tourist trade, the government has set aside about half of the valley as reserves and national parks. The other half remains communal land divided into tribal areas. Many of these communal lands border protected reserves. Because of this close proximity, poaching and illegal settlement has been a persistent problem on protected lands.

Many rural residents view the wildlife of Zambezi as, at best, a food source, and, at worst, an enemy. Local people receive little benefit from their proximity to parks and reserves. No compensation is provided when crops are trampled by elephants or when livestock is killed by lions. Buffalo, crocodiles, and hippopotamuses wander off reserve lands and endanger human lives in villages. Endemic poverty exacerbated by wildlife depredations and restricted hunting rights create serious problems for parks. As one tribal leader said, "African elephants have no chance until Africans have a chance."

In an attempt to alleviate both social and conservation issues around the parks and preserves, a new program was established entitled the Communal Areas Management Program for Indigenous Resources (CAMPFIRE). This program gives local people authority over wildlife on communal lands. Each village negotiates with safari companies to establish hunting and tourist concessions. A single hunter may spend more than $200,000 on a hunt. Half of this goes to the local community. The World Wildlife Fund estimates that CAMPFIRE has increased local household incomes up to 25 percent. At the end of the year, each village decides independently how it will use its income. Some villages divide it among households, giving each a cash dividend; others put it into community projects such as new schools or clinics or to pay children's school fees.

The communities involved in the CAMPFIRE program have gained a real sense of ownership and control over their local resources. They willingly take responsibility for the management and protection of wildlife. Poaching has declined and residents are actually starting to complain there are not enough animals. Wildlife is now viewed as a valuable commodity—a financial asset to the communities rather than a threat to their well-being.

Everyone wins in the CAMPFIRE program. Local communities receive funds and food from the wildlife in their region; the governmental costs of reserve management have declined; tourists get to view the unique animals and landscapes of the Zambezi River Valley; and illegal poaching has declined, allowing wildlife to thrive again in Zimbabwe.

Members of the Hikwaka, Zimbabwe Sewing and Bakery Co-Op display clothes made with a sewing machine financed by profits from ecotourism.

disturbances, a 1948 compromise allowed hunting in refuges in exchange for an agreement by hunters to purchase special duck stamps to raise money for wetland protection. Although a refuge that allows hunting seems like an oxymoron, and in spite of the fact that only 10 percent of refuge lands have been acquired with duck stamp funds, hunting has become firmly established in most units (Fig. 22.16). Critics charge that the U.S. Fish and Wildlife Service, which administers the refuge system, is so strongly oriented toward hunting that many units have become little more than duck and goose farms.

FIGURE 22.16 Hunters on their way to duck blinds in a wildlife refuge. Although it seems a contradiction, half of all refuges allow hunting.

Over the years, a number of improbable and incompatible uses have become accepted in wildlife refuges including oil drilling, cattle grazing, snowmobiling, motorboating, off-road vehicle use, timber harvesting, hay cutting, trapping, and camping. One Nevada refuge has a bombing and gunnery range, while another is the site of a brothel. A General Accounting Office report found that 60 percent of all refuges allow activities that are harmful to wildlife.

Refuges also face threats from external activities. More than three-quarters of all refuges in the United States have water pollution problems, two-thirds of which are serious enough to affect wildlife. A notorious example is the former Kesterson Wildlife Refuge in California, where selenium-contaminated irrigation water drained from farm fields turned the marsh into a death trap for wildlife rather than a sanctuary. Eventually, the marsh had to be drained and capped with clean soil to protect wildlife. Subsequent research has shown that at least 20 wildlife refuges in western states have toxic metal pollution caused by agriculture and industrial activities outside the refuge.

The biggest current battle over wildlife refuges concerns proposals for oil and gas drilling in the Arctic National Wildlife Refuge on the north slope of Alaska's Brooks Range.

International Wildlife Preserves

As we saw earlier in this chapter, most developing countries rarely have separate systems of parks and wildlife refuges. Many nature preserves are set up primarily to protect wildlife, however. An outstanding example of both the promise and the problems in managing parks in the less-developed countries is seen in the Serengeti ecosystem in Kenya and Tanzania. This area of savanna, thorn woodland, and volcanic highland lying between Lake Victoria and the Great Rift Valley in East Africa is home to the highest density of ungulates (hoofed grazing animals) in the world. Over 1.5 million wildebeests (or gnus) graze on the savanna in the wet season, when grass is available, and then migrate through the woodlands into the northern highlands during the dry season. The ecosystem also supports hundreds of thousands of zebras, gazelles, impalas, giraffes, and other beautiful and intriguing animals. The herbivores, in turn, support lions and a variety of predators and scavengers, such as leopards, hyenas, cheetahs, wild dogs, and vultures. This astounding diversity and abundance is surely one of the greatest wonders of the world.

Tanzania's Serengeti National Park was established in 1940 to protect 15,000 sq km, an area about the size of Prince Edward Island or twice as big as Yellowstone Park. It is bordered on the east by the much smaller Ngorongoro Conservation Area and Lake Manyara National Park. Kenya's Masa Mara National Reserve borders the Serengeti on the north. Rapidly growing human populations push against the boundaries of the park on all sides. Herds of domestic cattle compete with wild animals for grass and water. Agriculturalists clamor for farmland, especially in the temperate highlands along the Kenya-Tanzania border. So many tourists flock to these parks that the vegetation is ground to dust by hundreds of sight-seeing vans, and wildlife find it impossible to carry out normal lives.

Perhaps the worst problem in Africa is **poachers,** illegal hunters who massacre wildlife for valuable meat, horns, and tusks. Where there once were about 1 million rhinos in Africa, the population had dropped by the mid-1980s to less than 10,000 animals. Antipoaching efforts have allowed the population to recover to about 15,000 currently (Fig. 22.17).

Elephants are under a similar assault. Forty years ago there were no elephants in the Serengeti, but perhaps 3 million in all of Africa. Since then, about 80 percent of the African elephants have been killed—mainly for their ivory—at a rate of 100,000 each year. The 2,000 elephants now in Serengeti National Park have been driven there by hunting pressures elsewhere.

FIGURE 22.17 A guard protects black rhinoceroses from poachers. His ancient rifle is no match, however, for the powerful automatic weapons with which the poachers are now armed.

The poachers continue to pursue the elephants and rhinos, even in the park. Armed with high-powered rifles and even machine guns and bazookas from the many African wars in the last decade, the poachers take a terrible toll on the wildlife. Park rangers try to stop the carnage, but they often are outgunned by the poachers. The parks themselves are beginning to resemble war zones, with fierce, lethal firefights rather than peace and tranquility.

SUMMARY

Parks, wildlife refuges, wilderness areas, and nature preserves occupy a small percentage of our total land area but protect valuable cultural resources and representative samples of the earth's species and ecosystems. We can discern much about people's environmental ideals by examining the gardens, parks, and recreation areas that they create.

Parks are havens for wildlife and places for healthful outdoor recreation. Many are overcrowded, misused, and neglected, however. Pollution and incompatible uses outside parks threaten the values that we seek to protect. Wildlife is at the centre of many park controversies. Is the park's purpose to preserve wild nature or provide entertainment for visitors? How much management is acceptable? When should we intervene and when should we let nature take its course?

There are proposals to remove distracting or damaging uses from parks and to limit entrance permits. We may reach the point where each of us might be allowed to visit some of the more popular parks only once in a lifetime. One solution to congestion and overuse is to create additional parks in some of the many deserving but unprotected areas.

Worldwide, only about 4 percent of total land area has been protected in parks, wildlife refuges, and nature preserves. Some biomes such as dry tropical forests and tundra are well represented in this network, but others such as grassland and wetlands are underrepresented. The optimum size for nature preserves depends on the terrain and the values they are intended to protect, but—in general—the larger a reserve is, the more species it can protect. Establishing corridors to link separate areas can be a good way to increase effective space and to allow migration from one area to another. Economic development and nature protection can go hand in hand. Ecotourism may be the most lucrative and long-lasting way to use resources in many developing countries.

Areas chosen for preservation often are lands of indigenous people. Careful planning and zoning can protect nature and also allow sustainable use of resources. Man and Biosphere (MAB) reserves provide for multiple use in some areas but strict conservation in others. Wildlife refuges were intended to be sanctuaries for wildlife, but over the years many improbable and damaging uses have become established in them.

QUESTIONS FOR REVIEW

1. List some problems and threats from inside and outside our national parks.

2. Why is the reintroduction of wolves into Yellowstone a controversial issue?

3. Describe the IUCN categories of protected areas and the amount of human impact or intervention allowed in each. Can you name some parks or preserves in each category?

4. Which biomes or landscape types are best represented among the world's protected areas and which are least represented?

5. Draw a diagram of an ideal MAB reserve. What activities would be allowed in each zone?

QUESTIONS FOR CRITICAL THINKING

1. Is "contrived" naturalness a desirable feature in parks and nature preserves? How much human intervention do you think is acceptable in trying to make nature more beautiful, safe, comfortable, or attractive to human visitors? Think of some specific examples that you would or would not accept.

2. Suppose you were superintendent of Banff National Park. How would you determine the carrying capacity of the park for elk? How would you weigh having more elk or more ground squirrels? If there are too many elk, how would you thin the herd?

3. Why do you suppose that dry tropical forests and tundra are well represented in protected areas, whereas grasslands and wetlands are rarely protected? Consider social, cultural, and economic as well as biogeographical reasons in your answer.

4. Suppose that preserving healthy populations of grizzly bears and wolves requires that we set aside some large fraction of our national parks as a zone into which no humans will ever again be allowed to enter for any reason. Would you support protecting bears even if no one ever sees them or could even be sure that they still existed?

5. Suppose that you had trespassed into the bear sanctuary and were attacked by a bear. Should the rangers shoot the bear or let it eat you?

6. Are there any conditions under which you would permit oil drilling in the Arctic National Wildlife Refuge?

7. Why do you think some ecoregions in Canada have a very low proportion (sometimes 0 percent) of protected land? Describe the conflicts between protection of ecosystems and other human uses. See http://atlas.gc.ca/site/english/maps/environment/ecology/protecting/protectedareas.

KEY TERMS

biogeographical
   area   494

biosphere reserves   495

corridors   498

inholdings   493

Man and Biosphere (MAB)
   program   499

poachers   503

wilderness   500

wildlife refuges   501

world conservation
   strategy   495

Web Exercises

EXPLORE OUR NATIONAL PARKS

Explore the variety of national parks and reserves in Canada at the Parks Canada website at http://www.pc.gc.ca/progs/np-pn/index_E.asp. Which Canadian national park is closest to where you live? Find two contrasting park contexts (e.g., wilderness versus heavily used for recreation) and think about the challenges and opportunities faced by managers of these two parks. How do you think areas that should be national parks are identified?

PARKS AND PRESERVES IN AUSTRALIA

Many countries have national park and nature preserve systems. Australia provides information on these areas on the Web. Go to http://www.ea.gov.au/parks/, and choose a national park or nature preserve by clicking on a name in the list in the left frame. Look at the web page for that site. What is the site like? Where is it? What characteristics make it worth preserving? Is there anything like it near where you live, or have you ever been to a place like this one?

Spilling down a 2,000 metre icefall, the Kennecott Glacier in the U.S. flows through the heart of Wrangell-St. Elias National Park. Encompassing more than 5 million hectares, the park is larger than Switzerland and, together with adjacent parks in Canada, makes up the largest protected natural area in the world.

Right Whale Environmental Scientist

Moira Brown

Moira Brown's first degree was from McGill University in Education, Physical Education (1977) and she taught Phys Ed for four years in the West Island school board, three years in Special Education, and the fourth year in an elementary school. By 1981 her interest in wildlife and science was developing from a hobby to a desire to gain more education in the field. She resigned from teaching in June 1981 and returned to Macdonald College of McGill University to pursue a Bachelors degree in Science in the Department of Renewable Resources (B.Sc Agriculture, Renewable Resources, 1985). There her interests were primarily in the field of conservation and ornithology, particularly raptors.

Although contemplating a master's program in parasitology of raptors at MacDonald College, in 1984 she was working part time at the Arctic Biological Station, Department of Fisheries and Oceans in Ste. Anne de Bellevue. When her undergraduate studies finished in December 1984, two scientists, Drs. Ed Mitchell and Randall Reeves, were looking for a contractor to research the history of whales that were taken in the Canadian Arctic during the 1800s and 1900s. She spent her days pouring through logbooks of American whale ships and post journals from the Hudson's Bay Company compiling tables on what species of whales were hunted, how many, from where, and by whom. When that work ended in March, the two scientists she worked with suggested she try interning with a humpback field project in Maine and a right whale field project in the Bay of Fundy to gain field experience. She remembered meeting Scott Kraus at the New England Aquarium in Boston in March 1985; he agreed to take her on as a volunteer, but the funding was so limited he could only provide housing, no salary, and all the volunteers would chip in for food during the field season. Just after the meeting with Scott, she also met with an assistant for Steve Katona, director of Allied Whale at the College of the Atlantic in Bar Harbor Maine. They agreed to take her on for the month of July at their remote field station at a lighthouse on an island 21 miles south of Bar Harbor. Thus the summer plans were set for her first field season with marine mammals: July on an island and August and September in Lubec Maine for surveys for right whales in Canadian waters in the Bay of Fundy.

By 1988, she realized an advanced degree was essential to success and enrolled at the University of Guelph under the tutelage of Dr. David Gaskin. Her thesis involved developing a technique to obtain skin biopsy samples from right whales so they could use molecular techniques to determine the sex of right whales. This would seem like a very basic question but as she learned over the years, information on right whales was hard won and although the photo identification program had been carried out since 1980 and there was an extensive catalogue of individually identified right whales, many of the basic life history parameters easily determined with other species were difficult with right whales and by 1989, they knew the sex of less than 30 percent of the catalogued whales. Because this species was so endangered, such basic information was essential to figuring out the potential for recovery. The genetic work was carried out with Dr. Bradley White with whom they still collaborate today and the early struggles at figuring out how to get samples from right whales from which DNA could be extracted has now developed into one of the premier studies where over 70 percent of the living animals have been sampled and graduate students are in the process of defining the genealogy or family tree of the entire species.

In 1997, she moved to Cape Cod to work as a scientist with the Center for Coastal Studies in Provincetown, MA. Cape Cod Bay is a winter habitat for right whales and she was contracted by the State of Massachusetts to carry out their right whale conservation and monitoring program, which involved aerial and vessel surveys in the winter and spring months.

Perhaps the late Dr. Gaskin, who was very much a conservationist as well as a scientist, nurtured her interest in using science to promote stewardship of an endangered species. The more they learned about right whales the more they learned that there were primarily two human activities that were causing mortality to right whales and thus reducing the potential for population recovery. Right whales are a near shore species, an urban whale if you will. Their habitat areas parallel the east coast of Canada and the United States, an area with a great deal of shipping traffic and fishing activity. Every year there were a couple of dead right whales on the beach, victims of an accidental collision with a ship or entanglement in fixed fishing gear. The interaction with the shipping industry was the most likely to be fatal and the one issue she chose to concentrate on in Canada. While working on her doctorate they were successful at convincing DFO to designate right whale conservation areas in Canada for the two habitats where the right whales spend the summer and fall (1993). They were also working with the Canadian Coast Guard (Fundy Traffic) so that commercial vessels were warned when right whales were in the Bay of Fundy. Each large commercial vessel that enters the internationally mandated shipping lanes in the Bay of Fundy is in radio contact and tracked by radar through Fundy Traffic. Awareness was on the increase through communication with the shipping industry, articles in newspapers, documentaries, and with the summer visitors who went out to see right whales on the many whale watching vessels that depart from the shore of New Brunswick and Nova Scotia. Even though she was working full time in the U.S., she always kept attention on Canadian issues through East Coast Ecosystems. For example, in order to better educate the public, they wrote a code of ethics for the whale watchers, many of the companies signed on and would attend yearly meetings to keep up to date on the science and biology of this still rare whale.

Their most important conservation action for right whales to date was realized last year with the relocation of the shipping lanes in the Bay of Fundy from an area of high-density right whales to an area with lower densities of whales. The whole process took four years and started with the right whale data collected each summer, which was presented to a diverse vessel whale working group made up of representatives from the shipping industry, government agencies, conservation groups, shipping agents, fishermen, and port authorities.

The key to the success of this project and the recovery of the North Atlantic right whales is a group of very dedicated scientists working together to share the results of the science to further their understanding of what makes right whales tick. But it doesn't end there, the data are then shared with the shipping and fishing industry and government policymakers to use the science to make changes to shipping and fishing practices to reduce their affect on this species and help it recover. It is a huge undertaking by a large number of people working on a wide variety of topics of which Moira is just one member, but that is what it will take to recover this species.

Glossary

A GLOBAL CONCERN

A

abiotic Non-living.

abundance The number or amount of something.

acid precipitation Acidic rain, snow, or dry particles deposited from the air due to increased acids released by anthropogenic or natural resources.

acids Substances that release hydrogen ions (protons) in water.

active learner Someone who understands and remembers best by doing things physically.

active solar system A mechanical system that actively collects, concentrates, and stores solar energy.

acute effects Sudden, severe effects.

acute poverty Insufficient income or access to resources needed to provide the basic necessities for life such as food, shelter, sanitation, clean water, medical care, and education.

adaptive management A management plan designed from the outset to "learn by doing," and to actively test hypotheses and adjust treatments as new information becomes available.

aerosols Minute particles or liquid droplets suspended in the air.

aesthetic degradation Changes in environmental quality that offend our aesthetic senses.

albedo A description of a surface's reflective properties.

allergens Substances that activate the immune system.

ambient air The air immediately around us.

anthropocentric The belief that humans hold a special place in nature; being centred primarily on humans and human affairs.

antigens Chemical compounds to which antibodies bind.

appropriate technology Technology that can be made at an affordable price by ordinary people using local materials to do useful work in ways that do the least possible harm to both human society and the environment.

aquifers Porous, water-bearing layers of sand, gravel, and rock below the earth's surface; reservoirs for groundwater.

arbitration A formal process of dispute resolution in which there are stringent rules of evidence, cross-examination of witnesses, and a legally binding decision made by the arbitrator that all parties must obey.

arithmetic growth A pattern of growth that increases at a constant amount per unit time, such as 1, 2, 3, 4 or 1, 3, 5, 7.

artesian well The result of a pressurized aquifer intersecting the surface or being penetrated by a pipe or conduit, from which water gushes without being pumped; also called a spring.

atmospheric deposition Sedimentation of solids, liquids, or gaseous materials from the air.

atom The smallest unit of matter that has the characteristics of an element; consists of three main types of subatomic particles: protons, neutrons, and electrons.

atomic number The characteristic number of protons per atom of an element. Used as an identifying attribute.

B

barrier islands Low, narrow, sandy islands that form offshore from a coastline.

bases Substances that bond readily with hydrogen ions.

Batesian mimicry Evolution by one species to resemble the coloration, body shape, or behavior of another species that is protected from predators by a venomous stinger, bad taste, or some other defensive adaptation.

benthic/benthos The bottom of a sea or lake.

bioaccumulation The selective absorption and concentration of molecules by cells.

biocentric preservation A philosophy that emphasizes the fundamental right of living organisms to exist and to pursue their own goods.

biocentrism The belief that all creatures have rights and values; being centered on nature rather than humans.

biochemical oxygen demand (BOD) A standard test of water pollution measured by the amount of dissolved oxygen consumed by aquatic organisms over a given period.

biocide A broad-spectrum poison that kills a wide range of organisms.

biodegradable plastics Plastics that can be decomposed by microorganisms.

biodiversity The genetic, species, and ecological diversity of the organisms in a given area.

biogeochemical cycles Movement of matter within or between ecosystems; caused by living organisms, geological forces, or chemical reactions. The cycling of nitrogen, carbon, sulphur, oxygen, phosphorus, and water are examples.

biogeographical area An entire self-contained natural ecosystem and its associated land, water, air, and wildlife resources.

biological community The populations of plants, animals, and microorganisms living and interacting in a certain area at a given time.

biological controls Use of natural predators, patho gens, or competitors to regulate pest populations.

biological pests Organisms that reduce the availability, quality, or value of resources useful to humans.

biomagnification Increase in concentration of certain stable chemicals (for example, heavy metals or fat-soluble pesticides) in successively higher trophic levels of a food chain or web.

biomass The total mass or weight of all the living organisms in a given population or area.

biome A broad, regional type of ecosystem characterized by distinctive climate and soil conditions and a distinctive kind of biological community adapted to those conditions.

bioremediation Use of biological organisms to remove or detoxify pollutants from a contaminated area.

biosphere reserves World heritage sites identified by the IUCN as worthy for national park or wildlife refuge status because of high biological diversity or unique ecological features.

biotic Pertaining to life; environmental factors created by living organisms.

biotic potential The maximum reproductive rate of an organism, given unlimited resources and ideal environmental conditions. Compare with environmental resistance.

birth control Any method used to reduce births, including celibacy, delayed marriage, contraception; devices or medication that prevent implantation of fertilized zygotes, and induced abortions.

black lung disease Inflammation and fibrosis caused by accumulation of coal dust in the lungs or airways.

bog An area of waterlogged soil that tends to be peaty; fed mainly by precipitation; low productivity; some bogs are acidic.

boreal forest A broad band of mixed coniferous and deciduous trees that stretches across northern North America (and also Europe and Asia); its northernmost edge, the taiga, intergrades with the arctic tundra.

breeder reactor A nuclear reactor that produces fuel by bombarding isotopes of uranium and thorium with high-energy neutrons that convert inert atoms to fissionable ones.

brownfield An area of land, often former industrial sites, which has been contaminated but can be redeveloped if the site is cleaned up.

C

cancer Invasive, out-of-control cell growth that results in malignant tumours.

carbon cycle The circulation and reutilization of carbon atoms, especially via the processes of photosynthesis and respiration.

carbon dating Isotopes that decay on a known schedule used for establishing the age of organic matter.

carbon management Storing CO_2 or using it in ways that prevent its release into the air.

carbon monoxide (CO) Colourless, odourless, nonirritating but highly toxic gas produced by incomplete combustion of fuel, incineration of biomass or solid waste, or partially anaerobic decomposition of organic material.

carbon sink Places of carbon accumulation, such as in large forests (organic compounds) or ocean sediments (calcium carbonate); carbon is thus removed from the carbon cycle for moderately long to very long periods of time.

carcinogens Substances that cause cancer.

carnivores Organisms that mainly prey upon animals.

carrying capacity The maximum number of individuals of any species that can be supported by a particular ecosystem on a long-term basis.

catastrophic systems Dynamic systems that jump abruptly from one seemingly steady state to another without any intermediate stages.

cell Minute biological compartments within which the processes of life are carried out.

cellular respiration The process in which a cell breaks down sugar or other organic compounds to release energy used for cellular work; may be anaerobic or aerobic, depending on the availability of oxygen.

chain reaction A self-sustaining reaction in which the fission of nuclei produces subatomic particles that cause the fission of other nuclei.

chaotic systems Systems that exhibit variability, which may not be necessarily random, yet whose complex patterns are not discernible over a normal human time scale.

chaparral Thick, dense, thorny evergreen scrub found in Mediterranean climates.

chemical bond The force that holds atoms together in molecules and compounds.

chemical energy Potential energy stored in chemical bonds of molecules.

chlorofluorocarbons Chemical compounds with a carbon skeleton and one or more attached chlorine and fluorine atoms. Commonly used as refrigerants, solvents, fire retardants, and blowing agents.

chronic effects Long-lasting results of exposure to a toxin; can be a permanent change caused by a single, acute exposure or a continuous, low-level exposure.

city A differentiated community with a sufficient population and resource base to allow residents to specialize in arts, crafts, services, and professional occupations.

clear-cutting Cutting every tree in a given area, regardless of species or size; an appropriate harvest method for some species; can be destructive if not carefully controlled.

climate A description of the long-term pattern of weather in a particular area.

climax community A relatively stable, long-lasting community reached in a successional series; usually determined by climate and soil type.

closed canopy A forest where tree crowns spread over 20 percent of the ground; has the potential for commercial timber harvests.

cloud forests High mountain forests where temperatures are uniformly cool and fog or mist keeps vegetation wet all the time.

coevolution The process in which species exert selective pressure on each other and gradually evolve new features or behaviours as a result of those pressures.

cogeneration The simultaneous production of electricity and steam or hot water in the same plant.

cold front A moving boundary of cooler air displacing warmer air.

coliform bacteria Bacteria that live in the intestines (including the colon) of humans and other animals; used as a measure of the presence of feces in water or soil.

commensalism A symbiotic relationship in which one member is benefited and the second is neither harmed nor benefited.

complexity (ecological) The number of species at each trophic level and the number of trophic levels in a community.

composting The biological degradation of organic material under aerobic (oxygen-rich) conditions to produce compost, a nutrient-rich soil amendment and conditioner.

compound A molecule made up of two or more kinds of atoms held together by chemical bonds.

condensation The aggregation of water molecules from vapour to liquid or solid when the saturation concentration is exceeded.

condensation nuclei Tiny particles that float in the air and facilitate the condensation process.

conifers Needle-bearing trees that produce seeds in cones.

conservation development Consideration of landscape history, human culture, topography, and ecological values in subdivision design. Using cluster housing, zoning, covenants, and other design features, at least half of a subdivision can be preserved as open space, farmland, or natural areas.

conservation of matter In any chemical reaction, matter changes form; it is neither created nor destroyed.

consumer An organism that obtains energy and nutrients by feeding on other organisms or their remains. *See* also heterotroph.

consumption The fraction of withdrawn water that is lost in transmission or that is evaporated, absorbed, chemically transformed, or otherwise made unavailable for other purposes as a result of human use.

contour plowing Plowing along hill contours; reduces erosion.

control rods Neutron-absorbing material inserted into spaces between fuel assemblies in nuclear reactors to regulate fission reaction.

convection currents Rising or sinking air currents that stir the atmosphere and transport heat from one area to another. Convection currents also occur in water; see spring overturn.

conventional or criteria air contaminants The seven major pollutants (sulphur dioxide, carbon monoxide, nitrogen oxide, volatile organic compounds, and three categories of particulates) identified and regulated by Environment Canada.

core The dense, intensely hot mass of molten metal, mostly iron and nickel, thousands of kilometres in diameter at the earth's centre.

core region The primary industrial region of a country; usually located around the capital or largest port; has both the greatest population density and the greatest economic activity of the country.

Coriolis effect The influence of friction and drag on air layers near the earth; deflects air currents to the direction of the earth's rotation.

cornucopian fallacy The belief that nature is limitless in its abundance and that perpetual growth is not only possible but essential.

corridor A strip of natural habitat that connects two adjacent nature preserves to allow migration of organisms from one place to another.

covenants "Voluntary agreements" concluded between different levels of government and representatives of industry with the status of binding contracts in civil law.

cover crops Plants, such as rye, alfalfa, or clover, that can be planted immediately after harvest to hold and protect the soil.

crude birth rate The number of births in a year divided by the midyear population.

crude death rate The number of deaths per thousand persons in a given year; also called crude mortality rate.

crust The cool, lightweight, outermost layer of the earth's surface that floats on the soft, pliable underlying layers; similar to the "skin" on a bowl of warm pudding.

cultural eutrophication An increase in biological productivity and ecosystem succession caused by human activities.

D

debt-for-nature swap Forgiveness of international debt in exchange for nature protection in developing countries.

decay (radioactively) Spontaneous disintegration of a radioactive substance accompanied by emission of ionizing radiation in the form of particles and gamma rays.

deciduous Trees and shrubs that shed their leaves at the end of the growing season.

decomposers Fungi and bacteria that break complex organic material into smaller molecules.

deductive reasoning Deriving testable predictions about specific cases from general principles.

degradation (of water resource) Deterioration in water quality due to contamination or pollution; makes water unsuitable for other desirable purposes.

delta Fan-shaped sediment deposit found at the mouth of a river.

demanufacturing Disassembly of products so components can be reused or recycled.

demography Vital statistics about people: births, marriages, deaths, etc.; the statistical study of human populations relating to growth rate, age structure, geographic distribution, etc., and their effects on social, economic, and environmental conditions.

dependency ratio The number of non-working members compared to working members for a given population.

desalination Removal of salt from water by distillation, freezing, or ultrafiltration.

desert A type of biome characterized by low moisture levels and infrequent and unpredictable precipitation. Daily and seasonal temperatures fluctuate widely.

desertification Denuding and degrading a once-fertile land, initiating a desert-producing cycle that feeds on itself and causes long-term changes in soil, climate, and biota of an area.

detritivore Organisms that consume organic litter, debris, and dung.

detrital food chain A linked feeding series of organisms that consume dead matter.

dew point The temperature at which condensation occurs for a given concentration of water vapour in the air.

disability-adjusted life year (DALY) A measure of premature deaths and losses due to illnesses and disabilities in a population.

discharge The amount of water that passes a fixed point in a given amount of time; usually expressed as litres or cubic feet of water per second.

disclimax communities *See* equilibrium community.

disease A deleterious change in the body's condition in response to destabilizing factors, such as nutrition, chemicals, or biological agents.

dissolved oxygen (DO) content Amount of oxygen dissolved in a given volume of water at a given temperature and atmospheric pressure; usually expressed in parts per million (ppm).

diversity (species diversity, biological diversity) The number of species present in a community (species richness), as well as the relative abundance of each species.

doughnut Unit of energy that equals 199 kilocalories or 835 kilojoules.

downbursts Sudden, very strong, downdrafts of cold air associated with an advancing storm front.

dry alkali injection Spraying dry sodium bicarbonate into flue gas to absorb and neutralize acidic sulphur compounds.

E

earthquakes Sudden, violent movement of the earth's crust.

ecocentric (ecologically centered) A philosophy that claims moral values and rights for both organisms and ecological systems and processes.

ecofeminism A pluralistic, nonhierarchical, relationship-oriented philosophy that suggests how humans could reconceive themselves and their relationships to nature in nondominating ways as an alternative to patriarchal systems of domination.

ecological development A gradual process of environmental modification by organisms.

ecological niche The functional role and position of a species (population) within a community or ecosystem, including what resources it uses, how and when it uses the resources, and how it interacts with other populations.

ecology The scientific study of relationships between organisms and their environment. It is concerned with the life histories, distribution, and behaviour of individual species as well as the structure and function of natural systems at the level of populations, communities, and ecosystems.

economic thresholds In pest management, the point at which the cost of pest damage exceeds the costs of pest control.

ecosystem A specific biological community and its physical environment interacting in an exchange of matter and energy.

ecotone A boundary between two types of ecological communities.

edge effects A change in species composition, physical conditions, or other ecological factors at the boundary between two ecosystems.

effluent sewerage A low-cost alternative sewage treatment for cities in poor countries that combines some features of septic systems and centralized municipal treatment systems.

electrostatic precipitators The most common particulate controls in power plants; fly ash particles pick up an electrostatic surface charge as they pass between large electrodes in the effluent stream, causing particles to migrate to the oppositely charged plate.

element A molecule composed of one kind of atom; cannot be broken into simpler units by chemical reactions.

El Niño A climatic change marked by shifting of a large warm water pool from the western Pacific Ocean towards the east. Wind direction and precipitation patterns are changed over much of the Pacific and perhaps around the world.

emergent disease A new disease or one that has been absent for at least 20 years.

emigration The movement of members from a population.

endangered species A species considered to be in imminent danger of extinction.

energy The capacity to do work (that is, to change the physical state or motion of an object).

energy efficiency A measure of energy produced compared to energy consumed.

energy recovery Incineration of solid waste to produce useful energy.

environment The circumstances or conditions that surround an organism or group of organisms as well as the complex of social or cultural conditions that affect an individual or community.

environmental ethics A search for moral values and ethical principles in human relations with the natural world.

Environmental Impact Statement (EIS) An analysis required by government regulation or statute that determines the magnitude and significance of impacts on the environment arising from a resource development that is specified and/or promoted by an agency that is specified in the regulation or statute.

environmental indicators Organisms or physical factors that serve as a gauge for environmental changes. More specifically, organisms with these characteristics are called bioindicators.

environmentalism Active participation in attempts to solve environmental pollution and resource problems.

environmental policy The official rules or regulations concerning the environment adopted, implemented, and enforced by some governmental agency.

environmental resistance All the limiting factors that tend to reduce population growth rates and set the maximum allowable population size or carrying capacity of an ecosystem.

environmental science The systematic, scientific study of our environment as well as our role in it.

enzymes Molecules, usually proteins or nucleic acids, that act as catalysts in biochemical reactions.

equilibrium community Also called a **disclimax community;** a community subject to periodic disruptions, usually by fire, that prevent it from reaching a climax stage.

estuary A bay or drowned valley where a river empties into the sea.

eutectic chemicals Phase-changing chemicals used to store a large amount of energy in a small volume.

eutrophic Rivers and lakes rich in organisms and organic material (*eu* = truly; *trophic* = nutritious).

evaporation The process in which a liquid is changed to vapour (gas phase).

evolution A theory that explains how random changes in genetic material and competition for scarce resources cause species to change gradually.

evolutionary trap Habitat that was once beneficial is now a dangerous or degraded environment.

existence value The importance we place on just knowing that a particular species or a specific organism exists.

exotic organisms Alien species introduced by human agency into biological communities where they would not naturally occur.

exponential growth Growth at a constant rate of increase per unit of time; can be expressed as a constant fraction or exponent. *See* geometric growth.

externality The impact associated with a resource development/allocation decision or activity on people who were/are not involved in that decision or activity.

extinction The irrevocable elimination of species; can be a normal process of the natural world as species out-compete or kill off others or as environmental conditions change.

extirpated species Extinction of species caused by direct human action, such as hunting, trapping, etc.

F

family planning Controlling reproduction; planning the timing of birth and having as many babies as are wanted and can be supported.

famines Acute food shortages characterized by large-scale loss of life, social disruption, and economic chaos.

fecundity The physical ability to reproduce.

federal statute A law passed by the Parliament of Canada.

fen An area of waterlogged soil that tends to be peaty; fed mainly by upwelling water; low productivity.

fertility Measurement of actual number of offspring produced through sexual reproduction; usually described in terms of number of offspring of females, since paternity can be difficult to determine.

fetal alcohol syndrome A tragic set of permanent physical and mental and behavioural birth defects that result when mothers drink alcohol during pregnancy.

filters A porous mesh of cotton cloth, spun glass fibers, or asbestos-cellulose that allows air or liquid to pass through but holds back solid particles.

fire-climax community An equilibrium community maintained by periodic fires; examples include grasslands, chaparral shrubland, and some pine forests.

first law of thermodynamics States that energy is *conserved*; that is, it is neither created nor destroyed under normal conditions.

First World Industrialized, market-oriented democracies of Western Europe, North America, Japan, Australia, New Zealand, and their allies.

food chain A linked feeding series; in an ecosystem, the sequence of organisms through which energy and materials are transferred, in the form of food, from one trophic level to another.

food security The ability of individuals to obtain sufficient food on a day-to-day basis.

food web A complex, interlocking series of individual food chains in an ecosystem.

forest management Scientific planning and administration of forest resources for sustainable harvest, multiple use, regeneration, and maintenance of a healthy biological community.

fossil fuels Petroleum, natural gas, and coal created by geological forces from organic wastes and dead bodies of formerly living biological organisms.

fuel assembly A bundle of hollow metal rods containing uranium oxide pellets; used to fuel a nuclear reactor.

fuel cells Mechanical devices that use hydrogen or hydrogen-containing fuel such as methane to produce an electric current. Fuel cells are clean, quiet, and highly efficient sources of electricity.

fuelwood Branches, twigs, logs, wood chips, and other wood products harvested for use as fuel.

fugitive emissions Substances that enter the air without going through a smokestack, such as dust from soil erosion, strip mining, rock crushing, construction, and building demolition.

fungicide A chemical that kills fungi.

G

Gaia hypothesis A theory that the living organisms of the biosphere form a single, complex interacting system that creates and maintains a habitable Earth; named after Gaia, the Greek Earth mother goddess.

gap analysis A biogeographical technique of mapping biological diversity and endemic species to find gaps between protected areas that leave endangered habitats vulnerable to disruption.

garden city A new town with special emphasis on landscaping and rural ambience.

gasohol A mixture of gasoline and ethanol.

genetic assimilation The disappearance of a species as its genes are diluted through crossbreeding with a closely related species.

genetic engineering Laboratory manipulation of genetic material using molecular biology techniques to create desired characteristics in organisms.

genetically modified organisms (GMOs) Organisms whose genetic code has been altered by artificial means such as interspecies gene transfer.

geometric growth Growth that follows a geometric pattern of increase, such as 2, 4, 8, 16, etc. *See* exponential growth.

geothermal energy Energy drawn from the internal heat of the earth, either through geysers, fumaroles, hot springs, or other natural geothermal features, or through deep wells that pump heated groundwater.

ghettoization Polarization of income and spatial segregation between rich and poor. Those in poverty cluster to poor areas of a community.

government voluntary programs Initiatives undertaken by governments that allow an environmental agency to address environmental problems or allow the development of solutions in areas where its statutory authority is non-existent or weak.

grazing food chain A linked feeding series of growing grasses and herbage.

grasslands A biome dominated by grasses and associated herbaceous plants.

greenhouse gas/greenhouse effect Gases in the atmosphere are transparent to visible light but absorb infrared (heat)

waves that are reradiated from the earth's surface.

green plans Integrated national environmental plans for reducing pollution and resource consumption while achieving sustainable development and environmental restoration. Usually incorporate community roundtables for vision development.

green revolution Dramatically increased agricultural production brought about by "miracle" strains of grain; usually requires high inputs of water, plant nutrients, and pesticides.

groundwater Water held in gravel deposits or porous rock below the earth's surface; does not include water or crystallization held by chemical bonds in rocks or moisture in upper soil layers.

gully erosion Removal of layers of soil, creating channels or ravines too large to be removed by normal tillage operations.

H

habitat The place or set of environmental conditions in which a particular organism lives.

hazardous Describes chemicals that are dangerous, including flammables, explosives, irritants, sensitizers, acids, and caustics; may be relatively harmless in diluted concentrations.

hazardous waste Any discarded material containing substances known to be toxic, mutagenic, carcinogenic, or teratogenic to humans or other life-forms; ignitable, corrosive, explosive, or highly reactive alone or with other materials.

heads of power The division of powers or responsibilities between federal and provincial governments under sections 91, 92 and 92A of the Canadian Constitution.

health A state of physical and emotional well-being; the absence of disease or ailment.

heap-leach extraction A technique for separating gold from extremely low-grade ores. Crushed ore is piled in huge heaps and sprayed with a dilute

alkaline-cyanide solution, which percolates through the pile to extract the gold, which is separated from the effluent in a processing plant. This process has a high potential for water pollution.

heat A form of energy transferred from one body to another because of a difference in temperatures.

herbicide A chemical that kills plants.

herbivore An organism that eats only plants.

homeostasis Maintaining a dynamic, steady state in a living system through opposing, compensating adjustments.

horizontal fragmentation Division of responsibility that occurs within a level of government. For instance, one agency may be responsible for agricultural drainage, which often impacts wetlands. Another agency within the same level of government could have responsibility for wetland conservation. The mandates of these agencies conflict.

human development index (HDI) A measure of quality of life using life expectancy, child survival, adult literacy, childhood education, gender equity and access to clan water and sanitation as well as income.

humus Sticky, brown, insoluble residue from the bodies of dead plants and animals; gives soil its structure, coating mineral particles and holding them together; serves as a major source of plant nutrients.

hurricanes Large cyclonic oceanic storms with heavy rain and winds exceeding 119 km/hr.

hybrid gasoline-electric vehicles Automobiles that run on electric power and a small gasoline or diesel engine.

hypothesis A provisional explanation that can be tested scientifically.

I

igneous rocks Crystalline minerals solidified from molten magma from deep in the earth's interior; basalt, rhyolite, andesite, lava, and granite are examples.

individual company initiatives A type of voluntary initiative characterized by companies establishing their own environmental goals and programs such as, corporate environmental management systems, corporate environmental policy, guidelines, principles or codes of conduct, corporate environmental programs, and corporate environmental reports.

inductive reasoning Inferring general principles from specific examples.

industrial timber Trees used for lumber, plywood, veneer, particleboard, chipboard, and paper; also called roundwood.

industry initiatives A general type of voluntary program where industry has exclusive management responsibilities for all aspects of implementation. Government agencies may recognize the voluntary initiative. These inititives may be company-based or industry wide.

inertial confinement A nuclear fusion process in which a small pellet of nuclear fuel is bombarded with extremely high-intensity laser light.

infiltration The process of water percolation into the soil and pores and hollows of permeable rocks.

inherent value Ethical values or rights that exist as an intrinsic or essential characteristic of a particular thing or class of things simply by the fact of their existence.

inholdings Private lands within public parks, forests, or wildlife refuges.

insecticide A chemical that kills insects.

instrumental value Value or worth of objects that satisfy the needs and wants of moral agents. Objects that can be used as a means to some desirable end.

integrated pest management (IPM) An ecologically based pest-control strategy that relies on natural mortality factors, such as natural enemies, weather, cultural control methods, and carefully applied doses of pesticides.

interspecific competition In a community, competition for resources between members of *different* species.

intraspecific competition In a community, competition for resources among members of the *same* species.

ionosphere The lower part of the thermosphere.

ions Electrically charged atoms that have gained or lost electrons.

irruptive growth *See* Malthusian growth.

island biogeography The study of rates of colonization and extinction of species on islands or other isolated areas based on size, shape, and distance from other inhabited regions.

isotopes Forms of a single element that differ in atomic mass due to a different number of neutrons in the nucleus.

issue or interest fragmentation A government approach to environmental management characterized by issue-specific statutes—energy, navigation, fisheries, parks—in the absence of effective coordinating and/or conflict resolution mechanisms.

J

J curve A growth curve that depicts exponential growth; called a J curve because of its shape.

jet streams Powerful winds or currents of air that circulate in shifting flows; similar to oceanic currents in extent and effect on climate.

joule A unit of energy. One joule is the energy expended in 1 second by a current of 1 amp flowing through a resistance of 1 ohm.

K

keystone species A species whose impacts on its community or ecosystem are much larger and more influential than would be expected from mere abundance.

kinetic energy Energy contained in moving objects such as a rock rolling down a hill, the wind blowing through the trees, or water flowing over a dam.

Kyoto Protocol An international agreement to reduce greenhouse gas emissions.

L

land reform Democratic redistribution of landownership to recognize the rights of those who actually work the land to

a fair share of the products of their labour.

landscape ecology The study of the reciprocal effects of spatial pattern on ecological processes. A study of the ways in which landscape history shapes the features of the land and the organisms that inhabit it as well as our reaction to, and interpretation of, the land.

landslide The sudden fall of rock and earth from a hill or cliff. Often triggered by an earthquake or heavy rain.

LD50 A chemical dose lethal to 50 percent of a test population.

life expectancy The average age that a newborn infant can expect to attain in a particular time and place.

life span The longest period of life reached by a type of organism.

limiting factors Chemical or physical factors that limit the existence, growth, abundance, or distri bution of an organism.

logistic growth Growth rates regulated by internal and external factors that establish an equilibrium with environmental resources. *See* S curve.

low-head hydropower Small-scale hydro technology that can extract energy from small headwater dams; causes much less ecological damage.

M

magma Molten rock from deep in the earth's interior; called lava when it spews from volcanic vents.

magnetic confinement A technique for enclosing a nuclear fusion reaction in a powerful magnetic field inside a vacuum chamber.

Malthusian growth A population explosion followed by a population crash; also called irruptive growth.

Man and Biosphere (MAB) program A design for nature preserves that divides protected areas into zones with different purposes. A highly protected core is surrounded by a buffer zone and peripheral regions in which multiple-use resource harvesting is permitted.

mantle A hot, pliable layer of rock that surrounds the earth's core and underlies the cool, outer crust.

marsh Wetland without trees; in North America, this type of land is characterized by cattails and rushes.

mass burn Incineration of unsorted solid waste.

matter Anything that takes up space and has mass.

mediation An informal dispute resolution process in which parties are encouraged to discuss issues openly but in which all decisions are reached by consensus and any participant can withdraw at any time.

megacity *See* megalopolis.

megalopolis Also known as a megacity or supercity; megalopolis indicates an urban area with more than 10 million inhabitants.

mesosphere The atmospheric layer above the stratosphere and below the thermosphere; the middle layer; temperatures are usually very low.

metabolism All the energy and matter exchanges that occur within a living cell or organism; collectively, the life processes.

metamorphic rock Igneous and sedimentary rocks modified by heat, pressure, and chemical reactions.

metapopulation A collection of populations that have regular or intermittent gene flow between geographically separate units.

methane hydrate Small bubbles or individual molecules of methane (natural gas) trapped in a crystalline matrix of frozen water.

micro-hydro generators Small power generators that can be used in low-level rivers to provide economical power for four to six homes, freeing them from dependence on large utilities and foreign energy supplies.

Milankovitch cycles Periodic variations in tilt, eccentricity, and wobble in the earth's orbit; Milutin Milankovitch suggested that it is responsible for cyclic weather changes.

milpa agriculture An ancient farming system in which small patches of tropical forests are cleared and perennial polyculture agriculture practiced and is then followed by many years of fallow

to restore the soil; also called **swidden agriculture.**

mineral A naturally occurring, inorganic, crystalline solid with definite chemical composition and characteristic physical properties.

mixed perennial polyculture Growing a mixture of different perennial crop species (where the same plant persists for more than one year) together in the same plot.

molecule A combination of two or more atoms.

monoculture agroforestry Intensive planting of a single species; and efficient wood production approach, but one that encourages pests and disease infestations and conflicts with wildlife habitat or recreation uses.

monsoon A seasonal reversal of wind patterns caused by the different heating and cooling rates of the oceans and continents.

moral agents Beings capable of making distinctions between right and wrong and acting accordingly. Those whom we hold responsible for their actions.

moral extensionism Expansion of our understanding of inherent value or rights to persons, organisms, or things that might not be considered worthy of value or rights under some ethical philosophies.

moral subjects Beings that are not capable of distinguishing between right or wrong or that are not able to act on moral principles and yet are capable of being wronged by others.

morals A set of ethical principles that guide our actions and relationships.

morbidity Illness or disease.

mortality Death rate in a population; the probability of dying.

Müellerian mimicry Evolution of two species, both of which are unpalatable and, have poisonous stingers or some other defence mechanism, to resemble each other.

mulch Protective ground cover, including both natural products and synthetic materials that protect the soil, save water, and prevent weed growth.

mutagens Agents, such as chemicals or radiation, that damage or alter genetic material (DNA) in cells.

mutualism A symbiotic relationship between individuals of two different species in which both species benefit from the association.

N

natality The production of new individuals by birth, hatching, germination, or cloning.

National Environmental Policy Act (NEPA) The first legislation of its kind passed into law in the United States in 1969. It requires federal departments in the United States to consider the environmental aspects of their activities. NEPA forms the cornerstone of both U.S. environmental policy and law.

natural increase Crude death rate subtracted from crude birthrate.

natural selection The mechanism for evolutionary change in which environmental pressures cause certain genetic combinations in a population to become more abundant; genetic combinations best adapted for present environmental conditions tend to become predominant.

neo-Luddites People who reject technology as the cause of environmental degradation and social disruption. Named after the followers of Ned Ludd who tried to turn back the Industrial Revolution in England.

neo-Malthusian A belief that the world is characterized by scarcity and competition in which too many people fight for too few resources. Named for Thomas Malthus, who predicted a dismal cycle of misery, vice, and starvation as a result of human overpopulation.

net energy yield Total useful energy produced during the lifetime of an entire energy system minus the energy used, lost, or wasted in making useful energy available.

neurotoxins Toxic substances, such as lead or mercury, that specifically poison nerve cells.

new towns Experimental urban environments that seek to combine the best features of the rural village and the modern city.

nihilists Those who believe the world has no meaning or purpose other than a dark, cruel, unceasing struggle for power and existence.

nitrogen cycle The circulation and reutilization of nitrogen in both inorganic and organic phases.

nitrogen oxides Highly reactive gases formed when nitrogen in fuel or combustion air is heated to over 650°C in the presence of oxygen or when bacteria in soil or water oxidize nitrogen-containing compounds.

nonpoint sources Scattered, diffuse sources of pollutants, such as runoff from farm fields, golf courses, construction sites, etc.

North/South division A description of the fact that most of the world's wealthier countries tend to be in North America, Europe, and Japan while the poorer countries tend to be located closer to the equator.

nuclear fission The radioactive decay process in which isotopes split apart to create two smaller atoms.

nuclear fusion A process in which two smaller atomic nuclei fuse into one larger nucleus and release energy; the source of power in a hydrogen bomb.

O

ocean thermal electric conversion (OTEC) Energy derived from temperature differentials between warm ocean surface waters and cold deep waters. This differential can be used to drive turbines attached to electric generators.

oil sands Deposits of bitumen, a heavy black viscous oil. *See* tar sands.

oil shale A fine-grained sedimentary rock rich in solid organic material called kerogen. When heated, the kerogen liquefies to produce a fluid petroleum fuel.

old-growth forests Forests free from disturbance for long enough (generally 150 to 200 years) to have mature trees,

physical conditions, species diversity, and other characteristics of equilibrium ecosystems.

oligotrophic Condition of rivers and lakes that have clear water and low biological productivity (*oligo* = little; *trophic* = nutrition); are usually clear, cold, infertile headwater lakes and streams.

omnivore An organism that eats both plants and animals.

open canopy A forest where tree crowns cover less than 20 percent of the ground; also called woodland.

open range Unfenced, natural grazing lands; includes woodland as well as grassland.

organic compounds Complex molecules organized around skeletons of carbon atoms arranged in rings or chains; includes biomolecules, molecules synthesized by living organisms.

overnutrition Receiving too many calories.

overshoot The extent to which a population exceeds the carrying capacity of its environment.

oxygen sag Oxygen decline downstream from a pollution source that introduces materials with high biological oxygen demands.

ozone A highly reactive molecule containing three oxygen atoms; a dangerous pollutant in ambient air. In the stratosphere, however, ozone forms an ultraviolet absorbing shield that protects us from mutagenic radiation.

P

Pacific Decadal Oscillation (PDO) A large pool of warm water that moves north and south in the Pacific Ocean every 30 years or so and has large effects on North America's climate.

paradigm A model that provides a framework for interpreting observations.

parasite An organism that lives in or on another organism, deriving nourishment at the expense of its host, usually without killing it.

parsimony If two explanations appear equally plausible, choose the simpler one.

particulate material Atmospheric aerosols, such as dust, ash, soot, lint, smoke, pollen, spores, algal cells, and other suspended materials; originally applied only to solid particles but now extended to droplets of liquid.

passive heat absorption The use of natural materials or absorptive structures without moving parts to gather and hold heat; the simplest and oldest use of solar energy.

pasture Grazing lands suitable for domestic livestock.

pathogen An organism that produces disease in a host organism, disease being an alteration of one or more metabolic functions in response to the presence of the organism.

penal sanctions Penalties applied when a proponent fails to meet standards specified in a permit or as defined in government regulations.

perennial species Plants that grow for more than two years.

permanent retrievable storage Placing waste storage containers in a secure building, salt mine, or bedrock cavern where they can be inspected periodically and retrieved, if necessary.

persistent organic polutants (POPs) Chemical compounds that persist in the environment and retain biological activity for long times.

pesticide Any chemical that kills, controls, drives away, or modifies the behaviour of a pest.

pesticide treadmill A need for constantly increasing doses or new pesticides to prevent pest resurgence.

pest resurgence Rebound of pest populations due to acquired resistance to chemicals and nonspecific destruction of natural predators and competitors by broadscale pesticides.

pH A value that indicates the acidity or alkalinity of a solution on a scale of 0 to 14, based on the proportion of H^+ ions present.

phosphorus cycle The movement of phosphorus atoms from rocks through the biosphere and hydrosphere and back to rocks.

photochemical oxidants Products of secondary atmospheric reactions. *See* smog.

photodegradable plastics Plastics that break down when exposed to sunlight or to a specific wavelength of light.

photosynthesis The biochemical process by which green plants and some bacteria capture light energy and use it to produce chemical bonds. Carbon dioxide and water are consumed while oxygen and simple sugars are produced.

photovoltaic generation Converting energy by capturing solar energy and directly converting it to electrical current.

pioneer species In primary succession on a terrestrial site, the plants, lichens, and microbes that first colonize the site.

plankton Primarily microscopic organisms that occupy the upper water layers in both freshwater and marine ecosystems.

poachers Those who hunt wildlife illegally.

point sources Specific locations of highly concentrated pollution discharge, such as factories, power plants, sewage treatment plants, underground coal mines, and oil wells.

policy A societal plan or statement of intentions intended to accomplish some social good.

population A group of individuals of the same species occupying a given area.

population crash A sudden population decline caused by predation, waste accumulation, or resource depletion; also called a dieback.

population explosion Growth of a population at exponential rates to a size that exceeds environmental carrying capacity; usually followed by a population crash.

population momentum A potential for increased population growth as young members reach reproductive age.

potential energy Stored energy that is latent but available for use. A rock poised at the top of a hill or water stored behind a dam are examples of potential energy.

power The rate of energy delivery; measured in horsepower or watts.

precautionary principle Where there are threats of serious or irreversible damage, lack of full scientific uncertainty shall not be used as a reason for postponing cost-effective measures to prevent environmental degradation.

predator An organism that feeds directly on other organisms in order to survive; live-feeders, such as herbivores and carnivores.

primary pollutants Chemicals released directly into the air in a harmful form.

primary productivity Synthesis of organic materials (biomass) by green plants using the energy captured in photosynthesis.

primary succession An ecological succession that begins in an area where no biotic community previously existed.

primary treatment A process that removes solids from sewage before it is discharged or treated further.

prior approval Proponents of a development are required to seek the permission of a government agency or body before an activity takes place.

private law Also referred to as common law, and is intended to resolve disputes between individuals and where the outcome is of specific relevance to landowners and not society as a whole.

producer An organism that synthesizes food molecules from inorganic compounds by using an external energy source; most producers are photo synthetic.

productivity The synthesis of new organic material. That done by green plants using solar energy is called primary productivitiy.

Promethean environmentalism *See* technological optimists.

pronatalist pressures Influences that encourage people to have children.

proprietary rights Are associated with land ownership and allow landowners (governments or individuals) to reasonable use of their land and to set conditions for those interests wishing to develop resources (e.g., forests, minerals, recreational opportunities) on their land, or decide to leave their land in a natural state and forego development opportunities.

pull factors (in urbanization) Conditions that draw people from the countryside into the city.

push factors (in urbanization) Conditions that force people out of the countryside and into the city.

R

rainforest A forest with high humidity, constant temperature, and abundant rainfall (generally over 380 cm per year); can be tropical or temperate.

rain shadow Dry area on the downwind side of a mountain.

recharge zone Area where water infiltrates into an aquifer.

recycling Reprocessing of discarded materials into new, useful products; not the same as reuse of materials for their original purpose, but the terms are often used interchangeably.

red tide A population explosion or bloom of minute, single-celled marine organisms called dinoflagellates. Billions of these cells can accumulate in protected bays where the toxins they contain can poison other marine life.

reduced tillage systems Systems, such as minimum till, conserve-till, and no-till, that preserve soil, save energy and water, and increase crop yields.

reformer A device that strips hydrogen from fuels such as natural gas, methanol, ammonia, gasoline, or vegetable oil so they can be used in a fuel cell.

refuse-derived fuel Processing of solid waste to remove metal, glass, and other unburnable materials; organic residue is shredded, formed into pellets, and dried to make fuel for power plants.

regenerative farming Farming techniques and land stewardship that restore the health and productivity of the soil by rotating crops, planting ground cover, protecting the surface with crop residue, and reducing synthetic chemical inputs and mechanical compaction.

relative humidity At any given temperature, a comparison of the actual water content of the air with the amount of water that could be held at saturation.

GLOSSARY

relativists Those who believe moral principles are always dependent on the particular situation.

remineralize To convert matter to a mineral substance so nutrients can once again be taken up by primary producers.

renewable water supplies Annual freshwater surface runoff plus annual infiltration into underground freshwater aquifers that are accessible for human use.

residence time The length of time a component, such as an individual water molecule, spends in a particular compartment or location before it moves on through a particular process or cycle.

resilience The ability of a community or ecosystem to recover from disturbances.

resistance (inertia) The ability of a community to resist being changed by potentially disruptive events.

resource partitioning In a biological community, various populations sharing environmental resources through specialization, thereby reducing direct competition. *See also* ecological niche.

rill erosion The removing of thin layers of soil as little rivulets of running water gather and cut small channels in the soil.

risk Probability that something undesirable will happen as a consequence of exposure to a hazard.

rock A solid, cohesive, aggregate of one or more crystalline minerals.

rock cycle The process whereby rocks are broken down by chemical and physical forces; sediments are moved by wind, water, and gravity, sedimented and reformed into rock, and then crushed, folded, melted, and recrystallized into new forms.

run-of-the-river flow Ordinary river flow not accelerated by dams, flumes, etc. Some small, modern, high-efficiency turbines can generate useful power with run-of-the-river flow or with a current of only a few kilometres per hour.

rural area An area in which most residents depend on agriculture or the harvesting of natural resources for their livelihood.

S

S curve A curve that depicts logistic growth; called an S curve because of its shape.

salinization A process in which mineral salts accumulate in the soil, killing plants; occurs when soils in dry climates are irrigated profusely.

saltwater intrusion Movement of saltwater into freshwater aquifers in coastal areas where groundwater is withdrawn faster than it is replenished.

sanitary landfills A landfill in which garbage and municipal waste is buried every day under enough soil or fill to eliminate odours, vermin, and litter.

SARS Severe Acute Respiratory Syndrome

saturation point The maximum concentration of water vapour the air can hold at a given temperature.

scavenger An organism that feeds on the dead bodies of other organisms.

science An orderly, methodical approach to investigating ideas and phenomena.

scientific method A systematic, precise, objective study of a problem. Generally this requires observation, hypothesis development and testing, data gathering, and interpretation.

scientific theory An explanation supported by many tests and accepted by a general consensus of scientists.

secondary pollutants Chemicals modified to a hazardous form after entering the air or that are formed by chemical reactions as components of the air mix and interact.

secondary recovery technique Pumping pressurized gas, steam, or chemical-containing water into a well to squeeze more oil from a reservoir.

secondary succession Succession on a site where an existing community has been disrupted.

secondary treatment Bacterial decomposition of suspended particulates and dissolved organic compounds that remain after primary sewage treatment.

second law of thermodynamics States that, with each successive energy transfer or transformation in a system, less energy is available to do work.

secure landfill A solid waste disposal site lined and capped with an impermeable barrier to prevent leakage or leaching. Drain tiles, sampling wells, and vent systems provide monitoring and pollution control.

sedimentary rock Deposited material that remains in place long enough or is covered with enough material to compact into stone; examples include shale, sandstone, breccia, and conglomerates.

sedimentation The deposition of organic materials or minerals by chemical, physical, or biological processes.

selective cutting Harvesting only mature trees of certain species and size; usually more expensive than clear-cutting, but it is less disruptive for wildlife and often better for forest regeneration.

Second World Centrally planned, socialist countries, such as the former Soviet Union and its Eastern European allies.

shantytowns Settlements created when people move onto undeveloped lands and build their own shelter with cheap or discarded materials; some are simply illegal subdivisions where a landowner rents land without city approval; others are land invasions.

sinkholes A large surface crater caused by the collapse of an underground channel or cavern; often triggered by groundwater withdrawal.

slums Legal but inadequate multifamily tenements or rooming houses; some are custom built for rent to poor people, others are converted from some other use.

smart growth Efficient use of land resources and existing urban infrastructure.

smog The term used to describe the combination of smoke and fog in the stagnant air of London; now often applied to photochemical pollution products or urban air pollution of any kind.

social justice Equitable access to resources and the benefits derived from them; a system that recognizes inalienable rights and adheres to what is fair, honest, and moral.

soil A complex mixture of weathered mineral materials from rocks, partially decomposed organic molecules, and a host of living organisms.

soil horizons Horizontal layers that reveal a soil's history, characteristics, and usefulness.

soil profile All the vertical layers or horizons that make up a soil in a particular place.

spatial fragmentation Different interests that arise from a resource or environmental development or problem over space. Externalities are one form of spatial fragmentation. Upstream-downstream conflicts are another form of spatial fragmentation. Is closely related to interest fragmentation.

species A population of morphologically similar organisms that can reproduce sexually among themselves but that cannot produce fertile offspring when mated with other organisms.

species of special concern Wildlife species that may become threatened or endangered species because of a combination of biological characteristics and identified threats.

sprawl Unlimited outward extension of city boundaries that lowers population density, consumes open space, generates highway congestion, and causes decay in central cities.

squatter towns Shantytowns that occupy land without owner's permission; some are highly organized movements in defiance of authorities; others grow gradually.

stable isotope Isotopes that do not decay over time.

standing crop The amount of energy or biomass sitting in an ecosystem at a given time.

statute law Formal documents or decrees enacted by the legislative branch of government.

stewardship A philosophy that holds that humans have a unique responsibility to manage, care for, and improve nature.

strategic metals and minerals Materials a country cannot produce itself but that it uses for essential materials or processes.

stratosphere The zone in the atmosphere extending from the tropopause to about 50 km above the earth's surface; temperatures are stable or rise slightly with altitude; has very little water vapour but is rich in ozone.

stress-related diseases/stress shock Diseases caused or accentuated by social stresses such as crowding.

strip cutting Harvesting trees in strips narrow enough to minimize edge effects and to allow natural regeneration of the forest.

strip-farming Planting different kinds of crops in alternating strips along land contours; when one crop is harvested, the other crop remains to protect the soil and prevent water from running straight down a hill.

sublimation The process by which water can move between solid and gaseous states without ever becoming liquid.

subsidence A settling of the ground surface caused by the collapse of porous formations that result from withdrawal of large amounts of groundwater, oil, or other underground materials.

subsoil A layer of soil beneath the topsoil that has lower organic content and higher concentrations of fine mineral particles; often contains soluble compounds and clay particles carried down by percolating water.

sulphur cycle The chemical and physical reactions by which sulphur moves into or out of storage and through the environment.

sulphur dioxide A colourless, corrosive gas directly damaging to both plants and animals.

survivorship The percentage of a population reaching a given age or the proportion of the maximum life span of the species reached by any individual.

sustainable agriculture An ecologically sound, economically viable, socially just, and humane agricultural system. Stewardship, soil conservation, and integrated pest management are essential for sustainability.

sustainable development A real increase in well-being and standard of life for the average person that can be maintained over the long-term without degrading the environment or compromising the ability of future generations to meet their own needs.

swamp Wetland with trees, such as the extensive swamp forests of the southern United States.

swidden agriculture See milpa agriculture.

symbiosis The intimate living together of members of two different species; includes **mutualism, commensalism,** and, in some classifications, **parasitism.**

synergistic effects When an injury caused by exposure to two environmental factors together is greater than the sum of exposure to each factor individually.

T

taiga The northernmost edge of the boreal forest, including species-poor woodland and peat deposits; intergrading with the arctic tundra.

tar sands Sand deposits containing petroleum or tar.

technological optimists Those who believe that technology and human enterprise will find cures for all our problems. Also called **Promethean environmentalism.**

tectonic plates Huge blocks of the earth's crust that slide around slowly, pulling apart to open new ocean basins or crashing ponderously into each other to create new, larger landmasses.

temperate rainforest The cool, dense, rainy forest of the northern Pacific coast; enshrouded in fog much of the time; dominated by large conifers.

temperature A measure of the speed of motion of a typical atom or molecule in a substance.

10 percent rule Each trophic level in the grazing food chain has about 10 percent of the productivity of the next lower level.

teratogens Chemicals or other factors that specifically cause abnormalities during embryonic growth and development.

terracing Shaping the land to create level shelves of earth to hold water and soil; requires extensive hand labor or expensive machinery, but it enables farmers to farm very steep hillsides.

territoriality An intense form of intraspecific competition in which organisms define an area surrounding their home site or nesting site and defend it, primarily against other members of their own species.

tertiary treatment The removal of inorganic minerals and plant nutrients after primary and secondary treatment of sewage.

thermal plume A plume of hot water discharged into a stream or lake by a heat source, such as a power plant.

thermocline In water, a distinctive temperature transition zone that separates an upper layer that is mixed by the wind (the epilimnion) and a colder, deep layer that is not mixed (the hypolimnion).

thermodynamics, first law Energy can be transformed and transferred, but cannot be destroyed or created.

thermodynamics, second law With each successive energy transfer or transformation, less energy is available to do work.

thermosphere The highest atmospheric zone; a region of hot, dilute gases above the mesosphere extending out to about 1,600 km from the earth's surface.

Third World Less-developed countries that are not capitalistic and industrialized (First World) or centrally-planned socialist economies (Second World); not intended to be derogatory.

thorn scrub A dry, semi-desert dominated by acacias and other spiny shrubs.

threatened species Wildlife species that are likely to become endangered species if nothing is done to reverse the factors leading to their extirpation or extinction.

tidal station A dam built across a narrow bay or estuary traps tide water flowing both in and out of the bay. Water flowing through the dam spins turbines attached to electric generators.

tolerance limits *See* limiting factors.

topsoil The first true layer of soil; layer in which organic material is mixed with mineral particles; thickness ranges from a meter or more under virgin prairie to zero in some deserts.

tornado A violent storm characterized by strong swirling winds and updrafts; tornadoes form when a strong cold front pushes under a warm, moist air mass over the land.

total fertility rate The number of children born to an average woman in a population during her entire reproductive life.

total growth rate The net rate of population growth resulting from births, deaths, immigration, and emigration.

toxins Poisonous chemicals that react with specific cellular components to kill cells or to alter growth or development in undesirable ways; often harmful, even in dilute concentrations.

trophic level Step in the movement of energy through an ecosystem; an organism's feeding status in an ecosystem.

tropical rainforests Forests in which rainfall is abundant-more than 200 cm per year-and temperatures are warm to hot year-round.

tropical seasonal forest Semievergreen or partly deciduous forests tending toward open woodlands and grassy savannas dotted with scattered, drought-resistant tree species; distinct wet and dry seasons, hot year-round.

troposphere The layer of air nearest to the earth's surface; both temperature and pressure usually decrease with increasing altitude.

tsunami Giant seismic sea swells that move rapidly from the center of an earthquake; they can be 10 to 20 metres high when they reach shorelines hundreds or even thousands of kilometres from their source.

tundra Treeless arctic or alpine biome characterized by cold, harsh winters, a short growing season, and potential for frost any month of the year; vegetation includes low-growing perennial plants, mosses, and lichens.

U

unconventional or noncriteria air contaminants Toxic or hazardous substances, such as asbestos, benzene, beryllium, mercury, polychlorinated biphenyls, and vinyl chloride.

undernourished Those who receive less than 90 percent of the minimum dietary intake over a long-term time period; they lack energy for an active, productive life and are more susceptible to infectious diseases.

universalists Those who believe that some fundamental ethical principles are universal and un changing. In this vision, these principles are valid regardless of the context or situation.

urban area An area in which a majority of the people are not directly dependent on natural resource-based occupations.

urbanization An increasing concentration of the population in cities and a transformation of land use to an urban pattern of organization.

utilitarian conservation A philosophy that resources should be used for the greatest good for the greatest number for the longest time.

utilitarianism *See* utilitarian conservation.

V

values An estimation of the worth of things; a set of ethical beliefs and preferences that determine our sense of right and wrong.

vertical fragmentation Division of responsibility for an issue that occurs between levels of government. The heads of power establish a framework for vertical fragmentation in Canada.

village A collection of rural households linked by culture, custom, and association with the land.

volatile organic compounds Organic chemicals that evaporate readily and exist as gases in the air.

volcanoes Vents in the earth's surface through which gases, ash, or molten lava are ejected. Also a mountain formed by this ejecta.

W

warm front A long, wedge-shaped boundary caused when a warmer advancing air mass slides over neighboring cooler air parcels.

waste stream   The steady flow of varied wastes, from domestic garbage and yard wastes to industrial, commercial, and construction refuse.

water stress   A situation when residents of a country don't have enough accessible, high-quality water to meet their everyday needs.

water table   The top layer of the zone of saturation; undulates according to the surface topography and subsurface structure.

waterlogging   Water saturation of soil that fills all air spaces and causes plant roots to die from lack of oxygen; a result of overirrigation.

watershed   The land surface and groundwater aquifers drained by a particular river system.

weather   Description of the physical conditions of the atmosphere (moisture, temperature, pressure, and wind).

weathering   Changes in rocks brought about by exposure to air, water, changing temperatures, and reactive chemical agents.

wetlands   Ecosystems of several types in which rooted vegetation is surrounded by standing water during part of the year. *See also* swamp, marsh, bog, *fen*.

whistle-blower legislation   Laws, such as Ontario's Environmental Bill of Rights, which protect employees who report their employers who might be conducting their operations in an environmentally inappropriate and possibly illegal manner, from reprisal by their employer.

wicked problems   Problems with no simple right or wrong answer where there is no single, generally agreed-on definition of or solution for the particular issue.

wilderness   An area of undeveloped land affected primarily by the forces of nature; an area where humans are visitors who do not remain.

wildlife refuges   Areas set aside to shelter, feed, and protect wildlife; due to political and economic pressures, refuges often allow hunting, trapping, mineral exploitation, and other activities that threaten wildlife.

withdrawal   A description of the total amount of water taken from a lake, river, or aquifer.

woodfuel   Consists of three main commodities: fuelwood, charcoal, and black liquor.

woodland   A forest where tree crowns cover less than 20 percent of the ground; also called open canopy.

work   The application of force through a distance; requires energy input.

world conservation strategy   A proposal for maintaining essential ecological processes, preserving genetic diversity, and ensuring that utilization of species and ecosystems is sustainable.

worldviews   Sets of basic beliefs, images, and values that make up a way of looking at and making sense of the world around us.

Z

zero population growth (ZPG)   The number of births at which people are just replacing themselves; also called the replacement level of fertility.

zone of aeration   Upper soil layers that hold both air and water.

zone of saturation   Lower soil layers where all spaces are filled with water.

Photo Credits

Funkhouser/Peter Arnold, Inc.; 16.16 (Purple loosestrife): William P. Cunningham; 16.16 (Kudzu vine): © David Dennis/Tom Stack & Associates; 16.16 (Beetle): © Taina Litwak; 16.16 (Zebra mussels): © Ken Cole/Animals Animals/Earth Scenes; 16.17: William P. Cunningham; 16.23: © The McGraw-Hill Companies Inc./Barry W. Barker, photographer; 16.24: © Tom McHugh/Photo researchers, Inc.; 16.25: © The McGraw-Hill Companies Inc./Barry W. Barker, photographer; 16.26: Courtesy Dr. Ronald Tilson, Minnesota Zoo; p. 358: Courtesy of At-Bristol.

Chapter 17 Opener, 17.5, 17.6: William P. Cunningham; 17.8: © The McGraw Hill Companies, Inc./Barry W. Barker, photographer; 17.10: William P. Cunningham; 17.11: © The McGraw Hill Companies, Inc./Barry W. Barker, photographer; 17.12: Courtesy Andy White, University of Minnesota, College of Agriculture; 17.13: © David Muench/CORBIS/MAGMA; 17.14: © Greg Vaughn/Stone Images; 17.15: © Gary Braasch/Stone Images; 17.16: © The McGraw Hill Companies, Inc./Barry W. Barker, photographer; 17.17: Courtesy John McColgan, Alaska Fires Service/ Bureau of Land Management; 17.18: © The McGraw Hill Companies, Inc./Barry W. Barker, photographer; 17.20: © Eastcott/Momatiuk/The Image Works; 17.22: William P. Cunningham; 17.23: © The McGraw Hill Companies, Inc./Barry W. Barker, photographer; 17.24: © AP/Wide World Photos; 17.25: © G. Prance/Visuals Unlimited.

Chapter 18 Opener: © Ray Pfortner/Peter Arnold, Inc.; 18.3: © Fred McConnaughey/Photo Researchers, Inc.; 18.6: William P. Cunningham; 18.9: © Mike Brisson; 18.10: Courtesy Urban Ore, Inc., Berkeley, CA; 18.12: © Michael Greenlar/The Images Works; 18.16: © Piet Van Lier; p. 402: (both) Courtesy of Dr. Phil Whiting.

Chapter 19 Opener: William P. Cunningham; 19.1: © Royalty-Free/CORBIS/MAGMA; 19.2: © Charles Mauzy/CORBIS/MAGMA; p. 408, p. 410: © Royalty-Free/ CORBIS; p. 411: © Charles Mauzy/CORBIS/MAGMA; 19.4: © Royalty-Free/CORBIS/MAGMA; 19.14: © John Riley/Stone Images; 19.15: © Phil Degginger/Animals Animals/Earth Scenes; 19.16: © David L. Brown/Tom Stack & Associates.

Chapter 20 Opener: AP Photo/Anat Givon; 20.1: CP PHOTO/Aaron Harris; 20.2: CP PHOTO/Kevin Frayer; 20.4: Courtesy of Stanley Erlandsen, University of Minnesota; 20.5: William P. Cunningham; 20.6: Courtesy WHO/World Bank/R. Witlin; 20.10: © John Maier Jr./The Image Works.

Chapter 21 Opener: William P. Cunningham; 21.1: © Royalty-Free/CORBIS/ MAGMA; 21.2: Corbis Royalty Free/World Travel; 21.3: © Jim Sugar/CORBIS/MAGMA; 21.4: William P. Cunningham; 21.5: © The McGraw Hill Companies, Inc./Barry W. Barker, photographer; 21.9, 21.10: William P. Cunningham; 21.11: © Louis Psihoyos/Matrix International, Inc.; 21.12: © The McGraw Hill Companies, Inc./Barry W. Barker, photographer; p. 473: CP PHOTO/Len Wagg; 21.14, 21.16: William P. Cunningham; 21.17: © Jocelyn Vachet; 21.18: © Tom Myers/Tom Stack & Associates; 21.19: William P. Cunningham; 21.20: © Kevin R. Morris/Stone Images; 21.22: William P. Cunningham.

Chapter 22 Opener: © Michael DeYoung/CORBIS/ MAGMA; 22.1: © Galen Rowell, 1987 Mountain Light Photography; 22.2: William P. Cunningham; 22.3: Courtesy David E. Wieprecht, USGS/CVO; 22.4:© Lee Battaglia 1973/Photo Researchers, Inc; 22.5: © John Gerlack/Visuals Unlimited; 22.6: © American Heritage Centre, University of Wyoming; 22.7: William P. Cunningham; p. 497: © Bruce Paton/Panos Pictures; 22.11: Courtesy R.O. Bierregaard; 22.12: © The McGraw Hill Companies, Inc./Barry W. Barker, photographer; p. 502: © Bruce Paton/Panos Pictures; 22.14: William P. Cunningham; 22.16: © Jeff Foote Productions; 22.17: © Steve Raymer/National Geographic Society; p. 505: William P. Cunningham; p. 506 (both) Courtesy of Moira Brown.

Index

A

abandoned wells, 233
Abbey, Edward, 500
abiotic factors, 67
abundance, 103
acacia ants, 100–101
Accelerated Reduction/Elimination of Toxins (ARET), 423
Access to Information Act, 419
acclimation, 93
acid deposition. *See* acid precipitation
acid precipitation
    aquatic effects, 187, 227
    buildings, 188–189
    described, 187
    forest damage, 187–188
    monuments, 188–189
    pH and atmospheric acidity, 187
    problem, 7*f*
    sulphates, 187
acid rain. *See* acid precipitation
acids, 40, 227
activated sludge process, 237–238
active solar systems, 278–279
acute effects, 453–454
acute poverty, 9–10
adaptation
    evolution theory, 93
    and habitats, 92–94
adaptive management, 429, 429*t*
adaptive responses, 99
additive materials, 451
Adirondack Mountains (New York), 187, 227
adzuki bean plant, 53*f*
aeration tank digestion, 237–238
aerosol effects, 165
aerosol spray cans, 184
aerosols, 151, 179
aesthetic benefits of biodiversity, 343

aesthetic degradation, 180
Afghanistan, doubling time in, 76
Africa
    see also developing countries; specific countries
    AIDS, 74, 440, 443
    biodiversity in, 346
    blindness in, 441
    DALY losses, 441
    declining population in, 74
    deforestation rate in, 367
    desertification, susceptibility to, 376
    droughts, 160
    Ebola, 443
    fertility rates, 75
    firewood demand, 251
    indoor air pollution, 180
    "miracle" crops, 14
    monsoon season, 159
    peasant movements, 378
    poachers, 503–504
    population growth, 469
    poverty of, 6*f*
    rural areas, 465
    the Sahel, 119, 160, 211, 377
    shantytowns in, 471
    water availability in, 210–211
    water quality, 232
    water stress, 214
age structure of population, 66–67, 67*f*, 79*f*
Agenda 21, 463
agricultural resources
    energy, 305, 305*f*
    fertilizer, 304–305
    soil. *See* soil
    water, 304
agriculture
    see also food production
    intensive agriculture, 237, 303

    irrigation problems, 304
    milpa agriculture, 367
    and nonpoint pollution, 235
    pesticide treadmill, 323
    regenerative agriculture, 329
    shared government responsibility, 412
    sustainable agriculture, 308–313
Agriculture and Agri-Food Canada, 315
AIDS, 74, 440, 443
air currents, 154
air exchange, high rates of, 181
air pollution
    in Canada, 476
    and coal, use of, 252–254
    control of. *See* air pollution control
    conventional air contaminants, 176–180
    criteria air contaminants, 176–180
    current conditions, 193–195
    developing countries, 193, 470
    dust domes, 182
    effects of. *See* air pollution effects
    emissions in Canada, 194*t*
    fugitive emissions, 176
    future prospects, 193–195
    heat islands, 182
    human-caused, 176–181
    indoor, 180–181
    long-range transport of pollutants, 182–183
    natural sources of, 175–176
    noncriteria air contaminants, 180
    primary pollutants, 176
    secondary pollutants, 176
    smelting, 140
    Southeast Asia forest fires, 174

    statistics, 175
    stratospheric ozone, 183–185
    temperature inversions, 181–182, 182*f*
    unconventional pollutants, 180
air pollution control
    dilution, 189–190
    electrostatic precipitators, 190
    emission-control devices, 190*f*
    filters, 190
    hydrocarbon controls, 192–193
    nitrogen oxide control, 191–192
    particulate removal, 190
    remote areas, movement to, 189–190
    sulphur removal, 190–191
air pollution effects
    acid precipitation, 187–189
    bronchitis, 185, 186*f*
    emphysema, 185, 186*f*
    human health, 185
    plant pathology, 185–187
    synergistic effects, 186–187
air pressure variations, 151
air quality, improvement in, 175
air-to-air heat exchangers, 181
albedo, 153
Alcan, 169
alcohol
    from biomass, 283
    during pregnancy, 446
algae, 298
alkali-metal batteries, 280
allergens, 445
allergic reactions, and mold spores, 181
alligators in Lake Apopka (Florida), 325
alpine glaciers, 165, 165*f*

H

habitat diversification, 327–328
Habitat II Conference, 469
habitats
 see also biological
  communities; species
  distribution and
  characteristics
 defined, 94
 destruction of, 344
 distribution around the
  globe, 91–95
 evolutionary traps, 106
 human disturbance of,
  124–126, 124*f*
 Mediterranean climate, 125
 preservation of, 106
 protection of, 353–355
 winter habitat of songbirds,
  destruction of, 98
hail suppression, 160
Haiti
 colonialism, impact of, 82
 deforestation rate in, 367
 firewood use, 251
 fuel, lack of, 250
 human disturbance of
  natural world, 125
 and Philadelphia's
  waste, 384
 poverty in, 10
 soil degradation in, 304
halite, 139
halogens, 179
Hansen, James, 169
Hardin, Garret, 82
hazardous substances, 444–445,
 445*t*, 448*f*
 see also toxins
hazardous wastes
 alternatives to hazardous
  household chemicals, 398
 Basel Convention, 384
 bioremediation, 398
 brownfields, 395
 chemical processing, 398
 conversion to less
  hazardous substances,
  397–398
 defined, 394
 disposal of, 394–395
 export of, 384
 incineration, 397
 limit on international
  shipments of, 384
 permanent retrievable
  storage, 398–399
 physical treatments, 397

plants, and toxic waste
 cleanup, 396
producing less waste,
 396–397
"recycling" of, 387
secure landfills, 399
transportation, risk of, 399
waste management
 options, 395–399
heads of power, 411
Healing the Wounds (King), 28
health and illness
 see also environmental
  health hazards
 AIDS, 440
 air pollution effects, 185
 antibiotic resistance, 444,
  445*f*
 black lung disease, 252
 bronchitis, 185, 186*f*
 cancer, 447
 cardiovascular disease, 439
 diet, 447
 emergent diseases, 442–444
 emphysema, 185, 186*f*
 flu, 442–443
 in global village, 16
 health, defined, 439
 HIV, 440
 infectious lung
  diseases, 439
 leading causes of death
  worldwide, 440*t*
 Nauru inhabitants, 141
 onchocerciasis, 441
 pesticide-related health
  problems, 326–327
 Severe Acute Respiratory
  Syndrome (SARS), 439
 tips for staying healthy, 446
 tuberculosis, 439, 444
 in twentieth century, 77
Health Canada, 295*f*
Health Care Without Harm, 451
heap-leach extraction, 142, 142*f*
heat, 42
heat islands, 182
heavy water, 262
Helsinki, and CO_2 reduction, 169
Herbicide Trials. *See Palmer v.*
 Nova Scotia Forest Industries
herbicides, 259, 308, 319
herbivores, 47
herbivorous insects, 328
Herculaneum (Italy), 145
heterogeneity, 126–127
high-quality energy, 42
high-temperature solar energy, 279
Highway 407 (Ontario), 474
highway privatization, 474

The Historic Roots of Our
 Ecological Crisis (White), 26
HIV, 440, 443
homeostasis, 48
Honda Insight, 274
Hooke, Rodger, 136
horizontal fragmentation, 412
horse latitudes, 155
household chemicals, alternatives
 to, 398
household waste disposal
 guide, 397*f*
housing in developing countries,
 470–472
Howard, Ebenezer, 483
Huang He (China), 198
Hubbard Brook Experimental
 Forest (New Hampshire), 188
human demography. *See*
 demography
human development
 developmental
  discrepancies, 13
 and ecological footprint,
  14–15
 gender inequities, 13
 sustainable development, 13
human development index,
 12–13, 19
human disturbance
 and air pollution, 176–181
 biodiversity, threats to,
  344–347
 CO_2 releases, 178
 of ecosystems, 124–126,
  124*f*
 and global climate
  changes, 163–165
 types of, 126
human population growth. *See*
 population growth
human waste disposal
 activated sludge process,
  237–238
 aeration tank digestion,
  237–238
 effluent sewerage, 238–239
 low-cost waste treatment,
  238–239
 municipal sewage
  treatment, 237–239
 natural processes, 237
 wetlands, 239
human wastes, 223–224
humanism, 26
humans, carrying capacity for, 64
humidity, 201
humus, 298
Hungary
 "black triangle" region, 185

surface water in, 231
hunting, and species depletion,
 344–346
Hurricane Floyd, 210
Hurricane Mitch, 157, 157*f*
hurricanes, 157
Hutchinson, G.E., 94
hybrid gas-electric vehicles, 274
Hydro-Quebec, 217–218
hydrocarbon controls, 192–193
hydrocarbons, 40, 180
hydroelectric power, 258–260
hydroelectricity, 248
hydrogen ions, 40
hydrologic cycle, 200–201, 200*f*
Hynes, Noel, 3
hypolimnion, 122, 122*f*
hypothesis, 31

I

ice, 203
icebergs, towing, 216
Iceland
 water availability in, 211
 wealth of, 10–11
 whaling, 345
igneous rocks, 135
illegal dumping, 386
illness. *See* health and illness
immigration, 65, 78–79
immigration pull factors, 468–469
immigration push factors, 468
immune system depressants, 445
inbreeding, 352–353
incineration, 388–389, 397
income
 average annual income in
  global village, 16
 average national income
  increases, 14*f*
 gap, 13
Independent Commission on
 International Development
 Issues, 11*f*
India
 arsenic in drinking
  water, 229
 CO_2 production, 167
 colonialism, impact of, 82
 dams, and displacement of
  people, 218
 DDT and malaria, 444
 drought, reaction to,
  295–296
 grassroots forest
  protection, 369
 Gujarat earthquake, 132
 homeless people, 470
 industrial water use, 213

Jakarta, traffic in, 469
landownership in, 377
monsoon season, 159
Narmada Valley
project, 259
population in, 74
sewage treatment, lack
of, 470
shantytowns in, 471
slums, 470
squatter settlements, 471
surface waters, 232
indigenous people
ancestral lands of, 378–379
and biosphere reserves,
499–500
and dams, 218, 259
decimation of culture, 378
rights of, 379
individual company initiative, 423
Indonesia
forest fires, and air
pollution, 174
forest preservation, 369
integrated pest management
programs, 331, 331*f*
population control
success, 82
rice output, 293
squatter towns, 485
"transmigration" plan, 79
Indonesia, poverty levels in, 10
indoor air pollution, 180–181
inductive reasoning, 29–31
industrial timber, 363
industrial waste, 385
industrial water use, 213
industrialists, defensiveness of, 5
industry initiatives, 424
inertia, 105
inertial confinement, 266
infant mortality rates, 75, 82–83
infectious agents, 223–224
infectious diseases
and animals, 443–444
and climate change, 166
emergent diseases, 443–444
recent outbreaks of, 442*f*
reduction of, 9
infectious lung diseases, 439
infiltration, 204, 205*f*
influenza epidemic, 442–443
information sharing, 419–420
infrared radiation, 153
inherent value, 25
inholdings, 493
inorganic pesticides, 320
inorganic water pollutants, 226–229
insect herbivores, 96*f*
insecticides, 319

insects, 318–319
institutional resilience, 429–430
Instituto Nacional de
Biodiversidad (INBIO), 342
instrumental value, 25
integrated pest management
(IPM), 330–332
intensive agriculture, 237, 303
interest fragmentation, 412
Intergovernmental Panel on
Climate Change, 150, 163,
167, 178
international climate
negotiations, 167
International Nickel Company
(INCO), 185–186, 186*f*
International Rice Institute, 306
international treaties and
conventions, 426–428, 427*f*
International Union for
Conservation of Nature and
Natural Resources (IUCN),
351, 494*t*, 495, 496*t*
International Whaling Ban, 8
International Whaling
Commission, 345
international wildlife treaties, 355
Internet Consumer Recycling
Guide, 401
interpretive science, 31–32
*Interprovincial Co-operatives Ltd.
et al v. The Queen,* 405, 413
interspecific competition, 97
interspecific interactions, 67–68
intraspecific competition, 97
intraspecific interactions, 68
intrinsic factors, 67
the Inuit, 183
the Inupiat people, 247
invasive species, 112
inversions, 181–182
invertebrate species, and pollution
degradation, 92*f*
ionic bonds, 40
ionosphere, 152
ions, 40
Iran
declining fertility rates, 76
family planning in, 77
population control
success, 82
Ireland
renewable energy
resources, 168
surface water quality, 231
wealth of, 10–11
iron, and minimills, 143
irrigation problems, 304
island biogeography, 352
isotopes, 39

Israel, water use in, 212
issue fragmentation, 412
Itaipu Dam, 259
Italy
declining fertility rates, 76
life expectancy in, 78

J

J curve, 61, 63*f*
Jamaica, declining fertility rates
in, 82
James Bay project, 217–218
James Bay (Quebec), 14
Japan
air quality, improvement
in, 175
CO$_2$ production, 167
energy conservation in, 249
female-to-male wage
ratio, 13
Itai-Itai disease, 227
Kobe, and earthquake
damage, 144, 145*f*
life expectancy in, 78
megacities in, 464–465
methane hydrate,
extraction of, 258
Minamata Bay, 232
recycling program in, 390
surface water quality, 231
whaling, 345
wood, imports of, 364
Jasper National Park (Alberta), 165
Java
human disturbance of
natural world, 125
"transmigration" plan in, 79
tropical rainforests in, 120
Javanese rhinos, 356
jet streams, 155–156
Jintsu River (Japan), 227
Jordan, rapid population growth
in, 81
Joshua Tree National Park, 116
joules (J), 42
jumping genes, 32

K

Kant, Immanuel, 23
Karban, Richard, 100
kelp, 90, 104–105, 105*f*
Kenetech, 284
Kennedy, Matt, 129
Kenya
historical and projected
birth rates, 81*f*
rapid population
growth in, 81

Kesterson Marsh (California), 227
keystone species, 103–105
Khian Sea, 384
killer whales, 90
kinetic energy, 42
King, Ynestra, 28
Knox, Paul, 465
Koop, C. Everett, 451
kudzu vine, 347–348
Kuhn, Thomas, 32
Kuwait
desalination, 216
water availability in, 211
water use by sector, 213
Kyoto Protocol
Canadian ratification of, 167
described, 5, 167
voluntary initiatives, role
of, 423

L

La Niña, 162–163
Lake Apopka (Florida), 325
Lake Chad (West Africa), 217
Lake Erie, 4, 228*f*, 230
Lake Laberge (Yukon Territory), 38
Lake Nasser, 260
Lake Nyos (Cameroon), 169
Lake Superior, 223
lakes, 206
land degradation, 301, 302*f*
land disposal
and nonpoint pollution, 236
radioactive waste disposal,
263–264
land invasions, 368–369
land reform, 378
land resources, 300–301
land tenure, 377–378
land uses
forests, 361–366
and land reform, 378
landownership, 377–378
rangelands, 374–377
urban land-use
planning, 480
world land uses, 361
landfills, 386–387, 399
landownership, 377–378
landscape ecology
Amazonian rainforest, 126
conservation biology,
similarities with, 127
described, 126
dynamics, 127
geographic information
systems (GIS), use
of, 126*f*

hydroelectric power, 259
Newton, Isaac, 24
Nicaragua, land reform in, 378
niche specialization, 94–95, 95f
nickel, 227
Nietzsche, Friedrich, 23
Nigeria
    population in, 74
    poverty in, 10
    socioeconomic
        disparities, 13
nihilists, 23
Nile River, 198
nitric acid (HNO$_3$), 178
nitric oxide (NO), 178, 182
nitrogen cycle, 52–53, 52f
nitrogen dioxide (NO$_2$), 178
nitrogen oxide control, 191–192
nitrogen oxides (NO$_x$), 178
"no regrets" policy, 168
Noah, 22
Noah question, 356
Nobel Prize for environmental
    issue, 185
nomadic hunters, 3
non-governmental organizations
    (NGOs), 427
non-timber forest products, 373–374
noncriteria air contaminants, 180
nondomesticated land, 124f
nonhumans, rights of, 25
nonmetal mineral resources,
    138–139
nonmetallic salts, 227
nonpoint sources, 222, 235–236
nonsentient beings, 25
nontarget species, pesticide effects
    on, 323
North America
    see also Canada; United
        States
    air quality, improvement
        in, 175
    forest damage, and acid
        precipitation, 187–188
    grain, use of, 297
    human disturbance of
        natural world, 125
    parks in, 491–494
    protected lands, 495
    sanitary sewers, 238
    urbanization, 465
    wet sulphate
        deposition, 188f
North Carolina, hurricane damage
    in, 210
North/South division, 12
Northern Hemisphere
    Coriolis effect, 157
    polar air masses, 156

northern lights, 152
Northern Sun, 264
Northwest Falling Contractors
    Ltd. v. R., 405, 411
Norway
    acid deposition in, 187
    storage of CO2, 168–169
    wealth of, 10–11
    whaling, 345
Notestein, Frank, 80
NO$_x$, 178
NRTEE. *See* National Round
    Table on the Economy and the
    Environment (NRTEE)
nuclear fission, 261, 262f
nuclear fusion, 265–267, 267f
nuclear power
    breeder reactors, 262–263,
        262f
    chain reaction, 261
    control rods, 261
    fuel assembly, 261
    future of, 260
    greatest danger, 261–262
    meltdown, 261–262
    net energy yield, 275
    nuclear fission, 261, 262f
    nuclear reactors, 260–262
    opposition to, 264–265
    radioactive waste
        management, 263–264
    use of, in developed
        countries, 248
nucleic acids, 40–41
nuees ardentes, 145
nuisance, 408–409
Nunavut
    creation of, 379
    life expectancy in, 78
nutrition
    balanced diet, 294–295
    chronic hunger, 293
    countries at risk of
        inadequate nutrition, 294f
    and health, 447
    malnutrition in poorer
        countries, 441
    overnutrition, 293, 293f,
        447
    undernourishment, 293

O

Ocean Arks International, 241
ocean currents, 204f
ocean dumping, 386
ocean thermal electric conversion
    (OTEC), 286–287
Oceania, cropland increases in,
    301

oceans
    barrier islands, 123–124,
        123f
    beach pollution, 234f
    biogenic sulphur
        emissions, 54
    chlorophyll levels, 121
    coral reefs, 123–124,
        124f, 165
    desalination, 216
    described, 203
    dimethlysulphide (DMS)
        production, 54, 55f
    and El Niño, 162–163
    ocean thermal electric
        conversion (OTEC),
        286–287
    oil pollution, location
        of, 235f
    and Pacific Decadal
        Oscillation (PDO), 163
    pollution, 234
    radioactive waste
        disposal, 263
    residence time, 203
    saline ecosystems, 121–122
    shorelines, 123–124
    tidal energy, 286
    warming of, 150
    wave energy, 286
odour maskants, 180
oil
    in Caspian Sea, 256
    petroleum deposits, 254
    recovery processes, 254f
    reserves, 255, 255f
    resources, 255
    secondary recovery
        techniques, 254
    topsoil, 299
oil sands, 257
oil shale, 255
Okefenokee Swamp (Georgia), 122f
old-growth forests, 362, 363t
oligotrophic waters, 225
Oliver, Merlin, 309
Olmstead, Frederick Law, 491
Olson, Sigurd, 500
Ombudsman, 419–420
omnivores, 47
onchocerciasis, 441
One Tonne Challenge, 171
open canopy, 362
open communities, 105
open dumps, 385–386
open-pit mine, world's largest, 140f
open range, 374
open space designs, 484
optimism, 17
orange roughy, 348
orca, 90

organic chemicals in water, 228
organic compounds, 40–41, 41f
organisms
    aquatic organisms, 121
    biotic potential, 61
    carnivores, 47
    distribution of. *See* habitats
    exponential growth, 61
    food chain, 47–49, 47f
    herbivores, 47
    identification in
        ecosystem, 49f
    interactions of. *See* species
        interactions
    maturation, 107
    omnivores, 47
    trophic levels, 47
Organization for Economic
    Cooperation and Development
    (OECD), 12, 424, 425
organochlorines, 321
Our Common Future (Brundtland
    Commission), 13
outhouses, 237
overfertilization, 188, 304
overgrazing, and land degradation,
    375–376
overharvesting, 344–346
overnutrition, 293, 293f, 447
overshoot, 62
Oxfam, 366
oxygen-demanding wastes, 224
oxygen sag, 224, 225f
oyamel forests, 114
ozone, stratospheric, 183–185
ozone depletion, 183f
ozone "hole," 183
ozone layer, 152f
ozone (O$_3$), 180
Ozyer, Carl, 289

P

Pacific Decadal Oscillation
    (PDO), 163
Pacific Northwest, ancient forests
    of, 370–371
Pakistan
    air pollution in, 175
    squatter towns, 485
paleobiogeography, 48
*Palmer v. Nova Scotia Forest
    Industries,* 404, 410
Pamlico Sound (North Carolina),
    210, 226
Pangaea, 135f
parabolic mirrors, 279
paradigms, 32
parasites, 96
parasitism, 100

invertebrate species, and pollution degradation, 92f
rainfall and topography, 201–202
runoff water, slowing of, 123
water budget, balancing, 202
water-rich countries, 210–211
water shortages, 6–7, 211–212, 213–215
see also droughts
water stress, 213
water table, 204, 215f
water use
consumption, 212
degradation, 212
growth of global water use, 212f
quantities, 212
by sector, 212–213
types of, 212
typical household water use in Canada, 220f
withdrawal, 212
water vapour, 154, 201
waterlogging, 304
watershed, 219
watershed protection, 236
Watershed Stewardship, 208
wave energy, 286
The Wealth of Nations (Smith), 377
weather
and atmosphere, 152–154
and biomes and ecosystem distribution, 150–151
vs. climate, 160
cold front, 156, 156f
convection cells and prevailing winds, 155
cyclonic storms, 156–158
defined, 150
downbursts, 158
energy balance in atmosphere, 154–155
frontal weather, 156
hurricanes, 157
jet streams, 155–156
modification, 160
and Pacific Decadal Oscillation (PDO), 163
and population growth, 67
seasonal winds, 159–160
supercell frontal systems, 157
tornadoes, 157–158, 158f
warm front, 156, 156f
weathering, 135–136
weed control, 308
Welwyn Garden (London), 483

West Antarctic Ice Sheet, 165
West Nile virus, 443
wetlands
described, 122–123, 206
disturbance of, 206
loss of, 126
methane production, 179
waste treatment systems, 239
whales, 8, 90, 506–507
whaling, 8, 345
whaling nations, 8
wheat, 296
whirling disease, 350
whistle-blower legislation, 418, 420–421
White, Lynn Jr., 25, 26
Whiting, Phil, 402
wicked problems, 428–429
wilderness, 500
wildlife
see also animals
captive breeding, 355–356
demographic bottleneck, 352
diversity of. *See* biodiversity
endangered. *See* endangered species
existence value, 343
founder effect, 352
genetic drift, 352
harvesting of, 377
inbreeding, 352–353
international wildlife preserves, 503–504
international wildlife treaties, 355
List of Wildlife Species at Risk (SARA), 351–352
live specimens, 346–347
in parks, 493–494
as pets, 347
preserves. *See* nature preserves
protection of, in temperate forests, 371
small population sizes, effect of, 352
smuggling, 346–347
species survival plans, 355–356
trade in, 346–347
wildlife refuges, 501–503
wolves, in Yellowstone National Park, 497
wildlife refuges, 501–503
wildlife-related recreation, 343
Wilson, E.O., 352
wind energy, 283–285

wind ranching, 272
windmills, 284, 284f
winds
cyclones, 157
cyclonic storms, 156–158
downbursts, 158
general circulation patterns, 155f
horse latitudes, 155
hurricanes, 157
jet streams, 155–156
monsoon, 159, 159f
prevailing winds, 155
seasonal winds, 159–160
tornadoes, 157–158, 158f
trade winds, 155
winged beans, 306f
withdrawal, 212
Wittgenstein, Ludwig, 31
wolves, in Yellowstone National Park, 497
women
and ecofeminism, 27–28
and global poverty, 16
and indoor air pollution, 180–181
lack of status, 80
leading causes of death, 447
rights of, and children's welfare, 82–83, 83f
wood, 250–252
wood warblers, 96f
woodfuels, 251
woodland, 119, 362
woodland songbirds, 98
work, 42
world
acute poverty, 9–10
age structure of population, 79f
average fertility rate, 76
average national income increases, 14f
contraceptive use, 84
death, leading causes of, 440t
energy consumption, 248
estimated and projected population, 84f
extent of human disturbance, 124f
fires, 381f
forest distribution, 362, 362f
global village, 16
gross domestic product, increase in, 13
income gap, 13
land uses, 361
largest metropolitan regions, 467t

metal trade, 137f
population growth, 70t
urbanization, 465–466, 465t, 466f
wildlife preserves, 503–504
World Bank, 9, 19, 214, 229, 239, 485
world cities, 465
World Commission on Environment and Development, 13
world conservation strategy, 495–496
World Energy Council, 284
World Health Organization
acute pesticide poisoning estimates, 326
air pollution standards, 175
Health, definition of, 439
minimum life-sustaining water levels, 214
pollution from burning of organic fuels, 180
replacement rates, 75
waterborne infectious agents, and inadequate sanitation, 224
world conception rates, 84
World Meteorological Organization, 283
World Resources Institute, 19, 423
World Resources Report, 19
World Trade Organization, 428
World Wildlife Fund, 498, 502
worldviews, 27t, 28–29
Worldwatch Institute, 82, 304
Wright, Henry, 483

Y

Yamuna River, 232
Yellow River (China), 198
Yellowstone National Park wolves, 497
Yemen
poverty in, 10
rapid population growth in, 81
Yosemite National Park, 492–493, 493–494
Yukon Environment Act, 419
Yukon Placer Authorization, 138
Yukon placer mining, 138

Z

Zaire
forest preservation, 369